The Quadratic Formula

If $ax^2 + bx + c = 0$, for $a \neq 0$, then

$$x = \frac{-b \pm \sqrt{b^2 - 4ac}}{2a}$$

The Pythagorean Theorem

In any right triangle,

$$c^2 = a^2 + b^2$$

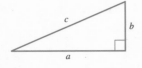

Formula for Midpoint

The midpoint M of points $A(x_1, y_1)$ and $B(x_2, y_2)$ is found with the formula

$$M = \left(\frac{x_1 + x_2}{2}, \frac{y_1 + y_2}{2} \right)$$

Formula for Distance

The distance between points (x_1, y_1) and (x_2, y_2) can be found using the formula

$$d = \sqrt{(x_2 - x_1)^2 + (y_2 - y_1)^2}$$

Slope of a Line

The slope of a line containing points $P(x_1, y_1)$ and $Q(x_2, y_2)$ is

$$m = \frac{y_2 - y_1}{x_2 - x_1}, \text{ where } x_2 \neq x_1$$

Slope-Intercept Form

The equation of a line with slope m and y intercept b is

$$y = mx + b$$

Geometric Formulas

Perimeter and Circumference

1. Rectangle
   $P = 2L + 2W$

2. Square
   $P = 4S$

3. Circle
   $C = \pi d$
   $C = 2\pi r$
   $(\pi \approx 3.14)$

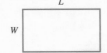  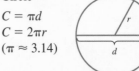

Area

4. Rectangle
   $A = L \cdot W$

5. Square
   $A = S^2$

6. Parallelogram
   $A = b \cdot h$

7. Triangle
   $A = \frac{1}{2} \cdot b \cdot h$

8. Circle
   $A = \pi r^2$

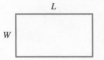 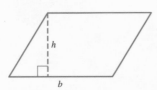

Volume

9. Rectangular Solid
   $V = L \cdot W \cdot H$

10. Cylinder
    $V = \pi r^2 h$

 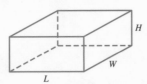

IMPORTANT:

HERE IS YOUR REGISTRATION CODE TO ACCESS
YOUR PREMIUM McGRAW-HILL ONLINE RESOURCES.

MCGRAW-HILL
ONLINE RESOURCES

For key premium online resources you need THIS CODE to gain access. Once the code is entered, you will be able to use the Web resources for the length of your course.

If your course is using **WebCT** or **Blackboard**, you'll be able to use this code to access the McGraw-Hill content within your instructor's online course.

Access is provided if you have purchased a new book. If the registration code is missing from this book, the registration screen on our Website, and within your WebCT or Blackboard course, will tell you how to obtain your new code.

Registering for McGraw-Hill Online Resources

TO gain access to your McGraw-Hill web resources simply follow the steps below:

1. USE YOUR WEB BROWSER TO GO TO: **www.mhhe.com/hutchison**
2. CLICK ON **FIRST TIME USER**.
3. ENTER THE REGISTRATION CODE* PRINTED ON THE TEAR-OFF BOOKMARK ON THE RIGHT.
4. AFTER YOU HAVE ENTERED YOUR REGISTRATION CODE, CLICK **REGISTER**.
5. FOLLOW THE INSTRUCTIONS TO SET-UP YOUR PERSONAL UserID AND PASSWORD.
6. WRITE YOUR UserID AND PASSWORD DOWN FOR FUTURE REFERENCE. KEEP IT IN A SAFE PLACE.

BT15-HLI4-SO3A-AZWS-QU8F

REGISTRATION CODE

TO GAIN ACCESS to the McGraw-Hill content in your instructor's **WebCT** or **Blackboard** course simply log in to the course with the UserID and Password provided by your instructor. Enter the registration code exactly as it appears in the box to the right when prompted by the system. You will only need to use the code the first time you click on McGraw-Hill content.

Thank you, and welcome to your McGraw-Hill online Resources!

Higher Education

0-07-292313-X T/A HUTCHISON: ELEMENTARY AND INTERMEDIATE ALGEBRA, 2/E

Elementary and Intermediate
Algebra

Elementary and Intermediate
Algebra
A Unified Approach
Second Edition

Donald Hutchison | **Barry Bergman** | **Louis Hoelzle**
Clackamas Community College | *Clackamas Community College* | *Bucks County Community College*

 **Higher Education**

Boston   Burr Ridge, IL   Dubuque, IA   Madison, WI   New York   San Francisco   St. Louis
Bangkok   Bogotá   Caracas   Kuala Lumpur   Lisbon   London   Madrid   Mexico City
Milan   Montreal   New Delhi   Santiago   Seoul   Singapore   Sydney   Taipei   Toronto

Higher Education

ELEMENTARY AND INTERMEDIATE ALGEBRA: A UNIFIED APPROACH, SECOND EDITION

1 2 3 4 5 6 7 8 9 0 VNH/VNH 0 9 8 7 6 5 4 3
1 2 3 4 5 6 7 8 9 0 VNH/VNH 0 9 8 7 6 5 4 3

ISBN 0–07–254671–9
ISBN 0–07–254674–3 (Annotated Instructor's Edition)

Publisher: *William K. Barter*
Director of development: *David Dietz*
Developmental editor: *Peter Galuardi*
Executive marketing manager: *Marianne C. P. Rutter*
Senior marketing manager: *Mary K. Kittell*
Lead project manager: *Susan J. Brusch*
Senior production supervisor: *Laura Fuller*
Media project manager: *Sandra M. Schnee*
Senior media technology producer: *Jeff Huettman*
Senior coordinator of freelance design: *Michelle D. Whitaker*
Cover/interior designer: *Rokusek Design*
Cover image: *Rokusek Design*
Senior photo research coordinator: *Lori Hancock*
Supplement producer: *Brenda A. Ernzen*
Compositor: *Interactive Composition Corporation*
Typeface: *10/12 Times Roman*
Printer: *Von Hoffmann Corporation*

PHOTO CREDITS
Opener 0: © Richard Lord/The Image Works; Opener 1: © Bob Dammrich/The Image Works; Opener 2: © Alan Levenson/ Getty Images; Opener 3: © Michelle Bridwell/Photo Edit; Opener 4: © Michael Newman/Photo Edit; Opener 5: © Mitch Wojnarowicz/The Image Works; Opener 6: © Lawrence Migdale/Getty Images; Opener R: © Vol. 41/PhotoDisc; Opener 7: © David Young-Wolf/Getty Images; Opener 8: © Frank Siteman/Picture Cube/Index Stock; Opener 9: © Hank Morgan/SS/ Photo Researchers, Inc.; Opener 10: © Bob Daemmrich/The Image Works; Opener 11: © David Burnett/The Stock Market/ Corbis; Opener 12: © Kathy McLaughlin/The Image Works; Opener 13: © Chris Jones/The Stock Market/Corbis

Library of Congress Cataloging-in-Publication Data

Hutchison, Donald, 1948–
    Elementary and intermediate algebra : a unified approach / Donald Hutchison, Barry Bergman, Louis Hoelzle. — 2nd ed.
      p. cm.
    Includes index.
    ISBN 0–07–254671–9 (hard copy : alk. paper)
    1. Algebra. I. Bergman, Barry. II. Hoelzle, Louis. III. Title.

QA152.3.H874   2004
512—dc22
                                       2003016024
                                           CIP

www.mhhe.com

Dedications

Donald Hutchison

This book is dedicated to my life's travel companion, Claudia. She has allowed me to grow as an author, a partner, and a parent.

Barry Bergman

This book is dedicated to my wife Marcia, who encouraged me to enter into this work and who, along with our wonderful boys, Joel and Adam, gave me support through the duration of the project.

Louis Hoelzle

This book is dedicated to my wife and children who have shared my joys, sorrows, successes, and failures over 38 years: Rosemary, my friend, my inspiration, and my companion on the journey of life; and Beth, Ray, Amy, Oscar, Meg, Jerry, Johanna, and Patrick, the joys of my life.

> This series is dedicated to the memory of
> James Arthur Streeter, an artisan with words,
> a genius with numbers, and a virtuoso with
> pictures from 1940 until 1989.

About the Authors

Donald Hutchison Don began teaching in a preschool while he was an undergraduate. He subsequently taught children with disabilities, adults with disabilities, high school mathematics, and college mathematics. Although all these positions were challenging and satisfying, the act of breaking a difficult lesson into teachable components was what he enjoyed most.

It was at Clackamas Community College that he found his professional niche. The community college allowed him to focus on teaching within a department that constantly challenged faculty and students to expect more. Under the guidance of Jim Streeter, Don learned to present his approach to teaching in the form of a textbook.

Don has also been an active member of many professional organizations. He has been President of ORMATYC, AMATYC committee chair, and ACM curriculum committee member. He has presented at AMATYC, ORMATYC, AACC, MAA, ICTCM, and numerous other conferences.

Barry Bergman Barry has enjoyed teaching mathematics to a wide variety of students over the years. He began in the field of adult basic education and moved into the teaching of high school mathematics in 1977. He taught at that level for 11 years, at which point he served as a K–12 mathematics specialist for his county. This work allowed him the opportunity to promote the emerging NCTM Standards in his region.

In 1990 Barry began the current stage of his career, having been hired to teach at Clackamas Community College. He maintains a strong interest in the appropriate use of technology and visual models in the learning of mathematics.

Throughout the past 26 years, Barry has played an active role in professional organizations. As a member of OCTM, he contributed several articles and activities to the group's journal. He has made presentations at OCTM, NCTM, ORMATYC, and ICTCM conferences. Barry also served as an officer of ORMATYC for four years, and participated on an AMATYC committee to provide feedback to revisions of NCTM's Standards.

Louis Hoelzle Lou has taught at Bucks County Community College for more than 30 years. In 1989, Lou became chair of the mathematics department at the college. He has taught the entire range of courses from arithmetic to calculus, giving him an excellent view of the current and future developmental needs of students.

Over the past 36 years, Lou has also taught physics courses at 4-year colleges, which has given him the perspective of the practical applications of mathematics. Lou has always focused his writing on the student.

Lou is also active in several professional organizations. He has served on the Placement and Assessment committee and the Grants committee for AMATYC. He was president of PSMATYC from 1997 to 1999.

Contents

To the Student

You are about to begin a course in algebra. We made every attempt to provide a text that will help you understand what algebra is about and how to use it effectively. We made no assumptions about your previous experience with algebra. Your progress through the course will depend on the amount of time and effort you devote to the course and your previous background in math. There are some specific features in this book that will aid you in your studies. Here are some suggestions about how to use this book. (Keep in mind that a review of *all* the chapter and summary material will further enhance your ability to grasp later topics and to move more effectively through the text).

1. If you are in a lecture class, make sure that you take the time to read the appropriate text section *before* your instructor's lecture on the subject. Then take careful notes on the examples that your instructor presents during class.

2. After class, work through similar examples in the text, making sure that you understand each of the steps shown. Examples are followed in the text by *Check Yourself* exercises. You can best learn algebra by being involved in the process, and that is the purpose of these exercises. Always have a pencil and paper at hand, and work out the problems presented and check your results immediately. If you have difficulty, go back and carefully review the previous examples. Make sure you understand what you are doing and why. The best test of whether you do understand a concept lies in your ability to explain that concept to one of your classmates. Try working together.

3. At the end of each chapter section you will find a set of exercises. Work these carefully to check your progress on the section you have just finished. You will find the answers for the odd-numbered exercises at the end of the book. If you have difficulties with any of the exercises, review the appropriate parts of the chapter section. If your questions are not completely cleared up, by all means do not become discouraged. Ask your instructor or an available tutor for further assistance. A word of caution: Work the exercises on a regular (preferably daily) basis. Again, learning algebra requires becoming involved. As is the case with learning any skill, the main ingredient is practice.

4. When you complete a chapter, review by using the *Summary*. You will find all the important terms and definitions in this section, along with examples illustrating all the techniques developed in the chapter. Following the summary are *Summary Exercises* for further practice. The exercises are keyed to chapter sections, so you will know where to turn if you are still having problems. Answers to odd-numbered *Summary Exercises* are located at the end of the book.

5. When you finish with the *Summary Exercises,* try the *Self-Test* that appears at the end of each chapter. It is an actual practice test you can work on as you review for in-class testing. Again, answers with section references are provided at the end of the book.

6. Finally, an important element of success in studying algebra is the process of regular review. We provide a series of *Cumulative Tests* throughout the textbook, beginning at the end of Chapter 2. These tests will help you review not only the concepts of the chapter that you have just completed but also those of previous chapters. Use these tests in preparation for any midterm or final exams. Answers to the *Cumulative Tests* are located at the end of the book. If it appears that you have forgotten some concepts that are being tested, don't worry. Go back and review the sections where the idea was initially explained, or the appropriate chapter summary. That is the purpose of the *Cumulative Tests*.

We hope that you will find our suggestions helpful as you work through this material, and we wish you the best of luck in the course.

Donald Hutchison
Barry Bergman
Louis Hoelzle

Preface

Our Philosophy

We believe the key to learning mathematics, at any level, is active participation. When students actively take part in the learning process, they have the opportunity to construct their own mathematical ideas and make connections to previously studied material. Such participation leads to understanding, success, and confidence. We developed this text with that philosophy in mind, and we integrated many features throughout the book to reflect that approach to learning and teaching.

The *Check Yourself* exercises are designed to keep the students active and involved with every page of exposition. Almost every end-of-section exercise set has applications, challenging exercises, writing exercises, and/or collaborative exercises. The application exercises are designed to awaken interest and insight within students. Our hope is that every student who uses this text will be a better mathematical thinker as a result.

Visual Approach

For students of elementary and intermediate algebra, one key to success is the visualization of a problem and its solution. One theme of this text is the teaching of concepts, aided by a visual approach. In some instances, the visualization is encouraged through use of a graphing calculator. At other times, it may be encouraged through a hand-drawn graph. In any event, the text provides students with an opportunity to see the relationship between symbolic manipulation and a visual, concrete model.

Features of This Text

In every discipline and college course, students must utilize critical-thinking skills. These skills can be divided into three areas: communication skills, pattern-recognition skills, and problem-solving skills. In mathematics, we stress the latter two skills, but all three are germane to what we do in our texts and in our classrooms. We will now examine each of the areas as it relates to *Elementary and Intermediate Algebra*.

Communication
Early Practice in Interpreting Expressions
Beginning in Chapter 1, we prepare students for the process of interpreting word problems.

An Abundance of Application Problems (See, for Instance, pp. 65 and 66)
No matter what language one is trying to learn, the secret is repeated exposure. This is certainly true of the language of algebra. We present many interesting applications that appeal to every interest and are applicable to everyday life.

Writing and Group Exercises (See, for Instance, p. 76)

Relating what one knows to peers and instructors is always daunting at first. Only with regular practice are students able to develop this important skill. We provide dozens of exercises designed to help hone these skills.

Pattern Recognition

Matching Exercises (See, for Instance, pp. 122 and 232)

Part of any top-down design requires the recognition of patterns. Students need to identify whether they are dealing with an expression or an equation. Likewise, students need to identify what kind of graph to expect (linear, quadratic, etc.) before they begin to plot it. We have provided exercises that emphasize equation and graph recognition.

Transition between Examples (See, for Instance, p. 238)

Students must have the opportunity to see why studying multiple examples is necessary. A good transition between similar examples helps students to discern a difference in the examples so that they can spot these differences in the problem sets.

Section Exercises

When doing homework exercises, students should expect to find similar worked examples in the preceding text. In *Elementary and Intermediate Algebra,* exercise sets are graded in a manner parallel to the examples. Exceptions are clearly marked; for example, challenge exercises are marked with a mountain icon.

Transition Material

The review material between Chapters 6 and 7, or Chapter R, is designed to help students who begin using this text in intermediate algebra, having studied elementary algebra using another book, possibly long ago.

Problem Solving

Consistent Use of Problem Solving (See, for Instance, p. 157)

The five-step problem-solving approach first introduced in Chapter 2 is used consistently throughout the text for all applications.

Problem-Solving Hints (See, for Instance, p. 162)

Students are encouraged, by way of margin notes, to use sketches and tables where appropriate. These techniques are consistently reinforced throughout the text.

Group Work (See, for Instance, p. 168)

Some designated exercises are best completed by small groups. Critical-thinking and problem-solving skills develop more fully when students interact with peers to solve problems. Students will then bring this skill to their other courses as well as to their careers.

Use of Technology (See, for Instance, pp. 743 and 744)

More and more often, instructors report that students place too much reliance upon calculators. At the same time, most instructors see the graphing calculator as a tool that can enhance students' understanding of algebra by enabling them to visualize concepts more effectively. This text includes exercises and examples that make use of graphing and scientific calculators. These exercises and examples may be considered optional. The graphing calculator exercises and examples are intended to enhance the experience of learning algebra and to actively engage students in mathematical learning. Furthermore, the graphing calculator supports a visual approach to learning algebra.

What's New for the Second Edition

At the request of users of the first edition and reviewers, a number of important changes were made for this second edition. We believe these will make the second edition a more effective teaching tool. The following are the major changes to the text.

New coverage:

- A new review chapter, Chapter R, was added between Chapters 6 and 7. This review chapter provides a concise review of elementary algebra topics.

- Two new sections, *Problem Solving with Factoring* (6.5) and *A General Strategy for Factoring* (6.6), were added to Chapter 6.

- A new section, *Examining Rational Functions* (7.7), was added to Chapter 7.

- A new section, *Solving Radical Equations* (10.4), was added to Chapter 10.

Revised material:

- Chapters 4 and 5 from the previous edition were swapped. *A Beginning Look at Functions* now follows *Exponents and Polynomials.*

- Negative exponents, scientific notation, and division of polynomials are now included in Chapter 4, *Exponents and Polynomials.*

- Several topics formerly covered in one section were split into two sections:
  Solving linear inequalities (now 2.5 and 2.6)
  Absolute value equations and inequalities (now 2.7 and 2.8)
  Rational equations and inequalities (now 7.5 and 7.6)

Pedagogical changes:

- **NEW** tips on overcoming math anxiety were added to Chapters 0, 1, and 2. These tips are designed to provide timely and useful advice.

- **NEW** chapter summaries were added at the end of each chapter. Each chapter summary contains brief descriptions or definitions of each concept and provides an example along with section and page references.

- **NEW** exercises, examples, and applications were added throughout the text.

Feature Walk-Through

This text contains a wealth of pedagogical features designed to enhance its usefulness to students and teachers. Some of the more important features are presented here.

Overcoming Math Anxiety

Throughout this text, we will present you with a series of class-tested techniques that are designed to improve your performance in this math class.

Hint 1: Become familiar with your textbook.

Perform each of the following tasks.

1. Use the Table of Contents to find the title of Section 5.1.

2. Use the index to find the earliest reference to the term *factor*.

3. Find the answer to the first Check Yourself exercise in Section 0.1.

4. Find the answers to the Self-Test for Chapter 1.

5. Find the answers to the odd-numbered exercises in Section 0.1.

Now you know where some of the most important features of the text are. When you have a moment of confusion, think about using one of these features to help you clear up that confusion.

Tips on Overcoming Math Anxiety Located within the first few chapters are suggestions on overcoming math anxiety. These suggestions are designed to be timely and useful. They are the same suggestions most instructors make in class, but sometimes those words are given extra weight when students see them in print.

CHAPTER

1

From Arithmetic to Algebra

LIST OF SECTIONS

1.1 Transition to Algebra

1.2 Evaluating Algebraic Expressions

1.3 Adding and Subtracting Algebraic Expressions

1.4 Sets

Cultures from all over the world have developed number systems and ways to record patterns in their natural surroundings. The Mayans in Central America had one of the most sophisticated number systems in the world in 12th century C.E. The Chinese numbering and recording system dates from around 1200 B.C.E. The oldest evidence of numerical record is in Africa, where a bone notched in numerical patterns and dating from about 35,000 B.C.E. was found in Southern Africa.

The roots of algebra developed among the Babylonians 4,000 years ago in an area now part of the country of Iraq. The Babylonians developed ways to record useful numerical relationships so that they were easy to remember, easy to record, and helpful in solving problems. Archaeologists have found instructions for solving problems in engineering, economics, city planning, and agriculture. The writing was on clay tablets. Some of these formulas developed by the Babylonians are still in use today.

1-1

57

Chapter-Opening Vignette Each chapter opens with a real-world vignette that showcases an example of how mathematics is used in our world. Exercise sets for each section then feature one or more applications that relate to the vignette.

1.3 OBJECTIVES

1. Combine like terms
2. Add algebraic expressions
3. Subtract algebraic expressions

Section Objectives Objectives for each section are clearly identified. These numbered objectives are used to organize lessons on the accompanying tutorial SMART-CD and Online Learning Center.

Example 4 **Applying the Properties of Equality with Like Terms**

Solve for x.

Note the like terms on the left and right sides of the equation.

$$8x + 2 - 3x = 8 + 3x + 2$$

Here we combine the like terms $8x$ and $-3x$ on the left and the like terms 8 and 2 on the right as our first step. We then have

$$5x + 2 = 3x + 10$$

We can now solve as before.

$$5x + 2 - 2 = 3x + 10 - 2 \quad \text{Subtract 2 from both sides}$$
$$5x = 3x + 8$$

Then

$$5x - 3x = 3x - 3x + 8 \quad \text{Subtract } 3x \text{ from both sides}$$
$$2x = 8$$
$$\frac{2x}{2} = \frac{8}{2} \quad \text{Divide both sides by 2}$$
$$x = 4$$

The solution set is {4}, which can be checked by returning to the *original equation.*

Examples Concise examples effectively illustrate the concepts and ideas presented. Student annotations and *margin notes* are used to further clarify or highlight important steps in the solution to the problem presented.

 CHECK YOURSELF 4

Solve for x.

$$7x - 3 - 5x = 10 + 4x + 3$$

Check Yourself Along with the readability, the *Check Yourself* exercises are the hallmark of this text; they are designed to actively involve students throughout the learning process. Each example is followed by a *Check Yourself* exercise that encourages students to solve a problem similar to the one just presented. Answers to these exercises are provided at the end of each section for immediate feedback.

Remember: The x coordinate gives the *horizontal* distance from the y axis. The y coordinate gives the *vertical* distance from the x axis.

Margin Notes *Margin notes* are provided throughout and are designed to help students focus on important topics and techniques. Some margin notes also point out potential trouble spots.

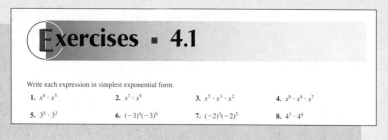

Exercises ■ 4.1

Write each expression in simplest exponential form.

1. $x^4 \cdot x^5$ **2.** $x^7 \cdot x^9$ **3.** $x^5 \cdot x^3 \cdot x^2$ **4.** $x^8 \cdot x^4 \cdot x^7$

5. $3^5 \cdot 3^2$ **6.** $(-3)^4(-3)^6$ **7.** $(-2)^3(-2)^5$ **8.** $4^3 \cdot 4^4$

Comprehensive Exercise Sets Each section ends with a comprehensive set of *Exercises* that are designed to reinforce basic skills and develop critical-thinking and communication abilities. Nearly every set of *Exercises* contains real-world application problems demonstrating the use and utility of mathematics in everyday life.

Writing, Group, and Challenge Exercises Included in the exercise sets are writing problems, collaborative and group exercises, and challenge exercises. Distinctive icons denote each type of exercise: an icon of a sheet of paper for writing problems, three people together in a group for group exercises, and a mountain climber for challenge exercises.

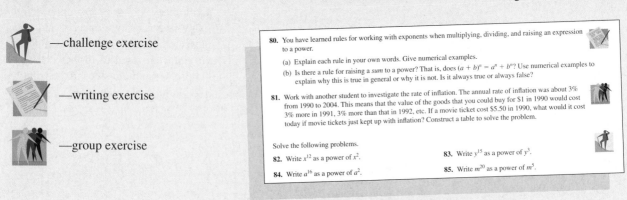

—challenge exercise

—writing exercise

—group exercise

80. You have learned rules for working with exponents when multiplying, dividing, and raising an expression to a power.
  (a) Explain each rule in your own words. Give numerical examples.
  (b) Is there a rule for raising a *sum* to a power? That is, does $(a + b)^n = a^n + b^n$? Use numerical examples to explain why this is true in general or why it is not. Is it always true or always false?

81. Work with another student to investigate the rate of inflation. The annual rate of inflation was about 3% from 1990 to 2004. This means that the value of the goods that you could buy for $1 in 1990 would cost 3% more in 1991, 3% more than that in 1992, etc. If a movie ticket cost $5.50 in 1990, what would it cost today if movie tickets just kept up with inflation? Construct a table to solve the problem.

Solve the following problems.

82. Write x^{12} as a power of x^2.   **83.** Write y^{15} as a power of y^3.

84. Write a^{16} as a power of a^2.   **85.** Write m^{20} as a power of m^5.

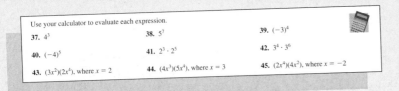

Use your calculator to evaluate each expression.

37. 4^3 **38.** 5^7 **39.** $(-3)^4$

40. $(-4)^5$ **41.** $2^3 \cdot 2^5$ **42.** $3^4 \cdot 3^6$

43. $(3x^2)(2x^4)$, where $x = 2$ **44.** $(4x^3)(5x^4)$, where $x = 3$ **45.** $(2x^4)(4x^2)$, where $x = -2$

Calculator Examples and Exercises Throughout the text are examples displaying how students may use a calculator to solve various problems. Consequently, exercises utilizing calculators are given in some exercise sets. These examples and exercises are denoted by a calculator icon. Calculator examples and exercises are designed to enhance, rather than replace, traditional mathematical topics.

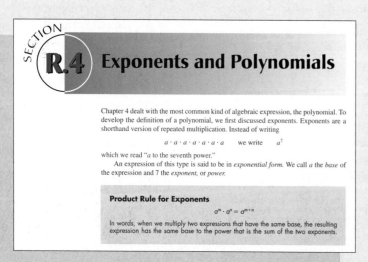

SECTION R.4 Exponents and Polynomials

Chapter 4 dealt with the most common kind of algebraic expression, the polynomial. To develop the definition of a polynomial, we first discussed exponents. Exponents are a shorthand version of repeated multiplication. Instead of writing

$$a \cdot a \cdot a \cdot a \cdot a \cdot a \cdot a \quad \text{we write} \quad a^7$$

which we read "a to the seventh power."

An expression of this type is said to be in *exponential form*. We call a the *base* of the expression and 7 the *exponent*, or *power*.

Product Rule for Exponents

$$a^m \cdot a^n = a^{m+n}$$

In words, when we multiply two expressions that have the same base, the resulting expression has the same base to the power that is the sum of the two exponents.

Review Chapter Appearing in the middle of the text, Chapter R, the review chapter, is a concise review of important topics from Chapters 1 through 6. Students entering intermediate algebra without having used this book for elementary algebra will find the review chapter an excellent resource to get them up to speed on important topics from the first half of the book. The review chapter is also a superb reference for elementary algebra students preparing for their final exam. Following the review chapter is a final exam for elementary algebra.

Summary for Chapter 7

Example	Topic	Reference
	Simplifying Rational Expressions	7.1
$\frac{x^2 - 5x}{x - 3}$ is a rational expression. The variable x cannot have the value 3.	**Rational expressions** have the form $\frac{P}{Q}$ in which P and Q are polynomials and $Q(x) \neq 0$ for all x.	p. 568
This uses the fact that $\frac{R}{R} = 1$ when $R \neq 0$.	**Fundamental Principle of Rational Expressions** For polynomials P, Q, and R, $$\frac{P}{Q} = \frac{PR}{QR} \qquad \text{when } Q \neq 0 \text{ and } R \neq 0$$ This principle can be used in two ways. We can multiply or divide the numerator and denominator of a rational expression by the same nonzero polynomial.	p. 570

Summaries   *Summaries* are given at the end of each chapter in a tabular format. These enable students to quickly review important concepts. *Summaries* contain brief descriptions or definitions of each concept and provide an example of each concept along with a section and page reference.

Summary Exercises ■ 7

This summary exercise set is provided to give you practice with each of the objectives in the chapter. Each exercise is keyed to the appropriate chapter section.

[7.1] For what value(s) of the variable will each of the following rational expressions be defined?

1. $\dfrac{x}{2}$

2. $\dfrac{3}{y}$

3. $\dfrac{2}{x - 5}$

4. $\dfrac{4x}{3x - 2}$

[7.1] Simplify each of the following rational expressions.

5. $\dfrac{18x^5}{24x^3}$

6. $\dfrac{15m^3n}{-5mn^2}$

7. $\dfrac{7y - 49}{y - 7}$

8. $\dfrac{5x - 20}{x^2 - 16}$

9. $\dfrac{9 - x^2}{x^2 + 2x - 15}$

10. $\dfrac{3w^2 + 8w - 35}{2w^2 + 13w + 15}$

11. $\dfrac{6a^2 - ab - b^2}{9a^2 - b^2}$

12. $\dfrac{6w - 3z}{8w^3 - z^3}$

Summary Exercises   *Summary Exercises* provide an opportunity for students to practice the important concepts they reviewed in the *Summary*. Answers to the odd-numbered *Summary Exercises* are located at the end of the book.

Self-Tests   Beginning with Chapter 1, each chapter contains a *Self-Test*. *Self-Tests* provide students with an opportunity to check their progress and review important skills, as well as building confidence in preparing for in-class exams. The answers to the questions in the *Self-Tests* are provided at the end of the book.

Self-Test   5

The purpose of this self-test is to help you check your progress and to review for a chapter test in class. Allow yourself about an hour to take the test. When you are done, check your answers in the back of the book. If you missed any questions, be sure to go back and review the appropriate sections in the chapter and the exercises that are provided.

1. For each of the following sets of ordered pairs, identify the domain and range.

   (a) $\{(1, 6), (-3, 5), (2, 1), (4, -2), (3, 0)\}$
   (b) $\{(United States, 101), (Germany, 65), (Russia, 63), (China, 50)\}$

2. For each of the following, determine if the given relation is a function. Identify the domain and range of the relation.

   (a) $\{(2, 5), (-1, 6), (0, -2), (-4, 5)\}$

   (b)

x	y
-3	2
0	4
1	7
2	0
-3	1

Cumulative Tests   Beginning with Chapter 2, a *Cumulative Test* follows the *Self-Test*. These tests help students build on what was previously covered and give them greater opportunity to reinforce skills necessary in preparing for midterm and final exams. Answers to the questions in these tests are provided at the end of the book.

Cumulative Test   0–5

This test is provided to help you in the process of reviewing previous chapters. Answers are provided in the back of the book. If you miss any answers, be sure to go back and review the appropriate chapter sections.

In Exercises 1 to 4, evaluate each expression if $x = -2$, $y = 4$, and $z = -1$.

1. $4x^2 + 3y - z$

2. $-z^2 + 4y + 3x$

3. $\dfrac{2z - 3x + 4y}{x^2 - z^2}$

4. $\dfrac{x^2 + y^2 + z^2}{x - y + z}$

Supplements for the Instructor

Annotated Instructor's Edition

This ancillary provides answers to problems and exercises in the text, including answers to all section *Exercises, Summary Exercises, Self-Tests,* and *Cumulative Tests.* These answers are printed in a second color for ease of use by the instructor and are located on the appropriate pages throughout the text.

Instructor's Testing and Resource CD

This cross-platform CD-ROM provides resources for the instructor. Supplements featured on this CD-ROM include a computerized test bank utilizing Brownstone's Diploma © testing software to quickly create customized exams. This user-friendly program allows instructors to search for questions by topic, format, or difficulty level; edit existing questions or add new ones; and scramble questions and answer keys for multiple versions of the same test.

Other assets on the Instructor's Testing and Resource CD-ROM include test item files, preformatted chapter tests, and PowerPoint slides.

Instructor's Solutions Manual

Prepared by Stefan Baratto, this supplement contains detailed solutions to all the exercises in the text. The methods used to solve the problems in the manual are the same as those used to solve the examples in the textbook.

Online Learning Center

Web-based interactive learning is available for your students on the Online Learning Center, located at www.mhhe.com/hutchison. Student resources are located in the **Student Edition** and include interactive applications, algorithmically generated practice exams and exercises, audiovisual tutorials, and web links. Instructor resources are located in the **Instructor's Edition** and include links to PageOut, ALEKS, NetTutor, and other recommended sites.

NetTutor

NetTutor is a revolutionary system that enables students to interact with a live tutor over the World Wide Web. Students can receive instruction from live tutors using NetTutor's Web-based, graphical chat capabilities. They can also submit questions and receive answers, browse previously answered questions, and view previous live chat sessions. NetTutor is available for free through this text's Online Learning Center.

PageOut

PageOut is McGraw-Hill's unique point-and-click course website tool, which enables you to create a full-featured, professional-quality course website without knowing

HTML coding. With PageOut you can post your course syllabus, assign McGraw-Hill Online Learning Center content, add links to important off-site resources, and maintain student results in the online grade book. You can send class announcements, copy your course site to share with colleagues, and upload original files. PageOut is free for every McGraw-Hill user, and if you're short on time, we even have a team ready to help you create your site!

ALEKS

ALEKS (Assessment and LEarning in Knowledge Spaces) is an artificial intelligence-based system for individualized math learning, available over the World Wide Web. ALEKS delivers precise, qualitative diagnostic assessments of students' math knowledge, guides them in the selection of appropriate new study material, and records their progress toward mastery of curricular goals in a robust classroom management system. See page xxvii for more details regarding ALEKS.

Supplements for the Student

Student's Solutions Manual

Prepared by Stefan Baratto, this manual provides complete worked-out solutions to all the odd-numbered section exercises from the text. The procedures followed in the Solutions Manual match those shown in worked examples in the text.

SMART Tutorial CD-ROM

This interactive CD-ROM is a self-paced tutorial linked directly to the text that reinforces topics through unlimited opportunities to review concepts and practice problem solving. The CD-ROM provides algorithmically generated, bookmarkable practice exercises (including hints), section-level and chapter-level testing with grade book capabilities, and a built-in graphing calculator. This product requires virtually no computer training on the part of students and is supported by Windows and Macintosh systems.

Video Series

The video series is composed of 14 videocassettes (one for each chapter of the text). An on-screen instructor introduces topics and works through examples, using the methods presented in the text. The video series is also available on video CD-ROMs.

Online Learning Center

Student resources are located in the **Student Edition** of the Online Learning Center (OLC) and include interactive applications; algorithmically generated, bookmarkable practice exercises (including hints); section-level and chapter-level testing; a built-in graphing calculator; a glossary; audiovisual tutorials; and links to PageOut, NetTutor, and other fun and useful algebra websites. The OLC is located at www.mhhe.com/hutchison, and a free password card is included with each new copy of this text.

NetTutor

NetTutor is a revolutionary system that enables students to interact with a live tutor over the World Wide Web. Students can receive instruction from live tutors using NetTutor's Web-based, graphical chat capabilities. They can also submit questions and receive answers, browse previously answered questions, and view previous live chat sessions.

Acknowledgments

Throughout the writing process, the list of contributors to this text has grown steadily. Foremost are the reviewers. Every person on this list has helped by questioning, correcting, or complimenting. Our thanks to each of them.

Reviewers of the second edition:

Mary Black, Illinois Valley Community College
Julie A. Brown, Yuba College
Bonni Buchmeier, University of Minnesota–Crookston
Reneé M. Cooper, Arizona Western College
Michael Dixon, Midland College
Jeanette G. Eggert, Concordia University
Susan Forman, Bronx Community College
Deborah D. Fries, Wor Wic Community College
Andrea Hendricks, Georgia Perimeter College–Clarkston
Ming-Hang Yun Her, Georgia Perimeter College–Decatur
Richard Hobbs, Mission College
Kathryn Hodge, Midland College
Everett House, Nashville State Tech Community College
Stefanie H. Hunt, Towson University
Glenn Jablonski, Triton College
Christine Heinecke Lehmann, Purdue University–Westville
Robert A. Maynard, Tidewater Community College–Virginia Beach
Linda G. Mudge, Illinois Valley Community College
Deloria Nanze-Davis, University of Texas at Brownsville
Kathy C. Nickell, College of DuPage
Dr. Ellen O'Connell, Triton College
Thomas S. Pomykalski, Madison Area Technical College
José Rico, Laredo Community College
Dr. Jo Temple, Texas Tech University
Bella Zamansky, University of Cincinnati

Reviewers of the first edition:

Mary Kay Abbey, Montgomery College
Victor Akatsa, Chicago State University
Dr. Vivian Morgan Alley, Middle Tennessee State University
Dr. John Barker, Oklahoma City Community College
Cynthia Broughton, Arizona Western College
Linda K. Buchanan, Howard College
Walter E. Burlage, Lake City Community College
Marc Campbell, Daytona Beach Community College
Dora S. Cantu, Laredo Community College
Connie Carruthers, Scottsdale Community College
Tim Chappell, Penn Valley Community College
John W. Coburn, St. Louis Community College at Florissant Valley

Patrick S. Cross, University of Oklahoma

Walter Czarnec, Framingham State College

Antonio David, Del Mar College

Yolanda F. Davis, Central Texas College

Andrea DeCosmo, Waubonsee Community College

Dr. Said Fariabi, University of Texas at San Antonio

Joan Finney, Dyersburg State Community College

Barb Gentry, Parkland College

David Graser, Yavapai College

Roberta Grenz, Community College of Southern Nevada

Garry Hart, California State University–Dominquez Hills

Margret M. Hathaway, Kansas City, Kansas, Community College

Alan T. Hayashi, Oxnard College

Dr. Jeannie Hollar, Lenoir-Rhyne College

Jefferson A. Humphries, San Antonio College

Linda Hunt, Marshall University

Marvin Johnson, College of Lake County

Maryann Justinger, Erie Community College

Eric Paul Kraus, Sinclair Community College

Ann Loving, J. Sargeant Reynolds College

Marva Lucas, Middle Tennessee State University

David P. MacAdam, Cape Cod Community College

Anthony P. Malone, Raymond Walters College

Sarah Martin, Virginia Western Community College

Jeff Mock, Diablo Valley College

Feridoon Moinian, Cameron University

Ann C. Mugavero, College of Staten Island

Carol Murphy, San Diego Miramar College

Linda J. Murphy, Northern Essex Community College

Pinder Naidu, Kennesaw State University

Barbara Napoli, Our Lady of the Lake College

Nancy K. Nickerson, Northern Essex Community College

Sergei Ovchinnikov, San Francisco State University

Dena Perkins, Oklahoma Christian University

Evelyn A. Puaa, Hawaii Pacific University

Dr. Atma Sahu, Coppin State College

Ellen Sawyer, College of DuPage

Martha W. Scarbrough, Motlow State Community College

Wade Sick, Southwestern Community College

Helen Smith, South Mountain Community College

Andrea Spratt, University of Maryland–Baltimore County

Eleanor Storey, Front Range Community College

Nancy Szen, Northern Essex Community College

Lana Taylor, Siena Heights College

Jo Anne Temple, Texas Tech University

Jane H. Theiling, Dyersburg State Community College

John B. Thoo, Yuba College

Victoria C. Wacek, Missouri Western College

Pansy Waycaster, Southwest Virginia Community College

Danny Whited, Virginia Intermont College

Mary Jane Wolfe, University of Rio Grande
Judith B. Wood, Central Florida Community College
Karl Zilm, Lewis and Clark Community College

Selecting the excellent list of reviewers was only one of the many contributions of our McGraw-Hill editors, David Dietz and Peter Galuardi. Thanks must go to everybody at McGraw-Hill with whom we worked. Their professional, caring attitude was always appreciated. We are especially appreciative for the keen eyes and good sense of Susan Brusch.

In this age of the high-technology mathematics lab, the supplements and the supplement authors have become as important as the text itself. We are proud that our names appear on the supplements together with these authors.

ALEKS is an artificial intelligence-based system for individualized math learning, available for Higher Education from McGraw-Hill over the World Wide Web.

ALEKS delivers precise assessments of math knowledge, guides the student in the selection of appropriate new study material, and records student progress toward mastery of goals.

ALEKS interacts with a student much as a skilled human tutor would, moving between explanation and practice as needed, correcting and analyzing errors, defining terms and changing topics on request. By accurately assessing a student's knowledge, ALEKS can focus clearly on what the student is ready to learn next, helping to master the course content more quickly and easily.

ALEKS is:

- **A comprehensive course management system**. It tells the instructor exactly what students know and don't know.
- **Artificial intelligence.** It totally individualizes assessment and learning.
- **Customizable.** ALEKS can be set to cover the material in your course.
- **Web-based.** It uses a standard browser for easy Internet access.
- **Inexpensive.** There are no setup fees or site license fees.

ALEKS 2.0 adds the following new features:

- **Automatic Textbook Integration**
- **New Instructor Module**
- **Instructor-Created Quizzes**
- **New Message Center**

ALEKS maintains the features that have made it so popular including:

- **Web-Based Delivery** No complicated network or lab setup
- **Immediate Feedback** for students in learning mode
- **Integrated Tracking of Student Progress and Activity**
- **Individualized Instruction** which gives students problems they are *Ready to Learn*

For more information please contact your McGraw-Hill Sales Representative or visit ALEKS at http://www.highedmath.aleks.com.

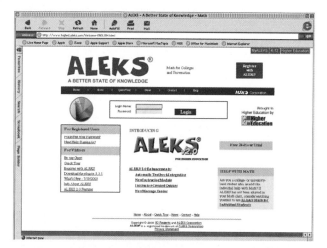

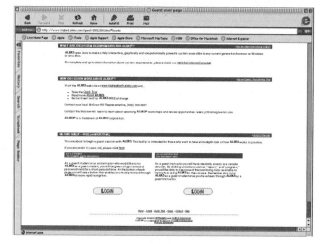

0 Prealgebra Review

Anthropologists and archaeologists sometimes investigate cultures that existed so long ago that their characteristics must be inferred from buried objects. When some interesting object is found, often the first question is "How old is this?" With methods such as carbon dating, it has been established that large, organized cultures existed around 3000 B.C.E. in Egypt, 2800 B.C.E. in India, no later than 1500 B.C.E. in China, and around 1000 B.C.E. in the Americas.

How long ago was 1500 B.C.E.? Using the Christian notation for dates, we have to count C.E. years and B.C.E. years differently. An object from 500 C.E. is $2000 = 500$ years old, or about 1500 years old. But an object from 1500 B.C.E. is $2000 + 1500$ years old, or about 3500 years old. Note that zero is used on this line only as a reference point. There was no year zero.

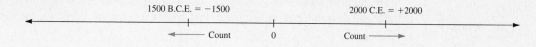

1500 B.C.E. = −1500   2000 C.E. = +2000

← Count   0   Count →

A Review of Fractions

1. Simplify a fraction
2. Multiply or divide two fractions
3. Add or subtract two fractions

Overcoming Math Anxiety

Throughout this text, we will present you with a series of class-tested techniques that are designed to improve your performance in this math class.

Hint 1: Become familiar with your textbook.

Perform each of the following tasks.

1. Use the Table of Contents to find the title of Section 5.1.

2. Use the index to find the earliest reference to the term *factor*.

3. Find the answer to the first Check Yourself exercise in Section 0.1.

4. Find the answers to the Self-Test for Chapter 1.

5. Find the answers to the odd-numbered exercises in Section 0.1.

Now you know where some of the most important features of the text are. When you have a moment of confusion, think about using one of these features to help you clear up that confusion.

This chapter provides a review of the basic operations—addition, subtraction, division, and multiplication—on signed numbers.

The numbers used for counting are called the **natural numbers.** We write them as 1, 2, 3, 4, ... The three dots indicate that the pattern continues in the same way.

If we include zero in this group of numbers, we then call them the **whole numbers.**

The **rational numbers** consist of all the whole numbers and all fractions, whether they are proper fractions such as $\frac{1}{2}$ and $\frac{2}{3}$ or improper fractions such as $\frac{7}{2}$ or $\frac{19}{5}$.

Every rational number can be written in fraction form $\frac{a}{b}$.

The number 1 has many different fractional forms. Any fraction in which the numerator and denominator are the same (and not zero) is another name for the number 1.

$$1 = \frac{2}{2} \qquad 1 = \frac{12}{12} \qquad 1 = \frac{257}{257}$$

The fundamental principle of fractions states that multiplying the numerator and denominator of a fraction by the same number is the same as multiplying the fraction by 1. We express the principle in symbols here.

The Fundamental Principle of Fractions

$$\frac{a}{b} = \frac{a \times c}{b \times c} \quad \text{or} \quad \frac{a \times c}{b \times c} = \frac{a}{b}$$

Example 1	**Rewriting Fractions**

Use the fundamental principle to write three fractional representations for each number.

Each representation is a numeral, or name, for the number. Each number has many names.

(a) $\dfrac{2}{3}$

Multiplying the numerator and denominator by the same number is the same as multiplying by 1.

Numerator and denominator are multiplied by 2.

$$\frac{2}{3} = \frac{2 \times 2}{3 \times 2} = \frac{4}{6}$$

Numerator and denominator are multiplied by 3.

$$\frac{2}{3} = \frac{2 \times 3}{3 \times 3} = \frac{6}{9}$$

Numerator and denominator are multiplied by 10.

$$\frac{2}{3} = \frac{2 \times 10}{3 \times 10} = \frac{20}{30}$$

(b) 5

$$5 = \frac{5 \times 2}{1 \times 2} = \frac{10}{2}$$

$$5 = \frac{5 \times 3}{1 \times 3} = \frac{15}{3}$$

$$5 = \frac{5 \times 100}{1 \times 100} = \frac{500}{100}$$

✔ *CHECK YOURSELF 1*

Use the fundamental principle to write three fractional representations for each number.

(a) $\dfrac{5}{8}$ (b) $\dfrac{4}{3}$ (c) 3

The simplest fractional representation for a number has the smallest numerator and denominator. Fractions written in this form are said to be **simplified.**

| Example 2 | **Simplifying Fractions** |

Use the fundamental principle to simplify each fraction.

(a) $\dfrac{22}{55}$ (b) $\dfrac{35}{45}$ (c) $\dfrac{24}{36}$

A prime number is any whole number greater than 1 that has only itself and 1 as factors.

In each case, we first write the numerator and denominator as a product of prime numbers.

(a) $\dfrac{22}{55} = \dfrac{2 \times 11}{5 \times 11}$

We then use the fundamental principle of fractions to "remove" the common factor of 11.

$$\dfrac{22}{55} = \dfrac{2 \times 11}{5 \times 11} = \dfrac{2}{5}$$

(b) $\dfrac{35}{45} = \dfrac{5 \times 7}{3 \times 3 \times 5}$

Removing the common factor of 5 yields

$$\dfrac{35}{45} = \dfrac{7}{3 \times 3} = \dfrac{7}{9}$$

(c) $\dfrac{24}{36} = \dfrac{2 \times 2 \times 2 \times 3}{2 \times 2 \times 3 \times 3}$

Removing the common factor $2 \times 2 \times 3$ yields

$$\dfrac{2}{3}$$

  CHECK YOURSELF 2

Use the fundamental principle to simplify each fraction.

(a) $\dfrac{21}{33}$ (b) $\dfrac{15}{30}$ (c) $\dfrac{12}{54}$

When multiplying fractions, we use the property

$$\dfrac{a}{b} \times \dfrac{c}{d} = \dfrac{a \times c}{b \times d}$$

We then write the numerator and denominator in factored form and simplify before multiplying.

Example 3

Multiplying Fractions

Find the product of the two fractions.

$$\frac{9}{2} \times \frac{4}{3}$$

A product is the result of multiplication.

$$\frac{9}{2} \times \frac{4}{3} = \frac{9 \times 4}{2 \times 3}$$

$$= \frac{3 \times 3 \times 2 \times 2}{2 \times 3}$$

$$= \frac{3 \times 2}{1}$$

$$= \frac{6}{1} \qquad \text{The denominator of 1 is not necessary.}$$

$$= 6$$

✔ *CHECK YOURSELF 3*

Multiply and simplify each pair of fractions.

(a) $\dfrac{3}{5} \times \dfrac{10}{7}$ (b) $\dfrac{12}{5} \times \dfrac{10}{6}$

To divide two fractions, the divisor is inverted, then the fractions are multiplied. The following property shows this process:

$$\frac{a}{b} \div \frac{c}{d} = \frac{a}{b} \times \frac{d}{c} = \frac{a \times d}{b \times c}$$

Example 4

Dividing Fractions

Find the quotient of the two fractions.

A quotient is the result of division.

$$\frac{7}{3} \div \frac{5}{6}$$

The divisor $\dfrac{5}{6}$ is inverted and becomes $\dfrac{6}{5}$.

$$\frac{7}{3} \div \frac{5}{6} = \frac{7}{3} \times \frac{6}{5} = \frac{7 \times 6}{3 \times 5}$$

The common factor of 3 is removed from the numerator and denominator.

$$= \frac{7 \times 2 \times 3}{3 \times 5} = \frac{7 \times 2}{5}$$

$$= \frac{14}{5}$$

CHECK YOURSELF 4

Find the quotient of the two fractions.

$$\frac{9}{2} \div \frac{3}{5}$$

When adding two fractions, we will need to find the **least common denominator (LCD)** first. The least common denominator is the smallest number that both denominators evenly divide. The process of finding the LCD is outlined below.

To Find the Least Common Denominator

Step 1.   Write the prime factorization for each of the denominators.

Step 2.   Find all the prime factors that appear in any one of the prime factorizations.

Step 3.   Form the product of those prime factors, using each factor the greatest number of times it occurs in any one factorization.

Example 5   Finding the Least Common Denominator (LCD)

Find the LCD of fractions with denominators 6 and 8.

Our first step in adding fractions with denominators 6 and 8 is to determine the least common denominator. Factor 6 and 8.

$$6 = 2 \times 3$$
$$8 = 2 \times 2 \times 2$$

Note that since 2 appears 3 times as a factor of 8, it is used 3 times in writing the LCD.

The LCD is $2 \times 2 \times 2 \times 3$, or 24.

  CHECK YOURSELF 5

Find the LCD of fractions with denominators 9 and 12.

The process is similar if more than two denominators are involved.

| **Example 6** | **Finding the Least Common Denominator** |

Find the LCD of fractions with denominators 6, 9, and 15.

To add fractions with denominators 6, 9, and 15, we need to find the LCD. Factor the three numbers.

$$6 = 2 \times 3 \qquad \text{2 and 5 appear only once in any one factorization.}$$

$$9 = 3 \times 3 \qquad \text{3 appears twice as a factor of 9.}$$

$$15 = 3 \times 5$$

The LCD is $2 \times 3 \times 3 \times 5$, or 90.

 CHECK YOURSELF 6

Find the LCD of fractions with denominators 5, 8, and 20.

To add two fractions, we will use the following property:

$$\frac{a}{b} + \frac{c}{b} = \frac{a + c}{b}$$

| **Example 7** | **Adding Fractions** |

Find the sum of the two fractions.

A sum is the result of addition.

$$\frac{5}{8} + \frac{7}{12}$$

The LCD of 8 and 12 is 24. Each fraction should be rewritten as a fraction with that denominator.

$$\frac{5}{8} = \frac{15}{24} \qquad\qquad \text{Multiply the numerator and denominator by 3.}$$

$$\frac{7}{12} = \frac{14}{24} \qquad\qquad \text{Multiply the numerator and denominator by 2.}$$

$$\frac{5}{8} + \frac{7}{12} = \frac{15}{24} + \frac{14}{24} = \frac{15 + 14}{24} = \frac{29}{24} \qquad \text{This fraction cannot be simplified.}$$

✔ CHECK YOURSELF 7

Find the sum for each fraction.

(a) $\dfrac{4}{5} + \dfrac{7}{9}$ (b) $\dfrac{5}{6} + \dfrac{4}{15}$

To subtract two fractions, the following rule is used:

$$\frac{a}{b} - \frac{c}{b} = \frac{a - c}{b}$$

Subtracting fractions is treated exactly like adding them, except the numerator becomes the difference of two numerators.

| Example 8 | **Subtracting Fractions** |

Find the difference.

The difference is the result of subtraction.

$$\frac{7}{9} - \frac{1}{6}$$

The LCD is 18. We rewrite the fractions with that denominator.

$$\frac{7}{9} = \frac{14}{18}$$

$$\frac{1}{6} = \frac{3}{18}$$

$$\frac{7}{9} - \frac{1}{6} = \frac{14}{18} - \frac{3}{18} = \frac{14 - 3}{18} = \frac{11}{18}$$ This fraction cannot be simplified.

✔ CHECK YOURSELF 8

Find the difference $\dfrac{11}{12} - \dfrac{5}{8}$.

 CHECK YOURSELF ANSWERS

1. Answers will vary.   **2.** (a) $\dfrac{7}{11}$; (b) $\dfrac{1}{2}$; (c) $\dfrac{2}{9}$ **3.** (a) $\dfrac{6}{7}$; (b) 4   **4.** $\dfrac{15}{2}$

5. 36   **6.** 40   **7.** (a) $\dfrac{71}{45}$; (b) $\dfrac{11}{10}$ **8.** $\dfrac{7}{24}$

Exercises ▪ 0.1

Use the fundamental principle of fractions to write three fractional representations for each number.

1. $\dfrac{3}{7}$

2. $\dfrac{2}{5}$

3. $\dfrac{4}{9}$

4. $\dfrac{7}{8}$

5. $\dfrac{5}{6}$

6. $\dfrac{11}{13}$

7. $\dfrac{10}{17}$

8. $\dfrac{2}{7}$

9. $\dfrac{9}{16}$

10. $\dfrac{6}{11}$

11. $\dfrac{7}{9}$

12. $\dfrac{15}{16}$

Use the fundamental principle of fractions to write each fraction in simplest form.

13. $\dfrac{10}{15}$

14. $\dfrac{12}{15}$

15. $\dfrac{10}{14}$

16. $\dfrac{18}{60}$

17. $\dfrac{12}{18}$

18. $\dfrac{28}{35}$

19. $\dfrac{35}{40}$

20. $\dfrac{28}{32}$

21. $\dfrac{11}{44}$

22. $\dfrac{10}{25}$

23. $\dfrac{11}{33}$

24. $\dfrac{18}{48}$

25. $\dfrac{24}{27}$

26. $\dfrac{27}{45}$

27. $\dfrac{32}{40}$

28. $\dfrac{17}{51}$

29. $\dfrac{75}{105}$

30. $\dfrac{62}{93}$

31. $\dfrac{24}{30}$

32. $\dfrac{48}{66}$

33. $\dfrac{105}{135}$

34. $\dfrac{39}{91}$

Multiply. Be sure to simplify each product.

35. $\dfrac{3}{7} \times \dfrac{4}{5}$

36. $\dfrac{2}{7} \times \dfrac{5}{9}$

37. $\dfrac{3}{4} \times \dfrac{7}{5}$

38. $\dfrac{3}{5} \times \dfrac{2}{7}$

39. $\dfrac{3}{5} \times \dfrac{5}{7}$

40. $\dfrac{6}{11} \times \dfrac{8}{6}$

41. $\dfrac{6}{13} \times \dfrac{4}{9}$

42. $\dfrac{5}{9} \times \dfrac{6}{11}$

43. $\dfrac{3}{11} \times \dfrac{7}{9}$

44. $\dfrac{7}{9} \times \dfrac{3}{5}$

45. $\dfrac{4}{21} \times \dfrac{7}{12}$

46. $\dfrac{5}{21} \times \dfrac{14}{25}$

Divide. Write each result in simplest form.

47. $\dfrac{1}{7} \div \dfrac{3}{5}$

48. $\dfrac{2}{5} \div \dfrac{1}{3}$

49. $\dfrac{2}{5} \div \dfrac{3}{4}$

50. $\dfrac{5}{8} \div \dfrac{3}{4}$

51. $\dfrac{8}{9} \div \dfrac{4}{3}$

52. $\dfrac{4}{7} \div \dfrac{6}{11}$

53. $\dfrac{7}{10} \div \dfrac{5}{9}$

54. $\dfrac{8}{9} \div \dfrac{11}{15}$

55. $\dfrac{8}{15} \div \dfrac{2}{5}$

56. $\dfrac{5}{27} \div \dfrac{15}{54}$

57. $\dfrac{8}{21} \div \dfrac{24}{35}$

58. $\dfrac{9}{28} \div \dfrac{27}{35}$

Find the least common denominator (LCD) for fractions with the given denominators.

59. 30 and 50

60. 36 and 48

61. 48 and 80

62. 60 and 84

63. 3, 4, and 5

64. 3, 4, and 6

65. 8, 10, and 15

66. 6, 22, and 33

67. 5, 10, and 25

68. 8, 24, and 48

Add. Write each result in simplest form.

69. $\dfrac{2}{5} + \dfrac{1}{4}$

70. $\dfrac{2}{3} + \dfrac{3}{10}$

71. $\dfrac{2}{5} + \dfrac{7}{15}$

72. $\dfrac{2}{3} + \dfrac{4}{5}$

73. $\dfrac{3}{8} + \dfrac{5}{12}$

74. $\dfrac{5}{36} + \dfrac{7}{24}$

75. $\dfrac{7}{30} + \dfrac{5}{18}$

76. $\dfrac{9}{14} + \dfrac{10}{21}$

77. $\dfrac{7}{15} + \dfrac{13}{18}$

78. $\dfrac{12}{25} + \dfrac{19}{30}$

79. $\dfrac{1}{5} + \dfrac{1}{10} + \dfrac{1}{15}$

80. $\dfrac{1}{3} + \dfrac{1}{5} + \dfrac{1}{10}$

Subtract. Write each result in simplest form.

81. $\dfrac{8}{9} - \dfrac{3}{9}$

82. $\dfrac{9}{10} - \dfrac{6}{10}$

83. $\dfrac{6}{7} - \dfrac{2}{7}$

84. $\dfrac{11}{12} - \dfrac{7}{12}$

85. $\dfrac{7}{8} - \dfrac{2}{3}$

86. $\dfrac{4}{9} - \dfrac{2}{5}$

87. $\dfrac{11}{18} - \dfrac{2}{9}$

88. $\dfrac{5}{6} - \dfrac{1}{4}$

89. $\dfrac{2}{3} - \dfrac{7}{11}$

90. $\dfrac{13}{18} - \dfrac{5}{12}$

91. $\dfrac{5}{42} - \dfrac{1}{36}$

92. $\dfrac{13}{18} - \dfrac{7}{15}$

93. Baking. If a recipe calls for $\frac{1}{3}$ cup of white flour, $\frac{1}{3}$ cup of wheat flour, and $\frac{1}{2}$ cup of soy flour, how much flour is in the pancakes?

94. Salary. Deductions from your paycheck are made roughly as follows: $\frac{1}{8}$ for federal tax, $\frac{1}{20}$ for state tax, $\frac{1}{20}$ for social security, and $\frac{1}{40}$ for a savings withholding plan. What portion of your pay is deducted?

95. Distance. Jose walked $\frac{3}{4}$ mile (mi) to the store, $\frac{1}{2}$ mi to a friend's house, and then $\frac{2}{3}$ mi home. How far did he walk?

96. Perimeter. Find the perimeter of, or the distance around, the accompanying figure.

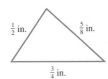

97. Cooking. A hamburger that weighed $\frac{1}{4}$ pound (lb) before cooking weighed $\frac{3}{16}$ lb after cooking. How much weight was lost in cooking?

98. Baking. Geraldo has $\frac{3}{4}$ cup of flour. Biscuits use $\frac{5}{8}$ cup. Will he have enough left over for a small pie crust that requires $\frac{1}{4}$ cup? Explain.

0.2 Real Numbers

 OBJECTIVES

1. Identify integers
2. Place integers on a number line
3. Find the opposite of a number
4. Find the absolute value of a number

In arithmetic, you learned to solve problems that involved working with numbers. In algebra, you will learn to use tools that will help you solve many new types of problems. The first tool we will provide you with involves the expansion of the numbers with which you do computation. Let us look at some important sets of numbers.

The **natural numbers** are all the counting numbers 1, 2, 3, ...

The **whole numbers** are the natural numbers together with zero.

We can represent whole numbers on a **number line.** Here is the number line.

And here is the number line with the whole numbers 0, 1, 2, and 3 plotted.

Now suppose you want to represent a temperature of 10 degrees below zero, a debt of $50, or an altitude 100 feet below sea level. These situations require a new set of numbers called *negative numbers*. We will expand the number line to include negative numbers.

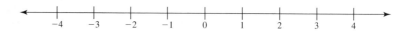

Numbers to the right of (greater than) 0 on the number line are called **positive numbers.** Numbers to the left of (less than) 0 are called **negative numbers.** Zero is neither positive nor negative.

To indicate a negative number, we use a minus sign (−) in front of the number. Positive numbers may be written with a plus sign (+) or with no sign at all, in which case the number is understood to be positive.

Since −3 is to the left of 0, it is a negative number. Read −3 as "negative three."

12

| Example 1 | ## Identifying Real Numbers |

$+6$ is a positive number.

-9 is a negative number.

If no sign appears, a number (other than 0) is positive.

5 is a positive number.

0 is neither positive nor negative.

✔ CHECK YOURSELF 1

Label each of the following as positive, negative, or neither.

(a) $+3$ (b) 7   (c) -5 (d) 0

The natural numbers, 0, and the negatives of natural numbers make up the set of integers.

The Set of Integers
The **set of integers** consists of the natural numbers, their negatives, and 0.

Here we have a graphical representation of the set of integers.

The arrowheads indicate that the number line extends forever in both directions.

Note that the integers occur at the hash marks on the number line. Any plotted point that would fall on one of the circles on the number line above is an integer. This will be true no matter how far in either direction we extend our number line.

| Example 2 | ## Identifying Integers |

Which of the following are integers?

$$-3, 5.3, \frac{2}{3}, 4$$

Of these four numbers, only -3 and 4 are integers.

✔ CHECK YOURSELF 2

Which of the following are integers?

$$7, 0, \frac{4}{7}, -5, 0.2$$

Any number that can be written as the ratio of two integers is called a **rational number.** Examples of rational numbers are $6, \frac{7}{3}, -\frac{15}{4}, 0, \frac{4}{1}$. On the number line, you can estimate the location of a rational number, as Example 3 illustrates.

| Example 3 | **Plotting Rational Numbers** |

Plot each of the following rational numbers on the number line provided.

Note that a decimal is really a "decimal fraction." So −1.445 is another way of writing

$$\frac{-1445}{1000}$$

$$\frac{2}{3}, -3\frac{1}{4}, \frac{27}{5}, -1.445$$

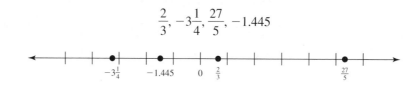

✔ *CHECK YOURSELF 3*

Plot each of the following rational numbers on the number line provided.

$$-2\frac{1}{3}, \frac{37}{11}, 5.66, -\frac{1}{4}$$

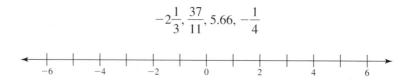

Not every number can be written as the ratio of two integers. Such numbers are called **irrational numbers.** Some examples of irrational numbers are $\sqrt{3}, \sqrt{7}$, and π. We will say more about this type of number later in the text. This diagram illustrates the relationships among the various sets of numbers

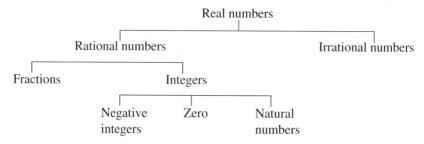

An important idea in our work with real numbers is the *opposite* of a number. Every number has an opposite.

Opposite of a Number

The **opposite** of a number corresponds to a point the same distance from 0 as the given number, but in the opposite direction.

| **Example 4** | **Writing the Opposite of a Real Number** |

The opposite of a *positive* number is *negative*.

(a) The opposite of 5 is -5.

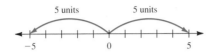

Both numbers are located 5 units from 0.

The opposite of a *negative* number is *positive*.

(b) The opposite of -3 is 3.

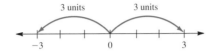

Both numbers correspond to points that are 3 units from 0.

✔ *CHECK YOURSELF 4*

(a) What is the opposite of 8?   (b) What is the opposite of -9?

To represent the opposite of a number, place a minus sign in front of the number.

We write the opposite of 5 as -5. You can now think of -5 in two ways: as negative 5 and as the opposite of 5.

Using the same idea, we can write the opposite of a negative number. The opposite of -3 is $-(-3)$. Since we know from looking at the number line that the opposite of -3 is 3, this means that

Again place a minus sign in front of the number to represent the opposite of that number.

$$-(-3) = 3$$

So the opposite of a negative number must be positive.

Let's summarize our results:

> **The Opposite of a Real Number**
>
> 1. The opposite of a positive number is negative.
>
> 2. The opposite of a negative number is positive.
>
> 3. The opposite of 0 is 0.

We also need to define the *absolute value,* or magnitude, of a real number.

> **Absolute Value**
>
> The **absolute value** of a real number is the distance (on the number line) between the number and 0.

Example 5

Finding the Absolute Value

(a) The absolute value of 5 is 5.

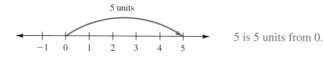

5 is 5 units from 0.

(b) The absolute value of -5 is 5.

-5 is also 5 units from 0.

We usually write the absolute value of a number by placing vertical bars before and after the number. We can write

|5| is read "the absolute value of 5."

$$|5| = 5 \qquad \text{and} \qquad |-5| = 5$$

✔ CHECK YOURSELF 5

Complete the following statements.

(a) The absolute value of 9 is _____.

(b) The absolute value of -12 is _____.

(c) $|-6| =$ (d) $|15| =$

✔ CHECK YOURSELF ANSWERS

1. (a) Positive; (b) positive; (c) negative; (d) neither   **2.** 7, 0, -5

3. ◄─┼─┼─┼●┼─┼●┼─┼─┼●┼─┼●┼─►
        $-2\frac{1}{3}$ $-\frac{1}{4}\,0$ $\frac{37}{11}$ 5.66   **4.** (a) -8; (b) 9

5. (a) 9; (b) 12; (c) 6; (d) 15

Exercises · 0.2

Indicate whether the following statements are true or false.

1. The opposite of -7 is 7.

2. The opposite of -10 is -10.

3. -9 is an integer.

4. 5 is an integer.

5. The opposite of -11 is 11.

6. The absolute value of -5 is 5.

7. $|-6| = -6$

8. $-(-30) = -30$

9. -12 is not an integer.

10. The opposite of -18 is 18.

11. $|-7| = -7$

12. The absolute value of -9 is -9.

13. $-(-8) = 8$

14. $\dfrac{2}{3}$ is not an integer.

15. $-|-15| = -15$

16. The absolute value of -3 is 3.

17. $\dfrac{3}{5}$ is an integer.

18. 0.7 is not an integer.

19. 0.15 is not an integer.

20. $|-9| = -9$

21. $\dfrac{5}{7}$ is not an integer.

22. 0.23 is not an integer.

23. $-(-7) = -7$

24. The opposite of 15 is -15.

Complete each of the following statements.

25. The absolute value of -10 is _____.

26. $-(-12) = $ _____

27. $|-20| = $ _____

28. The absolute value of -12 is _____.

29. The absolute value of -7 is _____.

30. The opposite of -9 is _____.

31. The opposite of 30 is _____.

32. $-|-15| = $ _____

33. $-(-6) = $ _____

34. The absolute value of 0 is _____.

35. $|50| = $ _____

36. The opposite of 18 is _____.

37. The absolute value of the opposite of 3 is _____.

38. The opposite of the absolute value of 3 is _____.

39. The opposite of the absolute value of -7 is _____.

40. The absolute value of the opposite of -7 is _____.

Complete each of the following statements, using the symbol < (less than), = (equal to), or > (greater than).

41. -5 _____ -9 **42.** -15 _____ -10 **43.** -20 _____ -10 **44.** -15 _____ -14

45. $|3|$ _____ 3 **46.** $|-5|$ _____ $-(-5)$ **47.** -4 _____ $|-4|$ **48.** 7 _____ $|7|$

For Exercises 49 to 52, use the following numbers: $-3, \dfrac{2}{3}, -1.5, 2,$ and 0.

49. Which of the numbers are integers?   **50.** Which of the numbers are natural numbers?

51. Which of the numbers are whole numbers?   **52.** Which of the numbers are negative numbers?

For Exercises 53 to 56, use the following numbers: $-2, -\dfrac{4}{3}, 3.5, 0,$ and 1.

53. Which of the numbers are integers?   **54.** Which of the numbers are natural numbers?

55. Which of the numbers are whole numbers?   **56.** Which of the numbers are negative numbers?

57. (a) Every number has an opposite. The opposite of 5 is -5. In English, a similar situation exists for words. For example, the opposite of *regular* is *irregular*. Write the opposite of these words:

irredeemable, uncomfortable, uninteresting, uninformed, irrelevant, immoral.

(b) Note that the idea of an opposite is usually expressed by a prefix such as *un-* or *ir-*. What other prefixes can be used to negate or change the meaning of a word to its opposite? List four words using these prefixes, and use the words in a sentence.

58. (a) What is the difference between positive integers and nonnegative integers?

(b) What is the difference between negative and nonpositive integers?

59. Simplify each of the following:

(a) $-(-3)$ (b) $-(-(-3))$ (c) $-(-(-(-3)))$

(d) Use the results of (a), (b), and (c) to create a rule for simplifying expressions of this type.

(e) Use the rule created in (d) to simplify $-(-(-(-(-(-(-7))))))$.

SECTION 0.3 Adding and Subtracting Real Numbers

0.3 OBJECTIVES

1. Add real numbers
2. Use the commutative property of addition
3. Use the associative property of addition
4. Subtract real numbers

"Imagination is more important than knowledge."

–Albert Einstein

"Don't play stupid with me. I'm better at it than you are."

–Colonel Flagg

The number line can be used to demonstrate the sum of two real numbers. To add a positive number, we will move to the right; to add a negative number, we will move to the left.

Example 1

Finding the Sum of Two Real Numbers

Find the sum $5 + (-2)$.

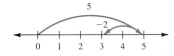

Since the first addend is positive, we start by moving right.

First move 5 units *to the right* of 0. Then, to add -2, move 2 units back *to the left*. We see that

$$5 + (-2) = 3$$

✔ CHECK YOURSELF 1

Find the sum.

$$9 + (-7)$$

We can also use the number line to picture addition when two negative numbers are involved. Example 2 illustrates this approach.

Example 2	**Finding the Sum of Two Real Numbers**

Find the sum $-2 + (-3)$.

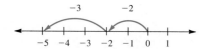

Move 2 units *to the left* of 0. Then move 3 more units *to the left* to add negative 3. We see that

$$-2 + (-3) = -5$$

 CHECK YOURSELF 2

Find the sum.

$$-7 + (-5)$$

You may have noticed some patterns in the previous examples. These patterns will let you do much of the addition mentally. Look at the following rule.

The sum of two positive numbers is positive, and the sum of two negative numbers is negative.

To Add Real Numbers

1. If two numbers have the same sign, add their absolute values. Give the sum the sign of the original numbers.

2. If two numbers have different signs, subtract the smaller absolute value from the larger. Attach the sign of the number with the larger absolute value to the result.

Example 3	**Finding the Sum of Two Real Numbers**

Find the sums.

(a) $5 + 2 = 7$ The sum of two positive numbers is positive.

(b) $-2 + (-6) = -8$ Add the absolute values of the two numbers ($2 + 6 = 8$). Give to the sum the sign of the original numbers.

 CHECK YOURSELF 3

Find the sums.

(a) $6 + 7$ (b) $-8 + (-7)$

There are three important pieces to the study of algebra. The first is the set of numbers, which we have discussed in this section. The second is the set of operations, such as addition and multiplication. The third is the set of rules, which we call **properties.** Example 4 enables us to look at an important property of addition.

Example 4   ## Finding the Sum of Two Real Numbers

Find the sums.

(a) $2 + (-7) = (-7) + 2 = -5$

(b) $-3 + (-4) = -4 + (-3) = -7$

Note that in both cases the order in which we add the numbers does not affect the sum. This leads us to the following property of real numbers.

> ### The Commutative Property of Addition
>
> The *order* in which we add two numbers does not change the sum. Addition is **commutative.** In symbols, for any numbers a and b,
>
> $$a + b = b + a$$

 CHECK YOURSELF 4

Find the sums $-8 + 2$ and $2 + (-8)$. How do the results compare?

What if we want to add more than two numbers? Another property of addition will be helpful. Look at Example 5.

Example 5   ## Finding the Sum of Three Real Numbers

Find the sum $2 + (-3) + (-4)$. First,

$$\underbrace{[2 + (-3)]} + (-4) \qquad \text{Add the first two numbers.}$$
$$= \quad -1 \quad + (-4) \qquad \text{Then add the third to that sum.}$$
$$= \quad -5$$

Now let's try a second approach.

$$2 + \underbrace{[(-3) + (-4)]} \qquad \text{This time, add the second and third numbers.}$$
$$= \quad 2 + \quad (-7) \qquad \text{Then add the first number to that sum.}$$
$$= -5$$

✔   *CHECK YOURSELF 5*

Show that $-2 + (-3 + 5) = [-2 + (-3)] + 5$

Do you see that it makes no difference which way we group numbers in addition? The final sum is not changed. We can state the following.

The Associative Property of Addition

The way we *group* numbers does not change the sum. Addition is **associative.** In symbols, for any numbers a, b, and c,

$$(a + b) + c = a + (b + c)$$

A number's opposite (or negative) is called its **additive inverse.** Use the following rule to add opposite numbers.

The Additive Inverse

The sum of any number and its additive inverse is 0. In symbols, for any number a,

$$a + (-a) = 0$$

| Example 6 | **Finding the Sum of Two Additive Inverses** |

Find the sums.

(a) $6 + (-6) = 0$

(b) $-8 + 8 = 0$

   CHECK YOURSELF 6

Find the sum.

$$9 + (-9)$$

So far we have looked only at the addition of integers. The process is the same if we want to add other types of real numbers.

| **Example 7** | **Finding the Sum of Two Real Numbers** |

Find the sums.

(a) $\dfrac{15}{4} + \left(-\dfrac{9}{4}\right) = \dfrac{6}{4} = \dfrac{3}{2}$ Subtract the absolute values $\dfrac{15}{4} - \dfrac{9}{4} = \dfrac{6}{4} = \dfrac{3}{2}$.

The sum is positive since $\dfrac{15}{4}$ has the larger absolute value.

(b) $-0.5 + (-0.2) = -0.7$ Add the absolute values $(0.5 + 0.2 = 0.7)$. The sum is negative.

  CHECK YOURSELF 7

Find the sums.

(a) $-\dfrac{5}{2} + \left(-\dfrac{7}{2}\right)$ (b) $5.3 + (-4.3)$

Now we turn our attention to the subtraction of real numbers. Subtraction is called the *inverse* operation to addition. This means that any subtraction problem can be written as a problem in addition. Let's see how it works with the following rule.

To find the difference $a - b$, we add a and the opposite of b.

> **To Subtract Real Numbers**
>
> To subtract real numbers, add the first number and the *opposite* of the number being subtracted. In symbols, by definition
>
> $$a - b = a + (-b)$$

Example 8 illustrates this property.

| **Example 8** | **Finding the Difference of Two Real Numbers** |

(a) Subtract $5 - 3$.

$$5 - 3 = 5 + (-3) = 2$$ To subtract 3, we can add the opposite of 3.

The opposite of 3

(b) Subtract $2 - 5$.

$$2 - 5 = 2 + (-5) = -3$$

The opposite of 5

By the definition, add the opposite of 4, -4, to the value -3.

(c) Subtract $-3 - 4$.

$$-3 - 4 = -3 + (-4) = -7$$ -4 is the opposite of 4.

(d) Subtract $-10 - 15$.

$$-10 - 15 = -10 + (-15) = -25 \qquad \text{−15 is the opposite of 15.}$$

 CHECK YOURSELF 8

Find each difference, using the definition of subtraction.

(a) $8 - 3$ (b) $7 - 9$ (c) $-5 - 9$ (d) $-12 - 6$

Real numbers other than integers are subtracted in exactly the same way, as Example 9 illustrates.

Example 9 | Finding the Difference of Two Fractions

Subtract.

$$-\frac{11}{4} - \left(-\frac{5}{4}\right) = -\frac{11}{4} + \frac{5}{4} = -\frac{6}{4} = -\frac{3}{2}$$

 CHECK YOURSELF 9

Find the difference.

$$\frac{47}{8} - \left(-\frac{19}{8}\right)$$

Caution

Your graphing calculator can be used to simplify the kinds of problems we've encountered in this section. The negation key is the $\boxed{(-)}$ or the $\boxed{+/-}$ found on the calculator. Do not confuse this with the subtraction key! The subtraction key is used between two numbers. The negation key is used in front of a number.

Example 10 | Using a Graphing Calculator to Subtract

Find each difference.

(a) $-4.567 - 3.92$

The keystrokes

$\boxed{(-)}$ 4.567 $\boxed{-}$ 3.92 $\boxed{\text{Enter}}$

yield -8.487.

(b) $-15.782 - (-19.4)$

The keystrokes

$\boxed{(-)}$ 15.782 $\boxed{-}$ $\boxed{(-)}$ 19.4 $\boxed{\text{Enter}}$

yield 3.618.

✔ *CHECK YOURSELF 10*

Find each difference.

(a) $-0.936 - 2.75$ (b) $-12.134 - (-1.656)$

✔ *CHECK YOURSELF ANSWERS*

1. 2   **2.** -12 **3.** (a) 13; (b) -15 **4.** $-6 = -6$ **5.** $0 = 0$ **6.** 0

7. (a) -6; (b) 1   **8.** (a) 5; (b) -2; (c) -14; (d) -18 **9.** $\dfrac{33}{4}$

10. (a) -3.686; (b) -10.478

Exercises · 0.3

Perform the indicated operation.

1. $-6 + (-5)$ **2.** $3 + 9$ **3.** $11 + (-7)$

4. $-6 + (-7)$ **5.** $4 + (-6)$ **6.** $9 + (-2)$

7. $7 + 9$ **8.** $-7 + 11$ **9.** $(-11) + 5$

10. $5 + (-8)$ **11.** $-8 + (-7)$ **12.** $8 + (-7)$

13. $-12 + 4$ **14.** $7 + (-7)$ **15.** $-9 + 10$

16. $-6 + 8$ **17.** $-4 + 4$ **18.** $5 + (-20)$

19. $7 + (-13)$ **20.** $0 + (-10)$ **21.** $-8 + 5$

22. $-7 + 3$ **23.** $6 + (-6)$ **24.** $-9 + 9$

25. $\dfrac{45}{16} - \dfrac{9}{16}$ **26.** $-\dfrac{35}{16} + \dfrac{17}{16}$ **27.** $\dfrac{29}{8} + \left(-\dfrac{17}{8}\right)$

28. $-\dfrac{81}{20} + \left(-\dfrac{107}{20}\right)$ **29.** $-\dfrac{73}{16} + \dfrac{119}{16}$ **30.** $-\dfrac{13}{8} - \left(-\dfrac{15}{4}\right)$

31. $4 + (-7) + (-5)$ **32.** $-7 + 8 + (-6)$ **33.** $-2 + (-6) + (-4)$

34. $12 + (-6) + (-4)$ **35.** $-3 + (-7) + 5 + (-2)$ **36.** $7 + (-8) + (-9) + 10$

37. $11 - 13$ **38.** $7 - 5$ **39.** $9 - 3$

40. $4 - 9$ **41.** $-8 - 3$ **42.** $-13 - 8$

43. $-12 - 8$ **44.** $9 - 15$ **45.** $-2 - (-3)$

46. $-9 - (-6)$ **47.** $-5 - (-5)$ **48.** $9 - (-7)$

49. $28 - (-22)$ **50.** $50 - (-25)$ **51.** $-15 - (-25)$

52. $-20 - (-30)$ **53.** $-25 - (-15)$ **54.** $-30 - (-20)$

55. $-(-20) - (-15)$ **56.** $18 - (-12)$ **57.** $48 - (-15)$

58. $-25 - (-30)$ **59.** $\dfrac{10}{2} - \left(-\dfrac{7}{2}\right)$ **60.** $-\dfrac{4}{2} - \dfrac{3}{2}$

26

61. $-\dfrac{7}{8} - \left(-\dfrac{19}{8}\right)$ **62.** $\dfrac{13}{4} - \left(-\dfrac{7}{4}\right)$ **63.** $-7 - (-5) - 6$

64. $-5 - (-8) - 10$ **65.** $-10 - 8 - (-7)$ **66.** $-5 - 8 - (-15)$

Find each difference using a graphing calculator.

67. $-11.392 - 13.491$ **68.** $-9.245 - 14.316$ **69.** $-7.259 - 4.235$

70. $-6.319 - 2.628$ **71.** $-18.271 - (-12.569)$ **72.** $-15.586 - (-9.874)$

73. $-17.346 - (-28.293)$ **74.** $-11.358 - (-23.145)$

Solve the following applications.

75. Temperature. The temperature in Chicago dropped from $18°F$ at 4 P.M. to $-9°F$ at midnight. What was the drop in temperature?

76. Banking. Charley's checking account had $175 deposited at the beginning of the month. After he wrote checks for the month, the account was $95 *overdrawn*. What amount of checks did he write during the month?

77. Elevator stops. Micki entered the elevator on the 34th floor. From that point the elevator went up 12 floors, down 27 floors, down 6 floors, and up 15 floors before she got off. On what floor did she get off the elevator?

78. Submarines. A submarine dives to a depth of 500 ft below the ocean's surface. It then dives another 217 ft before climbing 140 ft. What is the depth of the submarine?

79. Military vehicles. A helicopter is 600 ft above sea level, and a submarine directly below it is 325 ft below sea level. How far apart are they?

80. Bank balance. Tom has received an overdraft notice from the bank telling him that his account is overdrawn by $142. How much must he deposit in order to have $625 in his account?

81. Change in temperature. At 9:00 A.M., Jose had a temperature of $99.8°$. It rose another $2.5°$ before falling $3.7°$ by 1:00 P.M. What was his temperature at 1:00 P.M.?

82. Checking account. Olga has $250 in her checking account. She deposits $52 and then writes a check for $77. What is her new balance?

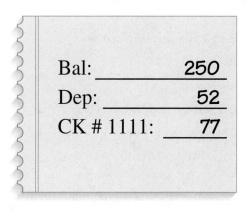

83. Education. Ezra's scores on five tests taken in a mathematics class were 87, 71, 95, 81, and 90. What was the difference between the highest and the lowest of his scores?

84. Bank balance. Aaron had $769 in his bank account on June 1. He deposited $125 and $986 during the month and wrote checks for $235, $529, and $712 during June. What was his balance at the end of the month?

85. Complete the following problem: "$4 - (-9)$ is the same as _____." Write an application problem that might be answered using this subtraction.

86. Explain the difference between the two phrases "a number less than 7" and "a number subtracted from 7." Use both algebra and English to explain the meanings of these phrases. Write some other ways of expressing subtraction in English.

87. Create an example to show that subtraction of real numbers is *not* commutative.

88. Create an example to show that subtraction of real numbers is *not* associative.

89. Do you think that the following statement is true?

$$|a + b| = |a| + |b| \quad \text{for all numbers } a \text{ and } b$$

When we don't know whether such a statement is true, we refer to the statement as a **conjecture.** We may "test" the conjecture by substituting specific numbers for the variables.

Test the conjecture, using two positive numbers for a and b.

Test again, using a positive number for a and 0 for b.

Test again, using two negative numbers.

Now try using one positive number and one negative number.

Summarize your results in a rule that you feel is true.

90. If a represents a positive number and b represents a negative number, determine whether the given expressions are positive or negative.

(a) $|b| + a$ (b) $b + (-a)$ (c) $(-b) + a$ (d) $-b + |-a|$

0.4 Multiplying and Dividing Real Numbers

0.4 OBJECTIVES

1. Multiply real numbers
2. Use the commutative property of multiplication
3. Use the associative property of multiplication
4. Use the distributive property
5. Divide real numbers

"Man's mind, once stretched by a new idea, never regains its original dimensions."

–Oliver Wendell Holmes

Multiplication can be seen as repeated addition. We can interpret

$$3 \times 4 = 4 + 4 + 4 = 12$$

We can use this interpretation together with the work of Section 0.3 to find the product of two real numbers.

Example 1

Finding the Product of Two Real Numbers

Multiply.

Note that we use parentheses ( ) to indicate multiplication when negative numbers are involved.

(a) $(3)(-4) = (-4) + (-4) + (-4) = -12$

(b) $(4)\left(-\dfrac{1}{3}\right) = \left(-\dfrac{1}{3}\right) + \left(-\dfrac{1}{3}\right) + \left(-\dfrac{1}{3}\right) + \left(-\dfrac{1}{3}\right) = -\dfrac{4}{3}$

 CHECK YOURSELF 1

Find the product by writing as repeated addition.

$$4(-3)$$

Looking at the products we found by repeated addition in Example 1 should suggest our first rule for multiplying real numbers.

> ### To Multiply Real Numbers
> 1. The product of two numbers with different signs is negative.

The rule is easy to use. To multiply two numbers with different signs, just multiply their absolute values and attach a minus sign to the product.

| Example 2 | ## Finding the Product of Two Real Numbers |

Find each product.

$$(5)(-6) = -30$$

$$(10)(-12) = -120$$

$$(-7)\left(\frac{1}{8}\right) = -\frac{7}{8}$$

The product must have two decimal places.

$$(1.5)(-0.3) = -0.45$$

The product is negative. You can simplify as before in finding the product.

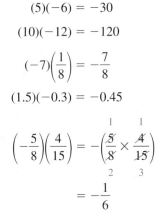

$$\left(-\frac{5}{8}\right)\left(\frac{4}{15}\right) = -\left(\frac{\overset{1}{\cancel{5}}}{\underset{2}{\cancel{8}}} \times \frac{\overset{1}{\cancel{4}}}{\underset{3}{\cancel{15}}}\right)$$

$$= -\frac{1}{6}$$

✔ *CHECK YOURSELF 2*

Find each product.

(a) $(15)(-5)$ (b) $(-0.8)(0.2)$ (c) $\left(-\frac{2}{3}\right)\left(\frac{6}{7}\right)$

The product of two negative numbers is harder to visualize. The following pattern may help you see how we can determine the sign of the product.

$$(3)(-2) = -6$$
$$(2)(-2) = -4$$
$$(1)(-2) = -2$$
$$(0)(-2) = 0$$
$$(-1)(-2) = 2$$
$$(-2)(-2) = 4$$

Do you see that the product is *increasing* by 2 each time the first number *decreases* by 1?

We already know that the product of two positive numbers is positive.

This suggests that the product of two negative numbers is positive, and this is in fact the case. To extend our multiplication rule, we have the following.

> **To Multiply Real Numbers**
> 2. The product of two numbers with the same sign is positive.

Example 3	**Finding the Product of Two Real Numbers**

Find each product.

$$8 \times 7 = 56$$

$$(-9)(-6) = 54$$

$$(-0.5)(-2) = 1$$

Since the numbers have the same sign, the product is positive.

Caution

Be Careful! $(-8)(-6)$ tells you to multiply. The parentheses are *next to* one another. The expression $-8 - 6$ tells you to subtract. The numbers are *separated* by the operation sign.

 CHECK YOURSELF 3

Find each product.

(a) 5×7 (b) $(-8)(-6)$ (c) $(9)(-6)$ (d) $(-1.5)(-4)$

To multiply more than two real numbers, apply the multiplication rule repeatedly.

Example 4	**Finding the Product of a Set of Real Numbers**

Multiply.

$$(5)(-7)(-3)(-2) \qquad (5)(-7) = -35$$

$$= (-35)(-3)(-2) \qquad (-35)(-3) = 105$$

$$= (105)(-2)$$

$$= -210$$

 CHECK YOURSELF 4

Find the product.

$$(-4)(3)(-2)(5)$$

We saw in Section 0.3 that the commutative and associative properties for addition could be extended to real numbers. The same is true for multiplication. What about the order in which we multiply? Look at the following examples.

| **Example 5** | ## Using the Commutative Property of Multiplication |

Find the products.

$$(-5)(7) = (7)(-5) = -35$$

$$(-6)(-7) = (-7)(-6) = 42$$

The order in which we multiply does not affect the product. This gives us the following rule.

The centered dot represents multiplication. This could have been written as

$$a \times b = b \times a$$

> ### The Commutative Property of Multiplication
> The order in which we multiply does not change the product. Multiplication is *commutative*. In symbols, for any a and b,
>
> $$a \cdot b = b \cdot a$$

  CHECK YOURSELF 5

Show that $(-8)(-5) = (-5)(-8)$.

What about the way we group numbers in multiplication? Look at Example 6.

| **Example 6** | ## Using the Associative Property of Multiplication |

Multiply.

The symbols [ ] are called *brackets* and are used to group numbers in the same way as parentheses.

$$[(3)(-7)](-2) \qquad \text{or} \qquad (3)[(-7)(-2)]$$

$$= (-21)(-2) \qquad\qquad\quad = (3)(14)$$

$$= 42 \qquad\qquad\qquad\qquad = 42$$

We group the first two numbers on the left and the second two numbers on the right. Note that the product is the same in either case.

The Associative Property of Multiplication

The way we *group* the numbers does not change the product. Multiplication is *associative*. In symbols, for any a, b, and c,

$$(a \cdot b) \cdot c = a \cdot (b \cdot c)$$

✔ *CHECK YOURSELF 6*

Show that $[(2)(-6)](-3) = (2)[(-6)(-3)]$.

Another important property in mathematics is the **distributive property.** The distributive property involves addition and multiplication together. We can illustrate the property with an application.

Remember: The area of a rectangle is the product of its length and width:

$$A = L \cdot W$$

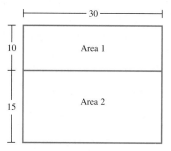

We can find the total area by multiplying the length by the overall width, which is found by adding the two widths.   or

We can find the total area as a sum of the two areas.

Length	Overall width
30	$\cdot (10 + 15)$
$= 30 \cdot 25$	
$= 750$	

	(Area 1) Length · Width		(Area 2) Length · Width
	$30 \cdot 10$	$+$	$30 \cdot 15$
	$= 300$	$+$	450
	$= 750$		

So

$$30 \cdot (10 + 15) = 30 \cdot 10 + 30 \cdot 15$$

This leads us to the following property.

Note the pattern.

$$a(b + c) = a \cdot b + a \cdot c$$

We "distributed" the multiplication "over" the addition.

The Distributive Property

If a, b, and c are any numbers,

$$a(b + c) = a \cdot b + a \cdot c \quad \text{and} \quad (b + c)a = b \cdot a + c \cdot a$$

| Example 7 | **Using the Distributive Property** |

Use the distributive property to simplify (remove the parentheses in) the following.

Note: It is also true that

$$5(3 + 4) = 5 \cdot 7 = 35$$

(a) $5(3 + 4)$

$$5(3 + 4) = 5 \cdot 3 + 5 \cdot 4 = 15 + 20 = 35$$

Note: It is also true that

$$\frac{1}{3}(9 + 12) = \frac{1}{3}(21) = 7$$

(b) $\dfrac{1}{3}(9 + 12) = \dfrac{1}{3} \cdot 9 + \dfrac{1}{3} \cdot 12$

$$= 3 + 4 = 7$$

✔ CHECK YOURSELF 7

Use the distributive property to simplify (remove the parentheses).

(a) $4(6 + 7)$

(b) $\dfrac{1}{5}(10 + 15)$

The distributive property can also be used to distribute multiplication over subtraction.

| Example 8 | **Distributing Multiplication over Subtraction** |

Use the distributive property to remove the parentheses and simplify the following.

(a) $4(-3 - 6) = 4(-3) - 4(6) = -12 - 24 = -36$

(b) $-7(-3 - 2) = -7(-3) - (-7)(2) = 21 - (-14) = 21 + 14 = 35$

✔ CHECK YOURSELF 8

Use the distributive property to remove the parentheses and simplify the following.

(a) $7(-3 - 4)$ (b) $-2(-4 - 3)$

A detailed explanation of why the product of two negative numbers must be positive concludes our discussion of multiplying real numbers.

The Product of Two Negative Numbers

The following argument shows why the product of two negative numbers must be positive.

From our earlier work, we know that a number added to its opposite is 0.	$5 + (-5) = 0$
Multiply both sides of the statement by -3.	$(-3)[5 + (-5)] = (-3)(0)$
A number multiplied by 0 is 0, so on the right we have 0.	$(-3)[5 + (-5)] = 0$
We can now use the distributive property on the left.	$(-3)(5) + (-3)(-5) = 0$
Since we know that $(-3)(5) = -15$, the statement becomes	$-15 + (-3)(-5) = 0$

We now have a statement of the form $-15 + \boxed{} = 0$. This asks, What number must we add to -15 to get 0, where $\boxed{}$ is the value of $(-3)(-5)$? The answer is, of course, 15. This means that

$$(-3)(-5) = 15 \qquad \text{The product must be positive.}$$

It doesn't matter what numbers we use in the argument. The product of two negative numbers will always be positive.

Multiplication and division are related operations. So every division problem can be stated as an equivalent multiplication problem.

$$8 \div 4 = 2 \qquad \text{since } 8 = 4 \cdot 2$$

$$\frac{12}{3} = 4 \qquad \text{since } 12 = 3 \cdot 4$$

Since the operations are related, the rules of signs for multiplication are also true for division.

To Divide Real Numbers

1. If two numbers have the same sign, the quotient is positive.

2. If two numbers have different signs, the quotient is negative.

As you would expect, division with fractions or decimals uses the same rules for signs. Example 9 illustrates this concept.

| Example 9 | **Dividing Two Real Numbers** |

Divide.

First note that the quotient is positive. Then invert the divisor and multiply.

$$\left(-\frac{3}{5}\right) \div \left(-\frac{9}{20}\right) = \frac{\overset{1}{\cancel{3}}}{\underset{1}{\cancel{5}}} \cdot \frac{\overset{4}{\cancel{20}}}{\underset{3}{\cancel{9}}} = \frac{4}{3}$$

✔ CHECK YOURSELF 9

Find each quotient.

(a) $-\dfrac{5}{8} \div \dfrac{3}{4}$ (b) $-4.2 \div (-0.6)$

Be very careful when 0 is involved in a division problem. Remember that 0 divided by any nonzero number is 0. However, division *by 0* is not allowed and will be described as *undefined*.

| Example 10 | **Dividing Real Numbers When Zero Is Involved** |

Divide.

(a) $0 \div 7 = 0$ (b) $\dfrac{0}{-4} = 0$

A statement like $-9 \div 0$ has no meaning. There is no answer to the problem. Just write "undefined."

(c) $-9 \div 0$ is undefined.   (d) $\dfrac{-5}{0}$ is undefined.

✔ CHECK YOURSELF 10

Find the quotient if possible.

(a) $\dfrac{0}{-7}$ (b) $\dfrac{-12}{0}$

The result of Example 10 can be confirmed on your calculator. That will be included in Example 11.

| Example 11 | **Dividing with a Calculator** |

Use your calculator to find each quotient.

(a) $\dfrac{-12.567}{0}$

The keystroke sequence on a graphing calculator

$\boxed{(-)}$ 12.567 $\boxed{\div}$ 0 $\boxed{\text{Enter}}$

results in a "Divide by 0" error message. The calculator recognizes that it cannot divide by zero.

On a scientific calculator, 12.567 $\boxed{+/-}$ $\boxed{\div}$ 0 $\boxed{=}$ results in an error message.

(b) $-10.992 \div -4.58$

The keystroke sequence

$\boxed{(-)}$ 10.992 $\boxed{\div}$ $\boxed{(-)}$ 4.58 $\boxed{\text{Enter}}$

or 10.992 $\boxed{+/-}$ $\boxed{\div}$ 4.58 $\boxed{+/-}$ $\boxed{=}$

yields 2.4.

✔ *CHECK YOURSELF 11*

Find each quotient.

(a) $\dfrac{-31.44}{6.55}$ (b) $-23.6 \div 0$

✔ *CHECK YOURSELF ANSWERS*

1. $(-3) + (-3) + (-3) + (-3) = -12$ **2.** (a) -75; (b) -0.16; (c) $-\dfrac{4}{7}$

3. (a) 35; (b) 48; (c) -54; (d) 6   **4.** 120   **5.** $40 = 40$ **6.** $36 = 36$

7. (a) 52; (b) 5   **8.** (a) -49; (b) 14   **9.** (a) $-\dfrac{5}{6}$; (b) 7

10. (a) 0; (b) undefined   **11.** (a) -4.8; (b) undefined

Exercises · 0.4

Multiply.

1. $7 \cdot 8$

2. $(6)(-12)$

3. $(4)(-3)$

4. $15 \cdot 5$

5. $(-8)(9)$

6. $(-8)(3)$

7. $(-5)\left(\dfrac{1}{3}\right)$

8. $(-12)(-2)$

9. $(-10)(0)$

10. $(10)(-10)$

11. $(-8)(-8)$

12. $(0)(-50)$

13. $(-4)\left(\dfrac{5}{7}\right)$

14. $(-25)(-8)$

15. $(-9)(-12)$

16. $(-3)(-27)$

17. $(-20)(1)$

18. $(1)(-30)$

19. $(-1.3)(6)$

20. $(-25)(5)$

21. $(-10)(-15)$

22. $(-2.4)(0.2)$

23. $\left(-\dfrac{7}{10}\right)\left(-\dfrac{5}{14}\right)$

24. $\left(-\dfrac{7}{20}\right)\left(\dfrac{10}{21}\right)$

25. $\left(\dfrac{3}{5}\right)\left(-\dfrac{10}{27}\right)$

26. $\left(-\dfrac{15}{4}\right)(0)$

27. $\left(-\dfrac{5}{8}\right)\left(-\dfrac{4}{15}\right)$

28. $\left(-\dfrac{8}{21}\right)\left(-\dfrac{7}{4}\right)$

29. $(-5)(3)(-8)$

30. $(4)(-3)(-5)$

31. $(-5)(-9)(-3)$

32. $(-7)(-5)(-2)$

33. $(2)(-5)(-3)(-5)$

34. $(-2)(-5)(-5)(-6)$

35. $(-4)(-3)(-6)(-2)$

36. $(-8)(3)(-2)(5)$

Use the distributive property to remove parentheses and simplify the following.

37. $5(-6 + 9)$

38. $12(-5 + 9)$

39. $-8(-9 + 15)$

40. $-11(-8 + 3)$

41. $-4(-5 - 3)$

42. $-2(-7 - 11)$

43. $-4(-6 - 3)$

44. $-6(-3 - 2)$

Divide.

45. $15 \div (-3)$

46. $\dfrac{90}{18}$

47. $\dfrac{54}{9}$

48. $-20 \div (-2)$

49. $\dfrac{-50}{5}$

50. $-36 \div 6$

51. $\dfrac{-24}{-3}$

52. $\dfrac{42}{-6}$

53. $\dfrac{90}{-6}$

38

54. $70 \div (-10)$

55. $18 \div (-1)$

56. $\dfrac{-250}{-25}$

57. $\dfrac{0}{-9}$

58. $\dfrac{-12}{0}$

59. $-180 \div (-15)$

60. $\dfrac{0}{-10}$

61. $-7 \div 0$

62. $\dfrac{-25}{-1}$

63. $\dfrac{-150}{6}$

64. $\dfrac{-80}{-16}$

65. $-45 \div (-9)$

66. $-\dfrac{2}{3} \div \dfrac{4}{9}$

67. $-\dfrac{8}{11} \div \dfrac{18}{55}$

68. $(-8) \div (-4)$

69. $\dfrac{7}{10} \div \left(-\dfrac{14}{25}\right)$

70. $\dfrac{6}{13} \div \left(-\dfrac{18}{39}\right)$

71. $\dfrac{-75}{15}$

72. $-\dfrac{5}{8} \div \left(-\dfrac{5}{16}\right)$

Divide by using a graphing calculator. Round answers to the nearest thousandth.

73. $-5.634 \div 2.398$

74. $-2.465 \div 7.329$

75. $-18.137 \div (-5.236)$

76. $-39.476 \div (-17.629)$

77. $32.245 \div (-48.298)$

78. $43.198 \div (-56.249)$

79. Dieting. A woman lost 42 pounds (lb). If she lost 3 lb each week, how long has she been dieting?

80. Mowing lawns. Patrick worked all day mowing lawns and was paid $9 per hour. If he had $125 at the end of a 9-hour day, how much did he have before he started working?

81. Unit pricing. A 4.5-lb can of food costs $8.91. What is the cost per pound?

82. Investment. Suppose that you and your two brothers bought equal shares of an investment for a total of $20,000 and sold it later for $16,232. How much did each person lose?

83. Temperature. Suppose that the temperature outside is dropping at a constant rate. At noon, the temperature is 70°F and it drops to 58°F at 5:00 P.M. How much did the temperature change each hour?

84. **Test tube count.** A chemist has 84 ounces (oz) of a solution. He pours the solution into test tubes. Each test tube holds $\frac{2}{3}$ oz. How many test tubes can he fill?

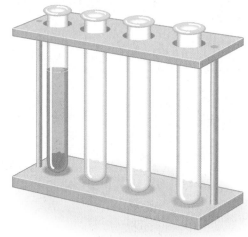

To evaluate an expression involving a fraction (indicating division), we evaluate the numerator and then the denominator. We then divide the numerator by the denominator as the last step. Using this approach, find the value of each of the following expressions.

85. $\dfrac{5 - 15}{2 + 3}$

86. $\dfrac{4 - (-8)}{2 - 5}$

87. $\dfrac{-6 + 18}{-2 - 4}$

88. $\dfrac{-4 - 21}{3 - 8}$

89. $\dfrac{(5)(-12)}{(-3)(5)}$

90. $\dfrac{(-8)(-3)}{(2)(-4)}$

91. Create an example to show that the division of real numbers is *not* commutative.

92. Create an example to show that the division of real numbers is *not* associative.

93. Here is another conjecture to consider:

$$|ab| = |a|\,|b| \qquad \text{for all numbers } a \text{ and } b$$

(See the discussion in Exercises 0.3, number 89, concerning testing a conjecture.) Test this conjecture for various values of a and b. Use positive numbers, negative numbers, and 0. Summarize your results in a rule.

94. Use a calculator (or mental calculations) to compute the following:

$$\frac{5}{0.1}, \frac{5}{0.01}, \frac{5}{0.001}, \frac{5}{0.0001}, \frac{5}{0.00001}$$

In this series of problems, while the numerator is always 5, the denominator is getting smaller (and is getting closer to 0). As this happens, what is happening to the value of the fraction?

Write an argument that explains why $\dfrac{5}{0}$ could not have any finite value.

0.5 Exponents and Order of Operations

1. Write a product of like factors in exponential form
2. Evaluate numbers with exponents
3. Use the order of operations

"The intelligent person is one who has successfully fulfilled many accomplishments, and yet is willing to learn more."

–Ed Parker

In Section 0.4, we mentioned that multiplication is a more compact, or "shorthand," form for repeated addition. For example, an expression with repeated addition, such as

$$3 + 3 + 3 + 3 + 3$$

can be rewritten as

$$5 \cdot 3$$

Thus, multiplication is shorthand for repeated addition.

In algebra, we frequently have a number or variable that is repeated in an expression several times. For instance, we might have

$$5 \cdot 5 \cdot 5$$

A factor is a number or a variable that is being multiplied by another number or variable.

To abbreviate this product, we write

$$5 \cdot 5 \cdot 5 = 5^3$$

Since an exponent represents repeated multiplication, 5^3 is an expression.

This is called **exponential notation** or **exponential form.** The exponent or power, here 3, indicates the number of times that the factor or base, here 5, appears in a product.

Caution

Be careful: 5^3 is *not* the same as $5 \cdot 3$. Notice that
$5^3 = 5 \cdot 5 \cdot 5 = 125$ and
$5 \cdot 3 = 15$.

$$5 \cdot 5 \cdot 5 = 5^3$$

Exponent or power

Factor or base

41

Example 1	Writing Expressions in Exponential Form

(a) Write $3 \cdot 3 \cdot 3 \cdot 3$, using exponential form. The number 3 appears 4 times in the product, so

$$3 \cdot 3 \cdot 3 \cdot 3 = 3^4 \qquad \text{Four factors of 3}$$

This is read "3 to the fourth power."

(b) Write $10 \cdot 10 \cdot 10$, using exponential form. Since 10 appears 3 times in the product, you can write

$$10 \cdot 10 \cdot 10 = 10^3$$

This is read "10 to the third power" or "10 cubed."

 CHECK YOURSELF 1

Write in exponential form.

(a) $4 \cdot 4 \cdot 4 \cdot 4 \cdot 4 \cdot 4$ (b) $10 \cdot 10 \cdot 10 \cdot 10$

When one is evaluating a number raised to a power, it is important to note whether there is a sign attached to the number. Note that

$$(-2)^4 = (-2)(-2)(-2)(-2) = 16$$

whereas

$$-2^4 = -(2)(2)(2)(2) = -16$$

Example 2	Evaluating Exponential Terms

Evaluate each expression.

(a) $(-3)^3 = (-3)(-3)(-3) = -27$ (b) $-3^3 = -(3)(3)(3) = -27$

(c) $(-3)^4 = (-3)(-3)(-3)(-3) = 81$ (d) $-3^4 = -(3)(3)(3)(3) = -81$

 CHECK YOURSELF 2

Evaluate each expression.

(a) $(-4)^3$ (b) -4^3 (c) $(-4)^4$ (d) -4^4

Your calculator can help you to evaluate expressions containing exponents. If you have a graphing calculator, the appropriate key is the carat, $\wedge$. Enter the base, followed by the carat, followed by the exponent. Other calculators use a key labeled y^x in place of the carat.

| **Example 3** | **Evaluating Expressions with Exponents** |

Use your calculator to evaluate each expression.

(a) $3^5 = 243$ Type 3 $\boxed{\wedge}$ 5 $\boxed{\text{Enter}}$

or 3 $\boxed{y^x}$ 5 $\boxed{=}$

(b) $2^{10} = 1024$ 2 $\boxed{\wedge}$ 10 $\boxed{\text{Enter}}$

or 2 $\boxed{y^x}$ 10 $\boxed{=}$

✔ *CHECK YOURSELF 3*

Use your calculator to evaluate each expression.

(a) 3^4 (b) 2^{16}

We have used the term *expression* with numbers taken to powers, such as 3^4. But what about something like $4 + 12 - 6$? We call *any* meaningful combination of numbers and operations an **expression.** When we evaluate an expression, we find a number that is equal to the expression. To evaluate an expression, we need to establish a set of rules that tell us the correct order in which to perform the operations. To see why, simplify the expression $5 + 2 \cdot 3$.

Caution

Only one of these results can be correct.

	Method 1	or	Method 2
	$5 + 2 \cdot 3$		$5 + 2 \cdot 3$
	Add first.		Multiply first.
	$= 7 \cdot 3$		$= 5 + 6$
	$= 21$		$= 11$

Since we get different answers depending on how we do the problem, the language of algebra would not be clear if there were no agreement on which method is correct. The following rules tell us the order in which operations should be done.

Parentheses and brackets are both grouping symbols. Fraction bars and radicals are also grouping symbols.

The Order of Operations

Step 1.   Evaluate all expressions inside grouping symbols.

Step 2.   Evaluate all expressions involving exponents.

Step 3.   Do any multiplication or division in order, working from left to right.

Step 4.   Do any addition or subtraction in order, working from left to right.

| Example 4 | Evaluating Expressions |

Evaluate $5 + 2 \cdot 3$.

There are no parentheses or exponents, so start with step 3: First multiply and then add.

$$5 + 2 \cdot 3$$

Note: Method 2 shown on the previous page is the correct one.

Multiply first.

$$= 5 + 6$$

Then add.

$$= 11$$

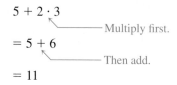

✔ CHECK YOURSELF 4

Evaluate the following expressions.

(a) $20 - 3 \cdot 4$ (b) $9 + 6 \div 3$

| Example 5 | Evaluating Expressions |

Evaluate $-5 \cdot 3^2$.

$$-5 \cdot 3^2$$

$$= -5 \cdot 9$$ Evaluate the exponent first.

$$= -45$$

✔ CHECK YOURSELF 5

Evaluate $-4 \cdot 2^4$.

Both scientific and graphing calculators correctly interpret the order of operations. This is demonstrated in Example 6.

| Example 6 | Using a Calculator to Evaluate Expressions |

Use your scientific or graphing calculator to evaluate each expression.

(a) $24.3 + 6.2 \cdot 3.5$

When evaluating expressions by hand, you must consider the order of operations. In this case, the multiplication must be done first, then the addition. With a modern calculator, you need only enter the expression correctly. The calculator is programmed to follow the order of operations.

Entering 24.3　$\boxed{+}$　6.2　$\boxed{\times}$　3.5　$\boxed{\text{Enter}}$

yields the evaluation 46.

(b)　$(2.45)^3 - 49 \div 8,000 + 12.2 \cdot 1.3$

As we mentioned earlier, some calculators use the carat ($\wedge$) to designate exponents. Others use the symbol x^y (or y^x).

Entering　$\boxed{(}$　2.45　$\boxed{)}$　$\boxed{\wedge}$　3　$\boxed{-}$　49　$\boxed{\div}$　8,000　$\boxed{+}$　12.2　$\boxed{\times}$　1.3

or　　　　　$\boxed{(}$　2.45　$\boxed{)}$　$\boxed{y^x}$　3　$\boxed{-}$　49　$\boxed{\div}$　8,000　$\boxed{+}$　12.2　$\boxed{\times}$　1.3

yields the evaluation 30.56.

✔ CHECK YOURSELF 6

Use your scientific or graphing calculator to evaluate each expression.

(a)　$67.89 - 4.7 \cdot 12.7$

(b)　$4.3 \cdot 55.5 - (3.75)^3 + 8,007 \div 1,600$

Operations inside grouping symbols are done first.

Example 7　Evaluating Expressions

Evaluate $(5 + 2) \cdot 3$.
　　Do the operation inside the parentheses as the first step.

$$(5 + 2) \cdot 3 = 7 \cdot 3 = 21$$
　　　　　　　　　　　　　　—— Add.

✔ CHECK YOURSELF 7

Evaluate $4(9 - 3)$.

The principle is the same when more than two "levels" of operations are involved.

Example 8	**Evaluating Expressions**

(a) Evaluate $4(\underbrace{-2 + 7})^3$.

Add inside the parentheses first.

$$4(-2 + 7)^3 = 4(5)^3$$

Evaluate the exponent.

$$= 4 \cdot 125$$

Multiply.

$$= 500$$

(b) Evaluate $5(7 - 3)^2 - 10$.

Evaluate the expression inside the parentheses.

$$5(7 - 3)^2 - 10 = 5(4)^2 - 10$$

Evaluate the exponent.

$$= 5 \cdot 16 - 10$$

Multiply.

$$= 80 - 10 = 70$$

Subtract.

✔ *CHECK YOURSELF 8*

Evaluate.

(a) $4 \cdot 3^3 + 8 \cdot (-11)$ (b) $12 + 4(2 + 3)^2$

✔ *CHECK YOURSELF ANSWERS*

1. (a) 4^6; (b) 10^4 **2.** (a) -64; (b) -64; (c) 256; (d) -256

3. (a) 81; (b) 65,536   **4.** (a) 8; (b) 11

5. -64 **6.** (a) 8.2; (b) 190.92

7. 24   **8.** (a) 20; (b) 112

Exercises · 0.5

Write each expression, using exponential form.

1. $3 \cdot 3 \cdot 3 \cdot 3 \cdot 3$

2. $2 \cdot 2 \cdot 2 \cdot 2 \cdot 2 \cdot 2 \cdot 2$

3. $7 \cdot 7 \cdot 7 \cdot 7 \cdot 7$

4. $10 \cdot 10 \cdot 10 \cdot 10 \cdot 10$

5. $8 \cdot 8 \cdot 8 \cdot 8 \cdot 8 \cdot 8$

6. $5 \cdot 5 \cdot 5 \cdot 5 \cdot 5 \cdot 5$

7. $(-2)(-2)(-2)$

8. $(-4)(-4)(-4)(-4)$

Evaluate.

9. 3^2

10. 2^3

11. 2^4

12. 2^5

13. $(-8)^3$

14. 3^5

15. -8^3

16. 4^4

17. -5^2

18. $(-5)^2$

19. $(-4)^2$

20. $(-3)^4$

21. $-(-2)^5$

22. $-(-6)^4$

23. 10^3

24. 10^2

25. 10^6

26. 10^7

27. 2×4^3

28. $(2 \times 4)^3$

29. 3×4^2

30. $(3 \times 4)^2$

31. $5 + 2^2$

32. $(5 + 2)^2$

33. $3^4 \times 2^4$

34. $(3 \times 2)^4$

Evaluate each of the following expressions.

35. $4 + 3 \cdot 5$

36. $10 - 4 \cdot 2$

37. $(7 + 2) \cdot 6$

38. $(9 - 5) \cdot 3$

39. $-12 - 8 \div 4$

40. $10 + 20 \div 5$

41. $(12 - 8) \div 4$

42. $(10 + 20) \div 5$

43. $8 \cdot 7 + 2 \cdot 2$

44. $48 \div 8 - 14 \div 2$

45. $(7 \cdot 5) + 3 \cdot 2$

46. $48 \div (8 - 4) \div 2$

47. $3 \cdot 5^2$

48. $5 \cdot 2^3$

49. $(3 \cdot 5)^2$

50. $(5 \cdot 2)^3$

51. $4 \cdot 3^2 - 2$

52. $3 \cdot 2^4 - 8$

53. $7(2^3 - 5)$

54. $3(7 - 3^2)$

55. $3 \cdot 2^4 - 26 \cdot 2$

56. $4 \cdot 2^3 - 15 \cdot 6$

57. $(2 \cdot 4)^2 - 8 \cdot 3$

58. $(3 \cdot 2)^3 - 7 \cdot 3$

59. $5(3 + 4)^2$

60. $3(8 - 4)^2$

61. $(5 \cdot 3 + 4)^2$

62. $(3 \cdot 8 - 4)^2$

63. $5[3(2 + 5) - 5]$

64. $\dfrac{11 - (-9) + 6(8 - 2)}{2 + 3 \cdot 4}$

65. $-2[(3-5)^2 - (-4+2)^3 \cdot (8 \div 4 \cdot 2)]$

66. $5 \cdot 4 - 2^3$

67. $4(2+3)^2 - 125$

68. $8 + 2(3+3)^2$

69. $(4 \cdot 2 + 3)^2 - 25$

70. $8 + (2 \cdot 3 + 3)^2$

71. $[-20 - 4^2 + (-4)^2 + 2] \div 9$

72. $14 + 3 \cdot 9 - 28 \div 7 \cdot 2$

73. $4 \cdot 8 \div 2 - 5^2$

74. $-12 - 8 \div 4 \cdot 2$

75. $15 + 5 - 3 \cdot 2 + (-2)^3$

76. $-8 + 14 \div 2 \cdot 4 - 3$

Evaluate by using your calculator. Round your answer to the nearest tenth.

77. $(1.2)^3 \div 2.0736 \cdot 2.4 + 1.6935 - 2.4896$

78. $(5.21 \cdot 3.14 - 6.2154) \div 5.12 - 0.45625$

79. $1.23 \cdot 3.169 - 2.05194 + (5.128 \cdot 3.15 - 10.1742)$

80. $4.56 + (2.34)^4 \div 4.7896 \cdot 6.93 \div 27.5625 - 3.1269 + (1.56)^2$

81. Population doubling. Over the last 2,000 years, the earth's population has doubled approximately 5 times. Write the phrase "doubled 5 times" in exponential form.

82. Volume of a cube. The volume of a cube with each edge of length 9 inches (in.) is given by $9 \cdot 9 \cdot 9$. Write the volume, using exponential notation.

83. Insert grouping symbols in the proper place so that the value of the expression $36 \div 4 + 2 - 4$ is 2.

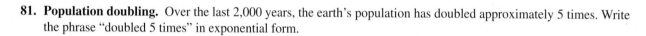

84. Work with a small group of students.

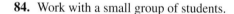

Part 1: Write the numbers 1 through 25 on slips of paper and put the slips in a pile, face down. Each of you randomly draws a slip of paper until each person has five slips. Turn the papers over and write down the five numbers. Put the five papers back in the pile, shuffle, and then draw one more. This last number is the answer. The first five numbers are the problem. Your task is to arrange the first five into a computation, using all you know about the order of operations, so that the answer is the last number. Each number must be used and may be used only once. If you cannot find a way to do this, pose it as a question to the whole class. Is this guaranteed to work?

Part 2: Use your five numbers in a problem, each number being used and used only once, for which the answer is 1. Try this 9 more times with the numbers 2 through 10. You may find more than one way to do each of these. Surprising, isn't it?

Part 3: Be sure that when you successfully find a way to get the desired answer by using the five numbers, you can then write your steps, using the correct order of operations. Write your 10 problems and exchange them with another group to see if they get these same answers when they do your problems.

Summary for Chapter 0

Example	Topic	Reference
	Fractions	0.1
$\dfrac{4 \cdot 3}{5 \cdot 3} = \dfrac{12}{15}$	**Equivalent Fractions** If the numerator and denominator of a fraction are both multiplied by some nonzero number, the result is a fraction that is equivalent to the original fraction.	p. 3
$\dfrac{9}{21} = \dfrac{3 \cdot 3}{3 \cdot 7} = \dfrac{3}{7}$	**Simplifying Fractions** A fraction is in simplest terms when the numerator and denominator have no common factor.	p. 3
$\dfrac{2}{3} \cdot \dfrac{5}{6} = \dfrac{10}{18} = \dfrac{5}{9}$	**Multiplying Fractions** To multiply two fractions, multiply the numerators, then multiply the denominators. Simplification can be done before or after the multiplication.	p. 4
$\dfrac{3}{5} \div \dfrac{2}{7} = \dfrac{3}{5} \cdot \dfrac{7}{2} = \dfrac{21}{10}$	**Dividing Fractions** To divide two fractions, invert the divisor (the second fraction), then multiply the fractions.	p. 5
$\dfrac{2}{5} + \dfrac{5}{8} = \dfrac{16}{40} + \dfrac{25}{40} = \dfrac{41}{40}$	**Adding Fractions** To add two fractions, find the LCD (least common denominator), rewrite the fractions with this denominator, then add the numerators.	p. 7
$\dfrac{2}{3} - \dfrac{1}{4} = \dfrac{8}{12} - \dfrac{3}{12} = \dfrac{5}{12}$	**Subtracting Fractions** To subtract two fractions, find the LCD, rewrite the fractions with this denominator, then subtract the numerators.	p. 8
	Real Numbers—The Terms	0.2
Negative / Positive numbers number line: $-3\ -2\ -1\ 0\ 1\ 2\ 3$; Zero is neither positive nor negative.	**Positive Numbers** Numbers used to name points to the right of 0 on the number line. **Negative Numbers** Numbers used to name points to the left of 0 on the number line.	p. 12
The natural numbers are $\{1, 2, 3, \ldots\}$	**Natural Numbers** The counting numbers.	p. 12
The integers are $\{\ldots -3, -2, -1, 0, 1, 2, 3, \ldots\}$	**Integers** The set consisting of the natural numbers, their opposites, and 0.	p. 13

Rational numbers are $\frac{2}{3}, \frac{5}{1}, 0.234$	**Rational Number** Any number that can be expressed as the ratio of two integers.	**p. 14**
Irrational numbers are $\sqrt{4}, e$	**Irrational Number** Any number that is not rational.	**p. 14**
All the numbers above are real numbers.	**Real Numbers** Rational and irrational numbers together.	**p. 14**
5 units 5 units ◄─┼───┼───┼─► −5 0 5 The opposite of 5 is −5.	**Opposites** Two numbers are opposites if the points name the same distance from 0 on the number line, but in opposite directions.	**p. 15**
	The opposite of a positive number is negative.	**p. 15**
3 units 3 units ◄─┼───┼───┼─► −3 0 3 The opposite of −3 is 3.	The opposite of a negative number is positive.	**p. 15**
	0 is its own opposite.	**p. 16**
The absolute value of a number a is written $\lvert a\rvert$. $\lvert 7\rvert = 7$ $\lvert -8\rvert = 8$	**Absolute Value** The distance on the number line between the point named by a number and 0. The absolute value of a number is always positive or 0.	**p. 16**

Operations on Real Numbers   0.3–0.4

	To Add Real Numbers	**p. 20**
$5 + 8 = 13$ $-3 + (-7) = -10$	1. If two numbers have the same sign, add their absolute values. Give to the sum the sign of the original numbers.	
$5 + (-3) = 2$ $7 + (-9) = -2$	2. If two numbers have different signs, subtract the smaller absolute value from the larger. Give to the result the sign of the number with the larger absolute value.	**p. 20**
$4 - (-2) = 4 + 2 = 6$ The opposite of −2	**To Subtract Real Numbers** To subtract real numbers, add the first number and the opposite of the number being subtracted.	**p. 23**
	To Multiply Real Numbers To multiply real numbers, multiply the absolute values of the numbers. Then attach a sign to the product according to the following rules:	**p. 30**
$5 \cdot 7 = 35$ $(-4)(-6) = 24$ $(8)(-7) = -56$	1. If the numbers have different signs, the product is negative. 2. If the numbers have the same sign, the product is positive.	

	The Properties of Addition and Multiplication	0.3–0.4
$3 + 4 = 4 + 3$ $7 = 7$	**The Commutative Properties** If a and b are any numbers, then **1.** $a + b = b + a$ **2.** $a \cdot b = b \cdot a$	**p. 21, 32**
$3 \cdot (4 \cdot 5) = (3 \cdot 4) \cdot 5$ $3 \cdot (20) = (12) \cdot 5$ $60 = 60$	**The Associative Properties** If a, b, and c are any numbers, then **1.** $a + (b + c) = (a + b) + c$ **2.** $a \cdot (b \cdot c) = (a \cdot b) \cdot c$	**p. 22, 33**
$2(5 + 3) = 2 \cdot 5 + 2 \cdot 3$ $2(8) = 10 + 6$ $16 = 16$	**The Distributive Property** If a, b, and c are any numbers, then $a(b + c) = a \cdot b + a \cdot c$.	**p. 33**
$\dfrac{-8}{-2} = 4$ $27 \div (-3) = -9$ $\dfrac{-16}{8} = -2$	**To Divide Real Numbers** To divide real numbers, divide the absolute values of the numbers. Then attach a sign to the quotient according to the following rules: **1.** If the numbers have the same sign, the quotient is positive. **2.** If the numbers have different signs, the quotient is negative.	**p. 35**
	Exponents and the Order of Operations	0.5
$5^3 = 5 \cdot 5 \cdot 5$ $\quad = 125$ $a^2 b^3 = a \cdot a \cdot b \cdot b \cdot b$ $6m^2 = 6 \cdot m \cdot m$	**The Notation** $$\overset{\text{Exponent}}{\underset{\text{Base}}{a^4}} = \underbrace{a \cdot a \cdot a \cdot a}_{\text{4 factors}}$$ The number or letter used as a factor, here a, is called the *base.* The *exponent,* which is written above and to the right of the base, tells us how many times the base is used as a factor.	**p. 41**

Operate inside
grouping symbols.

$5 + 3(6-4)^2$

Evaluate
the power.

$5 + 3 \cdot 2^2$

Multiply.

$= 5 + 3 \cdot 4$

Add.

$= 5 + 12$
$= 17$

The Order of Operations

1. Do any operations within grouping symbols.
2. Evaluate all expressions containing exponents.
3. Do any multiplication or division in order, working from left to right.
4. Do any addition or subtraction in order, working from left to right.

p. 43

Summary Exercises · 0

This summary exercise set is provided to give you practice with each of the objectives of the chapter. Each exercise is keyed to the appropriate chapter section. Your instructor will give you guidelines on how to best use these exercises in your instructional setting.

[0.1] In Exercises 1 to 3, write three fractional representations for each number.

1. $\dfrac{5}{7}$

2. $\dfrac{3}{11}$

3. $\dfrac{4}{9}$

4. Use the fundamental principle to write the fraction $\dfrac{24}{64}$ in simplest form.

[0.1] In Exercises 5 to 12, perform the indicated operations. Write each answer in simplest form.

5. $\dfrac{7}{15} \times \dfrac{5}{21}$

6. $\dfrac{10}{27} \times \dfrac{9}{20}$

7. $\dfrac{5}{17} \div \dfrac{15}{34}$

8. $\dfrac{7}{15} \div \dfrac{14}{25}$

9. $\dfrac{7}{8} + \dfrac{15}{24}$

10. $\dfrac{5}{18} + \dfrac{7}{12}$

11. $\dfrac{11}{18} - \dfrac{2}{9}$

12. $\dfrac{11}{27} - \dfrac{5}{18}$

[0.2] In Exercises 13 to 20, complete the statement.

13. The absolute value of 12 is _____.

14. The opposite of -8 is _____.

15. $-|-3| = $ _____

16. $-(-20) = $ _____

17. $-|-4| = $ _____

18. $|-(-5)| = $ _____

19. The absolute value of -16 is _____.

20. The opposite of the absolute value of -9 is _____.

[0.2] Complete each of the following statements, using the symbol $<$, $>$, or $=$.

21. -3 _____ -1

22. -6 _____ $-|-6|$

23. $-|-7|$ _____ $-(-2)$

24. $-|-5|$ _____ $|-(-5)|$

[0.3] In Exercises 25 to 32, simplify.

25. $15 + (-7)$

26. $4 + (-9)$

27. $-23 - (-12)$

28. $\dfrac{5}{2} + \left(-\dfrac{4}{2}\right)$

29. $-\dfrac{9}{13} - \dfrac{4}{39}$

30. $5 + (-6) + (-3)$

31. $7 + (-4) + 8 + (-7)$

32. $-6 + 9 + 9 + (-5)$

[0.3] In Exercises 33 to 42, simplify.

33. $-35 + 30$

34. $-10 - 5$

35. $3 - (-2)$

36. $-7 - (-3)$

37. $\dfrac{23}{4} - \left(-\dfrac{3}{4}\right)$

38. $-3 - 2$

39. $8 - 12 - (-5)$

40. $-6 - 7 - (-18)$

41. $7 - (-4) - 7 - 4$

42. $-9 - (-6) - 8 - (-11)$

[0.4] In Exercises 43 to 50, multiply.

43. $(-18)(-2)$

44. $(-10)(8)$

45. $(-5)(3)$

46. $\left(-\dfrac{3}{8}\right)\left(-\dfrac{4}{5}\right)$

47. $(-4)^2$

48. $(-2)(7)(-3)$

49. $(-6)(-5)(4)(-3)$

50. $(-9)(2)(-3)(1)$

[0.4] In Exercises 51 to 54, use the distributive property to remove parentheses and simplify.

51. $-4(8 - 7)$

52. $11(-15 + 4)$

53. $-8(5 - 2)$

54. $-4(-3 - 6)$

[0.4] In Exercises 55 to 62, divide.

55. $(-48) \div 12$

56. $\dfrac{-33}{-3}$

57. $-2 \div 0$

58. $-75 \div (-3)$

59. $-\dfrac{7}{9} \div \left(-\dfrac{2}{3}\right)$

60. $\left(-\dfrac{5}{11}\right) \div \dfrac{20}{33}$

61. $8 \div (-4)$

62. $(-12) \div (-1)$

[0.5] Write each expression in expanded form.

63. 3^3

64. 5^4

65. 2^6

66. 4^5

[0.5] Evaluate each of the following expressions.

67. $18 - 12 \div 2$

68. $(18 - 3) \cdot 5$

69. $6 \cdot 2^3$

70. $(5 \cdot 4)^2$

71. $5 \cdot 3^2 - 4$

72. $5(3^2 - 4)$

73. $5(4 - 2)^2$

74. $5 \cdot 4 - 2^2$

75. $(5 \cdot 4 - 2)^2$

76. $3(5 - 2)^2$

77. $3 \cdot 5 - 2^2$

78. $(3 \cdot 5 - 2)^2$

[0.1–0.5] Solve the following.

79. A kitchen measures $\dfrac{16}{3}$ by $\dfrac{16}{4}$ yd. If you purchase linoleum which costs \$9 per square yard, how much will it cost to cover the floor?

80. The scale on a map uses 1 in. to represent 80 mi. If two cities are $\dfrac{11}{4}$ in. apart on the map, what is the actual distance between the cities?

81. An 18-acre piece of land is to be subdivided into home lots that are each $\dfrac{3}{8}$ acre. How many lots can be formed?

82. Find the perimeter of the given figure.

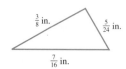

Self-Test ■ 0

The purpose of this self-test is to help you check your progress and to review for a chapter test in class. Allow yourself about an hour to take the test. When you are done, check the answers in the back of the book. If you missed any problems, go back and review the appropriate sections in the chapter and do the exercises provided.

In Exercises 1 and 2, use the fundamental principle to simplify each fraction.

1. $\dfrac{27}{99}$

2. $\dfrac{100}{64}$

In Exercises 3 to 6, perform the indicated operations. Write each answer in simplest form.

3. $\dfrac{4}{15} + \dfrac{7}{10}$

4. $\dfrac{9}{16} - \dfrac{3}{10}$

5. $\dfrac{3}{8} \times \dfrac{18}{33}$

6. $\dfrac{4}{5} \div \dfrac{2}{7}$

In Exercises 7 and 8, complete each statement, using the symbol $<$, $>$, or $=$.

7. -7 _____ -5

8. $8 + (-3)^2$ _____ $8 - (-3)$

In Exercises 9 to 12, add or subtract as indicated.

9. $13 + (-11) + (-5)$

10. $23 - 35$

11. $-9 - 8 - (-5)$

12. $7 - 11 + 15$

In Exercises 13 to 16, multiply or divide as indicated.

13. $(-7)(5)$

14. $(-9)(-6)$

15. $28 \div (-4)$

16. $(-44) \div (-11)$

In Exercises 17 and 18, use the distributive property to remove parentheses and simplify.

17. $5(-8 + 12)$

18. $-5(-9 - 14)$

In Exercises 19 and 20, evaluate each expression.

19. $23 - 4 \cdot 12 \div 3 + \left| -4 \right|$

20. $4 \cdot 5^2 - 35 + 21 - (-3)^3$

CHAPTER

$\left(1\right)$

From Arithmetic to Algebra

LIST OF SECTIONS

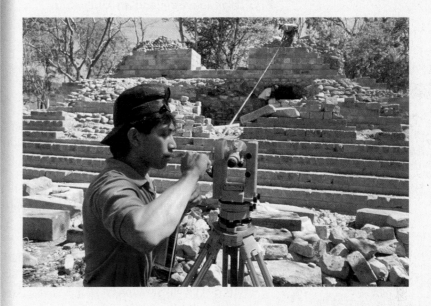

Cultures from all over the world have developed number systems and ways to record patterns in their natural surroundings. The Mayans in Central America had one of the most sophisticated number systems in the world in 12th century C.E. The Chinese numbering and recording system dates from around 1200 B.C.E. The oldest evidence of numerical record is in Africa, where a bone notched in numerical patterns and dating from about 35,000 B.C.E. was found in Southern Africa.

The roots of algebra developed among the Babylonians 4,000 years ago in an area now part of the country of Iraq. The Babylonians developed ways to record useful numerical relationships so that they were easy to remember, easy to record, and helpful in solving problems. Archaeologists have found instructions for solving problems in engineering, economics, city planning, and agriculture. The writing was on clay tablets. Some of these formulas developed by the Babylonians are still in use today.

1.1 Transition to Algebra

 OBJECTIVES

1. *Introduce the concept of variables*
2. *Identify algebraic expressions*
3. *Translate from English to algebra*

Overcoming Math Anxiety

Throughout this text, we will present you with a series of class-tested techniques that are designed to improve your performance in this math class.

Hint 2: Become familiar with your syllabus.

In the first class meeting, your instructor probably handed out a class syllabus. If you haven't done so already, you need to incorporate important information into your calendar and address book.

1. Write all important dates in your calendar. This includes homework due dates, quiz dates, test dates, and the date and time of the final exam. Never allow yourself to be surprised by any deadline!

2. Write your instructor's name, contact number, and office number in your address book. Also include the office hours. Make it a point to see your instructor early in the term. Although this is not the only person who can help clear up your confusion, it is the most important person.

3. Make note of other resources that are made available to you. This includes CDs, video tapes, web pages, and tutoring.

Given all of these resources, it is important that you never let confusion or frustration mount. If you can't "get it" from the text, try another resource. All the resources are there specifically for you, so take advantage of them!

In arithmetic, you learned how to do calculations with numbers by using the basic operations of addition, subtraction, multiplication, and division.

In algebra, you will still use numbers and the same four operations. However, you will also use letters to represent numbers. Letters such as x, y, L, or W are called **variables** when they represent numerical values.

Here we see two rectangles whose lengths and widths are labeled with numbers.

 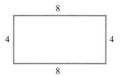

If we need to represent the length and width of *any* rectangle, we can use the variables *L* for length and *W* for width.

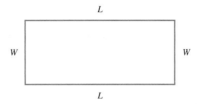

In arithmetic:
+ denotes addition
− denotes subtraction
× denotes multiplication
÷ denotes division.

You are familiar with the four symbols (+, −, ×, ÷) used to indicate the fundamental operations of arithmetic.

Let's look at how these operations are indicated in algebra. We begin by looking at addition.

Addition

$x + y$ means the *sum* of *x* and *y*, or *x plus y*.

Example 1

Writing Expressions That Indicate Addition

Some other words that tell you to add are *more than* and *increased by.*

(a) The *sum* of *a* and 3 is written as $a + 3$.

(b) *L plus W* is written as $L + W$.

(c) 5 *more than m* is written as $m + 5$.

(d) *x increased by* 7 is written as $x + 7$.

✔ CHECK YOURSELF 1

Write, using symbols.

(a) The sum of *y* and 4   (b) *a* plus *b*

(c) 3 more than *x*   (d) *n* increased by 6

Let's look at how subtraction is indicated in algebra.

Subtraction

$x - y$ means the *difference* of x and y, or x *minus* y.

Example 2	**Writing Expressions That Indicate Subtraction**

Some other words that mean to subtract are *decreased by* and *less than*.

(a) *r minus s* is written as $r - s$.

(b) The *difference* of m and 5 is written as $m - 5$.

(c) *x decreased by* 8 is written as $x - 8$.

(d) 4 *less than a* is written as $a - 4$.

   CHECK YOURSELF 2

Write, using symbols.

(a) *w* minus *z*   (b) The difference of *a* and 7

(c) *y* decreased by 3   (d) 5 less than *b*

You have seen that the operations of addition and subtraction are written exactly the same way in algebra as in arithmetic. This is not true in multiplication because the symbol × looks like the letter x. So in algebra we use other symbols to show multiplication to avoid any confusion. Here are some ways to write multiplication.

Multiplication

Note: x and y are called the **factors** of the product xy.

A raised dot	$x \cdot y$	These all indicate the *product* of x and y, or x *times* y.
Parentheses	$(x)(y)$	
Writing the letters next to each other	xy	

Example 3	**Writing Expressions That Indicate Multiplication**

(a) The product of 5 and a is written as $5 \cdot a$, $(5)(a)$, or $5a$. The last expression, $5a$, is the shortest and the most common way of writing the product.

(b) 3 times 7 can be written as $3 \cdot 7$ or $(3)(7)$.

Note: You can place letters next to each other or numbers and letters next to each other to show multiplication. But you *cannot* place numbers side by side to show multiplication: 37 means the number thirty-seven, not 3 times 7.

(c) Twice z is written as $2z$.

(d) The product of 2, s, and t is written as $2st$.

(e) 4 more than the product of 6 and x is written as $6x + 4$.

✔ **CHECK YOURSELF 3**

Write, using symbols.

(a) m times n

(b) The product of h and b

(c) The product of 8 and 9

(d) The product of 5, w, and y

(e) 3 more than the product of 8 and a

Before we move on to division, let's look at how we can combine the symbols we have learned so far.

Not every collection of symbols is an expression.

> An **expression** is a meaningful collection of numbers, variables, and symbols of operation.

Example 4   Identifying Expressions

(a) $2m + 3$ is an expression. It means that we multiply 2 and m, then add 3.

(b) $x + \cdot + 3$ is not an expression. The three operations in a row have no meaning.

(c) $y = 2x - 1$ is not an expression. The equals sign is not an operation sign.

(d) $3a + 5b - 4c$ is an expression.

✔ **CHECK YOURSELF 4**

Identify which are expressions and which are not.

(a) $7 - \cdot x$ (b) $6 + y = 9$

(c) $a + b - c$ (d) $3x - 5yz$

To write more complicated products in algebra, we need some "punctuation marks." Parentheses ( ) mean that an expression is to be thought of as a single quantity. Brackets [ ] are used in exactly the same way as parentheses in algebra. Look at the following example showing the use of these signs of grouping.

| Example 5 | **Expressions with More Than One Operation** |

(a) 3 times the sum of a and b is written as

This can be read as "3 times the quantity a plus b."

$$\underbrace{3(a + b)}$$
$$\uparrow$$

The sum of a and b is a single quantity, so it is enclosed in parentheses.

No parentheses are needed here since the 3 multiplies only a.

(b) The sum of 3 times a and b is written as $3a + b$.

(c) 2 times the difference of m and n is written as $2(m - n)$.

(d) The product of s plus t and s minus t is written as $(s + t)(s - t)$.

(e) The product of b and 3 less than b is written as $b(b - 3)$.

✔ CHECK YOURSELF 5

Write, using symbols.

(a) Twice the sum of p and q (b) The sum of twice p and q

(c) The product of a and the   (d) The product of x plus 2 and
     quantity $b - c$ x minus 2

(e) The product of x and 4 more than x

In algebra the fraction form is usually used.

Now let's look at the operation of division. In arithmetic, you see the division sign ÷, the long division symbol $\overline{)}$, and the fraction notation. For example, to indicate the quotient when 9 is divided by 3, you could write

$$9 \div 3 \qquad \text{or} \qquad 3\overline{)9} \qquad \text{or} \qquad \frac{9}{3}$$

Division

$\dfrac{x}{y}$ means x *divided* by y or the *quotient* of x and y.

| Example 6 | **Writing Expressions That Indicate Division** |

(a) m divided by 3 is written as $\dfrac{m}{3}$.

(b) The quotient of a plus b, divided by 5 is written as $\dfrac{a + b}{5}$.

(c) The quantity p plus q divided by the quantity p minus q is written as $\dfrac{p + q}{p - q}$.

✔ *CHECK YOURSELF 6*

Write, using symbols.

(a) *r* divided by *s*

(b) The quotient when *x* minus *y* is divided by 7

(c) The quantity *a* minus 2 divided by the quantity *a* plus 2

Notice that we can use many different letters to represent variables. In Example 6 the letters *m*, *a*, *b*, *p*, and *q* represented different variables. We often choose a letter that reminds us of what it represents, for example, *L* for *length* or *W* for *width*. These variables may be upper case or lower case letters, although lower case is used more often.

Example 7   Writing Geometric Expressions

(a) *Length* times *width* is written $L \cdot W$.

(b) One-half of *altitude* times *base* is written $\frac{1}{2} a \cdot b$.

(c) *Length* times *width* times *height* is written $L \cdot W \cdot H$.

(d) Pi (π) times *diameter* is written πd.

✔ *CHECK YOURSELF 7*

Write each geometric expression, using symbols.

(a) 2 times *length* plus two times *width*   (b) 2 times pi (π) times *radius*

✔ *CHECK YOURSELF ANSWERS*

1. (a) $y + 4$; (b) $a + b$; (c) $x + 3$; (d) $n + 6$

2. (a) $w - z$; (b) $a - 7$; (c) $y - 3$; (d) $b - 5$

3. (a) mn; (b) hb; (c) $8 \cdot 9$ or $(8)(9)$; (d) $5wy$; (e) $8a + 3$

4. (a) not an expression; (b) not an expression; (c) expression; (d) expression

5. (a) $2(p + q)$; (b) $2p + q$; (c) $a(b - c)$; (d) $(x + 2)(x - 2)$; (e) $x(x + 4)$

6. (a) $\dfrac{r}{s}$; (b) $\dfrac{x - y}{7}$; (c) $\dfrac{a - 2}{a + 2}$ **7.** (a) $2L + 2W$; (b) $2\pi r$

Exercises · 1.1

Write each of the following phrases, using symbols.

1. The sum of c and d

2. a plus 7

3. w plus z

4. The sum of m and n

5. x increased by 5

6. 3 more than b

7. 10 more than y

8. m increased by 4

9. a minus b

10. s less than 5

11. 7 decreased by b

12. r minus 3

13. 6 less than r

14. x decreased by 3

15. w times z

16. The product of 3 and c

17. The product of 5 and t

18. 8 times a

19. The product of 8, m, and n

20. The product of 7, r, and s

21. The product of 8 and the quantity m plus n

22. The product of 5 and the sum of a and b

23. Twice the sum of x and y

24. 3 times the sum of m and n

25. The sum of twice x and y

26. The sum of 3 times m and n

27. Twice the difference of x and y

28. 3 times the difference of c and d

29. The quantity a plus b times the quantity a minus b

30. The product of x plus y and x minus y

31. The product of m and 3 less than m

32. The product of a and 7 less than a

33. 5 divided by x

34. The quotient when b is divided by 8

35. The quotient of a plus b, divided by 7

36. The quantity x minus y, divided by 9

37. The difference of p and q, divided by 4

38. The sum of a and 5 divided by 9

39. The sum of a and 3, divided by the difference of a and 3

40. The difference of m and n, divided by the sum of m and n

Write each of the following phrases, using symbols. Use the variable x to represent the number in each case.

41. 5 more than a number

42. A number increased by 8

43. 7 less than a number

44. A number decreased by 8

64

45. 9 times a number

46. Twice a number

47. 6 more than 3 times a number

48. 5 times a number, decreased by the sum of the number and 3

49. Twice the sum of a number and 5

50. 3 times the difference of a number and 4

51. The product of 2 more than a number and 2 less than that same number

52. The product of 5 less than a number and 5 more than that same number

53. The quotient of a number and 7

54. A number divided by the sum of the number and 7

55. The sum of a number and 5, divided by 8

56. The quotient when 7 less than a number is divided by 3

57. 6 more than a number divided by 6 less than that same number

58. The quotient when 3 less than a number is divided by 3 more than that same number

Write each of the following geometric expressions, using symbols.

59. Four times the length of a side s

60. $\dfrac{4}{3}$ times π times the cube of the radius r

61. π times the radius r squared times the height h

62. Twice the length L plus twice the width W

63. One-half the product of the height h and the sum of two unequal sides b_1 and b_2

64. Six times the length of a side s squared

Identify which are expressions and which are not.

65. $2(x + 5)$

66. $4 - (x + 3)$

67. $4 + \div m$

68. $6 + a = 7$

69. $2b = 6$

70. $x(y + 3)$

71. $2a(3b + 5)$

72. $4x + \cdot 7$

73. **Number theory.** Two numbers have a sum of 35. If one number is x, express the other number in terms of x.

74. **Species extinction.** It is estimated that earth is losing 4,000 species of plants and animals every year. If S represents the number of species living last year, how many species are on earth this year?

75. **Interest.** The simple interest earned when a principal P is invested at a rate r for a time t is calculated by multiplying the principal by the rate by the time. Write an expression for the interest earned.

76. **Kinetic energy.** The kinetic energy of a particle of mass m is found by taking one-half of the product of the mass and the square of the velocity v. Write an expression for the kinetic energy of a particle.

77. Recreation. Four hundred tickets were sold for a school play. The tickets were of two types: general admission and student. There were x general admission tickets sold. Write a formula for the number of student tickets sold.

78. Geometry. Two angles are supplementary if their sum is 180°. One angle measures $d°$. Write a formula for the measurement of its supplement.

79. Money. Patrick has $375 in his bank account. He wrote a check for x dollars for a concert ticket. Write an expression that will represent the remaining money in his account.

80. Rewrite the following algebraic expressions in English phrases. Exchange papers with another student to edit your writing. Be sure the meaning in English is the same as in algebra. These expressions are not complete sentences, so your English does not have to be in complete sentences. Here is an example.

Algebra: $2(x - 1)$

English: We could write "1 less than a number doubled." Or we might write "a number diminished by 1 and then multiplied by 2."

(a) $n + 3$ (b) $\dfrac{x + 2}{5}$ (c) $3(5 + a)$ (d) $3 - 4n$ (e) $\dfrac{x + 6}{x - 1}$

1.2 Evaluating Algebraic Expressions

1.2 OBJECTIVES

1. Evaluate algebraic expressions given any real number value for the variable
2. Use a graphing calculator to evaluate algebraic expressions

"Black holes are where God divided by zero."

—Steven Wright

Overcoming Math Anxiety

Hint 3: Working Together

How many of your classmates do you know? Whether you are by nature gregarious or shy, you have much to gain by getting to know your classmates.

1. It is important to have someone to call when you have missed class or if you are unclear on an assignment.

2. Working with another person is almost always beneficial to both people. If you don't understand something, it helps to have someone to ask about it. If you do understand something, nothing will cement that understanding more than explaining the idea to another person.

3. Sometimes we need to commiserate. If an assignment is particularly frustrating, it is reassuring to find out that it is also frustrating for other students.

4. Have you ever thought you had the right answer, but it doesn't match the answer in the text? Frequently the answers are equivalent, but that's not always easy to see. A different perspective can help you see that. Occasionally there is an error in a textbook (here we are talking about *other* textbooks). In such cases it is wonderfully reassuring to find that someone else has the same answer as you do.

In applying algebra to problem solving, you will often want to find the value of an algebraic expression when you know certain values for the letters (or variables) in the expression. Finding the value of an expression is called *evaluating the expression* and uses the following steps.

To Evaluate an Algebraic Expression

Step 1.   Replace each variable with the given number value.

Step 2.   Do the necessary arithmetic operations, following the rules for order of operations.

Example 1   **Evaluating Algebraic Expressions**

Suppose that $a = 5$ and $b = 7$.

(a) To evaluate $a + b$, we replace a with 5 and b with 7.

$$a + b = 5 + 7 = 12$$

(b) To evaluate $3ab$, we again replace a with 5 and b with 7.

$$3ab = 3 \cdot 5 \cdot 7 = 105$$

  CHECK YOURSELF 1

If $x = 6$ and $y = 7$, evaluate.

(a) $y - x$ (b) $5xy$

We are now ready to evaluate algebraic expressions that require following the rules for the order of operations.

Example 2   **Evaluating Algebraic Expressions**

Evaluate the following expressions if $a = 2, b = 3, c = 4$, and $d = 5$.

(a) $5a + 7b$

$5a + 7b = 5 \cdot 2 + 7 \cdot 3$ Multiply first.

$\qquad = 10 + 21 = 31$ Then add.

Caution

This is different from

$(3c)^2 = (3 \cdot 4)^2$
$\qquad = 12^2 = 144$

(b) $3c^2$

$3c^2 = 3 \cdot 4^2$ Evaluate the power.

$\qquad = 3 \cdot 16 = 48$ Then multiply.

(c) $7(c + d)$

$$7(c + d) = 7(4 + 5) \qquad \text{Add inside the parentheses.}$$
$$= 7 \cdot 9 = 63 \qquad \text{Multiply.}$$

(d) $5a^4 - 2d^2$

$$5a^4 - 2d^2 = 5 \cdot 2^4 - 2 \cdot 5^2 \qquad \text{Evaluate the powers.}$$
$$= 5 \cdot 16 - 2 \cdot 25 \qquad \text{Multiply.}$$
$$= 80 - 50 = 30 \qquad \text{Subtract.}$$

✔ *CHECK YOURSELF 2*

If $x = 3$, $y = 2$, $z = 4$, and $w = 5$, evaluate the following expressions.

(a) $4x^2 + 2$ (b) $5(z + w)$

(c) $7(z^2 - y^2)$

To evaluate algebraic expressions when a fraction bar is used, do the following: Start by doing all the work in the numerator, then do the work in the denominator. Divide the numerator by the denominator as the last step.

Example 3

Evaluating Algebraic Expressions

If $p = 2$, $q = 3$, and $r = 4$, evaluate.

(a) $\dfrac{8p}{r}$

Replace p with 2 and r with 4.

$$\frac{8p}{r} = \frac{8 \cdot 2}{4} = \frac{16}{4} = 4 \qquad \text{Divide as the last step.}$$

(b) $\dfrac{7q + r}{p + q}$

$$\frac{7q + r}{p + q} = \frac{7 \cdot 3 + 4}{2 + 3} \qquad \text{Evaluate the top and bottom separately.}$$
$$= \frac{21 + 4}{2 + 3}$$
$$= \frac{25}{5} = 5$$

✔ CHECK YOURSELF 3

Evaluate the following if $c = 5$, $d = 8$, and $e = 3$.

(a) $\dfrac{6c}{e}$ (b) $\dfrac{4d + e}{c}$

(c) $\dfrac{10d - e}{d + e}$

Example 4 shows how a calculator can be used to evaluate algebraic expressions.

Example 4

Using a Calculator to Evaluate Expressions

Use a calculator to evaluate the following expressions.

(a) $\dfrac{4x + y}{z}$

$\dfrac{4x + y}{z}$ if $x = 2$, $y = 1$, and $z = 3$

Replace x with 2, y with 1, and z with 3:

$$\frac{4x + y}{z} = \frac{4 \cdot 2 + 1}{3}$$

Remember to use parentheses to group the numerator and denominator.

Now, use the following keystrokes:

$$\boxed{(}\ 4\ \boxed{\times}\ 2\ \boxed{+}\ 1\ \boxed{)}\ \boxed{\div}\ 3\ \boxed{\text{ENTER}}$$

The display will read 3.

(b) $\dfrac{7x - y}{3z - x}$

$\dfrac{7x - y}{3z - x}$ if $x = 2$, $y = 6$, and $z = 2$

$$\frac{7x - y}{3z - x} = \frac{7 \cdot 2 - 6}{3 \cdot 2 - 2}$$

Use the following keystrokes:

$$\boxed{(}\ 7\ \boxed{\times}\ 2\ \boxed{-}\ 6\ \boxed{)}\ \boxed{\div}\ \boxed{(}\ 3\ \boxed{\times}\ 2 - 2\ \boxed{)}\ \boxed{\text{ENTER}}$$

The display will read 2.

✔ *CHECK YOURSELF 4*

Use a calculator to evaluate the following if $x = 2$, $y = 6$, and $z = 5$.

(a) $\dfrac{2x + y}{z}$ (b) $\dfrac{4y - 2z}{x}$

| Example 5 | **Evaluating Expressions** |

Evaluate $5a + 4b$ if $a = -2$ and $b = \dfrac{3}{4}$.

Replace a with -2 and b with $\dfrac{3}{4}$.

Remember the rules for the order of operations. Multiply first, then add.

$$5a + 4b = 5(-2) + 4\left(\dfrac{3}{4}\right)$$
$$= -10 + 3$$
$$= -7$$

✔ *CHECK YOURSELF 5*

Evaluate $3x + 5y$ if $x = -2$ and $y = -\dfrac{4}{5}$.

We follow the same rules no matter how many variables are in the expression.

| Example 6 | **Evaluating Expressions** |

Evaluate the following expressions if $a = -4$, $b = 2$, $c = -5$, and $d = 6$.

(a) $7a - 4c$ This becomes $-(-20)$, or $+20$.

$$7a - 4c = 7(-4) - 4(-5)$$
$$= -28 + 20$$
$$= -8$$

(b) $7c^2$ Evaluate the power first, then multiply by 7.

$$7c^2 = 7(-5)^2 = 7 \cdot 25$$
$$= 175$$

Caution

When a squared variable is replaced by a negative number, square the negative.

$$(-5)^2 = (-5)(-5) = 25$$

The exponent applies to -5!

$$-5^2 = -(5 \cdot 5) = -25$$

The exponent applies only to 5!

(c) $b^2 - 4ac$

$$b^2 - 4ac = 2^2 - 4(-4)(-5)$$

$$= 4 - 4(-4)(-5)$$

$$= 4 - 80$$

$$= -76$$

(d) $b(a + d)$ ⟶ Add inside the parentheses first

$$b(a + d) = 2(-4 + 6)$$

$$= 2(2)$$

$$= 4$$

✔ CHECK YOURSELF 6

Evaluate if $p = -4$, $q = 3$, and $r = -2$.

(a) $5p - 3r$ (b) $2p^2 + q$ (c) $p(q + r)$

(d) $-q^2$ (e) $(-q)^2$

If an expression involves a fraction, remember that the fraction bar is a grouping symbol. This means that you should do the required operations first in the numerator and then in the denominator. Divide as the last step.

Example 7   **Evaluating Expressions**

Evaluate the following expressions if $x = 4$, $y = -5$, $z = 2$, and $w = -3$.

(a) $\dfrac{z - 2y}{x}$

$$\frac{z - 2y}{x} = \frac{2 - 2(-5)}{4} = \frac{2 + 10}{4}$$

$$= \frac{12}{4} = 3$$

(b) $\dfrac{3x - w}{2x + w}$

$$\frac{3x - w}{2x + w} = \frac{3(4) - (-3)}{2(4) + (-3)} = \frac{12 + 3}{8 + (-3)}$$

$$= \frac{15}{5} = 3$$

✔ *CHECK YOURSELF 7*

Evaluate if $m = -6$, $n = 4$, and $p = -3$.

(a) $\dfrac{m + 3n}{p}$ (b) $\dfrac{4m + n}{m + 4n}$

✔ *CHECK YOURSELF ANSWERS*

1. (a) 1; (b) 210   **2.** (a) 38; (b) 45; (c) 84   **3.** (a) 10; (b) 7; (c) 7

4. (a) 2; (b) 7   **5.** -10 **6.** (a) -14; (b) 35; (c) -4; (d) -9; (e) 9

7. (a) -2; (b) -2

Exercises · 1.2

Evaluate each of the expressions if $a = -2$, $b = 5$, $c = -4$, and $d = 6$.

1. $3c - 2b$

2. $4c - 2b$

3. $7c + 6b$

4. $7a - 2c$

5. $-b^2 + b$

6. $(-c)^2 + 5c$

7. $3a^2$

8. $6c^2$

9. $c^2 - 2d$

10. $3a^2 + 4c$

11. $2a^2 + 3b^2$

12. $4b^2 - 2c^2$

13. $2(c - d)$

14. $5(b - c)$

15. $4(2a - d)$

16. $6(3c - d)$

17. $a(b + 3c)$

18. $c(3a - d)$

19. $\dfrac{6d}{c}$

20. $\dfrac{3a}{5b}$

21. $\dfrac{3d + 2c}{b}$

22. $\dfrac{2b + 3d}{2a}$

23. $\dfrac{2b - 3a}{c + 2d}$

24. $\dfrac{3d - 2b}{5a + d}$

25. $b^2 - d^2$

26. $d^2 - b^2$

27. $(b - d)^2$

28. $(d - b)^2$

29. $(d - b)(d + b)$

30. $(c - a)(c + a)$

31. $c^3 - a^3$

32. $c^3 + a^3$

33. $(c - a)^3$

34. $(c + a)^3$

35. $(d - b)(d^2 + db + b^2)$

36. $(c + a)(c^2 - ac + a^2)$

37. $b^2 + a^2$

38. $d^2 - a^2$

39. $(b + a)^2$

40. $(d - a)^2$

41. $a^2 + 2ad + d^2$

42. $d^2 - 2ad + a^2$

Evaluate each of the expression if $x = -2$, $y = -3$, and $z = 4$.

43. $x^2 - 2y^2 + z^2$

44. $4yz + 6xy$

45. $2xy - (x^2 - 2yz)$

46. $3yz - 6xyz + x^2y^2$

47. $2y(z^2 - 2xy) + yz^2$

48. $-z - (-2x - yz)$

Use a calculator to evaluate each expression if $x = -2.34$, $y = -3.14$, and $z = 4.12$. Round your answer to the nearest tenth.

49. $x + yz$

50. $y - 2z$

51. $y^2 - 2x^2$

52. $x^2 + y^2$

53. $\dfrac{xy}{z - x}$

54. $\dfrac{y^2}{zy}$

55. $\dfrac{2x + y}{2x + z}$

56. $\dfrac{y^2z^2}{xy}$

Use a calculator to evaluate the expression $x^2 - 4x^3 + 3x$ for each given value.

57. $x = 3$

58. $x = 12$

59. $x = 27$

60. $x = 48$

Decide if the given numbers make the statement true or false.

61. $x - 7 = 2y + 5$; $x = 22$, $y = 5$

62. $3(x - y) = 6$; $x = 5$, $y = -3$

63. $2(x + y) = 2x + y$; $x = -4$, $y = -2$

64. $x^2 - y^2 = x - y$; $x = 4$, $y = -3$

74

65. Electrical resistance. The formula for the total resistance in a parallel circuit is $R_T = \dfrac{R_1 R_2}{R_1 + R_2}$. Find the total resistance if $R_1 = 9$ ohms (Ω) and $R_2 = 15\ \Omega$.

66. Area. The formula for the area of a triangle is given by $A = \dfrac{1}{2}\,ab$, where a is the altitude (or height) and b is the length of the base. Find the area of a triangle if $a = 4$ centimeters (cm) and $b = 8$ cm.

67. Perimeter. The perimeter of a rectangle of length L and width W is given by the formula $P = 2L + 2W$. Find the perimeter when $L = 10$ inches (in.) and $W = 5$ in.

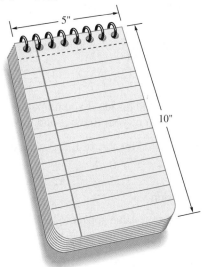

68. Simple interest. The simple interest I on a principal of P dollars at interest rate r for time t, in years, is given by $I = Prt$. Find the simple interest on a principal of $6,000 at 4% for 3 years. (**Note:** 4% = 0.04.)

69. Simple interest. Use the formula $P = \dfrac{I}{r \cdot t}$ to find the principal if the total interest earned was $150 and the rate of interest was 4% for 2 years.

70. Simple interest. Use the formula $r = \dfrac{I}{P \cdot t}$ to find the rate of interest if $5,000 earns $1,500 interest in 6 years.

71. Temperature conversion. The formula that relates Celsius and Fahrenheit temperatures is $F = \dfrac{9}{5}C + 32$. If the temperature is $-10°C$, what is the Fahrenheit temperature?

72. Geometry. If the area of a circle whose radius is r is given by $A = \pi r^2$, where $\pi = 3.14$, find the area when $r = 3$ meters (m).

73. Business. A local telephone company offers a long-distance telephone plan that charges $5.25 per month and $0.08 per minute of calling time. The expression $0.08t + 5.25$ represents the monthly long-distance bill for a customer who makes t minutes (min) of long-distance calling on this plan. Find the monthly bill for a customer who makes 173 min of long-distance calls on this plan.

74. Motion. The speed of a model car as it slows down is given by $v = 20 - 4t$, where v is the speed in meters per second (m/s) and t is the time in seconds (s) during which the car has slowed. Find the speed of the car 1.5 s after it has begun to slow.

75. Write an English interpretation of each of the following algebraic expressions.

    (a) $(2x^2 - y)^3$ (b) $3n = \dfrac{n - 1}{2}$ (c) $(2n + 3)(n - 4)$

76. Is $a^n + b^n = (a + b)^n$? Try a few numbers and decide if you think this is true for all numbers, for some numbers, or never true. Write an explanation of your findings and give examples.

77. (a) Evaluate the expression $4x(5 - x)(6 - x)$ for $x = 0, 1, 2, 3, 4,$ and 5. Complete the table below.

Value of x	0	1	2	3	4	5
Value of Expression						

    (b) For which value of x does the expression value appear to be largest?

    (c) Evaluate the expression for $x = 1.5, 1.6, 1.7, 1.8, 1.9, 2.0, 2.1, 2.2, 2.3, 2.4,$ and 2.5. Complete the table.

Value of x	1.5	1.6	1.7	1.8	1.9	2.0	2.1	2.2	2.3	2.4	2.5
Value of Expression											

    (d) For which value of x does the expression value appear to be largest?

    (e) Continue the search for the value of x that produces the greatest expression value. Determine this value of x to the nearest hundredth.

78. Work with other students on this exercise.

Part 1: Evaluate the three expressions $\dfrac{n^2 - 1}{2}$, n, $\dfrac{n^2 + 1}{2}$, using odd values of n: 1, 3, 5, 7, etc. Make a chart like the one below and complete it.

n	$a = \dfrac{n^2-1}{2}$	$b = n$	$c = \dfrac{n^2 + 1}{2}$	a^2	b^2	c^2
1						
3						
5						
7						
9						
11						
13						
15						

Part 2: The numbers a, b, and c that you get in each row have a surprising relationship to each other. Complete the last three columns and work together to discover this relationship. You may want to find out more about the history of this famous number pattern.

79. In problem 78 you investigated the numbers obtained by evaluating the following expressions for odd positive integer values of n: $\dfrac{n^2 - 1}{2}, n, \dfrac{n^2 + 1}{2}$. Work with other students to investigate what three numbers you get when you evaluate for a *negative* odd value. Does the pattern you observed before still hold? Try several negative odd numbers to test the pattern.

Have no fear of fractions—does the pattern work with fractions? Try even integers. Is there a pattern for the three numbers obtained when you begin with even integers?

80. Enjoyment of patterns in art, music, and language is common to all cultures, and many cultures also delight in and draw spiritual significance from patterns in numbers. One such set of patterns is that of the "magic" square. One of these squares appears in a famous etching by Albrecht Dürer, who lived from 1471 to 1528 in Europe. He was one of the first artists in Europe to use geometry to give perspective, a feeling of three dimensions, in his work. The magic square in his work is this one:

16	3	2	13
5	10	11	8
9	6	7	12
4	15	14	1

Why is this square "magic"? It is magic because every row, every column, and both diagonals add to the same number. In this square there are 16 spaces for the numbers 1 through 16.

Part 1: What number does each row and column add to?

Write the square that you obtain by adding -17 to each number. Is this still a magic square? If so, what number does each column and row add to? If you add 5 to each number in the original magic square, do you still have a magic square? You have been studying the operations of addition, multiplication, subtraction, and division with integers and with rational numbers. What operations can you perform on this magic square and still have a magic square? Try to find something that will not work. Use algebra to help you decide what will work and what won't. Write a description of your work and explain your conclusions.

Part 2: Here is the oldest published magic square. It is from China, about 250 B.C.E. Legend has it that it was brought from the River Lo by a turtle to Emperor Yii, who was a hydraulic engineer.

4	9	2
3	5	7
8	1	6

Check to make sure that this is a magic square. Work together to decide what operation might be done to every number in the magic square to make the sum of each row, column, and diagonal the *opposite* of what it is now. What would you do to every number to cause the sum of each row, column, and diagonal to equal zero?

1.3 Adding and Subtracting Algebraic Expressions

1.3 OBJECTIVES

1. Combine like terms
2. Add algebraic expressions
3. Subtract algebraic expressions

To find the perimeter of (or the distance around) a rectangle, we add 2 times the length and 2 times the width. In the language of algebra, this can be written as

$$\text{Perimeter} = 2L + 2W$$

We call $2L + 2W$ an **algebraic expression,** or more simply an **expression.** As we discussed in Section 1.1, an expression allows us to write a mathematical idea in symbols. It can be thought of as a meaningful collection of letters, numbers, and operation symbols.

Some expressions are

1. $5x^2$

2. $3a + 2b$

3. $4x^3 - 2y + 1$

4. $3(x^2 + y^2)$

In algebraic expressions, the addition and subtraction signs break the expressions into smaller parts called *terms.*

If a variable has no exponent, it is raised to the power 1.

> A **term** is a number, or the product of a number and one or more variables, raised to a power.

In an expression, each sign (+ or −) is a part of the term that follows the sign.

Example 1

Identifying Terms

(a) $5x^2$ has one term.

(b) $\underbrace{3a}_{\text{Term}} + \underbrace{2b}_{\text{Term}}$ has two terms: $3a$ and $2b$.

Note that each term "owns" the sign that precedes it.

(c) $\underbrace{4x^3}_{\text{Term}} \underbrace{- 2y}_{\text{Term}} \underbrace{+ 1}_{\text{Term}}$ has three terms: $4x^3$, $-2y$, and 1.

✔ CHECK YOURSELF 1

List the terms of each expression.

(a) $2b^4$ (b) $5m + 3n$ (c) $2s^2 - 3t - 6$

Note that a term in an expression may have any number of factors. For instance, $5xy$ is a term. It has factors of 5, x, and y. The number factor of a term is called the **numerical coefficient.** So for the term $5xy$, the numerical coefficient is 5.

Example 2

Identifying the Numerical Coefficient

(a) $4a$ has the numerical coefficient 4.

(b) $6a^3b^4c^2$ has the numerical coefficient 6.

(c) $-7m^2n^3$ has the numerical coefficient -7.

(d) x has the numerical coefficient 1 since $x = 1 \cdot x$.

✔ CHECK YOURSELF 2

Give the numerical coefficient for each of the following terms.

(a) $8a^2b$ (b) $-5m^3n^4$ (c) y

If terms contain exactly the *same letters* (or variables) raised to the *same powers,* they are called **like terms.**

Example 3	**Identifying Like Terms**

(a) The following are like terms.

$6a$ and $7a$ Each pair of terms has the same letters, with matching
$5b^2$ and b^2 letters raised to the same power—the numerical coefficients
$10x^2y^3z$ and $-6x^2y^3z$ can be any number.

(b) The following are *not* like terms.

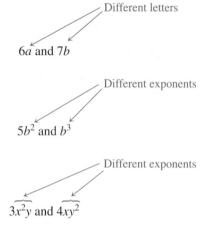

$6a$ and $7b$ Different letters

$5b^2$ and b^3 Different exponents

$3x^2y$ and $4xy^2$ Different exponents

✔ **CHECK YOURSELF 3**

Circle the like terms.

$$5a^2b \qquad ab^2 \qquad a^2b \qquad -3a^2 \qquad 4ab \qquad 3b^2 \qquad -7a^2b$$

Like terms of an expression can always be combined into a single term. Look at the following:

$$\underbrace{2x}_{x + x} + \underbrace{5x}_{x + x + x + x + x} = \underbrace{7x}_{x + x + x + x + x + x + x}$$

Rather than having to write out all those x's, try

Here we use the distributive property.

$$2x + 5x = (2 + 5)x = 7x$$

In the same way,

You don't have to write all this out—just do it mentally!

$$9b + 6b = (9 + 6)b = 15b$$

and

$$10a - 4a = (10 - 4)a = 6a$$

This leads us to the following rule.

> **Combining Like Terms**
>
> To combine like terms, do the following steps.
>
> Step 1.   Add or subtract the numerical coefficients.
>
> Step 2.   Attach the common variables.

| **Example 4** | **Combining Like Terms** |

Combine like terms.

(a) $8m + 5m = (8 + 5)m = 13m$

(b) $5pq^3 - 4pq^3 = 1pq^3 = pq^3$

(c) $7a^3b^2 - 7a^3b^2 = 0a^3b^2 = 0$

Remember that when any factor is multiplied by 0, the product is 0.

✔   *CHECK YOURSELF 4*

Combine like terms.

(a) $6b + 8b$ (b) $12x^2 - 3x^2$

(c) $8xy^3 - 7xy^3$ (d) $9a^2b^4 - 9a^2b^4$

Let's look at some expressions involving more than two terms. The idea is the same.

| **Example 5** | **Combining Like Terms** |

Combine like terms.

(a) $5ab - 2ab + 3ab$

   $= (5 - 2 + 3)ab = 6ab$

The distributive property can be used over any number of like terms.

Only like terms can be combined.

(b)  $8x - 2x + 5y$

   $= 6x + 5y$

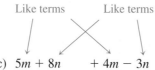

Like terms   Like terms

With practice you won't be
writing out these steps, but
doing them mentally.

(c) $5m + 8n \quad + 4m - 3n$ The order of the terms has been rearranged by using the
                                associative and commutative properties of addition.

$= (5m + 4m) + (8n - 3n)$

$= 9m \qquad + \qquad 5n$

Be careful when moving terms.
Remember that they own the
signs in front of them.

(d) $4x^2 + 2x - 3x^2 + x$

$= (4x^2 - 3x^2) + (2x + x)$

$= x^2 + 3x$

As these examples illustrate, combining like terms often means changing the
grouping and the order in which the terms are written. Again all this is possible because
of the properties of addition that we introduced in Section 0.3.

✔   *CHECK YOURSELF 5*

Combine like terms.

(a) $4m^2 - 3m^2 + 8m^2$ (b) $9ab + 3a - 5ab$

(c) $4p + 7q + 5p - 3q$

Addition is always a matter of combining like quantities (two apples plus three
apples, four books plus five books, and so on). If you keep that basic idea in mind,
adding expressions will be easy. It is just a matter of combining like terms. Suppose that
you want to add

$$5x^2 + 3x + 4 \qquad \text{and} \qquad 4x^2 + 5x - 6$$

Parentheses are sometimes used in adding, so for the sum of these expressions, we
can write

$$(5x^2 + 3x + 4) + (4x^2 + 5x - 6)$$

Now what about the parentheses? You can use the following rule.

Removing Grouping Symbols Case 1

If a plus sign (+) or nothing at all appears in front of parentheses, just remove the
parentheses. No other changes are necessary.

Now let's return to the addition.

$$(5x^2 + 3x + 4) + (4x^2 + 5x - 6)$$
$$= 5x^2 + 3x + 4 + 4x^2 + 5x - 6$$

Like terms

Like terms

Like terms

Just remove the parentheses. No other changes are necessary.

Note the use of the associative and commutative properties in reordering and regrouping.

Collect like terms. (*Remember:* Like terms have the same variables raised to the same power.)

$$= (5x^2 + 4x^2) + (3x + 5x) + (4 - 6)$$

Here we use the distributive property. For example,

$$5x^2 + 4x^2 = 9x^2$$

Combine like terms for the result:

$$= 9x^2 + 8x - 2$$

Much of this work can be done mentally. You can then write the sum directly by locating like terms and combining. Example 6 illustrates this property.

Example 6

Combining Like Terms

Add $3x - 5$ and $2x + 3$.

Write the sum.

$$(3x - 5) + (2x + 3)$$
$$= 3x - 5 + 2x + 3 = 5x - 2$$

Like terms　　　Like terms

✔ CHECK YOURSELF 6

Add $6x^2 + 2x$ and $4x^2 - 7x$.

Subtracting expressions requires another rule for removing signs of grouping.

Removing Grouping Symbols Case 2

If a minus sign (−) appears in front of a set of parentheses, the parentheses can be removed by changing the sign of each term inside the parentheses.

The use of this rule is illustrated in Example 7.

| Example 7 | **Removing Parentheses** |

In each of the following, remove the parentheses.

Note: This uses the distributive property, since

$$-(2x + 3y) = (-1)(2x + 3y)$$
$$= -2x - 3y$$

(a) $-(2x + 3y) = -2x - 3y$ Change each sign when removing the parentheses.

(b) $m - (5n - 3p) = m \underbrace{- 5n + 3p}_{\text{Sign changes}}$

(c) $2x - (-3y + z) = 2x \underbrace{+ 3y - z}_{\text{Sign changes}}$

✔ *CHECK YOURSELF 7*

Remove the parentheses.

(a) $-(3m + 5n)$ (b) $-(5w - 7z)$ (c) $3r - (2s - 5t)$ (d) $5a - (-3b - 2c)$

Subtracting expressions is now a matter of using the previous rule when removing the parentheses and then combining the like terms. Consider Example 8.

| Example 8 | **Subtracting Expressions** |

(a) Subtract $5x - 3$ from $8x + 2$.

Write

Note: The expression following *from* is written first in the problem.

$$(8x + 2) - (5x - 3)$$
$$= 8x + 2 \underbrace{- 5x + 3}_{\text{Sign changes}}$$
$$= 3x + 5$$

(b) Subtract $4x^2 - 8x + 3$ from $8x^2 + 5x - 3$.

Write

$$(8x^2 + 5x - 3) - (4x^2 - 8x + 3)$$
$$= 8x^2 + 5x - 3 \underbrace{- 4x^2 + 8x - 3}_{\text{Sign changes}}$$
$$= 4x^2 + 13x - 6$$

✔ CHECK YOURSELF 8

(a) Subtract $7x + 3$ from $10x - 7$.

(b) Subtract $5x^2 - 3x + 2$ from $8x^2 - 3x - 6$.

✔ CHECK YOURSELF ANSWERS

1. (a) $2b^4$; (b) $5m$, $3n$; (c) $2s^2$, $-3t$, -6 **2.** (a) 8; (b) -5; (c) 1

3. The like terms are $5a^2b$, a^2b, and $-7a^2b$.   **4.** (a) $14b$; (b) $9x^2$; (c) xy^3; (d) 0

5. (a) $9m^2$; (b) $4ab + 3a$; (c) $9p + 4q$ **6.** $10x^2 - 5x$

7. (a) $-3m - 5n$; (b) $-5w + 7z$; (c) $3r - 2s + 5t$; (d) $5a + 3b + 2c$

8. (a) $3x - 10$; (b) $3x^2 - 8$

Exercises · 1.3

List the terms of the following expressions.

1. $5a + 2$

2. $7a - 4b$

3. $5x^4$

4. $3x^2$

5. $3x^2 + 3x - 7$

6. $2a^3 - a^2 + a$

Circle the like terms in the following groups of terms.

7. $5ab, 3b, 3a, 4ab$

8. $9m^2, 8mn, 5m^2, 7m$

9. $4xy^2, 2x^2y, 5x^2, -3x^2y, 5y, 6x^2y$

10. $8a^2b, 4a^2, 3ab^2, -5a^2b, 3ab, 5a^2b$

Combine the like terms.

11. $6p + 9p$

12. $6a^2 + 8a^2$

13. $7b^3 + 10b^3$

14. $7rs + 13rs$

15. $21xyz + 7xyz$

16. $4n^2m + 11n^2m$

17. $9z^2 - 3z^2$

18. $7m - 6m$

19. $5a^3 - 5a^3$

20. $9xy - 13xy$

21. $16p^2q - 17p^2q$

22. $7cd - 7cd$

23. $6p^2q - 21p^2q$

24. $8r^3s^2 - 17r^3s^2$

25. $10x^2 - 7x^2 + 3x^2$

26. $13uv + 5uv - 12uv$

27. $-6c + 3d + 5c$

28. $5m^2 - 3m + 6m^2$

29. $4x + 4y - 7x - 5y$

30. $7a - 4a^2 - 13a + 9a^2$

31. $2a - 7b - 3 + 2a - 3b + 2$

32. $5p^2 - 2p - 8 - 7p^2 + 5p + 6$

Remove the parentheses in each of the following expressions, and simplify where possible.

33. $-(2a + 3b)$

34. $-(7x - 4y)$

35. $5a - (2b - 3c)$

36. $7x - (4y + 3z)$

37. $3x - (4y + 5x)$

38. $10m - (3m - 2n)$

39. $5p - (-3p + 2q)$

40. $8d - (-7c - 2d)$

Add.

41. $6a - 5$ and $3a + 9$

42. $9x + 3$ and $3x - 4$

43. $-7p^2 + 9p$ and $4p^2 - 5p$

44. $2m^2 + 3m$ and $6m^2 - 8m$

45. $3x^2 - 2x$ and $-5x^2 + 2x$

46. $3p^2 + 5p$ and $-7p^2 - 5p$

47. $2x^2 + 5x - 3$ and $3x^2 - 7x + 4$

48. $4d^2 - 8d + 7$ and $5d^2 - 6d - 9$

49. $2b^2 + 8$ and $5b + 8$

50. $5p - 2$ and $4p^2 - 7p$

51. $8y^3 - 5y^2$ and $5y^2 - 2y$

52. $9x^4 - 2x^2$ and $2x^2 + 3$

53. $3x^2 - 7x^3$ and $-5x^2 + 4x^3$ **54.** $9m^3 - 2m$ and $-6m - 4m^3$ **55.** $4x^2 - 2 + 7x$ and $5 - 8x - 6x^2$

56. $5b^3 - 8b + 2b^2$ and $3b^2 - 7b^3 + 5b$

Subtract.

57. $x + 2$ from $3x - 5$ **58.** $x - 2$ from $3x + 5$ **59.** $3m^2 - 2m$ from $4m^2 - 5m$

60. $9a^2 - 5a$ from $11a^2 - 10a$ **61.** $6y^2 + 5y$ from $4y^2 + 5y$ **62.** $9x^2 - 2x$ from $6x^2 - 2x$

63. $x^2 - 4x - 3$ from $3x^2 - 5x - 2$ **64.** $3x^2 - 2x + 4$ from $5x^2 - 8x - 3$ **65.** $3a + 7$ from $8a^2 - 9a$

66. $3x^3 + x^2$ from $4x^3 - 5x$ **67.** $2p - 5p^2$ from $-9p^2 + 4p$ **68.** $7y - 3y^2$ from $3y^2 - 2y$

69. $x^2 - 5 - 8x$ from $3x^2 - 8x + 7$ **70.** $4x - 2x^2 + 4x^3$ from $4x^3 + x - 3x^2$

Perform the indicated operations.

71. Subtract $3b + 2$ from the sum of $4b - 2$ and $5b + 3$.

72. Subtract $5m - 7$ from the sum of $2m - 8$ and $9m - 2$.

73. Subtract $5x^2 + 7x - 6$ from the sum of $2x^2 - 3x + 5$ and $3x^2 + 5x - 7$.

74. Subtract $4x^2 - 5x - 3$ from the sum of $x^2 - 3x - 7$ and $2x^2 - 2x + 9$.

75. Subtract $2x^2 - 3x$ from the sum of $4x^2 - 5$ and $2x - 7$.

76. Subtract $5a^2 - 3a$ from the sum of $3a - 3$ and $5a^2 + 5$.

77. Subtract the sum of $3y^2 - 3y$ and $5y^2 + 3y$ from $2y^2 - 8y$.

78. Subtract the sum of $3y^3 + 7y^2$ and $5y^3 - 7y^2$ from $4y^3 - 5y^2$.

Perform the indicated operations.

79. $[(9x^2 - 3x + 5) - (3x^2 + 2x - 1)] - (x^2 - 2x - 3)$

80. $[(5x^2 + 2x - 3) - (-2x^2 + x - 2)] - (2x^2 + 3x - 5)$

Using your calculator, evaluate each of the following for the given values of the variables. Round your answer to the nearest tenth.

81. $7x^2 - 5y^3$ for $x = 7.1695$ and $y = 3.128$ **82.** $2x^2 + 3y + 5x$ for $x = 3.61$ and $y = 7.91$

83. $4x^2y + 2xy^2 - 5x^3y$ for $x = 1.29$ and $y = 2.56$ **84.** $3x^3y - 4xy + 2x^2y^2$ for $x = 3.26$ and $y = 1.68$

85. Geometry. A rectangle has sides of $8x + 9$ and $6x - 7$. Find the expression that represents its perimeter.

86. Geometry. A triangle has sides $4x + 7$, $6x + 3$, and $2x - 5$. Find the expression that represents its perimeter.

87. Business. The cost of producing x units of an item is $C = 150 + 25x$. The revenue for selling x units is $R = 90x - x^2$. The profit is given by the revenue minus the cost. Find the expression that represents profit.

88. Business. The revenue for selling y units is $R = 3y^2 - 2y + 5$, and the cost of producing y units is $C = y^2 + y - 3$. Find the expression that represents profit.

89. Carpentry. A wooden beam is $3y^2 + 3y - 2$ meters (m) long. If a piece $y^2 - 8$ is cut, find an expression that represents the length of the remaining piece of beam.

90. Construction. A steel girder is $9y^2 + 6y - 4$ m long. Two pieces are cut from the girder. One has length $(3y^2 + 2y - 1)$ m and the other has length $(4y^2 + 3y - 2)$ m. Find the length of the remaining piece.

91. Geometry. Find an expression for the perimeter of the given triangle.

x ft $(3x + 3)$ ft $(2x^2 - 5x + 1)$ ft

92. Geometry. Find an expression for the perimeter of the given rectangle.

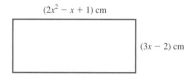

$(2x^2 - x + 1)$ cm $(3x - 2)$ cm

93. Geometry. Find an expression for the perimeter of the given figure.

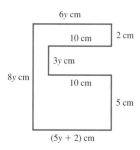

$6y$ cm 10 cm 2 cm $3y$ cm $8y$ cm 10 cm 5 cm $(5y + 2)$ cm

94. Geometry. Find the perimeter of the following figure.

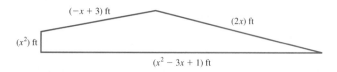

$(-x + 3)$ ft $(2x)$ ft (x^2) ft $(x^2 - 3x + 1)$ ft

95. Does replacing each occurrence of the variable y in $3y^5 - 7y^4 + 3y$ with its opposite result in the opposite of the polynomial? Why or why not?

96. Write a paragraph explaining the difference between n^2 and $2n$.

97. Complete the explanation: "x^3 and $3x$ are not the same because . . ."

98. Complete the statement: "$x + 2$ and $2x$ are different because . . ."

99. Write an English phrase for each algebraic expression.

    (a) $2x^3 + 5x$ (b) $(2x + 5)^3$ (c) $6(n + 4)^2$

100. Work with another student to complete this exercise. Place $>$, $<$, or $=$ in the blanks in these statements.

    1^2 _____ 2^1 What happens as the table of numbers is extended? Try more examples.

    2^3 _____ 3^2

                 What sign seems to occur the most in your table— $>$, $<$, or $=$?

    3^4 _____ 4^3

    4^5 _____ 5^4 Write an algebraic statement for the pattern of numbers in this table. Do you think this is a pattern that continues? Add more lines to the table and extend the pattern to the general case by writing the pattern in algebraic notation. Write a short paragraph stating your conjecture.

101. Compute and fill in the following blanks.

    Case 1: $1^2 - 0^2 =$ _____

    Case 2: $2^2 - 1^2 =$ _____

    Case 3: $3^2 - 2^2 =$ _____

    Case 4: $4^2 - 3^2 =$ _____

Based on the pattern you see in these four cases, predict the value of case 5: $5^2 - 4^2$. Compute case 5 to check your prediction. Write an expression for case n. Describe in words the pattern that you see in this exercise.

SECTION 1.4 · Sets

 1.4 OBJECTIVES

1. Write a set, using the roster method
2. Write a set, using set-builder notation
3. Write a set, using interval notation
4. Plot the elements of a set on a number line
5. Find the union and intersection of sets

"Whether you think that you can or you can't, you are usually right."

—Henry Ford

For his birthday, Jacob received a jacket, a ticket to a play, some candy, and a pen. We could call this collection of gifts "Jacob's presents." Such a collection is called a **set.** The things in the set are called **elements** of the set. We can write the set as {jacket, ticket, candy, pen}. The braces tell us where the set begins and ends. Every person could have a set that describes the presents she or he received on their last birthday.

What if I received no presents on my last birthday? What would my set look like? It would be the set { }, which we call the **empty set.** Sometimes the symbol ∅ is used to indicate the empty set.

Many sets can be written in *roster form,* as was the case with Jacob's presents. The set of prime numbers less than 15 can be written in roster form as {2, 3, 5, 7, 11, 13}. In Example 1, you will be asked to list sets in roster form. **Roster form** is a list enclosed in braces.

| Example 1 | Listing the Elements of a Set |

Use the roster form to list the elements of each set described.

(a) The set of all factors of 12.

The factors of 12 are {1, 2, 3, 4, 6, 12}.

(b) The set of all integers with an absolute value less than 4.

The integers are {−3, −2, −1, 0, 1, 2, 3}.

 CHECK YOURSELF 1

Use the roster form to list the elements of each set described.

(a) The set of all factors of 18.   (b) The set of all even prime numbers

---

Each set that we have examined has had a limited number of elements. If we need to indicate that a set continues in some pattern, we use three dots, called an *ellipsis,* to indicate that the set continues with the pattern it started.

| **Example 2** | **Listing the Elements of a Set** |

Use the roster form to list the elements of each set described.

(a) The set of all natural numbers less than 100.

The set $\{1, 2, 3, \ldots, 98, 99\}$ indicates that we continue increasing the numbers by 1 until we get to 99.

(b) The set of all positive multiples of 4.

The set $\{4, 8, 12, 16, \ldots\}$ indicates that we continue counting by fours forever. (There is no indicated stopping point.)

(c) The set of all integers.

$\{\ldots, -2, -1, 0, 1, 2, \ldots\}$ indicates that we continue forever in both directions.

✔ *CHECK YOURSELF 2*

Use the roster form to list the elements of each set described.

(a) The set of all natural numbers between 200 and 300.

(b) The set of all positive multiplies of 3.

(c) The set of all even numbers.

Not all sets can be described by using the roster form. What if we want to describe all the real numbers between 1 and 2? We could not list that set of numbers. Yet another way that we can describe the elements of a set is with **set-builder notation.** To describe the aforementioned set using this notation, we write

A statement such as

$$1 < x < 2$$

is called a *compound inequality.* It says that x is greater than 1 and also that x is less than 2.

$$\{x \mid 1 < x < 2\}$$

We read this as "the set of all x, where x is between 1 and 2." Note that neither 1 nor 2 is included in this set.

Example 3 further illustrates this idea.

| **Example 3** | **Using Set-Builder Notation** |

Use set-builder notation for each set described.

(a) The set of all real numbers less than 100.

We write $\{x \mid x < 100\}$.

(b) The set of all real numbers greater than -4 but less than or equal to 9.

$$\{x \mid -4 < x \le 9\}$$

The symbol $\le$ is a combination of the symbols $<$ and $=$. When we write $x \le 9$, we are indicating that either x is equal to 9 or it is less than 9.

✔ CHECK YOURSELF 3

Use set-builder notation for each set described.

(a) The set of all real numbers greater than -2.

(b) The set of all real numbers between 3 and 10 (inclusive).

Another notation that can be used to describe a set is called **interval notation.** For example, all the real numbers between 1 and 2 would be written as (1, 2). Note that the parentheses are used since neither 1 nor 2 is included in the set.

Example 4   Using Interval Notation

Use interval notation to represent each set described.

(a) The set of real numbers between 4 and 5.

We write (4, 5).

(b) The set of real numbers greater than -3 but less than or equal to 9.

We write $(-3, 9]$.

A square bracket is used at 9 to indicate that 9 is included in the interval while a parenthesis is used at -3 because -3 is not part of the interval.

(c) The set of all real numbers greater than 45.

We write $(45, \infty)$.

The positive infinity symbol ∞ does not indicate a number. It is used to show that the interval includes all real numbers greater than 45.

(d) The set of all real numbers less than 15.

We write $(-\infty, 15)$.

The negative infinity symbol $-\infty$ is used to show that the interval includes all real numbers less than 15.

✔ CHECK YOURSELF 4

Use interval notation to describe each set.

(a) The set of all real numbers less than 75.

(b) The set of all real numbers between 5 and 10.

(c) The set of all real numbers greater than 60.

(d) The set of all real numbers greater than or equal to 23 but less than or equal to 38.

Sets of numbers can also be represented graphically. In Example 5, we will look at the connection between sets and their graphs.

| Example 5 | **Plotting the Elements of a Set on a Number Line** |

Plot the points of each set on the number line.

(a) $\{-2, 1, 5\}$

(b) $\{x \mid x < 3\}$

Note that the blue line and blue arrow indicate that we continue forever in the negative direction. The parenthesis at 3 indicates that the 3 is not part of the graph.

(c) $\{x \mid -2 < x < 5\}$

The parentheses indicate that the numbers -2 and 5 are not part of the set.

(d) $\{x \mid x \geq 2\}$

The bracket indicates that 2 is part of the set that is graphed.

✔ *CHECK YOURSELF 5*

Plot the points of each set on a number line.

(a) $\{-5, -3, 0\}$

(b) $\{x \mid -3 < x < -1\}$

(c) $\{x \mid x \leq 5\}$

The following table summarizes the different ways of describing a set.

Basic Set Notation (*a* and *b* represent any real numbers)

Set	Set-Builder Notation	Interval Notation	Graph
All real numbers greater than b	$\{x \mid x > b\}$	(b, ∞)	
All real numbers less than or equal to b	$\{x \mid x \leq b\}$	$(-\infty, b]$	
All real numbers greater than a and less than b	$\{x \mid a < x < b\}$	(a, b)	
All real numbers greater than or equal to a and less than b	$\{x \mid a \leq x < b\}$	$[a, b)$	

There are occasions when we need to combine sets. There are two commonly used operations to accomplish this: *union* and *intersection*.

> The **union** of two sets A and B, written $A \cup B$, is the set of all elements that belong to either A or B or to both.
>
> The **intersection** of two sets A and B, written $A \cap B$, is the set of all elements that belong to both A and B.

Example 6

Finding Union and Intersection

Let $A = \{1, 3, 5\}$, $B = \{3, 5, 9\}$, and $C = \{9, 11\}$. List the elements in each of the following sets.

(a) $A \cup B$

This is the set of elements that are in A or B or in both.

$$A \cup B = \{1, 3, 5, 9\}$$

(b) $A \cap B$

This is the set of elements common to A and B.

$$A \cap B = \{3, 5\}$$

(c) $A \cup C$

This is the set of elements in A or C or in both.

$$A \cup C = \{1, 3, 5, 9, 11\}$$

(d) $A \cap C$

This is the set of elements common to A and C.

{ } or ∅ is the symbol for the empty set.

$A \cap C = \{ \; \}$ or $\varnothing$ since there are no elements in common.

✔ CHECK YOURSELF 6

Let $A = \{2, 4, 7\}$, $B = \{4, 7, 10\}$, and $C = \{8, 12\}$. Find

(a) $A \cup B$ (b) $A \cap B$

(c) $A \cup C$ (d) $A \cap C$

✔ CHECK YOURSELF ANSWERS

1. (a) $\{1, 2, 3, 6, 9, 18\}$; (b) $\{2\}$

2. (a) $\{201, 202, 203, \ldots, 298, 299\}$; (b) $\{3, 6, 9, 12, \ldots\}$;

  (c) $\{\ldots, -6, -4, -2, 0, 2, 4, 6, \ldots\}$

3. (a) $\{x \mid x > -2\}$; (b) $\{x \mid 3 \leq x \leq 10\}$

4. (a) $(-\infty, 75)$; (b) $(5, 10)$; (c) $(60, \infty)$; (d) $[23, 38]$

5. (a)

  (b)

  (c)

6. (a) $\{2, 4, 7, 10\}$; (b) $\{4, 7\}$; (c) $\{2, 4, 7, 8, 12\}$; (d) $\varnothing$

Exercises · 1.4

Use the roster method to list the elements of each set.

1. The set of all the days of the week

2. The set of all months of the year that have 31 days

3. The set of all factors of 18

4. The set of all factors of 24

5. The set of all prime numbers less than 30

6. The set of all prime numbers between 20 and 40

7. The set of all negative integers greater than -6

8. The set of all positive integers less than 6

9. The set of all even whole numbers less than 13

10. The set of all odd whole numbers less than 14

11. The set of integers greater than 2 and less than 7

12. The set of integers greater than 5 and less than 10

13. The set of integers greater than -4 and less than -1

14. The set of integers greater than -8 and less than -3

15. The set of integers between -5 and 2, inclusive

16. The set of integers between -1 and 4

17. The set of odd whole numbers

18. The set of even whole numbers

19. The set of all even whole numbers less than 100

20. The set of all odd whole numbers less than 100

21. The set of all positive multiples of 5

22. The set of all positive multiples of 6

Use set-builder notation and interval notation for each set described.

23. The set of all real numbers greater than 10

24. The set of all real numbers less than 25

25. The set of all real numbers greater than or equal to -5

26. The set of all real numbers less than or equal to -3

27. The set of all real numbers greater than or equal to 2 and less than or equal to 7

28. The set of all real numbers greater than -3 and less than -1

29. The set of all real numbers between -4 and 4, inclusive

30. The set of all real numbers between -8 and 3, inclusive

96

Plot the elements of each set on a number line.

31. $\{-2, -1, 0, 4\}$

32. $\{-5, -1, 2, 3, 5\}$

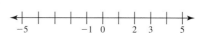

33. $\{x \mid x > 4\}$

34. $\{x \mid x > -1\}$

35. $\{x \mid x \geq -3\}$

36. $\{x \mid x \leq 6\}$

37. $\{x \mid 2 < x < 7\}$

38. $\{x \mid 4 < x < 8\}$

39. $\{x \mid -3 < x \leq 5\}$

40. $\{x \mid -6 < x \leq 1\}$

41. $\{x \mid -4 \leq x < 0\}$

42. $\{x \mid -5 \leq x < 2\}$

43. $\{x \mid -7 \leq x \leq -3\}$

44. $\{x \mid -1 \leq x \leq 4\}$

45. The set of all integers between -7 and -3, inclusive

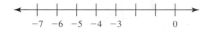

46. The set of all integers between -1 and 4, inclusive

Use set-builder notation and interval notation to describe each graphed set.

47.

48.

49.

50.

51.

52.

53.

54.

55.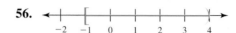

56.

57.

58.

59.

60.

In Exercises 61 to 70, $A = \{x \mid x$ is an even natural number less than 10$\}$, $B = \{1, 3, 5, 7, 9\}$, and $C = \{1, 2, 3, 4, 5\}$. List the elements in each set.

61. $A \cup B$ **62.** $A \cap B$ **63.** $B \cap \varnothing$ **64.** $C \cup A$

65. $A \cup \varnothing$ **66.** $B \cup C$ **67.** $B \cap C$ **68.** $C \cap A$

69. $(A \cup C) \cap B$ **70.** $A \cup (C \cap B)$

Summary for Chapter 1

Example	Topic	Reference
	From Arithmetic to Algebra	1.1
The sum of x and 5 is $x + 5$. 7 more than a is $a + 7$. b increased by 3 is $b + 3$.	**Addition** $x + y$ means the **sum** of x **and** y or x **plus** y. Some other words indicating addition are *more than* and *increased by*.	p. 59
The difference of x and 3 is $x - 3$. 5 less than p is $p - 5$. a decreased by 4 is $a - 4$.	**Subtraction** $x - y$ means the **difference** of x **and** y or x **minus** y. Some other words indicating subtraction are *less than* and *decreased by.*	p. 60
The product of m and n is mn. The product of 2 and the sum of a and b is $2(a + b)$.	**Multiplication** $\left. \begin{array}{l} x \cdot y \\ (x)(y) \\ xy \end{array} \right\}$ These all mean the **product** of x and y or x **times** y.	p. 60
n divided by 5 is $\dfrac{n}{5}$. The sum of a and b, divided by 3, is $\dfrac{a + b}{3}$.	**Division** $\dfrac{x}{y}$ means x **divided by** y or the **quotient** when x is divided by y.	p. 62
	Evaluating Algebraic Expressions	1.2
Evaluate $\dfrac{4a - b}{2c}$ if $a = -6$, $b = 8$, and $c = -4$. $\dfrac{4a - b}{2c} = \dfrac{4(-6) - 8}{2(-4)}$ $= \dfrac{-24 - 8}{-8}$ $= \dfrac{-32}{-8} = 4$	**To Evaluate an Algebraic Expression:** 1. Replace each variable by the given number value. 2. Do the necessary arithmetic operations. (Be sure to follow the rules for the order of operations.)	p. 68
	Adding and Subtracting Algebraic Expressions	1.3
$3x^2y$ is a term.	**Term** A number, or the product of a number and one or more variables, raised to a power.	p. 78

$4a^2$ and $3a^2$ are like terms. $5x^2$ and $2xy^2$ are not like terms.	**Like Terms** Terms that contain exactly the same variables raised to the same powers.	**p. 79**
$5a + 3a = 8a$ $7xy - 3xy = 4xy$	**Combining Like Terms** 1. Add or subtract the numerical coefficients. 2. Attach the common variables.	**p. 81**
	# Sets	**1.4**
$A = \{2, 3, 4, 5\}$ is a set.	**Set** A set is a collection of objects classified together.	**p. 90**
2 is an element of set A.	**Elements** The elements are the objects in a set.	**p. 90**
$S = \{2, 4, 6, 8\}$ is in roster form.	**Roster Form** A set is said to be in roster form if the elements are listed and enclosed in braces.	**p. 90**
$\{x \mid x < 4\}$ is written in set-builder notation.	**Set-Builder Notation** $\{x \mid x > a\}$ is read "the set of all x, where x is greater than a." $\{x \mid x < a\}$ is read "the set of all x, where x is less than a." $\{x \mid a < x < b\}$ is read "the set of all x, where x is greater than a and less than b."	**p. 91**
$(-4, 5]$ is written in interval notation.	**Interval Notation** (a, ∞) is read "all real numbers greater than a." $(-\infty, b)$ is read "all real numbers less than b." (a, b) is read "all real numbers greater than a and less than b." $[a, b]$ is read "all real numbers greater than or equal to a and less than or equal to b."	**p. 92**
$\{x \mid x < 4\}$ The parenthesis indicates every number below the marked value (here it is 4).	**Plotting the Elements of a Set on a Number Line** $\{x \mid x < a\}$ indicates the set of all points on the number line to the left of a. We plot those points by using a parenthesis at a (indicating that a is not included), then a bold line to the left.	**p. 93**
$\{x \mid x \geq -3\}$ The bracket indicates every number at or above the indicated value (-3).	$\{x \mid x \geq a\}$ indicates the set of all points on the number line to the right of, and including, a. We plot those points by using a bracket at a (indicating that a is included), then a bold line to the right.	**p. 93**

$\{x \mid 3 \le x < 10\}$ 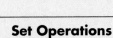 This notation indicates every number between 3 and 10, including 3 but not including 10.	$\{x \mid a \le x < b\}$ indicates the set of all points on the number line between a and b including a. We plot those points by using an opening bracket at a and a closing parenthesis at b, then a bold line in between.	**p. 93**
$A = \{2, 3, 4, 6\}$ and $B = \{3, 4, 7, 8\}$ $A \cup B = \{2, 3, 4, 6, 7, 8\}$ $A \cap B = \{3, 4\}$	**Set Operations** **Union**　$A \cup B$ is the set of elements in A or B or in both. **Intersection**　$A \cap B$ is the set of elements in both A and B.	**p. 94**

Summary Exercises · 1

This exercise set is provided to give you practice with each of the objectives of the chapter. Each exercise is keyed to the appropriate chapter section. Your instructor will give you guidelines on how to best use these exercises.

[1.1] Write, using symbols.

1. 8 more than y

2. c decreased by 10

3. The product of 8 and a

4. The quotient when y is divided by 3

5. 5 times the product of m and n

6. The product of x and 7 less than x

7. 3 more than the product of 17 and x

8. The quotient when a plus 2 is divided by a minus 2

9. The product of 6 more than a number and 6 less than the same number

10. The quotient of 9 and a number

11. The quotient when 4 less than a number is divided by 2 more than the same number

12. The product of a number and 3 more than twice the same number

[1.2] Evaluate the expressions if $x = -3$, $y = 6$, $z = -4$, and $w = 2$.

13. $3x + w$

14. $5y - 4z$

15. $x + y - 3z$

16. $5z^2$ 80

17. $3x^2 - 2w^2$

18. $3x^3$

19. $5(x^2 - w^2)$

20. $\dfrac{6z}{2w}$

21. $\dfrac{2x - 4z}{y - z}$

22. $\dfrac{3x - y}{w - x}$

23. $\dfrac{x(y^2 - z^2)}{(y + z)(y - z)}$

24. $\dfrac{y(x - w)^2}{x^2 - 2xw + w^2}$

25. $w^2y^2 - z^3x$

26. $-x^2 - 12xz^3$

27. $-4x^2 - 2zw^2 + 4z$

28. $3x^3w^2 + xy^2$

[1.3] List the terms of the expressions.

29. $4a^3 - 3a^2$

30. $5x^2 - 7x + 3$

[1.3] Circle like terms.

31. $5m^2, -3m, -4m^2, 5m^3, m^2$

32. $4ab^2, 3b^2, -5a, ab^2, 7a^2, -3ab^2, 4a^2b$

[1.3] Combine like terms.

33. $9x + 7x$

34. $2x + 5x$

35. $4a - 2a$

36. $-7y + 2y$

37. $9xy - 6xy$

38. $5ab^2 + 2ab^2$

39. $7a + 3b + 12a - 2b$

40. $-3x + 2y - 5x - 7y$

102

41. $5x^3 + 17x^2 - 2x^3 - 8x^2$

42. $3a^3 + 5a^2 + 4a - 2a^3 - 3a^2 - a$

43. Subtract $4a^3$ from the sum of $2a^3$ and $12a^3$.

44. Subtract the sum of $3x^2$ and $5x^2$ from $15x^2$.

[1.1–1.3] Write an expression for each of the following exercises.

45. Carpentry. If x feet (ft) is cut off the end of a board that is 37 ft long, how much is left?

46. Money. Sergei has 25 nickels and dimes in his pocket. If x of these are dimes, how many of the coins are nickels?

47. Age. Sam is 5 years older than Angela. If Angela is x years old now, how old is Sam?

48. Money. Margaret has $5 more than twice as much money as Pho. Write an expression for the amount of money that Margaret has.

49. Geometry. The length of a rectangle is 4 meters (m) more than the width. Write an expression for the length of the rectangle.

50. Number problem. A number is 7 less than 6 times the number n. Write an expression for the number.

51. Carpentry. A 25-ft plank is cut into two pieces. Write expressions for the length of each piece.

52. Money. Bernie has d dimes and q quarters in his pocket. Write an expression for the amount of money that Bernie has in his pocket.

53. Geometry. Find the perimeter of the given rectangle.

54. Geometry. If the length of a building is x m and the width is $\dfrac{x}{6}$ m, what is the perimeter of the building?

[1.4] Use the roster method to list the elements of each set.

55. The set of all factors of 3

56. The set of all positive integers less than 7

57. The set of integers greater than -2 and less than 4

58. The set of integers between -4 and 3 inclusive

59. The set of all odd whole numbers less than 4

60. The set of all integers greater than -2 and less than 3

[1.4] Use set-builder notation and interval notation to represent each set described.

61. The set of all real numbers greater than 9

62. The set of all real numbers greater than -2 and less than 4

63. The set of all real numbers less than or equal to -5

64. The set of all real numbers between -4 and 3 inclusive

[1.4] Plot the elements of each set on a number line.

65. $\{x \mid x \geq 1\}$

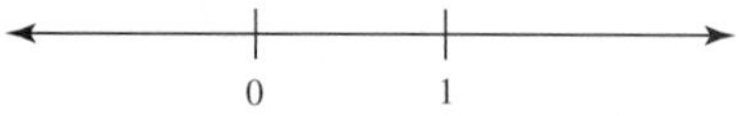

66. $\{x \mid x \leq -2\}$

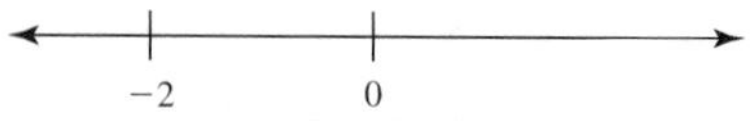

67. $\{x \mid -2 \leq x < 3\}$

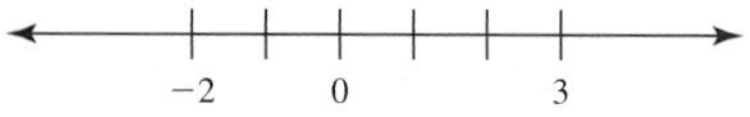

68. $\{x \mid -7 < x \leq -1\}$

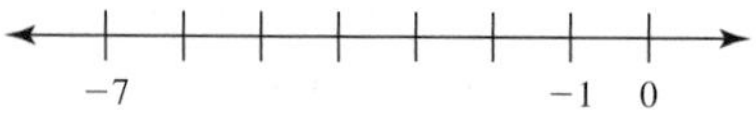

[1.4] Use set-builder notation and interval notation to describe each set.

69.

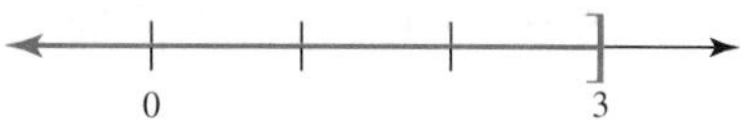

70.

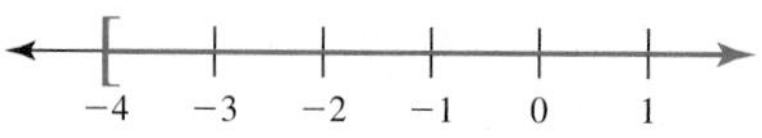

71.

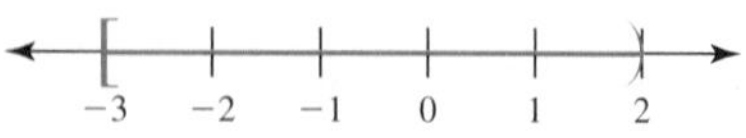

72.

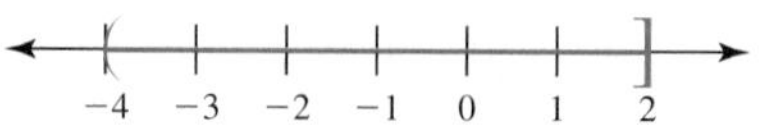

73.

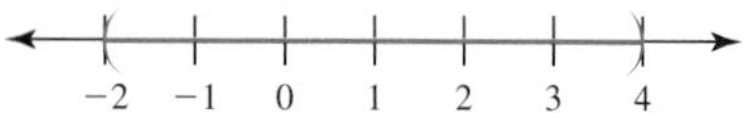

74.

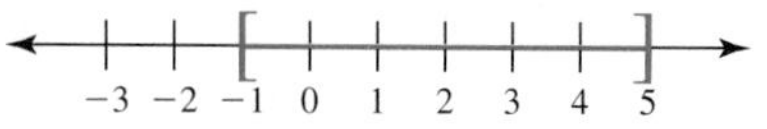

[1.4] In Exercises 75 to 78, $A = \{1, 5, 7, 9\}$ and $B = \{2, 5, 9, 11, 15\}$. List the elements in each set.

75. $A \cup B$

76. $A \cap B$

77. $B \cup \varnothing$

78. $A \cap \varnothing$

Self-Test   1

The purpose of this self-test is to help you check your progress and to review for a chapter test in class. Allow yourself about an hour to take the test. When you are done, check the answers in the back of the book. If you missed any answers, be sure to go back and review the appropriate sections in the chapter and their exercises.

Write in symbols.

1. The sum of x and y

2. The difference m minus n

3. The product of a and b

4. The quotient when p is divided by 3 less than q

5. 7 more than a

6. 5 less than c

7. The product of 3 and the quantity $2x$ minus $3y$

8. 3 times the difference of m and n

Evaluate where $x = -4$.

9. $-4x - 12$

10. $3x^2 + 2x - 4$

In Exercises 11 to 13, evaluate each expression if $a = -2$, $b = 6$, and $c = -4$.

11. $4a - c$

12. $6(2b - 3c)$

13. $\dfrac{3a - 4b}{a + c}$

In Exercises 14 and 15, combine like terms.

14. $8a - 3b - 5a + 2b$

15. $7x^2 - 3x + 2 - (5x^2 - 3x - 6)$

16. Plot the elements of the set $\{x \mid -5 < x \le 3\}$ on a number line.

17. (a) Use set-builder notation to describe the set

(b) Describe the set in interval notation.

$-3 \quad 0 \quad 2$

18. Patrick worked all day mowing lawns and was paid $7 per hour (h). If he had $85 at the end of an 8-h day, how much did he have before he started working?

19. A woman weighed 184 pounds (lb) when she started her diet. At the end of 12 weeks, she weighed 160 lb. How much did she lose each week, if weight loss was *equal* each week?

20. If $A = \{1, 2, 5\}$ and $B = \{3, 5, 7\}$, find

(a) $A \cup B$

(b) $A \cap B$

2 Equations and Inequalities

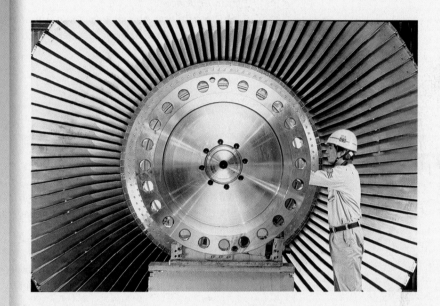

Many engineers, economists, and environmental scientists are working on the problem of meeting the increasing energy demands of a growing global population. One promising solution to this problem is power generated by wind-driven turbines.

The cost of wind-generated power has fallen from $0.25 per kilowatthour (kWh) in the early 1980s to about $0.043 per kilowatthour in 2000; thus, using this form of power production is becoming economically feasible. And compared to the cost of pollution from burning coal and oil, wind-generated power may be less expensive.

An economist for a city might use this equation to try to compute the cost for electricity for the city:

$$C = P(0.043)(1{,}000) \qquad \text{where } C = \text{cost and } P = \text{power}$$

The equation is an ancient tool for solving problems and writing numerical relationships clearly and accurately. In this chapter you will learn methods to solve linear equations and practice writing equations that accurately describe problem situations.

Solving Equations by Adding and Subtracting

SECTION 2.1

2.1 OBJECTIVES

1. Determine whether a given number is a solution for an equation
2. Use the addition property to solve equations
3. Determine whether a given number is a solution for an application
4. Translate words to equation symbols
5. Solve application problems

Overcoming Math Anxiety

Throughout this text, we will present you with a series of class-tested techniques that are designed to improve your performance in this math class.

Hint 4: Don't procrastinate!

1. Do your math homework while you're still fresh. If you wait until too late at night, your tired mind will have much greater difficulty understanding the concepts.

2. Do your homework the day it is assigned. The more recent the explanation is, the easier it is to recall.

3. When you've finished your homework, try reading the next section through one time. This will give you a sense of direction when you next encounter the material. This works whether you are in a lecture or lab setting.

Remember that, in a typical math class, you are expected to do 2 or 3 h of homework for each weekly class hour. This means 2 or 3 h per night. Schedule the time and stick to your schedule.

In this chapter you will begin working with one of the most important tools of mathematics—the equation. The ability to recognize and solve various types of equations is probably the most useful algebraic skill you will learn. We will continue to build upon the methods of this chapter throughout the remainder of the text. To start, let's describe what we mean by an *equation*.

An **equation** is a mathematical statement that two expressions are equal.

Some examples are $3 + 4 = 7$, $x + 3 = 5$, $P = 2L + 2W$. As you can see, an equals sign (=) separates the two equal expressions. These expressions are usually called the *left side* and the *right side* of the equation.

$$\underbrace{x + 3}_{\uparrow} = 5$$

Left side   Equals   Right side

An equation may be either true or false. For instance, $3 + 4 = 7$ is true because both sides name the same number. What about an equation such as $x + 3 = 5$ that has a letter or variable on one side? Any number can replace x in the equation. However, only one number will make this equation a true statement.

An equation such as
$$x + 3 = 5$$
is called a **conditional equation** because it can be either true or false depending on the value given to the variable.

We could also use set-builder notation. We write $\{x \mid x = 2\}$, which is read, "Every x such that x equals 2." We will use both notations throughout the text.

$$\text{If} \quad x = \begin{cases} 1 & 1 + 3 = 5 \text{ is false} \\ 2 & 2 + 3 = 5 \text{ is true} \\ 3 & 3 + 3 = 5 \text{ is false} \end{cases}$$

The number 2 is called the *solution* (or *root*) of the equation $x + 3 = 5$ because substituting 2 for x gives a true statement. The set $\{2\}$ is called the *solution set.*

A **solution** for an equation is any value for the variable that makes the equation a true statement.

The **solution set** for an equation is the set of all values for the variables that make the equation a true statement.

Example 1

Verifying a Solution

(a) Is 3 a solution for the equation $2x + 4 = 10$?

To find out, replace x with 3 and evaluate $2x + 4$ on the left.

Note that until the left side equals the right side, a question mark is placed over the equals sign.

Left side		Right side
$2 \cdot 3 + 4$	$\overset{?}{=}$	10
$6 + 4$	$\overset{?}{=}$	10
10	$=$	10

Since $10 = 10$ is a true statement, 3 is a solution of the equation. The solution set is $\{3\}$.

(b) Is 5 a solution of the equation $3x - 2 = 2x + 1$?

To find out, replace x with 5 and evaluate each side separately.

Remember the rules for the order of operations. Multiply first; then add or subtract.

Left side		Right side
$3 \cdot 5 - 2$	$\overset{?}{=}$	$2 \cdot 5 + 1$
$15 - 2$	$\overset{?}{=}$	$10 + 1$
13	$\neq$	11

Since the two sides do not name the same number, we do not have a true statement, and 5 is not a solution.

✔ *CHECK YOURSELF 1*

For the equation

$$2x - 1 = x + 5$$

(a) Is 4 a solution?   (b) Is 6 a solution?

You may be wondering whether an equation can have more than one solution. It certainly can. For instance,

The equation $x^2 = 9$ is an example of a **quadratic equation.** We will consider methods of solution in Chapter 11.

$$x^2 = 9$$

has two solutions. They are 3 and -3 because

$$3^2 = 9 \quad \text{and} \quad (-3)^2 = 9 \quad \text{The solution set is } \{-3, 3\}.$$

In this chapter, however, we will generally work with *linear equations*. These are equations that can be put into the form

$$ax + b = 0$$

where the variable is x, where a and b are any numbers, and where a is not equal to 0. In a linear equation, the variable can appear only to the first power. No other power (x^2, x^3, etc.) can appear. Linear equations are also called **first-degree equations.** The *degree* of an equation in one variable is the highest degree to which the variable appears.

Linear equations in one variable that can be written in the form

$$ax + b = 0 \qquad a \neq 0$$

will have exactly one solution.

Example 2

Identifying Expressions and Equations

Label each of the following as an expression, a linear equation, or a nonlinear equation.

(a) $4x + 5$ is an expression.

(b) $2x + 8 = 0$ is a linear equation.

(c) $3x^2 - 9 = 0$ is not a linear equation.

(d) $5x = 15$ is a linear equation.

There can be no variable in the denominator of a linear equation.

(e) $\dfrac{3}{x} + 2 = 0$ is not a linear equation.

✔ CHECK YOURSELF 2

Label each as an expression, a linear equation, or a nonlinear equation.

(a) $2x^2 = 8$ (b) $2x - 3 = 0$ (c) $5x - 10$

(d) $2x + 1 = 7$ (e) $5 - \dfrac{6}{x} = 2x$

One can find the solution for an equation such as $x + 3 = 8$ by guessing the answer to the question "What plus 3 is 8?" Here the answer to the question is 5, and that is also the solution for the equation. But for more complicated equations you will need something more than guesswork. A better method is to transform the given equation to an *equivalent equation* whose solution can be found by inspection. Let's make a definition.

> Equations that have the same solution set are called **equivalent equations.**

The following are all equivalent equations:

$$2x + 3 = 5 \qquad 2x = 2 \qquad \text{and} \qquad x = 1$$

They all have the same solution, 1. We say that a linear equation is *solved* when it is transformed to an equivalent equation of the form

Note: In some cases we'll write the equation in the form

$$\boxed{} = x$$

The number will be our solution when the equation has the variable isolated on the left or on the right.

$$x = \boxed{}$$

The variable is alone on the left side.   The right side is some number, the solution.

The addition property of equality is the first property you will need to transform an equation to an equivalent form.

The Addition Property of Equality

Remember: An equation is a statement that the two sides are equal. Adding the same quantity to both sides does not change the equality or "balance."

$$\text{If} \qquad a = b$$
$$\text{then} \qquad a + c = b + c$$

In words, adding the same quantity to both sides of an equation gives an equivalent equation.

Let's look at an example of applying this property to solve an equation.

Example 3   Using the Addition Property to Solve an Equation

Solve

$$x - 3 = 9$$

Remember that our goal is to isolate x on one side of the equation. Since 3 is being subtracted from x, we can add 3 to remove it. We must use the addition property to add 3 to both sides of the equation.

To check, replace x with 12 in the original equation:

$$x - 3 = 9$$
$$12 - 3 \overset{?}{=} 9$$
$$9 = 9 \quad \text{(True)}$$

Since we have a true statement, 12 is the solution.

$$
\begin{array}{rr}
x - 3 = & 9 \\
+\,3 & +3 \\
\hline
x \quad\;\; = & 12
\end{array}
$$

Adding 3 "undoes" the subtraction and leaves x alone on the left.

Since 12 is the solution for the equivalent equation $x = 12$, it is the solution for our original equation. For either equation, the solution set is $\{12\}$.

 CHECK YOURSELF 3

Solve and check.

$$x - 5 = 4$$

The addition property also allows us to add a negative number to both sides of an equation. This is really the same as subtracting the same quantity from both sides.

Example 4   Using the Addition Property to Solve an Equation

Solve

$$x + 5 = 9$$

Recall our comment that we could write an equation in the equivalent forms

$x = \Box$ or $\Box = x$,

where $\Box$ represents some number. Suppose we have an equation such as

$12 = x + 7$

Subtracting 7 will isolate x on the right:

$$\begin{array}{r} 12 = x + 7 \\ -7 \quad -7 \\ \hline 5 = x \end{array}$$

and the solution set is {5}.

In this case, 5 is *added* to x on the left. We can use the addition property to subtract 5 from both sides. This will "undo" the addition and leave the variable x alone on one side of the equation.

$$\begin{array}{r} x + 5 = \quad 9 \\ -5 \quad -5 \\ \hline x \quad = \quad 4 \end{array}$$

The solution set is {4}. To check, replace x with 4:

$$4 + 5 = 9 \qquad \text{True}$$

✔ CHECK YOURSELF 4

Solve and check.

$$x + 6 = 13$$

What if the equation has a variable term on both sides? You will have to use the addition property to add or subtract a term involving the variable to get the desired result.

Example 5

Using the Addition Property to Solve an Equation

Solve

$$5x = 4x + 7$$

We will start by subtracting $4x$ from both sides of the equation. Do you see why? Remember that an equation is solved when we have an equivalent equation of the form $x = \Box$.

$$\begin{array}{r} 5x = \quad 4x + 7 \\ -4x \quad -4x \\ \hline x = \quad \quad 7 \end{array}$$

Subtracting $4x$ from both sides *removes* $4x$ from the right.

To check: Since 7 is a solution for the equivalent equation $x = 7$, it should be a solution for the original equation. To find out, replace x with 7:

$$5 \cdot 7 \stackrel{?}{=} 4 \cdot 7 + 7$$

$$35 \stackrel{?}{=} 28 + 7$$

$$35 = 35 \qquad \text{True}$$

CHECK YOURSELF 5

Solve and check.

$$7x = 6x + 3$$

You may have to apply the addition property more than once to solve an equation. Look at Example 6.

| Example 6 | Using the Addition Property to Solve an Equation |

Solve

$$7x - 8 = 6x$$

We want all variables on *one* side of the equation. If we choose the left, we subtract $6x$ from both sides of the equation. This will remove $6x$ from the right:

$$
\begin{array}{rcl}
7x - 8 = & 6x \\
-6x & -6x \\
\hline
x - 8 = & 0
\end{array}
$$

We want the variable alone, so we add 8 to both sides. This isolates x on the left.

$$
\begin{array}{rcl}
x - 8 = & 0 \\
+8 & +8 \\
\hline
x = & 8
\end{array}
$$

The solution set is {8}. We'll leave it to you to check this result.

 CHECK YOURSELF 6

Solve and check.

$$9x + 3 = 8x$$

Often an equation will have more than one variable term *and* more than one number. You will have to apply the addition property twice in solving these equations.

| **Example 7** | **Using the Addition Property to Solve an Equation** |

Solve

$$5x - 7 = 4x + 3$$

We would like the variable terms on the left, so we start by subtracting $4x$ to remove that term from the right side of the equation:

$$
\begin{array}{rcr}
5x - 7 = & & 4x + 3 \\
-4x & & -4x \\
\hline
x - 7 = & & 3
\end{array}
$$

Now, to isolate the variable, we add 7 to both sides to undo the subtraction on the left:

$$
\begin{array}{rcr}
x - 7 = & & 3 \\
+7 & & +7 \\
\hline
x & = & 10
\end{array}
$$

You could just as easily have added 7 to both sides and then subtracted 4x. The result would have been the same. In fact, some students prefer to combine the two steps.

The solution set is $\{10\}$. To check, replace x with 10 in the original equation:

$$5 \cdot 10 - 7 \stackrel{?}{=} 4 \cdot 10 + 3$$

$$43 = 43 \qquad \text{True}$$

✔ *CHECK YOURSELF 7*

Solve and check.

(a) $4x - 5 = 3x + 2$ (b) $6x + 2 = 5x - 4$

Remember, by simplify we mean to combine all like terms.

In solving an equation, you should always simplify each side as much as possible before using the addition property.

| **Example 8** | **Combining Like Terms and Solving the Equation** |

Solve

$$\overset{\text{Like terms}}{\overbrace{5 + 8x - 2}} = \overset{\text{Like terms}}{\overbrace{2x - 3 + 5x}}$$

Since like terms appear on each side of the equation, we start by combining the numbers on the left (5 and -2). Then we combine the like terms ($2x$ and $5x$) on the right. We have

$$3 + 8x = 7x - 3$$

Now we can apply the addition property, as before:

$$
\begin{array}{rll}
3 + 8x = & 7x - 3 & \\
-7x = -7x & & \text{Subtract } 7x \\
\hline
3 + \ x = & -3 & \\
-3 \quad\quad & -3 & \text{Subtract } 3 \\
\hline
x = & -6 & \text{Isolate } x
\end{array}
$$

The solution set is $\{-6\}$. To check, always return to the original equation. That will catch any possible errors in simplifying. Replacing x with -6 gives

$$5 + 8(-6) - 2 \overset{?}{=} 2(-6) - 3 + 5(-6)$$

$$5 - 48 - 2 \overset{?}{=} -12 - 3 - 30$$

$$-45 = -45 \qquad \text{True}$$

✔ *CHECK YOURSELF 8*

Solve and check.

(a) $3 + 6x + 4 = 8x - 3 - 3x$ (b) $5x + 21 + 3x = 20 + 7x - 2$

We may have to apply some of the properties discussed in Section 0.4 in solving equations. Example 9 illustrates the use of the distributive property to clear an equation of parentheses.

Example 9 Using the Distributive Property and Solving Equations

Solve

$$2(3x + 4) = 5x - 6$$

Note:

$$2(3x + 4) = 2(3x) + 2(4)$$
$$= 6x + 8$$

Applying the distributive property on the left, we have

$$6x + 8 = 5x - 6$$

We can then proceed as before:

$$
\begin{array}{rll}
6x + 8 = & 5x - 6 & \\
-5x \quad = & -5x & \text{Subtract } 5x. \\
\hline
x + 8 = & -6 & \\
-8 \quad & -8 & \text{Subtract } 8. \\
\hline
x \quad = & -14 &
\end{array}
$$

Remember that

$$x = -14 \quad \text{and} \quad -14 = x$$

are equivalent equations.

The solution set is $\{-14\}$. We will leave the checking of this result to the reader.
Remember: Always return to the original equation to check.

✔ CHECK YOURSELF 9

Solve and check each of the following equations.

(a) $4(5x - 2) = 19x + 4$

(b) $3(5x + 1) = 2(7x - 3) - 4$

Given an expression such as

$$-2(x - 5)$$

the distributive property can be used to create the equivalent expression

$$-2x + 10$$

The distribution of a negative number is used in Example 10.

Example 10	**Distributing a Negative Number**

Solve each of the following equations.

(a) $-2(x - 5) = -3x + 2$ Distribute the -2.

$$
\begin{array}{rrl}
-2x + 10 &= -3x + 2 & \\
+3x & +3x & \text{Add } 3x. \\
\hline
x + 10 &= 2 & \\
- 10 &= -10 & \text{Add } -10. \\
\hline
x &= -8 & \text{The solution set is } \{-8\}.
\end{array}
$$

(b) $-3(3x + 5) = -5(2x - 2)$ Distribute the -3.

$-9x - 15 = -5(2x - 2)$ Distribute the -5.

$$
\begin{array}{rrl}
-9x - 15 &= -10x + 10 & \\
+10x & +10x & \text{Add } 10x. \\
\hline
x - 15 &= 10 & \\
+ 15 & + 15 & \text{Add } 15. \\
\hline
x &= 25 & \text{The solution set is } \{25\}.
\end{array}
$$

✔ CHECK YOURSELF 10

Solve each of the following.

(a) $-2(x - 3) = -x + 5$ (b) $-4(2x - 1) = -3(3x + 2)$

The main reason for learning how to set up and solve algebraic equations is so that we can use them to solve word problems. In fact, algebraic equations were *invented* to make solving word problems much easier. The first word problems that we know about are over 4,000 years old. They were literally "written in stone," on Babylonian tablets, about 500 years before the first algebraic equation made its appearance.

Before algebra, people solved word problems primarily by **substitution,** which is a method of finding unknown values by using trial and error in a logical way. Example 11 shows how to solve a word problem by using substitution.

Example 11	**Solving a Word Problem by Substitution**

Consecutive integers are integers that follow each other, such as 8 and 9.

The sum of two consecutive integers is 37. Find the two integers.

If the two integers were 20 and 21, their sum would be 41. Since that's more than 37, the integers must be smaller. If the integers were 15 and 16, the sum would be 31. More trials yield that the sum of 18 and 19 is 37.

   CHECK YOURSELF 11

The sum of two consecutive integers is 91. Find the two integers.

Most word problems are not so easily solved by substitution. For more complicated word problems, a five-step procedure is used. Using this step-by-step approach will, with practice, allow you to organize your work. Organization is the key to solving word problems. Here are the five steps.

To Solve Word Problems

Step 1.   Read the problem carefully. Then reread it to decide what you are asked to find.

Step 2.   Choose a letter to represent one of the unknowns in the problem. Then represent all other unknowns of the problem with expressions that use the same letter.

Step 3.   Translate the problem to the language of algebra to form an equation.

Step 4.   Solve the equation and answer the question of the original problem.

Step 5.   Check your solution by returning to the original problem.

We discussed these translations in Section 1.1. You might find it helpful to review that section before going on.

The third step is usually the hardest part. We must translate words to the language of algebra. Before we look at a complete example, the following table may help you review that translation step.

Translating Words to Algebra

Words	Algebra
The sum of x and y	$x + y$
3 plus a	$3 + a$ or $a + 3$
5 more than m	$m + 5$
b increased by 7	$b + 7$
The difference of x and y	$x - y$
4 less than a	$a - 4$
s decreased by 8	$s - 8$
The product of x and y	$x \cdot y$ or xy
5 times a	$5 \cdot a$ or $5a$
Twice m	$2m$
The quotient of x and y	$\dfrac{x}{y}$
a divided by 6	$\dfrac{a}{6}$
One-half of b	$\dfrac{b}{2}$ or $\dfrac{1}{2}b$

Now let's look at some typical examples of translating phrases to algebra.

Example 12	## Translating Statements

Translate each English language expression to an algebraic expression.

(a) The sum of a and 2 times b

$$a + 2b$$

Sum   2 times b

(b) 5 times m, increased by 1

$$5m + 1$$

5 times m Increased by 1

(c) 5 less than 3 times x

$$3x - 5$$

3 times x 5 less than

(d) The product of x and y, divided by 3

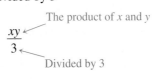

The product of x and y

$$\frac{xy}{3}$$

Divided by 3

CHECK YOURSELF 12

Translate to algebra.

(a) 2 more than twice x

(b) 4 less than 5 times n

(c) The product of twice a and b

(d) The sum of s and t, divided by 5

Now let's work through a complete example. Although this problem could be solved by substitution, it is presented here to help you practice the five-step approach.

| Example 13 | **Solving an Application** |

The sum of a number and 5 is 17. What is the number?

Step 1   *Read carefully.* You must find the unknown number.

Step 2   *Choose letters or variables.* Let x represent the unknown number. There are no other unknowns.

Step 3   *Translate.*

The sum of

$$x + 5 = 17$$

is

Step 4   *Solve.*

Always return to the *original problem* to check your result and *not* to the equation of step 3. This will prevent possible errors!

$$x + 5 = 17 \qquad \text{Subtract 5.}$$
$$x + 5 - 5 = 17 - 5$$
$$x = 12$$

So the number is 12.

Step 5   *Check.* Is the sum of 12 and 5 equal to 17? Yes ($12 + 5 = 17$). We have checked our solution.

 CHECK YOURSELF 13

The sum of a number and 8 is 35. What is the number?

✔ *CHECK YOURSELF ANSWERS*

1. (a) 4 is not a solution; (b) 6 is a solution.

2. (a) nonlinear equation; (b) linear equation; (c) expression; (d) linear equation; (e) a nonlinear equation

3. {9}   **4.** {7}   **5.** {3}   **6.** {−3}   **7.** (a) {7}; (b) {−6}

8. (a) {−10}; (b) {−3}   **9.** (a) {12}; (b) {−13}   **10.** (a) {1}; (b) {−10}

11. 45 and 46   **12.** (a) $2x + 2$; (b) $5n − 4$; (c) $2ab$; (d) $\dfrac{s + t}{5}$

13. The equation is $x + 8 = 35$. The number is 27.

Is the number shown in parentheses a solution for the given equation?

1. $x + 7 = 12$ (5)

2. $x + 2 = 11$ (8)

3. $x - 15 = 6$ (−21)

4. $x - 11 = 5$ (16)

5. $5 - x = 2$ (4)

6. $10 - x = 7$ (3)

7. $8 - x = 5$ (−3)

8. $5 - x = 6$ (−3)

9. $3x + 4 = 13$ (8)

10. $5x + 6 = 31$ (5)

11. $4x - 5 = 7$ (3)

12. $2x - 5 = 1$ (3)

13. $5 - 2x = 7$ (−1)

14. $4 - 5x = 9$ (−2)

15. $6 - \dfrac{3}{4}x = 4 + \dfrac{1}{4}x$ (2)

16. $5x + 4 = 2x + 10$ (4)

17. $x + 3 + 2x = 5 + x + 8$ (5)

18. $5x - 3 + 2x = 3 + x - 12$ (−2)

19. $\dfrac{3}{4}x = 18$ (20)

20. $\dfrac{3}{5}x = 24$ (40)

21. $\dfrac{3}{5}x + 5 = 11$ (10)

22. $\dfrac{2}{3}x + 8 = -12$ (−6)

Label each of the following as an expression, or a linear equation, or a nonlinear equation.

23. $2x + 1 = 9$

24. $5x - 11$

25. $7x + 2x + 8 - 3$

26. $x + 5 = 13$

27. $3x + 5 = 9$

28. $12x^2 - 5x + 2 = 5$

Solve each equation and check your results. Express each answer in set notation.

29. $x + 7 = 9$

30. $x - 4 = 6$

31. $x - 8 = 3$

32. $x + 11 = 15$

33. $x - 8 = -10$

34. $x - 8 = -11$

35. $11 = x + 5$

36. $x + 7 = 0$

37. $4x = 3x + 4$

38. $7x = 6x - 8$

39. $11x = 10x - 10$

40. $9x = 8x + 5$

41. $6x + 3 = 5x$

42. $12x - 6 = 11x$

43. $2x + 3 = x + 5$

44. $3x - 2 = 2x + 1$

45. $5x - 7 = 4x - 3$

46. $8x + 5 = 7x - 2$

47. $\dfrac{5}{3}x = 9 + \dfrac{2}{3}x$

48. $\dfrac{4}{7}x = 8 - \dfrac{3}{7}x$

49. $3 + 6x + 2 = 3x + 11 + 2x$

50. $6x - 3 + 2x = 7x + 8$

51. $4x + 7 + 3x = 5x + 13 + x$

52. $5x + 9 + 4x = 9 + 8x - 7$

53. $4(3x + 4) = 11x - 2$

54. $2(5x - 3) = 9x + 7$

55. $3(7x + 2) = 5(4x + 1) + 17$

56. $5(5x + 3) = 3(8x - 2) + 4$

57. $\dfrac{5}{4}x - 1 = \dfrac{1}{4}x + 7$

58. $\dfrac{7}{5}x + 3 = \dfrac{2}{5}x - 8$

122

59. $\dfrac{9}{2}x - \dfrac{3}{4} = \dfrac{7}{2}x + \dfrac{5}{4}$

60. $\dfrac{11}{3}x + \dfrac{1}{6} = \dfrac{8}{3}x + \dfrac{19}{6}$

61. $0.56x = 9 - 0.44x$

62. $8 - 0.37x = 5 + 0.63x$

63. $0.12x + 0.53x - 8 = -0.92x + 0.57x + 4$

64. $0.71x + 6 + 0.35x = 0.25x - 11 - 0.19x$

In Exercises 65 to 70, translate each English language statement to an algebraic equation. Let x represent the number in each case.

65. 3 more than a number is 7.

66. 5 less than a number is 12.

67. 7 less than 3 times a number is twice that same number.

68. 4 more than 5 times a number is 6 times that same number.

69. 2 times the sum of a number and 5 is 18 more than that same number.

70. 3 times the sum of a number and 7 is 4 times that same number.

71. Which of the following is equivalent to the equation $8x + 5 = 9x - 4$?

(a) $17x = -9$ (b) $x = -9$ (c) $8x + 9 = 9x$ (d) $9 = 17x$

72. Which of the following is equivalent to the equation $5x - 7 = 4x - 12$?

(a) $9x = 19$ (b) $9x - 7 = -12$ (c) $x = -18$ (d) $x - 7 = -12$

73. Which of the following is equivalent to the equation $12x - 6 = 8x + 14$?

(a) $4x - 6 = 14$ (b) $x = 20$ (c) $20x = 20$ (d) $4x = 8$

74. Which of the following is equivalent to the equation $7x + 5 = 12x - 10$?

(a) $5x = -15$ (b) $7x - 5 = 12x$ (c) $-5 = 5x$ (d) $7x + 15 = 12x$

True or false?

75. Every linear equation with one variable has exactly one solution.

76. Isolating the variable on the right side of the equation will result in a negative solution.

Solve the following word problems. Be sure to show the equation you use for the solution.

77. Number problem. The sum of a number and 7 is 33. What is the number?

78. Number problem. The sum of a number and 15 is 22. What is the number?

79. Number problem. The sum of a number and -15 is 7. What is the number?

80. Number problem. The sum of a number and -8 is 17. What is the number?

81. Number of votes cast. In an election, the winning candidate has 1,840 votes. If the total number of votes cast was 3,260, how many votes did the losing candidate receive?

82. Monthly earnings. Mike and Stefanie work at the same company and make a total of $2,760 per month. If Stefanie makes $1,400 per month, how much does Mike earn every month?

83. Appliance costs. A washer-dryer combination costs $650. If the washer costs $360, what does the dryer cost?

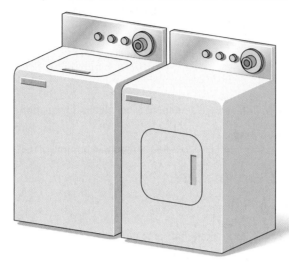

84. Computer costs. You have $2,350 saved for the purchase of a new computer that costs $3,675. How much more must you save?

85. Price increases. The price of an item has increased by $215 over last year. If the item is now selling for $675, what was the price last year?

86. An algebraic equation is a complete sentence. It has a subject, a verb, and a predicate. For example, $x + 2 = 5$ can be written in English as "Two more than a number is five" or "A number added to two is five." Write an English version of the following equations. Be sure to write complete sentences and that the sentences express the same idea as the equations. Exchange sentences with another student and see if your interpretations of each other's sentences result in the same equation.

     (a) $2x - 5 = x + 1$ (b) $2(x + 2) = 14$

     (c) $n + 5 = \dfrac{n}{2} - 6$ (d) $7 - 3a = 5 + a$

87. Complete the following explanation in your own words: "The difference between $3(x - 1) + 4 - 2x$ and $3(x - 1) + 4 = 2x$ is . . ."

88. "I make $2.50 an hour more in my new job." If x = the amount I used to make per hour and y = the amount I now make, which equations below say the same thing as the statement above? Explain your choices by translating the equation to English and comparing with the original statement.

     (a) $x + y = 2.50$ (b) $x - y = 2.50$ (c) $x + 2.50 = y$

     (d) $2.50 + y = x$ (e) $y - x = 2.50$ (f) $2.50 - x = y$

89. "The river rose 4 feet above flood stage last night." If a = the river's height at flood stage and b = the river's height now (the morning after), which equations below say the same thing as the statement? Explain your choices by translating the equations to English and comparing the meaning with the original statement.

     (a) $a + b = 4$ (b) $b - 4 = a$ (c) $a - 4 = b$

     (d) $a + 4 = b$ (e) $b + 4 = a$ (f) $b - a = 4$

90. "Surprising Results!" Work with other students to try this experiment. Each person should do the following six steps mentally, not telling anyone else what his or her calculations are.

(a) Think of a number   (b) Add 7

(c) Multiply by 3   (d) Add 3 more than the original number

(e) Divide by 4   (f) Subtract the original number

What number do you end up with? Compare your answer with everyone else's. Does everyone have the same answer? Make sure that everyone followed the directions accurately. How do you explain the results? Algebra makes the explanation clear. Work together to do the problem again, using a variable for the number. Make up another series of computations that give "surprising results."

91. (a) Do you think that the following is a linear equation in one variable?

$$3(2x + 4) = 6(x + 2)$$

(b) What happens when you use the properties of this section to solve the equation?

(c) Pick *any* number to substitute for x in this equation. Now try a different number to substitute for x in the equation. Try yet another number to substitute for x in the equation. Summarize your findings.

(d) Can this equation be called *linear in one variable?* Refer to the definition as you explain your answer.

92. (a) Do you think the following is a linear equation in one variable?

$$4(3x - 5) = 2(6x - 8) - 3$$

(b) What happens when you use the properties of this section to solve the equation?

(c) Do you think it is possible to find a solution for this equation?

(d) Can this equation be called *linear in one variable?* Refer to the definition as you explain your answer.

2.2 Solving Equations by Multiplying and Dividing

2.2 OBJECTIVES

1. *Determine whether a given number is a solution for an equation*
2. *Use the multiplication property to solve equations*

Let's look at a different type of equation. For instance, what if we want to solve an equation like the following?

$$6x = 18$$

Using the addition property of Section 2.1 won't help. We will need a second property for solving equations.

Again, as long as you do the *same* thing to *both* sides of the equation, the "balance" is maintained. But multiplying by 0 gives 0 = 0. We have lost the variable!

The Multiplication Property of Equality

$$\text{If } a = b \quad \text{then} \quad ac = bc \quad \text{where } c \neq 0$$

In words, multiplying both sides of an equation by the same nonzero number gives an equivalent equation.

Let's work through some examples, using this second rule.

Example 1

Solving Equations by Using the Multiplication Property

Solve

$$6x = 18$$

Here the variable x is multiplied by 6. So we apply the multiplication property and multiply both sides by $\frac{1}{6}$. Keep in mind that we want an equation of the form

$$x = \boxed{}$$

$$\frac{1}{6}(6x) = \left(\frac{1}{6} \cdot 6\right)x$$
$$= 1 \cdot x \quad \text{or} \quad x$$

$$\frac{1}{6}(6x) = \left(\frac{1}{6}\right)18$$

126

We can now simplify.

$$1 \cdot x = 3 \qquad \text{or} \qquad x = 3$$

The solution set is $\{3\}$. To check, replace x with 3:

$$6 \cdot 3 \stackrel{?}{=} 18$$
$$18 = 18 \qquad \text{True}$$

 CHECK YOURSELF 1

Solve and check.

$$8x = 32$$

In Example 1 we solved the equation by multiplying both sides by the reciprocal of the coefficient of the variable. Example 2 illustrates a slightly different approach to solving an equation by using the multiplication property.

Example 2

Solving Equations by Using the Multiplication Property

Solve

$$5x = -35$$

The variable x is multiplied by 5. We *divide* both sides by 5 to "undo" that multiplication:

Since division is defined in terms of multiplication, we can also divide both sides of an equation by the same nonzero number.

$$\frac{5x}{5} = \frac{-35}{5}$$
$$x = -7$$

Note that the right side reduces to -7. Be careful with the rules for signs.

The solution set is $\{-7\}$.
    We will leave it to you to check the solution.

 CHECK YOURSELF 2

Solve and check.

$$7x = -42$$

Example 3

Solving Equations by Using the Multiplication Property

Solve

$$-9x = 54$$

In this case, x is multiplied by -9, so we divide both sides by -9 to isolate x on the left:

$$\frac{-9x}{-9} = \frac{54}{-9}$$

$$x = -6$$

The solution set is $\{-6\}$. To check:

$$(-9)(-6) \stackrel{?}{=} 54$$

$$54 = 54 \qquad \text{True}$$

✔ *CHECK YOURSELF 3*

Solve and check.

$$-10x = -60$$

Example 4 illustrates the use of the multiplication property when there are fractions in an equation.

Example 4

Solving Equations by Using the Multiplication Property

(a) Solve

$$\frac{x}{3} = 6$$

Here x is *divided* by 3. We will use multiplication to isolate x.

$$3\left(\frac{x}{3}\right) = 3 \cdot 6$$

This leaves x alone on the left because

$$3\left(\frac{x}{3}\right) = \frac{3}{1} \cdot \frac{x}{3} = \frac{x}{1} = x$$

$$x = 18$$

The solution set is $\{18\}$.

To check:

$$\frac{18}{3} \stackrel{?}{=} 6$$

$$6 = 6 \qquad \text{True}$$

(b) Solve

$$\frac{x}{5} = -9$$

$$5\left(\frac{x}{5}\right) = 5(-9) \qquad \text{Since } x \text{ is divided by 5,}$$
$$\text{multiply both sides by 5.}$$

$$x = -45$$

The solution set is $\{-45\}$. To check, we replace x with -45:

$$\frac{-45}{5} \stackrel{?}{=} -9$$

$$-9 = -9 \qquad \text{True}$$

The solution is verified.

✔ *CHECK YOURSELF 4*

Solve and check.

(a) $\dfrac{x}{7} = 3$ (b) $\dfrac{x}{4} = -8$

When the variable is multiplied by a fraction that has a numerator other than 1, there are two approaches to finding the solution.

Example 5   Solving Equations by Using Reciprocals

Solve

$$\frac{3}{5}x = 9$$

One approach is to multiply by 5 as the first step.

$$5\left(\frac{3}{5}x\right) = 5 \cdot 9$$

$$3x = 45$$

Now we divide by 3.

$$\frac{3x}{3} = \frac{45}{3}$$

$$x = 15$$

To check the solution set $\{15\}$, substitute 15 for x.

$$\frac{3}{5} \cdot 15 \stackrel{?}{=} 9$$

$$9 = 9 \qquad \text{True}$$

A second approach combines the multiplication and division steps and is generally a bit more efficient. We multiply by $\frac{5}{3}$.

We multiply by $\frac{5}{3}$ because it is the reciprocal of $\frac{3}{5}$, and the product of a number and its reciprocal is 1! So $\left(\frac{5}{3}\right)\left(\frac{3}{5}\right) = 1$.

$$\frac{5}{3}\left(\frac{3}{5}x\right) = \frac{5}{3} \cdot 9$$

$$x = \frac{5}{\cancel{3}} \cdot \frac{\cancel{9}^{3}}{1} = 15$$

So $x = 15$, as before.

  CHECK YOURSELF 5

Solve and check.

$$\frac{2}{3}x = 18$$

You may sometimes have to simplify an equation before applying the methods of this section. Example 6 illustrates this property.

Example 6   Combining Like Terms and Solving Equations

Solve and check.

$$3x + 5x = 40$$

Using the distributive property, we can combine the like terms on the left to write

$$8x = 40$$

We can now proceed as before.

$$\frac{8x}{8} = \frac{40}{8} \qquad \text{Divide by 8}$$

$$x = 5$$

The solution set is {5}. To check, we return to the original equation. Substituting 5 for x yields

$$3 \cdot 5 + 5 \cdot 5 \stackrel{?}{=} 40$$

$$15 + 25 \stackrel{?}{=} 40$$

$$40 = 40 \qquad \text{True}$$

The solution is verified.

 CHECK YOURSELF 6

Solve and check.

$$7x + 4x = -66$$

| Example 7 | **Solving an Application** |

Elisa recently won an election by receiving 760 votes. If this total was four-fifths of the total votes cast, how many votes were cast?

Step 1   We want to find the number of votes cast.

Step 2   Let x be the total number of votes cast. Elisa received 760 votes, which represented four-fifths of the total.

Step 3 $$\frac{4}{5}x = 760$$

Step 4 $$\left(\frac{5}{4}\right)\frac{4}{5}x = \left(\frac{5}{4}\right)760$$

$$x = 950$$

Step 5   Elisa received 950 votes.

 CHECK YOURSELF 7

Peter has recently received an inheritance. He invests $10,000 of this money in his company's stock. If this investment represents five-eighths of his money, how much was his inheritance?

✔ *CHECK YOURSELF ANSWERS*

1. {4}   **2.** {−6}   **3.** {6}   **4.** (a) {21}; (b) {−32}

5. {27}   **6.** {−6}   **7.** $16,000

Exercises ▪ 2.2

Solve for x and check your result. Express your answer in set notation.

1. $5x = 20$

2. $6x = 30$

3. $7x = 42$

4. $6x = -42$

5. $63 = 9x$

6. $66 = 6x$

7. $4x = -16$

8. $-3x = 27$

9. $-9x = 72$

10. $-9x = 90$

11. $6x = -54$

12. $-7x = 49$

13. $-4x = -12$

14. $52 = -4x$

15. $-42 = 6x$

16. $-7x = -35$

17. $-6x = -54$

18. $-4x = -24$

19. $\dfrac{x}{4} = 2$

20. $\dfrac{x}{3} = 2$

21. $\dfrac{x}{5} = 3$

22. $\dfrac{x}{8} = 5$

23. $6 = \dfrac{x}{7}$

24. $6 = \dfrac{x}{3}$

25. $\dfrac{x}{5} = -4$

26. $\dfrac{x}{5} = -7$

27. $-\dfrac{x}{3} = 8$

28. $-\dfrac{x}{4} = -3$

29. $\dfrac{2}{3}x = 6$

30. $\dfrac{4}{5}x = 8$

31. $\dfrac{3}{4}x = -15$

32. $\dfrac{7}{8}x = -21$

33. $-\dfrac{2}{5}x = 10$

34. $-\dfrac{5}{6}x = -15$

35. $5x + 4x = 36$

36. $8x - 3x = -50$

37. $\dfrac{7}{9}x - 5 = \dfrac{3}{9}x + 11$

38. $\dfrac{4}{11}x - 9 + \dfrac{2}{11}x = 18 - \dfrac{3}{11}x$

39. $4(x + 5) + 7x = -3(x - 2)$

40. $-2(x - 3) + 10 = -4(5 - 4x)$

Once again, certain equations involving decimal fractions can be solved by the methods of this section. For instance, to solve $2.3x = 6.9$, we simply use the multiplication property to divide both sides of the equation by 2.3. This will isolate x on the left as desired. Use this idea to solve each of the following equations for x.

41. $3.2x = 12.8$

42. $5.1x = -15.3$

43. $-4.5x = 13.5$

44. $-8.2x = -32.8$

45. $1.3x + 2.8x = 12.3$

46. $2.7x + 5.4x = -16.2$

47. $9.3x - 6.2x = 12.4$

48. $12.5x - 7.2x = -21.2$

Translate each of the following statements to an equation. Let x represent the number in each case.

49. 6 times a number is 72.

50. Twice a number is 36.

51. A number divided by 7 is equal to 6.

52. A number divided by 5 is equal to -4.

53. $\dfrac{1}{3}$ of a number is 8.

54. $\dfrac{1}{5}$ of a number is 10.

55. $\dfrac{3}{4}$ of a number is 18.

56. $\dfrac{2}{7}$ of a number is 8.

57. Twice a number, divided by 5, is 12.

58. 3 times a number, divided by 4, is 36.

59. Customer satisfaction. Three-fourths of the theater audience left in disgust. If 87 angry patrons walked out, how many were there originally?

60. Auto repair. A mechanic charged $45 an hour to replace the ignition coil on a car plus $225 for parts. If the total bill was $450, how many hours did the repair job take?

61. Telephone charges. A call to Phoenix, Arizona, from Dubuque, Iowa, costs 55 cents for the first minute and 23 cents for each additional minute or portion of a minute. If Barry has $6.30 in change, how long can he talk?

62. Number problem. The sum of 4 times a number and 14 is 34. Find the number.

63. Number problem. If 6 times a number is subtracted from 42, the result is 24. Find the number.

64. Number problem. When a number is divided by −6, the result is 3. Find the number.

65. Geometry. Suppose that the circumference of a tree is measured to be 9 ft 2 in., or 110 in. To find the diameter of the tree at that point, we must solve the equation

$$110 = 3.14d$$

(*Note:* 3.14 is the approximation for π.) Find the diameter of the tree to the nearest inch.

66. Geometry. Suppose that the circumference of a circular swimming pool is 88 ft. Find the diameter of the pool by solving the equation to the nearest foot.

$$88 = 3.14d$$

SECTION 2.3 Combining the Rules to Solve Equations

2.3 OBJECTIVES

1. Use both addition and multiplication to solve equations
2. Solve equations that involve fractions
3. Solve applications

"Success usually comes to those who are too busy to be looking for it."

—David Henry Thoreau

In Section 2.1 we solved equations by using the addition property, which allowed us to solve equations such as $x + 3 = 9$. Then, in Section 2.2, we solved equations by using the multiplication property, which allowed us to solve equations such as $5x = 32$. Now, we will solve equations that require us to use both the addition and multiplication properties.

In Example 1, we will check to see that a given value for the variable is a solution to a given equation.

Example 1

Checking a Solution

Test to see if -3 is a solution for

$$5x + 6 = 2x - 3$$

The solution set includes -3 because replacing x with -3 gives

$$5(-3) + 6 \stackrel{?}{=} 2(-3) - 3$$

$$-15 + 6 \stackrel{?}{=} -6 - 3$$

$$-9 = -9 \qquad \text{A true statement}$$

 CHECK YOURSELF 1

Test to see if 7 is a solution for the equation

$$5x - 15 = 2x + 6$$

In Example 2, we will apply the addition and multiplication properties to find the solution of a linear equation.

| **Example 2** | **Applying the Properties of Equality** |

Solve for x.

$$3x - 5 = 4 \tag{1}$$

Why did we add 5? We added 5 because it is the *opposite* of -5, and the resulting equation will have the variable term on the left and the constant term on the right.

We start by using the addition property to add 5 to both sides of the equation.

$$3x - 5 + 5 = 4 + 5$$
$$3x = 9 \tag{2}$$

Now we want to get the x term alone on the left with a coefficient of 1 (we call this *isolating* the x). To do this, we use the multiplication property and multiply both sides by $\frac{1}{3}$.

We choose $\frac{1}{3}$ because $\frac{1}{3}$ is the *reciprocal* of 3 and

$$\frac{1}{3} \cdot 3 = 1$$

$$\frac{1}{3}(3x) = \frac{1}{3}(9)$$
$$\left(\frac{1}{3} \cdot 3\right)(x) = 3$$

So $x = 3$. The solution set is $\{3\}$. $\tag{3}$

Since any application of the addition or multiplication properties leads to an equivalent equation, equations (1), (2), and (3) all have the same solution, 3.

To check this result, we can replace x with 3 in the original equation:

$$3(3) - 5 \stackrel{?}{=} 4$$
$$9 - 5 \stackrel{?}{=} 4$$
$$4 = 4 \qquad \text{A true statement}$$

You may prefer a slightly different approach in the last step of the solution above. From equation (2),

$$3x = 9$$

The multiplication property can be used to *divide* both sides of the equation by 3. Then

$$\frac{3x}{3} = \frac{9}{3}$$
$$x = 3$$

The result is the same.

CHECK YOURSELF 2

Solve for x.

$$4x - 7 = 17$$

The steps involved in using the addition and multiplication properties to solve an equation are the same if more terms are involved in an equation.

| Example 3 | **Applying the Properties of Equality** |

Solve for x.

$$5x - 11 = 2x - 7$$

Our objective is to use the properties of equality to isolate x on one side of an equivalent equation. We begin by adding 11 to both sides.

Again, adding 11 leaves us with the constant term on the right.

$$5x - 11 + 11 = 2x - 7 + 11$$
$$5x = 2x + 4$$

We continue by adding $-2x$ to (or subtracting $2x$ from) both sides.

If you prefer, write

$$5x - 2x = 2x - 2x + 4$$

Again,

$$3x = 4$$

$$5x + (-2x) = 2x + (-2x) + 4$$
$$3x = 4$$

This is the same as dividing both sides by 3. So

$$\frac{3x}{3} = \frac{4}{3}$$

$$x = \frac{4}{3}$$

To isolate x, we now multiply both sides by $\frac{1}{3}$.

$$\frac{1}{3}(3x) = \frac{1}{3}(4)$$

$$x = \frac{4}{3}$$

In set notation, we write $\left\{\frac{4}{3}\right\}$. We leave it to you to check this result by substitution.

 CHECK YOURSELF 3

Solve for x.

$$7x - 12 = 2x - 9$$

Both sides of an equation should be simplified as much as possible *before* the addition and multiplication properties are applied. If like terms are involved on one side (or on both sides) of an equation, they should be combined before an attempt is made to isolate the variable. Example 4 illustrates this approach.

| **Example 4** | ## Applying the Properties of Equality with Like Terms |

Solve for x.

Note the like terms on the left and right sides of the equation.

$$8x + 2 - 3x = 8 + 3x + 2$$

Here we combine the like terms $8x$ and $-3x$ on the left and the like terms 8 and 2 on the right as our first step. We then have

$$5x + 2 = 3x + 10$$

We can now solve as before.

$$5x + 2 - 2 = 3x + 10 - 2 \quad \text{Subtract 2 from both sides}$$
$$5x = 3x + 8$$

Then

$$5x - 3x = 3x - 3x + 8 \quad \text{Subtract } 3x \text{ from both sides}$$
$$2x = 8$$
$$\frac{2x}{2} = \frac{8}{2} \quad \text{Divide both sides by 2}$$
$$x = 4$$

The solution set is {4}, which can be checked by returning to the *original equation*.

✔ CHECK YOURSELF 4

Solve for x.

$$7x - 3 - 5x = 10 + 4x + 3$$

If there are parentheses on one or both sides of an equation, the parentheses should be removed by applying the distributive property as the first step. Like terms should then be combined before an attempt is made to isolate the variable. Consider Example 5.

| **Example 5** | ## Applying the Properties of Equality with Parentheses |

Solve for x.

$$x + 3(3x - 1) = 4(x + 2) + 4$$

First, apply the distributive property to remove the parentheses on the left and right sides.

$$x + 9x - 3 = 4x + 8 + 4$$

Combine like terms on each side of the equation.

$$10x - 3 = 4x + 12$$

Recall that to isolate x, we must get x alone on the left side with a coefficient of 1.

Now, isolate the variable x on the left side.

$$10x - 3 + 3 = 4x + 12 + 3 \qquad \text{Add 3 to both sides}$$
$$10x = 4x + 15$$
$$10x - 4x = 4x - 4x + 15 \qquad \text{Subtract } 4x \text{ from both sides}$$
$$6x = 15$$
$$\frac{6x}{6} = \frac{15}{6} \qquad \text{Divide both sides by 6}$$
$$x = \frac{5}{2}$$

The solution set is $\left\{\dfrac{5}{2}\right\}$. Again, this can be checked by returning to the original equation.

 CHECK YOURSELF 5

Solve for x.

$$x + 5(x + 2) = 3(3x - 2) + 18$$

The LCM of a set of denominators is also called the **least common denominator (LCD).**

To solve an equation involving fractions, the first step is to multiply both sides of the equation by the **least common multiple (LCM)** of all denominators in the equation. This will clear the equation of fractions, and we can proceed as before.

Example 6

Applying the Properties of Equality with Fractions

Solve for x.

$$\frac{x}{2} - \frac{2}{3} = \frac{5}{6}$$

First, multiply each side by 6, the least common multiple of 2, 3, and 6.

$$6\left(\frac{x}{2}-\frac{2}{3}\right)=6\left(\frac{5}{6}\right)$$

$$6\left(\frac{x}{2}\right)-6\left(\frac{2}{3}\right)=6\left(\frac{5}{6}\right) \qquad \text{Apply the distributive property}$$

$$\overset{3}{6}\left(\frac{x}{2}\right)-\overset{2}{6}\left(\frac{2}{3}\right)=\overset{1}{6}\left(\frac{5}{6}\right) \qquad \text{Simplify}$$

The equation is now cleared of fractions.

$$3x-4=5$$

Next, isolate the variable x on the left side.

$$3x=9$$
$$x=3$$

The solution set, {3}, can be checked as before by returning to the original equation.

✔ **CHECK YOURSELF 6**

Solve for x.

$$\frac{x}{4}-\frac{4}{5}=\frac{19}{20}$$

Be sure that the distributive property is applied properly so that *every term* of the equation is multiplied by the LCM.

Example 7   ## Applying the Properties of Equality with Fractions

Solve for x.

$$\frac{2x-1}{5}+1=\frac{x}{2}$$

First, multiply each side by 10, the LCM of 5 and 2.

$$10\left(\frac{2x-1}{5}+1\right)=10\left(\frac{x}{2}\right) \qquad \text{Apply the distributive property on the left. Reduce.}$$

$$\overset{2}{10}\left(\frac{2x-1}{5}\right)+10(1)=\overset{5}{10}\left(\frac{x}{2}\right) \qquad \text{Next, isolate } x. \text{ Here we isolate } x \text{ on the right side.}$$

$$2(2x-1)+10=5x$$
$$4x-2+10=5x$$
$$4x+8=5x$$

$4x$ is subtracted from both sides of the equation.

$$8=x \qquad \text{The solution set for the original equation is {8}.}$$

✔ *CHECK YOURSELF 7*

Solve for *x*.

$$\frac{3x + 1}{4} - 2 = \frac{x + 1}{3}$$

Conditional Equations, Identities, and Contradictions

1. An equation that is true for only particular values of the variable is called a **conditional equation.** Here the equation can be written in the form

$$ax + b = 0$$

where $a \neq 0$. This case was illustrated in all our previous examples and exercises.

2. An equation that is true for all possible values of the variable is called an **identity.** In this case, *both a* and *b* are 0, so we get the equation $0 = 0$. This will be the case if both sides of the equation reduce to the same expression (a true statement).

3. An equation that is never true, no matter what the value of the variable, is called a **contradiction.** For example, if *a* is 0 but *b* is 4. This will be the case if the equation simplifies as a false statement.

Example 8 illustrates the second and third cases.

| Example 8 | ## Identities and Contradictions |

(a) Solve for *x*.

$$2(x - 3) - 2x = -6$$

Apply the distributive property to remove the parentheses.

$$2x - 6 - 2x = -6$$

$$-6 = -6 \qquad \text{A } \textit{true} \text{ statement}$$

See the definition of an identity, above. By adding 6 to both sides of this equation, we have $0 = 0$.

Because the two sides reduce to the true statement $-6 = -6$, the original equation is an *identity,* and the solution set is the set of all real numbers. This is sometimes written as $\mathbb{R}$, or $\{x \mid x \in \mathbb{R}\}$, which is read, "the set of all numbers *x* that are elements of the real number set."

(b) Solve for *x*.

$$3(x + 1) - 2x = x + 4$$

Again, apply the distributive property.

$$3x + 3 - 2x = x + 4$$

$$x + 3 = x + 4$$

$$3 = 4 \qquad \text{A } \textit{false} \text{ statement}$$

See the definition of a contradiction. Subtracting 3 from both sides, we have $0 = 1$.

Since the two sides reduce to the false statement $3 = 4$, the original equation is a contradiction. There are no values of the variable that can satisfy the equation. The solution set has nothing in it. We call this the **empty set** or **null set** and write $\{\ \}$ or $\varnothing$.

✔ *CHECK YOURSELF 8*

Determine whether each of the following equations is a conditional equation, an identity, or a contradiction. Write the solution set.

(a) $2(x + 1) - 3 = x$ (b) $2(x + 1) - 3 = 2x + 1$

(c) $2(x + 1) - 3 = 2x - 1$

An **algorithm** is a step-by-step process for problem solving.

An organized step-by-step procedure is the key to an effective equation-solving strategy. The following algorithm summarizes our work in this section and gives you guidance in approaching the problems that follow.

Solving Linear Equations in One Variable

Step 1. Remove any grouping symbols by applying the distributive property.

Step 2. Multiply both sides of the equation by the LCM of any denominators to clear the equation of fractions.

Step 3. Combine any like terms that appear on either side of the equation.

Step 4. Apply the addition property of equality to write an equivalent equation with the variable term on *one side* of the equation and the constant term on the *other side*.

Step 5. Apply the multiplication property of equality to write an equivalent equation with the variable isolated on one side of the equation with coefficient 1.

Step 6. Check the solution in the *original* equation.

Note: If the equation derived in step 5 is always true, the original equation is an *identity*. If the equation is always false, the original equation is a *contradiction*.

When you are solving an equation for which a calculator is recommended, it is often easiest to do all calculations as the last step. For more complex equations, it is usually best to calculate at each step.

Example 9	**Solving Equations by Using a Calculator**

Solve the following equation for x.

$$5(x - 3.25) + \frac{3}{4} = 2,110.75$$

Following the steps of the algorithm, we get

$$5x - 16.25 + \frac{3}{4} = 2,110.75 \qquad \text{Remove parentheses}$$

$$20x - 65 + 3 = 8,443 \qquad \text{Multiply by the LCM}$$

$$20x = 8,443 + 62 \qquad \text{Isolate the variable}$$

$$x = \frac{8,505}{20}$$

Now, we use a calculator to simplify the expression on the right.

$$x = 425.25 \qquad \text{or} \qquad \{425.25\}$$

  CHECK YOURSELF 9

Solve the following equation for x.

$$7(x + 4.3) - \frac{3}{5} = 467$$

Consecutive integers are integers that follow one another, such as 10, 11, and 12.

Consecutive Integers

If x is an integer, then $x + 1$ is the next consecutive integer, $x + 2$ is the next, and so on.

If x is an odd integer, the next **consecutive odd integer** is $x + 2$, and the next is $x + 4$.

If x is an even integer, the next **consecutive even integer** is $x + 2$, and the next is $x + 4$.

We'll need this idea in Example 10.

| **Example 10** | **Solving an Application** |

The sum of two consecutive integers is 41. What are the two integers?

Step 1   We want to find the two consecutive integers.

Step 2   Let x be the first integer. Then $x + 1$ must be the next.

Step 3   The first integer ⟶   The second integer

$$x + \overbrace{x + 1} = 41$$

The sum ⟶   Is

Solve the equation.

Step 4

$$x + x + 1 = 41$$
$$2x + 1 = 41$$
$$2x = 40$$
$$x = 20$$

The first integer (x) is 20, and the next integer ($x + 1$) is 21.

Check.

Step 5   The sum of the two integers 20 and 21 is 41.

✔ CHECK YOURSELF 10

The sum of three consecutive integers is 51. What are the three integers?

Sometimes algebra is used to reconstruct missing information. Example 11 does just that with some election information.

| **Example 11** | **Solving an Application** |

There were 55 more yes votes than no votes on an election measure. If 735 votes were cast in all, how many yes votes were there? How many no votes?

What do you need to find?

Step 1   We want to find the number of yes votes and the number of no votes.

Assign letters to the unknowns.

Step 2   Let x be the number of no votes. Then

$$\underline{x + 55}$$
↑
55 more than x

is the number of yes votes.

Write an equation.

Step 3

$$x + \underbrace{x + 55} = 735$$

No votes   Yes votes   Total votes cast

Solve the equation.   **Step 4**

$$x + x + 55 = 735$$

$$2x + 55 = 735$$

$$2x = 680$$

$$x = 340$$

$$\text{No votes } (x) = 340$$

$$\text{Yes votes } (x + 55) = 395$$

Check.   **Step 5**   Thus 340 no votes plus 395 yes votes equal 735 total votes. The solution
                                              checks.

   CHECK YOURSELF 11

Francine earns $120 per month more than Rob. If they earn a total of $2,680 per month,
what are their monthly salaries?

Similar methods will allow you to solve a variety of word problems. Example 12
includes three unknown quantities but uses the same basic solution steps.

Example 12   # Solving an Application

Juan worked twice as many hours as Jerry. Marcia worked 3 h more than Jerry. If they
worked a total of 31 h, find out how many hours each worked.

Step 1   We want to find the hours each worked, so there are three unknowns.

There are other choices for *x*,   **Step 2**   Let *x* be the hours that Jerry worked.
but choosing the smallest
quantity will usually give the   Juan worked twice Jerry's hours.
easiest equation to write and
solve.   Then $2x$ is Juan's hours worked.

                                           Marcia worked 3 h more than Jerry worked.

                                           And $x + 3$ is Marcia's hours.

Step 3

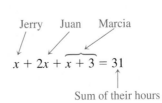

$$x + 2x + \overbrace{x + 3} = 31$$

Sum of their hours

Step 4 $$x + 2x + x + 3 = 31$$

$$4x + 3 = 31$$

$$4x = 28$$

$$x = 7$$

Jerry's hours $(x) = 7$

Juan's hours $(2x) = 14$

Marcia's hours $(x + 3) = 10$

Step 5   The sum of their hours $(7 + 14 + 10)$ is 31, and the solution is verified.

✔ *CHECK YOURSELF 12*

Lucy jogged twice as many miles as Paul but 3 mi less than Isaac. If the three ran a total of 23 mi, how far did each person run?

Many applied problems involve the use of *percents*. The idea of *percent* is a useful way of naming parts of a whole. We can think of a percent as a fraction whose denominator is 100. Thus 15% would be written as $\dfrac{15}{100}$ and represents 15 parts out of 100. A percentage can also be expressed as a decimal by converting the fractional representation to a decimal. So 15% is written as 0.15. Examples 13 and 14 illustrate the use of percent in applications.

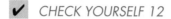

Example 13

Solving an Application

Marzenna inherits \$5,000 and invests part of her money in bonds at 4% and the remaining in savings at 3%. What amount has she invested at each rate if she receives \$180 in interest for 1 year?

Step 1   We want to find the amount invested at each rate, so there are two unknowns.

Step 2   Let x be the amount invested at 4%.

$5,000 was the total amount of money invested.

So $5,000 - x$ is the amount invested at 3%.

$0.04x$ is the amount of interest from the 4% investment.

$0.03(5,000 - x)$ is the amount of interest from the 3% investment.

$180 is the total interest for the year.

Step 3 $$0.04x + 0.03(5{,}000 - x) = 180$$

Step 4 $$0.04x + 0.03(5{,}000) - 0.03x = 180$$

$$0.04x + 150 - 0.03x = 180$$

$$0.04x - 0.03x = 180 - 150$$

$$0.01x = 30$$

$$x = \frac{30}{0.01}$$

$$x = 3{,}000$$

Amount invested at 4% (x) = \$3,000

The check (Step 5) is left to you.

Amount invested at 3% $(5000 - x)$ = \$2,000

 CHECK YOURSELF 13

José won \$8,000 in the lottery. He invested some of his winnings in bonds at 2% and the rest in savings at 5%. If he receives \$295 interest for 1 year, how much was invested at each rate?

Example 14   # Solving an Application

Tony Gonzalez earns a take-home pay of \$592 per week. If his deductions for taxes, retirement, union dues, and a medical plan amount to 26% of his wages, what is his weekly pay before the deductions?

Step 1   We want to find his weekly pay before deductions (gross pay).

Step 2   Let x = gross pay.

Since 26% of his gross pay is deducted from his weekly salary, the amount deducted is $0.26x$.

\$592 is Tony's take-home pay (net pay).

Step 3   Net pay = gross pay − deductions

$$\$592 = x - 0.26x$$

Step 4 $$592 = 0.74x$$

$$\frac{592}{0.74} = x$$

$$800 = x$$

So Tony's weekly pay before deductions is \$800.

✔ *CHECK YOURSELF 14*

Joan Grow gives 10% of her take-home pay to the church. This amounts to $90 per month. In addition, her paycheck deductions are 25% of her gross monthly income. What is her gross monthly income?

✔ *CHECK YOURSELF ANSWERS*

1. $5(7) - 15 \stackrel{?}{=} 2(7) + 6$ **2.** $\{6\}$ **3.** $\left\{\dfrac{3}{5}\right\}$ **4.** $\{-8\}$ **5.** $\left\{-\dfrac{2}{3}\right\}$ **6.** $\{7\}$
$35 - 15 \stackrel{?}{=} 14 + 6$
$20 = 20$
↑
A true statement

7. $\{5\}$ **8.** (a) conditional, $\{1\}$; (b) contradiction, $\{\ \}$; (c) identity, $\mathbb{R}$ **9.** $\{62.5\}$

10. The equation is $x + x + 1 + x + 2 = 51$. The integers are 16, 17, and 18.

11. The equation is $x + x + 120 = 2{,}680$. Rob's salary is $1,280, and Francine's is $1,400.

12. Paul: 4 mi; Lucy: 8 mi; Isaac: 11 mi

13. $3,500 invested at 2% and $4,500 at 5%   **14.** $1,200.00

Exercises · 2.3

Simplify and then solve each equation. Express your answer in set notation.

1. $3x + 1 = 13$

2. $3x - 1 = 17$

3. $3x - 2 = 7$

4. $5x + 3 = 23$

5. $4 - 7x = 18$

6. $7 - 4x = -5$

7. $3 - 4x = -9$

8. $5 - 4x = 25$

9. $\dfrac{x}{2} + 1 = 5$

10. $\dfrac{x}{3} - 2 = 3$

11. $\dfrac{3}{4}x + 8 = 32$

12. $\dfrac{5}{6}x - 9 = 16$

13. $5x = 2x + 9$

14. $7x = 18 - 2x$

15. $9x + 2 = 3x + 38$

16. $8x - 3 = 4x + 17$

17. $4x - 8 = x - 14$

18. $6x - 5 = 3x - 29$

19. $-\dfrac{2}{3}x + 5 = 9 - \dfrac{8}{3}x$

20. $\dfrac{4}{3}x - 7 = 11 - \dfrac{5}{3}x$

21. $5x + 4 = 7x - 8$

22. $2x + 23 = 6x - 5$

23. $6x + 7 - 4x = 8 + 7x - 26$

24. $7x - 2 - 3x = 5 + 8x + 13$

25. $6x - 3 + 5x + 11 = 8x - 12$

26. $3x + 3 + 8x - 9 = 7x + 5$

27. $5(8 - x) = 3x$

28. $7x = 7(6 - x)$

29. $7(2x - 1) - 5x = x + 25$

30. $9(3x + 2) - 10x = 12x - 7$

31. $2(2x - 1) = 3(x + 1)$

32. $3(3x - 1) = 4(3x + 1)$

33. $8x - 3(2x - 4) = 17$

34. $7x - 4(3x + 4) = 9$

35. $7(3x + 4) = 8(2x + 5) + 13$

36. $-4(2x - 1) + 3(3x + 1) = 9$

37. $9 - 4(3x + 1) = 3(6 - 3x) - 9$

38. $13 - 4(5x + 1) = 3(7 - 5x) - 15$

39. $5.3x - 7 = 2.3x + 5$

40. $9.8x + 2 = 3.8x + 20$

Clear fractions and then solve each equation. Express your answer in set notation.

41. $\dfrac{2x}{3} - \dfrac{5}{3} = 3$

42. $\dfrac{3x}{4} + \dfrac{1}{4} = 4$

43. $\dfrac{x}{6} + \dfrac{x}{5} = 11$

44. $\dfrac{x}{6} - \dfrac{x}{8} = 1$

45. $\dfrac{2x}{5} - \dfrac{x}{3} = \dfrac{7}{15}$

46. $\dfrac{2x}{7} - \dfrac{3x}{5} = \dfrac{6}{35}$

47. $\dfrac{x}{5} - \dfrac{x - 7}{3} = \dfrac{1}{3}$

48. $\dfrac{x}{6} + \dfrac{3}{4} = \dfrac{x - 1}{4}$

49. $\dfrac{5x - 3}{4} - 2 = \dfrac{x}{3}$

50. $\dfrac{6x - 1}{5} - \dfrac{2x}{3} = 3$

51. $\dfrac{3x - 2}{3} - \dfrac{2x - 5}{5} = \dfrac{7}{15}$

52. $\dfrac{4x}{7} - \dfrac{2x - 3}{3} = \dfrac{19}{21}$

Classify each equation as a conditional equation, an identity, or a contradiction. Write the solution set.

53. $3(x - 1) = 2x + 3$

54. $2(x + 3) = 2x + 6$

55. $3(x - 1) = 3x + 3$

56. $2(x + 3) = x + 5$

57. $3(x - 1) = 3x - 3$

58. $2(2x - 1) = 3x - 4$

59. $3x - (x - 3) = 2(x + 1) + 2$

60. $5x - (x + 4) = 4(x - 2) + 4$

61. $\dfrac{x}{2} - \dfrac{x}{3} = \dfrac{x}{6}$

62. $\dfrac{3x}{4} - \dfrac{2x}{3} = \dfrac{1}{6}$

Translate each of the following statements to an equation. Let x represent the number in each case.

63. 3 more than twice a number is 7.

64. 5 less than 3 times a number is 25.

65. 7 less than 4 times a number is 41.

66. 10 more than twice a number is 44.

67. 5 more than two-thirds a number is 21.

68. 3 less than three-fourths of a number is 24.

69. 3 times a number is 12 more than that number.

70. 5 times a number is 8 less than that number.

Solve the following word problems.

71. Number addition. The sum of twice a number and 16 is 24. What is the number?

72. Number addition. 3 times a number, increased by 8, is 50. Find the number.

73. Number subtraction. 5 times a number, minus 12, is 78. Find the number.

74. Number subtraction. 4 times a number, decreased by 20, is 44. What is the number?

75. Consecutive integers. The sum of two consecutive integers is 71. Find the two integers.

76. Consecutive integers. The sum of two consecutive integers is 145. Find the two integers.

77. Consecutive integers. The sum of three consecutive integers is 90. What are the three integers?

78. Consecutive integers. If the sum of three consecutive integers is 93, find the three integers.

79. Even integers. The sum of two consecutive even integers is 66. What are the two integers? (*Hint:* Consecutive even integers such as 10, 12, and 14 can be represented by x, $x + 2$, $x + 4$, and so on.)

80. Even integers. If the sum of two consecutive even integers is 110, find the two integers.

81. Odd integers. If the sum of two consecutive odd integers is 52, what are the two integers? (*Hint:* Consecutive odd integers such as 21, 23, and 25 can be represented by x, $x + 2$, $x + 4$, and so on.)

82. Odd integers. The sum of two consecutive odd integers is 88. Find the two integers.

83. Consecutive integers. 4 times an integer is 9 more than 3 times the next consecutive integer. What are the two integers?

84. Consecutive even integers. 4 times an even integer is 30 less than 5 times the next consecutive even integer. Find the two integers.

85. Election votes. In an election, the winning candidate had 160 more votes than the loser. If the total number of votes cast was 3,260, how many votes did each candidate receive?

86. Monthly salaries. Jody earns $140 more per month than Frank. If their monthly salaries total $2,760, what amount does each earn?

87. Appliance costs. A washer-dryer combination costs $950. If the washer costs $90 more than the dryer, what does each appliance cost?

88. Age. Yan Ling is 1 year less than twice as old as his sister. If the sum of their ages is 14 years, how old is Yan Ling?

89. Age. Diane is twice as old as her brother Dan. If the sum of their ages is 27 years, how old are Diane and her brother?

90. Interest investments. Patrick has invested $15,000 in two bonds; one bond yields 4% annual interest, and the other yields 3% annual interest. How much is invested in each certificate if the combined yearly interest from both bonds is $545?

91. Interest investments. Johanna deposited $21,000 in two banks. One bank gives $2\frac{1}{2}\%$ annual interest, and the other gives $3\frac{1}{4}\%$ annual interest. How much did she deposit in each bank if she received a total of $615 in annual interest?

92. Payroll. Tonya takes home $1,080 per week. If her deductions amount to 28% of her wages, what is her weekly pay before deductions?

93. Payroll. Sam donates 5% of his net income to charity. This amounts to $190 per month. His payroll deductions are 24% of his gross monthly income. What is Sam's gross monthly income?

94. Gallons of fuel. The Randolphs used 12 more gallons (gal) of fuel oil in October than in September and twice as much oil in November as in September. If they used 132 gal for the 3 months, how much was used during each month?

95. Complete this statement in your own words: "You can tell that an equation is a linear equation when . . ."

96. What is the common characteristic of equivalent equations?

97. What is meant by a *solution* to a linear equation?

98. Define (a) identity and (b) contradiction.

99. Why does the multiplication property of equality not include multiplying both sides of the equation by 0?

100. Maxine lives in Pittsburgh, Pennsylvania, and pays 8.33 cents per kilowatthour (kWh) for electricity. During the 6 months of cold winter weather, her household uses about 1,500 kWh of electric power per month. During the two hottest summer months, the usage is also high because the family uses electricity to run an air conditioner. During these summer months, the usage is 1,200 kWh per month; the rest of the year, usage averages 900 kWh per month.

    (a) Write an expression for the total yearly electric bill.

    (b) Maxine is considering spending $2,000 for more insulation for her home so that it is less expensive to heat and to cool. The insulation company claims that "with proper installation the insulation will reduce your heating and cooling bills by 25%." If Maxine invests the money in insulation, how long will it take her to get her money back in savings on her electric bill? Write to her about what information she needs to answer this question. Give her your opinion about how long it will take to save $2,000 on heating bills, and explain your reasoning. What is your advice to Maxine?

101. Solve each of the following equations. Express each solution as a fraction.

    (a) $2x + 3 = 0$ (b) $4x + 7 = 0$ (c) $6x - 1 = 0$

    (d) $5x - 2 = 0$ (e) $-3x + 8 = 0$ (f) $-5x - 9 = 0$

    (g) Based on these problems, express the solution to the equation

$$ax + b = 0$$

    where a and b represent real numbers and $a \neq 0$.

102. You are asked to solve the following equation, but one number is missing. It reads

$$\frac{5x - ?}{4} = \frac{9}{2}$$

    The solution is 4. What is the missing number?

SECTION 2.4

Literal Equations and Their Applications

2.4 OBJECTIVES

1. Solve a literal equation for any one of its variables
2. Solve applications involving geometric figures
3. Solve mixture problems
4. Solve motion problems

Formulas are extremely useful tools in any field in which mathematics is applied. Formulas are simply equations that express a relationship between more than one letter or variable. You are no doubt familiar with all kinds of formulas, such as

$$A = \frac{1}{2}bh \qquad \text{The area of a triangle}$$

$$I = Prt \qquad \text{Interest}$$

$$V = \pi r^2 h \qquad \text{The volume of a cylinder}$$

Actually a formula is also called a **literal equation** because it involves several letters or variables. For instance, our first formula or literal equation, $A = \frac{1}{2}bh$, involves the three letters A (for area), b (for base), and h (for height).

Unfortunately, formulas are not always given in the form needed to solve a particular problem. Then algebra is needed to change the formula to a more useful equivalent equation, which is solved for a particular letter or variable. The steps used in the process are very similar to those you used in solving linear equations. Let's consider an example.

Example 1

Solving a Literal Equation Involving a Triangle

Suppose that we know the area A and the base b of a triangle and want to find its height h.

We are given

$$A = \frac{1}{2}bh$$

Our job is to find an equivalent equation with h, the unknown, by itself on one side and everything else on the other side. We call $\frac{1}{2}b$ the **coefficient** of h.

We can remove the two *factors* of that coefficient, $\dfrac{1}{2}$ and b, separately.

Note:

$2\left(\dfrac{1}{2}bh\right) = \left(2 \cdot \dfrac{1}{2}\right)(bh)$

$\quad = 1 \cdot bh$

$\quad = bh$

or

$2A = 2\left(\dfrac{1}{2}bh\right)$ Multiply both sides by 2 to clear the equation of fractions.

or

$2A = bh$

$\dfrac{2A}{b} = \dfrac{bh}{b}$ Divide by b to isolate h.

$\dfrac{2A}{b} = h$

or

$h = \dfrac{2A}{b}$ Reverse the sides to write h on the left.

We now have the height h in terms of the area A and the base b. This is called **solving the equation for h** and means that we are rewriting the formula as an equivalent equation of the form

Here means an expression containing all the numbers or letters *other than* h.

$$h = \boxed{}$$

✔ *CHECK YOURSELF 1*

Solve $V = \dfrac{1}{3}Bh$ for h.

You have already learned the methods needed to solve most literal equations or formulas for some specified variable. As Example 1 illustrates, the rules of Sections 2.2 and 2.3 are applied in exactly the same way as they were applied to equations with one variable.

You may have to apply both the addition and the multiplication properties when solving a formula for a specified variable. Example 2 illustrates this situation.

Example 2

Solving a Literal Equation

This is a linear equation in two variables. You will see this again in Chapter 3.

Solve $y = mx + b$ for x.

Remember that we want to end up with x alone on one side of the equation. Let's start by subtracting b from both sides to undo the addition on the right.

$$y = mx + b$$

$$y - b = mx + b - b$$

$$y - b = mx$$

If we now divide both sides by m, then x with be alone on the right side.

$$\frac{y - b}{m} = \frac{mx}{m}$$

$$\frac{y - b}{m} = x$$

or

$$x = \frac{y - b}{m}$$

✔ *CHECK YOURSELF 2*

Solve $v = a + gt$ for t.

Let's summarize the steps illustrated by our examples.

Solving a Formula or Literal Equation

Step 1. If necessary, multiply both sides of the equation by the same term to clear the equation of fractions.

Step 2. Add or subtract the same term on both sides of the equation so that all terms involving the variable that you are solving for are on one side of the equation and all other terms are on the other side.

Step 3. Divide both sides of the equation by the coefficient of the variable that you are solving for.

Let's look at one more example, using the above steps.

| **Example 3** | **Solving a Literal Equation Involving Money** |

This is a formula for the *amount* of money in an account after interest has been earned.

Solve $A = P + Prt$ for r.

$$A = P + Prt$$

$$A - P = P - P + Prt$$ Subtracting P from both sides will leave the term involving r alone on the right.

$$A - P = Prt$$

$$\frac{A - P}{Pt} = \frac{Prt}{Pt}$$ Dividing both sides by Pt will isolate r on the right.

$$\frac{A - P}{Pt} = r$$

or

$$r = \frac{A - P}{Pt}$$

 CHECK YOURSELF 3

Solve $2x + 3y = 6$ for y.

Now let's look at an application of solving a literal equation.

| Example 4 | ## Solving a Literal Equation Involving Money |

Suppose that the amount in an account, 3 years after a principal of $5,000 was invested, is $6,050. What was the interest rate?

From Example 3,

$$A = P + Prt \tag{1}$$

where A is the amount in the account, P is the principal, r is the interest rate, and t is the time in years that the money has been invested. By the result of Example 3 we have

$$r = \frac{A - P}{Pt} \tag{2}$$

Do you see the advantage of having our equation solved for the desired variable?

and we can substitute the known values in equation (2):

$$r = \frac{6,050 - 5,000}{(5,000)(3)}$$

$$= \frac{1,050}{15,000} = 0.07 = 7\%$$

The interest rate was 7%.

 CHECK YOURSELF 4

Suppose that the amount in an account, 4 years after a principal of $3,000 was invested, is $3,720. What was the interest rate?

In our subsequent applications, we will use the five-step process first described in Section 2.1. As a reminder, here are those steps.

To Solve Word Problems

Step 1.   Read the problem carefully. Then reread it to decide what you are asked to find.

Step 2.   Choose a letter to represent one of the unknowns in the problem. Then represent all other unknowns of the problem with expressions that use the same letter.

Step 3.   Translate the problem to the language of algebra to form an equation.

Step 4.   Solve the equation and answer the question of the original problem.

Step 5.   Check your solution by returning to the original problem.

| Example 5 | Solving a Geometry Application |

Whenever you are working on an application involving geometric figures, you should draw a sketch of the problem, including the labels assigned in step 2.

The length of a rectangle is 1 centimeter (cm) less than 3 times the width. If the perimeter is 54 cm, find the dimensions of the rectangle.

Step 1 You want to find the dimensions (the width and length).

Step 2 Let x be the width.

Length $3x - 1$

Width
x

Then $3x - 1$ is the length.

3 times the width   1 less than

Step 3 To write an equation, we'll use this formula for the perimeter of a rectangle:

$$P = 2W + 2L \quad \text{or} \quad 2W + 2L = P$$

So

$$2x + 2(3x - 1) = 54$$

Twice the width   Twice the length   Perimeter

Step 4 Solve the equation.

$$2x + 2(3x - 1) = 54$$
$$2x + 6x - 2 = 54$$
$$8x = 56$$
$$x = 7$$

Be sure to return to the original statement of the problem when checking your result.

The width x is 7 cm, and the length, $3x - 1$, is 20 cm. We leave step 5, the check, to you.

✔ *CHECK YOURSELF 5*

The length of a rectangle is 5 inches (in.) more than twice the width. If the perimeter of the rectangle is 76 in., what are the dimensions of the rectangle?

You will also often use parentheses in solving *mixture problems.* Mixture problems involve combining things that have a different value, rate, or strength. Look at Example 6.

Example 6

Solving a Mixture Problem

Four hundred tickets were sold for a school play. General admission tickets were $4, while student tickets were $3. If the total ticket sales were $1,350, how many of each type of ticket were sold?

Step 1 You want to find the number of each type of ticket sold.

Step 2 Let x be the number of general admission tickets.

Then $\underline{400 - x}$ student tickets were sold.

400 tickets were sold in all.

We subtract x, the number of general admission tickets, from 400, the total number of tickets, to find the number of student tickets.

Step 3 The sales value for each kind of ticket is found by multiplying the price of the ticket by the number sold.

Value of general admission tickets: $4x$ $4 for each of the x tickets

Value of student tickets: $3(400 - x)$ $3 for each of the $400 - x$ tickets

So to form an equation, we have

$$4x + \underbrace{3(400 - x)} = 1,350$$

Value of general admission tickets   Value of student tickets   Total value

Step 4 Solve the equation.

$$4x + 3(400 - x) = 1,350$$
$$4x + 1,200 - 3x = 1,350$$
$$x + 1,200 = 1,350$$
$$x = 150$$

This shows that 150 general admission and 250 student tickets were sold. We leave the check to you.

✔ *CHECK YOURSELF 6*

Beth bought 40¢ stamps and 15¢ stamps at the post office. If she purchased 60 stamps at a cost of $19, how many of each kind did she buy?

| **Example 7** | ## Solving a Mixture Problem |

How much pure alcohol must be added to 9 quarts (qt) of water to obtain a mixture of 40% alcohol?

Step 1 We want the amount of pure alcohol (in quarts) that must be added to water. Pure alcohol is 100% alcohol.

Step 2 Let x be the amount of alcohol that must be added to the water. So $x + 9$ is the amount, in quarts, of the final mixture. Since the final mixture is 40% alcohol, the amount of alcohol in the final mixture is expressed as $0.40(x + 9)$.

Step 3 Since the amount of alcohol in the mixture will be equal to the amount added, we can write the equation

$$x = 0.40(x + 9)$$

A table helps summarize this information

	Quarts	**% Alcohol**	**Amount of Alcohol**
Alcohol	x	100	$1x$
Mixture	$x + 9$	40	$0.40(x + 9)$

Step 4 Solve the equation.

$$x = 0.40(x + 9)$$

$$x = 0.40x + (0.40)(9)$$

$$x = 0.40x + 3.6$$

$$x - 0.40x = 3.6$$

$$0.6x = 3.6$$

$$x = \frac{3.6}{0.6}$$

$$x = 6$$

Step 5 If we add 6 qt to 9 qt, there will be a total of 15 qt in the mixture, 40% of which is 6 qt.

✔ *CHECK YOURSELF 7*

How much pure alcohol must be added to 35 centiliters (cl) of a 20% solution to obtain a 30% solution?

The next group of applications we will look at in this section involves *motion problems.* They involve a distance traveled, a rate or speed, and time. To solve motion problems, we need a relationship among these three quantities.

Suppose you travel at a rate of 50 miles per hour (mi/h) on a highway for 6 hours (h). How far (what distance) will you have gone? To find the distance, you multiply:

Be careful to make your units consistent. If a rate is given in *miles per hour,* then the time must be given in *hours* and the distance in *miles.*

$$(50 \text{ mi/h})(6 \text{ h}) = 300 \text{ mi}$$

↑	↑	↑
Speed or rate	Time	Distance

In general, if *r* is the rate, *t* is the time, and *d* is the distance traveled, then

$$d = r \cdot t$$

This is the key relationship, and it will be used in all motion problems. Let's see how it is applied in Example 8.

Example 8 **Solving a Motion Problem**

On Friday morning Ricardo drove from his house to the beach in 4 h. In coming back on Sunday afternoon, heavy traffic slowed his speed by 10 mi/h, and the trip took 5 h. What was his average speed (rate) in each direction?

Step 1 We want the speed or rate in each direction.

Step 2 Let *x* be Ricardo's speed to the beach. Then $x - 10$ is his return speed.

It is always a good idea to sketch the given information in a motion problem. Here we would have

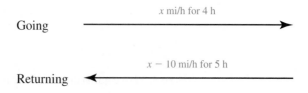

Going   *x* mi/h for 4 h

Returning $x - 10$ mi/h for 5 h

Step 3   Since we know that the distance is the same each way, we can write an equation, using the fact that the product of the rate and the time each way must be the same. So

$$\text{Distance (going)} = \text{distance (returning)}$$

$$\text{Time} \cdot \text{rate (going)} = \text{time} \cdot \text{rate (returning)}$$

$$\underbrace{4x}_{\substack{\uparrow \\ \text{Time} \cdot \text{rate} \\ \text{(going)}}} = \underbrace{5(x - 10)}_{\substack{\uparrow \\ \text{Time} \cdot \text{rate} \\ \text{(returning)}}}$$

A chart can help summarize the given information. We begin by filling in the information given in the problem.

	Rate	**Time**	**Distance**
Going	x	4	
Returning	$x - 10$	5	

Now we fill in the missing information. Here we use the fact that $d = rt$ to complete the chart.

	Rate	**Time**	**Distance**
Going	x	4	$4x$
Returning	$x - 10$	5	$5(x - 10)$

From here we set the two distances equal to each other and solve as before.

Step 4   Solve.

$$4x = 5(x - 10)$$
$$4x = 5x - 50$$
$$-x = -50$$
$$x = 50 \text{ mi/h}$$

x was his rate going; x − 10, his rate returning.

So Ricardo's rate going to the beach was 50 mi/h, and his rate returning was 40 mi/h.

Step 5   To check, you should verify that the product of the time and the rate is the same in each direction.

✔ CHECK YOURSELF 8

A plane made a flight (with the wind) between two towns in 2 h. In returning against the wind, the plane's speed was 60 mi/h slower, and the flight took 3 h. What was the plane's speed in each direction?

Example 9 illustrates another way of using the distance relationship.

Example 9	## Solving a Motion Problem

Katy leaves Las Vegas, Nevada, for Los Angeles, California, at 10 A.M., driving at 50 mi/h. At 11 A.M. Jensen leaves Los Angeles for Las Vegas, driving at 55 mi/h along the same route. If the cities are 260 mi apart, at what time will they meet?

Step 1   Let's find the time that Katy travels until they meet.

Step 2   Let x be Katy's time.

Then $x - 1$ is Jensen's time.   Jensen left 1 hr later!

Again, you should draw a sketch of the given information.

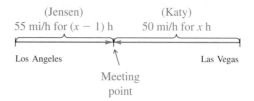

Step 3   To write an equation, we will again need the relationship $d = rt$. From this equation, we can write

$$\text{Katy's distance} = 50x$$

$$\text{Jensen's distance} = 55(x - 1)$$

As before, we can use a table to solve.

	Rate	**Time**	**Distance**
Katy	50	x	$50x$
Jensen	55	$x - 1$	$55(x - 1)$

From the original problem, the sum of those distances is 260 mi, so

$$50x + 55(x - 1) = 260$$

Step 4
$$50x + 55(x - 1) = 260$$
$$50x + 55x - 55 = 260$$
$$105x - 55 = 260$$
$$105x = 315$$
$$x = 3 \text{ h}$$

Be sure to answer the question asked in the problem.

Finally, since Katy left at 10 A.M., the two will meet at 1 P.M. We leave the check of this result to you.

✔ *CHECK YOURSELF 9*

At noon a jogger leaves one point, running at 8 mi/h. One hour later a bicyclist leaves the same point, traveling at 20 mi/h in the opposite direction. At what time will they be 36 mi apart?

✔ *CHECK YOURSELF ANSWERS*

1. $h = \dfrac{3V}{B}$ **2.** $t = \dfrac{v-a}{g}$ **3.** $y = \dfrac{6-2x}{3}$ or $y = -\dfrac{2}{3}x + 2$

4. The interest rate was 6%.   **5.** The width is 11 in.; the length is 27 in.

6. 40 at 40¢, and 20 at 15¢.   **7.** 5 cl

8. 180 mi/h with the wind and 120 mi/h against the wind

9. At 2 P.M.

Exercises · 2.4

Solve each literal equation for the indicated variable.

1. $P = 4s$ (for s)   Perimeter of a square

2. $V = Bh$ (for B)   Volume of a prism

3. $E = IR$ (for R)   Voltage in an electric circuit

4. $I = Prt$ (for r)   Simple interest

5. $V = LWH$ (for H)   Volume of a rectangular solid

6. $V = \pi r^2 h$ (for h)   Volume of a cylinder

7. $A + B + C = 180$ (for B)   Measure of angles in a triangle

8. $P = I^2 R$ (for R)   Power in an electric circuit

9. $ax + b = 0$ (for x)   Linear equation in one variable

10. $y = mx + b$ (for m)   Slope-intercept form for a line

11. $s = \dfrac{1}{2}gt^2$ (for g)   Distance

12. $K = \dfrac{1}{2}mv^2$ (for m)   Energy

13. $x + 5y = 15$ (for y)   Linear equation in two variables

14. $2x + 3y = 6$ (for x)   Linear equation in two variables

15. $P = 2L + 2W$ (for L)   Perimeter of a rectangle

16. $ax + by = c$ (for y)   Linear equation in two variables

17. $V = \dfrac{KT}{P}$ (for T)   Volume of a gas

18. $V = \dfrac{1}{3}\pi r^2 h$ (for h)   Volume of a cone

19. $x = \dfrac{a + b}{2}$ (for b)   Average of two numbers

20. $D = \dfrac{C - s}{n}$ (for s)   Depreciation

21. $F = \dfrac{9}{5}C + 32$ (for C)   Celsius/Fahrenheit conversion

22. $A = P + Prt$ (for t)   Amount at simple interest

23. $S = 2\pi r^2 + 2\pi rh$ (for h)   Total surface area of a cylinder

24. $A = \dfrac{1}{2}h(B + b)$ (for b)   Area of a trapezoid

25. Height of a solid. A rectangular solid has a base with length 8 centimeters (cm) and width 5 cm. If the volume of the solid is 120 cm^3, find the height of the solid. (See Exercise 5.)

26. Height of a cylinder. A cylinder has a radius of 4 inches (in.). If the volume of the cylinder is 144π in.^{3}, what is the height of the cylinder? (See Exercise 6.)

27. Interest rate. A principal of $2,000 was invested in a savings account for 4 years. If the interest earned for the period was $240, what was the interest rate? (See Exercise 4.)

28. Length of a rectangle. If the perimeter of a rectangle is 60 feet (ft) and the width is 12 ft, find its length. (See Exercise 15.)

29. Temperature conversion. The high temperature in New York for a particular day was reported at 77°F. How would the same temperature have been given in degrees Celsius? (See Exercise 21.)

30. Garden length. Rose's garden is in the shape of a trapezoid. If the height of the trapezoid is 16 meters (m), one base is 20 m, and the area is 224 m^2, find the length of the other base. (See Exercise 24.)

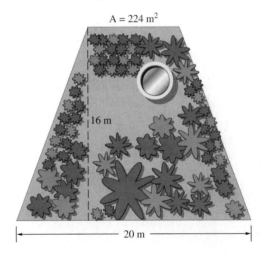

A = 224 m^2

16 m

20 m

Translate each of the following statements to equations. Let x represent the number in each case.

31. Twice the sum of a number and 6 is 18.

32. The sum of twice a number and 4 is 20.

33. 3 times the difference of a number and 5 is 21.

34. The difference of 3 times a number and 5 is 21.

35. The sum of twice an integer and 3 times the next consecutive integer is 48.

36. The sum of 4 times an odd integer and twice the next consecutive odd integer is 46.

Solve the following word problems.

37. Number problem. One number is 8 more than another. If the sum of the smaller number and twice the larger number is 46, find the two numbers.

38. Number problem. One number is 3 less than another. If 4 times the smaller number minus 3 times the larger number is 4, find the two numbers.

39. **Number problem.** One number is 7 less than another. If 4 times the smaller number plus 2 times the larger number is 62, find the two numbers.

40. **Number problem.** One number is 10 more than another. If the sum of twice the smaller number and 3 times the larger number is 55, find the two numbers.

41. **Consecutive integers.** Find two consecutive integers such that the sum of twice the first integer and 3 times the second integer is 28. (*Hint:* If x represents the first integer, $x + 1$ represents the next consecutive integer.)

42. **Consecutive integers.** Find two consecutive odd integers such that 3 times the first integer is 5 more than twice the second. (*Hint:* If x represents the first integer, $x + 2$ represents the next consecutive odd integer.)

43. **Dimensions of a rectangle.** The length of a rectangle is 1 inch (in.) more than twice its width. If the perimeter of the rectangle is 74 in., find the dimensions of the rectangle.

44. **Dimensions of a rectangle.** The length of a rectangle is 5 centimeters (cm) less than 3 times its width. If the perimeter of the rectangle is 46 cm, find the dimensions of the rectangle.

45. **Garden size.** The length of a rectangular garden is 5 m more than 3 times its width. The perimeter of the garden is 74 m. What are the dimensions of the garden?

46. **Size of a playing field.** The length of a rectangular playing field is 5 ft less than twice its width. If the perimeter of the playing field is 230 ft, find the length and width of the field.

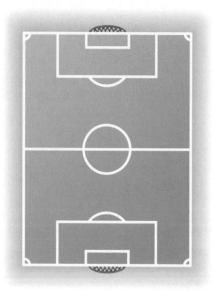

47. **Isosceles triangle.** The base of an isosceles triangle is 3 cm less than the length of the equal sides. If the perimeter of the triangle is 36 cm, find the length of each of the sides.

48. **Isosceles triangle.** The length of one of the equal legs of an isosceles triangle is 3 in. less than twice the length of the base. If the perimeter is 29 in., find the length of each of the sides.

49. **Ticket sales.** Tickets for a play cost $14 for the main floor and $9 in the balcony. If the total receipts from 250 tickets were $3,000, how many of each type of ticket were sold?

50. **Ticket sales.** Tickets for a basketball tournament were $6 for students and $9 for nonstudents. Total sales were $10,500, and 250 more student tickets were sold than nonstudent tickets. How many of each type of ticket were sold?

51. **Number of stamps.** Maria bought 80 stamps at the post office in 37¢ and 3¢ denominations. If she paid $17.70 for the stamps, how many of each denomination did she buy?

52. **Money denominations.** A bank teller had a total of 125 $10 bills and $20 bills to start the day. If the value of the bills was $1,650, how many of each denomination did he have?

53. **Ticket sales.** Tickets for a train excursion were $120 for a sleeping room, $80 for a berth, and $50 for a coach seat. The total ticket sales were $8,600. If there were 20 more berth tickets sold than sleeping room tickets and 3 times as many coach tickets as sleeping room tickets, how many of each type of ticket were sold?

54. **Baseball tickets.** Admission for a college baseball game is $6 for box seats, $5 for the grandstand, and $3 for the bleachers. The total receipts for one evening were $9,000. There were 100 more grandstand tickets sold than box seat tickets. Twice as many bleacher tickets were sold as box seat tickets. How many tickets of each type were sold?

55. **Driving speed.** Patrick drove 3 h to attend a meeting. On the return trip, his speed was 10 mi/h less, and the trip took 4 h. What was his speed each way?

56. **Bicycle speed.** A bicyclist rode into the country for 5 h. In returning, her speed was 5 mi/h faster and the trip took 4 h. What was her speed each way?

57. **Driving speed.** A car leaves a city and goes north at a rate of 50 mi/h at 2 P.M. One hour later a second car leaves, traveling south at a rate of 40 mi/h. At what time will the two cars be 320 mi apart?

58. **Bus distance.** A bus leaves a station at 1 P.M., traveling west at an average rate of 44 mi/h. One hour later a second bus leaves the same station, traveling east at a rate of 48 mi/h. At what time will the two buses be 274 mi apart?

59. **Traveling time.** At 8:00 A.M., Catherine leaves on a trip at 45 mi/h. One hour later, Max decides to join her and leaves along the same route, traveling at 54 mi/h. When will Max catch up with Catherine?

60. **Bicycling time.** Martina leaves home at 9 A.M., bicycling at a rate of 24 mi/h. Two hours later, John leaves, driving at the rate of 48 mi/h. At what time will John catch up with Martina?

61. Traveling time. Mika leaves Boston for Baltimore at 10:00 A.M., traveling at 45 mi/h. One hour later, Hiroko leaves Baltimore for Boston on the same route, traveling at 50 mi/h. If the two cities are 425 mi apart, when will Mika and Hiroko meet?

62. Traveling time. A train leaves town A for town B, traveling at 35 mi/h. At the same time, a second train leaves town B for town A at 45 mi/h. If the two towns are 320 mi apart, how long will it take for the two trains to meet?

63. Chemistry. How many liters of water must be added to 140 liters (l) of an 80% alcohol solution to obtain a 70% solution?

64. Chemistry. How many centiliters (cl) of water must be added to 500 cl of a 60% acid solution to obtain a 50% solution?

65. There is a universally agreed on *order of operations* used to simplify expressions. Explain how the order of operations is used in solving equations. Be sure to use complete sentences.

66. A common mistake in solving equations is the following:

The equation: $2(x - 2) = x + 3$
First step in solving: $2x - 2 = x + 3$

Write a clear explanation of what error has been made. What could be done to avoid this error?

67. Another very common mistake is seen in the equation below:

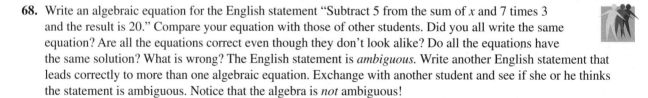

The equation: $6x - (x + 3) = 5 + 2x$
First step in solving: $6x - x + 3 = 5 + 2x$

Write a clear explanation of what error has been made and what could be done to avoid the mistake.

68. Write an algebraic equation for the English statement "Subtract 5 from the sum of x and 7 times 3 and the result is 20." Compare your equation with those of other students. Did you all write the same equation? Are all the equations correct even though they don't look alike? Do all the equations have the same solution? What is wrong? The English statement is *ambiguous*. Write another English statement that leads correctly to more than one algebraic equation. Exchange with another student and see if she or he thinks the statement is ambiguous. Notice that the algebra is *not* ambiguous!

2.5 Solving Linear Inequalities Using Addition

2.5 OBJECTIVES

1. Use the notation of inequalities
2. Graph the solution set of a linear inequality
3. Use the addition property to solve a linear inequality

"Few things are harder to put up with than a good example."

—Mark Twain

As pointed out earlier in this chapter, an equation is a statement that two expressions are equal. In algebra, an **inequality** is a statement that one expression is less than or greater than another. Four symbols, first introduced in Section 1.4, are used in writing inequalities. The use of two of them is illustrated in Example 1.

| **Example 1** | **Reading the Inequality Symbol** |

To help you remember, the "arrowhead" always points toward the smaller quantity.

$5 < 8$ is an inequality read "5 is less than 8."

$9 > 6$ is an inequality read "9 is greater than 6."

 CHECK YOURSELF 1

Fill in the blanks, using the symbols $<$ and $>$.

(a) 12 _____ 8   (b) 20 _____ 25

Just as was the case with equations, inequalities that involve variables may be either true or false depending on the value that we give to the variable. For instance, consider the inequality

$$x < 6$$

$$\text{If} \quad x = \begin{cases} 3 & 3 < 6 \text{ is true} \\ 5 & 5 < 6 \text{ is true} \\ -10 & -10 < 6 \text{ is true} \\ 8 & 8 < 6 \text{ is false} \end{cases}$$

Therefore 3, 5, and -10 are some *solutions* for the inequality $x < 6$; they make the inequality a true statement. You should see that 8 is *not* a solution. Recall the set of all solutions is the *solution set* for the inequality. Of course, there are many possible solutions.

Since there are so many solutions (an infinite number, in fact), we certainly do not want to try to list them all! A convenient way to show the solution set of an inequality is with the use of a number line.

| **Example 2** | ### Graphing Inequalities |

To graph the solution set for the inequality $x < 6$, we want to include all real numbers that are "less than" 6. This means all numbers *to the left* of 6 on the number line.

We then start at 6 and draw an arrow extending left, as shown:

The colored arrow indicates the direction of the *solution set*.

Note: The parenthesis at 6 means that we do not include 6 in the solution set (6 is not less than itself). The colored arrow shows all the numbers in the solution set, with the arrowhead indicating that the solution set continues infinitely to the left.

   CHECK YOURSELF 2

Graph the solution set of $x < -2$.

Two other symbols are used in writing inequalities. They are used with inequalities such as

$$x \geq 5 \qquad \text{and} \qquad x \leq 2$$

Here $x \geq 5$ is a combination of the two statements $x > 5$ and $x = 5$. It is read "x is greater than or equal to 5." The solution set includes 5 in this case.

The inequality $x \leq 2$ combines the statements $x < 2$ and $x = 2$. It is read "x is less than or equal to 2."

| **Example 3** | ### Graphing Inequalities |

Here the bracket means that we want to include 5 in the solution set.

The solution set for $x \geq 5$ is graphed as follows.

✔ CHECK YOURSELF 3

Graph the solution sets.

(a) $x \le -4$ (b) $x \ge 3$

We have looked at graphs of the solution sets of some simple inequalities, such as $x < 8$ or $x \ge 10$. Now we will look at more complicated inequalities, such as

$$2x - 3 < x + 4$$

This is called a **linear inequality in one variable.** Only one variable is involved in the inequality, and it appears only to the first power. Fortunately, the methods used to solve this type of inequality are very similar to those we used earlier in this chapter to solve linear equations in one variable. Here is our first property for inequalities.

> ### The Addition Property of Inequality
>
> If $a < b$ then $a + c < b + c$
>
> In words, adding the same quantity to both sides of an inequality gives an **equivalent inequality.**

Equivalent inequalities have exactly the same solution sets.

| Example 4 | # Solving Inequalities |

The inequality is solved when an equivalent inequality has the form

$x < \Box$ or $x > \Box$

Solve and graph the solution set for $x - 8 < 7$.

To solve $x - 8 < 7$, add 8 to both sides of the inequality by the addition property.

$$x - 8 < 7$$
$$x - 8 + 8 < 7 + 8$$
$$x < 15 \qquad \text{The inequality is solved.}$$

We read

$\{x \,|\, x < 15\}$

"every x such that x is less than 15."

The graph of the solution set, $\{x \,|\, x < 15\}$, is

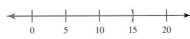

$$\begin{array}{ccccc} | & | & | &) & | \\ 0 & 5 & 10 & 15 & 20 \end{array}$$

✔ CHECK YOURSELF 4

Solve and graph the solution set for

$$x - 9 > -3$$

As with equations, the addition property allows us to *subtract* the same quantity from both sides of an inequality.

| Example 5 |

Solving Inequalities

Solve and graph the solution set for $4x - 2 \geq 3x + 5$. Write the solution set in set-builder notation.

First, we subtract $3x$ from both sides of the inequality.

$$4x - 2 \geq 3x + 5$$

$$4x - 3x - 2 \geq 3x - 3x + 5$$

$$x - 2 \geq 5 \qquad \text{Now we add 2 to both sides.}$$

$$x - 2 + 2 \geq 5 + 2$$

$$x \geq 7$$

We subtracted $3x$ and then added 2 to both sides. If these steps are done in the other order, the resulting inequality will be the same.

The graph of the solution set, $\{x \mid x \geq 7\}$, is

Using interval notation, the solution could be written as $[7, \infty)$.

✔ *CHECK YOURSELF 5*

Solve and graph the solution set. Write the solution set in set-builder notation.

$$7x - 8 \leq 6x + 2$$

Note that $x < 3$ is the same as $3 > x$. In our final example of this section, we graph an inequality in which the variable is on the right side.

| Example 6 |

Solving an Inequality

Solve and graph the solution set for the inequality

$$2x + 3 < 3x + 6$$

Note that the coefficient for x is larger on the right side of the inequality than it is the left. As a result, we will isolate the variable on the right.

$$2x - 2x + 3 < 3x - 2x + 6 \qquad \text{Subtract } 2x \text{ from both sides.}$$

$$3 < x + 6 \qquad \text{Add } -6 \text{ to both sides.}$$

$$3 - 6 < x + 6 - 6$$

$$-3 < x$$

The graph of the solution set, $\{x \mid -3 < x\}$, is

✔ *CHECK YOURSELF 6*

Solve and graph the solution set for the inequality

$$4x - 5 < 5x - 9$$

✔ *CHECK YOURSELF ANSWERS*

1. (a) $>$; (b) $<$ **2.**

3. (a)    (b)

4. $\{x \mid x > 6\}$

5. $\{x \mid x \leq 10\}$

6. $\{x \mid x > 4\}$

Complete the statements, using the symbol $<$ or $>$.

1. 5 _____ 10

2. 9 _____ 8

3. 7 _____ -2

4. 0 _____ -5

5. 0 _____ 4

6. -10 _____ -5

7. -2 _____ -5

8. -4 _____ -11

Write each inequality in words.

9. $x < 3$

10. $x \leq -5$

11. $x \geq -4$

12. $x < -2$

13. $-5 \leq x$

14. $2 < x$

Graph the solution set of each of the following inequalities.

15. $x > 2$

16. $x < -3$

17. $x < 6$

18. $x > 4$

19. $x > 1$

20. $x < -2$

21. $x < 8$

22. $x > 3$

23. $x > -5$

24. $x < -2$

25. $x \geq 9$

26. $x \geq 0$

27. $x < 0$

28. $x \leq -3$

Solve and graph the solution set of each of the following inequalities. Write the solution set in set-builder notation.

29. $x - 8 \leq 3$

30. $x + 5 \leq 4$

31. $x + 8 \geq 10$

32. $x - 11 > -14$

33. $5x < 4x + 7$

34. $8x \geq 7x - 4$

35. $6x - 8 \leq 5x$

36. $3x + 2 > 2x$

37. $8x + 1 \geq 7x + 9$

38. $5x + 2 \leq 4x - 6$

39. $7x + 5 < 6x - 4$

40. $8x - 7 > 7x + 3$

41. $\frac{3}{4}x - 5 \geq 7 - \frac{1}{4}x$

42. $\frac{7}{8}x + 6 < 3 - \frac{1}{8}x$

43. $11 + 0.63x > 9 - 0.37x$

44. $0.54x + 0.12x + 9 \leq 19 - 0.34x$

45. If an inequality simplifies to $7 > -5$, what is the solution set and why?

46. If an inequality simplifies to $7 < -5$, what is the solution set and why?

Match each inequality on the right with a statement on the left.

47. x is nonnegative.   (a) $x \geq 0$

48. x is negative.   (b) $x \geq 5$

49. x is no more than 5.   (c) $x \leq 5$

50. x is positive.   (d) $x > 0$

51. x is at least 5.   (e) $x < 5$

52. x is less than 5.   (f) $x < 0$

Solving Linear Inequalities Using Multiplication

2.6 OBJECTIVES

1. *Use the multiplication property to solve a linear inequality*
2. *Graph the solution set for a solved inequality*

In Section 2.5, we used the addition property to solve inequalities. For most inequalities, you will also need a rule for multiplying on both sides of an inequality. Here you'll have to be a bit careful. There is a difference between the multiplication property for inequalities and that for equations. Look at the following:

$$2 < 7 \qquad \text{A true inequality}$$

Let's multiply both sides by 3.

$$2 < 7$$
$$3 \cdot 2 < 3 \cdot 7$$
$$6 < 21 \qquad \text{A true inequality}$$

Now we multiply both sides by -3.

$$2 < 7$$
$$(-3)(2) < (-3)(7)$$
$$-6 < -21 \qquad \textit{Not} \text{ a true inequality}$$

Let's try something different.

$$2 < 7 \qquad \text{Change the "sense" of the inequality:}$$
$$(-3)(2) > (-3)(7) \qquad < \text{ becomes } >.$$
$$-6 > -21 \qquad \text{This is now a true inequality.}$$

This suggests that multiplying both sides of an inequality by a negative number changes the "sense" of the inequality.

When both sides of an inequality are multiplied by the same *negative* number, it is necessary to *reverse the sense* of the inequality to give an equivalent inequality.

The Multiplication Property of Inequality

$$\text{If} \quad a < b \quad \text{then} \quad ac < bc \quad \text{where } c > 0$$
$$\text{and} \quad ac > bc \quad \text{where } c < 0$$

In words, multiplying both sides of an inequality by the same *positive* number gives an equivalent inequality. Multiplying both sides of an inequality by the same *negative* number gives an equivalent inequality if we also reverse the direction of the inequality sign.

176

Overcoming Math Anxiety

Hint 5: Preparing for a test

Preparation for a test really begins on the first day of class. Everything you have done in class and at home has been part of that preparation. However, there are a few things that you should focus on in the last few days before a scheduled test.

1. Plan your test preparation to end at least 24 h before the test. The last 24 h is too late, and besides, you will need some rest before the test.

2. Go over your homework and class notes with pencil and paper in hand. Write down all the problem types, formulas, and definitions that you think might give you trouble on the test.

3. The day before the test, take the page(s) of notes from step 2, and transfer the most important ideas to a 3 × 5 in. card.

4. Just before the test, review the information on the card. You will be surprised at how much you remember about each concept.

5. Understand that if you have been successful at completing your homework assignments, you can be successful on the test. This is an obstacle for many students, but it is an obstacle that can be overcome. Truly anxious students are often surprised that they scored as well as they did on a test. They tend to attribute this to blind luck. It is not. It is the first sign that you really do "get it." Enjoy the success.

Example 1

Solving and Graphing Inequalities

(a) Solve and graph the solution set for $5x < 30$.
   Multiplying both sides of the inequality by $\frac{1}{5}$ gives

$$\frac{1}{5}(5x) < \frac{1}{5}(30)$$

Simplifying, we have

$$x < 6$$

The graph of the solution set, $\{x \mid x < 6\}$, is

Using interval notation, the solution set can be written as $(-\infty, 6)$.

(b) Solve and graph the solution set for $-4x \geq 28$.
   In this case we want to multiply both sides of the inequality by $-\frac{1}{4}$ to convert the coefficient of x to 1 on the left.

$$\left(-\frac{1}{4}\right)(-4x) \leq \left(-\frac{1}{4}\right)(28)$$

Reverse the sense of the inequality because you are multiplying by a negative number!

or $\qquad x \leq -7$

The graph of the solution set, $\{x \mid x \leq -7\}$, is

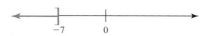

✔ CHECK YOURSELF 1

Solve and graph the solution sets.

(a) $7x > 35$ (b) $-8x \leq 48$

Example 2 illustrates the use of the multiplication property when fractions are involved in an inequality.

| Example 2 | **Solving and Graphing Inequalities** |

(a) Solve and graph the solution set for

$$\frac{x}{4} > 3$$

Here we multiply both sides of the inequality by 4. This will isolate x on the left.

$$4\left(\frac{x}{4}\right) > 4(3)$$

$$x > 12$$

The graph of the solution set, $\{x \mid x > 12\}$, is

(b) Solve and graph the solution set for

$$-\frac{x}{6} \geq -3$$

In this case, we multiply both sides of the inequality by -6:

$$(-6)\left(-\frac{x}{6}\right) \leq (-6)(-3)$$

Note that we reverse the sense of the inequality because we are multiplying by a negative number.

$$x \leq 18$$

The graph of the solution set, $\{x \mid x \leq 18\}$, is

✔ *CHECK YOURSELF 2*

Solve and graph the solution sets for the following inequalities.

(a) $\dfrac{x}{5} \leq 4$ (b) $-\dfrac{x}{3} < -7$

Example 3   **Solving and Graphing Inequalities**

(a) Solve and graph the solution set for $5x - 3 < 2x$.

First, add 3 to both sides to undo the subtraction on the left.

$$5x - 3 < 2x$$

$$5x - 3 + 3 < 2x + 3 \qquad \text{Add 3 to both sides to undo the subtraction.}$$

$$5x < 2x + 3$$

Now subtract $2x$, so that only the number remains on the right.

$$5x < 2x + 3$$

$$5x - 2x < 2x - 2x + 3 \qquad \text{Subtract } 2x \text{ to isolate the number on the right.}$$

$$3x < 3$$

Note that the multiplication property also allows us to divide both sides by a nonzero number.

Next *divide* both sides by 3.

$$\frac{3x}{3} < \frac{3}{3}$$

$$x < 1$$

The graph of the solution set, $\{x \mid x < 1\}$, is

(b) Solve and graph the solution set for $2 - 5x < 7$.

$$2 - 5x < 7$$

$$2 - 2 - 5x < 7 - 2 \qquad \text{Subtract 2.}$$

$$-5x < 5$$

$$\frac{-5x}{-5} > \frac{5}{-5} \qquad \text{Divide by } -5. \text{ Be sure to reverse the sense of the inequality.}$$

or $\qquad\qquad\qquad x > -1$

The graph of the solution set, $\{x \mid x > -1\}$, is

✔ *CHECK YOURSELF 3*

Solve and graph the solution sets.

(a) $4x + 9 \geq x$ (b) $5 - 6x < 41$

As with equations, when solving an inequality, we will collect all variable terms on one side and all constant terms on the other.

| **Example 4** | ## Solving and Graphing Inequalities |

Solve and graph the solution set for $5x - 5 \geq 3x + 4$.

$$5x - 5 \geq 3x + 4$$

$$5x - 5 + 5 \geq 3x + 4 + 5 \qquad \text{Add 5}$$

$$5x \geq 3x + 9$$

$$5x - 3x \geq 3x - 3x + 9 \qquad \text{Subtract } 3x$$

$$2x \geq 9$$

$$\frac{2x}{2} \geq \frac{9}{2} \qquad \text{Divide by 2}$$

$$x \geq \frac{9}{2}$$

The graph of the solution set, $\left\{x \mid x \geq \dfrac{9}{2}\right\}$, is

✔ *CHECK YOURSELF 4*

Solve and graph the solution set for

$$8x + 3 < 4x - 13$$

Be especially careful when negative coefficients occur in the process of solving.

| Example 5 | ## Solving and Graphing Inequalities |

Solve and graph the solution set for $x + 2 < \dfrac{5}{2}x - 1$.

$$2(x + 2) < 2\left(\dfrac{5}{2}x - 1\right)$$ Multiply by the LCD

$$2x + 4 < 5x - 2$$

$$2x + 4 - 4 < 5x - 2 - 4$$ Subtract 4

$$2x < 5x - 6$$

$$2x - 5x < 5x - 5x - 6$$ Subtract $5x$

$$-3x < -6$$ Divide by -3, and reverse the sense of the inequality.

$$\dfrac{-3x}{-3} > \dfrac{-6}{-3}$$

$$x > 2$$

The graph of the solution set, $\{x \mid x > 2\}$, is

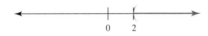

 CHECK YOURSELF 5

Solve and graph the solution set.

$$5x + 12 \geq 10x - 8$$

| Example 6 | ## Solving and Graphing Inequalities |

Solve and graph the solution set for

$$5(x - 2) \geq -8$$

Applying the distributive property on the left yields

$$5x - 10 \geq -8$$

Solving as before yields

$$5x - 10 + 10 \geq -8 + 10$$ Add 10.

$$5x \geq 2$$

or $$x \geq \dfrac{2}{5}$$ Divide by 5.

The graph of the solution set, $\left\{ x \mid x \geq \dfrac{2}{5} \right\}$, is

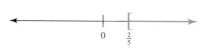

✔ *CHECK YOURSELF 6*

Solve and graph the solution set.

$$4(x + 3) < 9$$

| Example 7 | Solving and Graphing Inequalities |

Solve and graph the solution set for $-3(x + 2) \geq 5 - 3x$.

$$-3(x + 2) \geq 5 - 3x \qquad \text{Apply the distributive property.}$$

$$-3x - 6 \geq 5 - 3x \qquad \text{Add } 3x \text{ to both sides.}$$

$$0 - 6 \geq 5 \qquad \text{Add 6 to both sides.}$$

$$0 \geq 11$$

This is a false statement so no real number satisfies this inequality or the original inequality. Thus the solution set is the empty set, written { }. The graph of the solution set contains no points at all. (You might say that it is pointless!)

✔ *CHECK YOURSELF 7*

Solve and graph the solution set.

$$-4(x - 1) < 3 - 4x$$

| Example 8 | Solving and Graphing Inequalities |

Solve and graph the solution set for $3(x + 2) > 3x - 4$.

$$3(x + 2) > 3x - 4 \qquad \text{Apply the distributive property.}$$

$$3x + 6 > 3x - 4 \qquad \text{Subtract } 3x \text{ from both sides.}$$

$$6 > -4$$

This is a true statement for all values of x, so this inequality and the original inequality are true for all real numbers.

The graph of the solution set $\{x \mid x \in \mathbb{R}\}$ is every point on the number line.

✔ CHECK YOURSELF 8

Solve and graph the solution set.

$$-2(4 - x) \leq 5 + 2x$$

Some applications are solved by using an inequality instead of an equation. Example 9 illustrates such an application.

| Example 9 | ## Solving an Application with Inequalities |

Mohammed needs an average score of 92 or higher on four tests to get an A. So far his scores are 94, 89, and 88. What score on the fourth test will get him an A?

What do you need to find? **Step 1** We are looking for the score that will, when combined with the other scores, give Mohammed an A.

Assign a letter to the unknown. **Step 2** Let x represent a fourth-test score that will get him an A.

Write an inequality. **Step 3** The inequality will have the average (mean) on the left side, which must be greater than or equal to the 92 on the right.

$$\frac{94 + 89 + 88 + x}{4} \geq 92$$

Solve the equation for x. **Step 4** First, multiply both sides by 4:

$$94 + 89 + 88 + x \geq 368$$

Then add the test scores:

$$183 + 88 + x \geq 368$$
$$271 + x \geq 368$$

Subtracting 271 from both sides gives

$$x \geq 97 \qquad \text{Mohammed needs a test result of 97 or higher.}$$

Step 5 To check the solution, we find the mean of the four test scores 94, 89, 88, and 97.

$$\frac{94 + 89 + 88 + 97}{4} = \frac{368}{4} = 92$$

 CHECK YOURSELF 9

Felicia needs an average score of at least 75 on five tests to get a passing grade in her health class. On her first four tests she has scores of 68, 79, 71, and 70. What score on the fifth test will give her a passing grade?

The following outline (or algorithm) summarizes our work in this section.

Solving Linear Inequalities

Step 1.   Remove grouping symbols and fraction coefficients, and combine any like terms appearing on either side of the inequality.

Step 2.   Apply the addition property to write an equivalent inequality with the variable term on one side of the inequality and the number on the other.

Step 3.   Apply the multiplication property to write an equivalent inequality with the variable isolated on one side of the inequality. Be sure to reverse the sense of the inequality if you multiply or divide by a negative number. The solution set derived in step 3 can then be graphed on a number line.

✔ *CHECK YOURSELF ANSWERS*

1. (a) $\{x \mid x > 5\}$

**   ** (b) $\{x \mid x \geq -6\}$

2. (a) $\{x \mid x \leq 20\}$

**   ** (b) $\{x \mid x > 21\}$

3. (a) $\{x \mid x \geq -3\}$

**   ** (b) $\{x \mid x > -6\}$

4. $\{x \mid x < -4\}$ **5.** $\{x \mid x \leq 4\}$

6. $\left\{x \mid x < -\dfrac{3}{4}\right\}$ **7.** { }

8. $\{x \mid x \in \mathbb{R}\}$ **9.** 87 or greater

Exercises ▪ 2.6

Solve and graph the solution set of each of the following inequalities. Write the solution set in set-builder notation and interval notation.

1. $3x \le 9$

2. $5x > 20$

3. $5x > -35$

4. $6x \le -18$

5. $-6x \ge 18$

6. $-9x < 45$

7. $-2x < -12$

8. $-12x \ge -48$

9. $\dfrac{x}{4} > 5$

10. $\dfrac{x}{3} \le -3$

11. $-\dfrac{x}{2} \ge -3$

12. $-\dfrac{x}{4} < 5$

13. $\dfrac{2x}{3} < 6$

14. $\dfrac{3x}{4} \ge -9$

15. $5x > 3x + 8$

16. $4x \le x - 9$

17. $5x - 2 < 3x$

18. $7x + 3 \ge 2x$

19. $3 - 2x > 5$

20. $5 - 3x \le 17$

21. $2x \ge 5x + 18$

22. $3x < 7x - 28$

23. $\dfrac{1}{3}x - 5 \le \dfrac{5}{3}x + 11$

24. $\dfrac{3}{7}x + 6 \ge -\dfrac{12}{7}x - 9$

25. $0.34x + 21 \ge 19 - 1.66x$

26. $-1.57x - 15 \ge 1.43x + 18$

27. $7x - 5 < 3x + 2$

28. $5x - 2 \ge 2x - 7$

29. $5x + 7 > 8x - 17$

30. $4x - 3 \le 9x + 27$

31. $3x - 2 \le 5x + 3$

32. $2x + 3 > 8x - 2$

33. $-2(4 - x) \le 7 + 2x$

34. $6(x + 3) \ge 4 + 6x$

35. $-2(5 - x) \le -3(x + 2) + 5x$

36. $3(x + 5) \le 6(x + 2) - 3x$

Translate the following statements to inequalities. Let x represent the number in each case.

37. 5 more than a number is greater than 3.

38. 3 less than a number is less than or equal to 5.

39. 4 less than twice a number is less than or equal to 7.

40. 10 more than a number is greater than negative 2.

41. 4 times a number, decreased by 15, is greater than that number.

42. 2 times a number, increased by 28, is less than or equal to 6 times that number.

43. Panda population. There are fewer than 1,000 wild giant pandas left in the bamboo forests of China. Write an inequality expressing this relationship.

44. Forestry. Let C represent the amount of Canadian forest and M represent the amount of Mexican forest. Write an inequality showing the relationship of the forests of Mexico and Canada if Canada contains at least 9 times as much forest as Mexico.

45. Test scores. To pass a course with a grade of B or better, Liza must have an average of 80 or more. Her grades on three tests are 72, 81, and 79. Write an inequality representing the score that Liza must get on the fourth test to obtain a B average or better for the course.

46. Test scores. Sam must have an average of 70 or more in his summer course in order to obtain a grade of C. His first three test grades were 75, 63, and 68. Write an inequality representing the score that Sam must get on the last test in order to earn a C grade.

47. Commission. Juanita is a salesperson for a manufacturing company. She may choose to receive $500 or 5% commission on her sales as payment for her work. Write an inequality representing the amount she needs to sell to make the 5% offer a better deal.

48. Telephone costs. The cost for a long-distance telephone call is $0.24 for the first minute and $0.11 for each additional minute or portion thereof. The total cost of the call cannot exceed $3. Write an inequality representing the number of minutes a person could talk without exceeding $3.

49. College tuition. Samantha's financial aid stipulates that her tuition not exceed $1,500 per semester. If her local community college charges a $45 service fee plus $290 per course, what is the greatest number of courses for which Samantha can register?

50. Grade average. Nadia is taking a mathematics course in which five tests are given. To get a B, a student must average at least 80 on the five tests. Nadia scored 78, 81, 76, and 84 on the first four tests. What score on the last test will earn her at least a B?

51. Geometry. The width of a rectangle is fixed at 40 cm, and the perimeter can be no greater than 180 cm. Find the maximum length of the rectangle.

52. Entertainment. The women's soccer team can spend at most $900 for its annual awards banquet. If the restaurant charges a $75 setup fee and $24 per person, at most how many people can attend?

53. Consumer affairs. Joyce is determined to spend no more than $125 on clothes. She wants to buy two pairs of identical jeans and a blouse. If she spends $29 on the blouse, what is the maximum amount she can spend on each pair of jeans?

54. Commission. Ben earns $750 per month plus 4% commission on all his sales over $900. Find the minimum sales that will allow Ben to earn at least $2,500 per month.

55. You are the office manager for a small company. You need to acquire a new copier for the office. You find a suitable one that leases for $250 per month from the copy machine company. It costs 2.5¢ per copy to run the machine. You purchase paper for $3.50 per ream (500 sheets). If your copying budget is no more than $950 per month, is this machine a good choice? Write a brief recommendation to the purchasing department. Use equations and inequalities to explain your recommendation.

56. Nutritionists recommend that, for good health, no more than 30% of our daily intake of calories come from fat. Algebraically, we can write this as $f \le 0.30(c)$, where f = calories from fat and c = total calories for the day. But this does not mean that everything we eat must meet this requirement. For example, if you eat $\frac{1}{2}$ cup of Ben and Jerry's vanilla ice cream for dessert after lunch, you are eating a total of 250 calories, of which 150 calories is from fat. This amount is considerably more than 30% from fat, but if you are careful about what you eat the rest of the day, you can stay within the guidelines.

    Set up an inequality based on your normal caloric intake. Solve the inequality to find how many calories in fat you could eat over the day and still have no more than 30% of your daily calories from fat. The American Heart Association says that to maintain your weight, your daily caloric intake should be 15 calories for every pound. You can compute this number to estimate the number of calories a day you normally eat. Do some research in your grocery store or library to determine what foods satisfy the requirements for your diet for the rest of the day. There are 9 calories in every gram of fat; many food labels give the amount of fat only in grams.

57. Your aunt calls to ask your help in making a decision about buying a new refrigerator. She says that she found two that seem to fit her needs, and both are supposed to last at least 14 years, according to *Consumer Reports*. The initial cost for one refrigerator is $712, but it only uses 88 kilowatthours (kWh) per month. The other refrigerator costs $519 and uses an estimated 100 kWh/month. You do not know the price of electricity per kilowatthour where your aunt lives, so you will have to decide what, in cents per kilowatthour, will make the first refrigerator cheaper to run for its 14 years of expected usefulness. Write your aunt a letter, explaining what you did to calculate this cost, and tell her to make her decision based on how the kilowatthour rate she has to pay in her area compares with your estimation.

Solving Absolute Value Equations (Optional)

 2.7 OBJECTIVES

1. *Find the absolute value of an expression*
2. *Solve an absolute value equation*

Equations may contain absolute value notation in their statements. In this section, we will look at algebraic solutions to statements that include absolute values. First, we will review the concept of absolute value.

The **absolute value** of a real number is the distance from that real number to 0. Because absolute value is a distance, it is always positive. Formally, we say

The absolute value of a number *x* is given by

$$|x| = \begin{cases} -x & \text{if } x < 0 \\ x & \text{if } x \geq 0 \end{cases}$$

Example 1 ## Finding the Absolute Value of a Number

Find the absolute value for each expression.

(a) $|-3|$ (b) $|7 - 2|$ (c) $|-7 - 2|$

(a) Because $-3 < 0, |-3| = -(-3) = 3$.

(b) $|7 - 2| = |5|$ Because $5 \geq 0, |5| = 5$.

(c) $|-7 - 2| = |-9|$ Because $-9 < 0, |-9| = -(-9) = 9$.

 CHECK YOURSELF 1

Find the absolute value for each expression.

(a) $|12|$ (b) $|-9 + 5|$ (c) $|-3 - 4|$

Given an equation like

$$|x| = 5$$

there are two possible solutions. The value of x could be 5 or -5. In either case, the absolute value is 5. This can be generalized in the following property of absolute value equations.

Caution

Variable p must be positive because an equation such as $|x - 2| = -3$ has no solution. The absolute value of a quantity must always be equal to a nonnegative number.

Rules and Properties: Absolute Value Equations—Property 1

For any positive number p, if

$$|x| = p$$

then

$$x = p \quad \text{or} \quad x = -p$$

We'll use this property in the next several examples.

Example 2

Solving an Absolute Value Equation

Solve the equation

$$|x - 3| = 4$$

Caution

Be careful! A common mistake is to solve only the equation $x - 3 = 4$. You must solve *both* equations to find the **two** required solutions.

From property 1 above, we know that the expression inside the absolute value signs, $x - 3$, must equal either 4 or -4. We set up two equations and solve them both.

$$
\begin{array}{lll}
(x - 3) = 4 & \quad \text{or} \quad & (x - 3) = -4 \\
x - 3 = 4 & & x - 3 = -4 \qquad \text{Add 3 to both sides of the equation.} \\
x = 7 & & x = -1
\end{array}
$$

We arrive at the solution set, $\{-1, 7\}$.

 CHECK YOURSELF 2

Find the solution set for the equation.

$$|x - 2| = 3$$

We will use property 1 to solve subsequent examples in this section.

Example 3

Solving an Absolute Value Equation

Solve for x.

$$|3x - 2| = 4$$

From property 1, we know that $|3x - 2| = 4$ is equivalent to the equations

$$3x - 2 = 4 \quad \text{or} \quad 3x - 2 = -4 \qquad \text{Add 2}$$
$$3x = 6 \qquad\qquad\quad 3x = -2 \qquad \text{Divide by 3}$$
$$x = 2 \qquad\qquad\quad x = -\frac{2}{3}$$

The solution set is $\left\{-\dfrac{2}{3}, 2\right\}$. These solutions are easily checked by replacing x with $-\dfrac{2}{3}$ and 2 in the original absolute value equation.

✔ *CHECK YOURSELF 3*

Solve for x.

$$|4x + 1| = 9$$

An equation involving absolute value may have to be rewritten before you can apply property 1. Consider Example 4.

Example 4	## Solving an Absolute Value Equation

Solve for x.

$$|2 - 3x| + 5 = 10$$

To use property 1, we must first isolate the absolute value on the left side of the equation. This is easily done by subtracting 5 from both sides.

$$|2 - 3x| = 5$$

We can now proceed as before, by using property 1.

$$2 - 3x = 5 \quad \text{or} \quad 2 - 3x = -5 \qquad \text{Subtract 2.}$$
$$-3x = 3 \qquad\qquad -3x = -7 \qquad \text{Divide by } -3.$$
$$x = -1 \qquad\qquad\quad x = \frac{7}{3}$$

The solution set is $\left\{-1, \dfrac{7}{3}\right\}$.

✔ *CHECK YOURSELF 4*

Solve for x.

$$|5 - 2x| - 4 = 7$$

In some applications, there is more than one absolute value in an equation. Consider an equation of the form

$$|x| = |y|$$

Since the absolute values of x and y are equal, x and y are the same distance from 0, which means they are either *equal* or *opposite in sign*. This leads to a second general property of absolute value equations.

Absolute Value Equations—Property 2

If $|x| = |y|$

then $x = y$ or $x = -y$

Let's look at an application of this second property in Example 5.

Example 5	### Solving Equations with Two Absolute Value Expressions

Solve for x.

$$|3x - 4| = |x + 2|$$

By property 2, we can write

$3x - 4 = x + 2$	or	$3x - 4 = -(x + 2)$	
		$3x - 4 = -x - 2$	Add 4 to both sides.
$3x = x + 6$		$3x = -x + 2$	Isolate x.
$2x = 6$		$4x = 2$	Divide by 2.
$x = 3$		$x = \dfrac{1}{2}$	

The solution set is $\left\{ \dfrac{1}{2}, 3 \right\}$.

✔ *CHECK YOURSELF 5*

Solve for x.

$$|4x - 1| = |x + 5|$$

✔ *CHECK YOURSELF ANSWERS*

1. (a) 12; (b) 4; (c) 7

2. $|x - 2| = 3$

$x - 2 = 3$ or $x - 2 = -3$

$x = 5$ $x = -1$

$\{-1, 5\}$

3. $\left\{-\dfrac{5}{2}, 2\right\}$ **4.** $\{-3, 8\}$ **5.** $\left\{-\dfrac{4}{5}, 2\right\}$

Exercises ▪ 2.7

Find the absolute value for each expression.

1. $|15|$

2. $|21|$

3. $|-21|$

4. $|-18|$

5. $|8-3|$

6. $|35-23|$

7. $|-12+8|$

8. $|-23-11|$

9. $|-12-19|$

10. $|-19-27|$

11. $-|-19|+|-27|$

12. $|-13|-|-12|$

13. $|-13|+|12|$

14. $|-13-12|$

15. $-|-13|-|-12|$

16. $-|(-13)-(-12)|$

Solve the equations.

17. $|x|=5$

18. $|x|=6$

19. $|x+2|=6$

20. $|x-2|=6$

21. $|2x-1|=6$

22. $|3x-2|=8$

23. $3|2x-6|=9$

24. $2|3x-5|=12$

25. $|5x-2|=-3$

26. $|9-4x|=-1$

27. $|x-2|+3=5$

28. $|x+5|-2=5$

29. $8-|x-4|=5$

30. $10-|2x+1|=3$

31. $|2x-1|=|x+3|$

32. $|3x+1|=|2x-3|$

33. $|5x-2|=|2x-4|$

34. $|5x+2|=|x-3|$

35. $|4x-5|=|5-4x|$

36. $|x-2|=|2-x|$

37. $\left|3-\dfrac{1}{3}x\right|=\dfrac{5}{3}$

38. $\left|2+\dfrac{5}{8}x\right|=\dfrac{3}{8}$

Use absolute value notation to write an equation that represents each sentence.

39. x is exactly 5 units from 0 on the number line.

40. x is exactly 2 units from 0 on the number line.

41. x is exactly 4 units from 3 on the number line.

42. x is exactly 6 units from 4 on the number line.

43. x is exactly 4 units from -3 on the number line.

44. x is exactly 6 units from -4 on the number line.

SECTION 2.8

Solving Absolute Value Inequalities (Optional)

2.8 OBJECTIVES

1. Solve a compound inequality
2. Solve an absolute value inequality

In this section, we will solve absolute value inequalities.

First, we will look at two types of inequality statements that arise frequently in mathematics. Consider a statement such as

$$-2 < x < 5$$

It is called a **compound** inequality because it combines the two inequalities

$$-2 < x \quad \text{and} \quad x < 5$$

Because there are two inequality signs in a single statement, these are sometimes called double inequalities.

When we begin with a compound inequality such as

$$-3 \le 2x + 1 \le 7$$

we find an equivalent statement in which the variable is isolated in the middle. Example 1 illustrates.

| Example 1 | ## Solving a Compound Inequality |

Solve and graph the compound inequality.

$$-3 \le 2x + 1 \le 7$$

We are really applying the additive property to each of the two inequalities that make up the double-inequality statement.

First, we subtract 1 from each of the three members of the compound inequality.

$$-3 - 1 \le 2x + 1 - 1 \le 7 - 1$$

or

$$-4 \le 2x \le 6$$

We now divide by 2 to isolate the variable x.

When we divide by a positive number, the sense of the inequality is preserved.

$$-\frac{4}{2} \le \frac{2x}{2} \le \frac{6}{2}$$

$$-2 \le x \le 3$$

The solution set consists of all numbers between -2 and 3, including -2 and 3, and is written

In interval notation, we write $[-2, 3]$

$$\{x \mid -2 \leq x \leq 3\}$$

That set is graphed below.

Note: Our solution set is equivalent to

$$\{x \mid x \geq -2 \text{ and } x \leq 3\}$$

Look at the individual graphs.

$\{x \mid x \geq -2\}$

$\{x \mid x \leq 3\}$

$\{x \mid x \geq -2 \text{ and } x \leq 3\}$

Using set notation, we can write

$\{x \mid x \geq -2\} \cap \{x \mid x \leq 3\}$

Because the connecting word is *and*, we want the *intersection* of the sets, that is, those numbers common to both sets.

 CHECK YOURSELF 1

Solve and graph the double inequality.

$$-5 < 3 + 2x < 5$$

A compound inequality may also consist of two inequality statements connected by the word *or*. Example 2 illustrates how to solve that type of compound inequality.

Example 2   Simplifying a Compound Inequality

Solve and graph the inequality

$$2x - 3 < -5 \qquad \text{or} \qquad 2x - 3 > 5$$

In this case, we must work with each of the inequalities *separately*.

$2x - 3 < -5$	or	$2x - 3 > 5$	Add 3
$2x < -2$		$2x > 8$	Divide by 2
$x < -1$		$x > 4$	

The graph of the solution set, $\{x \mid x < -1 \text{ or } x > 4\}$, is shown.

$$\{x \mid x < -1 \text{ or } x > 4\}$$

In interval notation, we write $(-\infty, -1) \cup (4, \infty)$.

Note that since the connecting word is *or* in this case, the solution set of the original inequality is the *union* of the two sets, that is, those numbers that belong to either or both of the sets.

In set notation we can write $\{x \mid x < -1\} \cup \{x \mid x > 4\}$

✔ *CHECK YOURSELF 2*

Solve and graph the inequality.

$$3x - 4 \leq -7 \qquad \text{or} \qquad 3x - 4 \geq 7$$

Imagine that you've received a phone call from a friend who says that his car is stuck on the freeway, within 3 mi of mileage marker 255. Where is he? He must be somewhere between mileage marker 252 and marker 258. What you've actually solved here is an example of an **absolute value inequality.** Absolute value inequalities are in one of two forms

$$|x - a| < b \qquad \text{or} \qquad |x - a| > b$$

The mileage marker example is of the first form. We could say

$$|x - 255| < 3$$

To solve an equation of this type, use the following rule.

Rules and Properties: Absolute Value Inequalities—Property 3

For any positive number p, if

$$|x| < p$$

then

$$-p < x < p$$

Graphically, this is represented as

In our example, because $|x - 255| < 3$,

$$-3 < x - 255 < 3$$

Adding 255 to each member of the inequality, we get

$$-3 + 255 < x - 255 + 255 < 3 + 255$$

or $$252 < x < 258$$

| Example 3 | Solving an Absolute Value Inequality |

Solve the following inequality, then graph the solution set.

$$|x - 5| < 9$$

According to Property 3, we have

$$|x - 5| < 9$$
$$-9 < x - 5 < 9 \qquad \text{Add 5 to each member.}$$
$$-9 + 5 < x < 9 + 5$$
$$-4 < x < 14$$

In interval notation we would write $(-4, 14)$.

Graphing the solution set, $\{x \mid -4 < x < 14\}$, we get

✔ *CHECK YOURSELF 3*

Solve the following inequality, then graph the solution set.

$$|x - 25| < 136$$

What about inequalities of the form $|x - a| > b$? The following rule applies.

Rules and Properties: Absolute Value Inequalities—Property 4

For any positive number p, if

$$|x| > p$$

then

$$x < -p \qquad \text{or} \qquad x > p$$

Graphically, this is represented as

| Example 4 | **Solving an Absolute Value Inequality** |

Solve the following inequality, then graph the solution set.

$$|x - 7| > 19$$

By Property 4,

$$x - 7 > 19 \quad \text{or} \quad x - 7 < -19$$

Solving each inequality by adding 7 to each side, we get

$$x > 26 \quad \text{or} \quad x < -12$$

Now we can graph the solution set.

✔ *CHECK YOURSELF 4*

Solve the following inequality, then graph the solution set.

$$|x - 2| > 7$$

✔ *CHECK YOURSELF ANSWERS*

1. $\{x \mid -4 < x < 1\}$

2. $\left\{x \mid x \leq -1 \text{ or } x \geq \dfrac{11}{3}\right\}$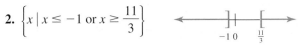

3. $\{x \mid -111 < x < 161\}$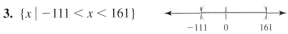

4. $\{x \mid x < -5 \text{ or } x > 9\}$

Exercises · 2.8

Solve each of the following compound inequalities. Then graph the solution set.

1. $4 \le x + 1 \le 7$

2. $-4 < x + 1 < 5$

3. $-8 < 2x < 4$

4. $-6 \le 3x \le 9$

5. $1 \le 2x - 3 \le 6$

6. $-2 < 3x - 5 < 4$

7. $-1 < 5 + 3x < 8$

8. $-7 \le 3 + 2x \le 8$

Solve each of the following compound inequalities. Then graph the solution set.

9. $x - 1 < -3$ or $x - 1 > 3$

10. $x + 2 < -5$ or $x + 2 > 5$

11. $3x - 2 \le -7$ or $3x - 2 \ge 7$

12. $2x + 5 < -3$ or $2x + 5 > 3$

13. $3x - 1 < -7$ or $3x - 1 > 7$

14. $4x + 3 < -5$ or $4x + 3 > 5$

Solve each inequality. Graph the solution set.

15. $|x| < 5$

16. $|x| > 3$

17. $|x| \ge 7$

18. $|x| \le 4$

200

19. $|x - 3| > 5$

20. $|x + 3| \leq 4$

21. $|x + 6| \leq 4$

22. $|x - 7| > 5$

23. $|3 - x| > 5$

24. $|5 - x| < 3$

25. $|x - 7| < 0$

26. $|x + 5| \geq 0$

27. $|2x - 5| < 3$

28. $|3x - 1| > 8$

29. $|3x + 4| \geq 5$

30. $|2x + 3| \leq 9$

31. $|5x - 3| > 7$

32. $|6x - 5| < 13$

33. $|2 - 3x| \leq 11$

34. $|3 - 2x| > 11$

Use absolute value notation to write an inequality that represents each sentence.

35. x is within 7 units of 0 on the number line.

36. x is within 4 units of 0 on the number line.

37. x is at least 5 units from 0 on the number line.

38. x is at least 2 units from 0 on the number line.

39. x is less than 7 units from -2 on the number line.

40. x is more than 6 units from 4 on the number line.

41. x is at least 3 units from -4 on the number line.

42. x is at most 3 units from 4 on the number line.

Summary for Chapter 2

Example	Topic	Reference
	Algebraic Equations	**2.1–2.3**
$3x - 5 = 7$ is an equation.	**Equation** A statement that two expressions are equal.	**p. 108**
4 is a solution for the equation because $$3 \cdot 4 - 5 = 7$$ $$12 - 5 = 7$$ $$7 = 7 \quad \text{True}$$	**Solution** A value for the variable that will make an equation a true statement.	**p. 109**
	Equivalent Equations Equations that have exactly the same solutions.	**p. 111**
If $x = y + 3$, then $x + 2 = y + 5$. $5x = 20$ and $x = 4$ are equivalent equations.	**Writing Equivalent Equations** There are two basic properties that will yield equivalent equations. 1. If $a = b$, then $a + c = b + c$. Adding (or subtracting) the same quantity on each side of an equation gives an equivalent equation. 2. If $a = b$, then $ac = bc$, $c \neq 0$. Multiplying (or dividing) both sides of an equation by the same number gives an equivalent equation.	**p. 112** **p. 126**
Solve: $$3(x - 2) + 4x = 3x + 14$$ $$3x - 6 + 4x = 3x + 14$$ $$7x - 6 = 3x + 14$$ $$\underline{+6 \qquad +6}$$ $$7x = 3x + 20$$ $$\underline{-3x \qquad -3x}$$ $$4x = 20$$ $$\frac{4x}{4} = \frac{20}{4}$$ $$x = 5$$	**Solving Linear Equations** We say that an equation is "solved" when we have an equivalent equation of the form $x = \Box$ or $\Box = x$ where $\Box$ is some number The steps of solving a linear equation are as follows: 1. Use the distributive property to remove any grouping symbols that appear. Then simplify by combining any like terms. 2. Add or subtract the same term on both sides of the equation until the term containing the variable is on one side and a number is on the other. 3. Multiply or divide both sides of the equation by the same nonzero number so that the variable is alone on one side of the equation. 4. Check the solution in the original equation.	**p. 142**

202

	Solving Literal Equations	**2.4**
$$a = \frac{2b + c}{3}$$ is a literal equation.	**Literal Equation** An equation that involves more than one letter or variable.	**p. 153**
Solve for b: $$a = \frac{2b + c}{3}$$ $$3a = 3\left(\frac{2b + c}{3}\right)$$ $$3a = 2b + c$$ $$\underline{-c \qquad -c}$$ $$3a - c = 2b$$ $$\frac{3a - c}{2} = b$$	**Solving Literal Equations** 1. Multiply both sides of the equation by the lowest common denominator (LCD) to clear of fractions. 2. Add or subtract the same term on both sides of the equation so that all terms containing the variable you are solving for are on one side. 3. Divide both sides by any numbers or letters multiplying the variable that you are solving for.	**p. 155**

	Applying Equations	**2.4**
	Using Equations to Solve Word Problems Follow these steps. 1. Read the problem carefully. Then reread it to decide what you are asked to find. 2. Choose a letter to represent one of the unknowns in the problem. Then represent each of the unknowns with an expression that uses the same letter. 3. Translate the problem to the language of algebra to form an equation. 4. Solve the equation and answer the question of the original problem. 5. Check your solution by returning to the original problem.	**p. 157**

	Inequalities	**2.5–2.6**
$4 < 9$ $-1 > -6$ $2 \leq 2$ $3 \geq -4$	**Inequality** A statement that one quantity is less than (or greater than) another. Four symbols are used: $a < b$ — a is less than b. $a > b$ — a is greater than b. $a \leq b$ — a is less than or equal to b. $a \geq b$ — a is greater than or equal to b.	**p. 169**

$x < 6$
is graphed

$x \geq 5$

Graphing Inequalities   To graph $x < a$, we use a parenthesis and an arrow pointing left.

To graph $x \geq b$, we use a bracket and an arrow pointing right.

p. 170

$$
\begin{array}{rr}
2x - 3 > & 5x + 6 \\
+ 3 & + 3 \\
\hline
2x > & 5x + 9 \\
-5x & -5x \\
\hline
-3x > & 9
\end{array}
$$

$$\frac{-3x}{-3} < \frac{9}{-3}$$

$$x < -3$$

Solving Inequalities   An inequality is "solved" when it is in the form $x < \square$ or $x > \square$.

Proceed as in solving equations by using the following properties.

1. If $a < b$, then $a + c < b + c$.

    Adding (or subtracting) the same quantity to both sides of an inequality gives an equivalent inequality.

2. If $a < b$, then $ac < bc$ when $c > 0$ and $ac > bc$ when $c < 0$.

    Multiplying both sides of an inequality by the same *positive number* gives an equivalent inequality. When both sides of an inequality are multiplied by the same *negative number, you must reverse the sense* of the inequality to give an equivalent inequality.

p. 171

p. 176

Absolute Value Equations (Optional)

2.7

$|2x - 5| = 7$ is equivalent
to $2x - 5 = -7$
or $2x - 5 = 7$
so $x = -1$ or $x = 6$

Property 1   For any positive number p, if

$$|x| = p$$

then $x = -p$ or $x = p$

p. 190

$|x| = |2x - 6|$
        $x = 2x - 6$
or $x = -(2x - 6)$
    $x = 6$ or $x = 2$

Property 2   If $|x| = |y|$, then

$$x = y \quad \text{or} \quad x = -y$$

p. 192

Absolute Value Inequalities (Optional)		**2.8**
$\|3x - 5\| < 7$ is equivalent to $$-7 < 3x - 5 < 7$$ This yields $$-2 < 3x < 12$$ $$-\frac{2}{3} < x < 4$$	**Property 3** For any positive number p, if $$\|x\| < p$$ then $$-p < x < p$$ To solve this form of inequality, translate to the equivalent compound inequality and solve as before.	**p. 197**
$\|2 - 5x\| \geq 12$ is equivalent to $$2 - 5x \leq -12$$ or $$2 - 5x \geq 12$$ This yields $$x \geq \frac{14}{5} \quad \text{or} \quad x \leq -2$$	**Property 4** For any positive number p, if $$\|x\| > p$$ then $$x < -p \quad \text{or} \quad x > p$$ To solve this form of inequality, translate to the equivalent compound inequality and solve as before.	**p. 198**

Summary Exercises ▪ 2

This summary exercise set is provided to give you practice with each of the objectives of the chapter. Each exercise is keyed to the appropriate chapter section. Your instructor will give you guidelines on how to best use these exercises in your instructional setting.

[2.1] Tell whether the number shown in parentheses is a solution for the given equation.

1. $5x - 3 = 7$ (2)

2. $5x - 8 = 3x + 2$ (4)

3. $7x - 2 = 2x + 8$ (2)

4. $4x + 3 = 2x - 11$ (-7)

5. $x + 5 + 3x = 2 + x + 23$ (6)

6. $\frac{2}{3}x - 2 = 10$ (21)

[2.1] Solve the following equations and check your results.

7. $x + 3 = 5$

8. $x - 9 = 3$

9. $5x = 4x - 5$

10. $4x - 9 = 3x$

11. $5x - 3 = 4x + 2$

12. $9x + 2 = 8x - 7$

13. $9x - 7 = 8x - 6$

14. $3 + 4x - 1 = x - 7 + 2x$

15. $4(2x + 3) = 7x + 5$

16. $5(5x - 3) = 6(4x + 1)$

[2.2–2.3] Solve the following equations and check your results.

17. $5x = 35$

18. $7x = -28$

19. $-9x = 36$

20. $-9x = -63$

21. $\frac{x}{4} = 8$

22. $-\frac{x}{5} = -3$

23. $\frac{2}{3}x = 18$

24. $\frac{7}{8}x = 28$

25. $5x - 3 = 12$

26. $4x + 3 = -13$

27. $7x + 8 = 3x$

28. $3 - 5x = -17$

29. $4x - 7 = 2x$

30. $2 - 4x = 5$

31. $\frac{x}{3} - 5 = 1$

32. $\frac{3}{4}x - 2 = 7$

33. $6x - 5 = 3x + 13$

34. $5x + 9 = 3x - 7$

35. $7x + 4 = 2x + 6$

36. $9x - 8 = 7x - 3$

37. $2x + 7 = 4x - 5$

38. $3x - 15 = 7x - 10$

39. $\frac{10}{3}x - 5 = \frac{4}{3}x + 7$

40. $\frac{11}{4}x - 15 = 5 - \frac{5}{4}x$

41. $3.7x + 8 = 1.7x + 16$

42. $5.4x - 3 = 8.4x + 9$

43. $2.4x + 6 - 1.2x = 9 - 1.8x + 12$

44. $3.6x - 7 - 2.1x = -21 - 2.5x + 12$

45. $5(3x - 1) - 6x = 3x - 2$

46. $5x + 2(3x - 4) = 14x - 7$

47. $4(2x - 1) = 6x + 5$

48. $7x - 3(x - 2) = 30$

49. $8x - 5(x + 3) = -10$

50. $7(3x + 1) - 13 = 8(2x + 3)$

51. $3(2x - 5) - 2(x - 3) = 11$

206

52. $\dfrac{2x}{3} - \dfrac{x}{4} = 5$

53. $\dfrac{3x}{4} - \dfrac{2x}{5} = 7$

54. $\dfrac{x}{2} - \dfrac{x+1}{3} = \dfrac{1}{6}$

55. $\dfrac{x+1}{5} - \dfrac{x-6}{3} = \dfrac{1}{3}$

[2.4] Solve for the indicated variable.

56. $V = LWH$ (for W)

57. $P = 2L + 2W$ (for L)

58. $ax + by = c$ (for y)

59. $A = \dfrac{1}{2}bh$ (for h)

60. $A = P + Prt$ (for t)

61. $m = \dfrac{n-p}{q}$ (for p)

[2.2–2.4] Solve the following word problems.

62. The sum of 3 times a number and 7 is 25. What is the number?

63. 5 times a number, decreased by 8, is 32. Find the number.

64. If the sum of two consecutive integers is 85, find the two integers.

65. The sum of three consecutive odd integers is 57. What are the three integers?

66. Salaries. Rafael earns $35 more per week than Andrew. If their weekly salaries total $715, what amount does each earn?

67. Age. Larry is 2 years older than Susan, while Nathan is twice as old as Susan. If the sum of their ages is 30 years, find each of their ages.

68. Speed. Lisa left Friday morning, driving on the freeway to visit friends for the weekend. Her trip took 4 h. When she returned on Sunday, heavier traffic slowed her average speed by 6 mi/h, and the trip took $4\dfrac{1}{2}$ h. What was her average speed in each direction, and how far did she travel each way?

69. Speed. A bicyclist started on a 132-mi trip and rode at a steady rate for 3 h. He began to tire at that point and slowed his speed by 4 mi/h for the remaining 2 h of the trip. What was his average speed for each part of the journey?

70. Motion. At noon, Jan left her house, jogging at an average rate of 8 mi/h. Two hours later, Stanley left on his bicycle along the same route, averaging 20 mi/h. At what time will Stanley catch up with Jan?

71. Motion. At 9 A.M., David left New Orleans, Louisiana, for Tallahassee, Florida, averaging 47 mi/h. Two hours later, Gloria left Tallahassee for New Orleans along the same route, driving 5 mi/h faster than David. If the two cities are 391 mi apart, at what time will David and Gloria meet?

72. Retail. A firm producing running shoes finds that its fixed costs are $3,900 per week, and its variable cost is $21 per pair of shoes. If the firm can sell the shoes for $47 per pair, how many pairs of shoes must be produced and sold each week for the company to break even?

[2.5] Graph the solution sets.

73. $x > 5$

74. $x < -3$

75. $x \le -4$

76. $x \ge 9$

77. $x \ge -6$

78. $x < 0$

[2.5–2.6] Solve the following inequalities.

79. $x - 2 \leq 9$

80. $x + 3 > -2$

81. $5x > 4x - 3$

82. $4x \geq -12$

83. $-12x < 36$

84. $-\dfrac{x}{3} \geq 5$

85. $2x \leq 8x - 3$

86. $2x + 3 \geq 9$

87. $7 - 6x > 15$

88. $5x - 2 \leq 4x + 5$

89. $7x + 13 \geq 3x + 19$

90. $4x - 2 < 7x + 16$

91. $-4(x + 5) \leq 7 - 4x$

92. $3(7 - x) \geq 2(x + 15) - 5x$

[2.7] Solve the following equations.

93. $|2x + 1| = 5$

94. $|3x - 2| = 7$

95. $|3x - 7| = 8$

96. $|4x + 1| = 9$

97. $|4x - 3| = 13$

98. $|6 - 2x| = 10$

99. $|-5x + 7| = 17$

100. $|-2x + 3| = 7$

101. $|7x - 1| = -3$

102. $|2x + 5| = -1$

103. $|2x + 1| - 3 = 6$

104. $7 - |x - 3| = 5$

105. $|3x - 1| = |x + 5|$

106. $|x - 5| = |x + 3|$

[2.8] Solve each of the following compound inequalities.

107. $4 \leq x + 3 \leq 5$

108. $-8 \leq 2x \leq 12$

109. $-3 < 5 + 4x \leq 17$

110. $-5 \leq 4 + 3x < 22$

111. $x - 2 < -4$ or $x - 2 > 4$

112. $3x + 2 \leq -11$ or $3x + 2 \geq 11$

[2.8] Solve the following inequalities.

113. $|x| \leq 3$

114. $|x + 3| > 5$

115. $|x - 7| > 4$

116. $|3 - x| < 6$

117. $|2x + 7| \geq 5$

118. $|3x - 1| \leq 4$

119. $|3x + 4| < 11$

120. $|5x + 2| \geq 12$

Self-Test 2

The purpose of this self-test is to help you check your progress and to review for a chapter test in class. Allow yourself about an hour to take the test. When you are done, check your answers in the back of the book. If you missed any answers, be sure to go back and review the appropriate sections in the chapter and the exercises that are provided.

Tell whether the number shown in parentheses is a solution for the given equation.

1. $7x - 3 = 25$ (5)

2. $8x - 3 = 5x + 9$ (4)

Solve the following equations and check your results. Express your answer in set-builder notation.

3. $7x - 12 = 6x$

4. $\frac{4}{5}x = 24$

5. $2x - 7 = 5x - 8$

6. $5x - 3(x - 5) = 19$

7. $\frac{x - 5}{3} = \frac{5}{4}$

8. $|2x - 5| = 9$

Solve for the indicated variable.

9. $V = \frac{1}{3}Bh$ (for B)

Solve the following word problems.

10. 5 times a number, decreased by 7, is 28. What is the number?

11. The sum of three consecutive integers is 66. Find the three integers.

12. Jan is twice as old as Juwan, while Rick is 5 years older than Jan. If the sum of their ages is 35 years, find each of their ages.

13. The perimeter of a rectangle is 62 inches (in.). If the length of the rectangle is 1 in. more than twice its width, what are the dimensions of the rectangle?

14. At 10 A.M., Sandra left her house on a business trip and drove an average of 45 mi/h. One hour later, Adam discovered that Sandra had left her briefcase behind, and he began driving at 55 mi/h along the same route. When will Adam catch up with Sandra?

Solve and graph the solution sets for the following inequalities.

15. $x - 5 \leq 9$

16. $5 - 3x > 17$

17. $6 \leq x + 5 \leq 9$

18. $3x + 7 < -8$ or $-4x + 5 < -11$

19. $|x - 1| < 4$

20. $|2x + 1| \geq 7$

Cumulative Test ▪ 0–2

This test is provided to help you in the process of reviewing the previous chapters. Answers are provided in the back of the book. If you missed any answers, be sure to go back and review the appropriate chapter sections.

Use the fundamental principle of fractions to simplify each fraction.

1. $\dfrac{56}{88}$

2. $\dfrac{132}{110}$

In Exercises 3 to 16, perform the indicated operations. Write each answer in simplest form.

3. $2 \cdot 3^2 - 8 \cdot 2$

4. $5(7 - 3)^2$

5. $|12 - 5|$

6. $|12| - |5|$

7. $(-7) + (-9)$

8. $\dfrac{17}{3} + \left(-\dfrac{5}{3}\right)$

9. $(-7)(-9)$

10. $(-3.2)(5)$

11. $\dfrac{0}{-13}$

12. $8 - 12 \div 2 \cdot 3 + 5$

13. $5 - 4^2 \div (-8) \cdot 2$

14. $\dfrac{4}{9} \times \dfrac{27}{36}$

15. $\dfrac{3}{4} + \dfrac{5}{6}$

16. $\dfrac{5}{6} \div \dfrac{25}{21}$

Evaluate each of the following expressions if $x = -2$, $y = 3$, and $z = 5$.

17. $3x - y$

18. $4x^2 - y$

19. $\dfrac{5z - 4x}{2y + z}$

20. $-y^2 - 8x$

Simplify and combine like terms.

21. $7x - 3y + 2(4x - 3y)$

22. $6x^2 - (5x - 4x^2 + 7) - 8x + 9$

Solve the following equations.

23. $12x - 3 = 10x + 5$

24. $|x - 3| = 5$

25. $\dfrac{x - 2}{3} - \dfrac{x + 1}{4} = 5$

26. $4(x - 1) - 2(x - 5) = 14$

Solve the following inequalities.

27. $7x + 5 \leq 4x - 7$ **28.** $-5 \leq 2x + 1 < 7$ **29.** $|x - 1| \leq 4$ **30.** $|x + 1| > 8$

Solve the following equations for the indicated variable.

31. $I = Prt$ (for r)   **32.** $A = \dfrac{1}{2}bh$ (for h)   **33.** $ax + by = c$ (for y)   **34.** $P = 2L + 2W$ (for W)

Solve the following word problems. Be sure to show the equation used for the solution.

35. If 4 times a number decreased by 7 is 45, find that number.

36. The sum of two consecutive integers is 85. What are those two integers?

37. If 3 times an odd integer is 12 more than the next consecutive odd integer, what is that integer?

38. Michelle earns $120 more per week than Dmitri. If their weekly salaries total $720, how much does Michelle earn?

39. The length of a rectangle is 2 centimeters (cm) more than 3 times its width. If the perimeter of the rectangle is 44 cm, what are the dimensions of the rectangle?

40. One side of a triangle is 5 in. longer than the shortest side. The third side is twice the length of the shortest side. If the triangle perimeter is 37 in., find the length of each leg.

3 Graphs and Linear Equations

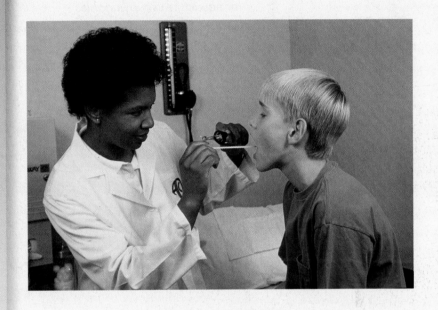

LIST OF SECTIONS

Graphs are used to discern patterns and trends that may be difficult to see when one is looking at a list of numbers or other kinds of data. The word *graph* comes from Latin and Greek roots and means "to draw a picture." This is what a graph does in mathematics: It draws a picture of a relationship between two or more variables.

In the field of pediatric medicine, there has been controversy about the use of somatotropin (human growth hormone) to help children whose growth has been impeded by various health problems. These children must be distinguished from children who are healthy and simply small of stature and thus should not be subjected to this treatment. Some of the measures used to distinguish between the two groups are blood tests and age and height measurements. The age and height measurements are graphed and monitored over several years of a child's life to monitor the rate of growth. If during a certain period the child's rate of growth slows to below 4.5 centimeters per year (cm/yr), this indicates that something may be seriously wrong. The graph can also indicate whether the child's size fits within a range considered normal at each age of the child's life.

3.1 Solutions of Equations in Two Variables

 OBJECTIVES

1. *Identify solutions for an equation in two variables*
2. *Use the ordered-pair notation to write solutions for equations in two variables*

We discussed finding solutions for equations in Section 2.1. Recall that a solution is a value for the variable that "satisfies" the equation, or makes the equation a true statement. For example, we know that 4 is a solution of the following equation.

$$2x + 5 = 13$$

We know this is true because when we replace x with 4, we have

$$2 \cdot 4 + 5 \stackrel{?}{=} 13$$

$$8 + 5 \stackrel{?}{=} 13$$

$$13 = 13 \qquad \text{A true statement}$$

We now want to consider **equations in two variables.** An example is

$$x + y = 5$$

> Recall that an equation is two expressions connected by an equal sign.

What will a solution look like? It is not going to be a single number, because there are two variables. Here a solution will be a pair of numbers—one value for each of the variables x and y. Suppose that x has the value 3. In the equation $x + y = 5$, you can substitute 3 for x.

$$3 + y = 5$$

Solving for y gives

$$y = 2$$

> An equation in two variables "pairs" two numbers, one for x and one for y.

So the pair of values $x = 3$ and $y = 2$ satisfies the equation because

$$3 + 2 = 5$$

That pair of numbers is then a *solution* for the equation in two variables.

> An **equation in two variables** is an equation for which *every* solution is a pair of values.

214

How many such pairs are there? Choose any value for *x* (or for *y*). You can always find the other *paired* or *corresponding* value in an equation of this form. We say that there are an *infinite* number of pairs that will satisfy the equation. Each of these pairs is a solution. We will find some other solutions for the equation $x + y = 5$ in Example 1.

Example 1

Solving for Corresponding Values

For the equation $x + y = 5$, find (a) *y* if $x = 5$ and (b) *x* if $y = 4$.

(a) If $x = 5$,

$$5 + y = 5, \qquad \text{so} \qquad y = 0$$

(b) If $y = 4$,

$$x + 4 = 5, \qquad \text{so} \qquad x = 1$$

So the pairs $x = 5$, $y = 0$ and $x = 1$, $y = 4$ are both solutions.

✔ *CHECK YOURSELF 1*

You are given the equation $2x + 3y = 26$.

(a) If $x = 4$, $y = ?$ (b) If $y = 0$, $x = ?$

To simplify writing the pairs that satisfy an equation, we use the **ordered-pair notation.** The numbers are written in parentheses and are separated by a comma. For example, we know that the values $x = 3$ and $y = 2$ satisfy the equation $x + y = 5$. So we write the pair as

$(3, 2)$

The *x* value   The *y* value

Caution

(3, 2) means $x = 3$ and $y = 2$. (2, 3) means $x = 2$ and $y = 3$. (3, 2) and (2, 3) are entirely different. That's why we call them *ordered pairs.*

The first number of the pair is *always* the value for *x* and is called the *x* **coordinate.** The second number of the pair is *always* the value for *y* and is the **y coordinate.**

Using this ordered-pair notation, we can say that (3, 2), (5, 0), and (1, 4) are all *solutions* for the equation $x + y = 5$. Each pair gives values for *x* and *y* that will satisfy the equation.

Example 2

Identifying Solutions of Two-Variable Equations

Which of the ordered pairs (2, 5), (5, −1), and (3, 4) are solutions for the equation $2x + y = 9$?

(a) To check whether $(2, 5)$ is a solution, let $x = 2$ and $y = 5$ and see if the equation is satisfied.

$$2x + y = 9 \qquad \text{Substitute 2 for } x \text{ and 5 for } y.$$

$$2 \cdot 2 + 5 \overset{?}{=} 9$$

$$4 + 5 \overset{?}{=} 9$$

$$9 = 9 \qquad \text{A true statement}$$

(2, 5) is a solution because a true statement results.

So $(2, 5)$ is a solution for the equation $2x + y = 9$.

(b) For $(5, -1)$, let $x = 5$ and $y = -1$.

$$2 \cdot 5 - 1 \overset{?}{=} 9$$

$$10 - 1 \overset{?}{=} 9$$

$$9 = 9 \qquad \text{A true statement}$$

So $(5, -1)$ is a solution for $2x + y = 9$.

(c) For $(3, 4)$, let $x = 3$ and $y = 4$. Then

$$2 \cdot 3 + 4 \overset{?}{=} 9$$

$$6 + 4 \overset{?}{=} 9$$

$$10 = 9 \qquad \textit{Not} \text{ a true statement}$$

So $(3, 4)$ is *not* a solution for the equation.

✔ **CHECK YOURSELF 2**

Which of the ordered pairs $(3, 4)$, $(4, 3)$, $(1, -2)$, and $(0, -5)$ are solutions for the following equation?

$$3x - y = 5$$

Equations such as those seen in Examples 1 and 2 are said to be in **standard form.**

A linear equation in two variables is in **standard form** if it is written as

$$Ax + By = C \qquad \text{where } A \text{ and } B \text{ are not both 0.}$$

Note, for example, that if $A = 1$, $B = 1$, and $C = 5$, we have

$$1 \cdot x + 1 \cdot y = 5$$

$$x + y = 5$$

which is the equation of Example 1.

It is therefore possible to view an equation in one variable as a two-variable equation. For example, if we have the equation $x = 2$, we can view this in standard form as

$$1 \cdot x + 0 \cdot y = 2$$

and we may search for ordered-pair solutions. The key is this: If the equation contains only one variable (in this case x), then the missing variable (in this case y) can take on any value. Consider Example 3.

| **Example 3** | ## Identifying Solutions of One-Variable Equations |

Which of the ordered pairs $(2, 0)$, $(0, 2)$, $(5, 2)$, $(2, 5)$, and $(2, -1)$ are solutions for the equation $x = 2$?

A solution is any ordered pair in which the x coordinate is 2. That makes $(2, 0)$, $(2, 5)$, and $(2, -1)$ solutions for the given equation.

✔ *CHECK YOURSELF 3*

Which of the ordered pairs $(3, 0)$, $(0, 3)$, $(3, 3)$, $(-1, 3)$, and $(3, -1)$ are solutions for the equation $y = 3$?

Remember that when an ordered pair is presented, the first number is always the x coordinate and the second number is always the y coordinate.

| **Example 4** | ## Completing Ordered-Pair Solutions |

Complete the ordered pairs $(9, \ )$, $(\ , -1)$, $(0, \ )$, and $(\ , 0)$ for the equation $x - 3y = 6$.

*The x coordinate is also called the **abscissa**, and the y coordinate is called the **ordinate**.*

(a) The first number, 9, appearing in $(9, \ )$ represents the x value. To complete the pair $(9, \ )$, substitute 9 for x and then solve for y.

$$9 - 3y = 6$$
$$-3y = -3$$
$$y = 1$$

The ordered pair $(9, 1)$ is a solution for $x - 3y = 6$.

(b) To complete the pair $(\ , -1)$, let y be -1 and solve for x.

$$x - 3(-1) = 6$$
$$x + 3 = 6$$
$$x = 3$$

The ordered pair $(3, -1)$ is a solution for the equation $x - 3y = 6$.

(c) To complete the pair $(0, \quad)$, let x be 0.

$$0 - 3y = 6$$
$$-3y = 6$$
$$y = -2$$

So $(0, -2)$ is a solution.

(d) To complete the pair $(\quad, 0)$, let y be 0.

$$x - 3 \cdot 0 = 6$$
$$x - 0 = 6$$
$$x = 6$$

Then $(6, 0)$ is a solution.

✔ *CHECK YOURSELF 4*

Complete the ordered pairs below so that each is a solution for the equation $2x + 5y = 10$.

$$(10, \quad), (\quad, 4), (0, \quad), \text{ and } (\quad, 0)$$

Example 5   **Finding Some Solutions of a Two-Variable Equation**

Find four solutions for the equation

$$2x + y = 8$$

Generally, you'll want to pick values for x (or for y) so that the resulting equation in one variable is easy to solve.

In this case the values used to form the solutions are *up to you.* You can assign any value for x (or for y). We'll demonstrate with some possible choices.

Solution with $x = 2$:

$$2x + y = 8$$
$$2 \cdot 2 + y = 8$$
$$4 + y = 8$$
$$y = 4$$

The ordered pair $(2, 4)$ is a solution for $2x + y = 8$.

Solution with $y = 6$:

$$2x + y = 8$$
$$2x + 6 = 8$$
$$2x = 2$$
$$x = 1$$

So $(1, 6)$ is also a solution for $2x + y = 8$.

Solution with $x = 0$:

$$2x + y = 8$$
$$2 \cdot 0 + y = 8$$
$$y = 8$$

The solutions $(0, 8)$ and $(4, 0)$ will have special significance later in graphing. They are also easy to find!

And $(0, 8)$ is a solution.

Solution with $y = 0$:

$$2x + y = 8$$
$$2x + 0 = 8$$
$$2x = 8$$
$$x = 4$$

So $(4, 0)$ is a solution.

✔ CHECK YOURSELF 5

Find four solutions for $x - 3y = 12$.

✔ CHECK YOURSELF ANSWERS

1. (a) $y = 6$; (b) $x = 13$ **2.** $(3, 4)$, $(1, -2)$, and $(0, -5)$ are solutions.

3. $(0, 3)$, $(3, 3)$, and $(-1, 3)$ are solutions.   **4.** $(10, -2)$, $(-5, 4)$, $(0, 2)$, and $(5, 0)$

5. $(6, -2)$, $(3, -3)$, $(0, -4)$, and $(12, 0)$ are four possibilities.

Exercises · 3.1

Determine which of the ordered pairs are solutions for the given equation.

1. $x + y = 6$ $(4, 2), (-2, 4), (0, 6), (-3, 9)$

2. $x - y = 10$ $(11, 1), (11, -1), (10, 0), (5, 7)$

3. $2x - y = 8$ $(5, 2), (4, 0), (0, 8), (6, 4)$

4. $x + 5y = 20$ $(10, -2), (10, 2), (20, 0), (25, -1)$

5. $4x + y = 8$ $(2, 0), (2, 3), (0, 2), (1, 4)$

6. $x - 2y = 8$ $(8, 0), (0, 4), (5, -1), (10, -1)$

7. $2x - 3y = 6$ $(0, 2), (3, 0), (6, 2), (0, -2)$

8. $6x + 2y = 12$ $(0, 6), (2, 6), (2, 0), (1, 3)$

9. $3x - 2y = 12$ $(4, 0), \left(\dfrac{2}{3}, -5\right), (0, 6), \left(5, \dfrac{3}{2}\right)$

10. $3x + 4y = 12$ $(-4, 0), \left(\dfrac{2}{3}, \dfrac{5}{2}\right), (0, 3), \left(\dfrac{2}{3}, 2\right)$

11. $3x + 5y = 15$ $(0, 3), \left(1, \dfrac{12}{5}\right), (5, 3)$

12. $y = 2x - 1$ $(0, -2), (0, -1), \left(\dfrac{1}{2}, 0\right), (3, -5)$

13. $x = 3$ $(3, 5), (0, 3), (3, 0), (3, 7)$

14. $y = 7$ $(0, 7), (3, 7), (-1, -4), (7, 7)$

Complete the ordered pairs so that each is a solution for the given equation.

15. $x + y = 12$ $(4, \ ), (\ , 5), (0, \ ), (\ , 0)$

16. $x - y = 7$ $(\ , 4), (15, \ ), (0, \ ), (\ , 0)$

17. $3x + y = 9$ $(3, \ ), (\ , 9), (\ , -3), (0, \ )$

18. $x + 4y = 12$ $(0, \ ), (\ , 2), (8, \ ), (\ , 0)$

19. $5x - y = 15$ $(\ , 0), (2, \ ), (4, \ ), (\ , -5)$

20. $x - 3y = 9$ $(0, \ ), (12, \ ), (\ , 0), (\ , -2)$

21. $4x - 2y = 16$ $(\ , 0), (\ , -6), (2, \ ), (\ , 6)$

22. $2x + 5y = 20$ $(0, \ ), (5, \ ), (\ , 0), (\ , 6)$

23. $y = 3x + 9$ $(\ , 0), \left(\dfrac{2}{3}, \ \right), (0, \ ), \left(-\dfrac{2}{3}, \ \right)$

24. $6x + 8y = 24$ $(0, \ ), \left(\ , \dfrac{3}{4}\right), (\ , 0), \left(-\dfrac{2}{3}, \ \right)$

25. $y = 3x - 4$ $(0, \ ), (\ , 5), (\ , 0), \left(\dfrac{5}{3}, \ \right)$

26. $y = -2x + 5$ $(0, \ ), (\ , 5), \left(\dfrac{3}{2}, \ \right), (\ , 1)$

220

Find four solutions for each of the following equations. *Note:* Your answers may vary from those shown in the answer section.

27. $x - y = 10$

28. $x + y = 18$

29. $2x - y = 6$

30. $4x - 2y = 8$

31. $x + 4y = 8$

32. $x + 3y = 12$

33. $5x - 2y = 10$

34. $2x + 7y = 14$

35. $y = 2x + 3$

36. $y = 5x - 8$

37. $x = -5$

38. $y = 8$

An equation in three variables has an ordered triple as a solution. For example, $(1, 2, 2)$ is a solution to the equation $x + 2y - z = 3$. Complete the ordered-triple solutions for each equation.

39. $x + y + z = 0$ $(2, -3, \ )$

40. $2x + y + z = 2$ $(, -1, 3)$

41. $x + y + z = 0$ $(1, \ , 5)$

42. $x + y - z = 1$ $(4, \ , 3)$

43. $2x + y + z = 2$ $(-2, \ , 1)$

44. $x + y - z = 1$ $(-2, 1, \ )$

45. Hourly wages. When an employee produces x units per hour, the hourly wage is given by $y = 0.75x + 8$. What are the hourly wages for the following number of units: 2, 5, 10, 15, and 20?

46. Temperature conversion. Celsius temperature readings can be converted to Fahrenheit readings by using the formula $F = \frac{9}{5}C + 32$. What is the Fahrenheit temperature that corresponds to each of the following Celsius temperatures: $-10, 0, 15, 100$?

47. Area. The area of a square is given by $A = s^2$. What is the area of the squares whose sides are 4 cm, 11 cm, 14 cm, and 17 cm?

48. Unit pricing. When x number of units is sold, the price of each unit is given by $p = 4x + 15$. Find the unit price in dollars when the following quantities are sold: 1, 5, 10, 12.

49. You now have had practice solving equations with one variable and equations with two variables. Compare equations with one variable to equations with two variables. How are they alike? How are they different?

50. Each of the following sentences describes pairs of numbers that are related. After completing the sentences in parts (a) to (g), write two of your own sentences in (h) and (i).

  (a) The *number of hours you work* determines the *amount you are* _____.

  (b) The *number of gallons of gasoline* you put in your car determines *the amount you* _____.

  (c) The *amount of the* _____ in a restaurant is related to *the amount of the tip*.

  (d) The *sales amount of a purchase in a store* determines _____.

  (e) The *age of an automobile* is related to _____.

  (f) The *amount of electricity you use in a month* determines _____.

  (g) The *cost of food for a family* is related to _____.

  Think of two more:

  (h) _____.

  (i) _____.

3.2 The Cartesian Coordinate System

1. Plot ordered pairs
2. Identify plotted points
3. Scale the axes
4. Find the midpoint

In Section 3.1, we saw that ordered pairs could be used to write solutions to equations in two variables. The next step is to graph those ordered pairs as points in a plane.

Since there are two numbers (one for x and one for y), we will need two number lines. One line is drawn horizontally, and the other is drawn vertically; their point of intersection (at their respective zero points) is called the *origin*. The horizontal line is called the ***x* axis,** while the vertical line is called the ***y* axis.** Together the lines form the **rectangular coordinate system.**

The axes (pronounced "axees") divide the plane into four regions called **quadrants,** which are numbered (usually by Roman numerals) counterclockwise from the upper right.

This system is also called the **Cartesian coordinate system,** named in honor of its inventor, René Descartes (1596–1650), a French mathematician and philosopher.

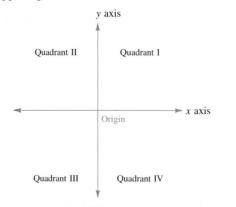

The origin is the point with coordinates $(0, 0)$.

We now want to establish correspondences between ordered pairs of numbers (x, y) and points in the plane.

For any ordered pair,

$$(x, y)$$

x coordinate y coordinate

the following are true:

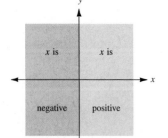

1. If the x coordinate is

   Positive, the point corresponding to that pair is located x units to the *right* of the y axis.

   Negative, the point is x units to the *left* of the y axis.

   Zero, the point is on the y axis.

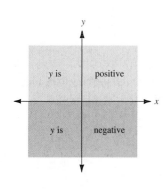

2. If the y coordinate is

    Positive, the point is y units *above* the x axis.

    Negative, the point is y units *below* the x axis.

    Zero, the point is on the x axis.

Example 1 illustrates how to use these guidelines to match coordinates with points in the plane.

Example 1   Identifying the Coordinates for a Given Point

Give the coordinates for the given points. Assume that each tick mark represents 1 unit.

Remember: The x coordinate gives the *horizontal* distance from the y axis. The y coordinate gives the *vertical* distance from the x axis.

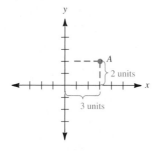

(a) Point A is 3 units to the *right* of the y axis and 2 units *above* the x axis. Point A has coordinates (3, 2).

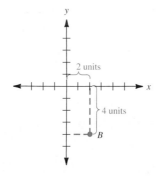

(b) Point B is 2 units to the *right* of the y axis and 4 units *below* the x axis. Point B has coordinates (2, −4).

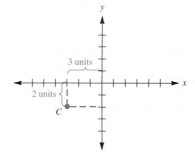

(c) Point C is 3 units to the *left* of the y axis and 2 units *below* the x axis. Point C has coordinates (−3, −2).

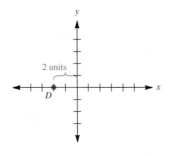

(d) Point *D* is 2 units to the *left* of the *y* axis and *on* the *x* axis. Point *D* has coordinates $(-2, 0)$.

✔ CHECK YOURSELF 1

Give the coordinates of points *P*, *Q*, *R*, and *S*.

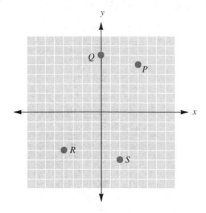

Reversing the process used in Example 1 will allow us to graph (or plot) a point in the plane, given the coordinates of the point. You can use the following steps.

The graphing of individual points is sometimes called **point plotting.**

To Graph a Point in the Plane

Step 1.   Start at the origin.

Step 2.   Move right or left according to the value of the *x* coordinate.

Step 3.   Move up or down according to the value of the *y* coordinate.

Example 2   Graphing Points

(a) Graph the point corresponding to the ordered pair (4, 3).

Move 4 units to the right on the *x* axis. Then move 3 units up from the point you stopped at on the *x* axis. This locates the point corresponding to (4, 3).

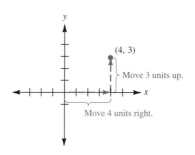

(b) Graph the point corresponding to the ordered pair $(-5, 2)$.

In this case move 5 units *left* (because the *x* coordinate is negative) and then 2 units *up*.

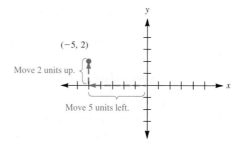

(c) Graph the point corresponding to $(-4, -2)$.

Here move 4 units *left* and then 2 units *down* (the *y* coordinate is negative).

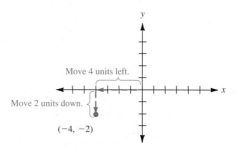

Any point on an axis will have 0 for one of its coordinates.

(d) Graph the point corresponding to $(0, -3)$.

There is *no* horizontal movement because the *x* coordinate is 0. Move 3 units *down*.

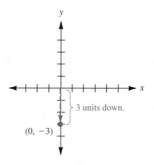

(e) Graph the point corresponding to (5, 0).

Move 5 units *right.* The desired point is on the *x* axis because the *y* coordinate is 0.

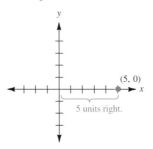

✔ *CHECK YOURSELF 2*

Graph the points corresponding to $M(4, 3)$, $N(-2, 4)$, $P(-5, -3)$, and $Q(0, -3)$.

Scaling the Axes

The same decisions must be made when you are using a graphing calculator. When graphing this kind of relation on a calculator, you must decide what the appropriate **viewing window** should be.

It is not necessary, or even desirable, to always use the same scale on both the *x* and *y* axes. For example, if we were plotting ordered pairs in which the first value represented the age of a used car and the second value represented the number of miles driven, it would be necessary to have a different scale on the two axes. If not, the following extreme cases could happen.

Assume that the cars range in age from 1 to 15 years. The cars have mileages from 2,000 to 150,000 miles (mi). If we used the same scale on both axes, 0.5 in. between each two counting numbers, how large would the paper have to be on which the points were plotted? The horizontal axis would have to be 15(0.5) = 7.5 in. The vertical axis would have to be 150,000(0.5) = 75,000 in. = 6,250 feet (ft) = almost 1.2 mi long!

So what do we do? We simply use a different, but clearly marked, scale on the axes. In this case, the horizontal axis could be marked in 10s, but the vertical axis would be marked in 50,000s. Additionally, all the numbers would be positive, so we really need only the first quadrant in which *x* and *y* are both always positive. We could scale the axes like this:

Every ordered pair is either in one of the quadrants or on one of the axes.

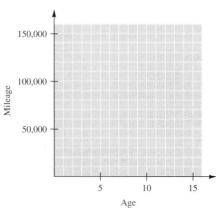

Example 3	**Scaling the Axes**

A survey of residents in a large apartment building was recently taken. The following points represent ordered pairs in which the first number is the number of years of education a person has had, and the second number is his or her year 2000 income (in thousands of dollars). Estimate, and interpret, each ordered pair represented.

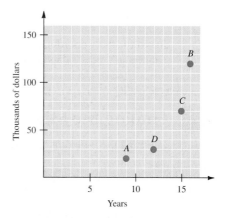

Point A is (9, 20), B is (16, 120), C is (15, 70), and D is (12, 30). Person A completed 9 years of education and made $20,000 in 2000. Person B completed 16 years of education and made $120,000 in 2000. Person C had 15 years' education and made $70,000. Person D had 12 years' and made $30,000.

Note that there is no obvious "relation" that would allow one to predict income from years of education, but you might suspect that in most cases, more education results in more income.

✔ CHECK YOURSELF 3

Each six months, Armand records his son's weight. The following points represent ordered pairs in which the first number represents his son's age and the second number represents his son's weight. For example, point A indicates that when his son was 1 year old, the boy weighed 14 pounds. Estimate each ordered pair represented.

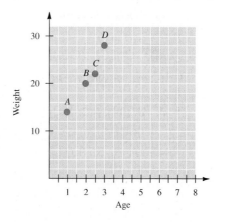

Changing the scales on the axes is definitely an important issue when you are using a graphing calculator. It is called *setting the viewing window* and is a skill that you will develop throughout this course.

Every two points have a *midpoint*. The midpoint of A and B is the point on line AB that is an equal distance from A and B. The following formula can be used to find the midpoint M.

$$M = \left(\frac{x_1 + x_2}{2}, \frac{y_1 + y_2}{2} \right)$$

In words, the x coordinate of the midpoint is the average of the two x values, and the y coordinate of the midpoint is the average of the two y values.

Example 4   **Finding the Midpoint**

(a) Find the midpoint of $(2, 0)$ and $(10, 0)$.

$$M = \left(\frac{2 + 10}{2}, \frac{0 + 0}{2} \right) = \left(\frac{12}{2}, \frac{0}{2} \right) = (6, 0)$$

(b) Find the midpoint of $(5, 7)$ and $(-1, -3)$.

$$M = \left(\frac{5 + (-1)}{2}, \frac{7 + (-3)}{3} \right) = \left(\frac{4}{2}, \frac{4}{2} \right) = (2, 2)$$

(c) Find the midpoint of $(-7, 3)$ and $(8, 6)$.

$$M = \left(\frac{-7 + 8}{2}, \frac{3 + 6}{2} \right) = \left(\frac{1}{2}, \frac{9}{2} \right)$$

✔ CHECK YOURSELF 4

Find the midpoint for each pair of points.

(a) $(2, 7)$ and $(-4, -1)$ (b) $(-3, -6)$ and $(5, -2)$ (c) $(-2, 3)$ and $(3, 8)$

✔ CHECK YOURSELF ANSWERS

1. $P(4, 5)$, $Q(0, 6)$, $R(-4, -4)$, and $S(2, -5)$ **2.**

3. $A(1, 14)$, $B(2, 20)$, $C\left(\frac{5}{2}, 22 \right)$, and $D(3, 28)$

4. (a) $(-1, 3)$; (b) $(1, -4)$; (c) $\left(\frac{1}{2}, \frac{11}{2} \right)$

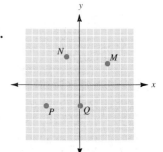

Exercises · 3.2

Give the coordinates of the points graphed below.

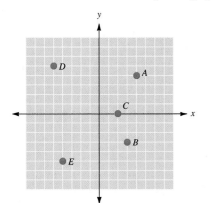

1. *A*
2. *B*
3. *C*
4. *D*
5. *E*

Give the coordinates of the points graphed below.

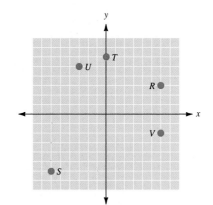

6. *R*
7. *S*
8. *T*
9. *U*
10. *V*

Plot the following points on the rectangular coordinate system.

11. $M(5, 3)$ **12.** $N(0, -3)$ **13.** $P(-4, 5)$ **14.** $Q(5, 0)$

15. $R(-4, -6)$ **16.** $S(-4, -3)$ **17.** $F(-3, -1)$ **18.** $G(4, 3)$

19. $H(4, -3)$ **20.** $I(-3, 0)$ **21.** $J(-5, 3)$ **22.** $K(0, 4)$

In Exercises 23 to 34 give the quadrant in which each of the following points is located or the axis on which the point lies.

23. $(4, 5)$ **24.** $(-3, 2)$ **25.** $(-6, -8)$ **26.** $(2, -4)$

27. $(5, 0)$ **28.** $(-1, 11)$ **29.** $(-4, 7)$ **30.** $(-3, -7)$

31. $(0, -4)$ **32.** $(-3, 0)$ **33.** $\left(5\frac{3}{4}, -3\right)$ **34.** $\left(-3, 4\frac{2}{3}\right)$

35. A company has kept a record of the number of items produced by an employee as the number of days on the job increases. In the following figure, points correspond to an ordered-pair relationship in which the first number represents days on the job and the second number represents the number of items produced. Estimate each ordered pair produced.

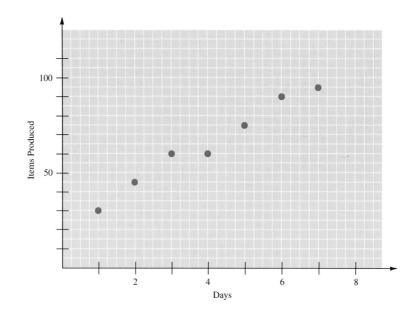

36. In the following figure, points correspond to an ordered-pair relationship between height and age in which the first number represents age and the second number represents height. Estimate each ordered pair represented.

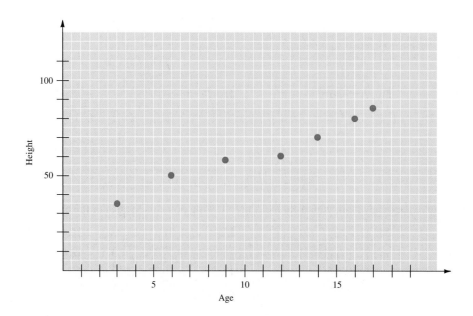

37. An unidentified company has kept a record of the number of hours devoted to safety training and the number of work hours lost due to on-the-job accidents. In the following figure, the points correspond to an ordered-pair relationship in which the first number represents hours in safety training and the second number represents hours lost by accidents. Estimate each ordered pair represented.

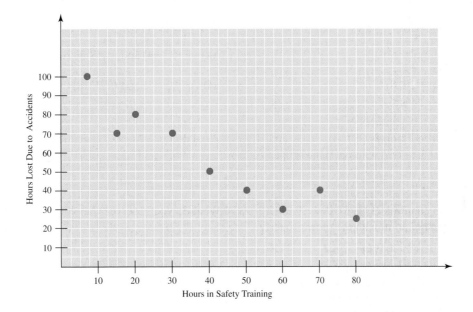

38. In the following figure, points correspond to an ordered-pair relationship between the age of a person and the annual average number of visits to doctors and dentists for a person that age. The first number represents the age, and the second number represents the number of visits. Estimate each ordered pair represented.

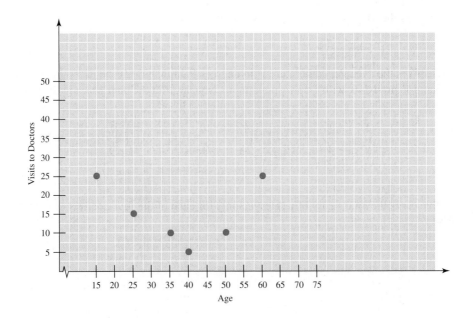

39. Find the midpoint for each pair of points.

    (a) $(3, 2)$ and $(11, 2)$

    (b) $(4, 5)$ and $(-4, -3)$

    (c) $(-5, -3)$ and $(7, -9)$

    (d) $(-4, 5)$ and $(7, -2)$

    (e) $(-1, -5)$ and $(-7, -10)$

    (f) $(-4, -9)$ and $(-20, 8)$

40. Find the midpoint for each pair of points.

    (a) $(4, 3)$ and $(2, 3)$

    (b) $(7, 5)$ and $(-7, -3)$

    (c) $(-8, -4)$ and $(6, -8)$

    (d) $(-5, 8)$ and $(6, -1)$

    (e) $(-6, -4)$ and $(-17, -20)$

    (f) $(-9, -1)$ and $(-2, 18)$

41. Graph points with coordinates $(-1, 3)$, $(0, 0)$, and $(1, -3)$. What do you observe? Can you give the coordinates of another point with the same property?

42. Graph points with coordinates $(1, 5)$, $(-1, 3)$, and $(-3, 1)$. What do you observe? Can you give the coordinates of another point with the same property?

43. Environment. A plastics company is sponsoring a plastics recycling contest for the local community. The focus of the contest is on collecting plastic milk, juice, and water jugs. The company will award $200, plus the current market price of the jugs collected, to the group that collects the most jugs in a single month. The number of jugs collected and the amount of money won can be represented as an ordered pair.

    (a) In April, group A collected 1,500 pounds (lb) of jugs to win first place. The prize for the month was $350. If x represents the pounds of jugs and y represents the amount of money that the group won, graph the point that represents the winner for April.

    (b) In May, group B collected 2,300 lb of jugs to win first place. The prize for the month was $430. Graph the point that represents the May winner on the same grid you used in part (a).

44. Science. The table gives the average temperature y (in degrees Fahrenheit) for the first 6 months of the year x. The months are numbered 1 through 6, with 1 corresponding to January. Plot the data given in the table.

x	1	2	3	4	5	6
y	4	14	26	33	42	51

45. Business. The table gives the total salary of a salesperson y for each of the four quarters of the year x. Plot the data given in the table.

x	1	2	3	4
y	$6,000	$5,000	$8,000	$9,000

46. Although high employment is a measure of a country's economic vitality, economists worry that periods of low unemployment will lead to inflation. Look at the following table.

Year	Unemployment Rate (%)	Inflation Rate (%)
1965	4.5	1.6
1970	4.9	5.7
1975	8.5	9.1
1980	7.1	13.5
1985	7.2	3.6
1990	5.5	5.4
1995	5.6	2.5
2000	3.8	3.2

Plot the figures in the table with unemployment rates on the x axis and inflation rates on the y axis. What do these plots tell you? Do higher inflation rates seem to be associated with lower unemployment rates? Explain.

47. We mentioned that the Cartesian coordinate system was named for the French philosopher and mathematician René Descartes. What philosophy book is Descartes most famous for? Use an encyclopedia as a reference.

48. What characteristic is common to all points on the x axis? On the y axis?

49. How would you describe a rectangular coordinate system? Explain what information is needed to locate a point in a coordinate system.

50. Some newspapers have a special day that they devote to automobile want ads. Use this special section or the Sunday classified ads from your local newspaper to find all the want ads for a particular automobile model. Make a list of the model year and asking price for 10 ads, being sure to get a variety of ages for this model. After collecting the information, make a plot of the age and the asking price for the car.

Describe your graph, including an explanation of how you decided which variable to put on the vertical axis and which on the horizontal axis. What trends or other information are given by the graph?

3.3

The Graph of a Linear Equation

 3.3 **OBJECTIVES**

1. Graph a linear equation by plotting points
2. Graph a linear equation by the intercept method
3. Graph a linear equation by solving the equation for y

"I think there is a world market for maybe five computers."

—Thomas Watson (IBM Chairman) in 1943

"640K ought to be enough for anybody."

—Bill Gates (Microsoft Chairman) in 1981

In Section 3.1, you learned to write the solutions of equations in two variables as ordered pairs. In Section 3.2, ordered pairs were graphed in the Cartesian plane. Putting these ideas together will help us graph certain equations. Example 1 illustrates one approach to finding the graph of a linear equation.

| Example 1 | Graphing a Linear Equation |

Graph $x + 2y = 4$.

Step 1 Find some solutions for $x + 2y = 4$. To find solutions, we choose any convenient values for x, say $x = 0$, $x = 2$, and $x = 4$. Given these values for x, we can substitute and then solve for the corresponding value for y.

We are going to find *three* solutions for the equation. We'll point out why shortly.

If $x = 0$, then $y = 2$, so $(0, 2)$ is a solution.

If $x = 2$, then $y = 1$, so $(2, 1)$ is a solution.

If $x = 4$, then $y = 0$, so $(4, 0)$ is a solution.

A handy way to show this information is in a table such as this:

The table is a convenient way to display the information. It is the same as writing $(0, 2)$, $(2, 1)$, and $(4, 0)$.

x	y
0	2
2	1
4	0

234

Step 2   We now graph the solutions found in step 1.

$$x + 2y = 4$$

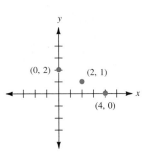

x	y
0	2
2	1
4	0

What pattern do you see? It appears that the three points lie on a straight line, and that is in fact the case.

Step 3   Draw a straight line through the three points graphed in step 2.

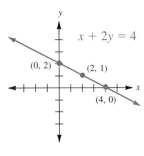

The arrowheads on the end of the line mean that the line extends infinitely in either direction.

The line shown is the **graph** of the equation $x + 2y = 4$. It represents *all* the ordered pairs that are solutions (an infinite number) for that equation.

The graph is a "picture" of the solutions for the given equation.

    Every ordered pair that is a solution will be plotted as a point on this line. Any point on the line will represent a pair of numbers that is a solution for the equation.

    Note: Why did we suggest finding *three* solutions in step 1? Two points determine a line, so technically you need only two. The third point that we find is a check to catch any possible errors.

   CHECK YOURSELF 1

Graph $2x - y = 6$, using the steps shown in Example 1.

As mentioned in Section 3.2, an equation that can be written in the form

$$Ax + By = C \qquad \text{where } A \text{ and } B \text{ are not both } 0$$

is called a linear equation in two variables **in standard form.** The graph of this equation is a *line.* That is why we call it a *linear* equation.

The steps of graphing follow.

> ### To Graph a Linear Equation
>
> Step 1.   Find at least three solutions for the equation, and put your results in tabular form.
>
> Step 2.   Graph the solutions found in step 1.
>
> Step 3.   Draw a straight line through the points determined in step 2 to form the graph of the equation.

Example 2	## Graphing a Linear Equation

Graph $y = 3x$.

Step 1   Some solutions are

Let $x = 0$, 1, and 2, and substitute to determine the corresponding y values. Again the choices for x are simply convenient. Other values for x would serve the same purpose.

x	y
0	0
1	3
2	6

Step 2   Graph the points.

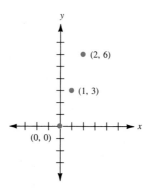

Notice that connecting any two of these points produces the same line.

Step 3   Draw a line through the points.

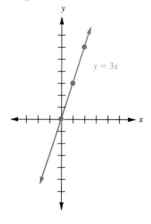

✔ *CHECK YOURSELF 2*

Graph the equation $y = -2x$ after completing the table of values.

x	y
0	
1	
2	

Let's work through another example of graphing a line from its equation.

Example 3	**Graphing a Linear Equation**

Graph $y = 2x + 3$.

Step 1   Some solutions are

x	y
0	3
1	5
2	7

Step 2   Graph the points corresponding to these values.

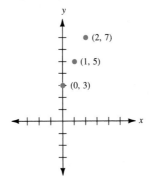

Step 3   Draw a line through the points.

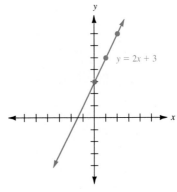

✔ CHECK YOURSELF 3

Graph the equation $y = 3x - 2$ after completing the table of values.

x	y
0	
1	
2	

In graphing equations, particularly when fractions are involved, a careful choice of values for x can simplify the process. Consider Example 4.

Example 4   Graphing a Linear Equation

Graph

$$y = \frac{3}{2}x - 2$$

As before, we want to find solutions for the given equation by picking convenient values for x. Note that in this case, choosing *multiples of 2*, the denominator of the x coefficient, will avoid fractional values for y and will make the plotting of those solutions much easier. For instance, here we might choose values of -2, 0, and 2 for x.

Step 1

If $x = -2$:

$$y = \frac{3}{2}x - 2$$
$$= \frac{3}{2}(-2) - 2$$
$$= -3 - 2 = -5$$

If $x = 0$:

$$y = \frac{3}{2}x - 2$$
$$= \frac{3}{2}(0) - 2$$
$$= 0 - 2 = -2$$

Suppose we do *not* choose a multiple of 2, say, $x = 3$. Then

$$y = \frac{3}{2}(3) - 2$$
$$= \frac{9}{2} - 2$$
$$= \frac{5}{2}$$

$\left(3, \frac{5}{2}\right)$ is still a valid solution, but we must graph a point with fractional coordinates.

If $x = 2$:

$$y = \frac{3}{2}x - 2$$
$$= \frac{3}{2}(2) - 2$$
$$= 3 - 2 = 1$$

In tabular form, the solutions are

x	y
-2	-5
0	-2
2	1

Step 2 Graph the points determined above.

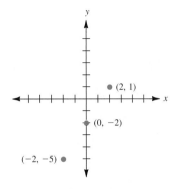

Step 3 Draw a line through the points.

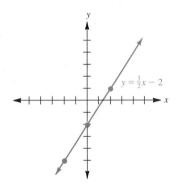

✔ *CHECK YOURSELF 4*

Graph the equation $y = -\dfrac{1}{3}x + 3$ after completing the table of values.

x	y
-3	
0	
3	

Some special cases of linear equations are illustrated in Examples 5 and 6.

| Example 5 | ## Graphing an Equation That Results in a Vertical Line |

Graph $x = 3$.

The equation $x = 3$ is equivalent to $1 \cdot x + 0 \cdot y = 3$. Let's look at some solutions.

If $y = 1$:	If $y = 4$:	If $y = -2$:
$x + 0 \cdot 1 = 3$	$x + 0 \cdot 4 = 3$	$x + 0(-2) = 3$
$x = 3$	$x = 3$	$x = 3$

In tabular form,

x	y
3	1
3	4
3	-2

What do you observe? The variable x has the value 3, regardless of the value of y. Look at the graph.

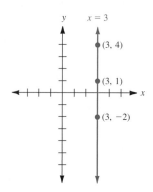

The graph of $x = 3$ is a vertical line crossing the x axis at $(3, 0)$.

Note that graphing (or plotting) points in this case is not really necessary. Simply recognize that the graph of $x = 3$ *must* be a vertical line (parallel to the y axis) which intercepts the x axis at $(3, 0)$.

 CHECK YOURSELF 5

Graph the equation $x = -2$.

Example 6 is a related example involving a horizontal line.

Example 6	**Graphing an Equation That Results in a Horizontal Line**

Graph $y = 4$.

Since $y = 4$ is equivalent to $0 \cdot x + 1 \cdot y = 4$, any value for x paired with 4 for y will form a solution. A table of values might be

x	y
-2	4
0	4
2	4

Here is the graph.

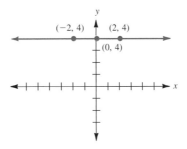

This time the graph is a horizontal line that crosses the y axis at $(0, 4)$. Again the graphing of points is not required. The graph of $y = 4$ *must* be horizontal (parallel to the x axis) and intercepts the y axis at $(0, 4)$.

  CHECK YOURSELF 6

Graph the equation $y = -3$.

The following box summarizes our work in Examples 5 and 6.

Vertical and Horizontal Lines

1. The graph of $x = a$ is a *vertical line* crossing the x axis at $(a, 0)$.

2. The graph of $y = b$ is a *horizontal line* crossing the y axis at $(0, b)$.

To simplify the graphing of certain linear equations, some students prefer the **intercept method** of graphing. This method makes use of the fact that the solutions that are easiest to find are those with an x coordinate or a y coordinate of 0. For instance, let's graph the equation

$$4x + 3y = 12$$

With practice, all this can be done mentally, which is the big advantage of this method.

First, let $x = 0$ and solve for y.

$$4x + 3y = 12$$
$$4 \cdot 0 + 3y = 12$$
$$3y = 12$$
$$y = 4$$

So $(0, 4)$ is one solution. Now let $y = 0$ and solve for x.

$$4x + 3y = 12$$
$$4x + 3 \cdot 0 = 12$$
$$4x = 12$$
$$x = 3$$

A second solution is $(3, 0)$.

The two points corresponding to these solutions can now be used to graph the equation.

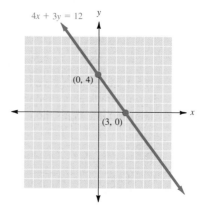

Remember, only two points are needed to graph a line. A third point is used only as a check.

The intercepts are the points where the line cuts the x and y axes. Here, the x intercept has coordinates (3, 0), and the y intercept has coordinates (0, 4).

The point $(3, 0)$ is called the **x intercept,** and the point $(0, 4)$ is the **y intercept** of the graph. Using these points to draw the graph gives the name to this method. Let's look at a second example of graphing by the intercept method.

| Example 7 | **Using the Intercept Method to Graph a Line** |

Graph $3x - 5y = 15$, using the intercept method.

To find the x intercept, let $y = 0$.

$$3x - 5 \cdot 0 = 15$$
$$x = 5$$

The x value of the intercept

To find the y intercept, let $x = 0$.

$$3 \cdot 0 - 5y = 15$$

$$y = -3$$

The y value of the intercept

So $(5, 0)$ and $(0, -3)$ are solutions for the equation, and we can use the corresponding points to graph the equation.

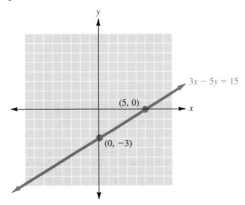

✔ *CHECK YOURSELF 7*

Graph $4x + 5y = 20$, using the intercept method.

Finding the third "checkpoint" is always a good idea.

This all looks quite easy, and for many equations it is. What are the drawbacks? For one, you don't have a third checkpoint, and it is possible for errors to occur. You can, of course, still find a third point (other than the two intercepts) to be sure your graph is correct. A second difficulty arises when the x and y intercepts are very close to each other (or are actually the same point—the origin). For instance, if we have the equation

$$3x + 2y = 1$$

the intercepts are $\left(\dfrac{1}{3}, 0\right)$ and $\left(0, \dfrac{1}{2}\right)$. It is hard to draw a line accurately through these intercepts, so choose other solutions farther away from the origin for your points.

Let's summarize the steps of graphing by the intercept method for appropriate equations.

Graphing a Line by the Intercept Method

Step 1.   To find the x intercept: Let $y = 0$, then solve for x.

Step 2.   To find the y intercept: Let $x = 0$, then solve for y.

Step 3.   Graph the x and y intercepts.

Step 4.   Draw a straight line through the intercepts.

A third method of graphing linear equations involves **solving the equation for y.** The reason we use this extra step is that often it will make finding solutions for the equation much easier. Let's look at an example.

| Example 8 | **Graphing a Linear Equation** |

Graph $2x + 3y = 6$.

Remember that solving for y means that we want to leave y isolated on the left.

Rather than finding solutions for the equation in this form, we solve for y.

$$2x + 3y = 6 \qquad \text{Subtract } 2x.$$

$$3y = 6 - 2x \qquad \text{Divide by 3.}$$

$$y = \frac{6 - 2x}{3}$$

We have solved for y. However, we will have reason in the coming sections to write this in a different form:

$$y = 2 - \frac{2}{3}x \qquad \text{We distributed the division.}$$

$$y = 2 + \left(-\frac{2}{3}x\right)$$

$$y = -\frac{2}{3}x + 2$$

Again, to pick convenient values for x, we suggest you look at the equation carefully. Here, for instance, picking multiples of 3 for x will make the work much easier.

Now find your solutions by picking convenient values for x.

If $x = -3$:

$$y = -\frac{2}{3}x + 2$$

$$= -\frac{2}{3}(-3) + 2$$

$$= 2 + 2 = 4$$

So $(-3, 4)$ is a solution.

If $x = 0$:

$$y = -\frac{2}{3}x + 2$$

$$= -\frac{2}{3}(0) + 2$$

$$= 0 + 2 = 2$$

So $(0, 2)$ is a solution.

If $x = 3$:

$$y = -\frac{2}{3}x + 2$$

$$= -\frac{2}{3}(3) + 2$$

$$= -2 + 2 = 0$$

So $(3, 0)$ is a solution.

We can now plot the points that correspond to these solutions and form the graph of the equation as before.

x	y
-3	4
0	2
3	0

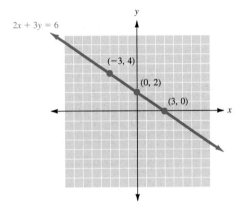

✔ CHECK YOURSELF 8

Graph the equation $5x + 2y = 10$. Solve for y to determine solutions.

x	y
0	
2	
4	

✔ CHECK YOURSELF ANSWERS

1.

x	y
1	-4
2	-2
3	0

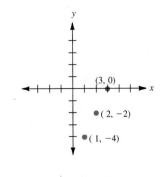

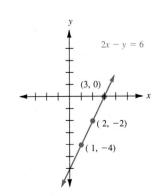

2.

x	y
0	0
1	−2
2	−4

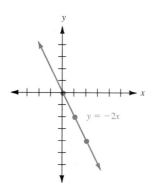

$y = -2x$

3.

x	y
0	−2
1	1
2	4

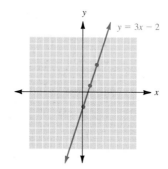

$y = 3x - 2$

4.

x	y
−3	4
0	3
3	2

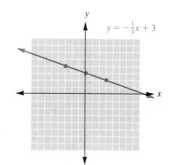

$y = -\frac{1}{3}x + 3$

5.

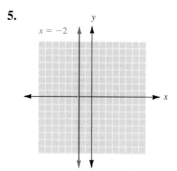

$x = -2$

6.

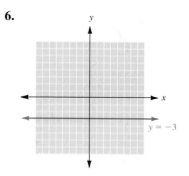

$y = -3$

7. $4x + 5y = 20$

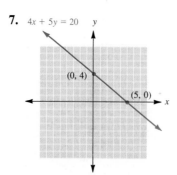

(0, 4)

(5, 0)

8. $y = -\frac{5}{2}x + 5$

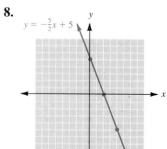

x	y
0	5
2	0
4	−5

Graph each of the following equations.

1. $x + y = 6$

2. $x - y = 5$

3. $x - y = -3$

4. $x + y = -3$

5. $3x + y = 6$

6. $x - 2y = 6$

7. $3x + y = 0$

8. $2x - y = 4$

9. $x + 4y = 8$

10. $2x - 3y = 6$

11. $y = 3x$

12. $y = -4x$

13. $y = 2x - 1$

14. $y = 2x + 5$

15. $y = -3x + 1$

16. $y = -3x - 3$

17. $y = \dfrac{1}{5}x$

18. $y = -\dfrac{1}{4}x$

19. $y = \dfrac{2}{3}x - 3$

20. $y = \dfrac{3}{4}x + 2$

21. $x = -3$

22. $y = -3$

23. $y = 1$

24. $x = 4$

25. $x - 2y = 4$ **26.** $6x + y = 6$ **27.** $5x + 2y = 10$

28. $2x + 3y = 6$ **29.** $3x + 5y = 15$ **30.** $4x + 3y = 12$

Graph each of the following equations by first solving for y.

31. $x + 3y = 6$ **32.** $x - 2y = 6$ **33.** $3x + 4y = 12$

34. $2x - 3y = 12$ **35.** $5x - 4y = 20$ **36.** $7x + 3y = 21$

Write an equation that describes the following relationships between x and y.

37. y is twice x.   **38.** y is 3 times x.

39. y is 3 more than x.   **40.** y is 2 less than x.

41. y is 3 less than 3 times x.   **42.** y is 4 more than twice x.

43. The difference of x and the product of 4 and y is 12.

44. The difference of twice x and y is 6.

Graph each pair of equations on the same grid. Give the coordinates of the point where the lines intersect.

45. $x + y = 4$ **46.** $x - y = 3$
     $x - y = 2$ $x + y = 5$

In each of the following exercises, graph both equations on the same set of axes and report what you observe about the graphs.

47. $y = 2x$ and $y = 2x + 1$ **48.** $y = 3x + 1$ and $y = 3x - 1$

49. $y = 2x$ and $y = -\dfrac{1}{2}x$ **50.** $y = \dfrac{1}{3}x + \dfrac{7}{3}$ and $y = -3x + 2$

51. Graph of winnings. The equation $y = 0.10x + 200$ describes the amount of winnings a group earns for collecting plastic jugs in the recycling contest described in Exercise 43 at the end of Section 3.2. Sketch the graph of the line.

52. Minimum values. The contest sponsor will award a prize only if the winning group in the contest collects 100 lb of jugs or more. Use your graph in Exercise 51 to determine the minimum prize possible.

53. Fund-raising. A high school class wants to raise some money by recycling newspapers. They decide to rent a truck for a weekend and to collect the newspapers from homes in the neighborhood. The market price for recycled newsprint is currently $15 per ton. The equation $y = 15x - 100$ describes the amount of money the class will make, where y is the amount of money made in dollars, x is the number of tons of newsprint collected, and 100 is the cost in dollars to rent the truck.

(a) Draw a graph that represents the relationship between newsprint collected and money earned.

(b) The truck is costing the class $100. How many tons of newspapers must the class collect to break even on this project?

(c) If the class members collect 16 tons of newsprint, how much money will they earn?

(d) Six months later the price of newsprint is $17 dollars per ton, and the cost to rent the truck has risen to $125. Write the equation that describes the amount of money the class might make at that time.

54. Production costs. The cost of producing a number of items x is given by $C = mx + b$, where b is the fixed cost and m is the marginal cost (the cost of producing one item).

(a) If the fixed cost is $40 and the marginal cost is $10, write the cost equation.

(b) Graph the cost equation.

(c) The revenue generated from the sale of x items is given by $R = 50x$. Graph the revenue equation on the same set of axes as the cost equation.

(d) How many items must be produced for the revenue to equal the cost (the break-even point)?

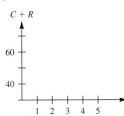

55. Consumer affairs. A car rental agency charges \$12 per day and 8¢ per mile for the use of a compact automobile. The cost of the rental C and the number of miles driven per day s are related by the equation

$$C = 0.08s + 12$$

Graph the relationship between C and s. Be sure to select appropriate scaling for the C and s axes.

56. Checking account charges. A bank has the following structure for charges on checking accounts. The monthly charges consist of a fixed amount of \$8 and an additional charge of 4¢ per check. The monthly cost of an account C and the number of checks written per month n are related by the equation

$$C = 0.04n + 8$$

Graph the relationship between C and n.

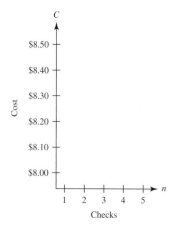

57. Tuition charges. A college has tuition charges based on the following pattern. Tuition is \$35 per credit-hour plus a fixed student fee of \$75.

(a) Write a linear equation that shows the relationship between the total tuition charge T and the number of credit-hours taken h.

(b) Graph the relationship between T and h.

58. Weekly salary. A salesperson's weekly salary is based on a fixed amount of \$200 plus 10% of the total amount of weekly sales.

(a) Write an equation that shows the relationship between the weekly salary S and the amount of weekly sales x (in dollars).

(b) Graph the relationship between S and x.

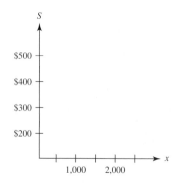

59. Consider the equation $y = 2x + 3$.

    (a) Complete the following table of values, and plot the resulting points.

Point	x	y
A	5	
B	6	
C	7	
D	8	
E	9	

    (b) As the x coordinate changes by 1 (for example, as you move from point A to point B), how much do the corresponding y coordinates change?

    (c) Is your answer to part (b) the same if you move from B to C? from C to D? from D to E?

    (d) Describe the "growth rate" of the line, using these observations. Complete the following statement: When the x value grows by 1 unit, the y value _____.

60. Describe how answers to parts (b), (c), and (d) would change if you were to repeat Exercise 59 using $y = 2x + 5$.

61. Describe how answers to parts (b), (c), and (d) would change if you were to repeat Exercise 59 using $y = 3x - 2$.

62. Describe how answers to parts (b), (c), and (d) would change if you were to repeat Exercise 59 using $y = 3x - 4$.

63. Describe how answers to parts (b), (c), and (d) would change if you were to repeat Exercise 59 using $y = -4x + 50$.

64. Describe how answers to parts (b), (c), and (d) would change if you were to repeat Exercise 59 using $y = -4x + 40$.

The Slope of a Line

3.4 OBJECTIVES

1. *Find the slope of a line*
2. *Find the slopes and y-intercepts of horizontal and vertical lines*
3. *Find the slope and y-intercepts of a line, given an equation*
4. *Write the equation of a line given the slope and y-intercept*
5. *Find the slope, given a graph*
6. *Graph linear equations, using the slope of a line*

Finding the Slope

On the coordinate system below, plot a point, any point.

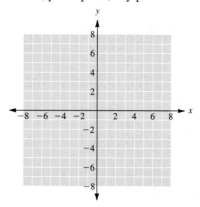

How many different lines can you draw through that point? Hundreds? Thousands? Millions? Actually, there is no limit to the number of different lines that pass through that point.

On the coordinate system below, plot two distinct points.

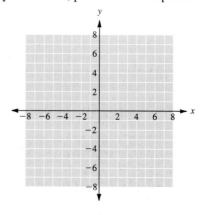

Now, how many different (straight) lines can you draw through those points? Only one! The two points were enough to define the line.

In Section 3.5, we will see how we can find the equation of a line if we are given two of its points. The first part of finding that equation is finding the **slope** of the line, which is a way of describing the *steepness* of a line.

Let us assume that the two points selected were $(-2, -3)$ and $(3, 7)$.

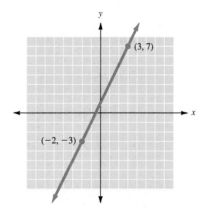

When moving between these two points, we go up 10 units and over 5 units.

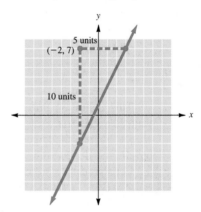

We refer to the 10 units as the *rise*. The 5 units is called the *run*. The slope is found by dividing the rise by the run. In this case, we have

$$\frac{\text{Rise}}{\text{Run}} = \frac{10}{5} = 2$$

The slope of this line is 2. This means that for any two points on the line, the rise (the change in the y value) is twice as much as the run (the change in the x value).

We will now proceed to a more formal look at the process for finding the slope when two points are given.

To define a formula for slope, choose any two distinct points on the line, say P with coordinates (x_1, y_1) and Q with coordinates (x_2, y_2). As we move along the line from P to Q, the x value, or coordinate, changes from x_1 to x_2. That change in x, also called the **horizontal change,** is $x_2 - x_1$. Similarly, as we move from P to Q, the corresponding change in y, called the **vertical change,** is $y_2 - y_1$. The *slope* is then defined as the ratio of the vertical change to the horizontal change. The letter m is used to represent the slope, which we now define.

Note: The difference $x_2 - x_1$ is called the **run.** The difference $y_2 - y_1$ is the **rise.**
Note that $x_1 \neq x_2$ or $x_2 - x_1 \neq 0$ ensures that the denominator is nonzero, so that the slope is defined. It also means the line cannot be vertical.

Slope of a Line

The *slope* of a line through two distinct points $P(x_1, y_1)$ and $Q(x_2, y_2)$ is given by

$$m = \frac{\text{change in } y}{\text{change in } x} = \frac{y_2 - y_1}{x_2 - x_1}$$

where $x_1 \neq x_2$.

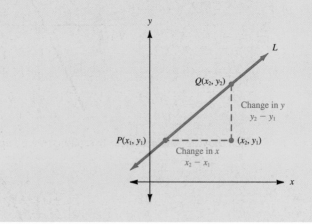

This definition provides exactly the numerical measure of "steepness" that we want. If a line "rises" as we move from left to right, the slope will be positive—the steeper the line, the larger the numerical value of the slope. If the line "falls" from left to right, the slope will be negative.

Let's proceed to some examples.

| Example 1 | ## Finding the Slope |

Find the slope of the line containing points with coordinates $(1, 2)$ and $(5, 4)$.

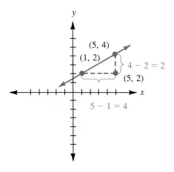

Let $P(x_1, y_1) = (1, 2)$ and $Q(x_2, y_2) = (5, 4)$. By the definition above, we have

$$m = \frac{y_2 - y_1}{x_2 - x_1} = \frac{4 - 2}{5 - 1} = \frac{2}{4} = \frac{1}{2}$$

Note: We would have found the same slope if we had reversed P and Q and subtracted in the other order. In that case, $P(x_1, y_1) = (5, 4)$ and $Q(x_2, y_2) = (1, 2)$, so

$$m = \frac{2 - 4}{1 - 5} = \frac{-2}{-4} = \frac{1}{2}$$

It makes no difference which point is labeled (x_1, y_1) and which is (x_2, y_2)—the resulting slope will be the same. You must simply stay with your choice once it is made and *not* reverse the order of the subtraction in your calculations.

✔ *CHECK YOURSELF 1*

Find the slope of the line containing points with coordinates $(2, 3)$ and $(5, 5)$.

By now you should be comfortable subtracting negative numbers. Let's apply that skill to finding a slope.

| **Example 2** | **Finding the Slope** |

Find the slope of the line containing points with the coordinates $(-1, -2)$ and $(3, 6)$.

Again, applying the definition, we have

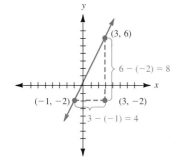

$$m = \frac{6 - (-2)}{3 - (-1)} = \frac{6 + 2}{3 + 1} = \frac{8}{4} = 2$$

The figure below compares the slopes found in Example 1 and in this example. Line l_1, from Example 1, had slope $\frac{1}{2}$. Line l_2, from this example, had slope 2. Do you see the idea of slope measuring steepness? The greater the value of a positive slope, the more steeply the line is inclined upward.

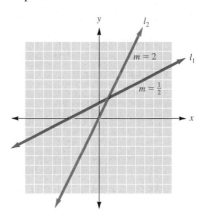

✔ *CHECK YOURSELF 2*

Find the slope of the line containing points with coordinates $(-1, 2)$ and $(2, 7)$. Draw a sketch of this line and the line of Check Yourself 1. Compare the lines and the two slopes.

Let's look at lines with a negative slope.

| Example 3 | ### Finding the Slope |

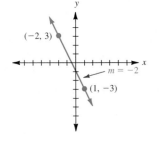

Find the slope of the line containing points with coordinates $(-2, 3)$ and $(1, -3)$.

By the definition,

$$m = \frac{-3 - 3}{1 - (-2)} = \frac{-6}{3} = -2$$

This line has a *negative* slope. The line *falls* as we move from left to right.

✔ *CHECK YOURSELF 3*

Find the slope of the line containing points with coordinates $(-1, 3)$ and $(1, -3)$.

We have seen that lines with positive slope rise from left to right and lines with negative slope fall from left to right. What about lines with a slope of zero? A line with a slope of 0 is especially important in mathematics.

| Example 4 | ### Finding the Slope |

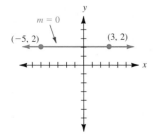

Find the slope of the line containing points with coordinates $(-5, 2)$ and $(3, 2)$.

By the definition,

$$m = \frac{2 - 2}{3 - (-5)} = \frac{0}{8} = 0$$

The slope of the line is 0. That will be the case for any horizontal line. Since any two points on the line have the same y coordinate, the vertical change $y_2 - y_1$ must always be 0, and so the resulting slope is 0.

✔ *CHECK YOURSELF 4*

Find the slope of the line containing points with coordinates $(-2, -4)$ and $(3, -4)$.

Since division by 0 is undefined, it is possible to have a line with an undefined slope.

Example 5

Finding the Slope

Find the slope of the line containing points with coordinates $(2, -5)$ and $(2, 5)$.

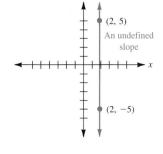

By the definition,

$$m = \frac{5 - (-5)}{2 - 2} = \frac{10}{0} \qquad \text{Remember that division by 0 is undefined.}$$

We say the vertical line has an undefined slope. On a vertical line, any two points have the same x coordinate. This means that the horizontal change $x_2 - x_1$ must always be 0, and since division by 0 is undefined, the slope of a vertical line will always be undefined.

✔ *CHECK YOURSELF 5*

Find the slope of the line containing points with the coordinates $(-3, -5)$ and $(-3, 2)$.

The following sketch summarizes the results of Examples 1 through 5.

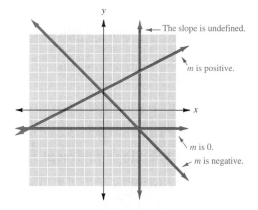

As the slope gets closer to 0, the line gets "flatter."

Four lines are illustrated in the figure. Note that

1. The slope of a line that rises from left to right is positive.

2. The slope of a line that falls from left to right is negative.

3. The slope of a horizontal line is 0.

4. A vertical line has an undefined slope.

We now want to consider finding the equation of a line when its slope and y intercept are known. Suppose that the y intercept is $(0, b)$. That is, the point at which the line crosses the y axis has coordinates $(0, b)$. Look at the sketch below.

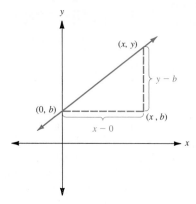

Now, using any other point (x, y) on the line and using our definition of slope, we can write

$$m = \frac{y - b}{x - 0}$$

Change in y

Change in x

or

$$m = \frac{y - b}{x}$$

Multiplying both sides by x, we have

$$mx = y - b$$

Finally, adding b to both sides gives

$$mx + b = y$$

or

$$y = mx + b$$

We can summarize the above discussion as follows:

In this form, the equation is *solved* for *y*. The coefficient of *x* will give you the slope of the line, and the constant term gives the *y* intercept.

The Slope-Intercept Form for a Line

An equation of the line with slope m and y intercept $(0, b)$ is

$$y = mx + b$$

Example 6   ## Finding the Slope and y Intercept

(a) Find the slope and y intercept for the graph of the equation

$$y = 3x + 4$$

m b

The graph has slope 3 and y intercept $(0, 4)$.

(b) Find the slope and y intercept for the graph of the equation

$$y = -\frac{2}{3}x - 5$$

$$\uparrow \qquad \uparrow$$
$$m \qquad b$$

The slope of the line is $-\frac{2}{3}$; the y intercept is $(0, -5)$.

 CHECK YOURSELF 6

Find the slope and y intercept for the graph of each of these equations.

(a) $y = -3x - 7$ (b) $y = \frac{3}{4}x + 5$

As Example 7 illustrates, we may have to solve for y as the first step in determining the slope and the y intercept for the graph of an equation.

Example 7 | Finding the Slope and y Intercept

Find the slope and y intercept for the graph of the equation

$$3x + 2y = 6$$

First, we must solve the equation for y.

If we write the equation as

$$y = \frac{-3x + 6}{2}$$

it is more difficult to identify the slope and the intercept.

$3x + 2y = 6$	Subtract $3x$ from both sides.
$2y = -3x + 6$	Divide each term by 2.
$y = -\frac{3}{2}x + 3$	

The equation is now in slope-intercept form. The slope is $-\frac{3}{2}$, and the y intercept is $(0, 3)$.

 CHECK YOURSELF 7

Find the slope and y intercept for the graph of the following equation.

$$2x - 5y = 10$$

As we mentioned earlier, knowing certain properties of a line (namely, its slope and *y* intercept) will also allow us to write the equation of the line by using the slope-intercept form. Example 8 illustrates this approach.

Example 8 | Writing the Equation of a Line

(a) Write the equation of a line with slope 3 and *y* intercept (0, 5).

We know that $m = 3$ and $b = 5$. Using the slope-intercept form, we have

$$y = 3x + 5$$

with arrows pointing to m and b

which is the desired equation.

(b) Write the equation of a line with slope $-\dfrac{3}{4}$ and *y* intercept (0, −3).

We know that $m = -\dfrac{3}{4}$ and $b = -3$. In this case,

$$y = -\frac{3}{4}x + (-3)$$

with arrows indicating m and b

or

$$y = -\frac{3}{4}x - 3$$

which is the desired equation.

✔ CHECK YOURSELF 8

Write the equation of a line with the following:

(a) Slope −2 and *y* intercept (0, 7)   (b) Slope $\dfrac{2}{3}$ and *y* intercept (0, −3)

We can also use the slope and *y* intercept of a line in drawing its graph. Consider Example 9.

Example 9 | Graphing a Line

(a) Graph the line with slope $\dfrac{2}{3}$ and *y* intercept (0, 2).

Because the *y* intercept is (0, 2), we begin by plotting this point. The horizontal change (or run) is 3, so we move 3 units to the right *from that y intercept*. The vertical change (or rise) is 2, so we move 2 units up to locate another point on the desired graph. Note that we will have located that second point at (3, 4). The final step is to draw a line through that point and the *y* intercept.

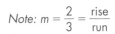

Note: $m = \dfrac{2}{3} = \dfrac{\text{rise}}{\text{run}}$

The line rises from left to right because the slope is positive.

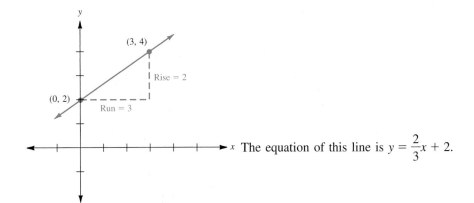

The equation of this line is $y = \dfrac{2}{3}x + 2$.

(b) Graph the line with slope -3 and *y* intercept (0, 3).

As before, first we plot the intercept point. In this case, we plot (0, 3). The slope is -3, which we interpret as $\dfrac{-3}{1}$. Because the rise is negative, we go down rather than up. We move 1 unit in the horizontal direction, then 3 units down in the vertical direction. We plot this second point, (1, 0), and connect the two points to form the line.

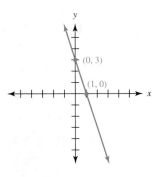

The equation of this line is $y = -3x + 3$.

✔ *CHECK YOURSELF 9*

Graph the equation of a line with slope $\dfrac{3}{5}$ and *y* intercept (0, -2).

A line can certainly pass through the origin, as Example 10 demonstrates. In such cases, the *y* intercept is (0, 0).

| Example 10 | # Graphing a Line |

Graph the line associated with the equation $y = -3x$ and the line associated with the equation $y = -3x - 3$.

In the first case, the slope is -3 and the y intercept is $(0, 0)$. We begin with the point $(0, 0)$. From there, we move down 3 units and to the right 1 unit, arriving at the point $(1, -3)$. Now we draw a line through those two points. On the same axes, we will draw the line with slope -3 through the intercept $(0, -3)$. Note that the two lines are parallel to each other.

Any two **parallel lines** have the same slope.

✔ *CHECK YOURSELF 10*

Graph the line associated with the equation $y = \dfrac{7}{2}x$.

We summarize the use of graphing by the slope-intercept form with the following algorithm.

Graphing by Using the Slope-Intercept Form

Step 1.   Write the original equation of the line in slope-intercept form $y = mx + b$.

Step 2.   Determine the slope m and the y intercept $(0, b)$.

Step 3.   Plot the y intercept at $(0, b)$.

Step 4.   Use m (the change in y over the change in x) to determine a second point on the desired line.

Step 5.   Draw a line through the two points determined above to complete the graph.

You have now seen two methods for graphing lines: the slope-intercept method (this section) and the intercept method (Section 3.3). When you graph a linear equation, you should first decide which is the appropriate method.

| **Example 11** | ## Selecting an Appropriate Graphing Method |

Decide which of the two methods for graphing lines—the intercept method or the slope-intercept method—is more appropriate for graphing equations (a), (b), and (c).

(a) $2x - 5y = 10$

Because both intercepts are easy to find, you should choose the intercept method to graph this equation.

(b) $2x + y = 6$

This equation can be quickly graphed by either method. As it is written, you might choose the intercept method. It can, however, be rewritten as $y = -2x + 6$. In that case the slope-intercept method is more appropriate.

(c) $y = \dfrac{1}{4}x - 4$

Since the equation is in slope-intercept form, that is the more appropriate method to choose.

✔ *CHECK YOURSELF 11*

Which would be more appropriate for graphing each equation, the intercept method or the slope-intercept method?

(a) $x + y = -2$ (b) $3x - 2y = 12$ (c) $y = -\dfrac{1}{2}x - 6$

✔ *CHECK YOURSELF ANSWERS*

1. $m = \dfrac{2}{3}$ **2.** $m = \dfrac{5}{3}$

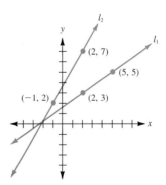

3. $m = -3$ **4.** $m = 0$ **5.** m is undefined

6. (a) $m = -3$, y intercept: $(0, -7)$; (b) $m = \dfrac{3}{4}$, y intercept: $(0, 5)$

7. $y = \dfrac{2}{5}x - 2$; $m = \dfrac{2}{5}$; y intercept: $(0, -2)$

8. (a) $y = -2x + 7$; (b) $y = \dfrac{2}{3}x - 3$

9.

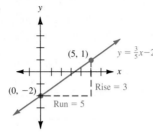

10.

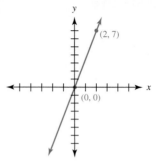

11. (a) either; (b) intercept; (c) slope-intercept

Exercises · 3.4

Find the slope of the line through the following pairs of points.

1. $(5, 7)$ and $(9, 11)$

2. $(4, 9)$ and $(8, 17)$

3. $(-3, -1)$ and $(2, 3)$

4. $(-3, 2)$ and $(0, 17)$

5. $(-2, 3)$ and $(3, 7)$

6. $(-2, -5)$ and $(1, -4)$

7. $(-3, 2)$ and $(2, -8)$

8. $(-6, 1)$ and $(2, -7)$

9. $(3, -2)$ and $(5, -5)$

10. $(-2, 4)$ and $(3, 1)$

11. $(5, -4)$ and $(5, 2)$

12. $(-2, 8)$ and $(6, 8)$

13. $(-4, -2)$ and $(3, 3)$

14. $(-5, -3)$ and $(-5, 2)$

15. $(-2, -6)$ and $(8, -6)$

16. $(-5, 7)$ and $(2, -2)$

17. $(-1, 7)$ and $(2, 3)$

18. $(-3, -5)$ and $(2, -2)$

Find the slope and y intercept of the line represented by each of the following equations.

19. $y = 3x + 5$

20. $y = -7x + 3$

21. $y = -3x - 6$

22. $y = 5x - 2$

23. $y = \dfrac{3}{4}x + 1$

24. $y = -5x$

25. $y = \dfrac{2}{3}x$

26. $y = -\dfrac{3}{5}x - 2$

27. $4x + 3y = 12$

28. $5x + 2y = 10$

29. $y = 9$

30. $2x - 3y = 6$

31. $3x - 2y = 8$

32. $x = -3$

Write the equation of the line with given slope and y intercept. Then graph each line, using the slope and y intercept.

33. Slope 3; y intercept: $(0, 5)$

34. Slope -2; y intercept: $(0, 4)$

35. Slope -4; y intercept: $(0, 5)$

36. Slope 5; y intercept: $(0, -2)$

37. Slope $\dfrac{1}{2}$; y intercept: $(0, -2)$

38. Slope $-\dfrac{2}{5}$; y intercept: $(0, 6)$

39. Slope $\dfrac{4}{3}$; y intercept: $(0, 0)$

40. Slope $\dfrac{2}{3}$; y intercept: $(0, -2)$

41. Slope $\dfrac{3}{4}$; y intercept: $(0, 3)$

42. Slope -3; y intercept: $(0, 0)$

In which quadrant(s) are there no solutions for each equation?

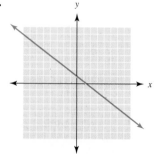

43. $y = 4x + 5$

44. $y = 3x + 2$

45. $y = -x + 1$

46. $y = -7x + 3$

47. $y = -2x - 5$

48. $y = -5x - 7$

49. $y = 5$

50. $x = -2$

In Exercises 51 to 58, match the graph with one of the equations below.

(a) $y = 2x$, (b) $y = x + 1$, (c) $y = -x + 3$, (d) $y = 2x + 1$, (e) $y = -3x - 2$, (f) $y = \dfrac{2}{3}x + 1$,

(g) $y = -\dfrac{3}{4}x + 1$, (h) $y = -4x$

51.

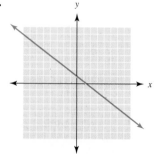

52.

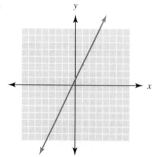

53.

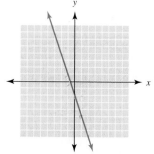

54.

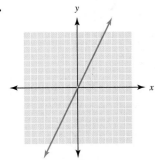

55.

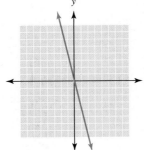

56.

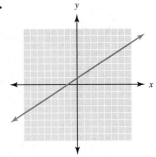

57.

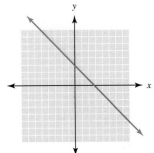

58.

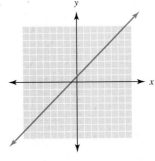

In Exercises 59 to 62, solve each equation for y, then graph each equation.

59. $2x + 5y = 10$ **60.** $5x - 3y = 12$ **61.** $x + 7y = 14$ **62.** $-2x - 3y = 9$

In Exercises 63 to 70, use the graph to determine the slope of the line.

63.

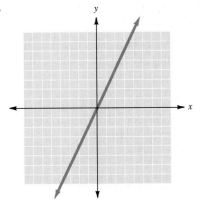

64.

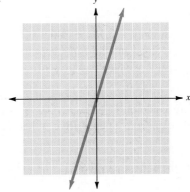

65.

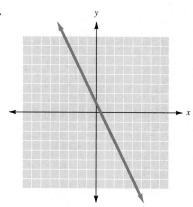

66.

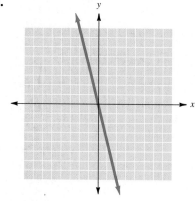

67.

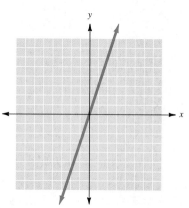

68.

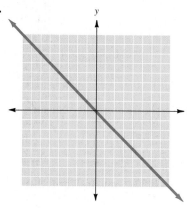

69.

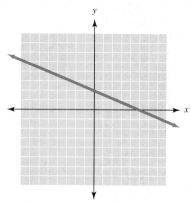

70.

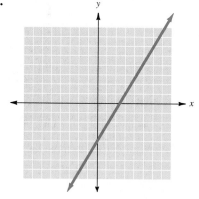

71. Recycling. The equation $y = 0.10x + 200$ was used in Section 3.3 to describe the award money in a recycling contest. What are the slope and the y intercept for this equation? What does the slope of the line represent in the equation? What does the y intercept represent?

72. Fund-raising. The equation $y = 15x - 100$ was used in Section 3.3 to describe the amount of money a high school class might earn from a paper drive. What are the slope and y intercept for this equation?

73. Fund-raising. In the equation in Exercise 72, what does the slope of the line represent? What does the y intercept represent?

74. Slope of a roof. A roof rises 8.75 feet (ft) in a horizontal distance of 15.09 ft. Find the slope of the roof to the nearest hundredth.

75. Slope of airplane descent. An airplane covered 15 miles (mi) of its route while decreasing its altitude by 24,000 ft. Find the slope of the line of descent that was followed. (1 mi = 5,280 ft) Round to the nearest hundredth.

76. Slope of road descent. Driving down a mountain, Tom finds that he has descended 1,800 ft in elevation by the time he is 3.25 mi horizontally away from the top of the mountain. Find the slope of his descent to the nearest hundredth.

77. Consumer affairs. In 1960, the cost of a soft drink was 20¢. By 2002, the cost of the same soft drink had risen to $1.50. During this time period, what was the annual rate of change of the cost of the soft drink?

78. Science. On a certain February day in Philadelphia, the temperature at 6:00 A.M. was 10°F. By 2:00 P.M. the temperature was up to 26°F. What was the hourly rate of temperature change?

79. Complete the following statement: "The difference between undefined slope and zero slope is . . . ."

80. Complete the following: "The slope of a line tells you . . . ."

81. On two occasions last month, Sam Johnson rented a car on a business trip. Both times it was the same model from the same company, and both times it was in San Francisco. Sam now has to fill out an expense account form and needs to know how much he was charged per mile and the base rate. On both occasions he dropped the car at the airport booth and just got the total charge, not the details. All Sam knows is that he was charged $210 for 625 mi on the first occasion and $133.50 for 370 mi on the second trip. Sam has called accounting to ask for help. Plot these two points on a graph, and draw the line that goes through them. What question does the slope of the line answer for Sam? How does the y intercept help? Write a memo to Sam, explaining the answers to his questions and how a knowledge of algebra and graphing has helped you find the answers.

82. On the same graph, sketch the following lines:

$$y = 2x - 1 \quad \text{and} \quad y = 2x + 3$$

What do you observe about these graphs? Will the lines intersect?

83. Repeat Exercise 82, using

$$y = -2x + 4 \quad \text{and} \quad y = -2x + 1$$

84. On the same graph, sketch the following lines:

$$y = \frac{2}{3}x \quad \text{and} \quad y = -\frac{3}{2}x$$

What do you observe concerning these graphs? Find the product of the slopes of these two lines.

85. Repeat Exercise 84, using

$$y = \frac{4}{3}x \quad \text{and} \quad y = -\frac{3}{4}x$$

86. Based on Exercises 84 and 85, write the equation of a line that is perpendicular to

$$y = \frac{3}{5}x$$

3.5 Forms of Linear Equations

3.5 OBJECTIVES

1. *Use the equations of lines to determine whether two lines are parallel, perpendicular, or neither*
2. *Write the equation of a line, given a slope and a point on the line*
3. *Write the equation of a line, given two points*
4. *Write the equation of a line satisfying given geometric conditions*

Recall that the form

$$Ax + By = C$$

where *A* and *B* cannot both be zero, is called the **standard form for a linear equation.** In Section 3.4 we determined the slope of a line from two ordered pairs. We also used the concept of slope to write the equation of a line.

In this section, we will see that the slope-intercept form clearly indicates whether the graphs of given equations will be parallel, intersecting, or perpendicular lines. We will make frequent use of the following properties.

Two lines are **parallel,** if and only if their slopes are equal:

$$m_1 = m_2$$

Two lines are **perpendicular,** if and only if their slopes are negative reciprocals:

$$m_1 = -\frac{1}{m_2}$$

Note that in this case $m_1 \cdot m_2 = -1$; the product of their slopes is -1.

Example 1 illustrates this concept.

Example 1   Verifying That Two Lines Are Perpendicular

Show that the graphs of $3x + 4y = 4$ and $-4x + 3y = 12$ are perpendicular lines.

First, we solve each equation for *y*.

$$3x + 4y = 4$$

$$4y = -3x + 4$$

$$y = -\frac{3}{4}x + 1$$

Noting that the slope of the line is $-\dfrac{3}{4}$

$$-4x + 3y = 12$$
$$3y = 4x + 12$$
$$y = \frac{4}{3}x + 4$$

The slope of the line is $\dfrac{4}{3}$.

We now look at the product of the two slopes: $-\dfrac{3}{4} \cdot \dfrac{4}{3} = -1$. Any two lines whose slopes have a product of -1 are perpendicular lines. These two lines are perpendicular.

✔ *CHECK YOURSELF 1*

Show that the graphs of the equations

$$-3x + 2y = 4 \qquad \text{and} \qquad 2x + 3y = 9$$

are perpendicular lines.

Let us review how the slope-intercept form can be used in graphing a line. Example 2 illustrates.

Example 2

Graphing the Equation of a Line

Graph the line $2x + 3y = 3$.

Solving for y, we find the slope-intercept form for this equation is

$$y = -\frac{2}{3}x + 1$$

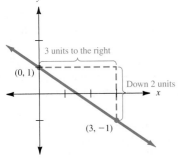

We treat $-\dfrac{2}{3}$ as $\dfrac{-2}{+3}$ to move to the right 3 units and down 2 units.

To graph the line, plot the y intercept at $(0, 1)$. Because the slope m is equal to $-\dfrac{2}{3}$, we move from $(0, 1)$ to the right 3 units and then *down* 2 units, to locate a second point on the graph of the line, here $(3, -1)$. We can now draw a line through the two points to complete the graph.

✔ *CHECK YOURSELF 2*

Graph the line with equation.

$$3x - 4y = 8$$

Hint: First rewrite the equation in slope-intercept form.

Assume that we were given the two points $(7, -2)$ and $(3, -1)$ and asked to find the equation of the line that goes through them. How might we proceed?

We first find the slope. Again, we divide the change in y by the change in x.

$$\frac{-2 - (-1)}{7 - 3} = \frac{-1}{4} = -\frac{1}{4}$$

The slope is $-\frac{1}{4}$.

We know that the equation of the line is of the form

$$y = -\frac{1}{4}x + b$$

We can find the value of b by substituting the values for either point in place of x and y in the equation. We will use $(7, -2)$.

$$-2 = -\frac{1}{4}(7) + b$$

$$-2 = -\frac{7}{4} + b$$

$$-\frac{1}{4} = b$$

The equation of the line is

$$y = -\frac{1}{4}x - \frac{1}{4}$$

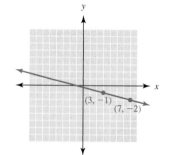

Graphing the line, we see that this is indeed the equation of the line that passes through the two points $(7, -2)$ and $(3, -1)$.

Often in mathematics it is useful to be able to write the equation of a line, given its slope and *any* point on the line. We will now derive a third special form for a line for this purpose.

Suppose a line has slope m and passes through the known point $P(x_1, y_1)$. Let $Q(x, y)$ be any other point on the line. Once again we can use the definition of slope and write

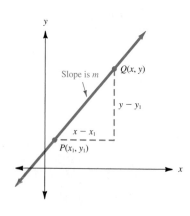

$$m = \frac{y - y_1}{x - x_1}$$

Multiplying both sides of the equation by $x - x_1$, we have

$$m(x - x_1) = y - y_1$$

or $$y - y_1 = m(x - x_1)$$

This last equation is called the **point-slope form** for the equation of a line, and all points lying on the line [including (x_1, y_1)] will satisfy this equation. We can state the following general result.

The equation of a line with undefined slope passing through the point (x_1, y_1) is given by $x = x_1$.

> ### Point-Slope Form for the Equation of a Line
>
> The equation of a line with slope m that passes through point (x_1, y_1) is given by
>
> $$y - y_1 = m(x - x_1)$$

Example 3   Finding the Equation of a Line

Write the equation for the line that passes through point $(3, -1)$ with a slope of 3.

Letting $(x_1, y_1) = (3, -1)$ and $m = 3$, we use point-slope form to get

$$y - (-1) = 3(x - 3)$$

or

$$y + 1 = 3x - 9$$

We can write the final result in slope-intercept form as

$$y = 3x - 10$$

 CHECK YOURSELF 3

Write the equation of the line that passes through point $(-2, 4)$ with a slope of $\frac{3}{2}$. Write your result in slope-intercept form.

Since we know that two points determine a line, it is natural that we should be able to write the equation of a line passing through two given points. Using the point-slope form together with the slope formula will allow us to write such an equation.

Example 4   Finding the Equation of a Line

Write the equation of the line passing through $(2, 4)$ and $(4, 7)$.

First, we find m, the slope of the line. Here

$$m = \frac{7 - 4}{4 - 2} = \frac{3}{2}$$

Note: We could just as well have chosen to let

$$(x_1, y_1) = (4, 7)$$

The resulting equation will be the same in either case. Take time to verify this for yourself.

Now we apply the point-slope form with $m = \dfrac{3}{2}$ and $(x_1, y_1) = (2, 4)$:

$$y - 4 = \frac{3}{2}(x - 2)$$

$$y - 4 = \frac{3}{2}x - 3$$

We write the result in slope-intercept form.

$$y = \frac{3}{2}x + 1$$

✔ CHECK YOURSELF 4

Write the equation of the line passing through $(-2, 5)$ and $(1, 3)$. Write your result in slope-intercept form.

A line with slope zero is a horizontal line. A line with an undefined slope is vertical. Example 5 illustrates the equations of such lines.

Example 5	**Finding the Equation of a Line**

(a) Find the equation of a line passing through $(7, -2)$ with a slope of 0.

We could find the equation by letting $m = 0$. Substituting into the slope-intercept form, we can solve for b.

$$y = mx + b$$
$$-2 = 0(7) + b$$
$$-2 = b$$

So $\qquad\qquad y = 0x - 2,\qquad$ so $\qquad y = -2$

It is far easier to remember that any line with a zero slope is a horizontal line and has the form

$$y = b$$

The value for b will always be the y coordinate for the given point.

(b) Find the equation of a line with undefined slope passing through $(4, -5)$.

A line with undefined slope is vertical. It will always be of the form $x = a$, where a is the x coordinate for the given point. The equation is

$$x = 4$$

CHECK YOURSELF 5

(a) Find the equation of a line with zero slope that passes through point $(-3, 5)$.

(b) Find the equation of a line passing through $(-3, -6)$ with undefined slope.

There are alternative methods for finding the equation of a line through two points. Example 6 shows such an alternate approach.

Example 6

Finding the Equation of a Line

Write the equation of the line through points $(-2, 3)$ and $(4, 5)$.

First, we find m, as before.

We could, of course, use the point-slope form as seen earlier.

$$m = \frac{5 - 3}{4 - (-2)} = \frac{2}{6} = \frac{1}{3}$$

We now make use of the slope-intercept form, but in a different manner.

Using $y = mx + b$, and the slope just calculated, we can immediately write

$$y = \frac{1}{3}x + b$$

We substitute these values because the line must pass through $(-2, 3)$.

Now, if we substitute a known point for x and y, we can solve for b. We may choose either of the two given points. Using $(-2, 3)$, we have

$$3 = \frac{1}{3}(-2) + b$$

$$3 = -\frac{2}{3} + b$$

$$3 + \frac{2}{3} = b$$

$$b = \frac{11}{3}$$

Therefore the equation of the desired line is

$$y = \frac{1}{3}x + \frac{11}{3}$$

 CHECK YOURSELF 6

Repeat the Check Yourself 4 exercise, using the technique illustrated in Example 6.

We now know that we can write the equation of a line once we have been given appropriate geometric conditions, such as a point on the line and the slope of that line. In some applications, the slope may be given not directly but through specified parallel or perpendicular lines.

| Example 7 | Finding the Equation of a Parallel Line |

Find the equation of the line passing through $(-4, -3)$ and parallel to the line determined by $3x + 4y = 12$.

First, we find the slope of the given parallel line, as before.

$$3x + 4y = 12$$
$$4y = -3x + 12$$
$$y = -\frac{3}{4}x + 3$$

The slope of the given line is $-\frac{3}{4}$, the coefficient of x.

The slopes of two parallel lines will be the same. Because the slope of the desired line must also be $-\frac{3}{4}$, we can use the point-slope form to write the required equation.

The line must pass through $(-4, -3)$, so let

$$(x_1, y_1) = (-4, -3)$$

$$y - (-3) = -\frac{3}{4}[x - (-4)]$$

This simplifies to

$$y = -\frac{3}{4}x - 6$$

and we have our equation in slope-intercept form.

✔ *CHECK YOURSELF 7*

Find the equation of the line passing through $(2, -5)$ and parallel to the line determined by $4x - y = 9$.

| Example 8 | Finding the Equation of a Perpendicular Line |

Find the equation of the line passing through $(3, -1)$ and perpendicular to the line $3x - 5y = 2$.

First, find the slope of the perpendicular line.

$$3x - 5y = 2$$
$$-5y = -3x + 2$$
$$y = \frac{3}{5}x - \frac{2}{5}$$

The slope of the perpendicular line is $\frac{3}{5}$. Recall that the slopes of perpendicular lines are negative reciprocals. The slope of our line is the negative reciprocal of $\frac{3}{5}$. It is therefore $-\frac{5}{3}$.

Using the point-slope form, we have the equation

$$y - (-1) = -\frac{5}{3}(x - 3)$$

$$y + 1 = -\frac{5}{3}x + 5$$

$$y = -\frac{5}{3}x + 4$$

✔ CHECK YOURSELF 8

Find the equation of the line passing through (5, 4) and perpendicular to the line with equation $2x - 5y = 10$.

There are many applications of our work with linear equations in various fields. The following is just one of many typical examples.

Example 9	**An Application of a Linear Equation**

In applications, it is common to use letters other than x and y. In this case, we use C to represent the **cost.**

In producing a new product, a manufacturer predicts that the number of items produced x and the cost in dollars C of producing those items will be related by a linear equation.

Suppose that the cost of producing 100 items will be $5,000 and the cost of producing 500 items will be $15,000. Find the linear equation relating x and C.

To solve this problem, we must find the equation of the line passing through points (100, 5,000) and (500, 15,000). Even though the numbers are considerably larger than we have encountered thus far in this section, the process is exactly the same.

First, we find the slope:

$$m = \frac{15000 - 5000}{500 - 100} = \frac{10000}{400} = 25$$

We can now use the point-slope form as before to find the desired equation.

$$C - 5,000 = 25(x - 100)$$

$$C - 5,000 = 25x - 2,500$$

$$C = 25x + 2,500$$

To graph the equation we have just derived, we must choose the scaling on the x and C axes carefully to get a "reasonable" picture. Here we choose increments of

100 on the x axis and 2,500 on the C axis since those seem appropriate for the given information.

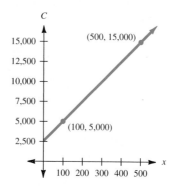

Note how the change in scaling "distorts" the slope of the line.

✔ CHECK YOURSELF 9

A company predicts that the value in dollars V and the time that a piece of equipment has been in use t are related by a linear equation. If the equipment is valued at \$1,500 after 2 years and at \$300 after 10 years, find the linear equation relating t and V.

✔ CHECK YOURSELF ANSWERS

1. $m_1 = \dfrac{3}{2}$ and $m_2 = -\dfrac{2}{3}$; $(m_1)(m_2) = -1$ **2.**

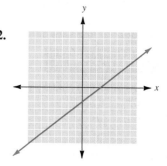

3. $y = \dfrac{3}{2}x + 7$ **4.** $y = -\dfrac{2}{3}x + \dfrac{11}{3}$ **5.** (a) $y = 5$; (b) $x = -3$

6. $y = -\dfrac{2}{3}x + \dfrac{11}{3}$ **7.** $y = 4x - 13$ **8.** $y = -\dfrac{5}{2}x + \dfrac{33}{2}$

9. $V = -150t + 1{,}800$

Exercises · 3.5

In Exercises 1 to 6, are the pairs of lines parallel, perpendicular, or neither?

1. L_1 through $(-2, -3)$ and $(4, 3)$; L_2 through $(3, 5)$ and $(5, 7)$

2. L_1 through $(-2, 4)$ and $(1, 8)$; L_2 through $(-1, -1)$ and $(-5, 2)$

3. L_1 through $(7, 4)$ and $(5, -1)$; L_2 through $(8, 1)$ and $(-3, -2)$

4. L_1 through $(-2, -3)$ and $(3, -1)$; L_2 through $(-3, 1)$ and $(7, 5)$

5. L_1 with equation $x - 3y = 6$; L_2 with equation $3x + y = 3$

6. L_1 with equation $2x + 4y = 8$; L_2 with equation $4x + 8y = 10$

7. Find the slope of any line parallel to the line through points $(-2, 3)$ and $(4, 5)$.

8. Find the slope of any line perpendicular to the line through points $(0, 5)$ and $(-3, -4)$.

9. A line passing through $(-1, 2)$ and $(4, y)$ is parallel to a line with slope 2. What is the value of y?

10. A line passing through $(2, 3)$ and $(5, y)$ is perpendicular to a line with slope $\dfrac{3}{4}$. What is the value of y?

In Exercises 11 to 22, write the equation of the line passing through each of the given points with the indicated slope. Give your results in slope-intercept form, where possible.

11. $(0, -5)$, slope $\dfrac{5}{4}$ **12.** $(0, -4)$, slope $-\dfrac{3}{4}$ **13.** $(1, 3)$, slope 5

14. $(-1, 2)$, slope 3   **15.** $(-2, -3)$, slope -3 **16.** $(1, 3)$, slope -2

17. $(5, -3)$, slope $\dfrac{2}{5}$ **18.** $(4, 3)$, slope 0   **19.** $(1, -4)$, slope undefined

20. $(2, -5)$, slope $\dfrac{1}{4}$ **21.** $(5, 0)$, slope $-\dfrac{4}{5}$ **22.** $(-3, 4)$, slope undefined

In Exercises 23 to 32, write the equation of the line passing through each of the given pairs of points. Write your result in slope-intercept form, where possible.

23. $(2, 3)$ and $(5, 6)$ **24.** $(3, -2)$ and $(6, 4)$ **25.** $(-2, -3)$ and $(2, 0)$

26. $(-1, 3)$ and $(4, -2)$ **27.** $(-3, 2)$ and $(4, 2)$ **28.** $(-5, 3)$ and $(4, 1)$

29. $(2, 0)$ and $(0, -3)$ **30.** $(2, -3)$ and $(2, 4)$ **31.** $(0, 4)$ and $(-2, -1)$

32. $(-4, 1)$ and $(3, 1)$

Write the equation of the line L satisfying the given geometric conditions.

33. L has slope 4 and y intercept $(0, -2)$.

34. L has slope $-\dfrac{2}{3}$ and y intercept $(0, 4)$.

35. L has x intercept $(4, 0)$ and y intercept $(0, 2)$.

36. L has x intercept $(-2, 0)$ and slope $\dfrac{3}{4}$.

37. L has y intercept $(0, 4)$ and a 0 slope.

38. L has x intercept $(-2, 0)$ and an undefined slope.

39. L passes through $(2, 3)$ with a slope of -2.

40. L passes through $(-2, -4)$ with a slope of $-\dfrac{3}{2}$.

41. L has y intercept $(0, 3)$ and is parallel to the line with equation $y = 3x - 5$.

42. L has y intercept $(0, -3)$ and is parallel to the line with equation $y = \dfrac{2}{3}x + 1$.

43. L has y intercept $(0, 4)$ and is perpendicular to the line with equation $y = -2x + 1$.

44. L has y intercept $(0, 2)$ and is parallel to the line with equation $y = -1$.

45. L has y intercept $(0, 3)$ and is parallel to the line with equation $y = 2$.

46. L has y intercept $(0, 2)$ and is perpendicular to the line with equation $2x - 3y = 6$.

47. L passes through $(-4, 5)$ and is parallel to the line $y = -4x + 5$.

48. L passes through $(-4, 3)$ and is parallel to the line with equation $y = -2x + 1$.

49. L passes through $(3, 2)$ and is parallel to the line with equation $y = \dfrac{4}{3}x + 4$.

50. L passes through $(-2, -1)$ and is perpendicular to the line with equation $y = 3x + 1$.

51. L passes through $(3, -1)$ and is perpendicular to the line with equation $y = -\dfrac{2}{3}x + 5$.

52. L passes through $(-4, 2)$ and is perpendicular to the line with equation $y = 4x + 5$.

53. L passes through $(-2, 1)$ and is parallel to the line with equation $x + 2y = 4$.

54. L passes through $(-3, 5)$ and is parallel to the x axis.

A four-sided figure (quadrilateral) is a parallelogram if the opposite sides have the same slope. If the adjacent sides are perpendicular, the figure is a rectangle. In Exercises 55 to 58, for each quadrilateral $ABCD$, determine whether it is a parallelogram; then determine whether it is a rectangle.

55. $A(0, 0)$, $B(2, 0)$, $C(2, 3)$, $D(0, 3)$

56. $A(-3, 2)$, $B(1, -7)$, $C(3, -4)$, $D(-1, 5)$

57. $A(0, 0)$, $B(4, 0)$, $C(5, 2)$, $D(1, 2)$

58. $A(-3, -5)$, $B(2, 1)$, $C(-4, 6)$, $D(-9, 0)$

59. Science. A temperature of 10°C corresponds to a temperature of 50°F. Also 40°C corresponds to 104°F. Find the linear equation relating F and C.

60. Business. In planning for a new item, a manufacturer assumes that the number of items produced x and the cost in dollars C of producing these items are related by a linear equation. Projections are that 100 items will cost $10,000 to produce and that 300 items will cost $22,000 to produce. Find the equation that relates C and x.

61. Business. Mike bills a customer at the rate of $65 per hour plus a fixed service call charge of $75.

    (a) Write an equation that will allow you to compute the total bill for any number of hours x that it takes to complete a job.

    (b) What will the total cost of a job be if it takes 3.5 hours to complete?

    (c) How many hours would a job have to take if the total bill were $247.25?

62. Business. Two years after an expansion, a company had sales of $42,000. Four years later (six years after the expansion) the sales were $102,000. Assuming that the sales in dollars S and the time t in years are related by a linear equation, find the equation relating S and t.

In Exercises 63 to 70, use the graph to determine the slope and y intercept of the line.

63.

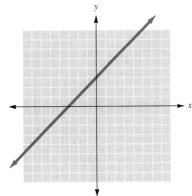

64.

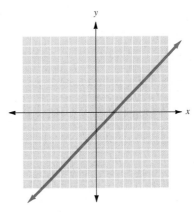

65.

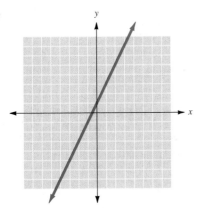

66.

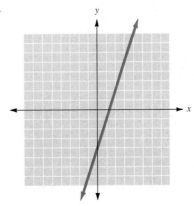

67.

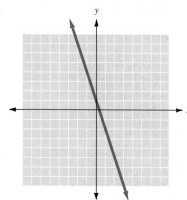

68.

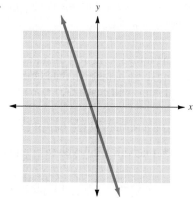

69.

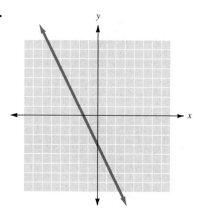

70.

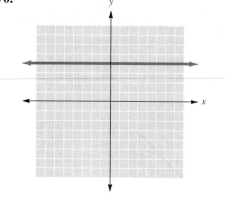

Summary for Chapter 3

Example	Topic	Reference
	Equations in Two Variables	3.1
If $2x - y = 10$, then $(6, 2)$ is a solution for the equation, because substituting 6 for x and 2 for y gives a true statement.	**Solutions of Linear Equations** Pairs of values that satisfy the equation. Solutions for linear equations in two variables are written as *ordered pairs*. An ordered pair has the form $$(x, y)$$ x coordinate y coordinate	p. 215
	The Cartesian Coordinate System	3.2
	The Rectangular Coordinate System A system formed by two perpendicular axes that intersect at a point called the **origin.** The horizontal line is called the x **axis.** The vertical line is called the y **axis.**	p. 222
To graph the point corresponding to $(2, 3)$: 	**Graphing Points from Ordered Pairs** The coordinates of an ordered pair allow you to associate a point in the plane with every ordered pair. To graph a point in the plane, 1. Start at the origin. 2. Move right or left according to the value of the x coordinate: to the right if x is positive or to the left if x is negative. 3. Then move up or down according to the value of the y coordinate: up if y is positive and down if y is negative.	p. 224
The midpoint of $(-4, 5)$ and $(2, 1)$ is $$\left(\frac{-4 + 2}{2}, \frac{5 + 1}{2}\right) = (-1, 3)$$	Midpoint formula $$M = \left(\frac{x_1 + x_2}{2}, \frac{y_1 + y_2}{2}\right)$$	p. 228

Graphing Linear Equations 3.3

$2x - 3y = 4$ is a linear equation.	**Linear Equation** An equation that can be written in the form $$Ax + By = C$$ where A and B are not both 0.	**p. 235**

	Graphing Linear Equations 1. Find at least three solutions for the equation, and put your results in tabular form. 2. Graph the solutions found in step 1. 3. Draw a straight line through the points determined in step 2 to form the graph of the equation.	**p. 235**

x	y
0	−6
3	−3
6	0

The Slope of a Line 3.4

To find the slope of the line through $(-2, -3)$ and $(4, 6)$, $$m = \frac{6 - (-3)}{4 - (-2)}$$ $$= \frac{6 + 3}{4 + 2}$$ $$= \frac{9}{6} = \frac{3}{2}$$	**Slope** The slope of a line gives a numerical measure of the steepness of the line. The slope m of a line containing the distinct points in the plane $P(x_1, y_1)$ and $Q(x_2, y_2)$ is given by $$m = \frac{y_2 - y_1}{x_2 - x_1} \qquad \text{where } x_2 \neq x_1.$$	**p. 253**

For the equation $$y = \frac{2}{3}x - 3$$ the slope m is $\frac{2}{3}$ and the y intercept is $(0, -3)$.	**Slope-Intercept Form** The slope-intercept form for the equation of a line is $$y = mx + b$$ where the line has slope m and y intercept $(0, b)$.	**p. 258**

Forms of Linear Equations 3.5

$y = 3x - 7$ and $y = 3x + \frac{2}{5}$ are parallel lines.	**Parallel Lines** Two lines are parallel if they have the same slope.	**p. 270**
$y = 2x + 1$ and $y = -\frac{1}{2}x + 5$ are perpendicular lines. $$(2)\left(-\frac{1}{2}\right) = -1$$	**Perpendicular Lines** Two lines are perpendicular if the product of the two slopes is -1. The slopes are said to be **negative reciprocals**.	**p. 270**

(S)ummary Exercises ▪ 3

This summary exercise set is provided to give you practice with each of the objectives of the chapter. Each exercise is keyed to the appropriate chapter section.

[3.1] Determine which of the ordered pairs are solutions for the given equations.

1. $x - y = 6$ $(6, 0), (3, 3), (3, -3), (0, -6)$

2. $2x + 3y = 6$ $(3, 0), (6, 2), (-3, 4), (0, 2)$

[3.2] Give the coordinates of the points graphed below.

3. A

4. B

5. E

6. F

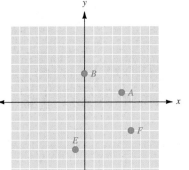

[3.2] Plot the points with the given coordinates.

7. $P(4, 0)$ **8.** $Q(5, 4)$ **9.** $T(-2, 4)$ **10.** $U(4, -2)$

[3.2] In Exercises 11 to 16, give the quadrant in which the following points are located or the axis on which the point lies.

11. $(3, 6)$ **12.** $(-7, 5)$ **13.** $(-1, -6)$

14. $(-7, 8)$ **15.** $(-5, 0)$ **16.** $(0, -5)$

[3.2] Plot the following points.

17. $(-1, 4)$ **18.** $(-1.25, 3.5)$ **19.** $(6, 3)$ **20.** $(-5, -2)$

[3.3] Graph each of the following equations.

21. $x + y = 5$ **22.** $x - y = 6$ **23.** $y = 5x$ **24.** $y = -3x$

25. $y = \dfrac{3}{2}x$ **26.** $y = 3x + 2$ **27.** $y = -2x + 4$ **28.** $y = -3x + 4$

29. $y = \dfrac{2}{3}x + 2$ **30.** $3x - y = 3$ **31.** $2x + y = 6$ **32.** $3x + 2y = 12$

33. $3x - 4y = 12$ **34.** $x = -5$ **35.** $y = -2$ **36.** $5x - 3y = 15$

37. $3x + 4y = 12$ **38.** $2x + y = 6$ **39.** $3x + 2y = 6$ **40.** $-4x - 5y = 20$

[3.4] Find the slope of the line through each of these pairs of points.

41. $(3, 4)$ and $(5, 8)$ **42.** $(-2, 3)$ and $(1, -6)$

43. $(-2, 5)$ and $(2, 3)$ **44.** $(-5, -2)$ and $(1, 2)$

45. $(-2, 6)$ and $(5, 6)$ **46.** $(-3, 2)$ and $(-1, -3)$

47. $(-3, -6)$ and $(5, -2)$ **48.** $(-6, -2)$ and $(-6, 3)$

[3.4] Find the slope and y intercept of the line represented by each of the following equations.

49. $y = 2x + 5$ **50.** $y = -4x - 3$

51. $y = -\dfrac{3}{4}x$ **52.** $y = \dfrac{2}{3}x + 3$

53. $2x + 3y = 6$

54. $5x - 2y = 10$

55. $y = -3$

56. $x = 2$

[3.4] Write the equation of the line with the given slope and y intercept.

57. Slope 2, y intercept $(0, 3)$

58. Slope $\dfrac{3}{4}$, y intercept $(0, -2)$

59. Slope $-\dfrac{2}{3}$, y intercept $(0, 2)$

[3.5] In Exercises 60 to 63, are the pairs of lines parallel, perpendicular, or neither?

60. L_1 through $(-3, -2)$ and $(1, 3)$
L_2 through $(0, 3)$ and $(4, 8)$

61. L_1 through $(-4, 1)$ and $(2, -3)$
L_2 through $(0, -3)$ and $(2, 0)$

62. L_1 with equation $x + 2y = 6$
L_2 with equation $x + 3y = 9$

63. L_1 with equation $4x - 6y = 18$
L_2 with equation $2x - 3y = 6$

[3.5] Write an equation of the line passing through the following points with the indicated slope. Give your result in slope-intercept form, where possible.

64. $(0, -5)$, $m = \dfrac{2}{3}$

65. $(0, -3)$, $m = 0$

66. $(2, 3)$, $m = 3$

67. $(4, 3)$, m is undefined

68. $(3, -2)$, $m = \dfrac{5}{3}$

69. $(-2, -3)$, $m = 0$

70. $(-2, -4)$, $m = -\dfrac{5}{2}$

71. $(-3, 2)$, $m = -\dfrac{4}{3}$

72. $\left(\dfrac{2}{3}, -5\right)$, $m = 0$

73. $\left(-\dfrac{5}{2}, -1\right)$, m is undefined

[3.5] Write an equation of the line L satisfying the following sets of conditions.

74. L passes through $(-3, -1)$ and $(3, 3)$.

75. L passes through $(2, 3)$ and $(-2, -5)$.

76. L has slope $\dfrac{3}{4}$ and y intercept $(0, 3)$.

77. L passes through $(4, -3)$ with a slope of $-\dfrac{5}{4}$.

78. L has y intercept $(0, -4)$ and is parallel to the line with equation $3x - y = 6$.

79. L passes through $(-5, 2)$ and is perpendicular to the line with equation $5x - 3y = 15$.

80. L passes through $(2, -1)$ and is perpendicular to the line with equation $3x - 2y = 5$.

81. L passes through the point $(-5, -2)$ and is parallel to the line with equation $4x - 3y = 9$.

The purpose of this self-test is to help you check your progress and to review for a chapter test in class. Allow yourself about an hour to take the test. When you are done, check your answers in the back of the book. If you missed any answers, be sure to go back and review the appropriate sections in the chapter and the exercises that are provided.

1. Determine which of the ordered pairs are solutions for the given equation.

$$4x - y = 16 \qquad (4, 0), (3, -1), (5, 4)$$

2. Complete the ordered pairs so that each is a solution for the given equation.

$$4x + 3y = 12 \qquad (3, \ ), (\ , 4), (\ , 3)$$

Give the coordinates of the points graphed below.

3. A

4. B

5. C

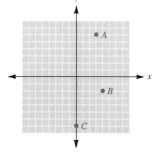

Plot points with the coordinates shown.

6. $S(1, -2)$

7. $T(0, 3)$

8. $U(-4, 5)$

Graph each of the following equations.

9. $x + y = 4$

10. $y = 3x$

11. $y = \dfrac{3}{4}x - 4$

12. $x + 3y = 6$

13. $2x + 5y = 10$

14. $y = -4$

Find the slope of the line through the following pairs of points.

15. $(-3, 5)$ and $(2, 10)$

16. $(4, 9)$ and $(-3, 6)$

Write the equation of the line with the given slope and y intercept. Then graph each line, *using* the slope and y intercept.

17. Slope -3, y intercept $(0, 6)$

18. Slope $\dfrac{2}{5}$, y intercept $(0, -3)$

Find the slope and y intercept of the line represented by each of the given equations.

19. $y = -5x - 9$

20. $6x + 5y = 30$

21. $y = 5$

Write an equation of the line L satisfying the given set of conditions.

22. L has slope 5 and y intercept $(0, -2)$.

23. L passes through $(-5, 4)$ and $(-2, -8)$.

24. L has y intercept $(0, 3)$ and is parallel to the line with the equation $4x - y = 9$.

25. L passes through the point $(-6, -2)$ and is perpendicular to the line with equation $2x - 5y = 10$.

288

Cumulative Test • 0–3

This test is provided to help you in the process of reviewing previous chapters. Answers are provided in the back of the book. If you missed any answers, be sure to go back and review the appropriate chapter sections.

In Exercises 1 to 4, perform the indicated operations. Write each answer in simplest form.

1. $\dfrac{5}{6} - \left(\dfrac{2}{3} + \dfrac{1}{2} \right)$

2. $\dfrac{7}{15} \times \left(\dfrac{5}{6} \div \dfrac{7}{12} \right)$

3. $2^3 - |-8| \div (-4) \cdot 2 + 5$

4. $4^2 + (-16 \div 4 \cdot 2)$

In Exercises 5 and 6, evaluate each expression if $x = -1$, $y = 3$, and $z = -2$.

5. $-4x^2 + 3y + 2z$

6. $\dfrac{2z - 3y^2 - 2}{y^2 + 2z}$

In Exercises 7 to 10, simplify the given expression.

7. $9x - 5y - (3x - 8y)$

8. $2x^2 - 4x - (-3x + x^2) - (4 + x^2)$

9. $-4x^2 + 7x - 4 - (-7x^2 + 11x) - (9x^2 - 5x - 6)$

10. $7 - 5x + 2x^2 + 2(9 - 5x^2)$

In Exercises 11 to 16, solve each equation.

11. $5x - 3(2x - 6) + 9 = -2(x - 5) + 6$

12. $\dfrac{4}{5}x - 2 = 3 + \dfrac{3}{4}x$

13. $|2x - 7| + 5 = 6$

14. $\dfrac{x + 1}{3} - \dfrac{2x + 3}{4} = \dfrac{1}{6}$

15. $|x + 1| = |2x - 3|$

16. $2x(x - 3) - 9 = 2x^2$

In Exercises 17 and 18, solve the equation for the indicated variable.

17. $F = \dfrac{9}{5}C + 32$ (for C)

18. $V = \dfrac{1}{3}\pi r^2 h$ (for h)

In Exercises 19 to 24, solve and graph the solution set for each inequality.

19. $4x - 7 < 9$

20. $6x + 4 > 3x - 8$

21. $4 \le 2x - 6 \le 12$

22. $x - 5 < -3$ or $x - 5 > 2$

23. $|2x + 5| \le 8$

24. $|x - 6| > 5$

In Exercises 25 and 26, find the slope of the line through the given pairs of points.

25. $(6, -4)$ and $(-2, 12)$

26. $(4, 5)$ and $(7, 5)$

Find the slope and the y intercept of the line represented by the given equation.

27. $y = -4x + 9$ **28.** $2x - 5y = 10$ **29.** $y = 9$ **30.** $x = 7$

Write an equation of the line L that satisfies the given conditions.

31. L has slope 5 and y intercept of $(0, -6)$.   **32.** L passes through $(-4, 9)$ and $(6, 8)$.

33. L has y intercept $(0, 6)$ and is parallel to the line with the equation $2x + 3y = 6$.

34. L passes through the point $(2, 4)$ and is perpendicular to the line with the equation $4x - 5y = 20$.

35. L has x intercept $(2, 0)$ and y intercept $(0, -3)$.

36. L has slope 3 and passes through the point $(-2, 4)$.

In Exercises 37 to 40, solve each problem.

37. If one-third of a number is added to 3 times the number, the result is 30. Find the number.

38. Two more than 4 times a number is 30. Find the number.

39. On a particular flight the cost of a coach ticket is one-half the cost of a first-class ticket. If the total cost of the tickets is $1,350, how much does each ticket cost?

40. The length of one side of a triangle is twice that of the second and 4 less than that of the third. If the perimeter is 64 meters (m), find the length of each of the sides.

4 Exponents and Polynomials

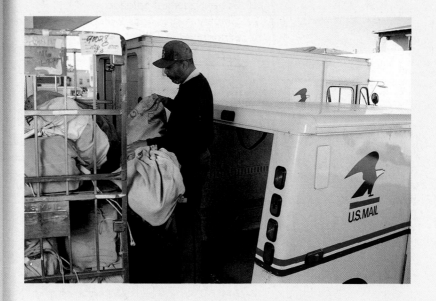

The U.S. Post Office limits the size of rectangular boxes it will accept for mailing. The regulations state the "length plus girth cannot exceed 108 inches." *Girth* means the distance around a cross section; in this case, this measurement is $2h + 2w$. Using the polynomial $l + 2w + 2h$ to describe the measurement required by the Post Office, the regulations say that $l + 2w + 2h \leq 108$ inches.

The volume of a rectangular box is expressed by another polynomial: $V = l \cdot w \cdot h$.

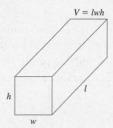

A company that wishes to produce boxes for use by postal patrons must use these formulas as well as do a statistical survey about the shapes that are useful to the largest number of customers. The surface area, expressed by another polynomial expression $2lw + 2wh + 2lh$, is also used so each box can be manufactured with the least amount of material, to help lower costs.

4.1 Positive Integer Exponents

1. *Use exponential notation*
2. *Simplify expressions with positive integer exponents*

Doubling

Take a typical sheet of paper, $8\frac{1}{2}$ by 11 inches (in.) and cut it in half. Stack the pieces together and then cut this stack again in half. You should now have four pieces. If you continue this process, stacking and cutting, you will double the number of pieces each time.

Cut Number	Number of Pieces
1	2
2	4
3	8
4	16
5	32
6	64
7	128
8	256

We can compute the height of the stack, for a given number of cuts, if we know the thickness of the original paper. Assuming the thickness to be $\frac{1}{500}$ in. = 0.002 in., the height after 8 cuts will be

$$(0.002)(256) = 0.512$$

which is a bit more than one-half inch.

If you have actually tried this, you have seen that it becomes *very* difficult to make even 8 cuts, and that the pieces become *very* small. But let's continue theoretically, for a surprising result. How high do you think the stack would be if we could make 16 cuts? The number of pieces would be

$$\underbrace{2 \cdot 2 \cdot 2 \cdots 2}_{16 \text{ times}}$$

which you can verify, with a calculator, to be 65,536 pieces. Then the height would be

$$(0.002)(65,536) \approx 131 \text{ in.}$$

which is almost 11 feet (ft) high!

292

Exponents

The calculations done in the paper-cutting exercise are best represented by using **exponents,** and the way in which the number of pieces (and the height of the stack!) grows is often called **exponential growth.**

Exponents were presented in Section 0.5, but we give a brief review here. Exponents are a shorthand form for writing repeated multiplication. Instead of writing

$$2 \cdot 2 \cdot 2 \cdot 2 \cdot 2 \cdot 2 \cdot 2$$

we write

$$2^7$$

Instead of writing

$$a \cdot a \cdot a \cdot a \cdot a$$

we write

$$a^5$$

which we read as "a to the fifth power."

We call a the **base** of the expression and 5 the **exponent,** or **power.**

> In general, for any real number a and any natural number n,
>
> $$a^n = \underbrace{a \cdot a \cdots a}_{n \text{ factors}}$$

An expression of this type is said to be in **exponential form.**

Example 1

Using Exponential Notation

Write each of the following, using exponential notation.

(a) $5y \cdot 5y \cdot 5y = (5y)^3$

(b) $w \cdot w \cdot w \cdot w = w^4$

 CHECK YOURSELF 1

Write each of the following, using exponential notation.

(a) $3z \cdot 3z \cdot 3z \cdot 3z$ (b) $x \cdot x \cdot x \cdot x \cdot x \cdot x$

Let's consider what happens when we multiply two expressions in exponential form with the same base.

We expand the expressions and then remove the parentheses.

$$a^4 \cdot a^5 = \underbrace{(a \cdot a \cdot a \cdot a)}_{4 \text{ factors}}\underbrace{(a \cdot a \cdot a \cdot a \cdot a)}_{5 \text{ factors}}$$

$$= \underbrace{a \cdot a \cdot a \cdot a \cdot a \cdot a \cdot a \cdot a \cdot a}_{9 \text{ factors}}$$

$$= a^9$$

Notice that the product is simply the base taken to the power that is the sum of the two original exponents. This leads us to our first property of exponents.

Product Rule for Exponents

For any real number a and positive integers m and n,

This is our *first property of exponents*

$a^m \cdot a^n = a^{m+n}$

$$a^m \cdot a^n = \underbrace{(a \cdot a \cdots a)}_{m \text{ factors}}\underbrace{(a \cdot a \cdots a)}_{n \text{ factors}}$$

$$= \underbrace{a \cdot a \cdots a}_{m + n \text{ factors}}$$

$$= a^{m+n}$$

Example 2 illustrates the product rule for exponents.

Example 2   Using the Product Rule

Simplify each expression.

(a) $b^4 \cdot b^6 = b^{4+6} = b^{10}$

(b) $(2a)^3 \cdot (2a)^4 = (2a)^{3+4} = (2a)^7$

(c) $(-2)^5(-2)^4 = (-2)^{5+4} = (-2)^9$

(d) $(10^7)(10^{11}) = (10)^{7+11} = (10)^{18}$

   ✔   CHECK YOURSELF 2

Simplify each expression.

(a) $(5b)^6(5b)^5$ (b) $(-3)^4(-3)^3$ (c) $10^8 \cdot 10^{12}$ (d) $(xy)^2(xy)^3$

Applying the commutative and associative properties of multiplication, we know that a product such as

$$2x^3 \cdot 3x^2$$

can be rewritten as

$$(2 \cdot 3)(x^3 \cdot x^2)$$

or as

$$6x^5$$

We expand on these ideas in Example 3.

| Example 3 | **Using Properties of Exponents** |

Using the product rule for exponents together with the commutative and associative properties, simplify each expression.

Multiply the coefficients and *add* the exponents by the product rule. With practice you will *not need to write* the regrouping step.

(a) $(x^4)(x^2)(x^3)(x) = x^{10}$

(b) $(3x^4)(5x^2) = (3 \cdot 5)(x^4 \cdot x^2) = 15x^6$

(c) $(2x^5y)(9x^3y^4) = (2 \cdot 9)(x^5 \cdot x^3)(y \cdot y^4) = 18x^8y^5$

(d) $(-3x^2y^2)(-2x^4y^3) = (-3)(-2)(x^2 \cdot x^4)(y^2 \cdot y^3) = 6x^6y^5$

✔ CHECK YOURSELF 3

Simplify each expression.

(a) $(x)(x^5)(x^3)$ (b) $(7x^5)(2x^2)$ (c) $(-2x^3y)(x^2y^2)$ (d) $(-5x^3y^2)(-x^2y^3)$

Now consider the quotient

$$\frac{a^6}{a^4}$$

If we write this in expanded form, we have

$$\frac{\overbrace{a \cdot a \cdot a \cdot a \cdot a \cdot a}^{6 \text{ factors}}}{\underbrace{a \cdot a \cdot a \cdot a}_{4 \text{ factors}}}$$

Divide the numerator and denominator by the four common factors of a.

This can be reduced to

$$\frac{\overset{1}{\cancel{a}} \cdot \overset{1}{\cancel{a}} \cdot \overset{1}{\cancel{a}} \cdot \overset{1}{\cancel{a}} \cdot a \cdot a}{\underset{1}{\cancel{a}} \cdot \underset{1}{\cancel{a}} \cdot \underset{1}{\cancel{a}} \cdot \underset{1}{\cancel{a}}} \quad \text{or} \quad a^2$$

Note that $\dfrac{a}{a} = 1$, where $a \neq 0$.

This means that

$$\frac{a^6}{a^4} = a^2$$

This leads us to our second property of exponents.

This is our *second property of exponents*. We write $a \neq 0$ to avoid division by 0.

Quotient Rule for Exponents

In general, for any real number a ($a \neq 0$) and positive integers m and n,

$$\frac{a^m}{a^n} = a^{m-n} \quad \text{where } m > n$$

Example 4 illustrates this rule.

Example 4   Using Properties of Exponents

Simplify each expression.

(a) $\dfrac{x^{10}}{x^4} = x^{10-4} = x^6$
           Subtract the exponents, applying the quotient rule.

Note that $a^1 = a$; there is no need to write the exponent 1 because it is understood.

(b) $\dfrac{a^8}{a^7} = a^{8-7} = a$

(c) $\dfrac{63w^8}{7w^5} = 9w^{8-5} = 9w^3$
           Divide the coefficients and subtract the exponents.

(d) $\dfrac{-32a^4b^5}{8a^2b} = -4a^{4-2}b^{5-1} = -4a^2b^4$
           Divide the coefficients and subtract the exponents for *each* variable.

(e) $\dfrac{10^{16}}{10^6} = 10^{16-6} = 10^{10}$

✔   *CHECK YOURSELF 4*

Simplify each expression.

(a) $\dfrac{y^{12}}{y^5}$ (b) $\dfrac{x^9}{x^8}$ (c) $\dfrac{45r^8}{-9r^6}$ (d) $\dfrac{49a^6b^7}{7ab^3}$ (e) $\dfrac{10^{13}}{10^5}$

What happens when a product such as xy is raised to a power? Consider, for example, $(xy)^4$:

$$(xy)^4 = (xy)(xy)(xy)(xy)$$
$$= (x \cdot x \cdot x \cdot x)(y \cdot y \cdot y \cdot y)$$
$$= x^4 y^4$$

Here we have used the commutative and associative properties.

This is expressed in the following product-power rule.

Product-Power Rule for Exponents

This is our *third property of exponents.*

In general, for any real numbers a and b and positive integer n,

$$(ab)^n = a^n b^n$$

Example 5 illustrates this rule.

Example 5

Using the Product–Power Rule

Simplify each expression.

(a) $(2x)^3 = 2^3 x^3 = 8x^3$

(b) $(-3x)^4 = (-3)^4 x^4 = 81x^4$

✔ CHECK YOURSELF 5

Simplify each expression.

(a) $(3x)^3$ (b) $(-2x)^4$

Now, consider the expression

$$(3^2)^3$$

This could be expanded to

$$(3^2)(3^2)(3^2)$$

and then expanded again to

$$(3)(3)(3)(3)(3)(3) = 3^6$$

This leads us to the power rule for exponents.

Power Rule for Exponents

This is our *fourth property of exponents.*

In general, for any real number a and positive integers m and n,

$$(a^m)^n = a^{mn}$$

Example 6	**Using the Power Rule for Exponents**

Simplify each expression.

(a) $(2^5)^3 = 2^{15}$

(b) $(x^2)^4 = x^8$

(c) $(2x^3)^3 = 2^3(x^3)^3 = 2^3x^9 = 8x^9$

 CHECK YOURSELF 6

Simplify each expression.

(a) $(3^4)^3$ (b) $(x^2)^6$ (c) $(3x^3)^4$

We have one final exponent property to develop. Suppose we have a quotient raised to a power. Consider the following:

$$\left(\frac{x}{3}\right)^3 = \frac{x}{3} \cdot \frac{x}{3} \cdot \frac{x}{3} = \frac{x \cdot x \cdot x}{3 \cdot 3 \cdot 3} = \frac{x^3}{3^3}$$

Note that the power, here 3, has been applied to the numerator x and to the denominator 3. This gives our fifth property of exponents.

Quotient-Power Rule for Exponents

This is our *fifth property of exponents.*

For any real numbers a and b, where b is not equal to 0, and positive integer m,

$$\left(\frac{a}{b}\right)^m = \frac{a^m}{b^m}$$

In words, to raise a quotient to a power, raise the numerator and denominator to that same power.

Example 7 illustrates the use of this property. Again note that the other properties may also have to be applied in simplifying an expression.

Example 7	# Using the Quotient–Power Rule for Exponents

Simplify each expression.

(a) $\left(\dfrac{3}{4}\right)^3 = \dfrac{3^3}{4^3} = \dfrac{27}{64}$

(b) $\left(\dfrac{x^3}{y^2}\right)^4 = \dfrac{(x^3)^4}{(y^2)^4}$

$= \dfrac{x^{12}}{y^8}$

(c) $\left(\dfrac{r^2 s^3}{t^4}\right)^2 = \dfrac{(r^2 s^3)^2}{(t^4)^2}$

$= \dfrac{(r^2)^2 (s^3)^2}{(t^4)^2}$

$= \dfrac{r^4 s^6}{t^8}$

✔ CHECK YOURSELF 7

Simplify each expression.

(a) $\left(\dfrac{2}{3}\right)^4$ (b) $\left(\dfrac{m^3}{n^4}\right)^5$ (c) $\left(\dfrac{a^2 b^3}{c^5}\right)^2$

The following table summarizes the five properties of exponents that were discussed in this section:

General Form	Example
1. $a^m a^n = a^{m+n}$	$x^2 \cdot x^3 = x^5$
2. $\dfrac{a^m}{a^n} = a^{m-n} \ (m > n)$	$\dfrac{5^7}{5^3} = 5^4$
3. $(a^m)^n = a^{mn}$	$(z^5)^4 = z^{20}$
4. $(ab)^m = a^m b^m$	$(4x)^3 = 4^3 x^3 = 64x^3$
5. $\left(\dfrac{a}{b}\right)^m = \dfrac{a^m}{b^m}$	$\left(\dfrac{2}{3}\right)^6 = \dfrac{2^6}{3^6} = \dfrac{64}{729}$

✔ *CHECK YOURSELF ANSWERS*

1. (a) $(3z)^4$; (b) x^6 **2.** (a) $(5b)^{11}$; (b) $(-3)^7$; (c) 10^{20}; (d) $(xy)^5$

3. (a) x^9; (b) $14x^7$; (c) $-2x^5y^3$; (d) $5x^5y^5$

4. (a) y^7; (b) x; (c) $-5r^2$; (d) $7a^5b^4$; (e) 10^8 **5.** (a) $27x^3$; (b) $16x^4$

6. (a) 3^{12}; (b) x^{12}; (c) $81x^{12}$ **7.** (a) $\dfrac{16}{81}$; (b) $\dfrac{m^{15}}{n^{20}}$; (c) $\dfrac{a^4b^6}{c^{10}}$

Exercises · 4.1

Write each expression in simplest exponential form.

1. $x^4 \cdot x^5$

2. $x^7 \cdot x^9$

3. $x^5 \cdot x^3 \cdot x^2$

4. $x^8 \cdot x^4 \cdot x^7$

5. $3^5 \cdot 3^2$

6. $(-3)^4(-3)^6$

7. $(-2)^3(-2)^5$

8. $4^3 \cdot 4^4$

9. $4 \cdot x^2 \cdot x^4 \cdot x^7$

10. $3 \cdot x^3 \cdot x^5 \cdot x^8$

11. $\left(\dfrac{1}{2}\right)^2\left(\dfrac{1}{2}\right)^3\left(\dfrac{1}{2}\right)$

12. $\left(-\dfrac{1}{3}\right)^4\left(-\dfrac{1}{3}\right)\left(-\dfrac{1}{3}\right)^5$

13. $(-2)^2(-2)^3(x^4)(x^5)$

14. $(-3)^4(-3)^2(x)^2(x)^6$

15. $(2x)^2(2x)^3(2x)^4$

16. $(-3x)^3(-3x)^5(-3x)^7$

Use the product rule of exponents together with the commutative and associative properties to simplify the products.

17. $(x^2y^3)(x^4y^2)$

18. $(x^4y)(x^2y^3)$

19. $(x^3y^2)(x^4y^2)(x^2y^3)$

20. $(x^2y^3)(x^3y)(x^4y^2)$

21. $(2x^4)(3x^3)(-4x^3)$

22. $(2x^3)(-3x)(-4x^4)$

23. $(5x^2)(3x^3)(x)(-2x^3)$

24. $(4x^2)(2x)(x^2)(2x^3)$

25. $(5xy^3)(2x^2y)(3xy)$

26. $(-3xy)(5x^2y)(-2x^3y^2)$

27. $(x^2yz)(x^3y^5z)(x^4yz)$

28. $(xyz)(x^8y^3z^6)(x^2yz)(xyz^4)$

Use the quotient rule of exponents to simplify each expression.

29. $\dfrac{x^{10}}{x^7}$

30. $\dfrac{b^{23}}{b^{18}}$

31. $\dfrac{x^7y^{11}}{x^4y^3}$

32. $\dfrac{x^5y^9}{xy^4}$

33. $\dfrac{x^5y^4z^2}{xy^2z}$

34. $\dfrac{x^8y^6z^4}{x^3yz^3}$

35. $\dfrac{21x^4y^5}{7xy^2}$

36. $\dfrac{48x^6y^6}{12x^3y}$

Use your calculator to evaluate each expression.

37. 4^3

38. 5^7

39. $(-3)^4$

40. $(-4)^5$

41. $2^3 \cdot 2^5$

42. $3^4 \cdot 3^6$

43. $(3x^2)(2x^4)$, where $x = 2$

44. $(4x^3)(5x^4)$, where $x = 3$

45. $(2x^4)(4x^2)$, where $x = -2$

46. $(3x^5)(2x^3)$, where $x = -3$

47. $(-2x^3)(-3x^5)$, where $x = 2$

48. $(-3x^2)(-4x^4)$, where $x = 4$

Simplify each expression.

49. $(-3x)(5x^5)$

50. $(-5x^2)(-2x^2)$

51. $(2x)^3$

52. $(-3x)^3$

53. $(x^3)^7$

54. $(-x^3)^5$

55. $(3x)(-2x)^3$

56. $(2x)(-3x)^3$

57. $(2x^3)^5$

58. $(-3x^2)^3$

59. $(-2x^2)^3(3x^2)^3$

60. $(-3x^2)^2(5x^2)^2$

4-11

61. $(2x^3)(5x^3)^2$

62. $(3x^3)^2(x^2)^4$

63. $(2x^3)^4(3x^4)^2$

64. $\left(\dfrac{3}{4}\right)^2$

65. $\left(\dfrac{2}{3}\right)^2$

66. $\left(\dfrac{x}{5}\right)^3$

67. $\left(\dfrac{a}{2}\right)^4$

68. $\left(\dfrac{m^3}{n^2}\right)^3$

69. $\left(\dfrac{a^4}{b^3}\right)^4$

70. $\left(\dfrac{a^3b^2}{c^4}\right)^2$

71. $\left(\dfrac{x^5y^2}{z^4}\right)^3$

72. $\left(\dfrac{2x^5}{y^3}\right)^2$

73. $\left(\dfrac{2x^5}{3x^3}\right)^3$

74. $(-8x^2y)(-3x^4y^5)^4$

75. $(5x^5y)^2(-3x^3y^4)^3$

76. $\left(\dfrac{3x^4y^9}{2x^2y^7}\right)\left(\dfrac{x^6y^3}{x^3y^2}\right)^2$

77. $\left(\dfrac{6x^5y^4}{5xy}\right)\left(\dfrac{x^3y^5}{xy^3}\right)^3$

78. Business. The value A of a savings account that compounds interest annually is given by the formula

$$A = P(1 + r)^t$$

where

$$P = \text{original amount (principle)}$$
$$r = \text{interest rate in decimal form}$$
$$t = \text{time in years}$$

Find the amount of money in the account after 8 years if $2,000 was invested initially at 5% compounded annually.

79. Business. Using the formula for compound interest in Exercise 78, determine the amount of money in the account if the original investment is doubled.

80. You have learned rules for working with exponents when multiplying, dividing, and raising an expression to a power.

   (a) Explain each rule in your own words. Give numerical examples.

   (b) Is there a rule for raising a *sum* to a power? That is, does $(a + b)^n = a^n + b^n$? Use numerical examples to explain why this is true in general or why it is not. Is it always true or always false?

81. Work with another student to investigate the rate of inflation. The annual rate of inflation was about 3% from 1990 to 2004. This means that the value of the goods that you could buy for $1 in 1990 would cost 3% more in 1991, 3% more than that in 1992, etc. If a movie ticket cost $5.50 in 1990, what would it cost today if movie tickets just kept up with inflation? Construct a table to solve the problem.

Solve the following problems.

82. Write x^{12} as a power of x^2.

83. Write y^{15} as a power of y^3.

84. Write a^{16} as a power of a^2.

85. Write m^{20} as a power of m^5.

86. Write each of the following as powers of 8 (remember that $8 = 2^3$): 2^{12}, 2^{18}, $(2^5)^3$, $(2^7)^6$.

87. Write each of the following as powers of 9: 3^8, 3^{14}, $(3^5)^8$, $(3^4)^7$.

88. What expression, raised to the third power, is $-8x^6y^9z^{15}$?

89. What expression, raised to the fourth power, is $81x^{12}y^8z^{16}$?

SECTION 4.2

Zero and Negative Exponents and Scientific Notation

4.2 OBJECTIVES

1. Define the zero exponent
2. Simplify expressions with negative exponents
3. Write a number in scientific notation
4. Solve an application of scientific notation

In Section 4.1, we examined properties of exponents, but all the exponents were positive integers. In this section, we look at zero and negative exponents. First we extend the quotient rule so that we can define an exponent of zero.

Recall that, in the quotient rule, to divide two expressions that have the same base, we keep the base and subtract the exponents.

$$\frac{a^m}{a^n} = a^{m-n} \qquad \text{where } a \neq 0$$

Now, suppose that we allow *m to equal n*. We then have

$$\frac{a^m}{a^m} = a^{m-m} = a^0 \qquad \text{where } a \neq 0$$

But we know that it is also true that

$$\frac{a^m}{a^m} = 1 \qquad \text{where } a \neq 0$$

Comparing these two equations we see that the following definition is reasonable.

We must have $a \neq 0$. The form 0^0 is called **indeterminate** and is considered in later mathematics classes.

The Zero Exponent

For any real number a where $a \neq 0$,

$$a^0 = 1$$

| Example 1 | The Zero Exponent |

Use the above definition to simplify each expression.

Note that in $6x^0$ the exponent 0 applied *only* to x.

(a) $17^0 = 1$ (b) $(a^3b^2)^0 = 1$ (c) $6x^0 = 6 \cdot 1 = 6$ (d) $-3y^0 = -3$

✔ CHECK YOURSELF 1

Simplify each expression.

(a) 25^0 (b) $(m^4n^2)^0$ (c) $8s^0$ (d) $-7t^0$

Recall that, in the product rule, to multiply expressions with the same base, we keep the base and add the exponents.

$$a^m \cdot a^n = a^{m+n}$$

Now, what if we allow one of the exponents to be negative and apply the product rule? Suppose, for instance, that $m = 3$ and $n = -3$. Then

$$a^m \cdot a^n = a^3 \cdot a^{-3} = a^{3+(-3)}$$
$$= a^0 = 1$$

Recall that if two numbers have a product of 1, they must be reciprocals of each other.

so $a^3 \cdot a^{-3} = 1$

If we divide both sides by a^3, where $a \neq 0$, we have

$$a^{-3} = \frac{1}{a^3}$$

This is the basis for the following.

Negative Integer Exponents

For any nonzero real number a and whole number n,

$$a^{-n} = \frac{1}{a^n}$$

and a^{-n} is the **multiplicative inverse** of a^n.

John Wallis (1616–1702), an English mathematician, was the first to fully discuss the meaning of 0, negative, and rational exponents (which we discuss in Section 10.3).

We can also say that a^{-n} is the reciprocal of a^n. Example 2 illustrates this definition.

Example 2   Using Properties of Exponents

From this point on, to *simplify* will mean to write the expression with *positive exponents only.*

Also, we will restrict all variables so that they represent nonzero real numbers.

Simplify the following expressions.

(a) $y^{-5} = \dfrac{1}{y^5}$

(b) $4^{-2} = \dfrac{1}{4^2} = \dfrac{1}{16}$

(c) $(-3)^{-3} = \dfrac{1}{(-3)^3} = \dfrac{1}{-27} = -\dfrac{1}{27}$

✔ CHECK YOURSELF 2

Simplify each of the following expressions.

(a) a^{-10} (b) 2^{-4} (c) $(-4)^{-2}$

Example 3 illustrates the case where coefficients are involved in an expression with negative exponents. As will be clear, some caution must be used. We must determine exactly what is included in the base of the exponent.

Example 3

Using Properties of Exponents

Simplify each of the following expressions.

(a) $2x^{-3} = 2 \cdot \dfrac{1}{x^3} = \dfrac{2}{x^3}$

The exponent -3 applies only to the variable x, and *not* to the coefficient 2.

Caution

The expressions

$4w^{-2}$ and $(4w)^{-2}$

are *not* the same. Do you see why?

(b) $4w^{-2} = 4 \cdot \dfrac{1}{w^2} = \dfrac{4}{w^2}$

(c) $(4w)^{-2} = \dfrac{1}{(4w)^2} = \dfrac{1}{16w^2}$

✔ CHECK YOURSELF 3

Simplify each of the following expressions.

(a) $3w^{-4}$ (b) $10x^{-5}$ (c) $(2y)^{-4}$ (d) $-5t^{-2}$

Suppose that a variable with a negative exponent appears in the denominator of an expression. For instance, if we wish to simplify

$$\frac{1}{a^{-2}}$$

we can multiply numerator and denominator by a^2:

$$\frac{1}{a^{-2}} = \frac{1 \cdot a^2}{a^{-2} \cdot a^2} = \frac{a^2}{a^0} = \frac{a^2}{1} = a^2$$

Negative exponent in denominator   Positive exponent in numerator

So

$$\frac{1}{a^{-2}} = a^2$$

We can write, in general:

For any nonzero real number a and integer n,

$$\frac{1}{a^{-n}} = a^n$$

| Example 4 | **Using Properties of Exponents** |

Simplify each of the following expressions.

(a) $\dfrac{1}{y^{-3}} = y^3$

(b) $\dfrac{1}{2^{-5}} = 2^5 = 32$

(c) $\dfrac{3}{4x^{-2}} = \dfrac{3x^2}{4}$ The exponent -2 applies only to x, not to 4.

(d) $\dfrac{a^{-3}}{b^{-4}} = \dfrac{b^4}{a^3}$

✔ CHECK YOURSELF 4

Simplify each of the following expressions.

(a) $\dfrac{1}{x^{-4}}$ (b) $\dfrac{1}{3^{-3}}$ (c) $\dfrac{2}{3a^{-2}}$ (d) $\dfrac{c^{-5}}{d^{-7}}$

To review these properties, return to Section 4.1.

The product and quotient rules for exponents apply to expressions that involve any integral exponent—positive, negative, or 0. Example 5 illustrates this concept.

| Example 5 | **Using Properties of Exponents** |

Simplify each of the following expressions, and write the result, using positive exponents only.

(a) $x^3 \cdot x^{-7} = x^{3+(-7)}$ Add the exponents by the product rule.

$$= x^{-4} = \frac{1}{x^4}$$

(b) $\dfrac{m^{-5}}{m^{-3}} = m^{-5-(-3)} = m^{-5+3}$ Subtract the exponents by the quotient rule.

$$= m^{-2} = \dfrac{1}{m^2}$$

(c) $\dfrac{x^5 x^{-3}}{x^{-7}} = \dfrac{x^{5+(-3)}}{x^{-7}} = \dfrac{x^2}{x^{-7}} = x^{2-(-7)} = x^9$ We apply first the product rule and then the quotient rule.

In simplifying expressions involving negative exponents, there are often alternate approaches. For instance, in Example 5(b), we could have made use of our earlier work to write

$$\dfrac{m^{-5}}{m^{-3}} = \dfrac{m^3}{m^5} = m^{3-5} = m^{-2} = \dfrac{1}{m^2}$$

Note that m^{-5} in the numerator becomes m^5 in the denominator, and m^{-3} in the denominator becomes m^3 in the numerator. We then simplify as before.

✔ CHECK YOURSELF 5

Simplify each of the following expressions.

(a) $x^9 \cdot x^{-5}$ (b) $\dfrac{y^{-7}}{y^{-3}}$ (c) $\dfrac{a^{-3} a^2}{a^{-5}}$

How do we simplify a rational expression raised to a negative power? As we have seen, the properties of exponents can be extended to include negative exponents.

Suppose for example, we wish to simplify $\left(\dfrac{x}{y}\right)^{-2}$:

$$\left(\dfrac{x}{y}\right)^{-2} = \dfrac{x^{-2}}{y^{-2}}$$ Use the quotient-power rule.

$$= \dfrac{y^2}{x^2} = \left(\dfrac{y}{x}\right)^2$$ Use the quotient-power rule again.

In general, we have:

Quotient Raised to a Negative Power

For any nonzero real numbers a and b and any integer n.

$$\left(\dfrac{a}{b}\right)^{-n} = \left(\dfrac{b}{a}\right)^{n}$$

| Example 6 | **Extending the Properties of Exponents** |

Simplify each expression.

(a) $\left(\dfrac{s^3}{t^2}\right)^{-2} = \left(\dfrac{t^2}{s^3}\right)^2 = \dfrac{t^4}{s^6}$

(b) $\left(\dfrac{m^2}{n^{-2}}\right)^{-3} = \left(\dfrac{n^{-2}}{m^2}\right)^3 = \dfrac{n^{-6}}{m^6} = \dfrac{1}{m^6 n^6}$

  CHECK YOURSELF 6

Simplify each expression.

(a) $\left(\dfrac{3t^2}{s^3}\right)^{-3}$ (b) $\left(\dfrac{x^5}{y^{-2}}\right)^{-3}$

| Example 7 | **Using Properties of Exponents** |

Simplify each of the following expressions.

(a) $\left(\dfrac{3}{q^5}\right)^{-2} = \left(\dfrac{q^5}{3}\right)^2$

$= \dfrac{q^{10}}{9}$

(b) $\left(\dfrac{x^3}{y^4}\right)^{-3} = \left(\dfrac{y^4}{x^3}\right)^3$

$= \dfrac{(y^4)^3}{(x^3)^3} = \dfrac{y^{12}}{x^9}$

  CHECK YOURSELF 7

Simplify each of the following expressions.

(a) $\left(\dfrac{r^4}{5}\right)^{-2}$ (b) $\left(\dfrac{a^4}{b^3}\right)^{-3}$

As you might expect, more complicated expressions require the use of more than one of the properties, for simplification. Example 8 illustrates such cases.

| Example 8 | **Using Properties of Exponents** |

Simplify each of the following expressions.

(a) $\dfrac{(a^2)^{-3}(a^3)^4}{(a^{-3})^3} = \dfrac{a^{-6} \cdot a^{12}}{a^{-9}}$ Apply the power rule to each factor.

$= \dfrac{a^{-6+12}}{a^{-9}} = \dfrac{a^6}{a^{-9}}$ Apply the product rule.

$= a^{6-(-9)} = a^{6+9} = a^{15}$ Apply the quotient rule.

It may help to separate the problem into three fractions—one for the coefficients and one for each of the variables.

(b) $\dfrac{8x^{-2}y^{-5}}{12x^{-4}y^3} = \dfrac{8}{12} \cdot \dfrac{x^{-2}}{x^{-4}} \cdot \dfrac{y^{-5}}{y^3}$

$= \dfrac{2}{3} \cdot x^{-2-(-4)} \cdot y^{-5-3}$

$= \dfrac{2}{3} \cdot x^2 \cdot y^{-8} = \dfrac{2x^2}{3y^8}$

Caution

Be Careful! Another possible first step (and generally an efficient one) is to rewrite an expression by using our earlier definitions

$a^{-n} = \dfrac{1}{a^n}$ and $\dfrac{1}{a^{-n}} = a^n$

(c) $\left(\dfrac{pr^3s^{-5}}{p^3r^{-3}s^{-2}}\right)^{-2} = (p^{1-3}r^{3-(-3)}s^{-5-(-2)})^{-2}$

$= (p^{-2}r^6s^{-3})^{-2}$ Apply the quotient rule inside the parentheses.

$= (p^{-2})^{-2}(r^6)^{-2}(s^{-3})^{-2}$ Apply the rule for a product to a power.

$= p^4r^{-12}s^6 = \dfrac{p^4s^6}{r^{12}}$ Apply the power rule.

Caution

A *common error* is to write

$\dfrac{8x^{-2}y^{-5}}{12x^{-4}y^3} = \dfrac{12x^4}{8x^2y^3y^5}$

This is *not* correct.

For instance, in Example 8(b), we would *correctly* write

$$\dfrac{8x^{-2}y^{-5}}{12x^{-4}y^3} = \dfrac{8x^4}{12x^2y^3y^5}$$

The coefficients should not have been moved along with the factors in *x*. Keep in mind that the negative exponents apply *only* to the variables. The coefficients remain *where they were* in the original expression when the expression is rewritten using this approach.

✔ CHECK YOURSELF 8

Simplify each of the following expressions.

(a) $\dfrac{(x^5)^{-2}(x^2)^3}{(x^{-4})^3}$ (b) $\dfrac{12a^{-3}b^{-2}}{16a^{-2}b^3}$ (c) $\left(\dfrac{xy^{-3}z^{-5}}{x^{-4}y^{-2}z^3}\right)^{-3}$

Let us now take a look at an important use of exponents, scientific notation.

We begin the discussion with a calculator exercise. On most scientific calculators, if you multiply 2.3 times 1,000, the display will read

$$2300$$

Multiply by 1,000 a second time. Now you will see

$$2300000$$

Multiplying by 1,000 a third time will result in the display

| 2.3   09   or   2.3   E09

And multiplying by 1,000 again yields

| 2.3   12   or   2.3   E12

This must equal 2,300,000,000.

Can you see what is happening? This is the way calculators display very large numbers. The number on the left is always between 1 and 10, and the number on the right indicates the number of places the decimal point must be moved to the right to put the answer in standard (or decimal) form.

Consider the following table:

$$2.3 = 2.3 \times 10^0$$
$$23 = 2.3 \times 10^1$$
$$230 = 2.3 \times 10^2$$
$$2{,}300 = 2.3 \times 10^3$$
$$23{,}000 = 2.3 \times 10^4$$
$$230{,}000 = 2.3 \times 10^5$$

This notation is used frequently in science. It is not uncommon in scientific applications of algebra to find yourself working with very large or very small numbers. Even in the time of Archimedes (287–212 B.C.E.), the study of such numbers was not unusual. Archimedes estimated that the universe was 23,000,000,000,000,000 m in diameter, which is the approximate distance light travels in $2\frac{1}{2}$ years. By comparison, Polaris (the North Star) is 680 light-years from the earth. Example 10 will discuss the idea of light-years.

In scientific notation, his estimate for the diameter of the universe would be

$$2.3 \times 10^{16} \text{ m}$$

In general, we can define scientific notation as follows.

Scientific Notation

Any number written in the form

$$a \times 10^n$$

where $1 \leq a < 10$ and n is an integer, is written in scientific notation.

Example 9   Using Scientific Notation

Write each of the following numbers in scientific notation.

Note the pattern for writing a number in scientific notation.

(a) $120{,}000. = 1.2 \times 10^5$
       5 places   The power is 5.

(b) $\underset{\text{7 places}}{88,000,000.} = 8.8 \times 10^7$ The power is 7.

The exponent on 10 shows the *number of places* we must move the decimal point so that the multiplier will be a number between 1 and 10. A positive exponent tells us to move right, while a negative exponent indicates to move left.

(c) $\underset{\text{8 places}}{520,000,000.} = 5.2 \times 10^8$

(d) $\underset{\text{9 places}}{4,000,000,000.} = 4 \times 10^9$

Note: To convert back to standard or decimal form, the process is simply reversed.

(e) $\underset{\text{4 places}}{0.0005} = 5 \times 10^{-4}$ If the decimal point is to be moved to the left, the exponent will be negative.

(f) $\underset{\text{9 places}}{0.0000000081} = 8.1 \times 10^{-9}$

✔ CHECK YOURSELF 9

Write in scientific notation.

(a) 212,000,000,000,000,000   (b) 0.00079

(c) 5,600,000   (d) 0.0000007

Example 10

An Application of Scientific Notation

(a) Light travels at a speed of 3.05×10^8 meters per second (m/s). There are approximately 3.15×10^7 s in a year. How far does light travel in a year?

We multiply the distance traveled in 1 s by the number of seconds in a year. This yields

$$(3.05 \times 10^8)(3.15 \times 10^7) = (3.05 \cdot 3.15)(10^8 \cdot 10^7)$$ Multiply the coefficients, and add the exponents.

Note that $9.6075 \times 10^{15} \approx 10 \times 10^{15} = 10^{16}$.

$$= 9.6075 \times 10^{15}$$

For our purposes we round the distance light travels in 1 year to 10^{16} m. This unit is called a **light-year,** and it is used to measure astronomical distances.

(b) The distance from earth to the star Spica (in Virgo) is 2.2×10^{18} m. How many light-years is Spica from earth?

We divide the distance (in meters) by the number of meters in 1 light-year.

$$\frac{2.2 \times 10^{18}}{10^{16}} = 2.2 \times 10^{18-16}$$

$$= 2.2 \times 10^2 = 220 \text{ light-years}$$

✔ CHECK YOURSELF 10

The farthest object that can be seen with the unaided eye is the Andromeda galaxy. This galaxy is 2.3×10^{22} m from earth. What is this distance in light-years?

✔ *CHECK YOURSELF ANSWERS*

1. (a) 1; (b) 1; (c) 8; (d) -7 **2.** (a) $\dfrac{1}{a^{10}}$; (b) $\dfrac{1}{16}$; (c) $\dfrac{1}{16}$

3. (a) $\dfrac{3}{w^4}$; (b) $\dfrac{10}{x^5}$; (c) $\dfrac{1}{16y^4}$; (d) $-\dfrac{5}{t^2}$ **4.** (a) x^4; (b) 27; (c) $\dfrac{2a^2}{3}$; (d) $\dfrac{d^7}{c^5}$

5. (a) x^4; (b) $\dfrac{1}{y^4}$; (c) a^4 **6.** (a) $\dfrac{s^9}{27t^6}$; (b) $\dfrac{1}{x^{15}y^6}$

7. (a) $\dfrac{25}{r^8}$; (b) $\dfrac{b^9}{a^{12}}$ **8.** (a) x^8; (b) $\dfrac{3}{4ab^5}$; (c) $\dfrac{y^3 z^{24}}{x^{15}}$

9. (a) 2.12×10^{17}; (b) 7.9×10^{-4}; (c) 5.6×10^6; (d) 7×10^{-7}

10. 2,300,000 light-years

Exercises · 4.2

Simplify each expression.

1. x^{-5}
2. 3^{-3}
3. 5^{-2}
4. x^{-8}

5. $(-5)^{-2}$
6. $(-3)^{-3}$
7. $(-2)^{-3}$
8. $(-2)^{-4}$

9. $\left(\dfrac{2}{3}\right)^{-3}$
10. $\left(\dfrac{3}{4}\right)^{-2}$
11. $3x^{-2}$
12. $4x^{-3}$

13. $-5x^{-4}$
14. $(-2x)^{-4}$
15. $(-3x)^{-2}$
16. $-5x^{-2}$

17. $\dfrac{1}{x^{-3}}$
18. $\dfrac{1}{x^{-5}}$
19. $\dfrac{2}{5x^{-3}}$
20. $\dfrac{3}{4x^{-4}}$

21. $\dfrac{x^{-3}}{y^{-4}}$
22. $\dfrac{x^{-5}}{y^{-3}}$

Use the properties of exponents to simplify the expressions.

23. $x^5 \cdot x^{-3}$
24. $y^{-4} \cdot y^5$
25. $a^{-9} \cdot a^6$
26. $w^{-5} \cdot w^3$

27. $z^{-2} \cdot z^{-8}$
28. $b^{-7} \cdot b^{-1}$
29. $a^{-5} \cdot a^5$
30. $x^{-4} \cdot x^4$

31. $\dfrac{x^{-5}}{x^{-2}}$
32. $\dfrac{x^{-3}}{x^{-6}}$

Use the properties of exponents to simplify.

33. $(x^5)^3$
34. $(w^4)^6$
35. $(2x^{-3})(x^2)^4$
36. $(p^4)(3p^3)^2$

37. $(3a^{-4})(a^3)(a^2)$
38. $(5y^{-2})(2y)(y^5)$
39. $(x^4y)(x^2)^3(y^3)^0$
40. $(r^4)^2(r^2s)(s^3)^2$

41. $(ab^2c)(a^4)^4(b^2)^3(c^3)^4$
42. $(p^2qr^2)(p^2)(q^3)^2(r^2)^0$
43. $(x^5)^{-3}$
44. $(x^{-2})^{-3}$

45. $(b^{-4})^{-2}$
46. $(a^0b^{-4})^3$
47. $(x^5y^{-3})^2$
48. $(p^{-3}q^2)^{-2}$

49. $(x^{-4}y^{-2})^{-3}$
50. $(3x^{-2}y^{-2})^3$
51. $(2x^{-3}y^0)^{-5}$
52. $\dfrac{a^{-6}}{b^{-4}}$

53. $\dfrac{x^{-2}}{y^{-4}}$
54. $\left(\dfrac{x^{-3}}{y^2}\right)^{-3}$
55. $\dfrac{x^{-4}}{y^{-2}}$
56. $\dfrac{(3x^{-4})^2(2x^2)}{x^6}$

57. $(4x^{-2})^2(3x^{-4})$
58. $(5x^{-4})^{-4}(2x^3)^{-5}$

Simplify each expression.

59. $(2x^5)^4(x^3)^2$

60. $(3x^2)^3(x^2)^4(x^2)$

61. $(2x^{-3})^3(3x^3)^2$

62. $(x^2y^3)^4(xy^3)^0$

63. $(xy^5z)^4(xyz^2)^8(x^6yz)^5$

64. $(x^2y^2z^2)^0(xy^2z)^2(x^3yz^2)$

65. $(3x^{-2})(5x^2)^2$

66. $(2a^3)^2(a^0)^5$

67. $(2w^3)^4(3w^{-5})^2$

68. $(3x^3)^2(2x^4)^5$

69. $\dfrac{3x^6}{2y^9} \cdot \dfrac{y^5}{x^3}$

70. $\dfrac{x^8}{y^6} \cdot \dfrac{2y^9}{x^3}$

71. $(-7x^2y)(-3x^5y^6)^4$

72. $\left(\dfrac{2w^5z^3}{3x^3y^9}\right)\left(\dfrac{x^5y^4}{w^4z^0}\right)^2$

73. $(2x^2y^{-3})(3x^{-4}y^{-2})$

74. $(-5a^{-2}b^{-4})(2a^5b^0)$

75. $\dfrac{(x^{-3})(y^2)}{y^{-3}}$

76. $\dfrac{6x^3y^{-4}}{24x^{-2}y^{-2}}$

77. $\dfrac{15x^{-3}y^2z^{-4}}{20x^{-4}y^{-3}z^2}$

78. $\dfrac{24x^{-5}y^{-3}z^2}{36x^{-2}y^3z^{-2}}$

79. $\dfrac{x^{-5}y^{-7}}{x^0y^{-4}}$

80. $\left(\dfrac{xy^3z^{-4}}{x^{-3}y^{-2}z^2}\right)^{-2}$

81. $\dfrac{x^{-2}y^2}{x^3y^{-2}} \cdot \dfrac{x^{-4}y^2}{x^{-2}y^{-2}}$

82. $\left(\dfrac{x^{-3}y^3}{x^{-4}y^2}\right)^3 \cdot \left(\dfrac{x^{-2}y^{-2}}{xy^4}\right)^{-1}$

83. $x^{2n} \cdot x^{3n}$

84. $x^{n+1} \cdot x^{3n}$

85. $\dfrac{x^{n+3}}{x^{n+1}}$

86. $\dfrac{x^{n-4}}{x^{n-1}}$

87. $(y^n)^{3n}$

88. $(x^{n+1})^n$

89. $\dfrac{x^{2n} \cdot x^{n+2}}{x^{3n}}$

90. $\dfrac{x^n \cdot x^{3n+5}}{x^{4n}}$

Express each number in scientific notation.

91. The distance from the earth to the sun: 93,000,000 mi.

92. The diameter of a grain of sand: 0.000021 m.

93. The diameter of the sun: 130,000,000,000 cm.

94. The number of oxygen atoms in 16 grams of oxygen of a gas: 602,000,000,000,000,000,000,000 (Avogadro's number).

95. The mass of the sun is approximately 1.98×10^{30} kg. If this were written in standard or decimal form, how many 0s would follow the digit 8?

96. Archimedes estimated the universe to be 2.3×10^{19} millimeters (mm) in diameter. If this number were written in standard or decimal form, how many 0s would follow the digit 3?

Write each expression in standard notation.

97. 8×10^{-3} **98.** 7.5×10^{-6} **99.** 2.8×10^{-5} **100.** 5.21×10^{-4}

Write each of the following in scientific notation.

101. 0.0005 **102.** 0.000003 **103.** 0.00037 **104.** 0.000051

Compute the expressions using scientific notation, and write your answer in that form.

105. $(4 \times 10^{-3})(2 \times 10^{-5})$ **106.** $(1.5 \times 10^{-6})(4 \times 10^2)$ **107.** $\dfrac{9 \times 10^3}{3 \times 10^{-2}}$ **108.** $\dfrac{7.5 \times 10^{-4}}{1.5 \times 10^2}$

Perform the indicated calculations. Write your result in scientific notation.

109. $(2 \times 10^5)(4 \times 10^4)$ **110.** $(2.5 \times 10^7)(3 \times 10^5)$ **111.** $\dfrac{6 \times 10^9}{3 \times 10^7}$ **112.** $\dfrac{4.5 \times 10^{12}}{1.5 \times 10^7}$

113. $\dfrac{(3.3 \times 10^{15})(6 \times 10^{15})}{(1.1 \times 10^8)(3 \times 10^6)}$ **114.** $\dfrac{(6 \times 10^{12})(3.2 \times 10^8)}{(1.6 \times 10^7)(3 \times 10^2)}$

115. Megrez, the nearest of the Big Dipper stars, is 6.6×10^{17} m from earth. Approximately how long does it take light, traveling at 10^{16} m/yr, to travel from Megrez to Earth?

116. Alkaid, the most distant star in the Big Dipper, is 2.1×10^{18} m from Earth. Approximately how long does it take light to travel from Alkaid to Earth?

117. The number of liters of water on Earth is 15,500 followed by 19 zeros. Write this number in scientific notation. Then use the number of liters of water on Earth to find out how much water is available for each person on earth. The population of Earth is 6.3 billion.

118. If there are 6.3×10^9 people on Earth and there is enough freshwater to provide each person with 7.40×10^5 l, how much freshwater is on Earth?

119. The United States uses an average of 2.6×10^6 l of water per person each year. The United States has 2.9×10^8 people. How many liters of water does the United States use each year?

120. Recall the paper "cut and stack" experiment described at the beginning of this chapter. If you had a piece of paper large enough to complete the following (and assuming that you *could* complete the following!), determine the height of the stack, given that the thickness of the paper is 0.002 in.

(a) After 30 cuts   (b) After 50 cuts

121. Can $(a + b)^{-1}$ be written as $\dfrac{1}{a} + \dfrac{1}{b}$ by using the properties of exponents? If not, why not? Explain.

122. Write a short description of the difference between $(-4)^{-3}$, -4^{-3}, $(-4)^3$, and -4^3. Are any of these equal?

123. If $n > 0$, which of the following expressions are negative?

$$-n^{-3}, n^{-3}, (-n)^{-3}, (-n)^3, -n^3$$

If $n < 0$, which of these expressions are negative? Explain what effect a negative in the exponent has on the sign of the result when an exponential expression is simplified.

124. Take the best offer. You are offered a 28-day job in which you have a choice of two different pay arrangements. Plan 1 offers a flat $4,000,000 at the end of the 28th day on the job. Plan 2 offers 1¢ the first day, 2¢ the second day, 4¢ the third day, and so on, with the amount doubling each day. Make a table to decide which offer is the best. Write a formula for the amount you make on the nth day and a formula for the total after n days. Which pay arrangement should you take? Why?

4.3 Introduction to Polynomials

4.3 OBJECTIVES

1. Identify types of polynomials
2. Find the degree of a polynomial
3. Write polynomials in descending-exponent form

Our work in this section deals with the most common kind of algebraic expression, a *polynomial*. To define a polynomial, let's recall our earlier definition of the word *term*.

> A **term** is a number or the product of a number and one or more variables, which may be raised to powers.

For example, x^5, $3x$, $-4xy^2$, 8, $\dfrac{5}{x}$, and $-14\sqrt{x}$ are terms. A **polynomial** consists of one or more terms in which the only allowable exponents are the whole numbers, 0, 1, 2, 3, and so on. These terms are connected by addition or subtraction signs. The variable is never used as a divisor in a term of a polynomial.

In a polynomial, terms are separated by + and − signs.

> In each term of a polynomial, the number is called the **numerical coefficient,** or more simply the **coefficient,** of that term.

Example 1 — Identifying Polynomials

Note: Each sign (+ or −) is attached to the term that *follows* that sign.

(a) $x + 3$ is a polynomial. The terms are x and 3. The coefficients are 1 and 3.

(b) $3x^2 - 2x + 5$ is also a polynomial. Its terms are $3x^2$, $-2x$, and 5. The coefficients are 3, −2, and 5.

(c) $5x^3 + 2 - \dfrac{3}{x}$ is *not* a polynomial because of the division by x in the third term.

 CHECK YOURSELF 1

Which of the following are polynomials?

(a) $5x^2$ (b) $3y^3 - 2y + \dfrac{5}{y}$ (c) $4x^2 - 2x + 3$

Certain polynomials are given special names because of the number of terms that they have.

The prefix *mono* means 1. The prefix *bi* means 2. The prefix *tri* means 3. There are no special names for polynomials with more than three terms.

> A polynomial with exactly one term is called a **monomial.**
>
> A polynomial with exactly two terms is called a **binomial.**
>
> A polynomial with exactly three terms is called a **trinomial.**

| Example 2 | **Identifying Types of Polynomials** |

(a) $3x^2y$ is a monomial. It has exactly one term.

(b) $2x^3 + 5x$ is a binomial. It has exactly two terms, $2x^3$ and $5x$.

(c) $5x^2 - 4x + 3$ is a trinomial. Its three terms are $5x^2$, $-4x$, and 3.

  CHECK YOURSELF 2

Classify each of these as a monomial, binomial, or trinomial.

(a) $5x^4 - 2x^3$ (b) $4x^7$ (c) $2x^2 + 5x - 3$

Remember, in a polynomial the allowable exponents are the whole numbers 0, 1, 2, 3, and so on. The degree will be a whole number.

We also classify polynomials by their *degree*. The **degree** of a polynomial that has only one variable is the highest power of that variable appearing in any one term.

| Example 3 | **Classifying Polynomials by Their Degree** |

The highest power

(a) $5x^3 - 3x^2 + 4x$ has degree 3.

The highest power

(b) $4x - 5x^4 + 3x^3 + 2$ has degree 4.

(c) $8x$ has degree 1 because $8x = 8x^1$.

In Section 4.2, we saw that $x^0 = 1$.

(d) 7 has degree 0 because $7 = 7 \cdot 1 = 7x^0$.

Note: Polynomials can have more than one variable, such as $4x^2y^3 + 5xy^2$. The degree is then the largest sum of the powers in any single term (here $2 + 3$, or 5). In general, we will be working with polynomials in a single variable, such as x.

✔ CHECK YOURSELF 3

Find the degree of each polynomial.

(a) $6x^5 - 3x^3 - 2$ (b) $5x$ (c) $3x^3 + 2x^6 - 1$ (d) 9

Working with polynomials is much easier if you get used to writing them in **descending-exponent form** (sometimes called *descending-power form*). When a polynomial has only one variable, this means that the term with the highest exponent is written first, then the term with the next-highest exponent, and so on.

| Example 4 | **Writing Polynomials in Descending Order** |

The exponents get smaller from left to right.

(a) $5x^7 - 3x^4 + 2x^2$ is in descending-exponent form.

(b) $4x^4 + 5x^6 - 3x^5$ is *not* in descending-exponent form. The polynomial should be written as

$$5x^6 - 3x^5 + 4x^4$$

Notice that the degree of the polynomial is the power of the *first*, or *leading*, term once the polynomial is arranged in descending-exponent form.

✔ CHECK YOURSELF 4

Write the following polynomials in descending-exponent form.

(a) $5x^4 - 4x^5 + 7$ (b) $4x^3 + 9x^4 + 6x^8$

A polynomial can represent any number. Its value depends on the value given to the variable.

Example 5

Evaluating Polynomials

Given the polynomial

$$3x^3 - 2x^2 - 4x + 1$$

(a) Find the value of the polynomial when $x = 2$.

Substituting 2 for x, we have

Again note how the rules for the order of operations are applied. See Section 0.5 for a review.

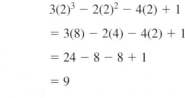

$$3(2)^3 - 2(2)^2 - 4(2) + 1$$
$$= 3(8) - 2(4) - 4(2) + 1$$
$$= 24 - 8 - 8 + 1$$
$$= 9$$

Caution

Be particularly careful when dealing with powers of negative numbers!

(b) Find the value of the polynomial when $x = -2$.

Now we substitute -2 for x.

$$3(-2)^3 - 2(-2)^2 - 4(-2) + 1$$
$$= 3(-8) - 2(4) - 4(-2) + 1$$
$$= -24 - 8 + 8 + 1$$
$$= -23$$

✔ CHECK YOURSELF 5

Find the value of the polynomial

$$4x^3 - 3x^2 + 2x - 1$$

when

(a) $x = 3$ (b) $x = -3$

✔ CHECK YOURSELF ANSWERS

1. (a) and (c) are polynomials.   **2.** (a) binomial; (b) monomial; (c) trinomial

3. (a) 5; (b) 1; (c) 6; (d) 0   **4.** (a) $-4x^5 + 5x^4 + 7$; (b) $6x^8 + 9x^4 + 4x^3$

5. (a) 86; (b) -142

Exercises ▪ 4.3

Which of the following expressions are polynomials?

1. $7x^3$

2. $4x^3 - \dfrac{2}{x}$

3. $2x^5y^3 - 4x^2y^4$

4. 7

5. -7

6. $4x^3 + x$

7. $\dfrac{3 + x}{x^2}$

8. $5a^2 - 2a + 7$

For each of the following polynomials, list the terms and the coefficients.

9. $2x^2 - 3x$

10. $5x^3 + x$

11. $4x^3 - 3x + 2$

12. $7x^2$

Classify each of the following as a monomial, binomial, or trinomial where possible.

13. $4x^3 - 2x^2$

14. $4x^7$

15. $7y^2 + 4y + 5$

16. $3x^3$

17. $2x^4 - 3x^2 + 5x - 2$

18. $x^4 + \dfrac{5}{x} + 7$

19. $7x^{10}$

20. $4x^4 - 2x^2 + 5x - 7$

21. $x^5 - \dfrac{3}{x^2}$

22. $25x^2 - 16$

Arrange in descending-exponent form and give the degree of each polynomial.

23. $4x^5 - 3x^2$

24. $5x^2 + 3x^3 + 4$

25. $7x^7 - 5x^9 + 4x^3$

26. $4 - x + x^2$

27. $-9x$

28. $x^{17} - 3x^4$

29. $5x^2 - 3x^5 + x^6 - 7$

30. 15

Find the values of each of the following polynomials for the given values of the variable.

31. $9x - 2$; $x = 1$ and $x = -1$

32. $5x - 5$; $x = 2$ and $x = -2$

33. $x^3 - 2x$; $x = 2$ and $x = -2$

34. $3x^2 + 7$; $x = 3$ and $x = -3$

35. $3x^2 + 4x - 2$; $x = 4$ and $x = -4$

36. $2x^2 - 5x + 1$; $x = 2$ and $x = -2$

37. $-x^2 - x + 12$; $x = 3$ and $x = -4$

38. $-x^2 - 5x - 6$; $x = -3$ and $x = -2$

Indicate whether each of the following statements is always true, sometimes true, or never true.

39. A monomial is a polynomial.

40. A binomial is a trinomial.

41. The degree of a trinomial is 3.

42. A trinomial has three terms.

43. A polynomial has four or more terms.

44. A binomial must have two coefficients.

45. If x equals 0, the value of a polynomial in x equals 0.

46. The coefficient of the leading term in a polynomial is the largest coefficient of the polynomial.

47. Cost of typing. The cost, in dollars, of typing a term paper is given as 3 times the number of pages plus $20. Use x as the number of pages to be typed and write a polynomial to describe this cost. Find the cost of typing a 50-page paper.

48. Manufacturing. The cost, in dollars, of making suits is described as 20 times the number of suits plus $150. Use x as the number of suits and write a polynomial to describe this cost. Find the cost of making seven suits.

49. Revenue. The revenue, in dollars, when x pairs of shoes are sold is given by $3x^2 - 95$. Find the revenue when 12 pairs of shoes are sold.

50. Manufacturing. The cost, in dollars, of manufacturing x wing nuts is given by $0.07x + 13.3$. Find the cost when 375 wing nuts are made.

4.4 Addition and Subtraction of Polynomials

4.4 OBJECTIVES

1. *Add two polynomials*
2. *Subtract two polynomials*

Addition is always a matter of combining like quantities (two apples plus three apples, four books plus five books, and so on). If you keep that basic idea in mind, adding polynomials will be easy. It is just a matter of combining like terms. Suppose that you want to add

$$5x^2 + 3x + 4 \qquad \text{and} \qquad 4x^2 + 5x - 6$$

Parentheses are sometimes used in adding, so for the sum of these polynomials, we can write

$$(5x^2 + 3x + 4) + (4x^2 + 5x - 6)$$

The plus sign between the parentheses indicates the addition.

Now what about the parentheses? You can use the following rule.

Removing Parentheses: Case 1

When you are adding or subtracting polynomials:

If a plus sign (+) or nothing at all appears in front of parentheses, just remove the parentheses. No other changes are necessary.

Now let's return to the addition.

Just remove the parentheses. No other changes are necessary.

$$(5x^2 + 3x + 4) + (4x^2 + 5x - 6)$$
$$= 5x^2 + 3x + 4 + 4x^2 + 5x - 6$$

Like terms   Like terms

Like terms

Note the use of the associative and commutative properties in reordering and regrouping.

Collect like terms. (*Remember:* Like terms have the same variables raised to the same power.)

$$= (5x^2 + 4x^2) + (3x + 5x) + (4 - 6)$$

Here we use the distributive property. For example,

$$5x^2 + 4x^2$$
$$= (5 + 4)x^2$$
$$= 9x^2$$

Combine like terms for the result:

$$= 9x^2 + 8x - 2$$

As should be clear, much of this work can be done mentally. You can then write the sum directly by locating like terms and combining.

Example 1 | Combining Like Terms

Add $3x - 5$ and $2x + 3$.

Write the sum.

$$(3x - 5) + (2x + 3)$$
$$= 3x - 5 + 2x + 3 = 5x - 2$$

Like terms   Like terms

✔ CHECK YOURSELF 1

Add $6x^2 + 2x$ and $4x^2 - 7x$.

The same technique is used to find the sum of two trinomials.

Example 2 | Adding Polynomials

Add $4a^2 - 7a + 5$ and $3a^2 + 3a - 4$.

Write the sum.

Remember: Only the like terms are combined in the sum.

$$(4a^2 - 7a + 5) + (3a^2 + 3a - 4)$$
$$= 4a^2 - 7a + 5 + 3a^2 + 3a - 4 = 7a^2 - 4a + 1$$

Like terms

Like terms

Like terms

 CHECK YOURSELF 2

Add $5y^2 - 3y + 7$ and $3y^2 - 5y - 7$.

Example 3	## Adding Polynomials

Add $2x^2 + 7x$ and $4x - 6$.

Write the sum.

$$(2x^2 + 7x) + (4x - 6)$$

$$= 2x^2 + \underbrace{7x + 4x} - 6$$

These are the only like terms;
$2x^2$ and -6 cannot be combined.

$$= 2x^2 + 11x - 6$$

 CHECK YOURSELF 3

Add $5m^2 + 8$ and $8m^2 - 3m$.

As we mentioned in Section 4.3, writing polynomials in descending-exponent form usually makes the work easier. Look at Example 4.

Example 4	## Adding Polynomials

Add $3x - 2x^2 + 7$ and $5 + 4x^2 - 3x$.

Write the polynomials in descending-exponent form; then add.

$$(-2x^2 + 3x + 7) + (4x^2 - 3x + 5)$$

$$= 2x^2 + 12$$

 CHECK YOURSELF 4

Add $8 - 5x^2 + 4x$ and $7x - 8 + 8x^2$.

Subtracting polynomials requires another rule for removing parentheses.

Removing Parentheses: Case 2

When you are adding or subtracting polynomials:

If a minus sign (−) appears in front of a set of parentheses, the parentheses can be removed by changing the sign of each item inside the parentheses.

The use of this rule is illustrated in Example 5.

| **Example 5** | **Removing Parentheses** |

In each of the following, remove the parentheses.

Note: This uses the distributive property, since

$$-(2x + 3y) = (-1)(2x + 3y)$$
$$= -2x - 3y$$

(a) $-(2x + 3y) = -2x - 3y$ Change each sign to remove the parentheses.

(b) $m - (5n - 3p) = m \underbrace{- 5n + 3p}_{\text{Sign changes}}$

(c) $2x - (-3y + z) = 2x \underbrace{+ 3y - z}_{\text{Sign changes}}$

✔ *CHECK YOURSELF 5*

Remove the parentheses.

(a) $-(3m + 5n)$ (b) $-(5w - 7z)$ (c) $3r - (2s - 5t)$ (d) $5a - (-3b - 2c)$

Subtracting polynomials is now a matter of using the previous rule to remove the parentheses and then combining like terms. Consider Example 6.

| **Example 6** | **Subtracting Polynomials** |

Note: The expression following *from* is written first in the problem.

(a) Subtract $5x - 3$ from $8x + 2$.

Write

$$(8x + 2) - (5x - 3)$$
$$= 8x + 2 \underbrace{- 5x + 3}_{\text{Sign changes}}$$
$$= 3x + 5$$

(b) Subtract $4x^2 - 8x + 3$ from $8x^2 + 5x - 3$.

Write

$$(8x^2 + 5x - 3) - (4x^2 - 8x + 3)$$
$$= 8x^2 + 5x - 3 \underbrace{- 4x^2 + 8x - 3}_{\text{Sign changes}}$$
$$= 4x^2 + 13x - 6$$

✔ *CHECK YOURSELF 6*

(a) Subtract $7x + 3$ from $10x - 7$.

(b) Subtract $5x^2 - 3x + 2$ from $8x^2 - 3x - 6$.

Again, writing all polynomials in descending-exponent form will make locating and combining like terms much easier. Look at Example 7.

| Example 7 | **Subtracting Polynomials** |

(a) Subtract $4x^2 - 3x^3 + 5x$ from $8x^3 - 7x + 2x^2$.

Write

$$(8x^3 + 2x^2 - 7x) - (-3x^3 + 4x^2 + 5x)$$
$$= 8x^3 + 2x^2 - 7x \underbrace{+ 3x^3 - 4x^2 - 5x}_{\text{Sign changes}}$$
$$= 11x^3 - 2x^2 - 12x$$

(b) Subtract $8x - 5$ from $-5x + 3x^2$.

Write

$$(3x^2 - 5x) - (8x - 5)$$
$$= 3x^2 \underbrace{- 5x - 8x}_{} + 5$$

Only the like terms can be combined.

$$= 3x^2 - 13x + 5$$

✔ *CHECK YOURSELF 7*

(a) Subtract $7x - 3x^2 + 5$ from $5 - 3x + 4x^2$.

(b) Subtract $3a - 2$ from $5a + 4a^2$.

If you think back to addition and subtraction in arithmetic, you'll remember that the work was arranged vertically. That is, the numbers being added or subtracted were placed under one another so that each column represented the same place value. This meant that in adding or subtracting columns you were always dealing with "like quantities."

It is also possible to use a vertical method for adding or subtracting polynomials. First rewrite the polynomials in descending-exponent form, then arrange them one under another, so that each column contains like terms. Then add or subtract in each column.

| Example 8 | Add by Using the Vertical Method |

(a) Add $3x - 5$ and $x^2 + 2x + 4$.

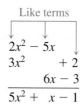

$$3x - 5$$
$$x^2 + 2x + 4$$
$$\overline{x^2 + 5x - 1}$$

(b) Add $2x^2 - 5x$, $3x^2 + 2$, and $6x - 3$.

Like terms

$$2x^2 - 5x$$
$$3x^2 \qquad + 2$$
$$\underline{\qquad 6x - 3}$$
$$5x^2 + x - 1$$

✔ *CHECK YOURSELF 8*

Add $3x^2 + 5$, $x^2 - 4x$, and $6x + 7$.

Example 9 illustrates subtraction by the vertical method.

| Example 9 | Subtract by Using the Vertical Method |

(a) Subtract $5x - 3$ from $8x - 7$.

Write

Since we are subtracting the entire quantity $5x - 3$, we place parentheses around $5x - 3$. Then we remove them according to the case 2 rule mentioned on page 325.

$$8x - 7$$
$$\underline{-(5x - 3)}$$

To subtract, change each sign of $5x - 3$ to get $-5x + 3$, then add.

$$8x - 7$$
$$\underline{-5x + 3}$$
$$3x - 4$$

(b) Subtract $5x^2 - 3x + 4$ from $8x^2 + 5x - 3$.

Write

$$8x^2 + 5x - 3$$
$$\underline{-(5x^2 - 3x + 4)}$$

To subtract, change each sign of $5x^2 - 3x + 4$ to get $-5x^2 + 3x - 4$, then add.

$$8x^2 + 5x - 3$$
$$\underline{-5x^2 + 3x - 4}$$
$$3x^2 + 8x - 7$$

Subtraction using the vertical method takes some practice. Take time to study the method carefully. You'll be using it in long division in Section 4.6.

✔ *CHECK YOURSELF 9*

Subtract, using the vertical method.

(a) $4x^2 - 3x$ from $8x^2 + 2x$ (b) $8x^2 + 4x - 3$ from $9x^2 - 5x + 7$

✔ *CHECK YOURSELF ANSWERS*

1. $10x^2 - 5x$ **2.** $8y^2 - 8y$ **3.** $13m^2 - 3m + 8$ **4.** $3x^2 + 11x$

5. (a) $-3m - 5n$; (b) $-5w + 7z$; (c) $3r - 2s + 5t$; (d) $5a + 3b + 2c$

6. (a) $3x - 10$; (b) $3x^2 - 8$ **7.** (a) $7x^2 - 10x$; (b) $4a^2 + 2a + 2$

8. $4x^2 + 2x + 12$ **9.** (a) $4x^2 + 5x$; (b) $x^2 - 9x + 10$

Exercises · 4.4

Add.

1. $5a - 7$ and $4a + 11$

2. $9x + 3$ and $3x - 4$

3. $8b^2 - 11b$ and $5b^2 - 7b$

4. $2m^2 + 3m$ and $6m^2 - 8m$

5. $3x^2 - 2x$ and $-5x^2 + 2x$

6. $9p^2 + 3p$ and $-13p^2 - 3p$

7. $2x^2 + 5x - 3$ and $3x^2 - 7x + 4$

8. $4d^2 - 8d + 7$ and $5d^2 - 6d - 9$

9. $3b^2 - 7$ and $2b - 7$

10. $4x - 3$ and $3x^2 - 9x$

11. $8y^3 - 5y^2$ and $5y^2 - 2y$

12. $9x^4 - 2x^2$ and $2x^2 + 3$

13. $2a^2 - 4a^3$ and $3a^3 + 2a^2$

14. $9m^3 - 2m$ and $-6m - 4m^3$

15. $7x^2 - 5 + 4x$ and $8 - 5x - 9x^2$

16. $5b^3 - 8b + 2b^2$ and $3b^2 - 7b^3 + 5b$

Remove the parentheses in each of the following expressions, and simplify where possible.

17. $-(4a + 5b)$

18. $-(7x - 4y)$

19. $5a - (2b - 3c)$

20. $8x - (2y - 5z)$

21. $9r - (3r + 5s)$

22. $10m - (3m - 2n)$

23. $5p - (-3p + 2q)$

24. $8d - (-7c - 2d)$

Subtract.

25. $x + 4$ from $2x - 3$

26. $5x + 1$ from $7x + 8$

27. $3m^2 - 2m$ from $4m^2 - 5m$

28. $9a^2 - 5a$ from $11a^2 - 10a$

29. $6y^2 + 5y$ from $4y^2 + 5y$

30. $9n^2 - 4n$ from $7n^2 - 4n$

31. $x^2 - 4x - 3$ from $3x^2 - 5x - 2$

32. $3x^2 - 2x + 4$ from $5x^2 - 8x - 3$

33. $3a + 7$ from $8a^2 - 9a$

34. $3x^3 + x^2$ from $4x^3 - 5x$

35. $4b^2 - 3b$ from $5b - 2b^2$

36. $7y - 3y^2$ from $3y^2 - 2y$

37. $5x^2 + 19 - 11x$ from $7x^2 - 11x + 31$

38. $4x - 2x^2 + 4x^3$ from $4x^3 + x - 3x^2$

Perform the indicated operations.

39. Subtract $2b + 5$ from the sum of $5b - 4$ and $3b + 8$.

40. Subtract $5m - 7$ from the sum of $2m - 8$ and $9m - 2$.

41. Subtract $3x^2 + 2x - 1$ from the sum of $x^2 + 5x - 2$ and $2x^2 + 7x - 8$.

330

42. Subtract $4x^2 - 5x - 3$ from the sum of $x^2 - 3x - 7$ and $2x^2 - 2x + 9$.

43. Subtract $2x^2 - 3x$ from the sum of $4x^2 - 5$ and $2x - 7$.

44. Subtract $7a^2 + 8a - 15$ from the sum of $14a - 5$ and $7a^2 - 8$.

45. Subtract the sum of $3y^2 - 3y$ and $5y^2 + 3y$ from $2y^2 - 8y$.

46. Subtract the sum of $7r^3 - 4r^2$ and $-3r^3 + 4r^2$ from $2r^3 + 3r^2$.

Add, using the vertical method.

47. $4w^2 + 11, 7w - 9$, and $2w^2 - 9w$ **48.** $3x^2 - 4x - 2, 6x - 3$, and $2x^2 + 8$

49. $3x^2 + 3x - 4, 4x^2 - 3x - 3$, and $2x^2 - x + 7$ **50.** $5x^2 + 2x - 4, x^2 - 2x - 3$, and $2x^2 - 4x - 3$

Subtract, using the vertical method.

51. $7a^2 - 9a$ from $9a^2 - 4a$ **52.** $6r^3 + 4r^2$ from $4r^3 - 2r^2$

53. $5x^2 - 6x + 7$ from $8x^2 - 5x + 7$ **54.** $8x^2 - 4x + 2$ from $9x^2 - 8x + 6$

55. $5x^2 - 3x$ from $8x^2 - 9$ **56.** $-13x^2 + 6x$ from $-11x^2 - 3$

Perform the indicated operations.

57. $[(9x^2 - 3x + 5) - (3x^2 + 2x - 1)] - (x^2 - 2x - 3)$

58. $[(5x^2 + 2x - 3) - (-2x^2 + x - 2)] - (2x^2 + 3x - 5)$

Find values for a, b, c, and d so that the following equations are true.

59. $3ax^4 - 5x^3 + x^2 - cx + 2 = 9x^4 - bx^3 + x^2 - 2d$

60. $(4ax^3 - 3bx^2 - 10) - 3(x^3 + 4x^2 - cx - d) = x^3 - 6x + 8$

61. Geometry. A rectangle has sides of $8x + 9$ and $6x - 7$. Find the polynomial that represents its perimeter.

62. Geometry. A triangle has sides $3x + 7$, $4x - 9$, and $5x + 6$. Find the polynomial that represents its perimeter.

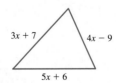

$3x + 7$ $4x - 9$

$5x + 6$

63. Business. The cost of producing x units of an item is $150 + 25x$. The revenue for selling x units is $90x - x^2$. The profit is given by the revenue minus the cost. Find the polynomial that represents the profit.

64. Business. The revenue for selling y units is $3y^2 - 2y + 5$, and the cost of producing y units is $y^2 + y - 3$. Find the polynomial that represents profit.

4.5 Multiplication of Polynomials and Special Products

OBJECTIVES

1. Find the product of a monomial and a polynomial
2. Find the product of two binomials
3. Square a binomial
4. Find the product of two binomials that differ only in sign

You have already had some experience in multiplying polynomials. In Section 4.1 we stated the product rule for exponents and used that rule to find the product of two monomials. Let's review briefly.

To Multiply Monomials

Step 1. Multiply the coefficients.

Step 2. Use the product rule for exponents to combine the variables:

$$ax^m \cdot bx^n = abx^{m+n}$$

Let's look at an example in which we multiply two monomials.

Example 1

Multiplying Monomials

Multiply $3x^2y$ and $2x^3y^5$.

Write

Once again we have used the commutative and associative properties to rewrite the problem.

$$(3x^2y)(2x^3y^5)$$

$$= (3 \cdot 2)(x^2 \cdot x^3)(y \cdot y^5)$$

Multiply the coefficients.   Add the exponents.

$$= 6x^5y^6$$

332

✔ CHECK YOURSELF 1

Multiply.

(a) $(5a^2b)(3a^2b^4)$ (b) $(-3xy)(4x^3y^5)$

You might want to review Section 0.5 before going on.

Our next task is to find the product of a monomial and a polynomial. Here we use the distributive property, which we introduced in Section 0.5. That property leads us to the following rule for multiplication.

Distributive property:

$$a(b + c) = ab + ac$$

To Multiply a Polynomial by a Monomial

Use the distributive property to multiply each term of the polynomial by the monomial and simplify the result.

Example 2

Multiplying a Monomial and a Binomial

(a) Multiply $2x + 3$ by x.

Write

Note: With practice you will do this step mentally.

$$x(2x + 3)$$

$$= x \cdot 2x + x \cdot 3$$

$$= 2x^2 + 3x$$

Multiply x by $2x$ and then by 3, the terms of the polynomial. That is, "distribute" the multiplication over the sum.

(b) Multiply $2a^3 + 4a$ by $3a^2$.

Write

$$3a^2(2a^3 + 4a)$$

$$= 3a^2 \cdot 2a^3 + 3a^2 \cdot 4a = 6a^5 + 12a^3$$

✔ CHECK YOURSELF 2

Multiply.

(a) $2y(y^2 + 3y)$ (b) $3w^2(2w^3 + 5w)$

The patterns of Example 2 extend to *any* number of terms.

| Example 3 | **Multiplying a Monomial and a Polynomial** |

Multiply the following.

(a) $3x(4x^3 + 5x^2 + 2)$

$= 3x \cdot 4x^3 + 3x \cdot 5x^2 + 3x \cdot 2 = 12x^4 + 15x^3 + 6x$

Again we have shown all the steps of the process. With practice you can write the product directly, and you should try to do so.

(b) $-5c(4c^2 - 8c)$

$= (-5c)(4c^2) - (-5c)(8c) = -20c^3 + 40c^2$

(c) $3c^2d^2(7cd^2 - 5c^2d^3)$

$= (3c^2d^2)(7cd^2) - (3c^2d^2)(5c^2d^3) = 21c^3d^4 - 15c^4d^5$

✔ *CHECK YOURSELF 3*

Multiply.

(a) $3(5a^2 + 2a + 7)$ (b) $4x^2(8x^3 - 6)$

(c) $-5m(8m^2 - 5m)$ (d) $9a^2b(3a^3b - 6a^2b^4)$

| Example 4 | **Multiplying Binomials** |

(a) Multiply $x + 2$ by $x + 3$.

We can think of $x + 2$ as a single quantity and apply the distributive property.

Note that this ensures that each term, x and 2, of the first binomial is multiplied by each term, x and 3, of the second binomial.

$(x + 2)(x + 3)$ Multiply $x + 2$ by x and then by 3.

$= (x + 2)x + (x + 2)3$

$= x \cdot x + 2 \cdot x + x \cdot 3 + 2 \cdot 3$

$= x^2 + 2x + 3x + 6$

$= x^2 + 5x + 6$

(b) Multiply $a - 3$ by $a - 4$.

$(a - 3)(a - 4)$ (Think of $a - 3$ as a single quantity and distribute.)

$= (a - 3)a - (a - 3)(4)$

$= a \cdot a - 3 \cdot a - [(a \cdot 4) - (3 \cdot 4)]$

$= a^2 - 3a - (4a - 12)$ Note that the parentheses are needed here because a *minus sign* precedes the binomial.

$= a^2 - 3a - 4a + 12$

$= a^2 - 7a + 12$

✔ *CHECK YOURSELF 4*

Multiply.

(a) $(x + 4)(x + 5)$ (b) $(y + 5)(y - 6)$

Fortunately, there is a pattern to this kind of multiplication that allows you to write the product of the two binomials directly without going through all these steps. We call it the **FOIL method** of multiplying. The reason for this name will be clear as we look at the process in greater detail.

To multiply $(x + 2)(x + 3)$:

Remember this by F.

1. $(x + 2)(x + 3)$

    $x \cdot x$ Find the product of the *first* terms of the factors.

Remember this by O.

2. $(x + 2)(x + 3)$

    $x \cdot 3$ Find the product of the *outer* terms.

Remember this by I.

3. $(x + 2)(x + 3)$

    $2 \cdot x$ Find the product of the *inner* terms.

Remember this by L.

4. $(x + 2)(x + 3)$

    $2 \cdot 3$ Find the product of the *last* terms.

Combining the four steps, we have

$$(x + 2)(x + 3)$$
$$= x^2 + 3x + 2x + 6$$
$$= x^2 + 5x + 6$$

It's called FOIL to give you an easy way of remembering the steps: First, Outer, Inner, and Last.

With practice, the FOIL method will let you write the products quickly and easily. Consider Example 5, which illustrates this approach.

| Example 5 | **Using the FOIL Method** |

Find the following products, using the FOIL method.

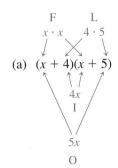

(a) $(x + 4)(x + 5)$

When possible, you should combine the outer and inner products mentally and write just the final product.

$$= x^2 + 5x + 4x + 20$$

$$\quad\quad\text{F}\quad\text{O}\quad\text{I}\quad\text{L}$$

$$= x^2 + 9x + 20$$

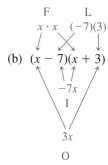

(b) $(x - 7)(x + 3)$

Combine the outer and inner products as $-4x$.

$$= x^2 - 4x - 21$$

✔ **CHECK YOURSELF 5**

Multiply.

(a) $(x + 6)(x + 7)$ (b) $(x + 3)(x - 5)$ (c) $(x - 2)(x - 8)$

Using the FOIL method, you can also find the product of binomials with leading coefficients other than 1 or with more than one variable.

| Example 6 | **Using the FOIL Method** |

Find the following products, using the FOIL method.

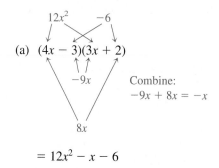

(a) $(4x - 3)(3x + 2)$

Combine:
$-9x + 8x = -x$

$$= 12x^2 - x - 6$$

$$6x^2 \qquad 35y^2$$

(b) $(3x - 5y)(2x - 7y)$

$-10xy$ Combine:
$\qquad\qquad -10xy + -21xy = -31xy$

$-21xy$

$$= 6x^2 - 31xy + 35y^2$$

The following rule summarizes our work in multiplying binomials.

To Multiply Two Binomials

Step 1. Multiply the first terms of the binomials (F).

Step 2. Then multiply the first term of the first binomial by the second term of the second binomial (O).

Step 3. Next multiply the second term of the first binomial by the first term of the second binomial (I).

Step 4. Finally multiply the second terms of the binomials (L).

Step 5. Form the sum of the four terms found above, combining any like terms.

✔ *CHECK YOURSELF 6*

Multiply.

(a) $(5x + 2)(3x - 7)$ (b) $(4a - 3b)(5a - 4b)$ (c) $(3m + 5n)(2m + 3n)$

The FOIL method works well for multiplying any two binomials. But what if one of the factors has three or more terms? The vertical format, shown in Example 7, works for factors with any number of terms.

Example 7

Using the Vertical Method

Multiply $x^2 - 5x + 8$ by $x + 3$.

Step 1
$$\begin{array}{r} x^2 - 5x + 8 \\ x + 3 \\ \hline 3x^2 - 15x + 24 \end{array}$$
Multiply each term of $x^2 - 5x + 8$ by 3.

Step 2

$$x^2 - 5x + 8$$
$$\underline{x + 3}$$
$$3x^2 - 15x + 24$$
$$\underline{x^3 - 5x^2 + 8x}$$

Now multiply each term by x.

Note that this line is shifted over so that like terms are in the same columns.

Note: Using this vertical method ensures that each term of one factor multiplies each term of the other. That's why it works!

Step 3

$$x^2 - 5x + 8$$
$$\underline{x + 3}$$
$$3x^2 - 15x + 24$$
$$\underline{x^3 - 5x^2 + 8x}$$
$$x^3 - 2x^2 - 7x + 24$$

Now add to combine like terms to write the product.

✔ *CHECK YOURSELF 7*

Multiply $2x^2 - 5x + 3$ by $3x + 4$.

Certain products occur frequently enough in algebra that it is worth learning special formulas for dealing with them. First, let's look at the **square of a binomial,** which is the product of two equal binomial factors.

$$(x + y)^2 = (x + y)(x + y) = x^2 + xy + xy + y^2$$
$$= x^2 + 2xy + y^2$$
$$(x - y)^2 = (x - y)(x - y) = x^2 - xy - xy + y^2$$
$$= x^2 - 2xy + y^2$$

The patterns above lead us to the following rule.

To Square a Binomial

Step 1. Find the first term of the square by squaring the first term of the binomial.

Step 2. Find the middle term of the square as twice the product of the two terms of the binomial.

Step 3. Find the last term of the square by squaring the last term of the binomial.

Example 8

Squaring a Binomial

(a) $(x + 3)^2 = x^2 + 2 \cdot x \cdot 3 + 3^2$

Square of first term   Twice the product of the two terms   Square of the last term

$$= x^2 + 6x + 9$$

Caution

A very common mistake in squaring binomials is to forget the middle term.

Again we have shown all the steps. With practice you can write just the square.

(b) $(3a + 4b)^2 = (3a)^2 + 2(3a)(4b) + (4b)^2$

$$= 9a^2 + 24ab + 16b^2$$

(c) $(y - 5)^2 = y^2 + 2 \cdot y \cdot (-5) + (-5)^2$

$$= y^2 - 10y + 25$$

(d) $(5c - 3d)^2 = (5c)^2 + 2(5c)(-3d) + (-3d)^2$

$$= 25c^2 - 30cd + 9d^2$$

✔ **CHECK YOURSELF 8**

Multiply.

(a) $(2x + 1)^2$ (b) $(4x - 3y)^2$

Example 9   ## Squaring a Binomial

Find $(y + 4)^2$.

$$(y + 4)^2 \quad \text{is } not \text{ equal to} \quad y^2 + 4^2 \quad \text{or} \quad y^2 + 16$$

The correct square is

You should see that $(2 + 3)^2 \neq 2^2 + 3^2$ because $5^2 \neq 4 + 9$.

$$(y + 4)^2 = y^2 + 8y + 16$$

The middle term is twice the product of y and 4.

✔ **CHECK YOURSELF 9**

Multiply.

(a) $(x + 5)^2$ (b) $(3a + 2)^2$

(c) $(y - 7)^2$ (d) $(5x - 2y)^2$

A second special product will be very important in Chapter 5, which deals with factoring. Suppose the form of a product is

$$(x + y)(x - y)$$

The two factors differ only in sign.

Let's see what happens when we multiply.

$$(x + y)(x - y)$$
$$= x^2 \underbrace{- xy + xy} - y^2$$
$$\qquad\qquad = 0$$
$$= x^2 - y^2$$

Special Product

The product of two binomials that differ only in the sign between the terms is the square of the first term minus the square of the second term.

Let's look at the application of this rule in Example 10.

Example 10	**Multiplying Polynomials**

Multiply each pair of factors.

(a) $(x + 5)(x - 5) = x^2 - 5^2$

    Square of   Square of
    the first term   the second term

$$= x^2 - 25$$

Note: $(2y)^2 = (2y)(2y)$
$\qquad\quad = 4y^2$

(b) $(x + 2y)(x - 2y) = x^2 - (2y)^2$

    Square of   Square of
    the first term   the second term

$$= x^2 - 4y^2$$

(c) $(3m + n)(3m - n) = 9m^2 - n^2$

(d) $(4a - 3b)(4a + 3b) = 16a^2 - 9b^2$

✔ *CHECK YOURSELF 10*

Find the products.

(a) $(a - 6)(a + 6)$ (b) $(x - 3y)(x + 3y)$

(c) $(5n + 2p)(5n - 2p)$ (d) $(7b - 3c)(7b + 3c)$

When you are finding the product of three or more factors, it is useful to first look for the pattern in which two binomials differ only in their sign. If you see it, finding this product first will make it easier to find the product of all the factors.

| **Example 11** | **Multiplying Polynomials** |

(a) $x(x - 3)(x + 3)$　　　　　　These binomials differ only in sign.

$= x(x^2 - 9)$

$= x^3 - 9x$

(b) $(x + 1)(x - 5)(x + 5)$　　　These binomials differ only in sign.

$= (x + 1)(x^2 - 25)$　　　　　With two binomials, use the FOIL method.

$= x^3 + x^2 - 25x - 25$

(c) $(2x - 1)(x + 3)(2x + 1)$

$(2x - 1)\,(x + 3)\,(2x + 1)$　　These two binomials differ only in the sign

$= (x + 3)(2x - 1)(2x + 1)$　　of the second term. We can use the commutative property to rearrange the terms.

$= (x + 3)(4x^2 - 1)$

$= 4x^3 + 12x^2 - x - 3$

✔ CHECK YOURSELF 11

Multiply.

(a) $3x(x - 5)(x + 5)$　　　　　　(b) $(x - 4)(2x + 3)(2x - 3)$

(c) $(x - 7)(3x - 1)(x + 7)$

✔ CHECK YOURSELF ANSWERS

1. (a) $15a^4b^5$; (b) $-12x^4y^6$　　**2.** (a) $2y^3 + 6y^2$; (b) $6w^5 + 15w^3$

3. (a) $15a^2 + 6a + 21$; (b) $32x^5 - 24x^2$; (c) $-40m^3 + 25m^2$; (d) $27a^5b^2 - 54a^4b^5$

4. (a) $x^2 + 9x + 20$; (b) $y^2 - y - 30$

5. (a) $x^2 + 13x + 42$; (b) $x^2 - 2x - 15$; (c) $x^2 - 10x + 16$

6. (a) $15x^2 - 29x - 14$; (b) $20a^2 - 31ab + 12b^2$; (c) $6m^2 + 19mn + 15n^2$

7. $6x^3 - 7x^2 - 11x + 12$　　**8.** (a) $4x^2 + 4x + 1$; (b) $16x^2 - 24xy + 9y^2$

9. (a) $x^2 + 10x + 25$; (b) $9a^2 + 12a + 4$; (c) $y^2 - 14y + 49$; (d) $25x^2 - 20xy + 4y^2$

10. (a) $a^2 - 36$; (b) $x^2 - 9y^2$; (c) $25n^2 - 4p^2$; (d) $49b^2 - 9c^2$

11. (a) $3x^3 - 75x$; (b) $4x^3 - 16x^2 - 9x + 36$; (c) $3x^3 - x^2 - 147x + 49$

Exercises · 4.5

Multiply.

1. $(5x^2)(3x^3)$

2. $(14a^4)(2a^7)$

3. $(-2b^2)(14b^8)$

4. $(14y^4)(-4y^6)$

5. $(-5p^7)(-8p^6)$

6. $(-6m^8)(9m^7)$

7. $(4m^5)(-3m)$

8. $(-5r^7)(-3r)$

9. $(4x^3y^2)(8x^2y)$

10. $(-3r^4s^2)(-7r^2s^5)$

11. $(-3m^5n^2)(2m^4n)$

12. $(3a^4b^2)(-14a^3b^4)$

13. $10(x + 3)$

14. $4(7b - 5)$

15. $3a(4a + 5)$

16. $5x(2x - 7)$

17. $3s^2(4s^2 - 7s)$

18. $3a^3(9a^2 + 15)$

19. $3x(5x^2 - 3x - 1)$

20. $5m(4m^3 - 3m^2 + 2)$

21. $3xy(2x^2y + xy^2 + 5xy)$

22. $5ab^2(ab - 3a + 5b)$

23. $6m^2n(3m^2n - 2mn + mn^2)$

24. $8pq^2(2pq - 3p + 5q)$

25. $(x + 3)(x + 2)$

26. $(a - 3)(a - 7)$

27. $(m - 5)(m - 9)$

28. $(b + 7)(b + 5)$

29. $(p - 8)(p + 7)$

30. $(x - 10)(x + 9)$

31. $(w - 7)(w + 6)$

32. $(s - 12)(s - 8)$

33. $(3x - 5)(x - 8)$

34. $(w + 5)(4w - 7)$

35. $(2x - 3)(3x + 4)$

36. $(7a - 2)(2a - 3)$

37. $(3a - b)(4a - 9b)$

38. $(7s - 3t)(3s + 8t)$

39. $(3p - 4q)(7p + 5q)$

40. $(5x - 4y)(2x - y)$

41. $(4x + 3y)(5x + 2y)$

42. $(4x - 5y)(4x + 3y)$

Find each of the following squares.

43. $(x + 3)^2$

44. $(y + 9)^2$

45. $(w - 6)^2$

46. $(a - 8)^2$

47. $(z + 12)^2$

48. $(p - 11)^2$

49. $(2a - 1)^2$

50. $(3x - 2)^2$

51. $(6m + 1)^2$

52. $(7b - 2)^2$

53. $(3x - y)^2$

54. $(5m + n)^2$

342

55. $(2r + 5s)^2$

56. $(3a - 4b)^2$

57. $(6a - 5b)^2$

58. $(7p + 6q)^2$

59. $\left(x + \dfrac{1}{2}\right)^2$

60. $\left(w - \dfrac{1}{4}\right)^2$

Find each of the following products.

61. $(x - 6)(x + 6)$

62. $(y + 8)(y - 8)$

63. $(m + 7)(m - 7)$

64. $(w - 10)(w + 10)$

65. $\left(x - \dfrac{1}{2}\right)\left(x + \dfrac{1}{2}\right)$

66. $\left(x + \dfrac{2}{3}\right)\left(x - \dfrac{2}{3}\right)$

67. $(p - 0.4)(p + 0.4)$

68. $(m - 1.1)(m + 1.1)$

69. $(a - 3b)(a + 3b)$

70. $(p + 4q)(p - 4q)$

71. $(3x + 2y)(3x - 2y)$

72. $(7x - y)(7x + y)$

73. $(8w + 5z)(8w - 5z)$

74. $(8c + 3d)(8c - 3d)$

75. $(5x - 9y)(5x + 9y)$

76. $(6s - 5t)(6s + 5t)$

Multiply.

77. $2x(3x - 2)(4x + 1)$

78. $3x(2x + 1)(2x - 1)$

79. $5a(4a - 3)(4a + 3)$

80. $6m(3m - 2)(3m - 7)$

81. $3s(5s - 2)(4s - 1)$

82. $7w(2w - 3)(2w + 3)$

83. $(x - 2)(x + 1)(x - 3)$

84. $(y + 3)(y - 2)(y - 4)$

85. $(a - 1)^3$

86. $(x + 1)^3$

87. $\left(\dfrac{x}{2} + \dfrac{2}{3}\right)\left(\dfrac{2x}{3} - \dfrac{2}{5}\right)$

88. $\left(\dfrac{x}{3} + \dfrac{3}{4}\right)\left(\dfrac{3x}{4} - \dfrac{3}{5}\right)$

89. $[x + (y - 2)][x - (y - 2)]$

90. $[x + (3 - y)][x - (3 - y)]$

Label Exercises 91 to 94 as true or false.

91. $(x + y)^2 = x^2 + y^2$

92. $(x - y)^2 = x^2 - y^2$

93. $(x + y)^2 = x^2 + 2xy + y^2$

94. $(x - y)^2 = x^2 - 2xy + y^2$

95. Length. The length of a rectangle is given by $(3x + 5)$ centimeters (cm), and the width is given by $(2x - 7)$ cm. Express the area of the rectangle in terms of x.

96. Area. The base of a triangle measures $(3y + 7)$ in., and the height is $(2y - 3)$ in. Express the area of the triangle in terms of y.

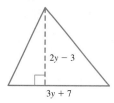

97. Revenue. The price of an item is given by $p = 12 - 0.05x$, where x is the number of items sold. If the revenue generated is found by multiplying the number of items sold by the price of an item, find the polynomial which represents the revenue.

98. Revenue. The price of an item is given by $p = 1{,}000 - 0.02x^2$, where x is the number of items sold. Find the polynomial that represents the revenue generated from the sale of x items.

99. Tree planting. Suppose an orchard is planted with trees in straight rows. If there are $5x - 4$ rows with $5x - 4$ trees in each row, find a polynomial that represents the number of trees in the orchard.

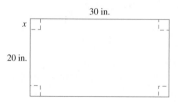

100. Area of a square. A square has sides of length $(3x - 2)$ cm. Express the area of the square as a polynomial.

101. Area of a rectangle. The length and width of a rectangle are given by two consecutive odd integers. Write an expression for the area of the rectangle.

102. Area of a rectangle. The length of a rectangle is 6 less than 3 times the width. Write an expression for the area of the rectangle.

103. Work with another student to complete this table and write the polynomial that represents the volume of the box. A paper box is to be made from a piece of cardboard 20 in. wide and 30 in. long. The box will be formed by cutting squares out of each of the four corners and folding up the sides to make a box.

If x is the dimension of the side of the square cut out of the corner, when the sides are folded up, the box will be x in. tall. You should use a piece of paper to try this to see how the box will be made. Complete the following chart.

Length of Side of Corner Square	Length of Box	Width of Box	Depth of Box	Volume of Box
1 in.				
2 in.				
3 in.				
n in.				

Write general formulas for the width, length, and height of the box and a general formula for the *volume* of the box, and simplify it by multiplying. The variable will be the height, the side of the square cut out of the corners. What is the highest power of the variable in the polynomial you have written for the volume? Extend the table to decide what the dimensions are for a box with maximum volume. Draw a sketch of this box and write in the dimensions.

104. Complete the following statement: $(a + b)^2$ is not equal to $a^2 + b^2$ because. . . . But wait! Isn't $(a + b)^2$ *sometimes* equal to $a^2 + b^2$? What do you think?

105. Is $(a + b)^3$ ever equal to $a^3 + b^3$? Explain.

106. In the following figures, identify the length and the width of the square:

Length = _____

Width = _____

Area = _____

Length = _____

Width = _____

Area = _____

107. The square shown is x units on a side. The area is _____.

Draw a picture of what happens when the sides are doubled. The area is _____.

Continue the picture to show what happens when the sides are tripled. The area is _____.

If the sides are quadrupled, the area is _____.

In general, if the sides are multiplied by n, the area is _____.

If each side is increased by 3, the area is increased by _____.

If each side is decreased by 2, the area is decreased by _____.

In general, if each side is increased by n, the area is increased by _____; and if each side is decreased by n, the area is decreased by _____.

For each of the following, let x represent the number, then write an expression for the product.

108. The product of 6 more than a number and 6 less than that number

109. The square of 5 more than a number

110. The square of 4 less than a number

111. The product of 5 less than a number and 5 more than that number

Note that $(28)(32) = (30 - 2)(30 + 2) = 900 - 4 = 896$. Use this pattern to find each of the following products.

112. $(49)(51)$ **113.** $(27)(33)$ **114.** $(34)(26)$

115. $(98)(102)$ **116.** $(55)(65)$ **117.** $(64)(56)$

4.6 Division of Polynomials

 OBJECTIVES

1. Find the quotient when a polynomial is divided by a monomial
2. Find the quotient of two polynomials

In Section 4.1, we introduced the quotient rule for exponents, which was used to divide one monomial by another monomial. Let's review that process.

To Divide a Monomial by a Monomial

Step 1. Divide the coefficients.

Step 2. Use the quotient rule for exponents to combine the variables.

Example 1

Dividing Monomials

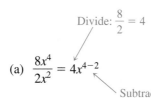

The quotient rule says: If x is not zero,

$$\frac{x^m}{x^n} = x^{m-n}$$

(a) $\dfrac{8x^4}{2x^2} = 4x^{4-2}$

Divide: $\dfrac{8}{2} = 4$

Subtract the exponents.

$$= 4x^2$$

(b) $\dfrac{45a^5b^3}{9a^2b} = 5a^3b^2$

 CHECK YOURSELF 1

Divide.

(a) $\dfrac{16a^5}{8a^3}$ (b) $\dfrac{28m^4n^3}{7m^3n}$

Now let's look at how this can be extended to divide any polynomial by a monomial. For example, to divide $12a^3 + 8a^2$ by $4a$, proceed as follows:

Technically, this step depends on the distributive property and the definition of division.

$$\frac{12a^3 + 8a^2}{4a} = \frac{12a^3}{4a} + \frac{8a^2}{4a}$$

Divide each term in the numerator by the denominator $4a$.

Now do each division.

$$= 3a^2 + 2a$$

The preceding work leads us to the following rule.

To Divide a Polynomial by a Monomial

Divide each term of the polynomial by the monomial. Then simplify the results.

Example 2   Dividing by Monomials

Divide each term by 2.

(a) $\dfrac{4a^2 + 8}{2} = \dfrac{4a^2}{2} + \dfrac{8}{2}$

$$= 2a^2 + 4$$

Divide each term by $6y$.

(b) $\dfrac{24y^3 - 18y^2}{6y} = \dfrac{24y^3}{6y} - \dfrac{18y^2}{6y}$

$$= 4y^2 - 3y$$

Remember the rules for signs in division.

(c) $\dfrac{15x^2 + 10x}{-5x} = \dfrac{15x^2}{-5x} + \dfrac{10x}{-5x}$

$$= -3x - 2$$

With practice you can write just the quotient.

(d) $\dfrac{14x^4 + 28x^3 - 21x^2}{7x^2} = \dfrac{14x^4}{7x^2} + \dfrac{28x^3}{7x^2} - \dfrac{21x^2}{7x^2}$

$$= 2x^2 + 4x - 3$$

(e) $\dfrac{9a^3b^4 - 6a^2b^3 + 12ab^4}{3ab} = \dfrac{9a^3b^4}{3ab} - \dfrac{6a^2b^3}{3ab} + \dfrac{12ab^4}{3ab}$

$$= 3a^2b^3 - 2ab^2 + 4b^3$$

✔ *CHECK YOURSELF 2*

Divide.

(a) $\dfrac{20y^3 - 15y^2}{5y}$ (b) $\dfrac{8a^3 - 12a^2 + 4a}{-4a}$

(c) $\dfrac{16m^4n^3 - 12m^3n^2 + 8mn}{4mn}$

We are now ready to look at dividing one polynomial by another polynomial (with more than one term). The process is very much like long division in arithmetic, as Example 3 illustrates.

| Example 3 | **Dividing by Binomials** |

Divide $x^2 + 7x + 10$ by $x + 2$.

The first term in the dividend x^2 is divided by the first term in the divisor x.

Step 1

$$x + 2 \overline{)\, x^2 + 7x + 10}$$
quotient x on top

Divide x^2 by x to get x.

Step 2

$$\begin{array}{r} x \phantom{{}+ 7x + 10} \\ x + 2 \overline{)\, x^2 + 7x + 10} \\ x^2 + 2x \phantom{{}+ 10} \end{array}$$

Multiply the divisor $x + 2$ by x.

Remember: To subtract $x^2 + 2x$, mentally change each sign to $-x^2 - 2x$ and then add. Take your time and be careful here. It's where most errors are made.

Step 3

$$\begin{array}{r} x \phantom{{}+ 7x + 10} \\ x + 2 \overline{)\, x^2 + 7x + 10} \\ x^2 + 2x \phantom{{}+ 10} \\ \hline 5x + 10 \end{array}$$

Subtract and bring down 10.

Step 4

$$\begin{array}{r} x + 5 \\ x + 2 \overline{)\, x^2 + 7x + 10} \\ x^2 + 2x \phantom{{}+ 10} \\ \hline 5x + 10 \end{array}$$

Divide $5x$ by x to get 5.

Note that we repeat the process until the degree of the remainder is less than that of the divisor or until there is no remainder.

Step 5

$$
\begin{array}{r}
x + 5 \\
x + 2 \overline{) x^2 + 7x + 10} \\
\underline{x^2 + 2x} \\
5x + 10 \\
\underline{5x + 10} \\
0
\end{array}
$$

Multiply $x + 2$ by 5 and then subtract.

The quotient is $x + 5$.

 CHECK YOURSELF 3

Divide $x^2 + 9x + 20$ by $x + 4$.

In Example 3, we showed all the steps separately to help you see the process. In practice, the work can be shortened.

Example 4   ## Dividing by Binomials

Divide $x^2 + x - 12$ by $x - 3$.

You might want to write out a problem like $408 \div 17$, to compare the steps.

$$
\begin{array}{r}
x + 4 \\
x - 3 \overline{) x^2 + x - 12} \\
\underline{x^2 - 3x} \\
4x - 12 \\
\underline{4x - 12} \\
0
\end{array}
$$

The Steps
1. Divide x^2 by x to get x, the first term of the quotient.
2. Multiply $x - 3$ by x.
3. Subtract and bring down -12. Remember to mentally change the signs to $-x^2 + 3x$ and add.
4. Divide $4x$ by x to get 4, the second term of the quotient.
5. Multiply $x - 3$ by 4 and subtract.

The quotient is $x + 4$.

 CHECK YOURSELF 4

Divide.

$$(x^2 + 2x - 24) \div (x - 4)$$

You may have a remainder in algebraic long division just as in arithmetic. Consider Example 5.

| Example 5 | ### Dividing by Binomials |

Divide $4x^2 - 8x + 11$ by $2x - 3$.

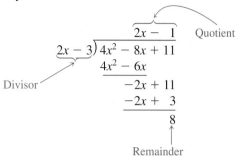

$$
\begin{array}{r}
2x - 1 \quad \text{Quotient} \\
2x - 3\overline{)\,4x^2 - 8x + 11} \\
\underline{4x^2 - 6x} \\
-2x + 11 \\
\underline{-2x + 3} \\
8
\end{array}
$$

Divisor

Remainder

This result can be written as

$$
\frac{4x^2 - 8x + 11}{2x - 3}
$$

$$
= \underbrace{2x - 1}_{\text{Quotient}} + \frac{8}{2x - 3} \quad
\begin{array}{l}
\leftarrow \text{Remainder} \\
\leftarrow \text{Divisor}
\end{array}
$$

✔ *CHECK YOURSELF 5*

Divide.

$$(6x^2 - 7x + 15) \div (3x - 5)$$

The division process shown in Examples 1 to 5 can be extended to dividends of a higher degree. The steps involved in the division process are exactly the same, as Example 6 illustrates.

| Example 6 | ### Dividing by Binomials |

Divide $6x^3 + x^2 - 4x - 5$ by $3x - 1$.

$$
\begin{array}{r}
2x^2 + x - 1 \\
3x - 1\overline{)\,6x^3 + x^2 - 4x - 5} \\
\underline{6x^3 - 2x^2} \\
3x^2 - 4x \\
\underline{3x^2 - x} \\
-3x - 5 \\
\underline{-3x + 1} \\
-6
\end{array}
$$

This result can be written as

$$\frac{6x^3 + x^2 - 4x - 5}{3x - 1} = 2x^2 + x - 1 + \frac{-6}{3x - 1}$$

✔ CHECK YOURSELF 6

Divide $4x^3 - 2x^2 + 2x + 15$ by $2x + 3$.

Suppose that the dividend is "missing" a term in some power of the variable. You can use 0 as the coefficient for the missing term. Consider Example 7.

Example 7   **Dividing by Binomials**

Divide $x^3 - 2x^2 + 5$ by $x + 3$.

Think of $0x$ as a placeholder. Writing it helps to align like terms.

$$
\begin{array}{r}
x^2 - 5x + 15 \\
x + 3 \overline{)x^3 - 2x^2 + 0x + 5} \\
\underline{x^3 + 3x^2} \\
-5x^2 + 0x \\
\underline{-5x^2 - 15x} \\
15x + 5 \\
\underline{15x + 45} \\
-40
\end{array}
$$

Write $0x$ for the "missing" term in x.

This result can be written as

$$\frac{x^3 - 2x^2 + 5}{x + 3} = x^2 - 5x + 15 + \frac{-40}{x + 3}$$

✔ CHECK YOURSELF 7

Divide.

$$(4x^3 + x + 10) \div (2x - 1)$$

You should always arrange the terms of the divisor and the dividend in descending-exponent form before starting the long division process, as illustrated in Example 8.

| Example 8 | **Dividing by Binomials** |

Divide $5x^2 - x + x^3 - 5$ by $-1 + x^2$.

Write the divisor as $x^2 - 1$ and the dividend as $x^3 + 5x^2 - x - 5$.

$$
\begin{array}{r}
x + 5 \\
x^2 - 1 \overline{)\,x^3 + 5x^2 - x - 5} \\
\underline{x^3 \qquad\quad - x} \\
5x^2 \qquad - 5 \\
\underline{5x^2 \qquad - 5} \\
0
\end{array}
$$

Write $x^3 - x$, the product of x and $x^2 - 1$, so that like terms fall in the same columns.

✔ *CHECK YOURSELF 8*

Divide.

$$(5x^2 + 10 + 2x^3 + 4x) \div (2 + x^2)$$

✔ *CHECK YOURSELF ANSWERS*

1. (a) $2a^2$; (b) $4mn^2$ **2.** (a) $4y^2 - 3y$; (b) $-2a^2 + 3a - 1$; (c) $4m^3n^2 - 3m^2n + 2$

3. $x + 5$ **4.** $x + 6$ **5.** $2x + 1 + \dfrac{20}{3x - 5}$ **6.** $2x^2 - 4x + 7 + \dfrac{-6}{2x + 3}$

7. $2x^2 + x + 1 + \dfrac{11}{2x - 1}$ **8.** $2x + 5$

Exercises · 4.6

Divide.

1. $\dfrac{22x^9}{11x^5}$

2. $\dfrac{20a^7}{5a^5}$

3. $\dfrac{35m^3n^2}{7mn^2}$

4. $\dfrac{42x^5y^2}{6x^3y}$

5. $\dfrac{3a + 6}{3}$

6. $\dfrac{3x - 6}{3}$

7. $\dfrac{9b^2 - 12}{3}$

8. $\dfrac{10m^2 + 5m}{5}$

9. $\dfrac{28a^3 - 42a^2}{7a}$

10. $\dfrac{9x^3 + 12x^2}{3x}$

11. $\dfrac{12m^2 + 6m}{-3m}$

12. $\dfrac{20b^3 - 25b^2}{-5b}$

13. $\dfrac{18a^4 - 45a^3 + 63a^2}{9a^2}$

14. $\dfrac{21x^5 - 28x^4 + 14x^3}{7x}$

15. $\dfrac{20x^4y^2 - 15x^2y^3 + 10x^3y}{5x^2y}$

16. $\dfrac{16m^3n^3 + 24m^2n^2 - 40mn^3}{8mn^2}$

Perform the indicated divisions.

17. $\dfrac{x^2 - x - 12}{x - 4}$

18. $\dfrac{x^2 + 8x + 15}{x + 3}$

19. $\dfrac{x^2 - x - 20}{x + 4}$

20. $\dfrac{x^2 - 2x - 35}{x + 5}$

21. $\dfrac{x^2 + x - 30}{x - 5}$

22. $\dfrac{3x^2 + 20x - 32}{3x - 4}$

23. $\dfrac{2x^2 - 3x - 5}{x - 3}$

24. $\dfrac{3x^2 + 17x - 12}{x + 6}$

25. $\dfrac{4x^2 - 18x - 15}{x - 5}$

26. $\dfrac{4x^2 - 11x - 24}{x - 5}$

27. $\dfrac{6x^2 - x - 10}{3x - 5}$

28. $\dfrac{4x^2 + 6x - 25}{2x + 7}$

29. $\dfrac{x^3 + x^2 - 4x - 4}{x + 2}$

30. $\dfrac{x^3 - 2x^2 + 4x - 21}{x - 3}$

31. $\dfrac{4x^3 + 7x^2 + 10x + 5}{4x - 1}$

32. $\dfrac{2x^3 - 3x^2 + 4x + 4}{2x + 1}$

33. $\dfrac{x^3 - x^2 + 5}{x - 2}$

34. $\dfrac{x^3 + 4x - 3}{x + 3}$

35. $\dfrac{49x^3 - 2x}{7x - 3}$

36. $\dfrac{8x^3 - 6x^2 + 2x}{4x + 1}$

37. $\dfrac{2x^2 - 8 - 3x + x^3}{x - 2}$ **38.** $\dfrac{x^2 - 18x + 2x^3 + 32}{x + 4}$ **39.** $\dfrac{x^4 - 1}{x - 1}$ **40.** $\dfrac{x^4 + x^2 - 16}{x + 2}$

41. $\dfrac{x^3 - 3x^2 - x + 3}{x^2 - 1}$ **42.** $\dfrac{x^3 + 2x^2 + 3x + 6}{x^2 + 3}$ **43.** $\dfrac{x^4 + 2x^2 - 2}{x^2 + 3}$ **44.** $\dfrac{x^4 + x^2 - 5}{x^2 - 2}$

45. $\dfrac{y^3 - 1}{y - 1}$ **46.** $\dfrac{y^3 - 8}{y - 2}$ **47.** $\dfrac{x^4 - 1}{x^2 - 1}$ **48.** $\dfrac{x^6 - 1}{x^3 - 1}$

49. Find the value of c so that $\dfrac{y^2 - y + c}{y + 1} = y - 2$.

50. Find the value of c so that $\dfrac{x^3 + x^2 + x + c}{x^2 + 1} = x + 1$.

51. Write a summary of your work with polynomials. Explain how a polynomial is recognized, and explain the rules for the arithmetic of polynomials—how to add, subtract, multiply, and divide. What parts of this chapter do you feel you understand very well, and what part(s) do you still have questions about, or feel unsure of? Exchange papers with another student and compare your answers.

52. An interesting (and useful) thing about division of polynomials: To find out about this interesting thing, do this division. Compare your answer with that of another student.

$(x - 2)\overline{\smash{)}\,2x^2 + 3x - 5}$ Is there a remainder?

Now, evaluate the polynomial $2x^2 + 3x - 5$ when $x = 2$. Is this value the same as the remainder?

Try $(x + 3)\overline{\smash{)}\,5x^2 - 2x + 1}.$ Is there a remainder?

Evaluate the polynomial $5x^2 - 2x + 1$ when $x = -3$. Is this value the same as the remainder?

What happens when there is no remainder?

Try $(x - 6)\overline{\smash{)}\,3x^3 - 14x^2 - 23x - 6}.$ Is the remainder zero?

Evaluate the polynomial $3x^3 - 14x^2 - 23x + 6$ when $x = 6$. Is this value zero? Write a description of the patterns you see. Make up several more examples and test your conjecture.

ⓢummary for Chapter 4

Example	Topic	Reference
	Positive Integer Exponents	**4.1**
	Properties of Exponents For any nonzero real numbers a and b and integers m and n:	
$x^5 \cdot x^7 = x^{5+7} = x^{12}$	**Product Rule** $$a^m \cdot a^n = a^{m+n}$$	**p. 294**
$\dfrac{x^7}{x^5} = x^{7-5} = x^2$	**Quotient Rule** $$\dfrac{a^m}{a^n} = a^{m-n} \qquad \text{where } m > n$$	**p. 296**
$(2y)^3 = 2^3 y^3 = 8y^3$	**Product-Power Rule** $$(ab)^n = a^n b^n$$	**p. 297**
$\left(\dfrac{2}{3}\right)^2 = \dfrac{2^2}{3^2} = \dfrac{4}{9}$	**Quotient-Power Rule** $$\left(\dfrac{a}{b}\right)^m = \dfrac{a^m}{b^m}$$	**p. 298**
$(2^3)^4 = 2^{12}$	**Power Rule** $$(a^m)^n = a^{mn}$$	**p. 298**
	Zero and Negative Exponents	**4.2**
$5x^0 = 5 \cdot 1 = 5$	**Zero Exponent** For any real number a where $a \neq 0$, $$a^0 = 1$$	**p. 303**
$x^{-3} = \dfrac{1}{x^3}$ $2y^{-5} = \dfrac{2}{y^5}$	**Negative Integer Exponents** For any nonzero real number a and whole number n, $$a^{-n} = \dfrac{1}{a^n}$$ and a^{-n} is the multiplicative inverse of a^n.	**p. 304**
$\left(\dfrac{x}{2}\right)^{-4} = \left(\dfrac{2}{x}\right)^4$ $= \dfrac{16}{x^4}$	**Quotient Raised to a Negative Power** For nonzero numbers a and b and a whole number n, $$\left(\dfrac{a}{b}\right)^{-n} = \left(\dfrac{b}{a}\right)^n$$	**p. 307**

$\underbrace{38,000,000.}_{\text{7 places}} = 3.8 \times 10^7$	**Scientific Notation** Scientific notation is a useful way of expressing very large or very small numbers through the use of powers of 10. Any number written in the form $$a \times 10^n$$ in which $1 \le a < 10$ and n is an integer, is said to be written in scientific notation.	**p. 310**

Introduction to Polynomials · 4.3

$4x^3 - 3x^2 + 5x$ is a polynomial.	**Polynomial** An algebraic expression made up of terms in which the exponents are whole numbers. These terms are connected by plus or minus signs. Each sign $(+$ or $-)$ is attached to the term following that sign.	**p. 317**
The terms of $4x^3 - 3x^2 + 5x$ are $4x^3$, $-3x^2$, and $5x$.	**Term** A number, or the product of a number and variables, raised to a power.	**p. 317**
The coefficients of $4x^3 - 3x^2$ are 4 and -3.	**Coefficient** In each term of a polynomial, the number that is multiplied by the variable(s) is called the *numerical coefficient* or, more simply, the *coefficient* of that term.	**p. 317**
	Types of Polynomials A polynomial can be classified according to the number of terms it has.	**p. 318**
$2x^3$ is a monomial. $3x^2 - 7x$ is a binomial. $5x^5 - 5x^3 + 2$ is a trinomial.	A *mono*mial has exactly one term. A *bi*nomial has exactly two terms. A *tri*nomial has exactly three terms.	
The degree of $4x^5 - 5x^3 + 3x$ is 5.	**Degree of a Polynomial with Only One Variable** The highest power of the variable appearing in any one term.	**p. 318**
$4x^5 - 5x^3 + 3x$ is written in descending-exponent form.	**Descending-Exponent Form** The form of a polynomial when it is written with the highest-degree term first, the next-highest-degree term second, and so on.	**p. 319**

Addition and Subtraction of Polynomials · 4.4

	Removing Signs of Grouping	
$+(3x - 5) = +3x - 5$	1. If a plus sign $(+)$ or no sign at all appears in front of parentheses, just remove the parentheses. No other changes are necessary.	**p. 323**
$-(3x - 5) = -3x + 5$	2. If a minus sign $(-)$ appears in front of parentheses, the parentheses can be removed by changing the sign of each term inside the parentheses.	**p. 325**

$(2x + 3) + (3x - 5)$ $= 2x + 3 + 3x - 5$ $= 5x - 2$	**Adding Polynomials** Remove the signs of grouping. Then collect and combine any like terms.	**p. 325**
$(3x^2 + 2x) - (2x^2 + 3x - 1)$ $= 3x^2 + 2x - 2x^2 - 3x + 1$ Sign changes $= 3x^2 - 2x^2 + 2x - 3x + 1$ $= x^2 - x + 1$	**Subtracting Polynomials** Remove the signs of grouping by changing the sign of each term in the polynomial being subtracted. Then combine any like terms.	**p. 326**

Multiplication of Polynomials

4.5

$(-2x^2y)(3x^3y)$ $= (-2)(3)(x^2 x^3)(yy)$ $= -6x^5y^2$	**To Multiply a Monomial by a Monomial** Multiply the coefficients, and use the product rule for exponents to combine the variables: $$ax^m \cdot bx^n = abx^{m+n}$$	**p. 332**
$2x(x^2 + 4)$ $= 2x^3 + 8x$	**To Multiply a Polynomial by a Monomial** Multiply each term of the polynomial by the monomial, and simplify the results.	**p. 333**
$(2x - 3)(3x + 5)$ $= 6x^2 + 10x - 9x - 15$ F O I L $= 6x^2 + x - 15$	**To Multiply a Binomial by a Binomial** Use the FOIL method: F O I L $(a + b)(c + d) = a \cdot c + a \cdot d + b \cdot c + b \cdot d$	**p. 335**
$\begin{array}{r} x^2 - 3x + 5 \\ 2x - 3 \\ \hline -3x^2 + 9x - 15 \\ 2x^3 - 6x^2 + 10x \\ \hline 2x^3 - 9x^2 + 19x - 15 \end{array}$	**To Multiply a Polynomial by a Polynomial** Arrange the polynomials vertically. Multiply each term of the upper polynomial by each term of the lower polynomial, and combine like terms.	**p. 337**

Special Products

4.5

$(2x - 5)^2$ $= 4x^2 + 2 \cdot 2x \cdot (-5) + 25$ $= 4x^2 - 20x + 25$	**The Square of a Binomial** $(a + b)^2 = a^2 + 2ab + b^2$ 1. The first term of the square is the square of the first term of the binomial. 2. The middle term is twice the product of the two terms of the binomial. 3. The last term is the square of the last term of the binomial.	**p. 338**

$(2x - 5y)(2x + 5y)$ $= (2x)^2 - (5y)^2$ $= 4x^2 - 25y^2$	**The Product of Binomials That Differ Only in Sign** Subtract the square of the second term from the square of the first term. $$(a + b)(a - b) = a^2 - b^2$$	**p. 340**
	## Division of Polynomials	**4.6**
$\dfrac{28m^4n^3}{7m^3n} = \left(\dfrac{28}{7}\right)m^{4-3}n^{3-1}$ $= 4mn^2$	**To Divide a Monomial by a Monomial** Divide the coefficients, and use the quotient rule for exponents to combine the variables.	**p. 347**
$\dfrac{9x^4 + 6x^3 - 15x^2}{3x}$ $= 3x^3 + 2x^2 - 5x$	**To Divide a Polynomial by a Monomial** Divide each term of the polynomial by the monomial.	**p. 348**
$\begin{array}{r} x + 5 \\ x - 3\overline{)x^2 + 2x - 7} \\ \underline{x^2 - 3x} \\ 5x - 17 \\ \underline{5x - 15} \\ 8 \end{array}$ The result is $x + 5 + \dfrac{8}{x - 3}$	**To Divide a Polynomial by a Polynomial** Use the long division method.	**p. 350**

Summary Exercises ▪ 4

This summary exercise set is provided to give you practice with each of the objectives in the chapter. Each exercise is keyed to the appropriate chapter section(s).

[4.1–4.2] Simplify each expression, using the properties of exponents.

1. $r^4 r^9$

2. $4x^{-5}$

3. $(2w)^{-3}$

4. $\dfrac{3}{m^{-4}}$

5. $y^{-5}y^2$

6. $\dfrac{w^{-7}}{w^{-3}}$

7. $\dfrac{x^{12}}{x^{15}}$

8. $(6c^0 d^4)(-3c^2 d^2)$

9. $(5a^2 b^3)(2a^{-2} b^{-6})$

10. $\left(\dfrac{3m^2 n^3}{p^4}\right)^3$

11. $\left(\dfrac{m^{-3} n^{-3}}{m^{-4} n^4}\right)^3$

12. $\left(\dfrac{r^{-5}}{s^4}\right)^{-2}$

13. $\left(\dfrac{x^3 y^4}{x^6 y^2}\right)^3$

14. $\left(\dfrac{a^8}{b^4}\right)\left(\dfrac{b^2}{2a^2}\right)^3$

15. $(2a^3)^0(-3a^4)^2$

16. $\left(\dfrac{2x^{-3} y^{-2}}{4x^{-5} y^{-4}}\right)^2 \left(\dfrac{8x^5 y^6}{4x^7 y^4}\right)^3$

17. Write 0.0000425 in scientific notation.

18. Write 3.1×10^{-4} in standard notation.

[4.3] Classify each of the following polynomials as a monomial, binomial, or trinomial.

19. $6x^4 - 3x$

20. $7x^5$

21. $4x^5 - 8x^3 + 5$

22. $x^3 + 2x^2 - 5x + 3$

23. $-7a^4 - 9a^3$

[4.3] Arrange in descending-exponent form and give the degree of each polynomial.

24. $5x^5 + 3x^2$

25. $13x^2$

26. $6x^2 + 4x^4 + 6$

27. $5 + x$

28. -8

29. $9x^4 - 3x + 7x^6$

[4.4] Add or subtract as indicated.

30. Add $7a^2 + 3a$ and $14a^2 - 5a$

31. Add $5x^2 + 3x - 5$ and $4x^2 - 6x - 2$

32. Add $5y^3 - 3y^2$ and $4y + 3y^2$

33. Subtract $7x^2 - 23x$ from $11x^2 - 15x$

34. Subtract $2x^2 - 5x - 7$ from $7x^2 - 2x + 3$

35. Subtract $5x^2 + 3$ from $9x^2 - 4x$

[4.4] Perform the indicated operations.

36. Subtract $5x - 3$ from the sum of $9x + 2$ and $-3x - 7$.

37. Subtract $5a^2 - 3a$ from the sum of $5a^2 + 2$ and $7a - 7$.

38. Subtract the sum of $16w^2 - 3w$ and $8w + 2$ from $7w^2 - 5w + 2$.

[4.4] Add, using the vertical method.

39. $x^2 + 5x - 3$ and $2x^2 + 4x - 3$ **40.** $9b^2 - 7$ and $8b + 5$ **41.** $x^2 + 7$, $3x - 2$, and $4x^2 - 8x$

[4.4] Subtract, using the vertical method.

42. $5x^2 - 3x + 2$ from $7x^2 - 5x - 7$ **43.** $8m - 7$ from $9m^2 - 7$

[4.5] Multiply.

44. $(3a^4)(2a^3)$ **45.** $(2x^2)(3x^5)$ **46.** $(-9p^3)(-6p^2)$

47. $(3a^2b^3)(-7a^3b^4)$ **48.** $5(3x - 8)$ **49.** $2a(5a - 3)$

50. $(-5rs)(2r^2s - 5rs)$ **51.** $7mn(3m^2n - 2mn^2 + 5mn)$ **52.** $(x + 5)(x + 4)$

53. $(w - 9)(w - 10)$ **54.** $(a - 9b)(a + 9b)$ **55.** $(p - 3q)^2$

56. $(a + 4b)(a + 3b)$ **57.** $(b - 8)(2b + 3)$ **58.** $(3x - 5y)(2x - 3y)$

59. $(5r + 7s)(3r - 9s)$ **60.** $(y + 2)(y^2 - 2y + 3)$ **61.** $(b + 3)(b^2 - 5b - 7)$

62. $(x - 2)(x^2 + 2x + 4)$ **63.** $(m^2 - 3)(m^2 + 7)$ **64.** $2x(x + 5)(x - 6)$

65. $a(2a - 5b)(2a - 7b)$

[4.5] Find the following products.

66. $(x + 7)^2$ **67.** $(a - 7)^2$ **68.** $(2w - 5)^2$

69. $(3p + 4)^2$ **70.** $(a + 7b)^2$ **71.** $(8x - 3y)^2$

72. $(x - 5)(x + 5)$ **73.** $(y + 9)(y - 9)$ **74.** $(2m + 3)(2m - 3)$

75. $(4r - 5)(4r + 5)$

76. $(5r - 2s)(5r + 2s)$

77. $(7a + 3b)(7a - 3b)$

78. $3x(x - 4)^2$

79. $3c(c + 5d)(c - 5d)$

80. $(y - 4)(y + 5)(y + 4)$

[4.6] Divide.

81. $\dfrac{9a^5}{3a^2}$

82. $\dfrac{24m^4n^2}{6m^2n}$

83. $\dfrac{15a - 10}{5}$

84. $\dfrac{32a^3 + 24a}{8a}$

85. $\dfrac{9r^2s^3 - 18r^3s^2}{-3rs^2}$

86. $\dfrac{35x^3y^2 - 21x^2y^3 + 14x^3y}{7x^2y}$

[4.6] Perform the indicated long division.

87. $\dfrac{x^2 - 2x - 15}{x + 3}$

88. $\dfrac{2x^2 + 9x - 35}{2x - 5}$

89. $\dfrac{x^2 - 8x + 17}{x - 5}$

90. $\dfrac{6x^2 - x - 10}{3x + 4}$

91. $\dfrac{6x^3 + 14x^2 - 2x - 6}{6x + 2}$

92. $\dfrac{4x^3 + x + 3}{2x - 1}$

93. $\dfrac{3x^2 + x^3 + 5 + 4x}{x + 2}$

94. $\dfrac{2x^4 - 2x^2 - 10}{x^2 - 3}$

Self-Test 4

The purpose of this self-test is to help you check your progress and to review for a chapter test in class. Allow yourself about 1 hour to take the test. When you are done, check your answers against the answers in the back of the book. If you missed any answers, be sure to go back and review the appropriate sections in the chapter and their exercises.

Simplify each expression, using the properties of exponents.

1. $(3x^2y)(-2xy^3)$

2. $\left(\dfrac{8m^2n^5}{2p^3}\right)^2$

3. $(x^4y^5)^2$

4. $\dfrac{9c^{-5}d^3}{18c^{-7}d^4}$

5. $(3x^2y)^3(-2xy^2)^2$

6. $(-3x^{-3}y)^2(4x^{-5}y^{-2})^{-1}$

7. $\dfrac{3x^0}{(2y)^0}$

Classify each of the following polynomials as a monomial, binomial, or trinomial.

8. $6x^2 + 7x$

9. $5x^2 + 8x - 8$

Arrange in descending-exponent form, and give the coefficients and degree.

10. $-3x^2 + 8x^4 - 7$

Add.

11. $3x^2 - 7x + 2$ and $7x^2 - 5x - 9$

12. $7a^2 - 3a$ and $7a^3 + 4a^2$

Subtract.

13. $5x^2 - 2x + 5$ from $8x^2 + 9x - 7$

14. $5a^2 + a$ from the sum of $3a^2 - 5a$ and $9a^2 - 4a$

Use the vertical method to add or subtract as indicated.

15. Add $x^2 + 3$, $5x - 9$, and $3x^2$.

16. Subtract $3x^2 - 5$ from $5x^2 - 7x$.

Multiply.

17. $5ab(3a^2b - 2ab + 4ab^2)$

18. $(x + 3y)(4x - 5y)$

19. $(3m + 2n)^2$

Divide.

20. $\dfrac{4x^3 - 5x^2 + 7x - 9}{x - 2}$

This test covers selected topics from the first five chapters. It is provided to help you in the process of reviewing previous chapters. Answers are provided in the back of the book. If you miss any answers, be sure to go back and review the appropriate chapter sections.

Evaluate each expression if $x = 3$, $y = -2$, and $z = 4$.

1. $-2x^2 - 3y^2 + 5z$

2. $\dfrac{-4y + 3x^2}{5z + x - y}$

Solve each equation.

3. $11x - 7 = 10x$

4. $-\dfrac{2}{3}x = 24$

5. $7x - 5 = 3x + 11$

6. $\dfrac{3}{5}x - 8 = 15 - \dfrac{2}{5}x$

7. $2(x - 3) + 5 = 2x - 1$

8. $4x - (2 - x) = 5x + 3$

9. $\dfrac{x + 1}{5} - \dfrac{2x - 3}{2} = 3$

10. $|x - 4| = 5$

11. $|3x + 5| - 5 = 9$

12. $|x + 1| = |2x - 3|$

Solve the following inequalities.

13. $7x - 5 > 8x + 10$

14. $6x - 9 < 3x + 6$

15. $|x - 8| < 5$

16. $|3x - 10| \geq 8$

17. Find the slope and y intercept of the line represented by the equation $4x + y = 9$.

18. Find the slope of the line perpendicular to the line represented by the equation $3x + 9y = 10$.

Write the equation of the line that satisfies the given conditions.

19. L has slope -2 and y intercept of $(0, 4)$.

20. L passes through the point $(3, 2)$ and is parallel to the line $4x - 5y = 20$.

21. L passes through the points $(1, -3)$ and $(3, 5)$ and has a y intercept of $(0, -7)$.

22. L is perpendicular to the line $x - 2y = 3$ and passes through the point $(1, -2)$.

Perform the indicated operations.

23. $(x^2 - 3x + 5) + (2x^2 + 5x - 9)$

24. $(3x^2 - 8x - 7) - (2x^2 - 5x + 11)$

25. $4x(3x - 5)$

26. $(2x - 5)(3x + 8)$

27. $(x + 2)(x^2 - 3x + 5)$ **28.** $(2x + 7)(2x - 7)$

29. $(3x - 5)^2$ **30.** $5x(2x - 5)^2$

31. $\dfrac{32x^2y^3 - 16x^4y^2 + 8xy^2}{8xy^2}$ **32.** $\dfrac{2x^3 - 15x - 7}{x - 3}$

Simplify each expression, using the properties of exponents.

33. $(3x^2)^2(2x^3)$ **34.** $(x^4y^{-3})^4$

35. $(2x^0)^3(-3x^2y)^2$ **36.** $\dfrac{6x^3y^{-5}}{3x^{-2}y^6}$

37. Calculate. Write your answer in scientific notation.

$$\frac{(4.2 \times 10^7)(6.0 \times 10^{-3})}{1.2 \times 10^{-5}}$$

Solve the following problems.

38. If 7 times a number decreased by 9 is 47, find the number.

39. The sum of two consecutive odd integers is 132. What are the two integers?

40. The length of a rectangle is 4 centimeters (cm) more than 5 times the width. If the perimeter is 56 cm, what are the dimensions of the rectangle?

CHAPTER 5

A Beginning Look at Functions

Economists are among the many professionals who use graphs to show connections between two sets of data. For example, an economist may use a graph to look for a connection between two different measures for the standard of living in various countries.

One measure of the standard of living in a country is the per capita gross domestic product (GDP). The GDP is the total value of all goods and services produced by all businesses and individuals over the course of 1 year. To find the per capita GDP, we divide that total value by the population of the country.

Other economists, including some who wrote an article in *Scientific American* in May 1993, question this method. Rather than compare GDP among countries, they use survival rate (life expectancy) to measure the quality of life.

5.1 Relations and Functions

1. Identify a function, using ordered pairs
2. Identify the domain and range of a function
3. Evaluate a function
4. Determine whether a relation is a function
5. Write an equation as a function

In Section 3.1, we introduced the concept of ordered pairs. We will now turn our attention to sets of ordered pairs.

Relation

A set of ordered pairs is called a **relation.**

We usually denote a relation with a capital letter. Given

$$A = \{(\text{Jane Trudameier, } 123\text{-}45\text{-}6789), (\text{Jacob Smith, } 987\text{-}65\text{-}4321),$$
$$(\text{Julia Jones, } 111\text{-}22\text{-}3333)\}$$

we have a relation, which we call A. In this case, there are three ordered pairs in the relation A.

Within this relation, there are two interesting sets. The first is the set of names, which happens to be the set of first elements. The second is the set of social security numbers (SSNs), which is the set of second elements. Each of these sets has a name.

Domain

The set of first elements in a relation is called the **domain** of the relation.

| Example 1 | ## Finding the Domain of a Relation |

Find the domain of each relation.

(a) $A = \{(\text{Ben Bender}, 58), (\text{Carol Clairol}, 32), (\text{David Duval}, 29)\}$

The domain of A is {Ben Bender, Carol Clairol, David Duval}.

(b) $B = \left\{\left(5, \dfrac{1}{2}\right), (-4, -5), (-12, 10), (-16, \pi)\right\}$

The domain of B is $\{5, -4, -12, -16\}$.

✔ CHECK YOURSELF 1

Find the domain of each relation.

(a) $A = \{(\text{Secretariat}, 10), (\text{Seattle Slew}, 8), (\text{Charismatic}, 5), (\text{Gallant Man}, 7)\}$

(b) $B = \left\{\left(-\dfrac{1}{2}, \dfrac{3}{4}\right), (0, 0), (1, 5), (\pi, \pi)\right\}$

Note: $X \to$ domain and $Y \to$ range. Many students find it helpful to remember that domain and range occur in alphabetical order.

Range

The set of second elements in a relation is called the **range** of the relation.

| Example 2 | ## Finding the Range of a Relation |

Find the range for each relation.

(a) $A = \{(\text{Ben Bender}, 58), (\text{Carol Clairol}, 32), (\text{David Duval}, 29)\}$

The range of A is $\{58, 32, 29\}$.

(b) $B = \left\{\left(5, \dfrac{1}{2}\right), (-4, -5), (-12, 10), (-16, \pi), (-16, 1)\right\}$

The range of B is $\left\{\dfrac{1}{2}, -5, 10, \pi, 1\right\}$.

✔ CHECK YOURSELF 2

Find the range of each relation.

(a) $A = \{(\text{Secretariat}, 10), (\text{Seattle Slew}, 8), (\text{Charismatic}, 5), (\text{Gallant Man}, 7)\}$

(b) $B = \left\{\left(-\dfrac{1}{2}, \dfrac{3}{4}\right), (0, 0), (1, 5), (\pi, \pi)\right\}$

For the remainder of this chapter, we will be interested in the special type of relation called a **function.** Before presenting a formal definition, we will note that a set of ordered pairs may be represented in a number of different ways, including a table. The set of ordered pairs in Example 2(b), for example, $B = \left\{ \left(5, \frac{1}{2}\right), (-4, -5), (-12, 10), (-16, \pi), (-16, 1) \right\}$, can be represented as

x	y
5	$\frac{1}{2}$
-4	-5
-12	10
-16	π
-16	1

Function

A **function** is a set of distinct ordered pairs (a relation) in which no two first coordinates are equal.

In Example 3, the sets of ordered pairs will be represented by tables.

Example 3	**Identifying a Function**

For each table of values below, decide whether the relation is a function.

(a)
x	y
-2	1
-1	1
1	3
2	3

(b)
x	y
-5	-2
-1	3
-1	6
2	9

(c)
x	y
-3	1
-1	0
0	2
2	4

Part (a) represents a function. No two first coordinates are equal. Part (b) is not a function because -1 appears as a first coordinate with two different second coordinates. Part (c) is a function.

✔ *CHECK YOURSELF 3*

For each table of values below, decide whether the relation is a function.

(a)
x	y
-3	0
-1	1
1	2
3	3

(b)
x	y
-2	-2
-1	-2
1	3
2	3

(c)
x	y
-2	0
-1	1
0	2
0	3

Next we will look at yet another way to represent relations, including the special relations called functions. Rather than being given a set of ordered pairs or a table, we may instead be given a rule or equation from which we must generate ordered pairs. To generate ordered pairs, we will need to recall how to evaluate an expression, first introduced in Section 1.2, and apply the order of operations rules of Section 0.5.

We have seen that variables can be used to represent numbers whose value is unknown. By using addition, subtraction, multiplication, division, and exponentiation, these numbers and variables form expressions such as

$$3 + 5 \qquad 7x - 4 \qquad x^2 - 3x - 4 \qquad x^4 - x^2 + 2$$

If a specific value is given for the variable, we **evaluate the expression.**

Example 4	**Evaluating Expressions**

Evaluate the expression $x^4 - 2x^2 + 3x + 4$ for the indicated value of x.

(a) $x = 0$

Substituting 0 for x in the expression yields

$$(0)^4 - 2(0)^2 + 3(0) + 4 = 0 - 0 + 0 + 4$$
$$= 4$$

(b) $x = 2$

Substituting 2 for x in the expression yields

$$(2)^4 - 2(2)^2 + 3(2) + 4 = 16 - 8 + 6 + 4$$
$$= 18$$

(c) $x = -1$

Substituting -1 for x in the expression yields

$$(-1)^4 - 2(-1)^2 + 3(-1) + 4 = 1 - 2 - 3 + 4$$
$$= 0$$

✔ *CHECK YOURSELF 4*

Evaluate the expression $2x^3 - 3x^2 + 3x + 1$ for the indicated value of x.

(a) $x = 0$

(b) $x = 1$

(c) $x = -2$

We could design a machine whose purpose would be to crank out the value of an expression for each given value of x. We could call this machine something simple such as f, our **function machine.** Our machine might look like this.

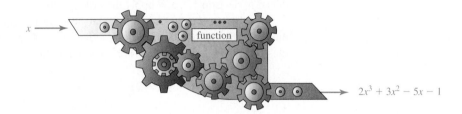

For example, if we put -1 into the machine, the machine would substitute -1 for x in the expression, and 5 would come out the other end because

$$2(-1)^3 + 3(-1)^2 - 5(-1) - 1 = -2 + 3 + 5 - 1 = 5$$

Note that, with this function machine, an input of -1 will always result in an output of 5. One of the most important aspects of a function machine is that each input has a unique output.

In fact, the idea of the function machine is very useful in mathematics. Your graphing calculator can be used as a function machine. You can enter the expression into the calculator as Y_1 and then evaluate Y_1 for different values of x.

Generally, in mathematics, we do not write $Y_1 = 2x^3 + 3x^2 - 5x - 1$. Instead, we write $f(x) = 2x^3 + 3x^2 - 5x - 1$, which is read "$f$ of x is equal to. . . ." Instead of calling f a function machine, we say that f is a function of x. The greatest benefit of this notation is that it lets us easily note the input value of x along with the output of the function. Instead of "the value of Y_1 is 155 when $x = 4$," we can write $f(4) = 155$.

> Two distinct input elements can have the same output. However, each input element can only be associated with exactly one output element.

Example 5	**Evaluating a Function**

Given $f(x) = x^3 + 3x^2 - x + 5$, find the following.

(a) $f(0)$

Substituting 0 for x in the above expression, we get

$$(0)^3 + 3(0)^2 - (0) + 5 = 5$$

(b) $f(-3)$

Substituting -3 for x in the above expression, we get

$$(-3)^3 + 3(-3)^2 - (-3) + 5 = -27 + 27 + 3 + 5$$

$$= 8$$

(c) $f\left(\dfrac{1}{2}\right)$

Substituting $\dfrac{1}{2}$ for x in the earlier expression, we get

$$\left(\dfrac{1}{2}\right)^3 + 3\left(\dfrac{1}{2}\right)^2 - \left(\dfrac{1}{2}\right) + 5 = \dfrac{1}{8} + 3\left(\dfrac{1}{4}\right) - \dfrac{1}{2} + 5$$

$$= \dfrac{1}{8} + \dfrac{3}{4} - \dfrac{1}{2} + 5$$

$$= \dfrac{1}{8} + \dfrac{6}{8} - \dfrac{4}{8} + 5$$

$$= \dfrac{3}{8} + 5$$

$$= 5\dfrac{3}{8} \text{ or } \dfrac{43}{8}$$

✔ CHECK YOURSELF 5

Given $f(x) = 2x^3 - x^2 + 3x - 2$, find the following.

(a) $f(0)$ (b) $f(3)$ (c) $f\left(-\dfrac{1}{2}\right)$

We can rewrite the relationship between x and $f(x)$ in Example 5 as a series of ordered pairs.

$$f(x) = x^3 + 3x^2 - x + 5$$

From this we found that

$$f(0) = 5, \qquad f(-3) = 8, \qquad \text{and} \qquad f\left(\dfrac{1}{2}\right) = \dfrac{43}{8}$$

Note: Because $y = f(x)$, $(x, f(x))$ is another way of writing (x, y).

There is an ordered pair, which we could write as $(x, f(x))$, associated with each of these. Those three ordered pairs are

$$(0, 5), \qquad (-3, 8), \qquad \text{and} \qquad \left(\dfrac{1}{2}, \dfrac{43}{8}\right)$$

| Example 6 | **Finding Ordered Pairs** |

Given the function $f(x) = 2x^2 - 3x + 5$, find the ordered pair $(x, f(x))$ associated with each given value for x.

(a) $x = 0$

$$f(0) = 2(0)^2 - 3(0) + 5 = 5$$

The ordered pair is $(0, 5)$.

(b) $x = -1$

$$f(-1) = 2(-1)^2 - 3(-1) + 5 = 10$$

The ordered pair is $(-1, 10)$.

(c) $x = \dfrac{1}{4}$

$$f\left(\frac{1}{4}\right) = 2\left(\frac{1}{4}\right)^2 - 3\left(\frac{1}{4}\right) + 5 = \frac{35}{8}$$

The ordered pair is $\left(\dfrac{1}{4}, \dfrac{35}{8}\right)$.

✔ CHECK YOURSELF 6

Given $f(x) = 2x^3 - x^2 + 3x - 2$, find the ordered pair associated with each given value of x.

(a) $x = 0$ (b) $x = 3$ (c) $x = -\dfrac{1}{2}$

We began this section by defining a relation as a set of ordered pairs. In Example 7, we will determine which relations can be modeled by a function machine.

Example 7   Modeling with a Function Machine

Determine which relations can be modeled by a function machine.

(a) The set of all possible ordered pairs in which the first element is a U.S. state and the second element is a U.S. Senator from that state.

We cannot model this relation with a function machine. Because there are two senators from each state, each input does not have a unique output. In the picture, New Jersey is the input, but New Jersey has two different senators.

(b) The set of all ordered pairs in which the input is the year and the output is the U.S. Open golf champion of that year.

Year 2000 →

Tiger Woods

This relation can be modeled with the function machine. Each input has a unique output. In the picture, an input of 2000 gives an output of Tiger Woods. Every time the input is 2000, the output will be Tiger Woods.

(c) The set of all ordered pairs in the relation R, when

$$R = \{(1, 3), (2, 5), (2, 7), (3 -4)\}$$

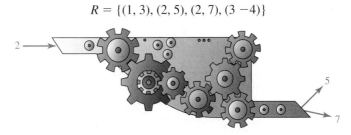

This relation cannot be modeled with a function machine. An input of 2 can result in two different outputs, either 5 or 7.

(d) The set of all ordered pairs in the relation S, when

$$S = \{(-1, 3), (0, 3), (3, 5), (5, -2)\}$$

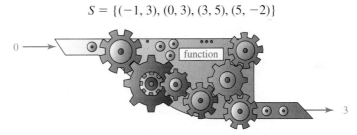

This relation can be modeled with a function machine. Each input has a unique output.

✔ *CHECK YOURSELF 7*

Determine which relations can be modeled by a function machine.

(a) The set of all ordered pairs in which the first element is a U.S. city and the second element is the mayor of that city

(b) The set of all ordered pairs in which the first element is a street name and the second element is a U.S. city in which a street of that name is found

(c) The relation $A = \{(-2, 3), (-4, 9), (9, -4)\}$

(d) The relation $B = \{(1, 2), (3, 4), (3, 5)\}$

If we are working with an equation in x and y, we may wish to rewrite the equation as a function of x. This is particularly useful if we want to use a graphing calculator to find y for a given x, or to view a graph of the equation.

| Example 8 | Writing Equations as Functions |

Rewrite each linear equation as a function of x. Use $f(x)$ notation in the final result.

(a) $y = 3x - 4$

We note that y is already isolated. Simply replace y with $f(x)$.

$$f(x) = 3x - 4$$

(b) $2x - 3y = 6$

We first solve for y.

$$-3y = -2x + 6$$

$$y = \frac{-2x + 6}{-3} \qquad \text{{\small y has been isolated. Using slope-intercept form gives}}$$

$$y = \frac{2}{3}x - 2 \qquad \text{{\small Now replace y with $f(x)$.}}$$

$$f(x) = \frac{2}{3}x - 2$$

✔ *CHECK YOURSELF 8*

Rewrite each linear equation as a function of x. Use $f(x)$ notation in the final result.

(a) $y = -2x + 5$ (b) $3x + 5y = 15$

The process of finding the graph of a linear function is identical to the process of finding the graph of a linear equation.

| Example 9 | Graphing a Linear Function |

Graph the function

$$f(x) = 3x - 5$$

We could use the slope and y intercept to graph the line, or we can find three points (the third is a checkpoint) and draw the line through them. We will do the latter.

$$f(0) = -5 \qquad f(1) = -2 \qquad f(2) = 1$$

We will use the three points $(0, -5)$, $(1, -2)$, and $(2, 1)$ to graph the line.

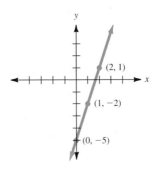

✔ CHECK YOURSELF 9

Graph the function.

$$f(x) = 5x - 3$$

One benefit of having a function written in $f(x)$ form is that it makes it fairly easy to substitute values for x. In Example 9, we substituted the values 0, 1, and 2. Sometimes it is useful to substitute nonnumeric values for x.

Example 10

Substituting Nonnumeric Values for x

Let $f(x) = 2x + 3$. Evaluate f as indicated.

(a) $f(a)$

Substituting a for x in the equation, we see that

$$f(a) = 2a + 3$$

(b) $f(2 + h)$

Substituting $2 + h$ for x in the equation, we get

$$f(2 + h) = 2(2 + h) + 3$$

Distributing the 2 and then simplifying, we have

$$f(2 + h) = 4 + 2h + 3$$

$$= 2h + 7$$

✔ CHECK YOURSELF 10

Let $f(x) = 4x - 2$. Evaluate f as indicated.

(a) $f(b)$ (b) $f(4 + h)$

The TABLE feature on a graphing calculator can also be used to evaluate a function. Example 11 illustrates this feature.

Example 11

Using a Graphing Calculator to Evaluate a Function

Evaluate the function $f(x) = 3x^3 + x^2 - 2x - 5$ for each x in the set $\{-6, -5, -4, -3, -2\}$.

1. Enter the function into a Y= screen.

2. Find the table setup screen.

3. Start the table at -6 with a change of 1.

4. View the table.

The table should look something like this

Although we have assumed that the graphing calculator was a TI, most such calculators have similar capability.

X	Y_1	
−6	−605	
−5	−345	
−4	−173	
−3	−71	
−2	−21	
−1	−5	
0	−5	
X=−6		

The Y_1 column is the function value for each value of x.

✔ CHECK YOURSELF 11

Evaluate the function $f(x) = 2x^3 - 3x^2 - x + 2$ for each x in the set $\{-5, -4, -3, -2, -1, 0, 1\}$.

✔ CHECK YOURSELF ANSWERS

1.   (a) The domain of A is {Secretariat, Seattle Slew, Charismatic, Gallant Man};

     (b) the domain of B is $\left\{-\dfrac{1}{2}, 0, 1, \pi\right\}$.

2. (a) The range of A is $\{10, 8, 5, 7\}$; (b) the range of B is $\left\{\dfrac{3}{4}, 0, 5, \pi\right\}$

3. (a) function; (b) function; (c) not a function

4. (a) 1; (b) 3; (c) -33 5. (a) -2; (b) 52; (c) -4

6. (a) $(0, -2)$; (b) $(3, 52)$; (c) $\left(-\dfrac{1}{2}, -4\right)$

7. (a) function; (b) not a function; (c) function; (d) not a function

8. (a) $f(x) = -2x + 5$; (b) $f(x) = -\dfrac{3}{5}x + 3$

9.

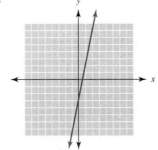

10. (a) $4b - 2$; (b) $4h + 14$

11.

X	Y₁	
-5	-318	
-4	-170	
-3	-76	
-2	-24	
-1	-2	
0	2	
1	0	
X=-5		

Exercises · 5.1

Find the domain and range of each relation.

1. $A = \{(\text{Colorado}, 21), (\text{Edmonton}, 5), (\text{Calgary}, 18), (\text{Vancouver}, 17)\}$

2. $F = \left\{ \left(\text{St. Louis}, \frac{1}{2}\right), \left(\text{Denver}, -\frac{3}{4}\right), \left(\text{Green Bay}, \frac{7}{8}\right), \left(\text{Dallas}, -\frac{4}{5}\right) \right\}$

3. $G = \left\{ (\text{Chamber}, \pi), (\text{Testament}, 2\pi), \left(\text{Rainmaker}, \frac{1}{2}\right), (\text{Street Lawyer}, 6) \right\}$

4. $C = \{(\text{John Adams}, -16), (\text{John Kennedy}, -23), (\text{Richard Nixon}, -5), (\text{Harry Truman}, -11)\}$

5. $\{(1, 2), (3, 4), (5, 6), (7, 8), (9, 10)\}$ **6.** $\{(2, 3), (3, 5), (4, 7), (5, 9), (6, 11)\}$

7. $\{(1, 2), (1, 3), (1, 4), (1, 5), (1, 6)\}$ **8.** $\{(3, 4), (3, 6), (3, 8), (3, 9), (3, 10)\}$

9. $\{(-1, 3), (-2, 4), (-3, 5), (4, 4), (5, 6)\}$ **10.** $\{(-2, 4), (1, 4), (-3, 4), (5, 4), (7, 4)\}$

11. The Dow Jones industrial averages over a 5-day period are displayed in a table. List this information as a set of ordered pairs, using the day of the week as the domain.

Day	1	2	3	4	5
Average	9,274	9,096	8,814	8,801	8,684

12. Food purchases. In the snack department of the local supermarket, candy costs $2.16 per pound. For 1 to 5 lb, write the cost of candy as a set of ordered pairs.

Bulk Candy
$2.16 per pound

Write a set of ordered pairs that describes each situation. Give the domain and range of each relation.

13. The first element is an integer between -3 and 3. The second coordinate is the cube of the first coordinate.

14. The first element is a positive integer less than 6. The second coordinate is the sum of the first coordinate and -2.

15. The first element is the number of hours worked—10, 20, 30, 40; the second coordinate is the salary at $9 per hour.

16. The first coordinate is the number of toppings on a pizza (up to four); the second coordinate is the price of the pizza, which is $9 plus $1 per topping.

In Exercises 17 to 24, determine which of the relations are also functions.

17. $\{(1, 6), (2, 8), (3, 9)\}$

18. $\{(2, 3), (3, 4), (5, 9)\}$

19. $\{(-1, 4), (-2, 5), (-3, 7)\}$

20. $\{(-2, 1), (-3, 4), (-4, 6)\}$

21. $\{(1, 3), (1, 2), (1, 1)\}$

22. $\{(2, 4), (2, 5), (3, 6)\}$

23. $\{(-3, 5), (6, 3), (6, 9)\}$

24. $\{(4, -4), (2, 8), (4, 8)\}$

Decide whether the relation, shown as a table of values, is a function.

25.

x	y
3	1
-2	4
5	3
-7	4

26.

x	y
-2	3
1	4
5	6
2	-1

27.

x	y
2	3
4	2
2	-5
-6	-3

28.

x	y
1	5
3	-6
1	-5
-2	-9

29.

x	y
-1	2
3	6
6	2
-9	4

30.

x	y
4	-6
2	3
-7	1
-3	-6

Evaluate each function for the values specified.

31. $f(x) = x^2 - x - 2$; find (a) $f(0)$, (b) $f(-2)$, and (c) $f(1)$.

32. $f(x) = x^2 - 7x + 10$; find (a) $f(0)$, (b) $f(5)$, and (c) $f(-2)$.

33. $f(x) = 3x^2 + x - 1$; find (a) $f(-2)$, (b) $f(0)$, and (c) $f(1)$.

34. $f(x) = -x^2 - x - 2$; find (a) $f(-1)$, (b) $f(0)$, and (c) $f(2)$.

35. $f(x) = x^3 - 2x^2 + 5x - 2$; find (a) $f(-3)$, (b) $f(0)$, and (c) $f(1)$.

36. $f(x) = -2x^3 + 5x^2 - x - 1$; find (a) $f(-1)$, (b) $f(0)$, and (c) $f(2)$.

37. $f(x) = -3x^3 + 2x^2 - 5x + 3$; find (a) $f(-2)$, (b) $f(0)$, and (c) $f(3)$.

38. $f(x) = -x^3 + 5x^2 - 7x - 8$; find (a) $f(-3)$, (b) $f(0)$, and (c) $f(2)$.

39. $f(x) = 2x^3 + 4x^2 + 5x + 2$; find (a) $f(-1)$, (b) $f(0)$, and (c) $f(1)$.

40. $f(x) = -x^3 + 2x^2 - 7x + 9$; find (a) $f(-2)$, (b) $f(0)$, and (c) $f(2)$.

Rewrite each equation as a function of x. Use $f(x)$ notation in the final result.

41. $y = -3x + 2$ **42.** $y = 5x + 7$ **43.** $y = 4x - 8$

44. $y = -7x - 9$ **45.** $3x + 2y = 6$ **46.** $4x + 3y = 12$

47. $-2x + 6y = 9$ **48.** $-3x + 4y = 11$

49. $-5x - 8y = -9$ **50.** $4x - 7y = -10$

Graph the functions.

51. $f(x) = 4x + 2$ **52.** $f(x) = -2x - 5$ **53.** $f(x) = -2x + 7$

54. $f(x) = -4x + 5$ **55.** $f(x) = x + 5$ **56.** $f(x) = 4x - 3$

If $f(x) = 5x - 1$, find the following:

57. $f(a)$ **58.** $f(2r)$ **59.** $f(x + 1)$

60. $f(a - 2)$ **61.** $f(x + h)$ **62.** $\dfrac{f(x + h) - f(x)}{h}$

If $g(x) = -3x + 2$, find the following:

63. $g(m)$ **64.** $g(5n)$ **65.** $g(x + 2)$ **66.** $g(s - 1)$

In Exercises 67 to 70, use your graphing calculator to evaluate the given function for each value in the given set.

67. $f(x) = 3x^2 - 5x + 7$; $\{-5, -4, -3, -2, -1, 0, 1, 2, 3, 4, 5\}$

68. $f(x) = 4x^3 - 7x^2 + 9$; $\{-3, -2, -1, 0, 1, 2, 3\}$

69. $f(x) = 2x^3 - 4x^2 + 5x - 9$; $\{-4, -3, -2, -1, 0, 1, 2, 3, 4\}$

70. $f(x) = -3x^4 + 5x^2 - 7x - 15$; $\{-3, -2, -1, 0, 1, 2, 3\}$

Solve the following application problems.

71. Profit. The marketing department of a company has determined that the profit for selling x units of a product is approximated by the function

$$f(x) = 50x - 600$$

Find the profit in selling 2,500 units.

72. Cost. The inventor of a new product believes that the cost of producing the product is given by the function

$$C(x) = 1.75x + 7{,}000 \qquad \text{where } x = \text{units produced}$$

What would be the cost of producing 2,000 units of the product?

73. Phone cost. A phone company has two different rates for calls made at different times of the day. These rates are given by the function

$$C(x) = \begin{cases} 24x + 33 & \text{between 5 P.M. and 11 P.M.} \\ 36x + 52 & \text{between 8 A.M. and 5 P.M.} \end{cases}$$

when x is the number of minutes of a call and C is the cost of a call in cents.

(a) What is the cost of a 10-minute call at 10:00 A.M.?

(b) What is the cost of a 10-minute call at 10:00 P.M.?

74. Accidents. The number of accidents in 1 month involving drivers x years of age can be approximated by the function

$$f(x) = 2x^2 - 125x + 3{,}000$$

Find the number of accidents in 1 month that involved (a) 17-year-olds and (b) 25-year-olds.

75. Stopping distance. The distance x (in feet) that a car will skid on a certain road surface after the brakes are applied is a function of the car's velocity v (in miles per hour). The function can be approximated by

$$x = f(v) = 0.017v^2$$

How far will the car skid if the brakes are applied at (a) 55 mi/h? (b) 70 mi/h?

76. Science. An object is thrown upward with an initial velocity of 128 ft/s. Its height h in feet after t seconds is given by the function

$$h(t) = -16t^2 + 128t$$

What is the height of the object at (a) 2 s? (b) 4 s? (c) 6 s?

5.2 Tables and Graphs

 OBJECTIVES

1. *Use the vertical line test*
2. *Identify the domain and range from the graph of a relation*
3. *Read function values from a table*
4. *Read function values from a graph*

In Section 5.1, we defined a function in terms of ordered pairs. A set of ordered pairs can be specified in several ways; here are the most common.

Rules and Properties: Ordered Pairs

1. We can present ordered pairs in a list or table.

2. We can give a rule or equation that will generate ordered pairs.

3. We can use a graph to indicate ordered pairs. The graph can show distinct ordered pairs, or it can show all the ordered pairs on a line or curve.

We have already seen functions presented as lists of ordered pairs, in tables, and as rules or equations. We now look at graphs of the ordered pairs from Example 3 in Section 5.1 to introduce the **vertical line test,** which is a graphical test for identifying a function.

(a) As a set of ordered pairs, the relation is $\{(-2, 1), (-1, 1), (1, 3), (2, 3)\}$. Recall that this relation did represent a function.

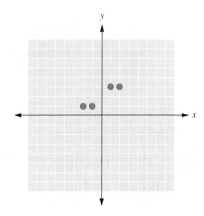

Recall: that, if a grid has no numeric labels, each mark represents one unit. In this text each such grid represents *x* values from −8 to 8 and *y* values from −8 to 8.

(b) As a set of ordered pairs, the relation is $\{(-5,-2), (-1, 3), (-1, 6), (2, 9)\}$. Recall that this relation did *not* represent a function.

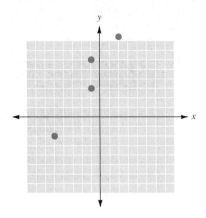

(c) As a set of ordered pairs, the relation is $\{(-3, 1), (-1, 0), (0, 2), (2, 4)\}$. Recall that this relation did represent a function.

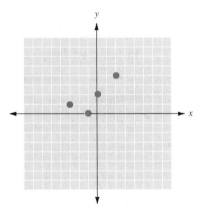

Notice that in the graphs of relations (a) and (c), there is no vertical line that can pass through two different points of the graph. In relation (b), a vertical line can pass through the two points that represent the ordered pairs $(-1, 3)$ and $(-1, 6)$. This leads to the following definition.

Vertical Line Test

A relation is a function if no vertical line can pass through two or more points on its graph.

Example 1	**Identifying a Function**

For each set of ordered pairs, plot the related points on the provided axes. Then use the vertical line test to determine which of the sets is a function.

(a) {(0, −1), (2, 3), (2, 6), (4, 2), (6, 3)}

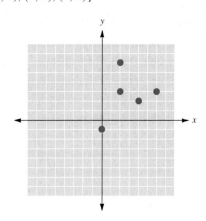

Because a vertical line can be drawn through the points (2, 3) and (2, 6), the relation does not pass the vertical line test. That is, if the input is 2, the output is *both* 3 and 6. This is not a function.

(b) {(1, 1), (2, 0), (3, 3), (4, 3), (5, 3)}

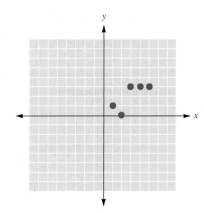

This is a function. Although a horizontal line can be drawn through several points, no vertical line passes through more than one point.

✔ *CHECK YOURSELF 1*

For each set of ordered pairs, plot the related points. Then use the vertical line test to determine which of the sets is a function.

(a) {(−2, 4), (−1, 4), (0, 4), (1, 3), (5, 5)}

(b) {(−3, −1), (−1, −3), (1, −3), (1, 3)}

By studying the graph of a relation, we can also determine the domain and range, as shown in Example 2. Recall that the domain is the set of *x* values that appear in the ordered pairs, while the range is the set of *y* values.

Example 2	**Identifying Functions, Domain, and Range**

Determine whether the given graph is the graph of a function. Also provide the domain and range in each case.

(a)

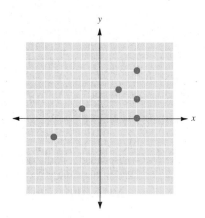

This is not a function. A vertical line at $x = 4$ passes through three points. The domain D of this relation is

$$D = \{-5, -2, 2, 4\}$$

and the range R is

$$R = \{-2, 0, 1, 2, 3, 5\}$$

(b)

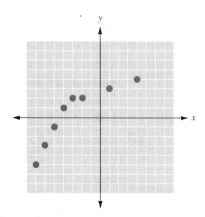

This is a function. No vertical line passes through more than one point. The domain is

$$D = \{-7, -6, -5, -4, -3, -2, 1, 4\}$$

and the range is

$$R = \{-5, -3, -1, 1, 2, 3, 4\}$$

✔ *CHECK YOURSELF 2*

Determine whether the given graph is the graph of a function. Also provide the domain and range in each case.

(a)   (b)

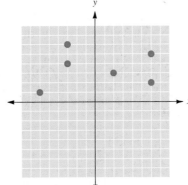   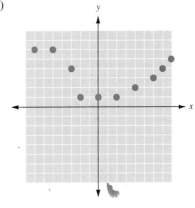

The graphs presented in this section have all depicted relations which are **finite** sets of ordered pairs. We now consider graphs composed of line segments, lines, or curves. Each such graph represents an **infinite** collection of points. The vertical line test can be used to decide if the relation is a function, and we can name the domain and range.

| Example 3 | **Identifying Functions, Domain, and Range** |

Determine whether the given graph is the graph of a function. Also provide the domain and range in each case.

(a)

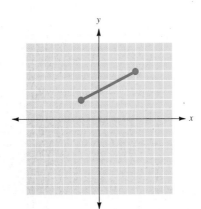

Note: When you see the statement −2 ≤ x ≤ 4 think "all real numbers between −2 and 4, including −2 and 4."

Since no vertical line will pass through more than point, this is a function. The *x* values that are used in the ordered pairs go from −2 to 4 inclusive. By using set-builder notation, the domain is then

$$D = \{x \mid -2 \le x \le 4\}$$

The y values that are used go from 2 to 5 inclusive. The range is

$$R = \{y \mid 2 \leq y \leq 5\}$$

(b)

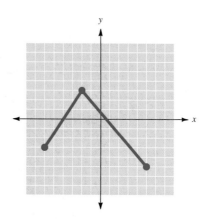

The relation graphed here is a function. The x values run from -6 to 5, so

$$D = \{x \mid -6 \leq x \leq 5\}$$

The y values go from -5 to 3, so

$$R = \{y \mid -5 \leq y \leq 3\}$$

✔ *CHECK YOURSELF 3*

Determine whether the given graph is the graph of a function. Also provide the domain and range in each case.

(a)

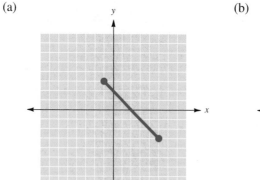

(b)

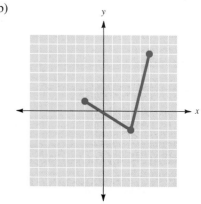

In Example 4, we consider the graphs of some common curves.

| Example 4 | **Identifying Functions, Domain, and Range** |

Determine whether the given graph is the graph of a function. Also provide the domain and range in each case.

(a)

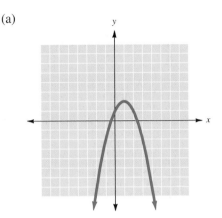

Note: Depicted here is a curve called a *parabola*.

Note: ℝ is the symbol for the set of all real numbers.

Since no vertical line will pass through more than one point, this is a function. Note that the arrows on the ends of the graph indicate that the pattern continues indefinitely. The x values that are used in this graph therefore consist of all real numbers. The domain is

$$D = \{x \mid x \text{ is a real number}\}$$

or simply $D = \mathbb{R}$.

The y values, however, are never higher than 2. The range is the set of all real numbers less than or equal to 2. So

$$R = \{y \mid y \leq 2\}$$

(b)

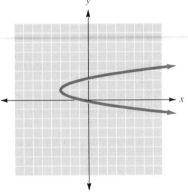

Note: This curve is also a parabola.

This relation is not a function. A vertical line drawn anywhere to the right of -3 will pass through two points. The x values which are used begin at -3 and continue indefinitely to the right, so

$$D = \{x \mid x \geq -3\}$$

The y values consist of all real numbers, so

$$R = \{y \mid y \text{ is a real number}\}$$

or simply $R = \mathbb{R}$.

(c)

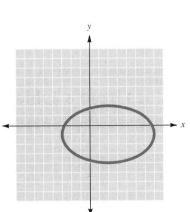

Note: This curve is called an *ellipse.*

This relation is not a function. A vertical line drawn anywhere between -3 and 7 will pass through two points. The x values which are used run from -3 to 7 inclusive. Thus

$$D = \{x \mid -3 \leq x \leq 7\}$$

The y values used in the ordered pairs go from -4 to 2 inclusive, so

$$R = \{y \mid -4 \leq y \leq 2\}$$

✔ *CHECK YOURSELF 4*

Determine whether the given graph is the graph of a function. Also provide the domain and range in each case.

(a)

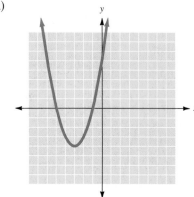

(b)

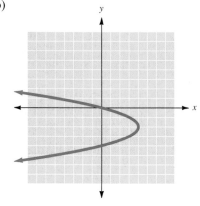

(c)

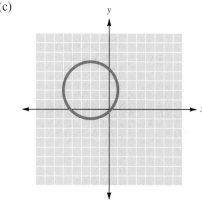

An important skill in working with functions is that of reading tables and graphs. If we are given a function f in either of these forms, our goals here are twofold:

1. Given x, we want to find $f(x)$.

2. Given $f(x)$, we want to find x.

Example 5 illustrates.

Example 5

Reading Values from a Table

Suppose we have the following functions f and g:

Note: Think of the x values as "input" values and the $f(x)$ values as "outputs."

x	$f(x)$		x	$g(x)$
-4	8		-2	5
0	6		1	0
2	-4		4	-4
1	-2		8	-2

(a) Find $f(0)$.

This means that 0 is the input value (a value for x). We want to know what f does to 0. Looking in the table, we see that the output value is 6. So $f(0) = 6$.

(b) Find $g(4)$.

We are given $x = 4$, and we want $g(x)$. In the table we find $g(4) = -4$.

(c) Find x, given that $f(x) = -4$.

Now we are given the output value of -4. We ask, what x value results in an output value of -4? The answer is 2. So $x = 2$.

(d) Find x, given that $g(x) = -2$.

Since the output is given as -2, we look in the table to find that when $x = 8$, $g(x) = -2$. So $x = 8$.

✔ CHECK YOURSELF 5

Use the functions in Example 5 to find

(a) $f(1)$ (b) $g(-2)$

(c) x, given that $f(x) = 8$ (d) x, given that $g(x) = 5$

In Example 6 we consider the same goals, given the graph of a function: (1) given x, find $f(x)$; and (2) given $f(x)$, find x.

Example 6   Reading Values from a Graph

Given the graph of f shown, find the desired values.

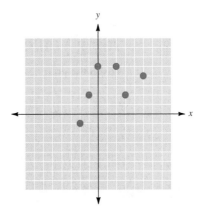

(a) Find $f(2)$.

Since 2 is the x value, we move to 2 on the x axis and then search vertically for a plotted point. We find $(2, 5)$, which tells us that an input of 2 results in an output of 5. Thus $f(2) = 5$.

(b) Find $f(-1)$.

Since $x = -1$, we move to -1 on the x axis. We note the point $(-1, 2)$, so $f(-1) = 2$.

(c) Find all x such that $f(x) = 2$.

Now we are told that the output value is 2, so we move up to 2 on the y axis and search horizontally for plotted points. There are two: $(-1, 2)$ and $(3, 2)$. So the desired x values are -1 and 3.

(d) Find all x such that $f(x) = 4$.

We move to 4 on the y axis and search horizontally. We find one point: $(5, 4)$. So $x = 5$.

✔ *CHECK YOURSELF 6*

Given the graph of f shown, find the desired values.

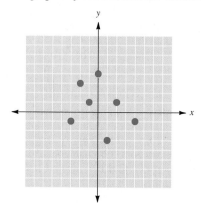

(a) Find $f(1)$.

(b) Find $f(-3)$.

(c) Find all x such that $f(x) = 1$.

(d) Find all x such that $f(x) = 4$.

Example 7 deals with graphs which represent **infinite** collections of points.

Example 7   Reading Values from a Graph

(a) Given the graph of f shown, find the desired values.

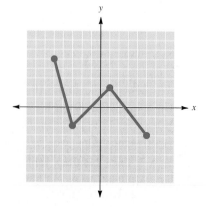

(i) Find $f(-1)$.

Since $x = -1$, we move to -1 on the x axis. There we find the point $(-1, 0)$. So $f(-1) = 0$.

(ii) Find all x such that $f(x) = -1$.

We are given that the output is -1, so we move to -1 on the y axis and search horizontally for plotted points. There are three of them, and we must estimate the coordinates for a couple of these. One point is exactly $(-2, -1)$, one is approximately $(-3.3, -1)$, and one is approximately $(3.5, -1)$. So the desired x values are -3.3, -2, and 3.5.

(b) Given the graph of f shown, find the desired values.

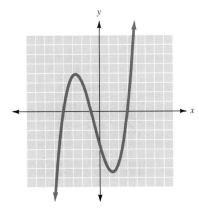

(i) Find $f(-3)$.

Since $x = -3$, we move to -3 on the x axis. We search vertically and estimate a plotted point at approximately $(-3, 3.7)$. So $f(-3) = 3.7$.

(ii) Find all x such that $f(x) = 0$.

Since the output (y value) is 0, we look for points with a y coordinate of 0. There are three: $(-4, 0)$, $(-1, 0)$, and $(3, 0)$. So the desired x values are -4, -1, and 3.

✔ *CHECK YOURSELF 7*

Given the graph of f shown, find the desired values.

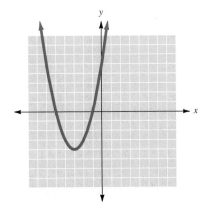

(a) Find $f(-4)$.

(b) Find all x such that $f(x) = 0$.

At this point, you may be wondering how the concept of function relates to anything outside the study of mathematics. A function is a relation that yields a single output (*y* value) each time a specific input (*x* value) is given. Any field in which predictions are made is building on the idea of functions. Here are a few examples:

- The physicist looks for the relationship that uses a planet's mass to predict its gravitational pull.

- The economist looks for the relationship that uses the tax rate to predict the employment rate.

- The business marketer looks for the relationship that uses an item's price to predict the number that will be sold.

- The college board looks for the relationship between tuition costs and the number of students enrolled at the college.

- The biologist looks for the relationship that uses temperature to predict a body of water's nutrient level.

In your future study of mathematics, you will see functions applied in areas such as these. In those applications, you'll find that you put to good use the basic skills developed here: (1) given *x*, find $f(x)$; and (2) given $f(x)$, find *x*.

✔ CHECK YOURSELF ANSWERS

1. (a) Is a function; (b) is not a function

2. (a) Is not a function; $D = \{-6, -3, 2, 6\}$; $R = \{1, 2, 3, 4, 5, 6\}$
    (b) Is a function; $D = \{-7, -5, -3, -2, 0, 2, 4, 6, 7, 8\}$; $R = \{1, 2, 3, 4, 5, 6\}$

3. (a) Is a function; $D = \{x \mid -1 \le x \le 5\}$; $R = \{y \mid -3 \le y \le 3\}$
    (b) Is a function; $D = \{x \mid -2 \le x \le 5\}$; $R = \{y \mid -2 \le y \le 6\}$

4. (a) Is a function; $D = \mathbb{R}$; $R = \{y \mid y \ge -4\}$
    (b) Is not a function; $D = \{x \mid x \le 4\}$; $R = \mathbb{R}$
    (c) Is not a function; $D = \{x \mid -5 \le x \le 1\}$; $R = \{y \mid -1 \le y \le 5\}$

5. (a) -2; (b) 5; (c) -4; (d) -2

6. (a) -3; (b) -1; (c) -1 and 2; (d) 0

7. (a) -3; (b) -5 and -1

Exercises ▪ 5.2

For each set of ordered pairs, plot the related points. Then use the vertical line test to determine which sets are functions.

1. $\{(-3, 1), (-1, 2), (-2, 3), (1, 4)\}$

2. $\{(2, 2), (1, 1), (3, 3), (4, 5)\}$

3. $\{(-1, 1), (2, 2), (3, 4), (5, 6)\}$

4. $\{(1, 4), (-1, 5), (0, 2), (2, 3)\}$

5. $\{(1, 2), (1, 3), (2, 1), (3, 1)\}$

6. $\{(-1, 1), (3, 4), (-1, 2), (5, 3)\}$

Determine whether the relation is a function. Also provide the domain and the range.

7.

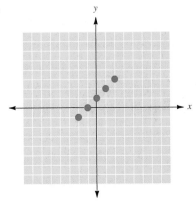

8.

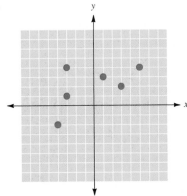

9.

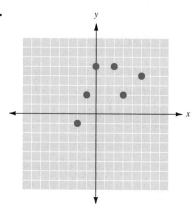

10.

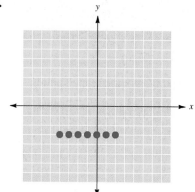

11.

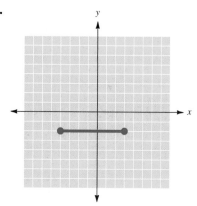

12.

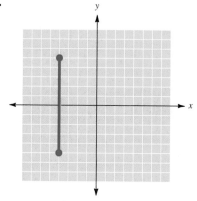

13.

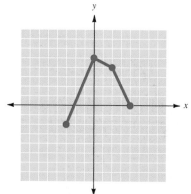

14.

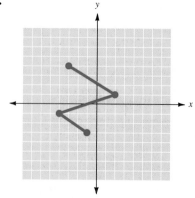

15.

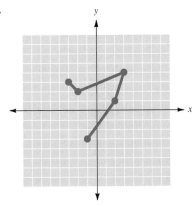

16.

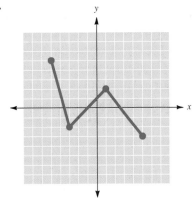

17.

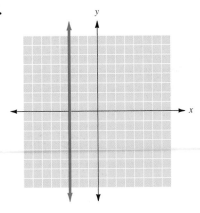

18.

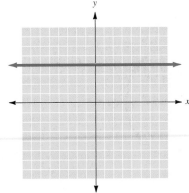

19.

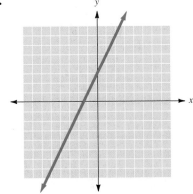

20.

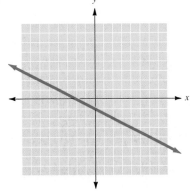

21.

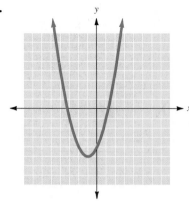

22.

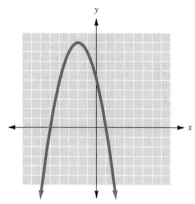

23.

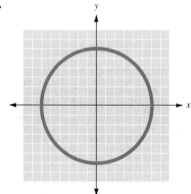

24.

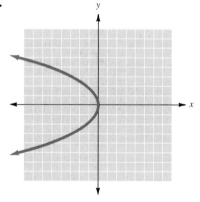

25.

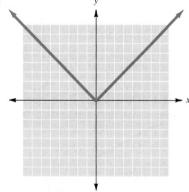

26.

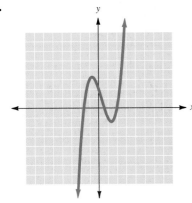

27.

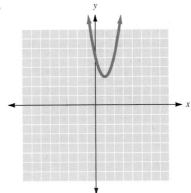

28.

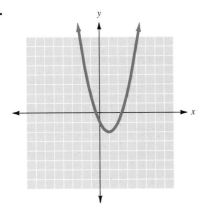

29.

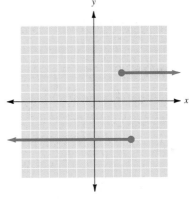

30.

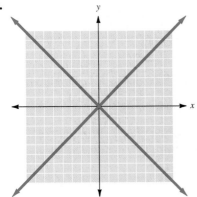

For Exercises 31 to 44, use the following tables to find the desired values.

x	f(x)	x	g(x)	x	h(x)	x	k(x)
−3	−8	−6	1	−4	3	−5	2
−1	2	0	3	−2	7	−3	−4
2	4	1	3	3	5	0	2
5	−3	4	5	7	−4	6	−4

31. $f(5)$

32. $g(-6)$

33. $h(3)$

34. $k(-5)$

35. All x such that $f(x) = -8$

36. All x such that $g(x) = 1$

37. All x such that $g(x) = 3$

38. All x such that $k(x) = 2$

39. $k(0)$

40. $g(4)$

41. $g(1)$

42. $h(-4)$

43. All x such that $k(x) = -4$

44. All x such that $h(x) = 3$

For Exercises 45 to 50, use the given graphs to find, or estimate, the desired values.

45.

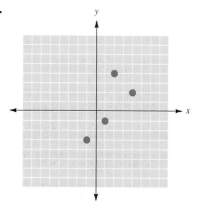

(a) Find $f(2)$.
(b) Find $f(-1)$.
(c) Find all x such that $f(x) = 2$.
(d) Find all x such that $f(x) = -1$.

46.

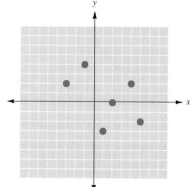

(a) Find $f(2)$.
(b) Find $f(-1)$.
(c) Find all x such that $f(x) = 2$.
(d) Find all x such that $f(x) = -3$.

47.

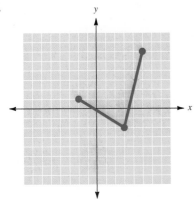

(a) Find $f(3)$.

(b) Find $f(4)$.

(c) Find all x such that $f(x) = 1$.

(d) Find all x such that $f(x) = 4$.

48.

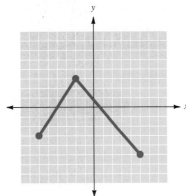

(a) Find $f(-2)$.

(b) Find $f(-5)$.

(c) Find all x such that $f(x) = 0$.

(d) Find all x such that $f(x) = -2$.

49.

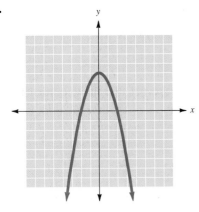

(a) Find $f(3)$.

(b) Find $f(0)$.

(c) Find all x such that $f(x) = 0$.

(d) Find all x such that $f(x) = -2$.

50.

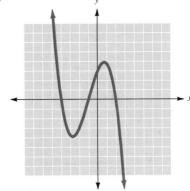

(a) Find $f(2)$.

(b) Find $f(-2)$.

(c) Find all x such that $f(x) = 3$.

(d) Find all x such that $f(x) = 5$.

SECTION 5.3 Algebra of Functions

5.3 OBJECTIVES

1. Find the sum or difference of two functions
2. Find the product of two functions
3. Find the quotient of two functions
4. Find the domain of the sum, difference, product, or quotient of two functions

The profit that a company makes on an item is determined by subtracting the cost of making the item from the total revenue the company receives from selling the item. This is an example of **combining functions.** It can be written as

$$P(x) = R(x) - C(x)$$

Many applications of functions involve the combination of two or more component functions. In this section, we will look at several properties that allow for the addition, subtraction, multiplication, and division of functions.

Sum of Two Functions

The **sum of two functions** f and g is written as $f + g$ and can be defined as

$$(f + g)(x) = f(x) + g(x)$$

for every value of x that is in the domain of both functions f and g.

Difference of Two Functions

The **difference of two functions** f and g is written as $f - g$ and can be defined as

$$(f - g)(x) = f(x) - g(x)$$

for every value of x that is in the domain of both functions f and g.

404

| Example 1 | **Finding the Sum or Difference of Two Functions** |

Suppose we have the following functions f and g.

x	$f(x)$		x	$g(x)$
-4	-8		-4	3
0	6		0	-5
2	5		2	7
1	-2		3	1

(a) Find $(f + g)(-4)$.

$$(f + g)(-4) = f(-4) + g(-4)$$
$$= -8 + 3$$
$$= -5$$

(b) Find $(f - g)(0)$.

$$(f - g)(0) = f(0) - g(0)$$
$$= 6 - (-5)$$
$$= 11$$

(c) Find $(f + g)(3)$.

$$(f + g)(3) = f(3) + g(3)$$

We can find $g(3)$, but not $f(3)$ since 3 is not in the domain of f. This means that 3 is not in the domain of $f + g$. So $(f + g)(3)$ does not exist.

(d) Find the domain of $f + g$.

We need to find all values of x that are in the domains of *both f and g*. Therefore,

$$D = \{-4, 0, 2\}$$

✔ *CHECK YOURSELF 1*

Suppose we have the following functions f and g.

x	$f(x)$		x	$g(x)$
-3	7		-3	-4
-1	0		2	8
5	-3		5	-6
6	2		7	0

(a) Find $(f + g)(5)$.   (b) Find $(f - g)(-3)$.

(c) Find $(f - g)(-1)$.   (d) Find the domain of $f - g$.

In Example 2, we will look at the sum and difference of two functions which are defined by equations rather than tables.

| **Example 2** | **Finding the Sum or Difference of Two Functions** |

You are given the functions $f(x) = 2x - 1$ and $g(x) = -3x + 4$.

(a) Find $(f + g)(x)$.

$$(f + g)(x) = f(x) + g(x)$$
$$= (2x - 1) + (-3x + 4) = -x + 3$$

(b) Find $(f - g)(x)$.

$$(f - g)(x) = f(x) - g(x)$$
$$= (2x - 1) - (-3x + 4) = 5x - 5$$

(c) Find $(f + g)(2)$.

If we use the definition of the sum of two functions, we find that

$$(f + g)(2) = f(2) + g(2) \qquad f(2) = 2(2) - 1 = 3$$
$$= 3 + (-2) = 1 \qquad g(2) = -3(2) + 4 = -2$$

As an alternative, we could use part (a) and say

$$(f + g)(x) = -x + 3$$

Therefore,

$$(f + g)(2) = -2 + 3 = 1$$

✔ CHECK YOURSELF 2

You are given the functions $f(x) = -2x - 3$ and $g(x) = 5x - 1$.

(a) Find $(f + g)(x)$.   (b) Find $(f - g)(x)$.   (c) Find $(f + g)(2)$.

In defining the sum of two functions, we indicated that the domain was determined by the domain of both functions. We will find the domain in Example 3.

| **Example 3** | **Finding the Domain of the Sum or Difference of Functions** |

You are given the functions $f(x) = 2x - 4$ and $g(x) = \dfrac{1}{x}$.

(a) Find $(f + g)(x)$.

$$(f + g)(x) = (2x - 4) + \frac{1}{x} = 2x - 4 + \frac{1}{x}$$

(b) Find the domain of $f + g$.

The domain of $f + g$ is the set of all numbers in the domain of f and also in the domain of g. The domain of f consists of all real numbers. The domain of g consists of all real numbers except 0 because we cannot divide by 0. The domain of $f + g$ is the set of all real numbers except 0. We write $D = \{x \mid x \neq 0\}$.

✔ **CHECK YOURSELF 3**

You are given $f(x) = -3x + 1$ and $g(x) = \dfrac{1}{x - 2}$.

(a) Find $(f + g)(x)$.   (b) Find the domain of $f + g$.

Product of Two Functions

The **product of two functions** f and g is written as $f \cdot g$ and can be defined as

$$(f \cdot g)(x) = f(x) \cdot g(x)$$

for every value of x that is in the domain of both functions f and g.

Quotient of Two Functions

The **quotient of two functions** f and g is written as $f \div g$ and can be defined as

$$(f \div g)(x) = f(x) \div g(x)$$

for every value of x that is in the domain of both functions f and g, such that $g(x) \neq 0$.

Example 4   ## Finding the Product or Quotient of Two Functions

Suppose we have the following functions f and g.

x	$f(x)$		x	$g(x)$
-3	7		-3	-4
-1	0		2	8
5	-3		5	-6
7	2		7	0

(a) Find $(f \cdot g)(-3)$.

$$
\begin{aligned}
(f \cdot g)(-3) &= f(-3) \cdot g(-3) \\
&= (7)(-4) \\
&= -28
\end{aligned}
$$

(b) Find $(f \div g)(5)$.

$$(f \div g)(5) = f(5) \div g(5)$$
$$= (-3) \div (-6)$$
$$= \frac{1}{2}$$

(c) Find $(f \div g)(7)$.

$(f \div g)(7) = f(7) \div g(7) = 2 \div 0$, which is undefined. Therefore 7 is not in the domain of $f \div g$. So we answer that $(f \div g)(7)$ does not exist.

(d) Find the domain of $f \cdot g$.

We want all values of x that are in *both* domains of f and of g. Then

$$D = \{-3, 5, 7\}$$

(e) Find the domain of $f \div g$.

Again we want all values of x that are in *both* domains of f and of g, but we must exclude any x value such that $g(x) = 0$. Since $g(7) = 0$, 7 cannot be in the domain of $f \div g$. Thus

$$D = \{-3, 5\}$$

✔ CHECK YOURSELF 4

Suppose we have the following functions f and g.

x	$f(x)$
-5	-1
-2	3
0	4
3	0

x	$g(x)$
-5	0
0	-6
1	7
3	2

(a) Find $(f \cdot g)(0)$.   (b) Find $(f \div g)(3)$.

(c) Find the domain of $f \cdot g$.   (d) Find the domain of $f \div g$.

Let us consider the product and quotient of two functions defined by equations.

| Example 5 | **Finding the Product of Two Functions** |

Note: From earlier work in algebra you should recall that the product of two binomials $(x + a)(x + b)$ is

$$x^2 + bx + ax + ab.$$

You are given $f(x) = x - 1$ and $g(x) = x + 5$, find $(f \cdot g)(x)$.

$$(f \cdot g)(x) = f(x) \cdot g(x) = (x - 1)(x + 5) = x^2 + 5x - x - 5 = x^2 + 4x - 5$$

✔ CHECK YOURSELF 5

Given $f(x) = x - 3$ and $g(x) = x + 2$, find $(f \cdot g)(x)$.

Example 6	# Finding the Quotient of Two Functions

You are given $f(x) = x - 1$ and $g(x) = x + 5$.

(a) Find $(f \div g)(x)$.

$$(f \div g)(x) = f(x) \div g(x) = (x - 1) \div (x + 5) = \frac{x - 1}{x + 5}$$

(b) Find the domain of $f \div g$.

The domain is the set of all real numbers except -5, because $g(-5) = 0$ and division by 0 is undefined. We write $D = \{x \mid x \neq -5\}$.

✔ *CHECK YOURSELF 6*

You are given $f(x) = x - 3$ and $g(x) = x + 2$.

(a) Find $(f \div g)(x)$.   (b) Find the domain of $f \div g$.

✔ *CHECK YOURSELF ANSWERS*

1. (a) -9; (b) 11; (c) does not exist; (d) $D = \{-3, 5\}$

2. (a) $3x - 4$; (b) $-7x - 2$; (c) 2

3. (a) $-3x + 1 + \dfrac{1}{x - 2}$; (b) $D = \{x \mid x \neq 2\}$

4. (a) -24; (b) 0; (c) $D = \{-5, 0, 3\}$; (d) $D = \{0, 3\}$

5. $x^2 - x - 6$ **6.** (a) $\dfrac{x - 3}{x + 2}$; (b) $D = \{x \mid x \neq -2\}$

Exercises · 5.3

Use the following tables to find the desired values.

x	$f(x)$		x	$g(x)$		x	$h(x)$		x	$k(x)$
-3	-5		0	8		-4	-1		-5	4
0	7		1	-3		0	-7		0	7
2	3		2	4		2	0		3	0
5	-3		7	-1		5	6		7	-3

1. $(f + g)(2)$

2. $(f + h)(5)$

3. $(k - g)(7)$

4. $(h - f)(5)$

5. $(f + k)(2)$

6. $(g - f)(0)$

7. Find the domain of $f + g$.

8. Find the domain of $h + k$.

9. Find the domain of $g - h$.

10. Find the domain of $k - f$.

Find (a) $(f + g)(x)$; (b) $(f - g)(x)$; (c) $(f + g)(3)$; and (d) $(f - g)(2)$.

11. $f(x) = -4x + 5$; $g(x) = 7x - 4$

12. $f(x) = 9x - 3$; $g(x) = -3x + 5$

13. $f(x) = 8x - 2$; $g(x) = -5x + 6$

14. $f(x) = -7x + 9$; $g(x) = 2x - 1$

15. $f(x) = x^2 + x - 1$; $g(x) = -3x^2 - 2x + 5$

16. $f(x) = -3x^2 - 2x + 5$; $g(x) = 5x^2 + 3x - 6$

17. $f(x) = -x^3 - 5x + 8$; $g(x) = 2x^2 + 3x - 4$

18. $f(x) = 2x^3 + 3x^2 - 5$; $g(x) = -4x^2 + 5x - 7$

Find (a) $(f + g)(x)$ and (b) the domain of $f + g$.

19. $f(x) = -9x + 11$; $g(x) = 15x - 7$

20. $f(x) = -11x + 3$; $g(x) = 8x - 5$

21. $f(x) = 3x + 2$; $g(x) = \dfrac{1}{x - 2}$

22. $f(x) = -2x + 5$; $g(x) = \dfrac{3}{x + 1}$

23. $f(x) = x^2 + x - 5$; $g(x) = \dfrac{2}{3x + 1}$

24. $f(x) = 3x^2 - 5x + 1$; $g(x) = \dfrac{-2}{2x - 3}$

410

Use the following tables to find the desired values.

x	f(x)	x	g(x)	x	h(x)	x	k(x)
−3	15	−4	18	−4	6	−2	2
0	−4	0	12	−3	−3	0	0
2	5	2	−3	−2	−4	2	3
4	9	5	4	5	9	4	18

25. $(f \cdot g)(2)$

26. $(h \cdot k)(-2)$

27. $(f \cdot k)(0)$

28. $(g \cdot h)(5)$

29. $(f \div h)(-3)$

30. $(g \div h)(-4)$

31. $(g \div k)(0)$

32. $(f \div k)(4)$

33. Find the domain of $f \div g$.

34. Find the domain of $h \div k$.

Find (a) $(f \cdot g)(x)$; (b) $(f \div g)(x)$; and (c) the domain of $f \div g$.

35. $f(x) = 2x - 1; g(x) = x - 3$

36. $f(x) = -x + 3; g(x) = x + 4$

37. $f(x) = 3x + 2; g(x) = 2x - 1$

38. $f(x) = -3x + 5; g(x) = -x + 2$

39. $f(x) = 2 - x; g(x) = 5 + 2x$

40. $f(x) = x + 5; g(x) = 1 - 3x$

In business, the profit $P(x)$ obtained from selling x units of a product is equal to the revenue $R(x)$ minus the cost $C(x)$. In Exercises 41 and 42, find the profit $P(x)$ for selling x units.

41. $R(x) = 25x; C(x) = x^2 + 4x + 50$ **42.** $R(x) = 20x; C(x) = x^2 + 2x + 30$

Let $V(t)$ be the velocity of an object that has been thrown in the air. It can be shown that $V(t)$ is the combination of three functions: the initial velocity V_0 (this is a constant), the acceleration due to gravity g (this is also a constant), and the time that has elapsed t. We have

$$V(t) = V_0 + g \cdot t$$

Find the velocity as a function of time t.

43. $V_0 = 10$ m/s; $g = -9.8$ m/s^2 **44.** $V_0 = 64$ ft/s; $g = -32$ ft/s^2

The revenue produced from the sale of an item can be found by multiplying the price $p(x)$ by the quantity sold x. In Exercises 45 and 46, find the revenue produced from selling x items.

45. $p(x) = 119 - 6x$ **46.** $p(x) = 1,190 - 36x$

SECTION 5.4

Composition of Functions

 OBJECTIVES

1. *Evaluate the composition of two functions given in table form*
2. *Evaluate the composition of two functions given in equation form*
3. *Write a function as a composition of two simpler functions*
4. *Solve an application involving function composition*

In Section 5.3, we learned that two functions could be combined by using any of the standard operations of arithmetic. There is still another way to combine two functions, called **composition.**

Composition of Functions

The **composition** of functions f and g is the function $f \circ g$, where

$$(f \circ g)(x) = f(g(x))$$

The domain of the composition is the set of all elements x in the domain of g for which $g(x)$ is in the domain of f.

Composition may be thought of as a chaining together of functions. To understand the meaning of $(f \circ g)(x)$, note that first the function g acts on x, producing $g(x)$, and then the function f acts on $g(x)$.

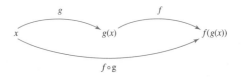

Consider Example 1, involving functions given in table form.

| Example 1 | **Composing Two Functions** |

Suppose we have the following functions f and g.

x	$g(x)$		x	$f(x)$
-2	5		-4	8
1	0		0	6
3	-4		2	5
8	2		1	-2

Think: What does g do to 1? And then, what does f do to the result?

(a) Find $(f \circ g)(1)$.

Since $(f \circ g)(1) = f(g(1))$, we first find $g(1)$. In the table for g we see that $g(1) = 0$. We then find $f(0)$. In the table for f we see that $f(0) = 6$. We then have

$$(f \circ g)(1) = f(g(1)) = f(0) = 6$$

or $$(f \circ g)(1) = 6$$

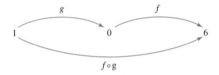

Caution

When you are evaluating $(f \circ g)(x)$, the first function to act is g, not f.

(b) Find $(f \circ g)(8)$.

Since $(f \circ g)(8) = f(g(8))$, we first note that $g(8) = 2$. Next we note that $f(2) = 5$. So

$$(f \circ g)(8) = f(g(8)) = f(2) = 5$$

or $$(f \circ g)(8) = 5$$

Note: -2 is not in the domain of $f \circ g$.

(c) Find $(f \circ g)(-2)$.

Since $(f \circ g)(-2) = f(g(-2))$, we see that $g(-2) = 5$. But we cannot compute $f(5)$ because 5 is not in the domain of f. So $(f \circ g)(-2)$ does not exist.

(d) Find $(f \circ g)(3)$.

Since $(f \circ g)(3) = f(g(3))$, we first find $g(3) = -4$. We then find that $f(-4) = 8$. Altogether

$$(f \circ g)(3) = f(g(3)) = f(-4) = 8$$

or $$(f \circ g)(3) = 8$$

✔ *CHECK YOURSELF 1*

Using the functions given in Example 1, find

(a) $(g \circ f)(-4)$ (b) $(g \circ f)(1)$ (c) $(g \circ f)(2)$

Typically, we encounter the composition of functions given in equation form. Example 2 again demonstrates how f and g are chained together to form the new function $f \circ g$.

| Example 2 | ## Composing Two Functions |

Suppose we have the functions $f(x) = x^2 - 2$ and $g(x) = x + 3$.

(a) Find $(f \circ g)(0)$.

The function g turns 0 into 3. The function f then turns 3 into 7.

Since $(f \circ g)(0) = f(g(0))$, we first find $g(0)$:

$$g(0) = 0 + 3 = 3$$

Then

$$f(3) = 3^2 - 2 = 7$$

Altogether

$$f(g(0)) = f(3) = 7$$

So

$$(f \circ g)(0) = 7$$

(b) Find $(f \circ g)(4)$.

Since $(f \circ g)(4) = f(g(4))$, we first find $g(4)$:

$$g(4) = 4 + 3 = 7$$

Then

$$f(7) = 7^2 - 2 = 47$$

So

$$(f \circ g)(4) = f(g(4)) = f(7) = 47 \qquad \text{or} \qquad (f \circ g)(4) = 47$$

(c) Find $(f \circ g)(x)$.

Since $(f \circ g)(x) = f(g(x))$, we first note that $g(x) = x + 3$.
Then

$$f(g(x)) = f(x + 3) = (x + 3)^2 - 2$$
$$= x^2 + 6x + 9 - 2 = x^2 + 6x + 7$$

So

$$(f \circ g)(x) = x^2 + 6x + 7$$

✔ *CHECK YOURSELF 2*

Suppose that $f(x) = x^2 + x$ and $g(x) = x - 1$. Find each of the following.

(a) $(f \circ g)(0)$ (b) $(f \circ g)(-2)$ (c) $(f \circ g)(x)$

In our next example of composed functions, we will need to pay attention to the domains of the functions involved.

| **Example 3** | **Composing Two Functions** |

Suppose we have the following functions: $f(x) = \sqrt{x}$ and $g(x) = 3 - x$. Find each of the following.

(a) $(f \circ g)(1) = f(g(1))$ Note that $g(1) = 2$.

$\qquad\qquad = f(2)$

$\qquad\qquad = \sqrt{2}$

(b) $(f \circ g)(-1) = f(g(-1))$ Note that $g(-1) = 4$.

$\qquad\qquad = f(4)$

$\qquad\qquad = \sqrt{4}$

$\qquad\qquad = 2$

(c) $(f \circ g)(7) = f(g(7))$ Note that $g(7) = -4$.

$\qquad\qquad = f(-4)$

$\qquad\qquad = \sqrt{-4}$

Since this is not a real number, we say that $(f \circ g)(7)$ does not exist. And 7 is not in the domain of $f \circ g$.

Note: Graph the function $Y = \sqrt{3 - x}$ in your graphing calculator. Note that the graph only exists for $x \leq 3$.

(d) $(f \circ g)(x) = f(g(x))$

$\qquad\qquad = f(3 - x)$

$\qquad\qquad = \sqrt{3 - x}$

Note that this function produces real number values only if $3 - x \geq 0$, that is, only if $x \leq 3$. This is in fact the domain of $f \circ g$.

✔ *CHECK YOURSELF 3*

Suppose that $f(x) = \dfrac{1}{x}$ and $g(x) = x^2 - 4$. Find each of the following.

(a) $(f \circ g)(0)$ (b) $(f \circ g)(-2)$ (c) $(f \circ g)(x)$

The order of composition is important. In general, $(f \circ g)(x) \neq (g \circ f)(x)$. Using the functions given in Example 2, we saw that $(f \circ g)(x) = x^2 + 6x + 7$ whereas

$$(g \circ f)(x) = g(f(x)) = g(x^2 - 2) = x^2 - 2 + 3 = x^2 + 1$$

Often it is convenient to write a given function as the composition of two simpler functions. While the choice of simpler functions is not unique, a good choice makes many applications easier to solve.

Example 4	## Writing a Function as the Composition of Two Functions

Use the functions $f(x) = x + 3$ and $g(x) = x^2$ to express the given function h as a composition of f and g.

(a) $h(x) = (x + 3)^2$

Note: Can you see how we are using the order of operations here?

Note that when a value is substituted for x, the first action that occurs is that 3 is added to the input. The function f does exactly that. The second action that occurs is that of squaring. The function g does this. Since f acts first and then g (to carry out the total action of h), we propose

$$h(x) = g(f(x)) = (g \circ f)(x)$$

It is easily checked that $(g \circ f)(x) = h(x)$:

$$(g \circ f)(x) = g(f(x)) = g(x + 3) = (x + 3)^2 = h(x)$$

(b) $h(x) = x^2 + 3$

Now when a value is substituted for x, the first action that occurs is a squaring action (which is what g does). The second action is that of adding 3 (which is what f does). We propose

$$h(x) = f(g(x)) = (f \circ g)(x)$$

Check:

$$(f \circ g)(x) = f(g(x)) = f(x^2) = x^2 + 3 = h(x)$$

✔ *CHECK YOURSELF 4*

Use the functions $f(x) = \sqrt{x}$ and $g(x) = x + 2$ to express the given function h as a composition of f and g.

(a) $h(x) = \sqrt{x + 2}$ (b) $h(x) = \sqrt{x} + 2$

There are many examples that involve the composition of functions, as illustrated in Example 5.

Example 5	## Solving an Application Involving Function Composition

At Kinky's Duplication Salon, customers pay $2 plus 4¢ per page copied. Duplication consultant Vinny makes a commission of 5% of the bill for each job he sends to Kinky's.

(a) Express a customer's bill B as a function of the number of pages copied p.

$$B(p) = 0.04p + 2$$

(b) Express Vinny's commission V as a function of each referral B.

$$V(B) = 0.05B$$

(c) Use function composition to express Vinny's commission V as a function of the number a customer has copied p.

$$V(B(p)) = 0.05(0.04p + 2) = 0.002p + 0.1$$

(d) Use the function in part (c) to find Vinny's commission on a job consisting of 2,000 pages.

$$V(2,000) = 0.002(2,000) + 0.1 = 4 + 0.1 = 4.1$$

Therefore, Vinny's commission is $4.10.

✔ *CHECK YOURSELF 5*

On his regular route, Gonzalo averages 62 mi/h between Charlottesville and Lawrenceville. His van averages 24 mi/gal, and his gas tank holds 12 gal of fuel. Assume that his tank is full when he starts the trip from Charlottesville to Lawrenceville.

(a) Express the fuel left in the tank as a function of n, the number of gallons used.

(b) Express the number of gallons used as a function of m, the number of miles driven.

(c) Express the fuel left in the tank as a function of m, the number of miles driven.

✔ *CHECK YOURSELF ANSWERS*

1. (a) 2; (b) 5; (c) does not exist   **2.** (a) 0; (b) 6; (c) $x^2 - x$

3. (a) $-\dfrac{1}{4}$; (b) does not exist; (c) $\dfrac{1}{x^2 - 4}$ **4.** (a) $(f \circ g)(x)$; (b) $(g \circ f)(x)$

5. (a) $f(n) = 12 - n$; (b) $g(m) = \dfrac{m}{24}$; (c) $(f \circ g)(m) = 12 - \dfrac{m}{24}$

Exercises ▪ 5.4

For Exercises 1 to 12, use the following tables to find the desired values.

x	$f(x)$	x	$g(x)$	x	$h(x)$	x	$k(x)$
-3	-1	-2	4	-2	5	-2	0
-1	7	1	-2	0	0	0	4
2	6	4	3	1	-2	2	3
3	-3	6	2	3	6	3	-4

1. $(f \circ g)(4)$

2. $(g \circ f)(2)$

3. $(h \circ g)(1)$

4. $(g \circ h)(1)$

5. $(g \circ h)(3)$

6. $(k \circ h)(0)$

7. $(h \circ k)(0)$

8. $(k \circ g)(1)$

9. $(f \circ h)(3)$

10. $(k \circ g)(4)$

11. $(k \circ k)(2)$

12. $(f \circ f)(-3)$

In Exercises 13 to 20, f and g are given. Evaluate the composite functions in each part.

13. $f(x) = x - 3$ and $g(x) = 2x + 1$

    (a) $(f \circ g)(0)$ (b) $(f \circ g)(-2)$ (c) $(f \circ g)(3)$ (d) $(f \circ g)(x)$

14. $f(x) = x - 1$ and $g(x) = 3x + 4$

    (a) $(f \circ g)(0)$ (b) $(f \circ g)(-2)$ (c) $(f \circ g)(3)$ (d) $(f \circ g)(x)$

15. $f(x) = 3x + 1$ and $g(x) = 4x - 3$

    (a) $(f \circ g)(0)$ (b) $(f \circ g)(-2)$ (c) $(g \circ f)(3)$ (d) $(g \circ f)(x)$

16. $f(x) = 4x - 2$ and $g(x) = -2x + 5$

    (a) $(f \circ g)(0)$ (b) $(f \circ g)(-2)$ (c) $(g \circ f)(3)$ (d) $(g \circ f)(x)$

17. $f(x) = x^2$ and $g(x) = x + 3$

    (a) $(f \circ g)(0)$ (b) $(f \circ g)(-2)$ (c) $(g \circ f)(3)$ (d) $(g \circ f)(x)$

18. $f(x) = x^2 + 3$ and $g(x) = 3x$

    (a) $(f \circ g)(0)$ (b) $(f \circ g)(-2)$ (c) $(g \circ f)(3)$ (d) $(g \circ f)(x)$

19. $f(x) = 2x^2 - 1$ and $g(x) = -2x$

    (a) $(g \circ f)(0)$ (b) $(g \circ f)(-2)$ (c) $(f \circ g)(3)$ (d) $(f \circ g)(x)$

20. $f(x) = x^2 + 3$ and $g(x) = 3x$

    (a) $(g \circ f)(0)$ (b) $(g \circ f)(-2)$ (c) $(f \circ g)(3)$ (d) $(f \circ g)(x)$

In Exercises 21 to 30, rewrite the function h as a composite of functions f and g.

21. $f(x) = 3x$ $g(x) = x + 2$ $h(x) = 3x + 2$

22. $f(x) = x - 4$ $g(x) = 7x$ $h(x) = 7x - 4$

23. $f(x) = x + 5$ $g(x) = \sqrt{x}$ $h(x) = \sqrt{x + 5}$

24. $f(x) = x + 5$ $g(x) = \sqrt{x}$ $h(x) = \sqrt{x} + 5$

25. $f(x) = x^2$ $g(x) = x - 5$ $h(x) = x^2 - 5$

26. $f(x) = x^2$ $g(x) = x - 5$ $h(x) = (x - 5)^2$

27. $f(x) = x - 3$ $g(x) = \dfrac{2}{x}$ $h(x) = \dfrac{2}{x - 3}$

28. $f(x) = x - 3$ $g(x) = \dfrac{2}{x}$ $h(x) = \dfrac{2}{x} - 3$

29. $f(x) = x - 1$ $g(x) = x^2 + 2$ $h(x) = x^2 + 1$

30. $f(x) = x - 1$ $g(x) = x^2 + 2$ $h(x) = x^2 - 2x + 3$

31. Carine and Jacob are getting married at East Fork Estates. The wedding will cost $1,000 plus $40 per guest. They have read that typically 80% of the people invited actually attend a wedding.

(a) Write a function to represent the number of people N expected to attend if v are invited.

(b) Write a function to represent the cost C of the wedding for N guests.

(c) Write a function to represent the cost C of the wedding if v people are invited.

32. On his regular route, Gonzalo averages 62 mi/h between Charlottesville and Lawrenceville. His van averages 24 mi/gal, and his gas tank holds 12 gal of fuel. Assume that his tank is full when he starts the trip from Charlottesville to Lawrenceville.

(a) Express the number of gallons used as a function of m, the miles driven.

(b) Express the number of miles driven as a function of t, the time on the road.

(c) Express the number of gallons used as a function of t, the time on the road.

33. When she arrives in London, Bichvan receives an exchange rate of 0.69 British pound for each U.S. dollar. In Reykjavik, she receives an exchange rate of 146.41 Icelandic kronas for each British pound. When she returns to the United States, how much (in U.S. dollars) should she expect to receive in exchange for 12,000 Icelandic kronas?

34. If the exchange rate for Japanese yen is 127.3 and the exchange rate for Indian rupees is 48.37 (both from U.S. dollars), then what is the exchange rate from Japanese yen to Indian rupees?

Summary for Chapter 5

Example	Topic	Reference
	Relations and Functions	**5.1**
(1, 4) is an ordered pair.	**Ordered Pair** Given two related values x and y, we write the pair of values as (x, y).	**p. 368**
The set {(1, 4), (2, 5), (1, 6)} is a relation.	**Relation** A relation is a set of ordered pairs.	**p. 368**
The domain is {1, 2}.	**Domain** The domain is the set of all first elements of a relation.	**p. 368**
The range is {4, 5, 6}.	**Range** The range is the set of all second elements of a relation.	**p. 369**
{(1, 2), (2, 3), (3, 4)} is a function. {(1, 2), (2, 3), (2, 4)} is *not* a function.	**Function** A function is a set of ordered pairs (a relation) in which no two first elements are equal.	**p. 370**
	Tables and Graphs	**5.2**
	Graph The graph of a relation is the set of points in the plane that correspond to the ordered pairs of the relation.	**p. 385**

A relation—*not* a function	**Vertical Line Test** The vertical line test is used to determine, from the graph, whether a relation is a function.
	    If a vertical line meets the graph of a relation in two or more points, the relation is *not* a function.
	    If no vertical line passes through two or more points on the graph of a relation, it is the graph of a function.
		p. 386

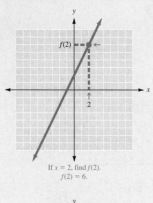

If $x = 2$, find $f(2)$.
$f(2) = 6$.

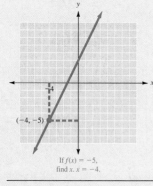

$(-4, -5)$

If $f(x) = -5$,
find x. $x = -4$.

Reading Values from Graphs   For a specific value of x, let's call it a, we can find $f(a)$ with the following algorithm:

1.   Draw a vertical line through a on the x axis.

2.   Find the point of intersection of that line with the graph.

3.   Draw a horizontal line through the graph at that point.

4.   Find the intersection of the horizontal line with the y axis.

5. $f(a)$ is that y value.

If given the function value, one finds the x value associated with it as follows:

1.   Find the given function value on the y axis.

2.   Draw a horizontal line through that point.

3.   Find every point on the graph that intersects the horizontal line.

4.   Draw a vertical line through each of those points of intersection.

5.   The x values are each point of intersection of the vertical lines and the x axis.

p. 393

p. 394

Algebra of Functions

5.3

Let $f(x) = 2x + 1$ and $g(x) = 3x^2$.

$(f + g)(x) = (2x + 1) + (3x^2)$
$= 3x^2 + 2x + 1$

The **sum of two functions** f and g is written $f + g$. It is defined as

$$(f + g)(x) = f(x) + g(x)$$

p. 404

$(f - g)(x) = (2x + 1) - (3x^2)$
$= -3x^2 + 2x + 1$

The **difference of two functions** f and g is written $f - g$. It is defined as

$$(f - g)(x) = f(x) - g(x)$$

p. 404

$(f \cdot g)(x) = (2x + 1)(3x^2)$
$= 6x^3 + 3x^2$

The **product of two functions** f and g is written $f \cdot g$. It is defined as

$$(f \cdot g)(x) = f(x) \cdot g(x)$$

p. 407

$(f \div g)(x) = (2x + 1) \div (3x^2)$
$= \dfrac{2x + 1}{3x^2}$

The **quotient of two functions** f and g is written $f \div g$. It is defined as

$$(f \div g)(x) = f(x) \div g(x) \qquad g(x) \neq 0$$

p. 407

Composition of Functions

5.4

$(f \circ g)(x) = 2(3x^2) + 1$
$= 6x^2 + 1$

The composition of two functions f and g is written $f \circ g$. It is defined as

$$(f \circ g)(x) = f(g(x))$$

p. 413

Summary Exercises ▪ 5

This summary exercise set is provided to give you practice with each of the objectives of the chapter. Each exercise is keyed to an appropriate chapter section.

[5.1] Find the domain and range of each relation.

1. $A = \{(\text{Maine}, 5), (\text{Massachusetts}, 13), (\text{Vermont}, 7), (\text{Connecticut}, 11)\}$

2. $B = \{(\text{John Wayne}, 1969), (\text{Art Carney}, 1974), (\text{Peter Finch}, 1976), (\text{Marlon Brando}, 1972)\}$

3. $C = \{(\text{Dean Smith}, 65), (\text{John Wooden}, 47), (\text{Denny Crum}, 42), (\text{Bob Knight}, 41)\}$

4. $E = \{(\text{Don Shula}, 328), (\text{George Halas}, 318), (\text{Tom Landry}, 250), (\text{Chuck Noll}, 193)\}$

5. $\{(3, 5), (4, 6), (1, 2), (8, 1), (7, 3)\}$

6. $\{(-1, 3), (-2, 5), (3, 7), (1, 4), (2, -2)\}$

7. $\{(1, 3), (1, 5), (1, 7), (1, 9), (1, 10)\}$

8. $\{(2, 4), (-1, 4), (-3, 4), (1, 4), (6, 4)\}$

Determine which relations are also functions.

9. $\{(1, 3), (2, 4), (5, -1), (-1, 3)\}$

10. $\{(-2, 4), (3, 6), (1, 5), (0, 1)\}$

11. $\{(1, 2), (0, 4), (1, 3), (2, 5)\}$

12. $\{(1, 3), (2, 3), (3, 3), (4, 3)\}$

13.

x	y
−3	2
−1	1
0	3
1	4
3	5

14.

x	y
−1	3
0	2
1	3
2	4
3	5

15.

x	y
−2	3
−1	4
0	−1
1	5
−2	13

16.

x	y
−3	−4
1	0
2	3
1	5
5	2

Evaluate each function for the value specified.

17. $f(x) = x^2 - 3x + 5$; find (a) $f(0)$, (b) $f(-1)$, and (c) $f(1)$.

18. $f(x) = -2x^2 + x - 7$; find (a) $f(0)$, (b) $f(2)$, and (c) $f(-2)$.

19. $f(x) = x^3 - x^2 - 2x + 5$; find (a) $f(-1)$, (b) $f(0)$, and (c) $f(2)$.

20. $f(x) = -x^2 + 7x - 9$; find (a) $f(-3)$, (b) $f(0)$, and (c) $f(1)$.

21. $f(x) = 3x^2 - 5x + 1$; find (a) $f(-1)$, (b) $f(0)$, and (c) $f(2)$.

22. $f(x) = -x^3 + 3x - 5$; find (a) $f(2)$, (b) $f(0)$, and (c) $f(1)$.

Rewrite each equation as a function of x. Use $f(x)$ notation in the final result.

23. $y = -2x + 5$ **24.** $y = 3x + 2$ **25.** $2x + 3y = 6$

26. $4x + 2y = 8$ **27.** $-3x + 4y = 12$ **28.** $-2x - 5y = -10$

In Exercises 29 to 32, graph the functions.

29. $f(x) = 2x + 3$ **30.** $f(x) = -3x + 4$ **31.** $f(x) = -4x - 3$ **32.** $f(x) = x - 3$

If $f(x) = 3x - 2$, find the following:

33. $f(a)$ **34.** $f(x - 1)$ **35.** $f(x + h)$ **36.** $\dfrac{f(x + h) - f(x)}{h}$

[5.2] For each set of ordered pairs, plot the related points. Then use the vertical line test to determine which sets are functions.

37. $\{(-3, -3), (-2, -2), (2, 2), (3, 3)\}$

38. $\{(-4, -4), (-2, 4), (2, 3), (0, 4)\}$

39. $\{(-2, 1), (-2, 3), (0, 1), (1, 2)\}$

40. $\{(0, 5), (1, 6), (1, -2), (3, 4)\}$

Use the vertical line test to determine whether the given graph represents a function. Find the domain and range of the relation.

41.

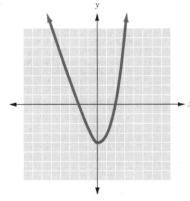

42.

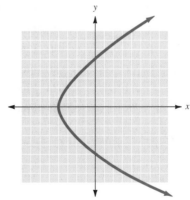

43.

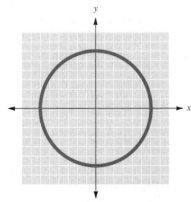

44.

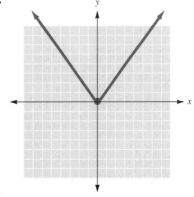

45. Use the given graph to answer parts (a) through (f). Estimate values where necessary.

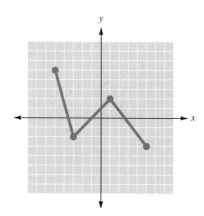

(a) Find $f(-1)$.
(b) Find $f(5)$.
(c) Find all x such that $f(x) = 5$.
(d) Find all x such that $f(x) = -3$.
(e) Find all x such that $f(x) = -1$.
(f) Find all x such that $f(x) = 2$.

46. Use the given graph to answer parts (a) through (f). Estimate values where necessary.

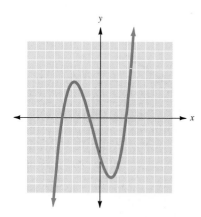

(a) Find $f(-3)$.
(b) Find $f(2)$.
(c) Find all x such that $f(x) = 7$.
(d) Find all x such that $f(x) = -8$.
(e) Find all x such that $f(x) = 3$.
(f) Find all x such that $f(x) = -4$.

47. Use the given table to answer parts (a) through (f).

x	$f(x)$
-5	-2
-2	0
4	-2
8	5
12	12

(a) Find $f(-5)$.
(b) Find $f(12)$.
(c) Find all x such that $f(x) = -2$.
(d) Find all x such that $f(x) = 0$.
(e) Find all x such that $f(x) = 5$.
(f) Find all x such that $f(x) = 12$.

48. Use the given table to answer parts (a) through (f).

x	$f(x)$
-3	0
-2	-3
0	-3
3	-2
7	0

(a) Find $f(-3)$.
(b) Find $f(0)$.
(c) Find $f(3)$.
(d) Find all x such that $f(x) = -3$.
(e) Find all x such that $f(x) = -2$.
(f) Find all x such that $f(x) = 0$.

[5.3] In Exercises 49 to 56, use the following tables to find the desired values.

x	$f(x)$
-2	3
-1	8
3	0
7	-6
8	-4

x	$g(x)$
-4	5
-2	-1
3	2
5	-3
7	0

49. $(f + g)(-2)$

50. $(g - f)(7)$

51. $(f - g)(-1)$

52. $(f \cdot g)(3)$

53. $(f \div g)(7)$

54. $(g \div f)(-2)$

55. Find the domain of $g - f$.

56. Find the domain of $g \div f$.

[5.3] In Exercises 57 to 60, $f(x)$ and $g(x)$ are given. Find $(f + g)(x)$.

57. $f(x) = 4x^2 + 5x - 3$ and $g(x) = -2x^2 + x - 5$

58. $f(x) = -3x^3 + 2x^2 - 5$ and $g(x) = 4x^3 - 4x^2 + 5x + 6$

59. $f(x) = 2x^4 + 4x^2 + 5$ and $g(x) = x^3 - 5x^2 + 6x$

60. $f(x) = 3x^3 + 5x - 5$ and $g(x) = -2x^3 + 2x^2 + 5x$

In Exercises 61 to 64, $f(x)$ and $g(x)$ are given. Find $(f - g)(x)$.

61. $f(x) = 7x^2 - 2x + 3$ and $g(x) = 2x^2 - 5x - 7$

62. $f(x) = 9x^2 - 4x$ and $g(x) = 5x^2 + 3$

63. $f(x) = 8x^2 + 5x$ and $g(x) = 4x^2 - 3x$

64. $f(x) = -2x^2 - 3x$ and $g(x) = -3x^2 + 4x - 5$

[5.3] In Exercises 65 to 68, find the product $(f \cdot g)(x)$.

65. $f(x) = 2x$ and $g(x) = 3x - 5$

66. $f(x) = x + 1$ and $g(x) = -3x$

67. $f(x) = 3x$ and $g(x) = x^2$

68. $f(x) = 2x$ and $g(x) = x^2 - 5$

In Exercises 69 to 72, find the quotient $(f \div g)(x)$ and state the domain of the resulting function.

69. $f(x) = 2x$ and $g(x) = x - 3$

70. $f(x) = x + 1$ and $g(x) = 2x - 4$

71. $f(x) = 3x$ and $g(x) = x^2$

72. $f(x) = 2x^2$ and $g(x) = x - 5$

[5.4] In Exercises 73 to 78, use the following tables to find the desired values.

x	$f(x)$		x	$g(x)$
-2	3		-4	5
-1	8		-2	-1
3	0		3	2
7	-6		5	-3
8	-4		7	0

73. $(f \circ g)(-2)$

74. $(g \circ f)(8)$

75. $(g \circ f)(-2)$

76. $(f \circ g)(-4)$

77. $(g \circ g)(-4)$

78. $(f \circ f)(-2)$

[5.4] In Exercises 79 to 82, evaluate the indicated composite functions in each part.

79. $f(x) = x - 3$ and $g(x) = 3x + 1$

    (a) $(f \circ g)(0)$ (b) $(f \circ g)(-2)$ (c) $(f \circ g)(3)$ (d) $(f \circ g)(x)$

80. $f(x) = 5x - 1$ and $g(x) = -4x + 5$

    (a) $(f \circ g)(0)$ (b) $(f \circ g)(-2)$ (c) $(g \circ f)(3)$ (d) $(g \circ f)(x)$

81. $f(x) = x^2$ and $g(x) = x - 5$

    (a) $(f \circ g)(0)$ (b) $(f \circ g)(-2)$ (c) $(g \circ f)(3)$ (d) $(g \circ f)(x)$

82. $f(x) = x^2 + 3$ and $g(x) = -2x$

    (a) $(g \circ f)(0)$ (b) $(g \circ f)(-2)$ (c) $(f \circ g)(3)$ (d) $(f \circ g)(x)$

In Exercises 83 and 84, rewrite the function h as a composite of functions f and g.

83. $f(x) = -2x, g(x) = x + 2, h(x) = -2x - 4$

84. $f(x) = 6x, g(x) = x^2 + 5, h(x) = 36x^2 + 5$

Self-Test 5

The purpose of this self-test is to help you check your progress and to review for a chapter test in class. Allow yourself about an hour to take the test. When you are done, check your answers in the back of the book. If you missed any questions, be sure to go back and review the appropriate sections in the chapter and the exercises that are provided.

1. For each of the following sets of ordered pairs, identify the domain and range.

   (a) $\{(1, 6), (-3, 5), (2, 1), (4, -2), (3, 0)\}$

   (b) $\{(\text{United States}, 101), (\text{Germany}, 65), (\text{Russia}, 63), (\text{China}, 50)\}$

2. For each of the following, determine if the given relation is a function. Identify the domain and range of the relation.

   (a) $\{(2, 5), (-1, 6), (0, -2), (-4, 5)\}$

   (b)
x	y
-3	2
0	4
1	7
2	0
-3	1

3. If $f(x) = x^2 - 5x + 6$, find (a) $f(0)$, (b) $f(-1)$, and (c) $f(1)$.

4. If $f(x) = -3x^2 - 2x + 3$, find (a) $f(-1)$ and (b) $f(-2)$.

5. Graph the function $f(x) = -2x + 3$.

6. If $f(x) = 6x - 3$, find $\dfrac{f(x + h) - f(x)}{h}$.

7. Plot the given points on a graph and then use the vertical line test to determine if it is a function.

$$\{(-1, 2), (0, 1), (2, 2), (3, -4)\}$$

Use the vertical line test to determine whether the following graphs represent functions.

8.

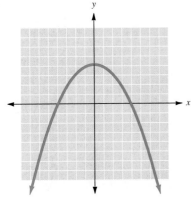

9.

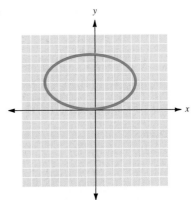

10. Use the given table to find the desired values.

(a) $f(-5)$ (b) Values of x such that $f(x) = 9$

(c) $f(4)$ (d) Values of x such that $f(x) = 3$

x	$f(x)$
-5	3
0	-1
1	9
4	2
5	3

Use the graph to determine the domain and range of the relation.

11.

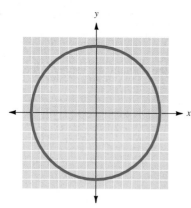

12.

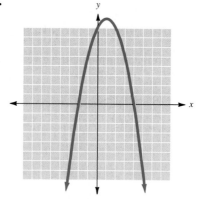

In Exercises 13 and 14, $f(x)$ and $g(x)$ are given. Find (a) $(f + g)(x)$, (b) $(f - g)(x)$.

13. $f(x) = 4x^2 - 3x + 7$ and $g(x) = 2x^2$

14. $f(x) = -3x^3 + 5x^2 - 2x - 7$ and $g(x) = -2x^2 + 7x - 2$

In Exercises 15 to 20, use $f(x) = x^2 - 1$ and $g(x) = 3x - 2$.

15. Find $(f \cdot g)(x)$ and state the domain of the resulting function.

16. Find $(f \div g)(x)$ and state the domain of the resulting function.

17. Find $(f \circ g)(x)$.   **18.** Find $(g \circ f)(x)$.

19. Find $(f \circ g)(3)$.   **20.** Find $(g \circ f)(3)$.

Cumulative Test ▪ 0–5

This test is provided to help you in the process of reviewing previous chapters. Answers are provided in the back of the book. If you miss any answers, be sure to go back and review the appropriate chapter sections.

In Exercises 1 to 4, evaluate each expression if $x = -2$, $y = 4$, and $z = -1$.

1. $4x^2 + 3y - z$

2. $-z^2 + 4y + 3x$

3. $\dfrac{2z - 3x + 4y}{x^2 - z^2}$

4. $\dfrac{x^2 + y^2 + z^2}{x - y + z}$

In Exercises 5 and 6, simplify each expression.

5. $4x^2 - 3x + 5 - (2x^2 - 2x - 6)$

6. $4x(3x - 4) - 5(2x - 3) - 4x + 6$

In Exercises 7 to 12, solve each equation. Express your answer in set notation.

7. $3x - 2(x - 5) + 8 = -3(4 - x)$

8. $\dfrac{2}{3}x - 5 = 4 + \dfrac{3}{4}x$

9. $\dfrac{x - 2}{3} - \dfrac{2x + 1}{5} = 1$

10. $|3x - 5| = 10$

11. $|2x - 1| - 4 = 7$

12. $|x - 5| = |3x + 1|$

13. Solve the equation $\dfrac{1}{R} = \dfrac{1}{R_1} + \dfrac{1}{R_2}$ for R.

In Exercises 14 to 18, solve and graph the solution set for each inequality.

14. $3x + 5 \le 8$

15. $2x - 9 > 4x - 5$

16. $|2x - 3| < 9$

17. $|x - 5| > 8$

18. $3x - 6 < -5$ or $2x - 1 \ge 7$

In Exercises 19 and 20, find the slope of the line through the given pairs of points.

19. $(2, -1)$ and $(4, -5)$

20. $(5, 6)$ and $(-3, 6)$

In Exercises 21 and 22, find the slope and the y intercept of the line represented by the given equation.

21. $y = 3x + 6$

22. $6x - 4y = 24$

In Exercises 23 to 26, write the equation of the line L that satisfies the given conditions.

23. L has y intercept of $(0, -3)$ and slope of 2.   **24.** L passes through $(1, 3)$ and $(-2, 1)$.

25. L has y intercept of $(0, -2)$ and is perpendicular to the line with equation $4x - 5y = 20$.

26. L passes through $(-3, 1)$ and is parallel to the line with equation $y = -x + 5$.

In Exercises 27 and 28, evaluate the given function for the indicated value.

27. $f(x) = -4x + 18$ $f(-1)$ **28.** $f(x) = -x^2 + 2x - 3$ $f(2)$

29. Graph the function $f(x) = -3x + 5$.

In Exercises 30 and 31, determine which relations are functions.

30. $\{(1, 2), (-1, 2), (3, 4), (5, 6)\}$

31.

x	y
-3	0
-2	1
-1	5
-1	3

32. Use the vertical line test to determine whether each of the following graphs represents a function.

(a)

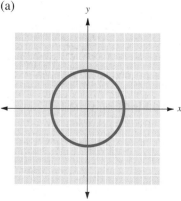

(b)

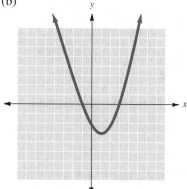

33. Use the given graph to determine

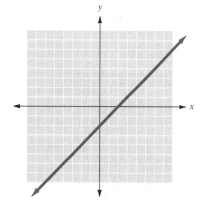

(a) $f(-3)$

(b) $f(0)$

(c) Value of x for which $f(x) = 3$

In Exercises 34 and 35, $f(x)$ and $g(x)$ are given. Find (a) $f(x) + g(x)$ and (b) $f(x) - g(x)$.

34. $f(x) = 3x - 5$ and $g(x) = x^2 - 7x + 4$

35. $f(x) = 2x^2 + 5x - 9$ and $g(x) = -2x^2 - 7x + 11$

In Exercises 36 and 37, $f(x)$ and $g(x)$ are given. Find (a) $(f \cdot g)(x)$ and (b) $(f \div g)(x)$.

36. $f(x) = -3x$ and $g(x) = 4x - 5$

37. $f(x) = 2x - 5$ and $g(x) = x^2$

In Exercises 38 and 39, $f(x)$ and $g(x)$ are given. Find (a) $(f \circ g)(1)$; (b) $(f \circ g)(x)$; (c) $(g \circ f)(1)$; and (d) $(g \circ f)(x)$.

38. $f(x) = 2x - 1$ and $g(x) = -3x + 2$

39. $f(x) = x^2$ and $g(x) = 4x$

40. The length of a rectangle is 2 centimeters (cm) more than 3 times its width. If the perimeter of the rectangle is 44 cm, what are the dimensions of the rectangle?

6 Factoring Polynomials

Civil engineers use "polynomial splines" when designing tunnels and highways. These splines help in the design of a roadway, ensuring that changes in direction and altitude occur smoothly and gradually. For example, a road passes through a valley at 50 meters (m) altitude and then climbs through some hills, reaching an altitude of 350 m before descending again. The graph on the next page shows the change in altitude for 19 kilometers (km) of the roadway.

The road seems very steep in this graph, but remember that the *y* axis is the altitude measured in meters, and the *x* axis is horizontal distance measured in kilometers. To get a true feeling for the vertical change, the horizontal axis would have to be stretched by a factor of 1,000.

Based on measurements of the distance, altitude, and change in the slope taken at intervals along the planned path of the road, formulas are developed that model the roadway for short sections. The formulas are then pieced together to form a model of the road over several kilometers. Such formulas are found by using algebraic methods to solve systems of equations. Here are some of the splines

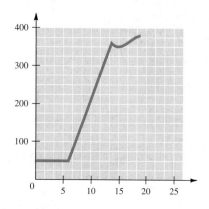

that could be fit together to create the roadway needed in our graph; y is the altitude measured in meters and x is the horizontal distance, measured in kilometers.

First 6 km: $y = 50$

Next 7.5 km: $y = 40x - 190$

Final 5.5 km: $y = -0.8x^3 + 41x^2 - 690x + 4{,}176$

By entering these equations in a graphing calculator, you can see part of the design for the road. Be certain that you adjust your graphing window appropriately.

Polynomials are useful in many areas. In this chapter, you will learn how to solve problems involving polynomials.

6.1 An Introduction to Factoring

In Chapter 4 you were given factors and asked to find a product. We are now going to reverse the process. You will be given a polynomial and asked to find its factors. This is called **factoring.**

Let's start with an example from arithmetic. To *multiply* $5 \cdot 7$, you write

$$5 \cdot 7 = 35$$

To *factor* 35, you write

$$35 = 5 \cdot 7$$

Factoring is the *reverse* of multiplication.

Now let's look at factoring in algebra. You have used the distributive property as

$$a(b + c) = ab + ac$$

3 and $x + 5$ are the factors of $3x + 15$.

For instance,

$$3(x + 5) = 3x + 15$$

To use the distributive property in factoring, we apply that property in the opposite fashion, as in

$$ab + ac = a(b + c)$$

The property lets us remove the common monomial factor a from the terms of $ab + ac$. To use this in factoring, the first step is to see whether each term of the polynomial has a common monomial factor. In our earlier example,

$$3x + 15 = 3 \cdot x + 3 \cdot 5$$

Common factor

So, by the distributive property,

$$3x + 15 = 3(x + 5)$$

The original terms are each divided by the greatest common factor to determine the terms in parentheses.

Again, factoring is the reverse of multiplication.

To check this, multiply $3(x + 5)$.

Here is a diagram that will relate the idea of multiplication and factoring.

Multiplying

$$3(x + 5) = 3x + 15$$

Factoring

The first step in factoring is to identify the *greatest common factor* (GCF) of a set of terms. This is the monomial with the largest common numerical coefficient and the largest power common to both variables.

In fact, we will see that factoring out the GCF is the *first* method to try in any of the factoring problems we will discuss.

The **greatest common factor (GCF)** of a polynomial is the monomial with the highest degree and the largest numerical coefficient that is a factor of each term of the polynomial.

Example 1   ## Finding the GCF

Find the GCF for each list of terms.

(a) 9 and 12

The largest number that is a factor of both is 3.

(b) 10, 25, 150

The GCF is 5.

(c) x^4 and x^7

The largest power common to the two variables is x^4.

(d) $12a^3$ and $18a^2$

The GCF is $6a^2$.

 CHECK YOURSELF 1

Find the GCF for each list of terms.

(a) 14, 24   (b) 9, 27, 81   (c) a^9, a^5 (d) $10x^5, 35x^4$

To Factor a Monomial from a Polynomial

Checking your answer is always important and perhaps is never easier than after you have factored.

Step 1. Find the *greatest* common factor for all the terms.

Step 2. Factor the GCF from each term, then apply the distributive property.

Step 3. Mentally check your factoring by multiplication.

Example 2 | **Finding the GCF of a Binomial**

(a) Factor $8x^2 + 12x$.

The largest common numerical factor of 8 and 12 is 4, and x is the variable factor with the largest common power. So $4x$ is the GCF. Write

$$8x^2 + 12x = 4x \cdot 2x + 4x \cdot 3$$

GCF

It is always a good idea to check your answer by multiplying to make sure that you get the original polynomial. Try it here. Multiply $4x$ by $2x + 3$.

Now, by the distributive property, we have

$$8x^2 + 12x = 4x(2x + 3)$$

(b) Factor $6a^4 - 18a^2$.

The GCF in this case is $6a^2$. Write

$$6a^4 - 18a^2 = 6a^2 \cdot a^2 - 6a^2 \cdot 3$$

GCF

It is also true that

$6a^4 - 18a^2 = 3a(2a^3 - 6a)$

However, this is *not completely factored*. Do you see why? You want to find the common monomial factor with the *largest possible* coefficient and the *largest* exponent, in this case $6a^2$.

Again, using the distributive property yields

$$6a^4 - 18a^2 = 6a^2(a^2 - 3)$$

You should check this by multiplying.

✔ *CHECK YOURSELF 2*

Factor each of the following polynomials.

(a) $5x + 20$ (b) $6x^2 - 24x$ (c) $10a^3 - 15a^2$

The process is exactly the same for polynomials with more than two terms. Consider Example 3.

| **Example 3** | **Finding the GCF of a Polynomial** |

(a) Factor $5x^2 - 10x + 15$.

The GCF is 5.

$$5x^2 - 10x + 15 = 5 \cdot x^2 - 5 \cdot 2x + 5 \cdot 3$$

$$\text{GCF}$$

$$= 5(x^2 - 2x + 3)$$

(b) Factor $6ab + 9ab^2 - 15a^2$.

The GCF is $3a$.

$$6ab + 9ab^2 - 15a^2 = 3a \cdot 2b + 3a \cdot 3b^2 - 3a \cdot 5a$$

$$\text{GCF}$$

$$= 3a(2b + 3b^2 - 5a)$$

(c) Factor $4a^4 + 12a^3 - 20a^2$.

The GCF is $4a^2$.

$$4a^4 + 12a^3 - 20a^2 = 4a^2 \cdot a^2 + 4a^2 \cdot 3a - 4a^2 \cdot 5$$

$$\text{GCF}$$

$$= 4a^2(a^2 + 3a - 5)$$

(d) Factor $\underbrace{6a^2b + 9ab^2 + 3ab}$.

Mentally note that 3, a, and b are factors of each term, so

$$6a^2b + 9ab^2 + 3ab = 3ab(2a + 3b + 1)$$

In each of these examples, you will want to check the result by multiplying the factors.

✔ *CHECK YOURSELF 3*

Factor each of the following polynomials.

(a) $8b^2 + 16b - 32$ (b) $4xy - 8x^2y + 12x^3$

(c) $7x^4 - 14x^3 + 21x^2$ (d) $5x^2y^2 - 10xy^2 + 15x^2y$

Factoring by Grouping

A related factoring method is called **factoring by grouping.** We introduce this method in Example 4.

Example 4	**Finding a Common Binomial Factor**

(a) Factor $3x(x + y) + 2(x + y)$.

We see that the *binomial x + y* is a common factor and can therefore be factored out.

Because of the commutative property, the factors can be written in either order.

$$3x(x + y) + 2(x + y)$$
$$= (x + y) \cdot 3x + (x + y) \cdot 2$$
$$= (x + y)(3x + 2)$$

(b) Factor $3x^2(x - y) + 6x(x - y) + 9(x - y)$.

We note that here the GCF is $3(x - y)$. Factoring as before, we have

$$3(x - y)(x^2 + 2x + 3)$$

✔ *CHECK YOURSELF 4*

Completely factor each of the polynomials.

(a) $7a(a - 2b) + 3(a - 2b)$ (b) $4x^2(x + y) - 8x(x + y) - 16(x + y)$

If the terms of a polynomial have no common factor (other than 1), factoring by grouping is the preferred method, as illustrated in Example 5.

Example 5	**Factoring by Grouping Terms**

Suppose we want to factor the polynomial

$$ax - ay + bx - by$$

As you can see, the polynomial has no common factors. However, look at what happens if we separate the polynomial into *two groups* of *two terms*.

Note that our example has *four* terms. That is the clue for trying the factoring by grouping method.

$$ax - ay + bx - by$$
$$= \underbrace{ax - ay}_{(1)} + \underbrace{bx - by}_{(2)}$$

Now *each* group has a common factor, and we can write the polynomial as

$$a(x - y) + b(x - y)$$

In this form, we can see that $x - y$ is the GCF. Factoring out $x - y$, we get

$$a(x - y) + b(x - y) = (x - y)(a + b)$$

✔ *CHECK YOURSELF 5*

Use the factoring by grouping method.

$$x^2 - 2xy + 3x - 6y$$

Be particularly careful of your treatment of algebraic signs when you apply the factoring by grouping method. Consider Example 6.

Example 6

Factoring by Grouping Terms

Factor $2x^3 - 3x^2 - 6x + 9$.

We group the polynomial as follows.

$$\underbrace{2x^3 - 3x^2}_{(1)} \underbrace{- 6x + 9}_{(2)}$$ Factor out the common factor of -3 from the second two terms.

Note that $9 = (-3)(-3)$.

$$= x^2(2x - 3) - 3(2x - 3)$$
$$= (2x - 3)(x^2 - 3)$$

✔ *CHECK YOURSELF 6*

Factor by grouping.

$$3y^3 + 2y^2 - 6y - 4$$

It may also be necessary to change the order of the terms as they are grouped. Look at Example 7.

Example 7

Factoring by Grouping Terms

Factor $x^2 - 6yz + 2xy - 3xz$.

Grouping the terms as before, we have

$$\underbrace{x^2 - 6yz}_{(1)} + \underbrace{2xy - 3xz}_{(2)}$$

Do you see that we have accomplished nothing because there are no common factors in the first group?

We can, however, rearrange the terms to write the original polynomial as

$$\underbrace{x^2 + 2xy}_{(1)} \; \underbrace{-\, 3xz - 6yz}_{(2)}$$

$= x(x + 2y) - 3z(x + 2y)$ We can now factor out the common
                                factor of $x + 2y$ in group (1)
$= (x + 2y)(x - 3z)$ and group (2).

Note: It is often true that the grouping can be done in more than one way. The factored form will be the same.

✔ CHECK YOURSELF 7

We can write the polynomial of Example 7 as

$$x^2 - 3xz + 2xy - 6yz$$

Factor, and verify that the factored form is the same in either case.

✔ CHECK YOURSELF ANSWERS

1. (a) 2; (b) 9; (c) a^5; (d) $5x^4$ **2.** (a) $5(x + 4)$; (b) $6x(x - 4)$; (c) $5a^2(2a - 3)$

3. (a) $8(b^2 + 2b - 4)$; (b) $4x(y - 2xy + 3x^2)$; (c) $7x^2(x^2 - 2x + 3)$;
  (d) $5xy(xy - 2y + 3x)$

4. (a) $(a - 2b)(7a + 3)$; (b) $4(x + y)(x^2 - 2x - 4)$

5. $(x - 2y)(x + 3)$ **6.** $(3y + 2)(y^2 - 2)$ **7.** $(x - 3z)(x + 2y)$

Find the greatest common factor for each of the following lists of terms.

1. 20, 22

2. 15, 35

3. 16, 32, 88

4. 44, 66, 143

5. x^2, x^5

6. y^7, y^9

7. a^3, a^6, a^9

8. b^4, b^6, b^8

9. $5x^4, 10x^5$

10. $8y^9, 24y^3$

11. $4a^4, 10a^7, 12a^{14}$

12. $9b^3, 6b^5, 12b^4$

13. $9x^2y, 12xy^2, 15x^2y^2$

14. $12a^3b^2, 18a^2b^3, 6a^4b^4$

15. $15ab^3, 10a^2bc, 25b^2c^3$

16. $12x^3, 15x^2y^2, 21y^5$

17. $15a^2bc^2, 9ab^2c^2, 6a^2b^2c^2$

18. $18x^3y^2z^3, 27x^4y^2z^3, 81xy^2z$

19. $(x + y)^2, (x + y)^5$

20. $12(a + b)^4, 4(a + b)^3$

Factor each of the following polynomials.

21. $10x + 5$

22. $5x - 15$

23. $24m - 32n$

24. $7p - 21q$

25. $12m^2 + 8m$

26. $30n^2 - 35n$

27. $10s^2 + 5s$

28. $12y^2 - 6y$

29. $24x^2 - 60x$

30. $14b^2 - 28b$

31. $15a^3 - 25a^2$

32. $36b^4 + 24b^2$

33. $6pq + 18p^2q$

34. $9xy - 18xy^2$

35. $7m^3n - 21mn^3$

36. $36p^2q^2 - 9pq$

37. $6x^2 - 18x + 30$

38. $7a^2 + 21a - 42$

39. $5a^3 - 15a^2 + 25a$

40. $5x^3 - 15x^2 + 25x$

41. $12x + 8xy - 28xy^2$

42. $4s + 6st - 14st^2$

43. $10x^2y + 15xy - 5xy^2$

44. $3ab^2 + 6ab - 15a^2b$

45. $10r^3s^2 + 25r^2s^2 - 15r^2s^3$ **46.** $28x^2y^3 - 35x^2y^2 + 42x^3y$ **47.** $9a^5 - 15a^4 + 21a^3 - 27a$

48. $8p^6 - 40p^4 + 24p^3 + 16p^2$ **49.** $15m^3n^2 - 20m^2n + 35mn^3 - 10mn$

50. $14ab^4 + 21a^2b^3 - 35a^3b^2 + 28ab^2$ **51.** $x(x - 9) + 5(x - 9)$

52. $y(y + 5) - 3(y + 5)$ **53.** $p(p - 2q) - q(p - 2q)$ **54.** $x(3x - 4y) - y(3x - 4y)$

55. $x(y - z) + 3(y - z)$ **56.** $2a(c - d) - b(c - d)$

Factor each polynomial by grouping. (*Hint:* You may have to rearrange terms.)

57. $ab - ac + b^2 - bc$ **58.** $ax + 2a + bx + 2b$ **59.** $6r^2 + 12rs - r - 2s$

60. $2mn - 4m^2 + 3n - 6m$ **61.** $ab^2 - 2b^2 + 3a - 6$ **62.** $r^2s^2 - 3s^2 - 2r^2 + 6$

63. $x^2 + 3x - 4xy - 12y$ **64.** $a^2 - 12b + 3ab - 4a$

65. $m^2 - 6n^3 + 2mn^2 - 3mn$ **66.** $r^2 - 3rs^2 - 12s^3 + 4rs$

Determine if the factoring in each of the following is correct.

67. $x^2 - x - 6 = (x - 3)(x + 2)$ **68.** $x^2 - x - 12 = (x - 4)(x + 3)$ **69.** $x^2 + x - 12 = (x + 6)(x - 2)$

70. $x^2 + 2x - 8 = (x + 8)(x - 1)$ **71.** $2x^2 - 5x - 3 = (2x + 1)(x - 3)$ **72.** $6x^2 - 13x + 6 = (3x - 2)(2x - 3)$

73. The GCF of $2x - 6$ is 2. The GCF of $5x + 10$ is 5. Find the greatest common factor of the product $(2x - 6)(5x + 10)$.

74. The GCF of $3z + 12$ is 3. The GCF of $4z + 8$ is 4. Find the GCF of the product $(3z + 12)(4z + 8)$.

75. The GCF of $2x^3 - 4x$ is $2x$. The GCF of $3x + 6$ is 3. Find the GCF of the product $(2x^3 - 4x)(3x + 6)$.

76. State, in a sentence, the rule that Exercises 73 to 75 illustrate.

77. For the monomials x^4y^2, x^8y^6, and x^9y^4, explain how you can determine the GCF by inspecting exponents.

78. It is not possible to use the grouping method to factor $2x^3 + 6x^2 + 8x + 4$. Is it correct to conclude that the polynomial is prime? Justify your answer.

79. Area of a rectangle. The area of a rectangle with width t is given by $33t - t^2$. Factor the expression and determine the length of the rectangle in terms of t.

80. Area of a rectangle. The area of a rectangle of length x is given by $3x^2 + 5x$. Find the width of the rectangle.

81. For centuries, mathematicians have found factoring numbers into prime factors a fascinating subject. A prime number is a number that cannot be written as a product of any whole numbers but 1 and itself. The list of primes begins with 2 because 1 is not considered a prime number and then goes on: 3, 5, 7, 11, ... What are the first 10 primes? What are the primes less than 100? If you list the numbers from 1 to 100 and then cross out all numbers that are multiples of 2, 3, 5, and 7, what is left? Are all the numbers not crossed out prime? Write a paragraph to explain why this might be so. You might want to investigate the Sieve of Eratosthenes, a system from 230 B.C.E. for finding prime numbers.

82. If we could make a list of all the prime numbers, what number would be at the end of the list? Because there are an infinite number of prime numbers, there is no "largest prime number." But is there some formula that will give us all the primes? Here are some formulas proposed over the centuries:

$$n^2 + n + 17 \qquad 2n^2 + 29 \qquad n^2 - n + 11$$

In all these expressions, $n = 1, 2, 3, 4, \ldots$, that is, a positive integer beginning with 1. Investigate these expressions with a partner. Do the expressions give prime numbers when they are evaluated for these values of n? Do the expressions give *every* prime in the range of resulting numbers? Can you put in *any* positive number for n?

83. How are primes used in coding messages and for security? Work together to decode the messages. The messages are coded using this code: After the numbers are factored into prime factors, the power of 2 gives the number of the letter in the alphabet. This code would be easy for a code breaker to figure out, but you might make up a code that would be more difficult to break.

(a) 1310720, 229376, 1572864, 1760, 460, 2097152, 336

(b) 786432, 142, 4608, 278528, 1344, 98304, 1835008, 352, 4718592, 5242880

(c) Code a message, using this rule. Exchange your message with a partner to decode it.

Factoring Special Polynomials

6.2 OBJECTIVES

1. Factor the difference of two squares
2. Factor the sum and difference of two cubes

In Section 4.5, we introduced some special products. Recall the following formula for the product of a sum and difference of two terms:

$$(a + b)(a - b) = a^2 - b^2$$

This also means that a binomial of the form $a^2 - b^2$, called a **difference of two squares,** has as its factors $a + b$ and $a - b$.

To use this idea for factoring, we can write

$$a^2 - b^2 = (a + b)(a - b)$$

The exponent must be even.

To apply this pattern, we look for **perfect squares.** Perfect square numbers are 1, 4, 9, 16, 25, 36, etc., since $1^2 = 1$, $2^2 = 4$, $3^2 = 9$, $4^2 = 16$, and so on. Since $(x^1)^2 = x^2$, $(x^2)^2 = x^4$, $(x^3)^2 = x^6$, $(x^4)^2 = x^8$, $(x^5)^2 = x^{10}$, and so on, variables that have exponents like 2, 4, 6, 8, 10, etc., are perfect squares.

Example 1

Factoring the Difference of Two Squares

Factor $x^2 - 16$.

Think $x^2 - 4^2$

You could also write $(x - 4)(x + 4)$. The order doesn't matter because multiplication is commutative.

Since $x^2 - 16$ is a difference of squares, we have

$$x^2 - 16 = (x + 4)(x - 4)$$

 CHECK YOURSELF 1

Factor $m^2 - 49$.

Any time an expression is a difference of two squares, it can be factored.

Example 2

Factoring the Difference of Two Squares

Factor $4a^2 - 9$.

Think $(2a)^2 - 3^2$

So
$$4a^2 - 9 = (2a)^2 - (3)^2$$
$$= (2a + 3)(2a - 3)$$

✔ *CHECK YOURSELF 2*

Factor $9b^2 - 25$.

The process for factoring a difference of squares does not change when more than one variable is involved.

Example 3

Think $(5a)^2 - (4b^2)^2$

Factoring the Difference of Two Squares

Factor $25a^2 - 16b^4$.

$$25a^2 - 16b^4 = (5a)^2 - (4b^2)^2 = (5a + 4b^2)(5a - 4b^2)$$

✔ *CHECK YOURSELF 3*

Factor $49c^4 - 9d^2$.

We will now consider an example that combines common-term factoring with difference-of-squares factoring. Note that the common factor is always factored out as the *first step*.

Example 4

Removing the GCF First

Factor $32x^2y - 18y^3$.

Note that $2y$ is a common factor, so

Step 1
Factor out the GCF.
Step 2
Factor the remaining binomial.

$$32x^2y - 18y^3 = 2y(\underbrace{16x^2 - 9y^2}_{\text{Difference of squares}})$$

$$= 2y(4x + 3y)(4x - 3y)$$

✔ *CHECK YOURSELF 4*

Factor $50a^3 - 8ab^2$.

You may also have to apply the difference of two squares method *more than once* to completely factor a polynomial.

| **Example 5** | ### Factoring the Difference of Two Squares |

Factor $m^4 - 81n^4$.

$$m^4 - 81n^4 = (m^2 + 9n^2)(m^2 - 9n^2)$$

Do you see that we are not done in this case? Since $m^2 - 9n^2$ is still factorable, we can continue to factor as follows.

$$m^4 - 81n^4 = (m^2 + 9n^2)(m + 3n)(m - 3n)$$

Note: The other binomial factor, $m^2 + 9n^2$, is a *sum of two squares,* which cannot be factored further.

✔ *CHECK YOURSELF 5*

Factor $x^4 - 16y^4$.

Factoring the Sum or Difference of Two Cubes

Two additional factoring patterns include the sum or difference of two cubes.

Be sure you take the time to expand the product on the right-hand side to confirm the formula.

> **The Sum or Difference of Two Cubes**
>
> $$a^3 + b^3 = (a + b)(a^2 - ab + b^2)$$
> $$a^3 - b^3 = (a - b)(a^2 + ab + b^2)$$

The exponent must be a multiple of 3.

We are now looking for **perfect cubes.** Perfect cube numbers are 1, 8, 27, 64, etc. Since $(x^1)^3 = x^3$, $(x^2)^3 = x^6$, $(x^3)^3 = x^9$, $(x^4)^3 = x^{12}$, $(x^5)^3 = x^{15}$, and so on, variables that have exponents like 3, 6, 9, 12, etc., are perfect cubes.

| **Example 6** | ### Factoring the Sum or Difference of Two Cubes |

We are now looking for perfect cubes—the exponents must be multiples of 3 and the coefficients perfect cubes—1, 8, 27, 64, and so on.

(a) Factor $x^3 + 27$.

The first term is the cube of x, and the second is the cube of 3, so we can apply the $a^3 + b^3$ equation. Letting $a = x$ and $b = 3$, we have

$$x^3 + 27 = (x)^3 + (3)^3 = (x + 3)(x^2 - 3x + 9)$$

(b) Factor $8w^3 - 27z^3$.

This is a difference of cubes, so use the $a^3 - b^3$ equation

$$8w^3 - 27z^3 = (2w)^3 - (3z)^3 = (2w - 3z)[(2w)^2 + (2w)(3z) + (3z)^2]$$
$$= (2w - 3z)(4w^2 + 6wz + 9z^2)$$

Again, looking for a *common factor* should be your first step. Remember to write the GCF as a part of the final factored form.

(c) Factor $5a^3b - 40b^4$.

First note the common factor of $5b$.

$$5a^3b - 40b^4 = 5b(a^3 - 8b^3)$$

The binomial is the difference of cubes, so:

$$= 5b[(a)^3 - (2b)^3]$$
$$= 5b(a - 2b)(a^2 + 2ab + 4b^2)$$

✔ CHECK YOURSELF 6

Factor completely.

(a) $27x^3 + 8y^3$ (b) $3a^4 - 24ab^3$

In each example in this section, we factored a polynomial expression. If we are given a polynomial function to factor, there is no change in the ordered pairs represented by the function after it is factored.

Example 7   **Factoring a Polynomial Function**

Given the function $f(x) = 9x^2 + 15x$, complete the following.

(a) Find $f(1)$.

$$f(1) = 9(1)^2 + 15(1)$$
$$= 9 + 15$$
$$= 24$$

(b) Factor $f(x)$.

$$f(x) = 9x^2 + 15x$$
$$= 3x(3x + 5)$$

(c) Find $f(1)$ from the factored form of $f(x)$.

$$f(1) = 3(1)(3(1) + 5)$$
$$= 3(8)$$
$$= 24$$

✔ *CHECK YOURSELF 7*

Given the function $f(x) = 16x^5 + 10x^2$, complete the following.

(a) Find $f(1)$.   (b) Factor $f(x)$.   (c) Find $f(1)$ from the factored form of $f(x)$.

✔ *CHECK YOURSELF ANSWERS*

1. $(m + 7)(m - 7)$ **2.** $(3b + 5)(3b - 5)$ **3.** $(7c^2 + 3d)(7c^2 - 3d)$

4. $2a(5a + 2b)(5a - 2b)$ **5.** $(x^2 + 4y^2)(x + 2y)(x - 2y)$

6. (a) $(3x + 2y)(9x^2 - 6xy + 4y^2)$; (b) $3a(a - 2b)(a^2 + 2ab + 4b^2)$

7. (a) 26; (b) $2x^2(8x^3 + 5)$; (c) 26

Exercises · 6.2

For each of the following binomials, state whether the binomial is a difference of squares.

1. $25x^2 + 9y^2$ **2.** $5x^2 - 7y^2$ **3.** $16a^2 - 25b^2$ **4.** $9n^2 - 16m^2$

5. $16r^2 + 4$ **6.** $9p^2 - 54$ **7.** $16a^2 - 12b^3$ **8.** $9a^2b^2 - 16c^2d^2$

9. $a^2b^2 - 25$ **10.** $8x^6 - 27y^3$

Factor the following binomials completely.

11. $m^2 - n^2$ **12.** $r^2 - 9$ **13.** $x^2 - 169$ **14.** $c^2 - d^2$

15. $49 - y^2$ **16.** $196 - y^2$ **17.** $9b^2 - 16$ **18.** $36 - x^2$

19. $16w^2 - 49$ **20.** $4x^2 - 25$ **21.** $4s^2 - 9r^2$ **22.** $64y^2 - x^2$

23. $9w^2 - 49z^2$ **24.** $25x^2 - 81y^2$ **25.** $49a^2 - 9b^2$ **26.** $64m^2 - 9n^2$

27. $x^4 - 36$ **28.** $y^6 - 49$ **29.** $x^2y^2 - 16$ **30.** $m^2n^2 - 64$

31. $25 - a^2b^2$ **32.** $49 - w^2z^2$ **33.** $r^4 - 4s^2$ **34.** $p^2 - 9q^4$

35. $81a^2 - 100b^6$ **36.** $64x^4 - 25y^4$ **37.** $18x^3 - 2xy^2$ **38.** $50a^2b - 2b^3$

39. $12m^3n - 75mn^3$ **40.** $63p^4 - 7p^2q^2$ **41.** $16a^4 - 81b^4$ **42.** $81x^4 - y^4$

43. $y^3 + 125$ **44.** $y^3 - 8$ **45.** $m^3 - 125$ **46.** $b^3 + 27$

47. $a^3b^3 - 27$ **48.** $p^3q^3 - 64$ **49.** $8w^3 + z^3$ **50.** $c^3 - 27d^3$

51. $r^3 - 64s^3$ **52.** $125x^3 + y^3$ **53.** $8x^3 - 27y^3$ **54.** $64m^3 + 27n^3$

55. $3a^3 + 81b^3$ **56.** $m^6 - 27n^3$ **57.** $4x^3 - 32y^3$

In each of the following, (a) find $f(1)$; (b) factor $f(x)$; and (c) find $f(1)$ from the factored form.

58. $f(x) = 12x^5 + 21x^2$

59. $f(x) = -6x^3 - 10x$

60. $f(x) = -8x^5 + 20x$

61. $f(x) = 5x^5 - 35x^3$

62. $f(x) = x^5 + 3x^2$

63. $f(x) = 6x^6 - 16x^5$

Factor each expression.

64. $x^2(x + y) - y^2(x + y)$

65. $a^2(b - c) - 16b^2(b - c)$

66. $2m^2(m - 2n) - 18n^2(m - 2n)$

67. $3a^3(2a + b) - 27ab^2(2a + b)$

68. Find the value for k so that $kx^2 - 25$ will have the factors $2x + 5$ and $2x - 5$.

69. Find the value for k so that $9m^2 - kn^2$ will have the factors $3m + 7n$ and $3m - 7n$.

70. Find the value for k so that $2x^3 - kxy^2$ will have the factors $2x$, $x - 3y$, and $x + 3y$.

71. Find the value for k so that $20a^3b - kab^3$ will have the factors $5ab$, $2a - 3b$, and $2a + 3b$.

72. Complete this statement: "To factor a number, you . . . ."

73. Complete this statement: "To factor an algebraic expression into prime factors means . . . ."

74. What binomial multiplied by $25x^2 - 15xy + 9y^2$ gives the sum of two cubes? What is the result of the multiplication?

75. What binomial when multiplied by $9x^2 + 6xy + 4y^2$ gives the difference of two cubes? What is the result of the multiplication?

76. What are the characteristics of a monomial that is a perfect cube?

77. Suppose you factored the polynomial $4x^2 - 16$ as follows:

$$4x^2 - 16 = (2x + 4)(2x - 4)$$

Would this be in completely factored form? If not, what would be the final form?

6.3 Factoring Trinomials: The *ac* Method

6.3 OBJECTIVES

1. Factor a trinomial of the form $ax^2 + bx + c$
2. Completely factor a trinomial

The product of two binomials of the form

$$(\_\_x + \_\_)(\_\_x + \_\_)$$

will always be a trinomial. In Section 4.5, we used the FOIL method to find the product of two binomials. In this section we will use the factoring-by-grouping method to find the binomial factors for a trinomial.

First let's look at some factored trinomials.

Example 1

Matching Trinomials and Their Factors

Determine which of the following are true statements.

(a) $x^2 - 2x - 8 = (x - 4)(x + 2)$

This is a true statement. Using the FOIL method, we see that

$$(x - 4)(x + 2) = x^2 + 2x - 4x - 8 = x^2 - 2x - 8$$

(b) $x^2 - 6x + 5 = (x - 2)(x - 3)$

This is not a true statement.

$$(x - 2)(x - 3) = x^2 - 3x - 2x + 6 = x^2 - 5x + 6$$

(c) $x^2 + 5x - 14 = (x - 2)(x + 7)$

This is true:

$$(x - 2)(x + 7) = x^2 + 7x - 2x - 14 = x^2 + 5x - 14$$

(d) $x^2 - 8x - 15 = (x - 5)(x - 3)$

This is false:

$$(x - 5)(x - 3) = x^2 - 3x - 5x + 15 = x^2 - 8x + 15$$

454

✔ *CHECK YOURSELF 1*

Determine which of the following are true statements.

(a) $2x^2 - 2x - 3 = (2x - 3)(x + 1)$

(b) $3x^2 + 11x - 4 = (3x - 1)(x + 4)$

(c) $2x^2 - 7x + 3 = (x - 3)(2x - 1)$

The first step in learning to factor a trinomial is to identify its coefficients. So that we are consistent, we first write the trinomial in standard $ax^2 + bx + c$ form, then label the three coefficients as a, b, and c.

Example 2   **Identifying the Coefficients of $ax^2 + bx + c$**

First, where necessary, rewrite the trinomial in $ax^2 + bx + c$ form. Then give the values for a, b, and c, where a is the coefficient of the x^2 term, b is the coefficient of the x term, and c is the constant.

(a) $x^2 - 3x - 18$

$$a = 1 \qquad b = -3 \qquad c = -18$$

(b) $x^2 - 24x + 23$

Note that the minus sign is attached to the coefficients.

$$a = 1 \qquad b = -24 \qquad c = 23$$

(c) $x^2 + 8 - 11x$

First rewrite the trinomial in descending order:

$$x^2 - 11x + 8$$

$$a = 1 \qquad b = -11 \qquad c = 8$$

✔ *CHECK YOURSELF 2*

First, where necessary, rewrite the trinomials in $ax^2 + bx + c$ form. Then label a, b, and c, where a is the coefficient of the x^2 term, b is the coefficient of the x term, and c is the constant.

(a) $x^2 + 5x - 14$ (b) $x^2 - 18x + 17$ (c) $x - 6 + 2x^2$

Not all trinomials can be factored. To discover if a trinomial is factorable, we try the *ac* **test.**

The *ac* Test

A trinomial of the form $ax^2 + bx + c$ is factorable if (and only if) there are two integers *m* and *n* such that

$$ac = mn \quad \text{and} \quad b = m + n$$

In Example 3 we will look for *m* and *n* to determine whether each trinomial is factorable.

Example 3

Using the *ac* Test

Use the *ac* test to determine which of the following trinomials can be factored. Find the values of *m* and *n* for each trinomial that can be factored.

(a) $x^2 - 3x - 18$

First, we find the values of *a*, *b*, and *c*, so that we can find *ac*.

$$a = 1 \qquad b = -3 \qquad c = -18$$

$$ac = 1(-18) = -18 \qquad \text{and} \qquad b = -3$$

Then we look for two integers *m* and *n* such that $mn = ac$ and $m + n = b$. In this case, that means

$$mn = -18 \qquad \text{and} \qquad m + n = -3$$

We now look at all pairs of integers with a product of -18. We then look at the sum of each pair of integers.

mn	*m + n*	
$1(-18) = -18$	$1 + (-18) = -17$	
$2(-9) = -18$	$2 + (-9) = -7$	
$3(-6) = -18$	$3 + (-6) = -3$	We need look no further than 3 and -6.
$6(-3) = -18$		
$9(-2) = -18$		
$18(-1) = -18$		

The two integers with a product of *ac* and a sum of *b* are 3 and -6. We can say that

$$m = 3 \qquad \text{and} \qquad n = -6$$

We could have chosen $m = -6$ and $n = 3$ as well.

Because we found values for *m* and *n*, we know that $x^2 - 3x - 18$ is factorable.

(b) $x^2 - 24x + 23$

We find that

$$a = 1 \qquad b = -24 \qquad c = 23$$

$$ac = 1(23) = 23 \qquad \text{and} \qquad b = -24$$

So $\qquad\qquad\qquad mn = 23 \qquad \text{and} \qquad m + n = -24$

We now calculate integer pairs, looking for two numbers with a product of 23 and a sum of -24.

mn	*m + n*
$1(23) = 23$	$1 + 23 = \quad 24$
$-1(-23) = 23$	$-1 + (-23) = -24$

$$m = -1 \qquad \text{and} \qquad n = -23$$

So $x^2 - 24x + 23$ is factorable.

(c) $x^2 - 11x + 8$

We find that $a = 1$, $b = -11$, and $c = 8$. Therefore, $ac = 8$ and $b = -11$. Thus $mn = 8$ and $m + n = -11$. We calculate integer pairs:

mn	*m + n*
$1(8) = 8$	$1 + 8 = \quad 9$
$2(4) = 8$	$2 + 4 = \quad 6$
$-1(-8) = 8$	$-1 + (-8) = -9$
$-2(-4) = 8$	$-2 + (-4) = -6$

There are no other pairs of integers with a product of 8, and none of these pairs has a sum of -11. The trinomial $x^2 - 11x + 8$ is not factorable.

(d) $2x^2 + 7x - 15$

We find that $a = 2$, $b = 7$, and $c = -15$. Therefore, $ac = 2(-15) = -30$ and $b = 7$. Thus $mn = -30$ and $m + n = 7$. We calculate integer pairs:

mn	*m + n*
$1(-30) = -30$	$1 + (-30) = -29$
$2(-15) = -30$	$2 + (-15) = -13$
$3(-10) = -30$	$3 + (-10) = \quad -7$
$5(-6) \ = -30$	$5 + (-6) \ = \quad -1$
$6(-5) \ = -30$	$6 + (-5) \ = \quad 1$
$10(-3) \ = -30$	$10 + (-3) \ = \quad 7$

There is no need to go any further. We see that 10 and -3 have a product of -30 and a sum of 7, so

$$m = 10 \quad \text{and} \quad n = -3$$

Therefore, $2x^2 + 7x - 15$ is factorable.

It is not always necessary to evaluate all the products and sums to determine whether a trinomial is factorable. You may have noticed patterns and shortcuts that make it easier to find m and n. By all means, use them to help you find m and n. This is essential in mathematical thinking. You are taught a mathematical process that will always work for solving a problem. Such a process is called an **algorithm.** It is very easy to teach a computer to use an algorithm. It is very difficult (some would say impossible) for a computer to have insight. Shortcuts that you discover are *insights*. They may be the most important part of your mathematical education.

✔ CHECK YOURSELF 3

Use the *ac* test to determine which of the following trinomials can be factored. Find the values of m and n for each trinomial that can be factored.

(a) $x^2 - 7x + 12$ (b) $x^2 + 5x - 14$

(c) $3x^2 - 6x + 7$ (d) $2x^2 + x - 6$

So far we have used the results of the *ac* test only to determine whether a trinomial is factorable. The results can also be used to help factor the trinomial.

Example 4

Using the Results of the *ac* Test to Factor

Rewrite the middle term as the sum of two terms, then factor by grouping.

(a) $x^2 - 3x - 18$

We find that $a = 1, b = -3$, and $c = -18$, so $ac = -18$ and $b = -3$. We are looking for two numbers m and n where $mn = -18$ and $m + n = -3$. In Example 3(a), we looked at every pair of integers whose product (mn) was -18, to find a pair that had a sum ($m + n$) of -3. We found the two integers to be 3 and -6, because $3(-6) = -18$ and $3 + (-6) = -3$, so $m = 3$ and $n = -6$. We now use that result to rewrite the middle term as the sum of $3x$ and $-6x$.

$$x^2 + 3x - 6x - 18$$

We then factor by grouping:

$$x^2 + 3x - 6x - 18 = x(x + 3) - 6(x + 3)$$
$$= (x + 3)(x - 6)$$

After factoring, immediately check your work. Here, multiply $(x + 3)(x - 6)$ and confirm that the product is $x^2 - 3x - 18$.

(b) $x^2 - 24x + 23$

We use the results from Example 3(b), in which we found $m = -1$ and $n = -23$, to rewrite the middle term of the equation.

$$x^2 - 24x + 23 = x^2 - x - 23x + 23$$

Then we factor by grouping:

$$\begin{aligned} x^2 - x - 23x + 23 &= (x^2 - x) - (23x - 23) \\ &= x(x - 1) - 23(x - 1) \\ &= (x - 1)(x - 23) \end{aligned}$$

Again, check by multiplying.

(c) $2x^2 + 7x - 15$

From Example 3(d), we know that this trinomial is factorable, and $m = 10$ and $n = -3$. We use that result to rewrite the middle term of the trinomial.

$$\begin{aligned} 2x^2 + 7x - 15 &= 2x^2 + 10x - 3x - 15 \\ &= (2x^2 + 10x) - (3x + 15) \\ &= 2x(x + 5) - 3(x + 5) \\ &= (x + 5)(2x - 3) \end{aligned}$$

Check!

Careful readers will note that we did not ask you to factor Example 3(c), $x^2 - 11x + 8$. Recall that, by the *ac* method, we determined that this trinomial was not factorable.

✔ *CHECK YOURSELF 4*

Use the results of Check Yourself 3 to rewrite the middle term as the sum of two terms, then factor by grouping.

(a) $x^2 - 7x + 12$ (b) $x^2 + 5x - 14$ (c) $2x^2 + x - 6$

Let's look at some examples that require us to first find m and n and then factor the trinomial.

Example 5	**Rewriting Middle Terms to Factor**

Rewrite the middle term as the sum of two terms, and then factor by grouping.

(a) $2x^2 - 13x - 7$

We find that $a = 2$, $b = -13$, and $c = -7$, so $mn = ac = -14$ and $m + n = b = -13$. Therefore,

mn	*m + n*
$1(-14) = -14$	$1 + (-14) = -13$

So $m = 1$ and $n = -14$. We rewrite the middle term of the trinomial as follows:

$$2x^2 - 13x - 7 = 2x^2 + x - 14x - 7$$
$$= (2x^2 + x) - (14x + 7)$$
$$= x(2x + 1) - 7(2x + 1)$$
$$= (2x + 1)(x - 7)$$

Check that

$(2x + 1)(x - 7)$
$= 2x^2 - 13x - 7$

(b) $6x^2 - 5x - 6$

We find that $a = 6, b = -5,$ and $c = -6,$ so $mn = ac = -36$ and $m + n = b = -5.$

mn	$m + n$
$1(-36) = -36$	$1 + (-36) = -35$
$2(-18) = -36$	$2 + (-18) = -16$
$3(-12) = -36$	$3 + (-12) = -9$
$4(-9) = -36$	$4 + (-9) = -5$

So $m = 4$ and $n = -9.$ We rewrite the middle term of the trinomial:

$$6x^2 + 5x - 6 = 6x^2 + 4x - 9x - 6$$
$$= (6x^2 + 4x) - (9x + 6)$$
$$= 2x(3x + 2) - 3(3x + 2)$$
$$= (3x + 2)(2x - 3)$$

Multiply to check.

✔ CHECK YOURSELF 5

Rewrite the middle term as the sum of two terms and then factor by grouping.

(a) $2x^2 - 7x - 15$ (b) $6x^2 - 5x - 4$

Be certain to check trinomials and binomial factors for any common monomial factor. (There is no common factor in the binomial unless it is also a common factor in the original trinomial.) Example 6 shows the factoring out of monomial factors.

Example 6   **Factoring Out Common Factors**

Completely factor the trinomial

$$3x^2 + 12x - 15$$

We could first factor out the common factor of 3:

$$3x^2 + 12x - 15 = 3(x^2 + 4x - 5)$$

Finding m and n for the trinomial $x^2 + 4x - 5$ yields $mn = -5$ and $m + n = 4$.

mn	$m + n$
$1(-5) = -5$	$1 + (-5) = -4$
$5(-1) = -5$	$5 + (-1) = \;\;\;4$

So $m = 5$ and $n = -1$. This gives us

$$
\begin{aligned}
3x^2 + 12x - 15 &= 3(x^2 + 4x - 5) \\
&= 3(x^2 + 5x - x - 5) \\
&= 3[(x^2 + 5x) - (x + 5)] \\
&= 3[x(x + 5) - (x + 5)] \\
&= 3[(x + 5)(x - 1)] \\
&= 3(x + 5)(x - 1)
\end{aligned}
$$

Again, multiply this result to check.

 CHECK YOURSELF 6

Completely factor the trinomial.

$$
6x^3 + 3x^2 - 18x
$$

Not all possible product pairs need to be tried to find m and n. A look at the sign pattern of the trinomial will eliminate many of the possibilities. Assuming the leading coefficient is positive, there are four possible sign patterns.

Pattern	Example	Conclusion
1. b and c are both positive.	$2x^2 + 13x + 15$	m and n must both be positive.
2. b is negative and c is positive.	$x^2 - 7x + 12$	m and n must both be negative.
3. b is positive and c is negative.	$x^2 + 3x - 10$	m and n are of opposite signs. (The value with the larger absolute value is positive.)
4. b and c are both negative.	$x^2 - 3x - 10$	m and n are of opposite signs. (The value with the larger absolute value is negative.)

Trial and Error

Sometimes the factors of a trinomial seem obvious. At other times you might be certain that there are only a couple of possible sets of factors for a trinomial. It is perfectly acceptable to check these proposed factors to see if they work. If you find the factors in this manner, we say that you have used the trial-and-error method. The difficulty with using this method exclusively is found in the reason that we call this method trial and *error.*

Section 6.3* offers instruction on the trial-and-error technique. Even if it has not been assigned, you may find it useful to examine that section.

✔ *CHECK YOURSELF ANSWERS*

1. (a) False; (b) true; (c) true

2. (a) $a = 1, b = 5, c = -14$; (b) $a = 1, b = -18, c = 17$; (c) $a = 2, b = 1, c = -6$

3. (a) Factorable, $m = -3, n = -4$; (b) factorable, $m = 7, n = -2$; (c) not factorable; (d) factorable, $m = 4, n = -3$

4. (a) $x^2 - 3x - 4x + 12 = (x - 3)(x - 4)$; (b) $x^2 + 7x - 2x - 14 = (x + 7)(x - 2)$; (c) $2x^2 + 4x - 3x - 6 = (2x - 3)(x + 2)$

5. (a) $2x^2 - 10x + 3x - 15 = (2x + 3)(x - 5)$; (b) $6x^2 - 8x + 3x - 4 = (3x - 4)(2x + 1)$

6. $3x(2x - 3)(x + 2)$

Exercises · 6.3

State whether each of the following is true or false.

1. $x^2 + 2x - 3 = (x + 3)(x - 1)$

2. $y^2 + 3y + 18 = (y - 6)(y + 3)$

3. $x^2 - 10x - 24 = (x - 6)(x + 4)$

4. $a^2 + 9a - 36 = (a - 12)(a + 4)$

5. $x^2 - 16x + 64 = (x - 8)(x - 8)$

6. $w^2 - 12w - 45 = (w + 9)(w - 5)$

7. $25y^2 - 10y + 1 = (5y - 1)(5y + 1)$

8. $6x^2 + 5xy - 4y^2 = (6x - 2y)(x + 2y)$

9. $10p^2 - pq - 3q^2 = (5p - 3q)(2p + q)$

10. $6a^2 + 13a + 6 = (2a - 3)(3a - 2)$

For each of the following trinomials, label a, b, and c.

11. $x^2 + 7x - 5$

12. $x^2 + 5x + 11$

13. $x^2 - 3x + 8$

14. $x^2 + 7x - 15$

15. $3x^2 + 5x - 8$

16. $3x^2 - 5x + 7$

17. $4x^2 + 8x + 11$

18. $5x^2 + 7x - 9$

19. $-7x^2 - 5x + 2$

20. $-7x^2 + 9x - 18$

Use the ac test to determine which of the following trinomials can be factored. Find the values of m and n for each trinomial that can be factored.

21. $x^2 + x - 6$

22. $x^2 - x - 6$

23. $x^2 + 3x - 1$

24. $x^2 - 3x + 7$

25. $x^2 - 5x + 6$

26. $x^2 - x + 2$

27. $2x^2 + 5x - 3$

28. $3x^2 - 14x - 5$

29. $6x^2 - 19x + 10$

30. $4x^2 + 5x + 6$

Rewrite the middle term as the sum of two terms and then factor by grouping.

31. $x^2 + 6x + 8$

32. $x^2 + 3x - 10$

33. $x^2 - 9x + 20$

34. $x^2 - 8x + 15$

35. $x^2 - 2x - 63$

36. $x^2 + 6x - 55$

Rewrite the middle term as the sum of two terms and then factor completely.

37. $x^2 + 10x + 24$ **38.** $x^2 - 11x + 24$ **39.** $x^2 - 11x + 28$ **40.** $y^2 - y - 20$

41. $s^2 + 13s + 30$ **42.** $b^2 + 11b + 28$ **43.** $a^2 - 2a - 48$ **44.** $x^2 - 17x + 60$

45. $x^2 - 8x + 7$ **46.** $x^2 + 7x - 18$ **47.** $x^2 - 7x - 18$ **48.** $x^2 - 11x + 10$

49. $x^2 - 14x + 49$ **50.** $s^2 - 4s - 32$ **51.** $p^2 - 10p - 24$ **52.** $x^2 - 11x - 60$

53. $x^2 + 5x - 66$ **54.** $a^2 - 5a - 24$ **55.** $c^2 + 19c + 60$ **56.** $t^2 - 4t - 60$

57. $n^2 + 5n - 50$ **58.** $x^2 - 16x + 63$ **59.** $x^2 + 7xy + 10y^2$ **60.** $x^2 - 8xy + 12y^2$

61. $x^2 + xy - 12y^2$ **62.** $m^2 - 8mn + 16n^2$ **63.** $x^2 - 13xy + 40y^2$ **64.** $r^2 - 9rs - 36s^2$

65. $6x^2 + 19x + 10$ **66.** $6x^2 - 7x - 3$ **67.** $15x^2 + x - 6$ **68.** $12w^2 + 19w + 4$

69. $6m^2 + 25m - 25$ **70.** $12x^2 + x - 20$ **71.** $9x^2 - 12x + 4$ **72.** $20x^2 - 23x + 6$

73. $12x^2 - 8x - 15$ **74.** $16a^2 + 40a + 25$ **75.** $3y^2 + 7y - 6$ **76.** $12x^2 + 11x - 15$

77. $8x^2 - 27x - 20$ **78.** $24v^2 + 5v - 36$ **79.** $2x^2 + 3xy + y^2$ **80.** $3x^2 - 5xy + 2y^2$

81. $5a^2 - 8ab - 4b^2$ **82.** $5x^2 + 7xy - 6y^2$ **83.** $9x^2 + 4xy - 5y^2$ **84.** $16x^2 + 32xy + 15y^2$

85. $6m^2 - 17mn + 12n^2$ **86.** $15x^2 - xy - 6y^2$ **87.** $36a^2 - 3ab - 5b^2$ **88.** $10q^2 + 14qr - 12r^2$

89. $x^2 + 4xy + 4y^2$ **90.** $25b^2 - 80bc + 64c^2$ **91.** $20x^2 - 20x - 15$ **92.** $24x^2 - 18x - 6$

93. $8m^2 + 12m + 4$ **94.** $14x^2 - 20x + 6$ **95.** $15r^2 - 21rs + 6s^2$ **96.** $10x^2 + 5xy - 30y^2$

97. $2x^3 - 2x^2 - 4x$ **98.** $2y^3 + y^2 - 3y$ **99.** $2y^4 + 5y^3 + 3y^2$ **100.** $4z^3 - 18z^2 - 10z$

101. $36a^3 - 66a^2 + 18a$ **102.** $20n^4 - 22n^3 - 12n^2$ **103.** $9p^2 + 30pq + 21q^2$ **104.** $12x^2 + 2xy - 24y^2$

Find a positive integer value for k for which each of the following can be factored.

105. $x^2 + kx + 8$ **106.** $x^2 + kx + 9$ **107.** $x^2 - kx + 16$ **108.** $x^2 - kx + 17$

109. $x^2 - kx - 5$ **110.** $x^2 - kx - 7$ **111.** $x^2 + 3x + k$ **112.** $x^2 + 5x + k$

113. $x^2 + 2x - k$ **114.** $x^2 + x - k$

In Exercises 115 to 118, (a) factor the given function; (b) identify the values of x for which $f(x) = 0$; (c) graph $f(x)$ on a graphing calculator and determine where the graph of $f(x)$ crosses the x axis; and (d) compare the results of (b) and (c).

115. $f(x) = x^2 - 2x - 8$ **116.** $f(x) = x^2 - 3x - 10$

117. $f(x) = 2x^2 - x - 3$ **118.** $f(x) = 3x^2 - x - 2$

SECTION 6.3* Factoring Trinomials: Trial and Error

 OBJECTIVES

> 1. Factor a trinomial of the form $ax^2 + bx + c$
> 2. Completely factor a trinomial

Recall that the product of two binomials may be a trinomial of the form

$$ax^2 + bx + c$$

This suggests that some trinomials may be factored as the product of two binomials. And, in fact, factoring trinomials in this way is probably the most common type of factoring that you will encounter in algebra. One process for factoring a trinomial into a product of two binomials is called *trial and error*.

As before, let's introduce the factoring technique with an example from multiplication. Consider

$$(x + 3)(x + 4) = x^2 + 4x + 3x + 12$$
$$= x^2 + 7x + 12$$

Product of first terms, x and x

Sum of inner and outer products, $3x$ and $4x$

Product of last terms, 3 and 4

To reverse the multiplication process to one of factoring, we see that the product of the *first* terms of the binomial factors is the *first* term of the given trinomial, the product of the *last* terms of the binomial factors is the *last* term of the trinomial, and the *middle* term of the trinomial must equal the sum of the *outer* and *inner* products. That leads us to the following sign patterns in factoring a trinomial.

Factoring Trinomials

		Factoring Sign Pattern
$x^2 + bx + c$	Both signs are positive.	$(x + \ )(x + \ )$
$x^2 - bx + c$	The constant is positive, and the x coefficient is negative.	$(x - \ )(x - \ )$
$x^2 + bx - c$ or $x^2 - bx - c$	The constant is negative.	$(x + \ )(x - \ )$

466

6-32

Given the above information, let's work through an example.

Example 1

Factoring Trinomials, $a = 1$

To factor

$$x^2 + 7x + 10$$

the desired sign pattern is

$$(x + \underline{\ \ })(x + \underline{\ \ })$$

From the constant 10 and the x coefficient of our original trinomial 7, for the second terms of the binomial factors, we want two integers whose product is 10 and whose sum is 7.

With practice, you will do much of this work mentally. We show the factors and their sums here, and in later examples, to emphasize the process.

Consider the following:

Factors of 10	Sum
1, 10	11
2, 5	7

We can see that the correct factorization is

Note: To check, multiply the factors, using the method of Section 4.5.

$$x^2 + 7x + 10 = (x + 2)(x + 5)$$

  CHECK YOURSELF 1

Factor $x^2 + 8x + 15$.

Example 2

Factoring Trinomials, $a = 1$

Factor $x^2 - 9x + 14$. Do you see that the sign pattern must be as follows?

$$(x - \underline{\ \ })(x - \underline{\ \ })$$

We then want two factors of 14 whose sum is -9.

Factors of 14	Sum
$-1, -14$	-15
$-2, -7$	-9

Here we use two negative factors of 14 since the coefficient of the x term is negative while the constant is positive.

Since the desired middle term is $-9x$, the correct factors are

$$x^2 - 9x + 14 = (x - 2)(x - 7)$$

CHECK YOURSELF 2

Factor $x^2 - 12x + 32$.

Let's turn now to applying our factoring technique to a trinomial whose constant term is negative. Consider Example 3.

| Example 3 | **Factoring Trinomials, $a = 1$** |

Factor

$$x^2 + 4x - 12$$

In this case, the sign pattern is

Since the constant is now negative, the signs in the binomial factors must be *opposite*.

$$(x - \underline{\ \ })(x + \underline{\ \ })$$

Here we want two integers whose product is -12 and whose sum is 4. Again let's look at the possible factors:

Factors of -12	Sum
$1, -12$	-11
$-1, 12$	11
$3, -4$	-1
$-3, 4$	1
$2, -6$	-4
$-2, 6$	4

From the information above, we see that the correct factors are

$$x^2 + 4x - 12 = (x - 2)(x + 6)$$

✔ *CHECK YOURSELF 3*

Factor $x^2 - 7x - 18$.

Thus far we have considered only trinomials of the form $x^2 + bx + c$. Suppose that the leading coefficient is *not* 1. In general, to factor the trinomial $ax^2 + bx + c$ (with $a \neq 1$), we must consider binomial factors of the form

$$(\underline{\ \ } x + \underline{\ \ })(\underline{\ \ } x + \underline{\ \ })$$

where one or both of the coefficients of x in the binomial factors are greater than 1. Again let's look at a multiplication example for some clues to the technique. Consider

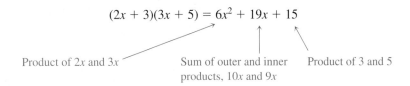

$$(2x + 3)(3x + 5) = 6x^2 + 19x + 15$$

Product of $2x$ and $3x$ Sum of outer and inner   Product of 3 and 5
products, $10x$ and $9x$

Now, to reverse the process to factoring, we can proceed as in Example 4.

| Example 4 | **Factoring Trinomials, $a \neq 1$** |

To factor $5x^2 + 9x + 4$, we must have the pattern

$$(\_x + \_)(\_x + \_)$$ This product must be 4.

This product must be 5.

	Factors of 5	Factors of 4
	1, 5	1, 4
		4, 1
		2, 2

Now that the lead coefficient is no longer 1, we must be prepared to try both 1, 4 and 4, 1.

Therefore the possible binomial factors are

$$(x + 1)(5x + 4)$$

$$(x + 4)(5x + 1)$$

$$(x + 2)(5x + 2)$$

Checking the middle terms of each product, we see that the proper factorization is

$$5x^2 + 9x + 4 = (x + 1)(5x + 4)$$

✔ *CHECK YOURSELF 4*

Factor $6x^2 - 17x + 7$.

The sign patterns discussed earlier remain the same when the leading coefficient is not 1. Look at Example 5 involving a trinomial with a negative constant.

| Example 5 | Factoring Trinomials, $a \neq 1$ |

(a) Factor $6x^2 + 7x - 3$. The sign patterns are

$$(\_\_x + \_\_)(\_\_x - \_\_)$$

Factors of 6	Factors of -3
1, 6	1, -3
2, 3	-1, 3

Again, as the number of factors for the first coefficient and the constant increase, the number of possible factors becomes larger. Can we reduce the search? One clue: If the trinomial has no common factors (other than 1), then a binomial factor can have no common factor. This means that $6x - 3$, $6x + 3$, $3x - 3$, and $3x + 3$ need not be considered. They are shown here to completely illustrate the possibilities.

There are eight possible binomial factors:

$$(x + 1)(6x - 3)$$
$$(x - 1)(6x + 3)$$
$$(x + 3)(6x - 1)$$
$$(x - 3)(6x + 1)$$
$$(2x + 1)(3x - 3)$$
$$(2x - 1)(3x + 3)$$
$$(3x + 1)(2x - 3)$$
$$(3x - 1)(2x + 3)$$

Again, checking the middle terms, we have the correct factors:

$$6x^2 + 7x - 3 = (3x - 1)(2x + 3)$$

Factoring certain trinomials in more than one variable involves similar techniques, as illustrated below.

(b) Factor

$$4x^2 - 16xy + 7y^2$$

From the first term of the trinomial we see that possible first terms for our binomial factors are $4x$ and x or $2x$ and $2x$. The last term of the trinomial tells us that the only choices for the last terms of our binomial factors are y and $7y$. So given the sign of the middle and last terms, the only possible factors are

Find the middle term of each product.

$$(4x - 7y)(x - y)$$
$$(4x - y)(x - 7y)$$
$$(2x - 7y)(2x - y)$$

From the middle term of our original trinomial we see that $2x - 7y$ and $2x - y$ are the proper factors.

✔   *CHECK YOURSELF 5*

Factor $6a^2 + 11ab - 10b^2$.

Recall our earlier comment that the *first step* in any factoring problem is to remove any existing common factors. As before, it may be necessary to combine common-term factoring with other methods (such as factoring a trinomial into a product of binomials) to completely factor a polynomial. Look at Example 6.

Example 6	## Factoring Trinomials with a Common Factor

(a) Factor

$$2x^2 - 16x + 30$$

First note the common factor of 2. So we can write

Factor out the common factor of 2.

$$2x^2 - 16x + 30 = 2(x^2 - 8x + 15)$$

Now, as the second step, examine the trinomial factor. By our earlier methods we know that

$$x^2 - 8x + 15 = (x - 3)(x - 5)$$

and we have

$$2x^2 - 16x + 30 = 2(x - 3)(x - 5)$$

in completely factored form.

(b) Factor

$$6x^3 + 15x^2y - 9xy^2$$

As before, note the common factor of $3x$ in each term of the trinomial. Factoring out that common factor, we have

$$6x^3 + 15x^2y - 9xy^2 = 3x(2x^2 + 5xy - 3y^2)$$

Again, considering the trinomial factor, we see that $2x^2 + 5xy - 3y^2$ has factors of $2x - y$ and $x + 3y$. And our original trinomial becomes

$$3x(2x - y)(x + 3y)$$

in completely factored form.

✔ *CHECK YOURSELF 6*

Factor.

(a) $9x^2 - 39x + 36$ (b) $24a^3 + 4a^2b - 8ab^2$

One final note. When factoring, we require that all coefficients be integers. Given this restriction, not all polynomials are factorable over the integers. The following illustrates.

To factor $x^2 - 9x + 12$, we know that the only possible binomial factors (using integers as coefficients) are

$$(x - 1)(x - 12)$$
$$(x - 2)(x - 6)$$
$$(x - 3)(x - 4)$$

You can easily verify that *none* of these pairs gives the correct middle term of $-9x$. We then say that the original trinomial is not factorable using integers as coefficients.

✔ *CHECK YOURSELF ANSWERS*

1. $(x + 3)(x + 5)$ **2.** $(x - 4)(x - 8)$ **3.** $(x - 9)(x + 2)$

4. $(2x - 1)(3x - 7)$ **5.** $(3a - 2b)(2a + 5b)$

6. (a) $3(x - 3)(3x - 4)$; (b) $4a(3a + 2b)(2a - b)$

Exercises · 6.3*

State whether each of the following is true or false.

1. $x^2 + 2x - 3 = (x + 3)(x - 1)$

2. $y^2 + 3y + 18 = (y - 6)(y + 3)$

3. $x^2 - 10x - 24 = (x - 6)(x + 4)$

4. $a^2 + 9a - 36 = (a - 12)(a + 4)$

5. $x^2 - 16x + 64 = (x - 8)(x - 8)$

6. $w^2 - 12w - 45 = (w + 9)(w - 5)$

7. $25y^2 - 10y + 1 = (5y - 1)(5y + 1)$

8. $6x^2 + 5xy - 4y^2 = (6x - 2y)(x + 2y)$

9. $10p^2 - pq - 3q^2 = (5p - 3q)(2p + q)$

10. $6a^2 + 13a + 6 = (2a - 3)(3a - 2)$

For each of the following trinomials, label a, b, and c.

11. $x^2 + 7x - 5$

12. $x^2 + 5x + 11$

13. $x^2 - 3x + 8$

14. $x^2 + 7x - 15$

15. $3x^2 + 5x - 8$

16. $3x^2 - 5x + 7$

17. $4x^2 + 8x + 11$

18. $5x^2 + 7x - 9$

19. $-7x^2 - 5x + 2$

20. $-7x^2 + 9x - 18$

Factor completely.

21. $x^2 + 10x + 24$

22. $x^2 + 3x - 10$

23. $x^2 - 9x + 20$

24. $x^2 - 8x + 15$

25. $x^2 - 2x - 63$

26. $x^2 + 6x - 55$

27. $x^2 + 8x + 14$

28. $x^2 - 11x + 24$

29. $x^2 - 11x + 28$

30. $y^2 - y - 21$

31. $s^2 + 13s + 30$

32. $b^2 + 11b + 28$

33. $a^2 - 2a - 48$

34. $x^2 - 17x + 60$

35. $x^2 - 8x + 7$

36. $x^2 + 7x - 18$

37. $x^2 + 11x + 24$

38. $x^2 - 11x + 10$

39. $x^2 - 14x + 49$

40. $s^2 - 4s - 32$

41. $p^2 - 10p - 28$

42. $x^2 - 11x - 60$

43. $x^2 + 5x - 66$

44. $a^2 - 5a - 24$

45. $c^2 + 19c + 60$

46. $t^2 - 4t - 60$

47. $n^2 + 5n - 50$

48. $x^2 - 16x + 65$

49. $x^2 + 7xy + 10y^2$

50. $x^2 - 8xy + 12y^2$

51. $x^2 + xy - 12y^2$

52. $m^2 - 8mn + 16n^2$

53. $x^2 - 13xy + 40y^2$

54. $r^2 - 9rs - 36s^2$

55. $6x^2 + 19x + 10$

56. $6x^2 - 7x - 3$

57. $15x^2 + x - 6$ **58.** $12w^2 + 19w + 4$ **59.** $6m^2 + 25m - 25$ **60.** $12x^2 + x - 20$

61. $9x^2 - 12x + 4$ **62.** $20x^2 - 23x + 6$ **63.** $12x^2 - 8x - 15$ **64.** $16a^2 + 40a + 25$

65. $3y^2 + 7y - 6$ **66.** $12x^2 + 11x - 15$ **67.** $8x^2 - 27x - 20$ **68.** $24v^2 + 5v - 36$

69. $2x^2 + 3xy + y^2$ **70.** $3x^2 - 5xy + 2y^2$ **71.** $5a^2 - 8ab - 4b^2$ **72.** $5x^2 + 7xy - 6y^2$

73. $9x^2 + 4xy - 5y^2$ **74.** $16x^2 + 32xy + 15y^2$ **75.** $6m^2 - 17mn + 12n^2$ **76.** $15x^2 - xy - 6y^2$

77. $36a^2 - 3ab - 5b^2$ **78.** $10q^2 + 14qr - 12r^2$ **79.** $x^2 + 4xy + 4y^2$ **80.** $25b^2 - 80bc + 64c^2$

81. $20x^2 - 20x - 15$ **82.** $24x^2 - 18x - 6$ **83.** $8m^2 + 12m + 4$ **84.** $14x^2 - 20x + 6$

85. $15r^2 - 21rs + 6s^2$ **86.** $10x^2 + 5xy - 30y^2$ **87.** $2x^3 - 2x^2 - 4x$ **88.** $2y^3 + y^2 - 3y$

89. $2y^4 + 5y^3 + 3y^2$ **90.** $4z^3 - 18z^2 - 10z$ **91.** $36a^3 - 66a^2 + 18a$ **92.** $20n^4 - 22n^3 - 12n^2$

93. $9p^2 + 30pq + 21q^2$ **94.** $12x^2 + 2xy - 24y^2$

Find a positive integer value for k for which each of the following can be factored.

95. $x^2 + kx + 8$ **96.** $x^2 + kx + 9$ **97.** $x^2 - kx + 16$ **98.** $x^2 - kx + 17$

99. $x^2 - kx - 5$ **100.** $x^2 - kx - 7$ **101.** $x^2 + 3x + k$ **102.** $x^2 + 5x + k$

103. $x^2 + 2x - k$ **104.** $x^2 + x - k$

In each of the following, (a) factor the given function; (b) identify the values of x for which $f(x) = 0$; (c) graph $f(x)$ on a graphing calculator and determine where the graph crosses the x axis; and (d) compare the results of (b) and (c).

105. $f(x) = x^2 - 2x - 8$ **106.** $f(x) = x^2 - 3x - 10$

107. $f(x) = 2x^2 - x - 3$ **108.** $f(x) = 3x^2 - x - 2$

Solving Quadratic Equations by Factoring

OBJECTIVES

1. Solve quadratic equations by factoring
2. Find the zeros of a quadratic function

The factoring techniques you have learned provide us with tools for solving equations that can be written in the form

$$ax^2 + bx + c = 0 \qquad a \neq 0$$

This is a quadratic equation in one variable, here x. You can recognize such a quadratic equation by the fact that the highest power of the variable x is the second power.

where a, b, and c are constants.

An equation written in the form $ax^2 + bx + c = 0$ is called a **quadratic equation in standard form.** Using factoring to solve quadratic equations requires the **zero-product principle,** which says that if the product of two factors is 0, then one or both of the factors must be equal to 0. In symbols:

Zero-Product Principle

If $a \cdot b = 0$, then $a = 0$ or $b = 0$ or $a = b = 0$.

Let's see how the principle is applied to solving quadratic equations.

| Example 1 | **Solving Equations by Factoring** |

Solve

$$x^2 - 3x - 18 = 0$$

Factoring on the left, we have

To use the zero-product principle, 0 must be on one side of the equation.

$$(x - 6)(x + 3) = 0$$

By the zero-product principle, we know that one or both of the factors must be zero. We can then write

$$x - 6 = 0 \qquad \text{or} \qquad x + 3 = 0$$

Graph the function

$$y = x^2 - 3x - 18$$

on your graphing calculator. The solutions to the equation $0 = x^2 - 3x - 18$ will be those x values of the points on the curve at which $y = 0$. Those are the points at which the graph intercepts the x axis.

Solving each equation gives

$$x = 6 \qquad \text{or} \qquad x = -3$$

The two solutions are 6 and -3, and the solution set is written as $\{x \mid x = -3, 6\}$ or simply $\{-3, 6\}$.

The solutions are sometimes called the **zeros,** or **roots,** of the equation. They represent the x coordinates of the points where the graph of the equation $y = x^2 - 3x - 18$ crosses the x axis. In this case, the graph crosses the x axis at $(-3, 0)$ and $(6, 0)$.

Quadratic equations can be checked in the same way as linear equations were checked: by substitution. For instance, if $x = 6$, we have

$$6^2 - 3 \cdot 6 - 18 \overset{?}{=} 0$$
$$36 - 18 - 18 \overset{?}{=} 0$$
$$0 = 0$$

which is a true statement. We leave it to you to check the solution of -3.

 CHECK YOURSELF 1

Solve $x^2 - 9x + 20 = 0$.

Other factoring techniques are also used in solving quadratic equations. Example 2 illustrates this concept.

Example 2

Solving Equations by Factoring

(a) Solve $x^2 - 5x = 0$.

Caution

A *common mistake* is to forget the statement $x = 0$ when you are solving equations of this type. Be sure to include the *two* values of x that satisfy the equation: $x = 0$ and $x = 5$.

Again, factor the left side of the equation and apply the zero-product principle.

$$x(x - 5) = 0$$

Now

$$x = 0 \qquad \text{or} \qquad x - 5 = 0$$
$$x = 5$$

The two solutions are 0 and 5. The solution set is $\{0, 5\}$.

(b) Solve $x^2 - 9 = 0$.

Factoring yields

$$(x + 3)(x - 3) = 0$$
$$x + 3 = 0 \qquad \text{or} \qquad x - 3 = 0$$
$$x = -3 \qquad\qquad x = 3$$

The symbol $\pm$ is read "plus or minus."

The solution set is $\{-3, 3\}$, which may be written as $\{\pm 3\}$.

✔ CHECK YOURSELF 2

Solve by factoring.

(a) $x^2 + 8x = 0$ (b) $x^2 - 16 = 0$

Example 3 illustrates a crucial point. Our solving technique depends on the zero-product principle, which means that the product of factors *must be equal to 0*. The importance of this is shown in Example 3.

Example 3

Caution

Consider the equation

$x(2x - 1) = 3$

Students are sometimes tempted to write

$x = 3$ or $2x - 1 = 3$

This is *not correct*.

If $a \cdot b = 0$, then a or b **must** be 0. But if $a \cdot b = 3$, we do not know that a or b is 3; for example, it could be that $a = 6$ and $b = 1/2$ (or many other possibilities).

Solving Equations by Factoring

Solve $2x^2 - x = 3$.

The first step in solving is to write the equation in standard form (i.e., when one side of the equation is 0). So start by adding -3 to both sides of the equation. Then

$$2x^2 - x - 3 = 0$$ Make sure all terms are on one side of the equation. The other side will be 0.

You can now factor and solve using the zero-product principle.

$$(2x - 3)(x + 1) = 0$$

$$2x - 3 = 0 \quad \text{or} \quad x + 1 = 0$$

$$2x = 3 \qquad\qquad x = -1$$

$$x = \frac{3}{2}$$

The solution set is $\left\{ \dfrac{3}{2}, -1 \right\}$.

 CHECK YOURSELF 3

Solve $3x^2 = 5x + 2$.

In all the previous examples, the quadratic equations had two distinct real number solutions. That may or may not always be the case, as we shall see.

Example 4

Solving Equations by Factoring

Solve $x^2 - 6x + 9 = 0$.

Factoring, we have

$$(x - 3)(x - 3) = 0$$

and

$$x - 3 = 0 \quad \text{or} \quad x - 3 = 0$$
$$x = 3 \qquad\qquad x = 3$$

The solution set is $\{3\}$.

A quadratic (or second-degree) equation always has *two* solutions. When an equation such as this one has two solutions that are the same number, we call 3 the **repeated** (or **double**) **solution** of the equation.

Even though a quadratic equation will always have two solutions, they may not always be real numbers. More is said about this in a later section.

 CHECK YOURSELF 4

Solve $x^2 + 6x + 9 = 0$.

Always examine the quadratic member of an equation for common factors. It will make your work much easier, as Example 5 illustrates.

Example 5

Solving Equations by Factoring

Solve $3x^2 - 3x - 60 = 0$.

First, note the common factor 3 in the quadratic member of the equation. Factoring out the 3, we have

$$3(x^2 - x - 20) = 0$$

Now divide both sides of the equation by 3.

Note the advantage of dividing both members by 3. The coefficients in the quadratic member become smaller, and that member is much easier to factor.

$$\frac{3(x^2 - x - 20)}{3} = \frac{0}{3}$$

or

$$x^2 - x - 20 = 0$$

We can now factor and solve as before.

$$(x - 5)(x + 4) = 0$$
$$x - 5 = 0 \quad \text{or} \quad x + 4 = 0$$
$$x = 5 \qquad\qquad x = -4 \quad \text{or} \quad \{-4, 5\}$$

 CHECK YOURSELF 5

Solve $2x^2 - 10x - 48 = 0$.

In Chapter 5, we introduced the concept of a function and expressed the equation of a line in function form. Another type of function is called a quadratic function.

> A **quadratic function** is a function that can be written in the form
>
> $$f(x) = ax^2 + bx + c$$
>
> where a, b, and c are real numbers and $a \neq 0$.

For example, $f(x) = 3x^2 - 2x - 1$ and $g(x) = x^2 - 2$ are quadratic functions. In working with functions, we often want to find the values of x for which $f(x) = 0$. These values are called the **zeros of the function.** They represent the x coordinates of the points where the graph of the function crosses the x axis. To find the zeros of a quadratic function, a quadratic equation must be solved.

Example 6

Finding the Zeros of a Function

Find the zeros of $f(x) = x^2 - x - 2$.

To find the zeros of $f(x) = x^2 - x - 2$, we must solve the quadratic equation $f(x) = 0$.

$$f(x) = 0$$
$$x^2 - x - 2 = 0$$
$$(x - 2)(x + 1) = 0$$

$$x - 2 = 0 \quad \text{or} \quad x + 1 = 0$$
$$x = 2 \qquad\qquad x = -1 \quad \text{or} \quad \{-1, 2\}$$

The zeros of the function are -1 and 2.

The graph of

$$f(x) = x^2 - x - 2$$

intercepts the x axis at the points $(2, 0)$ and $(-1, 0)$, so 2 and -1 are the *zeros* of the function.

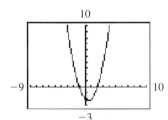

✔ *CHECK YOURSELF 6*

Find the zeros of $f(x) = 2x^2 - x - 3$.

✔ *CHECK YOURSELF ANSWERS*

1. $\{4, 5\}$ **2.** (a) $\{-8, 0\}$; (b) $\{-4, 4\}$ **3.** $\left\{-\dfrac{1}{3}, 2\right\}$ **4.** $\{-3\}$

5. $\{-3, 8\}$ **6.** $\left\{-1, \dfrac{3}{2}\right\}$

Solve the quadratic equations by factoring.

1. $x^2 - 3x - 10 = 0$ **2.** $x^2 - 5x + 4 = 0$ **3.** $x^2 - 2x - 15 = 0$

4. $x^2 + 4x - 32 = 0$ **5.** $x^2 - 11x + 30 = 0$ **6.** $x^2 + 13x + 36 = 0$

7. $x^2 - 4x - 21 = 0$ **8.** $x^2 + 5x - 36 = 0$ **9.** $x^2 - 5x = 50$

10. $x^2 + 14x = -33$ **11.** $x^2 = 5x + 84$ **12.** $x^2 = 6x + 27$

13. $x^2 - 8x = 0$ **14.** $x^2 + 7x = 0$ **15.** $x^2 + 10x = 0$

16. $x^2 - 9x = 0$ **17.** $x^2 = 5x$ **18.** $x^2 = 11x$

19. $x^2 - 25 = 0$ **20.** $x^2 - 49 = 0$ **21.** $9x^2 = 25$

22. $x^2 = 169$ **23.** $4x^2 + 12x + 9 = 0$ **24.** $9x^2 - 30x + 25 = 0$

25. $2x^2 - 17x + 36 = 0$ **26.** $5x^2 + 17x - 12 = 0$ **27.** $5x^2 + 9x = 18$

28. $12x^2 = 25x - 12$ **29.** $6x^2 = 7x - 2$ **30.** $4x^2 - 3 = x$

31. $2m^2 = 12m + 54$ **32.** $5x^2 - 55x = 60$ **33.** $7x^2 - 63x = 0$

34. $6x^2 - 9x = 0$ **35.** $5x^2 = 15x$ **36.** $7x^2 = -49x$

37. $x(x + 2) = 15$ **38.** $x(x + 3) = 28$ **39.** $x(2x - 3) = 9$

40. $x(3x + 1) = 52$ **41.** $2x(3x + 1) = 28$ **42.** $3x(2x - 1) = 30$

43. $(x - 3)(x - 1) = 15$ **44.** $(x + 3)(x - 2) = 14$ **45.** $(x - 5)(x + 2) = 18$

46. $(3x - 5)(x + 2) = 14$

Find the zeros of the functions.

47. $f(x) = 3x^2 - 24x + 36$ **48.** $f(x) = 2x^2 - 6x - 56$ **49.** $f(x) = 4x^2 + 16x - 20$

50. $f(x) = 3x^2 - 33x + 54$

51. Explain the differences between solving the equations $3(x - 2)(x + 5) = 0$ and $3x(x - 2)(x + 5) = 0$.

52. How can a graphing calculator be used to determine the zeros of a quadratic function?

Write an equation that has the given solutions. (*Hint:* Write the binomial factors and then find their product.)

53. $\{-4, 5\}$ **54.** $\{0, 5\}$ **55.** $\{2, 6\}$ **56.** $\{-4, 4\}$

The zero-product rule can be extended to three or more factors. If $a \cdot b \cdot c = 0$, then at least one of these factors is 0. In Exercises 57 to 60, use this information to solve the equations.

57. $x^3 - 3x^2 - 10x = 0$ **58.** $x^3 + 8x^2 + 15x = 0$ **59.** $x^3 - 9x = 0$ **60.** $x^3 = 16x$

In Exercises 61 to 64, extend the ideas in the previous exercises to find solutions for the following equations. (*Hint:* Apply factoring by grouping in Exercises 61 and 62.)

61. $x^3 + x^2 - 4x - 4 = 0$ **62.** $x^3 - 5x^2 - x + 5 = 0$ **63.** $x^4 - 10x^2 + 9 = 0$ **64.** $x^4 - 5x^2 + 4 = 0$

The net productivity of a forested wetland as related to the amount of water moving through the wetland can be expressed by a quadratic equation. In Exercises 65 to 68, if y represents the amount of wood produced and x represents the amount of water present, in cubic centimeters (cm^3), determine where the productivity is zero in each wetland represented by the equations.

65. $y = -3x^2 + 300x$ **66.** $y = -4x^2 + 500x$ **67.** $y = -6x^2 + 792x$ **68.** $y = -7x^2 + 1,022x$

69. Break-even analysis. The manager of a bicycle shop knows that the cost of selling x bicycles is $C = 20x + 60$ and the revenue from selling x bicycles is $R = x^2 - 8x$. Find the break-even value of x. (Recall that break-even occurs when cost equals revenue.)

70. Break-even analysis. A company that produces computer games has found that its daily operating cost in dollars is $C = 40x + 150$ and its daily revenue in dollars is $R = 65x - x^2$. For what value(s) of x will the company break even?

Graphing calculator. In Exercises 71 and 72, use your calculator to graph $f(x)$.

71. $f(x) = 3x^2 - 24x + 36$. Find the x values at which the graph crosses the x axis. Compare your answers to the solution for Exercise 47.

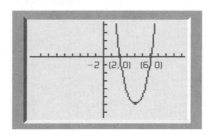

72. $f(x) = 2x^2 - 6x - 56$. Find the x values at which the graph crosses the x axis. Compare your answer to the solution for Exercise 48.

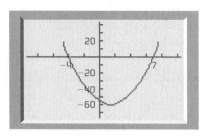

73. Work with another student and use your calculator to solve this exercise. Assume that each graph is the graph of a quadratic function. Find an equation for each graph given. There could be more than one equation for some of the graphs. Remember the connection between the x intercepts and the zeros.

(a)
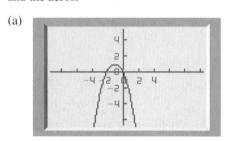

(b)
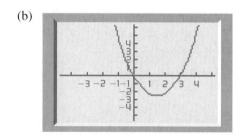

(c)
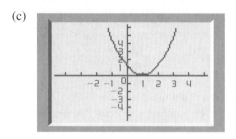

(d)
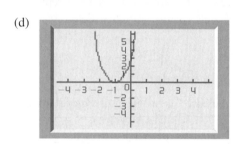

(e)
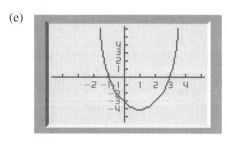

(f)
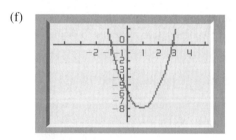

(g)
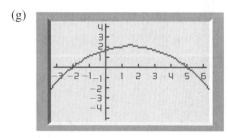

(h)
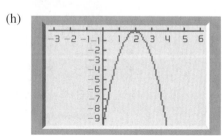

Problem Solving with Factoring

1. Solve a variety of applications, using factoring

With the techniques introduced in this chapter for solving equations by factoring, we can now examine a new group of applications. Recall that the key to problem solving lies in a step-by-step, organized approach to the process. You might want to take time now to review the five-step process introduced in Section 2.1. All the examples in this section make use of that model.

You will find that these applications typically lead to quadratic equations that may be solved by factoring. Remember that such quadratic equations will have two, one, or no distinct real number solutions. Step 5, the process of verifying or checking your solutions, is particularly important here. By checking solutions, you may find that both, only one, or none of the derived solutions satisfy the physical conditions stated in the original problem.

We will begin with a numerical application.

Example 1

Solving a Number Application

One integer is 3 less than twice another. If their product is 35, find the two integers.

Step 1 The unknowns are the two integers.

Step 2 Let x represent the first integer. Then

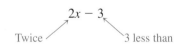

$$2x - 3$$

Twice ⟋   ⟍ 3 less than

represents the second.

Step 3 Form an equation:

$$\underbrace{x(2x - 3)}_{\text{Product of the two integers}} = 35$$

Step 4 Remove the parentheses and solve as before.

$$2x^2 - 3x = 35$$

$$2x^2 - 3x - 35 = 0$$

Factor on the left:

$$(2x + 7)(x - 5) = 0$$

$$2x + 7 = 0 \qquad \text{or} \qquad x - 5 = 0$$

$$2x = -7 \qquad\qquad\qquad x = 5$$

$$x = -\frac{7}{2}$$

Note: These represent solutions to the equation, but not the answer to the original problem.

Step 5   From step 4, we have the two solutions: $-\dfrac{7}{2}$ and 5. Since the original problem asks for *integers,* we must consider only 5 as a solution.

Using 5 for x, we see that the other integer is $2x - 3 = 2(5) - 3 = 7$. So the desired integers are 5 and 7.

To verify, note that (a) 7 is 3 less than 2 times 5, and (b) the product of 5 and 7 is 35.

 CHECK YOURSELF 1

One integer is 2 more than 3 times another. If their product is 56, what are the two integers?

Problems involving consecutive integers may also lead to quadratic equations. Recall that consecutive integers can be represented by x, $x + 1$, $x + 2$, and so on. Consecutive even (or odd) integers are represented by x, $x + 2$, $x + 4$, and so on.

Example 2

Solving a Number Application

The sum of the squares of two consecutive integers is 85. What are the two integers?

Step 1   The unknowns are the two consecutive integers.

Step 2   Let x be the first integer and $x + 1$ the second integer.

Step 3   Form the equation.

$$\underbrace{x^2 + (x + 1)^2}_{\text{Sum of squares}} = 85$$

Step 4   Solve.

$$x^2 + (x + 1)^2 = 85 \qquad \text{Remove parentheses.}$$

$$x^2 + x^2 + 2x + 1 = 85$$

$$2x^2 + 2x - 84 = 0 \qquad \text{Note the common factor 2, and divide}$$

$$x^2 + x - 42 = 0 \qquad \text{both sides of the equation by 2.}$$

$$(x + 7)(x - 6) = 0$$

$$x + 7 = 0 \qquad \text{or} \qquad x - 6 = 0$$

$$x = -7 \qquad\qquad\qquad x = 6$$

Step 5 The solution in step 4 leads to two possibilities, -7 or 6. Since *both numbers are integers,* both meet the conditions of the original problem. There are then two pairs of consecutive integers that work:

$$-7\ (x) \qquad \text{and} \qquad -6\ (x+1)$$

or $\qquad\qquad\qquad 6\ (x) \qquad \text{and} \qquad 7\ (x+1)$

You should confirm that in both cases the sum of the squares is 85.

✔ *CHECK YOURSELF 2*

The sum of the squares of two consecutive even integers is 100. Find the two integers.

We proceed now to applications involving geometry.

Example 3	## Solving a Geometric Application

The length of a rectangle is 3 cm greater than its width. If the area of the rectangle is 108 cm^2, what are the dimensions of the rectangle?

Step 1 You are asked to find the dimensions (the length and the width) of the rectangle.

Step 2 Whenever geometric figures are involved in an application, start by drawing, and *then labeling,* a sketch of the problem. Let x represent the width and $x+3$ the length.

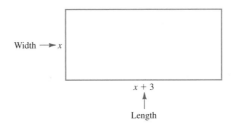

Step 3 Once the drawing is correctly labeled, this step should be easy. The area of a rectangle is the product of its length and width, so

$$x(x+3) = 108$$

Step 4

$$x(x+3) = 108 \qquad \text{Multiply and write in standard form.}$$

$$x^2 + 3x - 108 = 0 \qquad \text{Factor and solve as before.}$$

$$(x+12)(x-9) = 0$$

$$x + 12 = 0 \qquad \text{or} \qquad x - 9 = 0$$

$$x = -12 \qquad\qquad\qquad x = 9$$

Step 5   We reject -12 (cm) as a solution. A length cannot be negative, and so we must consider only 9 (cm) in finding the required dimensions.

   The width x is 9 cm, and the length $x + 3$ is 12 cm. Since this gives a rectangle of area 108 cm², the solution is verified.

✔ *CHECK YOURSELF 3*

In a triangle, the base is 4 in. less than its height. If its area is 30 in.², find the length of the base and the height of the triangle. (*Note:* The formula for the area of a triangle is $A = \dfrac{1}{2}bh$.)

We look at another geometric application.

Example 4

Solving a Rectangular Box Application

An open box is formed from a rectangular piece of cardboard, whose length is 2 in. more than its width, by cutting 2-in. squares from each corner and folding up the sides. If the volume of the box is to be 96 in.³, what must be the size of the original piece of cardboard?

Step 1   We are asked for the dimensions of the sheet of cardboard.

Step 2   Again, sketch the problem.

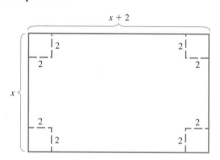

Step 3   To form an equation for volume, we sketch the completed box.

Note: The original width of the cardboard was x. Removing two 2-in. squares leaves $x - 4$ for the width of the box. Similarly, the length of the box is $x - 2$. Do you see why?

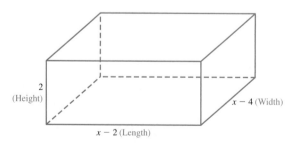

Since volume is the product of height, length, and width,

$$2(x - 2)(x - 4) = 96$$

Step 4

$$2(x - 2)(x - 4) = 96 \qquad \text{Divide both sides by 2.}$$
$$(x - 2)(x - 4) = 48 \qquad \text{Multiply on the left.}$$
$$x^2 - 6x + 8 = 48 \qquad \text{Write in standard form.}$$
$$x^2 - 6x - 40 = 0 \qquad \text{Solve as before.}$$
$$(x - 10)(x + 4) = 0$$
$$x = 10 \text{ in.} \qquad \text{or} \qquad x = -4 \text{ in.}$$

Step 5 Again, we need consider only the positive solution. The width x of the original piece of cardboard is 10 in., and its length $x + 2$ is 12 in. The dimensions of the completed box will be 6 in. by 8 in. by 2 in., which gives the required volume of 96 in.^{3}

✔ *CHECK YOURSELF 4*

A similar box is to be made by cutting 3-in. squares from a piece of cardboard that is 4 in. longer than it is wide. If the required volume is 180 in.^{3}, find the dimensions of the original piece of cardboard.

We now turn to another field for an application that leads to solving a quadratic equation. Many equations of motion in physics involve quadratic equations.

Example 5 | Solving a Thrown Ball Application

Suppose that a person throws a ball directly upward, releasing the ball at a height of 5 ft above the ground. If the ball is thrown with an initial velocity of 80 ft/s, the height of the ball, in feet, above the ground after t seconds(s) is given by

$$h = -16t^2 + 80t + 5$$

When is the ball at a height of 69 ft?

Step 1 The height h of the ball, and the time t since the ball was released are the unknowns.

Step 2 We want to know the value(s) of t for which $h = 69$.

Step 3 Write $69 = -16t^2 + 80t + 5$.

Step 4 Solve for t:

$$69 = -16t^2 + 80t + 5 \qquad \text{We need a 0 on one side.}$$
$$0 = -16t^2 + 80t - 64 \qquad \text{Divide through by } -16.$$
$$0 = t^2 - 5t + 4 \qquad \text{Factor.}$$
$$0 = (t - 1)(t - 4)$$
$$t - 1 = 0 \qquad \text{or} \qquad t - 4 = 0$$
$$t = 1 \qquad\qquad\qquad t = 4$$

Step 5 By substituting 1 for t, we confirm that $h = 69$:

$$h = -16(1)^2 + 80(1) + 5 = -16 + 80 + 5 = 69$$

You should also check that $h = 69$ when $t = 4$.

The ball is at a height of 69 ft at $t = 1$ s (on the way up) and at $t = 4$ s (on the way down).

✔ *CHECK YOURSELF 5*

A ball is thrown vertically upward from the top of a building 100 m high with an initial velocity of 25 m/s. After t s, the height h, in meters, is given by

$$h = -5t^2 + 25t + 100$$

When is the ball at a height of 130 m?

In Example 6, we must pay particular attention to the solutions that satisfy the given physical conditions.

Example 6	**Solving a Thrown Ball Application**

A ball is thrown vertically upward from the top of a building 60 m high with an initial velocity of 20 m/s. After t s, the height h is given by

$$h = -5t^2 + 20t + 60$$

(a) When is the ball at a height of 35 m?

Steps 1 and 2 We want to know the value(s) of t for which $h = 35$.

Step 3 Write $35 = -5t^2 + 20t + 60$.

Step 4 Solve for t:

$$35 = -5t^2 + 20t + 60 \qquad \text{We need a 0 on one side.}$$
$$0 = -5t^2 + 20t + 25 \qquad \text{Divide through by } -5.$$
$$0 = t^2 - 4t - 5 \qquad \text{Factor.}$$
$$0 = (t - 5)(t + 1)$$
$$t - 5 = 0 \qquad \text{or} \qquad t + 1 = 0$$
$$t = 5 \qquad\qquad\qquad t = -1$$

Step 5 Note that $t = -1$ does not make sense in the context of this problem. (It represents 1 s *before* the ball is released!) You should check that when $t = 5$, $h = 35$. The ball is at a height of 35 m at 5 s after release.

(b) When does the ball hit the ground?

**Steps 1
and 2**   When the ball hits the ground, the height of the ball is 0 m.

Step 3   Write $0 = -5t^2 + 20t + 60$.

Step 4   Solve for t:

$$0 = -5t^2 + 20t + 60 \qquad \text{Divide through by } -5.$$
$$0 = t^2 - 4t - 12 \qquad \text{Factor.}$$
$$0 = (t - 6)(t + 2)$$
$$t - 6 = 0 \qquad \text{or} \qquad t + 2 = 0$$
$$t = 6 \qquad\qquad\qquad t = -2$$

Step 5   As before, we reject the negative solution $t = -2$. The ball hits the ground after 6 s. You should verify that when $t = 6$, $h = 0$.

✔ *CHECK YOURSELF 6*

A ball is thrown vertically upward from the top of a building 150 ft high with an initial velocity of 64 ft/s. After t s, the height h is given by

$$h = -16t^2 + 64t + 150$$

When is the ball at a height of 70 ft?

An important application that leads to a quadratic equation involves the stopping distance of a car and its relation to the speed of the car. This is illustrated in Example 7.

| Example 7 | **Solving a Stopping Distance Application** |

The stopping distance d, in feet, of a car that is traveling at x mi/h is approximated by the equation

$$d = \frac{x^2}{20} + x$$

If the stopping distance of a car is 240 ft, at what speed is the car traveling?

Step 1   The stopping distance d, in feet, and the car's speed x, in miles per hour, are the unknowns.

Step 2   We want to know the value of x such that $d = 240$.

Step 3   Write $240 = \dfrac{x^2}{20} + x$.

Step 4   Solve for x:

$$240 = \frac{x^2}{20} + x \qquad\qquad \text{Multiply through by 20.}$$

$$4{,}800 = x^2 + 20x \qquad\qquad \text{We need a 0 on one side.}$$

$$0 = x^2 + 20x - 4{,}800 \qquad \text{Factor.}$$

$$0 = (x + 80)(x - 60)$$

$$x + 80 = 0 \qquad \text{or} \qquad x - 60 = 0$$

$$x = -80 \qquad\qquad\qquad x = 60$$

Step 5   Because x represents the speed of the car, we reject the value $x = -80$. You should verify that if $x = 60$, $d = 240$. The speed of the car is 60 mi/h.

✔ *CHECK YOURSELF 7*

If the stopping distance of a car is 175 ft, at what speed is the car traveling? Use the equation

$$d = \frac{x^2}{20} + x$$

✔ *CHECK YOURSELF ANSWERS*

1. 4, 14   **2.** −8, −6 or 6, 8   **3.** Base 6 in.; height 10 in.   **4.** 12 in. by 16 in.
5. 2 s or 3 s   **6.** 5 s   **7.** 50 mi/h

Exercises ▪ 6.5

Solve each of the applications.

1. One integer is 3 more than twice another. If the product of those integers is 65, find the two integers.

2. One positive integer is 5 less than 3 times another, and their product is 78. What are the two integers?

3. The sum of two integers is 10, and their product is 24. Find the two integers.

4. The sum of two integers is 12. If the product of the two integers is 27, what are the two integers?

5. The product of two consecutive integers is 72. What are the two integers?

6. If the product of two consecutive odd integers is 63, find the two integers.

7. The sum of the squares of two consecutive whole numbers is 61. Find the two whole numbers.

8. If the sum of the squares of two consecutive even integers is 100, what are the two integers?

9. The sum of two integers is 9, and the sum of the squares of those two integers is 41. Find the two integers.

10. The sum of two whole numbers is 12. If the sum of the squares of those numbers is 74, what are the two numbers?

11. The sum of the squares of three consecutive integers is 50. Find the three integers.

12. If the sum of the squares of three consecutive odd positive integers is 83, what are the three integers?

13. Twice the square of a positive integer is 12 more than 5 times that integer. What is the integer?

14. Find an integer such that if 10 is added to the integer's square, the result is 40 more than that integer.

15. The width of a rectangle is 3 ft less than its length. If the area of the rectangle is 70 ft^2, what are the dimensions of the rectangle?

16. The length of a rectangle is 5 cm more than its width. If the area of the rectangle is 84 cm^2, find the dimensions of the rectangle.

17. The length of a rectangle is 2 cm more than 3 times its width. If the area of the rectangle is 85 cm^2, find the dimensions of the rectangle.

18. If the length of a rectangle is 3 ft less than twice its width and the area of the rectangle is 54 ft^2, what are the dimensions of the rectangle?

19. The length of a rectangle is 1 cm more than its width. If the length of the rectangle is doubled, the area of the rectangle is increased by 30 cm^2. What were the dimensions of the original rectangle?

20. The height of a triangle is 2 in. more than the length of the base. If the base is tripled in length, the area of the new triangle is 48 in.^{2} more than the original. Find the height and base of the original triangle.

21. A box is to be made from a rectangular piece of tin that is twice as long as it is wide. To accomplish this, a 10-cm square is cut from each corner, and the sides are folded up. The volume of the finished box is to be 4,000 cm^3. Find the dimensions of the original piece of tin.

    Hint 1: To solve this equation, you will want to use the given sketch of the piece of tin. Note that the original dimensions are represented by x and $2x$. Do you see why? Also recall that the volume of the resulting box will be the product of the length, width, and height.

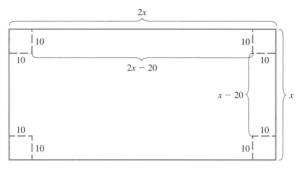

    Hint 2: From this sketch, you can see that the equation that results from $V = LWH$ will be

$$(2x - 20)(x - 20)10 = 4,000$$

22. An open box is formed from a square piece of material by cutting 2-in. squares from each corner of the material and folding up the sides. If the volume of the box that is formed is to be 72 in.^{3}, what was the size of the original piece of material?

23. An open carton is formed from a rectangular piece of cardboard that is 4 ft longer than it is wide, by removing 1-ft squares from each corner and folding up the sides. If the volume of the carton is then 32 ft^3, what were the dimensions of the original piece of cardboard?

24. A box that has a volume of 1,936 in.^{3} was made from a square piece of tin. The square piece cut from each corner had sides of length 4 in. What were the original dimensions of the square?

25. A square piece of cardboard is to be formed into a box. After 5-cm squares are cut from each corner and the sides are folded up, the resulting box will have a volume of 4,500 cm^3. Find the length of a side of the original piece of cardboard.

26. A rectangular piece of cardboard has a length that is 2 cm longer than twice its width. If 2-cm squares are cut from each of its corners, it can be folded into a box that has a volume of 280 cm^3. What were the original dimensions of the piece of cardboard?

27. If a ball is thrown vertically upward from the ground, with an initial velocity of 64 ft/s, its height h after t s is given by

$$h = -16t^2 + 64t$$

How long does it take the ball to return to the ground?

28. If a ball is thrown vertically upward from the ground, with an initial velocity of 64 ft/s, its height h after t s is given by

$$h = -16t^2 + 64t$$

How long does it take the ball to reach a height of 48 ft on the way up?

29. If a ball is thrown vertically upward from the ground, with an initial velocity of 96 ft/s, its height h after t s is given by

$$h = -16t^2 + 96t$$

How long does it take the ball to return to the ground?

30. If a ball is thrown vertically upward from the ground, with an initial velocity of 96 ft/s, its height h after t s is given by

$$h = -16t^2 + 96t$$

How long does it take the ball to pass through a height of 128 ft on the way back down to the ground?

31. If a ball is thrown vertically upward from the roof of a building 192 ft high with an initial velocity of 64 ft/s, its approximate height h after t s is given by

$$h = -16t^2 + 64t + 192$$

How long does it take the ball to fall back to the ground?

32. If a ball is thrown vertically upward from the roof of a building 192 ft high with an initial velocity of 64 ft/s, its approximate height h after t s is given by

$$h = -16t^2 + 64t + 192$$

When will the ball reach a height of 240 ft?

33. If a ball is thrown vertically upward from the roof of a building 192 ft high with an initial velocity of 96 ft/s, its approximate height h after t s is given by

$$h = -16t^2 + 96t + 192$$

How long does it take the ball to return to the thrower?

34. If a ball is thrown vertically upward from the roof of a building 192 ft high with an initial velocity of 96 ft/s, its approximate height h after t s is given by

$$h = -16t^2 + 96t + 192$$

When will the ball reach a height of 272 ft?

35. If a ball is thrown upward from the roof of a 105-m building with an initial velocity of 20 m/s, its approximate height h after t s is given by

$$h = -5t^2 + 20t + 105$$

How long will it take the ball to fall to the ground?

36. If a ball is thrown upward from the roof of a 105-m building with an initial velocity of 20 m/s, its approximate height h after t s is given by

$$h = -5t^2 + 20t + 105$$

When will the ball reach a height of 80 m?

Use the equation $d = \dfrac{x^2}{20} + x$, which relates the speed of a car x, in miles per hour, to the stopping distance d, in feet.

37. If the stopping distance of a car is 120 ft, how fast is the car traveling?

38. How fast is a car going if it requires 75 ft to stop?

39. Marcus is driving at high speed on the freeway when he spots a vehicle stopped ahead of him. If Marcus's car is 315 ft from the vehicle, what is the maximum speed at which Marcus can be traveling and still stop in time?

40. Juliana is driving at high speed on a country highway when she sees a deer in the road ahead. If Juliana's car is 400 ft from the deer, what is the maximum speed at which Juliana can be traveling and still stop in time?

41. Suppose that the cost C, in dollars, of producing x chairs is given by

$$C = 2x^2 - 40x + 2{,}400$$

How many chairs can be produced for $5,400?

42. Suppose that the profit P, in dollars, of producing and selling x appliances is given by

$$P = -3x^2 + 240x - 1{,}800$$

How many appliances must be produced and sold to achieve a profit of $3,000?

43. The relationship between the number x of calculators that a company can sell per month and the price of each calculator p is given by $x = 1{,}700 - 100p$. Find the price at which a calculator should be sold to produce a monthly revenue of $7,000. (*Hint:* Revenue = xp.)

44. A small manufacturer's weekly profit in dollars is given by

$$P = -3x^2 + 270x$$

Find the number of items x that must be produced to realize a profit of $4,200.

45. Suppose that a manufacturer's weekly profit in dollars is given by

$$P = -2x^2 + 240x$$

How many items x must be produced to realize a profit of $5,400?

SECTION 6.6 A General Strategy for Factoring

 OBJECTIVE

1. *Apply factoring techniques to a variety of polynomials*

In this chapter, we have studied a variety of factoring techniques. Our objective at this point is to summarize these and provide an overall plan for factoring. The types of exercises that we encounter will be mixed, so that our first tasks will involve recognizing patterns and polynomial types. This is very similar to what you'll face in your future algebra work.

The following is our general factoring plan.

Step 1. Check for a common factor in the terms of the polynomial. Factor out the greatest common factor (GCF).

Step 2. If the polynomial is a binomial, check for one of the following special patterns:

  a. Difference of squares: $a^2 - b^2 = (a - b)(a + b)$

  b. Difference of cubes: $a^3 - b^3 = (a - b)(a^2 + ab + b^2)$

  c. Sum of cubes: $a^3 + b^3 = (a + b)(a^2 - ab + b^2)$

Step 3. If the polynomial is a trinomial, apply the trial-and-error method or the *ac* method studied earlier.

Step 4. If the polynomial consists of four terms, try to factor by grouping.

Step 5. After successfully factoring a polynomial, check the factors to see if any polynomial factor can be further factored.

> Recall that a "sum of squares" such as $a^2 + b^2$ will not generally factor.

Remember that we can get a partial check of our work by multiplying the factors back together. We should obtain the original polynomial. We say "partial check" because we may have missed something: We may not have *completely* factored the polynomial. For example, suppose we wish to factor $2x^2 + 14x + 24$. If we go directly to step 3 above, we might obtain

$$2x^2 + 14x + 24 = (2x + 6)(x + 4)$$

There is no mistake in this factorization. It simply is not complete. The correct complete factorization is

$$2x^2 + 14x + 24 = 2(x + 3)(x + 4)$$

One final thought: Remember that some polynomials are simply not factorable! If we try all steps in the above plan and we are unable to break down the polynomial, we will answer "not factorable."

Example 1   ## Factoring Polynomials

Factor $2xy + 10x + 6y + 30$.

$$2xy + 10x + 6y + 30 \qquad \text{We note a GCF of 2, and factor.}$$
$$= 2(xy + 5x + 3y + 15)$$

Since we have a polynomial with four terms, we move to step 4 and try grouping:

$$= 2[(xy + 5x) + (3y + 15)]$$
$$= 2[x(y + 5) + 3(y + 5)]$$
$$= 2[(y + 5)(x + 3)]$$
$$= 2(x + 3)(y + 5)$$

Check by multiplying the factors together. You should get the original polynomial.

The binomial factors are first-degree and therefore cannot be further factored. We are done!

   CHECK YOURSELF 1

Factor $4mn - 12m - 20n + 60$.

Be sure to keep your eyes open for factors that can be further factored. This is illustrated in Example 2.

Example 2   ## Factoring Polynomials

Factor $3mn^4 - 48m$.

$$3mn^4 - 48m \qquad \text{Factor out the GCF.}$$
$$= 3m(n^4 - 16) \qquad \text{Note the difference-of-squares binomial.}$$
$$= 3m(n^2 - 4)(n^2 + 4) \qquad \text{Note another difference of squares.}$$
$$= 3m(n - 2)(n + 2)(n^2 + 4)$$

The only binomial that could possibly factor is $n^2 + 4$, but since this is a sum of squares, it will not factor. We are done.

   CHECK YOURSELF 2

Factor $3x^2y - 75y$.

Remember to always start with step 1: Factor out the GCF.

Example 3

Factoring Polynomials

Factor $6x^2y - 18xy - 60y$.

$$6x^2y - 18xy - 60y$$ We find a GCF of $6y$.

$$= 6y(x^2 - 3x - 10)$$ We factor the trinomial using the trial-and-error or the *ac* method.

$$= 6y(x - 5)(x + 2)$$

 CHECK YOURSELF 3

Factor $5xy^2 + 15xy - 90x$.

Do not become frustrated if your factoring attempts do not seem to produce results. You may have a polynomial that won't factor.

Example 4

Factoring Polynomials

Factor $9m^2 - 8$.

We cannot find a GCF greater than 1, so we proceed to step 2. We do have a binomial, but it does not fit any of our special patterns: $9m^2$ is a perfect square, but 8 is not. And 8 is a perfect cube, but $9m^2$ is not. So we conclude that the given binomial is not factorable.

 CHECK YOURSELF 4

Factor $2m^3 - 16$.

(*Caution:* This one may or may not be factorable!)

Example 5

Factoring Polynomials

Factor $x^2y^2 - 9x^2 + y^2 - 9$.

There is no GCF greater than 1. Since we have four terms, we'll try grouping:

$$x^2y^2 - 9x^2 + y^2 - 9$$

$$= x^2(y^2 - 9) + (y^2 - 9)$$

$$= (y^2 - 9)(x^2 + 1)$$ We note the difference of squares.

$$= (y - 3)(y + 3)(x^2 + 1)$$

Since $x^2 + 1$ is a sum of squares, we are done.

✔ *CHECK YOURSELF 5*

Factor $x^2y^2 - 3x^2 - 4y^2 + 12$.

✔ *CHECK YOURSELF ANSWERS*

1. $4(m - 5)(n - 3)$ **2.** $3y(x - 5)(x + 5)$ **3.** $5x(y + 6)(y - 3)$

4. $2(m - 2)(m^2 + 2m + 4)$ **5.** $(x - 2)(x + 2)(y^2 - 3)$

Completely factor each of the following polynomials.

1. $3x^2 + 30x + 75$

2. $4y^2 - 81$

3. $5n^2 + 20n + 30$

4. $8a^3 + 27$

5. $3ay^2 - 48a$

6. $4x^2 - 24x + 36$

7. $xy - 2x + 6y - 12$

8. $yz + 5y - 3z - 15$

9. $2abx - 14ax - 8bx + 56x$

10. $3ax + 12x + 3a + 12$

11. $2x^2 + 5x + 4$

12. $4m^4 - 4$

13. $3y^3 - 81$

14. $3x^2y + 9xy + 9y$

15. $18mn^2 - 15mn - 12m$

16. $3x^2 - 2x + 4$

17. $9x^3 - 4x$

18. $2n^4 - 16n$

19. $x^2y^2 - 4x^2 - 9y^2 + 36$

20. $40x^2 + 12x - 72$

21. $8x^2 + 8x + 2$

22. $m^2n^2 - 4n^2 - 25m^2 + 100$

23. $x^3 + 5x^2 - 4x - 20$

24. $x^3 - 2x^2 - 9x + 18$

Summary for Chapter 6

Example	Topic	Reference
	An Introduction to Factoring	**6.1**
$4x^2$ is the greatest common monomial factor of $8x^4 - 12x^3 + 16x^2$.	**Common Monomial Factor** A single term that is a factor of every term of the polynomial. The greatest common factor (GCF) is the common monomial factor that has the largest possible numerical coefficient and the largest possible exponents.	**p. 437**
$8x^4 - 12x^3 + 16x^2$ $= 4x^2(2x^2 - 3x + 4)$	**Factoring a Monomial from a Polynomial** 1. Determine the greatest common factor. 2. Apply the distributive law in the form $$ab + ac = a(b + c)$$ $\uparrow$ The greatest common factor	**p. 439**
$4x^2 - 6x + 10x - 15$ $= 2x(2x - 3) + 5(2x - 3)$ $= (2x - 3)(2x + 5)$	**Factoring by Grouping** When there are four terms of a polynomial, factor the first pair and factor the last pair. If these two pairs have a common binomial factor, factor that out. The result will be the product of two binomials.	**p. 441**
	Factoring Special Polynomials	**6.2**
To factor: $16x^2 - 25y^2$: $\downarrow$ $\downarrow$ Think: $(4x)^2 - (5y)^2$ so $16x^2 - 25y^2$ $= (4x + 5y)(4x - 5y)$	**Factoring a Difference of Squares** Use the following form: $$a^2 - b^2 = (a + b)(a - b)$$	**p. 447**
To factor: $x^3 - 64$ $x^3 - 64 = x^3 - 4^3$ $= (x - 4)(x^2 + 4x + 16)$	**Factoring a Difference of Cubes** Use the following form: $$a^3 - b^3 = (a - b)(a^2 + ab + b^2)$$	**p. 449**
To factor: $x^3 + 8y^3$ $x^3 + 8y^3 = x^3 + (2y)^3$ $= (x + 2y)(x^2 - 2xy + 4y^2)$	**Factoring a Sum of Cubes** Use the following form: $$a^3 + b^3 = (a + b)(a^2 - ab + b^2)$$	**p. 449**

	Factoring Trinomials	**6.3**
Given $2x^2 - 5x - 3$ $a = 2 \quad b = -5 \quad c = -3$ $ac = -6 \quad b = -5$ $m = -6 \quad n = 1$ $mn = -6 \quad m + n = -5$ $\quad 2x^2 - 6x + x - 3$ $\quad 2x(x - 3) + 1(x - 3)$ $\quad (x - 3)(2x + 1)$	**The ac Test** A trinomial of the form $ax^2 + bx + c$ is factorable if (and only if) there are two integers m and n such that $$ac = mn \quad b = m + n$$ **To Factor a Trinomial** In general, you can apply the following steps. 1. Write the trinomial in standard $ax^2 + bx + c$ form. 2. Label the three coefficients a, b, and c. 3. Find two integers m and n such that $$ac = mn \quad \text{and} \quad b = m + n$$ 4. Rewrite the trinomial as $$ax^2 + mx + nx + c$$ 5. Factor by grouping. Not all trinomials are factorable. To discover if a trinomial is factorable, try the **ac test.**	**p. 456**
	Solving Quadratic Equations by Factoring	**6.4**
To solve: $$x^2 + 7x = 30$$ $x^2 + 7x - 30 = 0$ $(x + 10)(x - 3) = 0$ $x + 10 = 0 \ \text{ or } \ x - 3 = 0$ $x = -10$ and $x = 3$ are solutions.	1. Add or subtract the necessary terms on both sides of the equation so that the equation is in standard form (set equal to 0). 2. Factor the quadratic expression. 3. Set each factor equal to 0. 4. Solve the resulting equations to find the solutions. 5. Check each solution by substituting in the original equation.	**p. 475**
	A General Strategy for Factoring	**6.6**
To factor: $3x^3 - 27x$ $\quad 3x^3 - 27x$ $\quad = 3x(x^2 - 9)$ $\quad = 3x(x - 3)(x + 3)$	**Step 1** Factor out the greatest common factor. **Step 2** If the polynomial is a binomial, check for a special pattern: (a) Difference of squares (b) Difference of cubes (c) Sum of cubes **Step 3** If the polynomial is a trinomial, use trial-and-error method or the ac method. **Step 4** If the polynomial consists of four terms, factor by grouping. **Step 5** After successfully factoring, check to see if any polynomial factor can be further factored.	**p. 495**

Summary Exercises · 6

This summary exercise set is provided to give you practice with each of the objectives of the chapter. Each exercise is keyed to the appropriate chapter section. Your instructor will give you guidelines on how to best use these exercises in your instructional setting.

[6.1] Factor each of the following polynomials.

1. $14a - 35$

2. $9m^2 - 21m$

3. $24s^2t - 16s^2$

4. $18a^2b + 36ab^2$

5. $27s^4 + 18s^3$

6. $3x^3 - 6x^2 + 15x$

7. $18m^2n^2 - 27m^2n + 36m^2n^3$

8. $81x^7y^6 + 63x^4y^6$

9. $8a^2b + 24ab - 16ab^2$

10. $3x^2y - 6xy^3 + 9x^3y - 12xy^2$

11. $2x(3x + 4y) - y(3x + 4y)$

12. $5(w - 3z) - w(w - 3z)$

[6.2] Factor each of the following binomials completely.

13. $p^2 - 49$

14. $25a^2 - 16$

15. $9n^2 - 25m^2$

16. $16r^2 - 49s^2$

17. $25 - z^2$

18. $a^4 - 16b^2$

19. $25a^2 - 36b^2$

20. $x^{10} - 9y^2$

21. $3w^3 - 12wz^2$

22. $16a^4 - 49b^2$

23. $2m^2 - 72n^4$

24. $3w^3z - 12wz^3$

[6.2] Factor the following polynomials completely.

25. $x^2 - 4x + 5x - 20$

26. $x^2 + 7x - 2x - 14$

27. $8x^2 + 6x - 20x - 15$

28. $12x^2 - 9x - 28x + 21$

29. $6x^3 + 9x^2 - 4x^2 - 6x$

30. $3x^4 + 6x^3 + 5x^3 + 10x^2$

[6.2] For each of the following functions (a) find $f(1)$; (b) factor $f(x)$; and (c) find $f(1)$ from the factored form.

31. $f(x) = 3x^3 + 15x^2$ **32.** $f(x) = -6x^3 - 2x^2$ **33.** $f(x) = 12x^6 - 8x^4$

34. $f(x) = 2x^5 - 2x$

[6.2] Factor each of the following binomials completely.

35. $x^2 - 81$ **36.** $25a^2 - 16$ **37.** $16m^2 - 49n^2$

38. $50m^3 - 18mn^2$ **39.** $a^4 - 16b^4$ **40.** $m^3 - 64$

41. $8x^3 + 1$ **42.** $8c^3 - 27d^3$ **43.** $125m^3 + 64n^3$

44. $2x^4 + 54x$

[6.3] Use the *ac* test to determine which of the following trinomials can be factored. Find the values of *m* and *n* for each trinomial that can be factored.

45. $x^2 - x - 30$ **46.** $x^2 + 3x + 2$ **47.** $2x^2 - 11x + 12$

48. $4x^2 - 23x + 15$

[6.3–6.3*] Completely factor each of the following trinomials.

49. $x^2 + 12x + 20$ **50.** $a^2 + a - 2$ **51.** $w^2 - 15w + 54$

52. $r^2 - 9r - 36$ **53.** $x^2 - 8xy - 48y^2$ **54.** $a^2 + 17ab + 30b^2$

55. $14x^2 + 43x - 21$ **56.** $2a^2 + 3a - 35$

57. $x^2 + 9x + 20$ **58.** $x^2 - 10x + 24$ **59.** $a^2 - 7a + 12$

60. $w^2 + 13w + 40$

61. $x^2 + 16x + 64$

62. $r^2 - 15r + 36$

63. $b^2 - 4bc - 21c^2$

64. $m^2n + 4mn - 32n$

65. $m^3 + 2m^2 - 35m$

66. $2x^2 - 2x - 40$

67. $3y^3 - 48y^2 + 189y$

68. $3b^3 - 15b^2 - 42b$

69. $3x^2 + 8x + 5$

70. $5w^2 + 13w - 6$

71. $2b^2 - 9b + 9$

72. $8x^2 + 2x - 3$

73. $10x^2 - 11x + 3$

74. $4a^2 + 7a - 15$

75. $16y^2 - 8xy - 15x^2$

76. $8x^2 + 14xy - 15y^2$

77. $8x^3 - 36x^2 - 20x$

78. $9x^2 - 15x - 6$

79. $6x^3 - 3x^2 - 9x$

80. $3x^2 - 3xy - 18y^2$

[6.4] Solve each of the following equations by factoring.

81. $x^2 + 5x - 6 = 0$

82. $x^2 + 2x - 8 = 0$

83. $x^2 + 7x = 30$

84. $x^2 - 6x = 40$

85. $x^2 + x = 20$

86. $x^2 = 28 - 3x$

87. $x^2 - 10x = 0$

88. $x^2 = 12x$

89. $x^2 - 25 = 0$

90. $x^2 = 225$

91. $2x^2 - x - 3 = 0$

92. $3x^2 - 4x = 15$

93. $3x^2 + 9x - 30 = 0$

94. $4x^2 + 24x = -32$

95. $x(x - 5) = 36$

96. $(x - 2)(2x + 1) = 33$

97. $x^3 - 2x^2 - 15x = 0$

98. $x^3 + x^2 - 4x - 4 = 0$

[6.5] Solve the following applications.

99. One integer is 15 more than another. If the product of the integers is -54, find the two integers.

100. The sum of the squares of two consecutive odd integers is 130. What are the two integers?

101. The length of a rectangle is 6 ft more than its width. If the area of the rectangle is 216 ft^2, find the dimensions of the rectangle.

102. An open carton is formed from a rectangular piece of cardboard that is 5 in. longer than it is wide, by removing 4-in. squares from each corner and folding up the sides. If the volume of the carton is then 200 in.^{3}, what were the dimensions of the original piece of cardboard?

103. If a ball is thrown vertically upward from the roof of a building 20 ft high with an initial velocity of 80 ft/s, its approximate height h after t s is given by

$$h = -16t^2 + 80t + 20$$

How long does it take the ball to pass through a height of 84 ft on the way back down to the ground?

104. A rectangular garden has dimensions 15 ft by 20 ft. The garden is to be enlarged with a strip of equal width surrounding it. If the resulting garden is to be 294 ft^2 larger than before, how wide should the strip be?

105. Suppose that the cost, in dollars, of producing x stereo systems is given by

$$C(x) = 3,000 - 60x + 3x^2$$

How many systems can be produced for $7,500?

106. The demand equation for a certain type of computer paper is predicted to be

$$D = -3p + 69$$

The supply equation is predicted to be

$$S = -p^2 + 24p - 3$$

Find the equilibrium price.

[6.6] Factor each polynomial completely.

107. $5x^4 - 5x^3 - 210x^2$

108. $81n^4 - 3n$

109. $72y - 2y^3$

110. $xy - 8x - 5y + 40$

111. $10x^5 + 80x^4 + 160x^3$

112. $3x^2 + 8x - 15$

113. $x^3 + 2x^2 - 25x - 50$

114. $54m + 16m^4$

Self-Test ■ 6

The purpose of this self-test is to help you check your progress and to review for a chapter test in class. Allow yourself about 1 hour to take the test. When you are done, check your answers in the back of the book. If you missed any answers, be sure to go back and review the appropriate sections in the chapter and the exercises that are provided.

Factor each of the following polynomials.

1. $7b + 42$ **2.** $5x^2 - 10x + 20$

Factor each of the polynomials completely.

3. $16y^2 - 49x^2$ **4.** $32a^2b - 50b^3$ **5.** $27y^3 - 8x^3$

6. $a^2 - 5a - 14$ **7.** $y^2 + 12yz + 20z^2$ **8.** $9x^2 - 12xy + 4y^2$

9. $x^2 + 2x - 5x - 10$ **10.** $3w^2 + 10w + 7$ **11.** $6x^2 + 4x - 15x - 10$

12. $8x^2 - 2xy - 3y^2$ **13.** $6x^3 + 3x^2 - 30x$ **14.** $x^4 - 81$

Solve each equation for the variable x.

15. $x^2 - 2x - 3 = 0$ **16.** $x^2 - 11x = -30$

17. $6x^2 - 7x = 3$ **18.** $2x^2 + 16x + 30 = 0$

Solve the following applications.

19. The length of a rectangle is 4 cm less than twice the width. If the area is 240 cm^2, what is the length of the rectangle?

20. If a ball is thrown upward from the roof of an 18-m building with an initial velocity of 20 m/s, its approximate height h after t s is given by

$$h = -5t^2 + 20t + 18$$

When is the ball at a height of 38 m?

This test covers selected topics from the first seven chapters. It is provided to help you in the process of reviewing previous chapters. Answers are provided in the back of the book. If you miss any answers, be sure to go back and review the appropriate chapter sections.

1. If $x = -3$, $y = 2$, and $z = -4$, find the value of the expression $\dfrac{3x^2 - 2z + 1}{5y + 2x}$.

In Exercises 2 to 5, solve each equation.

2. $7a - 3 = 6a + 8$ **3.** $\dfrac{2}{3}x = -22$ **4.** $|x - 6| = 9$ **5.** $|2x + 9| - 5 = 6$

Solve each inequality.

6. $2x - 7 < 5 + 4x$ **7.** $2x - 9 \le 7x - 3(x - 1)$ **8.** $|2x - 7| < 6$ **9.** $|x + 5| > 3$

10. Find the slope and y intercept of the line represented by the equation $2x + 5y = 10$.

In Exercises 11 and 12, write the equation of the line that satisfies the given conditions.

11. L passes through the points $(-2, 1)$ and $(1, 7)$.

12. L has y intercept $(0, -2)$ and is parallel to the line with equation $3x + y = 5$.

Perform the indicated operations.

13. $7x^2y + 3xy - 5x^2y + 2xy$

14. $(-3x^2 + 5x - 6) + (2x^2 - 3x + 5)$

15. $(5x^2 + 4x - 3) - (4x^2 - 5x - 1)$

16. $3x(x^2 - 2x + 2)$

17. $(2x - 5)(x + 1)$

18. $(x + 3)(x^2 - 3x + 9)$

19. $(4x - 3)(2x + 5)$

20. $(a - 3b)(a + 3b)$

21. $(x - 2y)^2$

22. $2x(5x + 3y)(5x - 3y)$

In Exercises 23 and 26, simplify each expression, using the laws of exponents.

23. $(-2x^{-2}y^3)^3$ **24.** $\dfrac{27a^5b^7}{9ab}$ **25.** $(2x^3y^{-2})^{-2}(x^{-3}y^{-1})$ **26.** $\dfrac{16x^2y^{-3}z^{-4}}{8x^{-1}y^2z^{-5}}$

Let $f(x) = 3x + 5$ and $g(x) = -x^2$.

27. Find $(f + g)(x)$.   **28.** Find $(f \cdot g)(x)$.   **29.** Find $(f \circ g)(x)$.   **30.** Find $(g \circ f)(2)$.

Factor the given expression.

31. $12x + 20$ **32.** $25x^2 - 49y^2$ **33.** $12x^2 - 15x + 8x - 10$ **34.** $2x^2 - 13x + 15$

Solve the given quadratic equation.

35. $x^2 + 2x = 15$ **36.** $x^2 - 9 = 0$

Perform the indicated division.

37. $\dfrac{24x^2y^5 - 15x^4y^2 + 9xy}{3xy}$ **38.** $\dfrac{x^2 + 3x + 2}{x - 1}$

Solve the following word problems. Show the equation used for the solution.

39. Three times a number decreased by 5 is 46. Find the number.

40. Juan's biology text cost $5 more than his mathematics text. Together they cost $81. Find the cost of the biology text.

41. The equation $d = \dfrac{x^2}{20} + x$ relates the speed of a car x, in miles per hour, to the stopping distance d, in feet. How fast is a car going if it requires 240 ft to stop?

42. Suppose that a manufacturer's weekly profit in dollars is given by

$$P = -5x^2 + 300x$$

How many items x must be produced to realize a profit of $2,500?

R

A Review of Elementary Algebra

C hapters 1–6 of this text covered topics that are commonly grouped under the title *Elementary Algebra*. This transitional portion of the text provides an opportunity to prepare for the topics of the second half of the text. These topics are commonly referred to as *Intermediate Algebra*.

There are two elements in this transitional portion of the text. This chapter provides a thorough review of Chapters 1 through 6. Each of the six sections of this chapter covers the material from one of those six chapters. This review is useful both for students who have recently completed these chapters and for students using this text to cover the intermediate algebra topics.

The second element of the transitional material provides a practice final. After taking that practice final, you may refer to either the original chapter or this abbreviated review to clarify topics with which you had difficulty.

In any case, it is important that you demonstrate mastery of the material in the first part of this text before moving on to the later topics.

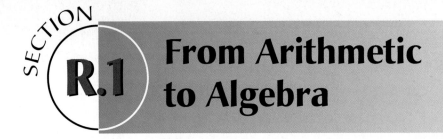

From Arithmetic to Algebra

R.1

In Chapter 1, we worked with algebraic expressions. An *expression* is a meaningful collection of numbers, variables, and symbols of operation.

The collection

$$3x - 2y + 5$$

is an example of an expression. The collection

$$2 + -3x \div \cdot \div 2x$$

is not an expression because the symbols are not presented in a meaningful way. In Example 1, we will translate expressions from words to symbols.

Example 1	**Translating Expressions into Mathematical Symbols**

Write each expression in mathematical symbols.

(a) The sum of a number and 5, divided by the number is written

$$\frac{x + 5}{x}$$

(b) One-half of the base times the height is written

$$\frac{1}{2}b \cdot h \qquad \text{or} \qquad \frac{1}{2}bh$$

(c) A number times the number decreased by 5 is written

$$x(x - 5)$$

(d) Pi (π) times the radius squared is written

$$\pi \cdot r^2 \qquad \text{or} \qquad \pi r^2$$

(e) Two times length plus 2 times width is written

$$2L + 2W$$

In our next example, we will evaluate an algebraic expression.

> **To Evaluate an Algebraic Expression**
>
> Step 1. Replace each variable with the assigned value.
>
> Step 2. Do the necessary arithmetic operations, following the rules for order of operations.

Example 2   Evaluating an Algebraic Expression

Use $x = 3$ and $y = -2$ and evaluate each expression.

(a) $2xy = 2(3)(-2)$

$\qquad\quad = -12$

(b) $3x - 2y + 3 = 3(3) - 2(-2) + 3$

$\qquad\qquad\qquad\quad = 9 + 4 + 3$

$\qquad\qquad\qquad\quad = 16$

(c) $\dfrac{2x}{3y} = \dfrac{2(3)}{3(-2)}$

$\qquad\quad = \dfrac{6}{-6}$

$\qquad\quad = -1$

(d) $3x^2 - 5y^2 = 3(3)^2 - 5(-2)^2$

$\qquad\qquad\qquad = 27 - 20$

$\qquad\qquad\qquad = 7$

A *term* is a number, or the product of a number and one or more variables, raised to a power. If terms contain exactly the same variables raised to the same powers, they are called *like terms*.

Like terms in an expression can always be combined into a single term.

> **Combining Like Terms**
>
> To combine like terms, use the following steps.
>
> Step 1. Add or subtract the numerical coefficients.
>
> Step 2. Attach the common variable(s).

Example 3	**Combining Like Terms**

Simplify the following expressions by combining like terms.

(a) $5m + 3n + 2m + 6n = 5m + 2m + 3n + 6n = 7m + 9n$

(b) $3x^2y + 5xy^2 + x^2y + 4xy^2 = 3x^2y + x^2y + 5xy^2 + 4xy^2 = 4x^2y + 9xy^2$

(c) $3a - 2b - a + 4b = 3a - a + 4b - 2b = 2a + 2b$

Removing Parentheses

Case 1. If a plus sign (+) or nothing at all appears in front of parentheses, just remove the parentheses. No other changes are necessary.

Case 2. If a minus sign (−) appears in front of a set of parentheses, the parentheses can be removed by changing the operation in front of each term inside the parentheses. Each addition becomes subtraction and each subtraction becomes addition.

Example 4	**Removing Parentheses**

Remove the parentheses and then simplify each expression.

(a) $(3x - 5) + (2x + 3) = 3x - 5 + 2x + 3$
$$= 3x + 2x - 5 + 3$$
$$= 5x - 2$$

(b) $(3x^2 + 2x - 1) + (x^2 - 5x + 3) = 3x^2 + 2x - 1 + x^2 - 5x + 3$
$$= 3x^2 + x^2 - 5x + 2x - 1 + 3$$
$$= 4x^2 - 3x + 2$$

(c) $(3x - 5) - (2x + 3) = 3x - 5 - 2x - 3$
$$= 3x - 2x - 5 - 3$$
$$= x - 8$$

(d) $(3x^2 + 2x - 1) - (x^2 - 5x + 3) = 3x^2 + 2x - 1 - x^2 + 5x - 3$
$$= 3x^2 - x^2 + 5x + 2x - 1 - 3$$
$$= 2x^2 + 7x - 4$$

A *set* is a collection of objects. Objects that belong to a set are called the *elements* of the set. Sets for which the elements are listed are said to be in *roster form*.

| **Example 5** | **Listing the Elements of a Set** |

Write each of the following sets in roster form.

(a) The set of all factors of 18

$$\{1, 2, 3, 6, 9, 18\}$$

(b) The set of all integers with an absolute value that is equal to 5

$$\{-5, 5\}$$

(c) The set of all even integers

$$\{\ldots, -4, -2, 0, 2, 4, \ldots\}$$

Not all sets can be described by using the roster form. For example, the set of real numbers between 0 and 1 cannot be put in a list. Instead, we use *set-builder notation*. We would write the set of real numbers greater than 0 but less than, or equal to, 1 as

$$\{x \mid 0 < x \leq 1\}$$

We can also describe a set by using *interval notation*. We would write the above set, using this notation, as

$$(0, 1]$$

The parenthesis indicates that 0 is not included in the set, and the square bracket indicates that 1 is included.

Such a set could also be plotted on a number line. Our next example combines sets and their graphs.

| **Example 6** | **Plotting the Elements of a Set on a Number Line** |

Plot the points of each set on the number line.

(a) $\{-2, 1, 5\}$

(b) $\{x \mid -3 < x < -1\}$

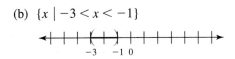

(c) $\{x \mid x \le 5\}$

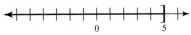

We can combine sets using the operations of *union* and *intersection*. The union of two sets contains all the elements in one or both of the sets. The intersection of two sets contains the elements that are common to both sets. Our final example illustrates these operations.

Example 7   **Finding Union and Intersection**

Let $A = \{2, 4, 6, 8, 9\}$ and $B = \{1, 3, 4, 6, 7, 9\}$.

$$A \cup B = \{1, 2, 3, 4, 6, 7, 8, 9\} \qquad \text{and} \qquad A \cap B = \{4, 6, 9\}$$

Write each phrase, using symbols.

1. 5 less than a number x

2. A number x increased by 10

3. Twice a number x, decreased by 5

4. The quantity $x + y$ times the quantity x minus y

5. The product of p and 6 more than p

6. The sum of m and 5, divided by 3 less than n

7. $\dfrac{4}{3}$ times π times the cube of the radius r

8. $\dfrac{1}{3}$ times the product of the base b and the height h

9. One-half the product of the height h and the sum of two unequal sides b_1 and b_2

10. Twice the sum of x and y minus 3 times the product of x and y

Evaluate each expression if $x = -4$ and $y = 2$.

11. $3x - 2y + 20$

12. $\dfrac{7x + 5y}{9y + 3x}$

13. $\dfrac{8y}{-2x}$

14. $(x - y)^2$

15. $x^2 - y^2$

16. $3xy - 5x$

Simplify each expression by combining like terms.

17. $11x^2 + 7x^2$

18. $-4p + 7q + 11p - 15q$

19. $4xy + 13y - 7xy - 5y$

20. $8r^3s^2 - 7r^2s^3 + 5r^3s^2 - 3r^2s^3$

21. $5x - (3x + 8)$

22. $9x + 11 - (-5x - 6)$

Write each set in roster form.

23. The set of all factors of 24

24. The set of all odd whole numbers

25. The set of even integers less than 6

26. The set of integers between -4 and 3, inclusive

In Exercises 27 to 30, use set-builder and interval notation to represent each set described.

27. The set of all real numbers between -2 and 5 inclusive

28. The set of all real numbers less than or equal to -4

29. The set of all real numbers greater than -4 and less than 4

30. The set of all real numbers greater than -7

In Exercises 31 to 34, plot the points of each set on a number line.

31. $\{-5, -1, 2, 3, 5\}$

32. $\{x \mid x \leq 6\}$

33. $\{x \mid -5 \leq x \leq 2\}$

34. $\{x \mid -3 \leq x < 4\}$

In Exercises 35 to 38, use set-builder and interval notation to describe each set.

35.

36.

37.

38.

In Exercises 39 and 40, find $A \cup B$ and $A \cap B$.

39. $A = \{1, 4, 7, 10\}; B = \{2, 3, 7, 10\}$

40. $A = \{3, 5, 7, 9, 11\}; B = \{1, 3, 7, 8, 12\}$

R.2 Equations and Inequalities

An *equation* is a mathematical statement that two expressions are equal. The statements

$$3x - 5 = 2x - 4 \qquad y = 2x + 1 \qquad \text{and} \qquad x + 3 = 0$$

are each an example of an equation.

A *solution* for an equation is any value for the variable that makes an equation a true statement.

Given the equation $3x - 5 = 2x - 4$, the solution is the value 1, because substituting 1 for x results in the true statement $-2 = -2$.

If we placed every solution to a particular equation in set braces, in this case we would have only $\{1\}$; we refer to it as the *solution set.*

Equations that have the same solution set are called *equivalent equations.* We can obtain equivalent equations by using the *addition property of equality.*

The Addition Property of Equality

If $\quad a = b \quad$ Then $\quad a + c = b + c$

In words, adding the same quantity to both sides of an equation gives an equivalent equation.

In Example 1, the addition property of equality is used to solve several equations.

Example 1	**Solving Equations**

Use the addition property to solve each equation.

(a) $x + 3 = 1$ Add -3 to each side of the equation.

$\qquad\quad x = -2$ The equation is solved. We could write the solution set as $\{-2\}$.

(b) $3x - 5 = 2x - 4$ Add 5 to each side of the equation.

$\qquad\quad 3x = 2x + 1$ Add $-2x$ to each side of the equation.

$\qquad\qquad x = 1$ The equation is solved. We could write the solution set as $\{1\}$.

(c) $4(5x - 2) = 19x + 4$ Use the distributive property to remove parentheses.

$\qquad 20x - 8 = 19x + 4$ Add 8 to each side.

$\qquad\qquad 20x = 19x + 12$ Add $-19x$ to each side.

$\qquad\qquad\quad x = 12$ The equation is solved. The solution set is $\{12\}$.

Perhaps the most important reason for learning mathematics is so that it can be applied. Such problems are called *word problems*. Below is a five-step approach that helps organize the solution to a word problem.

Solving Word Problems

Step 1.   Read the problem carefully. Then reread it to determine what you are asked to find.

Step 2.   Choose a letter to represent one of the unknowns in the problem. Other unknowns should be written as an expression using that same variable.

Step 3.   Translate the problem to the language of algebra to form an equation.

Step 4.   Solve the equation and answer the question of the original problem.

Step 5.   Check your solution by returning to the original problem.

In Example 2, we will solve a word problem. Although it is a problem that you can solve easily through trial and error, we present it to model the steps above. Subsequent word problems will be more challenging.

Example 2   ## Solving a Word Problem

The sum of a number and 13 is 29. Find the number.

Step 1   We are looking for a number.

Step 2   We will call the number x.

Step 3   The equation is $x + 13 = 29$.

Step 4   To solve the equation, we add -13 to each side of the equation, so $x = 16$.

Step 5   Check to make sure that $16 + 13 = 29$.

Given an equation such as $4x = 24$, the addition property is not enough to find a solution. We need a second property.

The Multiplication Property of Equality

If $a = b$, then $ac = bc$, where $c \neq 0$.

In words, multiplying both sides of an equation by the same nonzero number yields an equivalent equation.

To solve an equation of the form

$$ax = b$$

we multiply both sides by the reciprocal of a, or $1/a$.

Example 3 illustrates this process.

| **Example 3** | ## Using the Multiplication Property |

Solve the following equations.

(a) $5m = 15$ Multiply each side of the equation by $\frac{1}{5}$.

   $m = 3$

(b) $-\frac{1}{3}x = 2$ Multiply each side of the equation by -3.

   $x = -6$

In most cases, we must combine the addition and multiplication rules to solve an equation. In such cases, we use the following steps.

Solving Linear Equations in One Variable

Step 1.   Remove any grouping symbols by applying the distributive property.

Step 2.   Multiply both sides by the least common multiple (LCM) of any denominators.

Step 3.   Combine like terms on each side of the equation.

Step 4.   Apply the addition property of equality.

Step 5.   Apply the multiplication property of equality.

Step 6.   Check the solution.

| **Example 4** | ## Combining the Properties to Solve an Equation |

Solve each equation.

(a) $2(3x - 5) = 8$

   Step 1 $6x - 10 = 8$

   Step 2   No denominators

   Step 3   No like terms

   Step 4 $6x = 18$

Step 5 $x = 3$

Step 6 Check to see that $2[3(3) - 5] = 8$ is a true statement.

(b) $\dfrac{3x + 1}{2} = \dfrac{2}{5}$

Step 1 No parentheses

Step 2 Multiply through by the LCM of 2 and 5, which is 10.

$$15x + 5 = 4$$

Step 3 No like terms

Step 4 $15x = -1$

Step 5 $x = -\dfrac{1}{15}$

Step 6 Check to see that

$$\dfrac{3\left(-\dfrac{1}{15}\right) + 1}{2} = \dfrac{2}{5}$$

Every equation can be classified as one of the following types:

Conditional equations are true for some variable values and not true for others.

Identities are true for every possible value of the variable.

Contradictions are never true.

| Example 5 | **Classifying Equations** |

Decide whether each equation is a conditional equation, an identity, or a contradiction.

(a) $2x = x + x$

This statement is always true. The equation is an **identity.**

(b) $2x = 2x + 3$

This statement is never true. The equation is a **contradiction.**

(c) $2x + 1 = 3x - 4$

This equation is true only when $x = 5$. It is a **conditional equation.**

Inequalities are solved much as equations are. There are two important properties associated with solving inequalities. First, we have the addition property.

The Addition Property of Inequalities

If $a < b$ Then $a + c < b + c$

In words, adding the same quantity to both sides of an inequality gives an equivalent inequality.

Example 6   **Using the Addition Property to Solve an Inequality**

Solve and graph the solution set for $7x - 8 \leq 6x + 2$.

$$7x - 8 \leq 6x + 2 \qquad \text{Add 8 to each side.}$$

$$7x \leq 6x + 10 \qquad \text{Add } -6x \text{ to each side.}$$

$$x \leq 10$$

The graph indicates the set of real values equal to or less than 10.

Given an inequality such as $4x < 24$, the addition property is not enough to find a solution. We need a second property.

The Multiplication Property of Inequalities

If $a < b$,

then $ac < bc$ where $c > 0$

and $ac > bc$ where $c < 0$

In words, multiplying both sides of an equation by the same positive number yields an equivalent inequality. Multiplying by a negative number requires reversing the direction of the inequality to yield an equivalent inequality.

Example 7	## Solving and Graphing Inequalities

Solve each inequality and then graph the solution set.

(a) $4x + 9 \geq x$ Add -9 to each side.

$\qquad 4x \geq x - 9$ Add $-x$ to each side.

$\qquad 3x \geq -9$ Multiply by $\dfrac{1}{3}$ on each side.

$\qquad x \geq -3$ Graph the solution set.

(b) $5 - 6x < 41$ Add -5 to each side.

$\qquad -6x < 36$ Divide by -6. Reverse the direction of the inequality.

$\qquad x > -6$ Graph the solution set.

Solving an absolute value equation requires the following property.

Absolute Value Equations

If $|x| = p$ where $p > 0$,   then $x = p$ or $x = -p$

Example 8	## Solving an Absolute Value Equation

Solve for x.

$\qquad |5 - 2x| - 4 = 7$ Add 4 to each side.

$\qquad |5 - 2x| = 11$ Use the property of absolute value equations.

$\qquad 5 - 2x = 11 \quad$ or $\quad 5 - 2x = -11$

$\qquad -2x = 6 \quad$ or $\quad -2x = -16$

$\qquad x = -3 \quad$ or $\qquad x = 8$

Our final example uses the two properties of absolute value inequalities.

Solving Absolute Value Inequalities—Property 1

If $|x - a| < b$ where $b > 0$,   then $-b < x - a < b$

Solving Absolute Value Inequalities—Property 2

If $|x - a| > b$ where $b > 0$,   then $x - a > b$ or $x - a < -b$

| Example 9 | **Solving Absolute Value Inequalities** |

Solve each inequality and graph the solution set.

(a) $|x - 25| < 136$ Property 1 applies.

$-136 < x - 25 < 136$ Add 25.

$-111 < x < 161$ Graph the solution set.

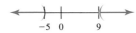

(b) $|x - 2| > 7$ Property 2 applies.

$x - 2 > 7$ or $x - 2 < -7$

$x > 9$ or $x < -5$

Exercises · R.2

Is the number shown in parentheses a solution for the given equation?

1. $5x + 3 = -16$ (-3)

2. $-7x + 9 = 30$ (-3)

3. $-5(x - 7) = 6(5 - x)$ (-5)

4. $\dfrac{5}{6}x = 25$ (6)

In Exercises 5 to 10, solve each equation.

5. $7x - 9 = 15 + 6x$

6. $5x + 3 = 2(3x - 4)$

7. $3x + 2(x - 5) = 11 - (x + 3)$

8. $4 - \dfrac{x + 2}{3} = 1$

9. $5 - (2x - 7) = (9 - 4x) + 6$

10. $3x - 4 = 3(x + 2)$

Solve and graph the solution set for the given inequalities.

11. $4x + 5 < 3x - 7$

12. $9 - 7x > 13 - 8x$

13. $5x - 1 > 3(x + 1)$

14. $-7x + 3 \leq 2x + 21$

15. $2(x + 3) > 7(x + 2) + 2$

16. $7x + 5 < 2(x - 1) - 23$

17. $\dfrac{2}{3}\left(x + \dfrac{3}{4}\right) < \dfrac{5}{6}$

18. $4x - 3 \leq 9(x + 3)$

524

Solve the given equation for *x*.

19. $|x - 3| = 8$

20. $|2x + 1| = 5$

21. $|2x + 5| - 7 = 8$

22. $12 - |3x + 2| = 4$

23. $|2x - 1| - 5 = 6$

24. $|3x - 1| = |x + 5|$

25. $|3x - 4| + 5 = 0$

26. $9 - |x + 1| = 5$

In Exercises 27 to 32, solve the given inequalities and graph the solution set.

27. $|x - 5| \le 6$

28. $|x + 3| > 4$

29. $|2x + 1| - 3 < 4$

30. $|4x + 3| \ge 0$

31. $\left|\dfrac{4x + 3}{5}\right| \le 3$

32. $|3x - 5| < 0$

In Exercises 33 to 40, solve the given problem.

33. The sum of a number and 14 is 33. Find the number.

34. The sum of two consecutive odd integers is 20. Find the integers.

35. A certain number is 8 more than a second number. The sum of the two numbers is 22. Find the numbers.

36. Michelle earns $90 more per week than Dan. If their weekly salaries total $950, how much does Dan earn?

37. A customer purchases a radio for $306.80 including a 4% sales tax. Find the price of the radio and the sales tax.

38. Town A has a population 10% greater than that of town B. The total population of the two towns is 49,245. Find the population of each.

39. The length of a rectangle is 3 centimeters (cm) more than 4 times its width. If the perimeter of the rectangle is 46 cm, find the dimensions of the rectangle.

40. The Amazon River is 3 times as long as the Ohio—Allegheny river. Find the length of each if the difference in their lengths is 2,610 mi.

SECTION R.3 Graphs and Linear Equations

A linear equation in two variables is said to be in **standard form** if it can be written as

$$Ax + By = C \qquad \text{where } A \text{ and } B \text{ are not both } 0$$

A **solution** for an equation in two variables is an **ordered pair** which, when substituted into the equation, results in a true statement.

Example 1 Finding Solutions for a Two-Variable Equation

Given the equation $3x - y = 5$:

(a) Determine if the ordered pair $(-2, -11)$ is a solution.

Substituting -2 for x and -11 for y, we have

$$3(-2) - (-11) \stackrel{?}{=} 5$$
$$-6 + 11 = 5$$

Since this is a true statement, $(-2, -11)$ is a solution.

(b) Find y if $x = 3$.

Substituting 3 for x, we have

$$3(3) - y = 5$$
$$9 - y = 5$$
$$-y = -4$$
$$y = 4$$

So, if $x = 3$, then $y = 4$.

(c) Complete the ordered pair $(\ , 7)$ so that it is a solution.

Substituting 7 for y, we have

$$3x - 7 = 5$$
$$3x = 12$$
$$x = 4$$

So $(4, 7)$ is a solution.

526

The **rectangular coordinate system** consists of two number lines, one drawn horizontally (called the ***x* axis**) and one drawn vertically (called the ***y* axis**). Their point of intersection is called the **origin,** and the axes divide the plane into four **quadrants.**

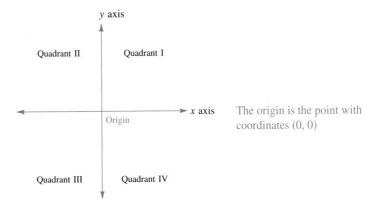

The two-dimensional picture shown above is referred to as the **plane.** There is a correspondence between **ordered pairs** of real numbers and **points in the plane.**

Example 2

Working with Ordered Pairs and Points in the Plane

(a)

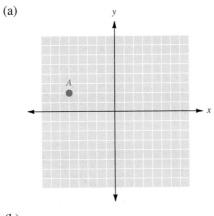

Name the ordered pair that corresponds to the given point *A*.

To move from the origin $(0, 0)$ to *A*, we see that *A* is 5 units to the *left* of the *y* axis and 2 units *above* the *x* axis. So point *A* has coordinates $(-5, 2)$.

(b)

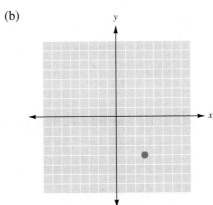

Graph the point corresponding to the ordered pair $(3, -4)$.

From the origin, move 3 units to the right on the *x* axis, and then move 4 units down.

We show *all* the solutions of a linear equation in two variables by plotting points that correspond to solutions of the equation. The set of all ordered pairs that satisfy

$$Ax + By = C \qquad \text{where } A \text{ and } B \text{ are not both } 0$$

will produce a **straight-line graph** when the points corresponding to solutions are plotted. One method of creating such a graph is to solve for y and make a table.

Example 3	## Graphing a Linear Equation

Draw the graph of $x - 2y = 8$.

Solve for y:

$$x - 2y = 8$$
$$-2y = -x + 8$$
$$y = \frac{1}{2}x - 4$$

Make a table of values, where the x values are multiples of 2.

x	y
-2	-5
0	-4
2	-3

Plot these points and draw a line through them.

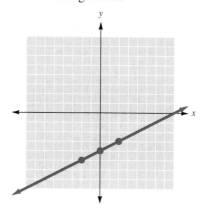

Another useful method for graphing linear equations is the **intercept method.**

Graphing a Line by the Intercept Method

Step 1. To find the x intercept: Let $y = 0$, then solve for x.

Step 2. To find the y intercept: Let $x = 0$, then solve for y.

Step 3. Graph the x and y intercepts.

Step 4. Draw a straight line through the intercepts.

Example 4   ## Using the Intercept Method to Graph a Line

Draw the graph of $3x - 4y = 12$.

Find the x intercept. Let $y = 0$:

$$3x - 4(0) = 12$$
$$3x = 12$$
$$x = 4$$

So the x intercept is $(4, 0)$.

     Find the y intercept. Let $x = 0$:

$$3(0) - 4y = 12$$
$$-4y = 12$$
$$y = -3$$

So the y intercept is $(0, -3)$.

     Plot the intercepts and draw a line through them.

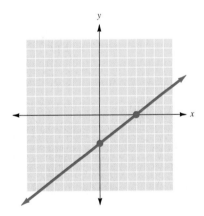

It is possible for the graph of a linear equation to be horizontal or vertical.

Vertical and Horizontal Lines

1. The graph of $x = a$ is a *vertical line* crossing the x axis at $(a, 0)$.

2. The graph of $y = b$ is a *horizontal line* crossing the y axis at $(0, b)$.

Example 5   ## Creating Graphs That Are Horizontal or Vertical

(a) Draw the graph of $x = -4$.

The line is vertical and crosses the x axis at $(-4, 0)$.

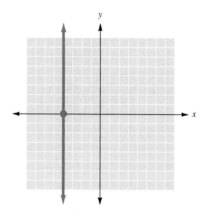

(b) Draw the graph of $y = 2$.

The line is horizontal and crosses the y axis at $(0, 2)$.

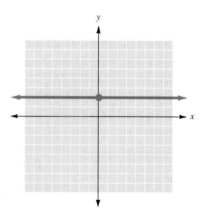

The **slope** of a line through two points $P(x_1, y_1)$ and $Q(x_2, y_2)$ is given by

$$m = \frac{\text{change in } y}{\text{change in } x}$$

$$= \frac{y_2 - y_1}{x_2 - x_1} \qquad \text{where } x_1 \neq x_2$$

| Example 6 | ### Finding the Slope |

Find the slope of the line which passes through the given points.

(a) $(-3, 2)$ and $(4, -1)$

Letting $(-3, 2) = (x_1, y_1)$ and $(4, -1) = (x_2, y_2)$, we have

$$m = \frac{y_2 - y_1}{x_2 - x_1}$$

$$= \frac{-1 - 2}{4 - (-3)}$$

$$= \frac{-3}{7}$$

$$= -\frac{3}{7}$$

(b) $(-4, 5)$ and $(2, 5)$

$$m = \frac{5 - 5}{2 - (-4)}$$

$$= \frac{0}{6}$$

$$= 0$$

(c) $(1, 6)$ and $(1, -3)$

$$m = \frac{6 - (-3)}{1 - 1}$$

$$= \frac{9}{0} \qquad \text{which is undefined}$$

The concepts of slope and y intercept are very useful for graphing linear equations.

The Slope-Intercept Form for a Line

A linear equation is said to be in **slope-intercept form** when it is written as $y = mx + b$, where m is the slope and the y intercept is $(0, b)$.

Example 7	## Using Slope-Intercept Form to Graph an Equation

Given the equation $4x - 3y = 6$:

(a) Find the slope and y intercept. Solve for y:

$$4x - 3y = 6$$

$$-3y = -4x + 6$$

$$y = \frac{4}{3}x - 2$$

Since $m = \frac{4}{3}$, the slope is $\frac{4}{3}$.

Since $b = -2$, the y intercept is $(0, -2)$.

(b) Draw the graph. Plot the y intercept $(0, -2)$. Then, using the slope of $\frac{4}{3}$, move from the y intercept with a rise of 4 and a run of 3 to plot a new point at $(3, 2)$. Draw a line through the plotted points.

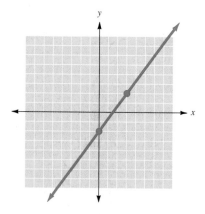

Parallel and Perpendicular Lines

Two lines are **parallel** if their slopes are equal: $m_1 = m_2$

Two lines are **perpendicular** if their slopes are negative reciprocals: $m_1 = -\dfrac{1}{m_2}$

Note that in this case $m_1 \cdot m_2 = -1$; the product of their slopes is -1.

There is another useful form for a linear equation: the **point-slope form.**

> **Point-Slope Form for a Linear Equation**
>
> Given that a line has slope m and passes through a point (x_1, y_1), the equation of the line is
>
> $$y - y_1 = m(x - x_1)$$

| Example 8 | ## Finding the Equation of a Line |

Write the equation of the line that passes through $(-3, 5)$ and $(6, 2)$.

First, find the slope:

$$m = \frac{y_2 - y_1}{x_2 - x_1} = \frac{5 - 2}{-3 - 6} = \frac{3}{-9} = -\frac{1}{3}$$

Then, choosing a given point, say $(6, 2)$, use the point-slope form to write

$$y - y_1 = m(x - x_1)$$

$$y - 2 = -\frac{1}{3}(x - 6)$$

$$y - 2 = -\frac{1}{3}x + 2$$

$$y = -\frac{1}{3}x + 4$$

| Example 9 | ## Finding the Equation of a Line |

Write the equation of the line that passes through $(-1, -5)$ and is perpendicular to the line whose equation is $x + 2y = 6$.

First, find the slope of the line with equation $x + 2y = 6$:

$$x + 2y = 6$$

$$2y = -x + 6$$

$$y = -\frac{1}{2}x + 3$$

The slope of this line is $-\dfrac{1}{2}$, so the slope of our desired line must be 2.

Using $m = 2$ and the given point $(-1, -5)$, we have

$$y - (-5) = 2[x - (-1)]$$

$$y + 5 = 2(x + 1)$$

$$y + 5 = 2x + 2$$

$$y = 2x - 3$$

Every two points have a midpoint. The midpoint of A and B is the point on the line AB that is an equal distance from A and B. The following formula can be used to find the midpoint M.

$$M = \left(\frac{x_1 + x_2}{2}, \frac{y_1 + y_2}{2}\right)$$

In words, the x coordinate of the midpoint is the average of the two x values, and the y coordinate of the midpoint is the average of the two y values.

Example 10   Finding the Midpoint

(a) Find the midpoint of $(2, 0)$ and $(10, 0)$.

$$M = \left(\frac{2 + 10}{2}, \frac{0 + 0}{2}\right) = (6, 0)$$

(b) Find the midpoint of $(5, 7)$ and $(-1, -3)$.

$$M = \left(\frac{5 + (-1)}{2}, \frac{7 + (-3)}{2}\right) = \left(\frac{4}{2}, \frac{4}{2}\right) = (2, 2)$$

Exercises · R.3

Determine which of the ordered pairs are solutions for the given equation.

1. $4x - 2y = 8$ $(0, 4), (1, -2), (-2, 0), (2, 0)$

2. $3x - y = 5$ $(0, 5), (2, 0), (0, -5), (2, 1)$

3. $x = 4$ $(4, 0), (0, 4), (4, -1), (2, 3)$

4. $y = 5$ $(5, 0), (0, 5), (1, 5), (-2, 4)$

Complete the ordered pairs so that each is a solution for the given equation.

5. $x + y = 9$ $(4, \ \ ), (0, \ \ ), (\ \ , 0), (\ \ , -1)$

6. $5x - 3y = 15$ $(6, \ \ ), (0, \ \ ), (\ \ , 0), (\ \ , 10)$

Find four solutions for each of the following equations. *Note:* Your answers may vary from those shown in the answer section.

7. $2x - y = 8$

8. $9x + 3y = 27$

Give the coordinates of the points graphed below.

9. A   **10.** B

11. C   **12.** D

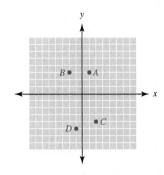

Plot the given points on the rectangular coordinate system.

13. $A(2, -1)$ **14.** $B(-3, -4)$

15. $C(-2, 5)$ **16.** $D(4, 3)$

Graph the given equation.

17. $x + y = 5$ **18.** $2x + y = 6$ **19.** $3x - y = 5$ **20.** $-4x - y = 7$

21. $y = 3x$ **22.** $y = -4x$ **23.** $2x + 5y = 10$ **24.** $3x - 5y = 15$

25. $x = 4$ **26.** $y = -2$

Find the slope of the line passing through the given pair of points.

27. $(4, -2)$ and $(-1, 3)$ **28.** $(2, 7)$ and $(-1, -5)$

29. $(-3, 2)$ and $(7, 2)$ **30.** $(2, 5)$ and $(2, -3)$

In Exercises 31 to 34, find the slope and y intercept of the line represented by the given equation. Draw the graph.

31. $y = 3x + 2$ **32.** $3y + 7x = 21$

33. $y + 5x = 3$ **34.** $4x - 3y = 12$

Determine if the lines are parallel, perpendicular, or neither.

35. L_1 through $(3, 4)$ and $(5, -6)$
    L_2 through $(5, 4)$ and $(10, 5)$

36. L_1 with equation $5x - 4y = 20$
    L_2 passes through $(4, 7)$ and $(8, 12)$

In Exercises 37 to 40, write an equation of the line satisfying the given geometric conditions.

37. L passes through the points $(-6, 2)$ and $(4, -3)$.

38. L passes through $(3, 4)$ and is perpendicular to the line with equation $y = -\dfrac{3}{5}x + 2$.

39. L has y intercept $(0, 6)$ and is parallel to the line with equation $y - 3x = 5$.

40. L has x intercept of $(-2, 0)$ and y intercept of $(0, 8)$.

Find the midpoint for each pair of points.

41. $(3, 0)$ and $(11, 0)$

42. $(2, 5)$ and $(-4, -3)$

43. $(-3, -3)$ and $(7, -9)$

44. $(-2, 5)$ and $(7, -2)$

R.4 Exponents and Polynomials

Chapter 4 dealt with the most common kind of algebraic expression, the polynomial. To develop the definition of a polynomial, we first discussed exponents. Exponents are a shorthand version of repeated multiplication. Instead of writing

$$a \cdot a \cdot a \cdot a \cdot a \cdot a \cdot a \qquad \text{we write} \qquad a^7$$

which we read "*a* to the seventh power."

An expression of this type is said to be in *exponential form*. We call *a* the *base* of the expression and 7 the *exponent*, or *power*.

Product Rule for Exponents

$$a^m \cdot a^n = a^{m+n}$$

In words, when we multiply two expressions that have the same base, the resulting expression has the same base to the power that is the sum of the two exponents.

| Example 1 | ## Using the Product Rule |

Simplify each product, first as a base to a single power and then, if possible, as a decimal.

(a) $a^5 \cdot a = a^6$ (b) $2^3 \cdot 2^5 = 2^8 = 256$

In Example 2, we will use the **quotient rule for exponents.**

Quotient Rule for Exponents

$$\frac{a^m}{a^n} = a^{m-n}$$

In words, when we divide two expressions that have the same base, the resulting expression has the same base to the power that is the difference of the two exponents.

| Example 2 | ## Using the Quotient Rule |

Simplify each quotient, first as a base to a single power and then, if possible, as a decimal.

(a) $\dfrac{a^9}{a^3} = a^6$ (b) $\dfrac{2^{12}}{2^3} = 2^9 = 512$

538

What does the quotient rule yield when m is equal to n? Let's take a look.

$$\frac{a^m}{a^m} = a^{m-m} = a^0$$

But we know that

$$\frac{a^m}{a^m} = 1$$

This implies that $a^0 = 1$. In fact, any base (other than zero) raised to the 0 power equals 1.

The Zero Exponent

For any real number $a \neq 0$,

$$a^0 = 1$$

What if we allow one of the exponents to be negative and apply the product rule? Suppose, for instance, that $m = 3$ and $n = -3$. Then

$$a^m \cdot a^n = a^3 \cdot a^{-3} = a^{3+(-3)} = a^0 = 1$$

so

$$a^3 \cdot a^{-3} = 1$$

Dividing both sides by a^3, we get

$$a^{-3} = \frac{1}{a^3}$$

This leads to the following definition.

Negative Integer Exponents

For any nonzero real number a and whole number n,

$$a^{-n} = \frac{1}{a^n}$$

Let's look at some examples.

Example 3   Working with Integer Exponents

Simplify the following expressions. Each answer should include no negative exponents.

(a) $y^{-3} = \dfrac{1}{y^3}$

(b) $2^{-5} = \dfrac{1}{2^5} = \dfrac{1}{32}$ (or, as a decimal number, 0.03125)

(c) $2(ab)^0 = 2(1) = 2$

When working with exponents, you must also keep three other properties in mind.

Power Rule

For any nonzero real number a and integers m and n,

$$(a^m)^n = a^{mn}$$

Product-Power Rule

For any nonzero real numbers a and b and integer n,

$$(ab)^n = a^n b^n$$

Quotient-Power Rule

For any nonzero real numbers a and b and integer n,

$$\left(\frac{a}{b}\right)^n = \frac{a^n}{b^n}$$

Example 4   **Using the Properties of Exponents**

Simplify each expression.

(a) $(a^3)^5 = a^{15}$

(b) $(2^4)^5 = 2^{20} = 1{,}048{,}576$

(c) $(ab)^5 = a^5 b^5$

(d) $(2a)^5 = 2^5 a^5 = 32a^5$

(e) $\left(\dfrac{2}{a}\right)^5 = \dfrac{2^5}{a^5} = \dfrac{32}{a^5}$

Scientific notation is a useful way of expressing very large or very small numbers through the use of powers of 10. Any number written in the form

$$a \times 10^n$$

in which $1 \le a < 10$ and n is an integer, is said to be written in scientific notation.

| Example 5 | ## Using Scientific Notation |

Write each number in scientific notation.

(a) $903{,}000{,}000{,}000. = 9.03 \times 10^{11}$

        11 places

(b) $0.000892 = 8.92 \times 10^{-4}$

     4 places

A *term* is a number, or the product of a number and one or more variables, which may be raised to powers. A polynomial consists of one or more terms in which the only allowable exponents are whole numbers. In each term of a polynomial, the number is called the *coefficient*.

A polynomial with exactly one term is called a *monomial*. A polynomial with exactly two terms is called a *binomial*. A polynomial with exactly three terms is called a *trinomial*.

Polynomials are also classified by their *degree*. The **degree** of a polynomial that has only one variable is the highest power of that variable appearing in any one term.

The value of a polynomial depends on the value given to its variable(s).

| Example 6 | ## Identifying and Evaluating Polynomials |

Classify each polynomial; then evaluate it for the given value of the variable. Also give the degree of the polynomial.

(a) $3x^2 + 5x - 4$ where $x = -2$

This is a trinomial. Its degree is 2. At $x = -2$, it has a value of

$$3(-2)^2 + 5(-2) - 4$$
$$= 12 - 10 - 4$$
$$= -2$$

(b) $-5x^4 + 7x$ where $x = 3$

This is a binomial. Its degree is 4. At $x = 3$, it has a value of

$$-5(3)^4 + 7(3)$$
$$= -5(81) + 21$$
$$= -405 + 21 = -384$$

When polynomials are added, it is a matter of adding like terms.

| Example 7 | **Adding Polynomials** |

Add the two polynomials.

$$(4a^2 + 3a - 5) + (3a^2 - 5a + 3)$$

To add two polynomials, remove the parentheses and combine like terms.

$$4a^2 + 3a - 5 + 3a^2 - 5a + 3$$

$$= 4a^2 + 3a^2 + 3a - 5a - 5 + 3$$

$$= 7a^2 - 2a - 2$$

When subtracting a polynomial, take care to distribute the subtraction sign to every term in the second polynomial.

| Example 8 | **Subtracting Polynomials** |

Subtract the two polynomials.

$$(5a^2 + 6a - 5) - (3a^2 - 3a + 1)$$

Note that distributing the subtraction sign changes addition to subtraction and subtraction to addition.

$$= 5a^2 + 6a - 5 - 3a^2 + 3a - 1$$

$$= 5a^2 - 3a^2 + 6a + 3a - 5 - 1$$

$$= 2a^2 + 9a - 6$$

When multiplying a polynomial by a monomial, distribute the monomial to each term of the polynomial and simplify the result.

| Example 9 | ## Multiplying a Polynomial by a Monomial |

Multiply $2x^2 + 3x - 1$ by $2x^2$.

$$2x^2(2x^2 + 3x - 1) = 2x^2(2x^2) + 2x^2(3x) - 2x^2(1)$$
$$= 4x^4 + 6x^3 - 2x^2$$

To multiply two binomials, we can use the FOIL method.

To Multiply Two Binomials

Step 1.   Multiply the first terms of the binomials (F).

Step 2.   Multiply the first term of the first binomial by the second term of the second binomial (O).

Step 3.   Multiply the second term of the first binomial by the first term of the second binomial (I).

Step 4.   Multiply the second terms of the binomials (L).

Step 5.   Form the sum of the four terms found above, combining any like terms.

| Example 10 | ## Multiplying Two Binomials |

Multiply $3x - 1$ by $2x + 3$.

$$(3x - 1)(2x + 3)$$

First: $(3x)(2x) = 6x^2$
Outer: $(3x)(3) = 9x$
Inner: $(-1)(2x) = -2x$
Last: $(-1)(3) = -3$

$$= 6x^2 + 9x - 2x - 3$$ Now combine like terms.
$$\quad\ \ \text{F} \quad\ \ \text{O} \quad\ \ \text{I} \quad\ \ \text{L}$$

$$= 6x^2 + 7x - 3$$

To Square a Binomial

Step 1.   Find the first term of the square by squaring the first term of the binomial.

Step 2.   Find the middle term of the square as twice the product of the two terms of the binomial.

Step 3.   Find the last term of the square by squaring the last term of the binomial.

In symbols:

$$(a + b)^2 = a^2 + 2ab + b^2$$

| Example 11 | **Squaring a Binomial** |

Square each binomial.

(a) $(x + 5)^2$

     Step 1   Square x, which gives us: x^2

     Step 2   Twice the product of the terms: $2 \cdot 5 \cdot x = 10x$

     Step 3   Square the final term: $5^2 = 25$

     We have

$$(x + 5)^2 = x^2 + 10x + 25$$

(b) $(2x - 3y)^2$

     Step 1   Square $2x$, which gives us: $4x^2$

     Step 2   Twice the product of the terms: $2 \cdot 2x \cdot (-3y) = -12xy$

     Step 3   Square the final term: $(-3y)^2 = 9y^2$

     We have

$$(2x - 3y)^2 = 4x^2 - 12xy + 9y^2$$

The product of two binomials that differ only in the sign between the terms is the square of the first term minus the square of the second term.

$$(a - b)(a + b) = a^2 - b^2$$

This pattern emerges because when we use the FOIL technique, we find that the sum of the inner and outer products is 0.

| Example 12 | **Special Patterns in Multiplication** |

Multiply.

$$(2x + 3y)(2x - 3y)$$
$$= (2x)^2 - (3y)^2$$
$$= 4x^2 - 9y^2$$

To divide a monomial by a monomial, divide the coefficients and use the quotient rule for exponents to combine the variables.

To divide a polynomial by a monomial, divide each term of the polynomial by the monomial.

| **Example 13** | **Dividing Polynomials** |

Divide.

(a) $\dfrac{36x^5y^2}{9x^2y} = \left(\dfrac{36}{9}\right)x^{5-2}\,y^{2-1} = 4x^3y$

(b) $\dfrac{24x^4 - 4x^3 + 8x^2}{4x^2} = \dfrac{24x^4}{4x^2} - \dfrac{4x^3}{4x^2} + \dfrac{8x^2}{4x^2}$

$$= 6x^2 - x + 2$$

To find the quotient when one polynomial is divided by another, we use long division.

| **Example 14** | **Dividing a Polynomial by a Binomial** |

Divide $2x^2 + x - 6$ by $x + 1$.

$$
\begin{array}{r}
2x \\
x + 1 \overline{)\,2x^2 + x - 6} \\
\underline{2x^2 + 2x}
\end{array}
$$

Start by dividing the lead term of the binomial (x) into the lead term of the trinomial ($2x^2$) and multiply the result ($2x$) by the binomial ($x + 1$).

$$
\begin{array}{r}
2x - 1 \\
x + 1 \overline{)\,2x^2 + x - 6} \\
\underline{2x^2 + 2x} \\
-x - 6 \\
\underline{-x - 1} \\
-5
\end{array}
$$

Subtract the product ($2x^2 + 2x$) to get $-x - 6$. Then divide again by $x + 1$.

The remainder is -5.

Rewriting the result, we have

$$(2x^2 + x - 6) \div (x + 1) = 2x - 1 - \dfrac{5}{x + 1}$$

Exercises · R.4

Simplify each expression.

1. x^8x^{10}

2. $\dfrac{x^{-2}x^{-6}}{x^7}$

3. $\dfrac{x^2y^3}{x^3y}$

4. $(3xy^2)^3$

5. $\dfrac{35x^{-4}y^{-5}z^3}{49x^{-6}y^{-8}z^4}$

6. $\left(\dfrac{x^{-1}y^{-2}}{y^{-3}}\right)^{-5}$

7. $(2x^{-2}y^3)^{-3}(-2x^{-3}y^{-2})^2$

8. $(3x^2y^{-5})^0(2x^{-3}y^{-2})^{-2}$

9. $(x^2y^3)^3(2xy^{-3})^2$

10. $(-2^0x^4)^{-1}(3x^{-3})^{-2}$

Write each number in scientific notation.

11. 0.00507

12. 80,630,000,000

In Exercises 13 and 14, write each number in decimal notation.

13. 4.28×10^7

14. 5.6×10^{-4}

Classify each polynomial and then evaluate it for the given value of the variable.

15. $-3x^4$ where $x = -1$

16. $4x^2 - 7x + 9$ where $x = -3$

17. $-2x^2 + 7x$ where $x = 2$

18. $4x^3 - 3x - 5$ where $x = 3$

Perform the indicated operations.

19. $(5x^2 + 3) + (x^2 - 6)$

20. $(-3w^2 + 15w) + (-4 + w^2) + (-11w + 5)$

21. $(14n^2 - 13n - 8) + (-19n + 7n^2 - 3) + (6n^2 + 5)$

22. $(23r^2 - 6r + 5) - (-9r^2 + 7r - 8)$

23. $(-6x^2 - 2x - 32) - (9x^2 + 13x - 19)$

24. $(2p^2 + 4p - 8) - (7p^2 - 8p + 5) + (14p^2 + 16p - 3)$

25. $-12y(-7y^2 - 9y)$

26. $2x(3x^2 + 4x - 3)$

27. $(7x + 2)(3x + 2)$

28. $(5t + 4)(7t - 1)$

29. $(3r - 2t)(4r - t)$

30. $(2m - 5)(2m + 5)$

31. $(3x - 4y)(3x + 4y)$

32. $(x + 8)^2$

33. $(x - 3)^2$

34. $(6x - 2y)^2$

35. $3x(x + 2)(x - 4)$

36. $-2y(3y - 1)(2y + 3)$

37. $x(5x - 2y)(5x + 2y)$

38. $-4x(5 - 3x) + 6x(3 - x)$

39. $(-2x^2)^3(x + 3)(2x - 1)$

40. $\dfrac{18x^2y^2 - 6xy^3}{2xy}$

41. $\dfrac{15n^4 - 5n^3 + 20n^2}{5n^2}$

42. $\dfrac{x^2 + 3x + 2}{x + 1}$

43. $\dfrac{y^2 - 5y - 6}{y + 2}$

44. $\dfrac{x^2 - 13x - 14}{x - 1}$

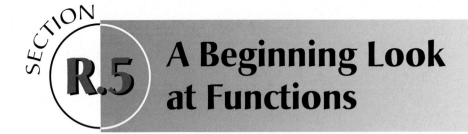
R.5 A Beginning Look at Functions

A **relation** is any set of ordered pairs. The set of first elements of the ordered pairs is called the **domain** of the relation, and the set of second elements of the ordered pairs is called the **range** of the relation.

The ordered pairs of a relation may be presented in a variety of forms. Among these are complete listings of the ordered pairs in set form or in table form.

| **Example 1** | **Finding the Domain and Range of a Relation** |

For each relation, find the domain and the range.

(a) $A = \{(-3, 5), (-1, 4), (2, 5)\}$

The domain is $D = \{-3, -1, 2\}$

The range is $R = \{4, 5\}$

(b) Suppose B contains the ordered pairs presented in the following table.

x	y
-2	3
0	2
4	-1
4	-2

The domain is $D = \{-2, 0, 4\}$

The range is $R = \{-1, -2, 2, 3\}$

A **function** is a set of ordered pairs (a relation) in which no two first elements are equal.

| **Example 2** | **Identifying a Function** |

For each relation given in Example 1, determine whether the relation is a function.

(a) A is a function since no two first coordinates are equal.

(b) Since $(4, -1)$ and $(4, -2)$ are both ordered pairs of B, we see that B is *not* a function.

Another way of expressing a function is through the use of $f(x)$ **notation.**

Example 3	**Evaluating a Function**

Given $f(x) = x^2 - 3x + 2$, find the following.

(a) $f(-5)$

    Substituting -5 for x, we have

$$f(x) = x^2 - 3x + 2$$
$$f(-5) = (-5)^2 - 3(-5) + 2$$
$$= 25 + 15 + 2$$
$$= 42$$

(b) $f(2)$

$$f(2) = (2)^2 - 3(2) + 2$$
$$= 4 - 6 + 2$$
$$= 0$$

(c) $f(4 + h)$

$$f(4 + h) = (4 + h)^2 - 3(4 + h) + 2$$
$$= 16 + 8h + h^2 - 12 - 3h + 2$$
$$= h^2 + 5h + 6$$

A set of ordered pairs can also be specified by means of a graph. In addition to finding the domain and range, we can determine from a graph whether a relation is a function. For this last purpose, we need the **vertical line test.**

Vertical Line Test

A relation is a function if no vertical line can pass through two or more points on its graph.

Example 4	**Identifying Functions, Domain, and Range**

Determine whether the given graph is the graph of a function. Also provide the domain and range.

(a)

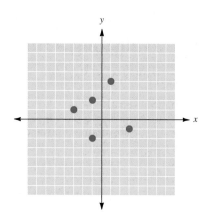

This is not a function. A vertical line at $x = -1$ passes through two points $(-1, 2)$ and $(-1, -2)$.

The domain is $\{-3, -1, 1, 3\}$

The range is $\{-2, -1, 1, 2, 4\}$

(b)

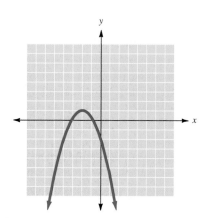

This is a function. The graph represents an infinite collection of ordered pairs, but since no vertical line will pass through more than one point, it passes the vertical line test.

The x values that are used in this graph consist of all real numbers, so

$$D = \{x \mid x \text{ is a real number}\} \qquad \text{or simply} \qquad D = \mathbb{R}$$

The y values never go higher than 1, so the range is the set of all real numbers less than or equal to 1.

$$R = \{y \mid y \leq 1\}$$

It is frequently necessary to read function values from a graph. This will generally involve one of the two following exercises: given x, find $f(x)$; or given $f(x)$, find x.

Example 5

Reading Values from a Graph

Given the graph of f, find the desired values.

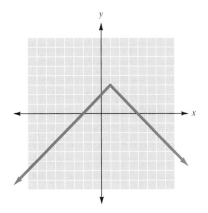

(a) Find $f(1)$.

Since $x = 1$, we move to 1 on the x axis. Searching vertically, we find the point $(1, 3)$. So $f(1) = 3$.

(b) Find all x such that $f(x) = 1$.

We are given that the output, or y value, is 1. So we move to 1 on the y axis and search horizontally. There are two points with a y value of 1: $(-1, 1)$ and $(3, 1)$. Since we want x values, we report $x = -1$ and 3.

We often find it useful to combine functions in mathematics. The following indicates how the usual operations of arithmetic are used to create new functions.

Algebra of Functions

1. The **sum of two functions** f and g is written as $f + g$ and is defined as

    $(f + g)(x) = f(x) + g(x)$ for every value of x in the domain of both f and g

2. The **difference of two functions** f and g is written as $f - g$ and is defined as

    $(f - g)(x) = f(x) - g(x)$ for every value of x in the domain of both f and g

3. The **product of two functions** f and g is written as $f \cdot g$ and is defined as

    $(f \cdot g)(x) = f(x) \cdot g(x)$ for every value of x in the domain of both f and g

4. The **quotient of two functions** f and g is written as $f \div g$ and is defined as

    $(f \div g)(x) = f(x) \div g(x)$ for every value of x in the domain of both f and g such that $g(x) \neq 0$

| Example 6 | **Combining Functions** |

Given $f(x) = x + 3$ and $g(x) = x - 4$:

(a) Find $(f \cdot g)(x)$.

Since $(f \cdot g)(x) = f(x) \cdot g(x)$, we have

$$(f \cdot g)(x) = f(x) \cdot g(x) = (x + 3)(x - 4) = x^2 - x - 12$$

(b) Find $(f \div g)(x)$.

Since $(f \div g)(x) = f(x) \div g(x)$, we have

$$(f \div g)(x) = f(x) \div g(x) = (x + 3) \div (x - 4) = \frac{x + 3}{x - 4}$$

(c) Find the domain of $f \div g$.

The domain is the set of all real numbers except 4 since $g(4) = 0$. So

$$D = \{x \mid x \neq 4\}$$

There is yet another way to combine functions, called **composition.**

Composition of Functions

The **composition** of functions f and g is the function $f \circ g$ where

$$(f \circ g)(x) = f(g(x))$$

The domain of the composition is the set of all elements x in the domain of g for which $g(x)$ is in the domain of f.

| Example 7 | **Composing Two Functions** |

Suppose we have the following functions f and g.

x	$g(x)$	x	$f(x)$
-2	3	-1	6
0	-1	0	-2
1	0	3	5
2	5	4	2

(a) Find $(f \circ g)(0)$.

$$
\begin{aligned}
(f \circ g)(0) &= f(g(0)) \qquad \text{We note that } g(0) = -1.\\
&= f(-1)\\
&= 6
\end{aligned}
$$

(b) Find $(g \circ f)(0)$.

$$
\begin{aligned}
(g \circ f)(0) &= g(f(0)) \qquad \text{We note that } f(0) = -2.\\
&= g(-2)\\
&= 3
\end{aligned}
$$

(c) Find $(f \circ g)(2)$.

$$
\begin{aligned}
(f \circ g)(2) &= f(g(2)) \qquad \text{We note that } g(2) = 5.\\
&= f(5)
\end{aligned}
$$

But $f(5)$ does not exist, because 5 is not in the domain of f. Therefore, 2 is not in the domain of $f \circ g$, and $(f \circ g)(2)$ does not exist.

More commonly, we deal with composition of functions represented in equation form.

Example 8   Composing Two Functions

Given $f(x) = x - 4$ and $g(x) = x^2 + 5$:

(a) Find $(f \circ g)(2)$.

$$
\begin{aligned}
(f \circ g)(2) &= f(g(2)) \qquad \text{We compute: } g(2) = 2^2 + 5 = 9.\\
&= f(9)\\
&= 9 - 4 = 5
\end{aligned}
$$

(b) Find $(g \circ f)(-1)$

$$
\begin{aligned}
(g \circ f)(-1) &= g(f(-1)) \qquad \text{We compute: } f(-1) = -1 - 4 = -5.\\
&= g(-5)\\
&= (-5)^2 + 5\\
&= 25 + 5 = 30
\end{aligned}
$$

(c) Find $(g \circ f)(x)$.

$$
\begin{aligned}
(g \circ f)(x) &= g(f(x))\\
&= g(x - 4)\\
&= (x - 4)^2 + 5\\
&= x^2 - 8x + 16 + 5\\
&= x^2 - 8x + 21
\end{aligned}
$$

Exercises · R.5

In Exercises 1 to 4, find the domain and range of each relation.

1. $\{(-1, 3), (0, 5), (2, 4), (5, 7)\}$

2. $\{(-2, 4), (1, 0), (2, 5), (3, 7)\}$

3. $\{(-1, 3), (0, 3), (2, 3), (4, 3)\}$

4. $\{(1, -2), (1, 1), (1, 2), (1, 5)\}$

In Exercises 5 and 6, find the domain and range of the relation given in the table.

5.

x	y
-3	1
0	5
1	3
2	-4

6.

x	y
0	0
1	3
2	0
4	5

In Exercises 7 to 12, determine which relations are also functions.

7. $\{(-2, 1), (0, 3), (1, 4), (2, 5)\}$

8. $\{(-3, -2), (-1, 0), (2, -1), (3, 0)\}$

9. $\{(-1, 2), (-1, -2), (2, 2), (4, 2)\}$

10. $\{(-2, 1), (-2, 3), (4, 5), (5, 6)\}$

11.

x	y
-3	0
-2	1
0	1
2	4

12.

x	y
-2	-2
0	5
-2	3
4	1

In Exercises 13 to 16, evaluate each function for the value specified.

13. $f(x) = x^2 + 3x - 1$ $f(-1)$

14. $f(x) = -2x^2 - 5x + 7$ $f(2)$

15. $f(x) = -2x^2 + 5x + 15$ $f(-3)$

16. $f(x) = 7x - 2$ $f(x + h)$

In Exercises 17 to 20, determine if the given graph is the graph of a function. Provide the domain and range.

17.

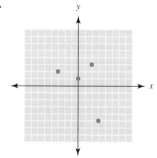

18.

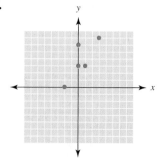

554

19.

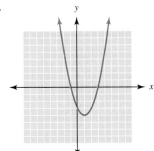

20.

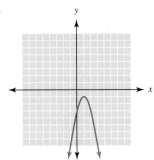

In Exercises 21 to 24, the graph of a function is shown. Find the indicated values

21.

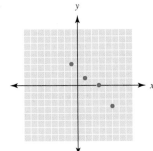

(a) $f(-1)$

(b) $f(5)$

(c) All x such that $f(x) = 3$

(d) All x such that $f(x) = 0$

22.

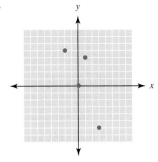

(a) $f(0)$

(b) $f(3)$

(c) All x such that $f(x) = 4$

(d) All x such that $f(x) = 5$

23.

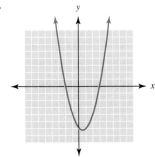

(a) $f(1)$

(b) $f(-2)$

(c) All x such that $f(x) = 0$

(d) All x such that $f(x) = -4$

24.

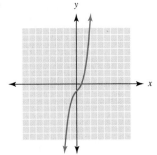

(a) $f(1)$

(b) $f(-1)$

(c) All x such that $f(x) = 8$

(d) All x such that $f(x) = -2$

In Exercises 25 to 28, find (a) $(f + g)(x)$; (b) $(f - g)(x)$; (c) $(f + g)(2)$; (d) $(f - g)(-1)$.

25. $f(x) = -3x + 5$ $g(x) = 7x - 2$ **26.** $f(x) = 4x - 9$ $g(x) = -11x + 7$

27. $f(x) = x^2 + 3x - 2$ $g(x) = -2x^2 - 5x + 7$ **28.** $f(x) = -2x^3 - 3x^2 + 5$ $g(x) = 4x^2 - 5x + 7$

In Exercises 29 and 30, find $(f + g)(x)$ and the domain of $f + g$.

29. $f(x) = 5x - 2$ $g(x) = \dfrac{1}{x - 3}$ **30.** $f(x) = x^2 + x - 1$ $g(x) = \dfrac{4}{x + 2}$

In Exercises 31 to 34, find (a) $(f \cdot g)(x)$; (b) $(f \div g)(x)$; (c) $(f \cdot g)(2)$; (d) $(f \div g)(-2)$.

31. $f(x) = x + 2$ $g(x) = 2x - 5$ **32.** $f(x) = 2x - 1$ $g(x) = x + 1$

33. $f(x) = 2 - 5x$ $g(x) = x - 3$ **34.** $f(x) = 7 - x$ $g(x) = x + 7$

In Exercises 35 and 36, use the following tables to find the desired values.

x	$f(x)$
-2	3
1	2
3	4
4	-1

x	$g(x)$
-2	5
0	3
1	2
3	5

35. $(g \circ f)(-2)$ **36.** $(f \circ g)(0)$

In Exercises 37 to 40, evaluate the composite function in each part.

37. $f(x) = x - 2$ $g(x) = 3x + 1$
Find (a) $(f \circ g)(0)$; (b) $(f \circ g)(-1)$; (c) $(g \circ f)(2)$; (d) $(g \circ f)(-1)$.

38. $f(x) = 2x + 1$ $g(x) = -x + 2$
Find (a) $(f \circ g)(-1)$; (b) $(f \circ g)(2)$; (c) $(g \circ f)(-1)$; (d) $(g \circ f)(-3)$.

39. $f(x) = x^2 + 3$ $g(x) = 3x$
Find (a) $(f \circ g)(0)$; (b) $(f \circ g)(-2)$; (c) $(g \circ f)(3)$; (d) $(g \circ f)(x)$.

40. $f(x) = 3x^2 + 1$ $g(x) = 3x$
Find (a) $(f \circ g)(-1)$; (b) $(f \circ g)(2)$; (c) $(g \circ f)(2)$; (d) $(g \circ f)(x)$.

R.6 Factoring Polynomials

In an earlier section, you were given polynomial factors and asked to find their product. In this section, you reverse that process. Here you will break a polynomial expression into its factors. First, we will look at an example of factoring integers.

Recall that we could write

$$18 = 3 \cdot 6$$

or we could write

$$18 = 2 \cdot 9$$

In either case, we have written 18 as a product of factors. On the other hand, there is only one correct *prime factorization* of 18. We write

$$18 = 2 \cdot 3 \cdot 3$$

It is important to keep this distinction in mind as we examine the topic of factoring polynomials.

Perhaps the most important property related to factoring is the distributive property. To this point, we have generally stated it as follows:

The Distributive Property

$$a(b + c) = ab + ac$$

When factoring, we are more interested in seeing this property applied in the opposite direction (it is actually mathematically identical).

$$ab + ac = a(b + c)$$

To factor a monomial from a polynomial, we find the GCF (greatest common factor) and use the distributive property.

Example 1	**Factoring a Monomial from a Polynomial**

Use the distributive property to factor each polynomial.

Note: To check, multiply the factors together. You should obtain the original polynomial.

(a) $6x^4 + 3x^3 + 12x^2$ The GCF is $3x^2$.

$= 3x^2(2x^2 + x + 4)$

(b) $12x^3y + 3x^2y - 3xy$ The GCF is $3xy$.

$= 3xy(4x^2 + x - 1)$

When we encounter four or more terms, it is sometimes possible to use the method of grouping terms to find the prime factorization of an expression.

| **Example 2** | **Factoring by Grouping Terms** |

Use the method of grouping terms to factor the expression

$$x^2 + 2xy - 3xz - 6yz$$

Grouping the two pairs of terms, we get

$$x(x + 2y) - 3z(x + 2y)$$

We now note a common factor of $x + 2y$.

$$(x + 2y)(x - 3z)$$

Several special products, introduced in Chapter 4, are used frequently in factoring.

$$(a + b)(a - b) = a^2 - b^2 \qquad \text{Difference of squares}$$
$$(a + b)(a + b) = a^2 + 2ab + b^2 \qquad \text{Perfect square trinomial}$$
$$(a + b)(a^2 - ab + b^2) = a^3 + b^3 \qquad \text{Sum of cubes}$$
$$(a - b)(a^2 + ab + b^2) = a^3 - b^3 \qquad \text{Difference of cubes}$$

Example 3 uses these special products.

| **Example 3** | **Factoring Using Special Products** |

Factor each expression.

(a) $4x^2 - y^4$ Note that $4x^2 = (2x)^2$ and $y^4 = (y^2)^2$.

  $= (2x + y^2)(2x - y^2)$

(b) $9m^2 - 24m + 16$ Note: $9m^2 = (3m)^2$, $16 = (-4)^2$, and $2(3m)(-4) = -24m$

  $= (3m - 4)^2$

(c) $x^3 - 8y^3$ Here, note that $8y^3 = (2y)^3$.

  $= (x - 2y)(x^2 + 2xy + 4y^2)$

The *ac* Test

A trinomial of the form $ax^2 + bx + c$ is factorable if (and only if) there are two integers m and n such that

$$ac = mn \quad \text{and} \quad b = m + n$$

Example 4   **Using the *ac* Test**

Use the *ac* test to determine if each of the following trinomials is factorable.

(a) $x^2 - 7x + 12$

$$a = 1 \qquad b = -7 \qquad c = 12$$

We are looking for m and n such that

$$mn = 12 \quad \text{and} \quad m + n = -7$$

We find that $m = -3$ and $n = -4$. The trinomial is factorable.

(b) $3x^2 - 6x + 7$

$$a = 3 \qquad b = -6 \qquad c = 7$$

We are looking for m and n such that

$$mn = 21 \quad \text{and} \quad m + n = -6$$

After trying every possible pair of factors for 21 (1 and 21, 3 and 7, -1 and -21, -3 and -7), we find that no such values exist. The trinomial is not factorable.

(c) $2x^2 + x - 6$

$$a = 2 \qquad b = 1 \qquad c = -6$$

We are looking for m and n such that

$$mn = -12 \quad \text{and} \quad m + n = 1$$

We find that $m = -3$ and $n = 4$. The trinomial is factorable.
    Having found that a trinomial is factorable, we can use the values of m and n to complete the factorization.

Example 5   **Using the Results of an *ac* Test to Factor**

Factor each of the following.

(a) $x^2 - 7x + 12$

We found that $m = -3$ and $n = -4$. Using those values, we rewrite the middle term as

$$x^2 - 3x - 4x + 12$$

Using the method of grouping, we get

$$x(x - 3) - 4(x - 3) = (x - 3)(x - 4)$$

So,

$$x^2 - 7x + 12 = (x - 3)(x - 4)$$

(b) $2x^2 + x - 6$

We found that $m = -3$ and $n = 4$.

$$
\begin{aligned}
2x^2 + x - 6 &= 2x^2 + (-3x) + 4x - 6 \\
&= x(2x - 3) + 2(2x - 3) \\
&= (2x - 3)(x + 2)
\end{aligned}
$$

An equation of the form

$$ax^2 + bx + c = 0$$

is called a quadratic equation in standard form. Suppose we have such an equation. If the trinomial $ax^2 + bx + c$ is factorable, the equation can be solved by using the following principle.

Zero-Product Principle

If $ab = 0$, then either $a = 0$ or $b = 0$ (or both equal zero).

Example 6 uses this principle to solve a quadratic equation.

Example 6   Using the Zero-Product Principle to Solve an Equation

Solve each of the following equations.

(a) $x^2 - 7x + 12 = 0$

From Example 5, we have

$$x^2 - 7x + 12 = (x - 3)(x - 4)$$

Therefore

$$(x - 3)(x - 4) = 0$$

By the zero-product principle, one of the two factors must be equal to zero.

You can (and should!) check
each solution by substituting
the value back into the
original equation.

$$x - 3 = 0 \quad \text{or} \quad x - 4 = 0$$
$$x = 3 \quad \text{or} \quad x = 4$$

The solution set is $\{3, 4\}$.

(b) $2x^2 + x - 6 = 0$

We found that

$$2x^2 + x - 6 = (x + 2)(2x - 3)$$

So $\qquad (x + 2)(2x - 3) = 0$

Therefore

Note that if $2x - 3 = 0$, then
$2x = 3$, and so $x = \dfrac{3}{2}$.

$$x + 2 = 0 \quad \text{or} \quad 2x - 3 = 0$$
$$x = -2 \quad \text{or} \quad x = \frac{3}{2}$$

The solution set is $\left\{-2, \dfrac{3}{2}\right\}$.

(c) $2x^2 - 10x - 48 = 0$

Be certain that all common factors are found before any other techniques are used.

$$2x^2 - 10x - 48 = 2(x^2 - 5x - 24) \qquad \text{We factor the trinomial by using}$$
$$= 2(x - 8)(x + 3) \qquad\qquad\quad \text{the } ac \text{ method or trial and error.}$$

Returning to the original equation, we have

$$2(x - 8)(x + 3) = 0$$

The zero-product principle tells us that one of the factors must equal zero. We know that $2 \neq 0$, so we have two other possibilities:

$$x - 8 = 0 \quad \text{or} \quad x + 3 = 0$$
$$x = 8 \quad \text{or} \quad x = -3$$

The solution set is $\{8, -3\}$.

The techniques for solving quadratic equations by factoring can be applied to many problems. Example 7 illustrates an application involving geometry.

| Example 7 | **Solving a Geometric Application** |

The length of a rectangle is 3 ft more than the width. If the area of the rectangle is 70 ft^2, what is the length of the rectangle?

Step 1   We identify the unknowns: the length and the width of the rectangle.

Step 2   We give the unknowns algebraic names: Let x represent the width. Then $x + 3$ represents the length. It's a good idea to draw a sketch.

Step 3   Form an equation. Since the area of a rectangle is the length times the width,

$$x(x + 3) = 70$$

Step 4   Solve.

$$x(x + 3) = 70 \qquad \text{Remove parentheses.}$$
$$x^2 + 3x = 70 \qquad \text{Create a "0" on one side.}$$
$$x^2 + 3x - 70 = 0 \qquad \text{Factor.}$$
$$(x + 10)(x - 7) = 0 \qquad \text{Use the zero-product principle.}$$
$$x + 10 = 0 \qquad \text{or} \qquad x - 7 = 0$$
$$x = -10 \qquad \text{or} \qquad x = 7$$

Step 5   We reject -10, since x (the width) must be positive. So the width is 7 ft. However, we were asked to find the length. Since length $= x + 3$, the length is 10 ft. We verify that these dimensions are correct: 7 ft $\times$ 10 ft $= 70$ ft^2, which is the given area.

Exercises · R.6

Factor by removing the greatest common factor (GCF).

1. $x^2 - xy$

2. $10a^2b - 2ab^2 + 6ab$

3. $36x^4y^2 - 18x^3y^3 - 4xy$

4. $16rt^2 + 24r^2t$

5. $3y(y - 1) + y^2(y - 1)$

6. $4x(7x - 8) + 3(7x - 8)$

Use the method of grouping to factor the expression.

7. $15x^2 - 10x + 9x - 6$

8. $10x^2 - 35x - 2x + 7$

9. $12m^2 - 20m + 40 - 24m$

10. $7y^2 - 7y + 35 - 35y$

Factor each expression.

11. $4y^2 - 25$

12. $5x^2 - 125$

13. $(x - 3)^2 - y^2$

14. $y^2 - 18y + 81$

15. $50x^2 + 60x + 18$

16. $-3t^2 + 48$

17. $27x^3 - 8y^3$

18. $8x^6 + y^3$

Use the *ac* test to determine if the given trinomial is factorable.

19. $x^2 - 2x - 8$

20. $x^2 + x + 6$

Factor each trinomial completely.

21. $5x^2 - 14x + 8$

22. $9y^2 + 28y + 20$

23. $12m^2 + 2m - 30$

24. $8y^2 - 68y + 84$

25. $63w^2 - 38w - 16$

26. $-20p^2 - 28p - 8$

Solve each equation.

27. $x^2 - 5x = 0$

28. $x^2 + 10x + 9 = 0$

29. $6x^2 + x = 15$

30. $y^2 - 13y = 14$

31. $3x^2 = 30 - 9x$

32. $8x^2 = 60 - 4x$

Solve each of the following applications.

33. One integer is 9 more than another. The sum of the squares of those two integers is 53. Find the two integers.

34. The length of a rectangle is 2 cm more than twice the width. If the area of the rectangle is 60 cm^2, find the length.

35. If a ball is thrown vertically upward from a height of 6 ft with an initial velocity of 80 ft/s, its approximate height h after t seconds is given by

$$h = -16t^2 + 80t + 6$$

When will the ball reach a height of 102 ft on the way down?

Final Exam   0–6

Evaluate each expression.

1. $4 - 5 + 3 \cdot 2^2 - 9$

2. $16 \div 8 + 4 \cdot 2$

3. $\dfrac{7}{8} + \dfrac{5}{12} - \dfrac{2}{3}$

4. $\dfrac{3}{5} \cdot \dfrac{6}{7} \div \dfrac{9}{10}$

Evaluate each expression where $x = 3$ and $y = -5$.

5. $-12xy$

6. $x^5 + y^3$

7. $(x + y)(x - y)$

8. $3x^2y - 2xy^2$

9. Plot the elements of the set $\{x \mid -2 \le x \le 5\}$.

10. Use set-builder notation to describe this set.

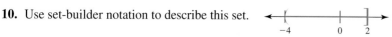

Simplify the expression.

11. $7x^2y + 3x - 5x^2y + 2xy$

12. $(5x^2 + 4x - 3) - (4x^2 - 5x - 3)$

Solve each equation and check your result.

13. $7a - 3 = 6a + 8$

14. $\dfrac{2}{3}x = -22$

15. $12 - 5x = 4$

16. $5x - 2(x - 3) = 9$

17. $|x - 6| = 9$

18. $|2x + 9| - 5 = 6$

19. $\dfrac{x + 2}{3} - \dfrac{2x + 5}{4} = 2$

Solve each inequality.

20. $2x - 7 < 5 + 4x$

21. $|2x - 7| < 6$

22. $|x + 5| > 3$

23. $|3x - 4| \ge 10$

24. $-3 < 2x + 5 < 9$

25. $2x + 1 < -5$ or $2x + 1 > 5$

564

Graph each of the functions.

26. $f(x) = 3x$

27. $f(x) = \dfrac{4}{5}x + 4$

28. $f(x) = -3x - 3$

29. $f(x) = 4x + \dfrac{9}{2}$

30. Find the slope of the line through the points $(3, -4)$ and $(-1, 4)$.

31. Find the slope and y intercept of the line represented by the equation $4x + y = 9$.

Find an equation of the line L that satisfies the given set of conditions.

32. L passes through $(-2, 5)$ and is parallel to the line with equation $y + 3x = 5$.

33. L has a y intercept of $(0, -1)$ and passes through the points $(2, 3)$ and $(4, 7)$.

In Exercises 34 to 37, determine whether each relation represents a function.

34. $\{(1, 1), (2, 1), (3, 1), (4, 1)\}$

35. $\{(1, 1), (2, 2), (1, -1)\}$

36.

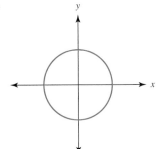

37.

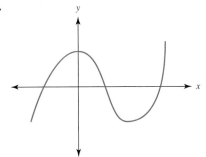

In Exercises 38 to 43, $f(x) = -4x + 7$, $g(x) = -2x^2$, and $h(x) = 3x - 5$. Find the following.

38. $(f + h)(x)$

39. $(f - h)(x)$

40. $(f \cdot g)(x)$

41. $(f \div g)(2)$

42. $(f \circ g)(x)$

43. $(g \circ f)(-1)$

Multiply or divide as indicated.

44. $(a - 3b)(a + 3b)$

45. $(x - 2y)^2$

46. $(x - 2)(x + 5)$

47. $(a - 3)(a + 4)$

48. $(9x^2 + 12x + 4) \div (3x + 2)$

49. $(3x^2 - 2) \div (x - 1)$

Factor each expression.

50. $12x + 20$

51. $25x^2 - 36y^2$

52. $12x^2 - 15x + 8x - 10$

53. $2x^2 - 13x + 15$

Solve each equation by factoring.

54. $x^2 - 3x = 28$

55. $x^2 - 10 = 3x$

56. $49x^2 - 9 = 0$

57. $2x^2 + 15x = 8$

In Exercises 58 to 64, solve the given problem.

58. The perimeter of a rectangle is 124 in. If the length of the rectangle is 1 in. more than its width, what are the dimensions of the rectangle?

59. Juan's biology text cost $5 more than his mathematics text. Together they cost $141. Find the cost of each text.

60. Boats can be rented for $34.50 for the first 3 h and $4.50 for each additional hour. How much will it cost you if you rent the boat for 5 h?

61. One integer is 8 more than another integer. The sum of the squares of those two integers is 34. Find the two integers.

62. The length of a rectangle is 2 ft less than 3 times the width. If the area of the rectangle is 65 ft², find the dimensions of the rectangle.

63. If a ball is thrown vertically upward from a height of 5 ft with an initial velocity of 80 ft/s, its approximate height h after t seconds is given by

$$h = -16t^2 + 80t + 5$$

When will the ball reach a height of 69 ft?

64. The cost C, in dollars, of producing x flashlights is given by

$$C = x^2 - 20x + 300$$

How many flashlights can be produced for $425?

In Exercises 65 to 68, factor each expression.

65. $8m^3 - n^3$

66. $27x - 3x^3$

67. $5xy^2 + 10xy + 15y^2 + 30y$

68. $3m^3 + 24m^2 + 48m$

LIST OF SECTIONS

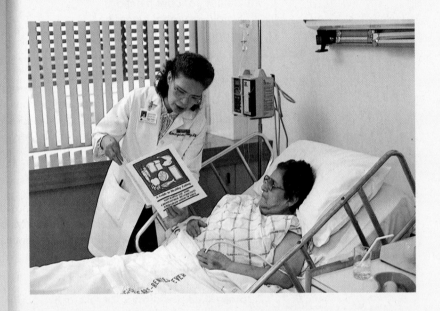

In the United States, disorders of the heart and circulatory system kill more people than all other causes combined. The major risk factors for heart disease are smoking, high blood pressure, obesity, cholesterol over 240, and a family history of heart problems. Although nothing can be done about family history, everyone can affect the first four risk factors by diet and exercise.

One quick way to check your risk of heart problems is to compare your waist and hip measurements. Measure around your waist at the naval and around your hips at the largest point. These measures may be in inches or centimeters. Use the ratio $\frac{w}{h}$ to assess your risk. For women, $\frac{w}{h} \geq 0.8$ indicates an increased health risk, and for men, $\frac{w}{h} \geq 0.95$ is the indicator of an increased risk.

7.1 Simplifying Rational Expressions

7.1 OBJECTIVES

1. Simplify rational expressions
2. Identify rational functions
3. Simplify rational functions
4. Graph rational functions

The word *rational* comes from "ratio."

Our work in this chapter will expand your experience with algebraic expressions to include algebraic fractions or **rational expressions.** We consider the four basic operations of addition, subtraction, multiplication, and division in the next sections. Fortunately, you will observe many parallels to your previous work with arithmetic fractions.

Evaluating Rational Expressions

First, let's define what we mean by a rational expression. Recall that a rational number is the ratio of two integers. Similarly, a rational expression can be written as the ratio of two polynomials, in which the denominator cannot have the value 0.

A **rational expression** is the ratio of two polynomials. It can be written as

$$\frac{P}{Q}$$

where P and Q are polynomials and Q cannot have the value 0.

$$\frac{x-3}{x+1} \qquad \frac{x^2+5}{x-3} \qquad \text{and} \qquad \frac{x^2-2x}{x^2+3x+1}$$

are all rational expressions.

| Example 1 | **Precluding Division by Zero** |

(a) For what values of x is the following expression undefined?

$$\frac{x}{x-5}$$

A fraction is undefined when its denominator is equal to 0. Note that when $x = 5$, $\dfrac{x}{x-5}$ becomes $\dfrac{5}{5-5}$, or $\dfrac{5}{0}$.

To answer this question, we must find where the denominator is 0. Set

$$x - 5 = 0$$
$$x = 5$$

The expression $\dfrac{x}{x-5}$ is undefined for $x = 5$.

(b) For what values of x is the following expression undefined?

$$\frac{3}{x+5}$$

Again, set the denominator equal to 0:

$$x + 5 = 0$$
$$x = -5$$

The expression $\dfrac{3}{x+5}$ is undefined for $x = -5$.

 CHECK YOURSELF 1

For what values of the variable are the following expressions undefined?

(a) $\dfrac{1}{r+7}$ (b) $\dfrac{5}{2x-9}$

Scientific calculators are often used to evaluate rational expressions for values of the variable. The *parentheses keys* help in this process.

| **Example 2** | **Evaluating a Rational Expression** |

Using a calculator, evaluate the following expressions for the given value of the variable.

(a) $\dfrac{3x}{2x-5}$ for $x = 4$

Enter the expression in your calculator as follows:

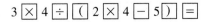

$$3 \boxed{\times} 4 \boxed{\div} \boxed{(} 2 \boxed{\times} 4 \boxed{-} 5 \boxed{)} \boxed{=}$$

Caution

Be sure to use the parentheses keys before the 2 and after the 5.

The display will read the value 4.

(b) $\dfrac{2x + 7}{4x - 11}$ for $x = 4$

Enter the expression as follows:

$\boxed{(}\ \boxed{2}\ \boxed{\times}\ \boxed{4}\ \boxed{+}\ \boxed{7}\ \boxed{)}\ \boxed{\div}\ \boxed{(}\ \boxed{4}\ \boxed{\times}\ \boxed{4}\ \boxed{-}\ \boxed{11}\ \boxed{)}\ \boxed{=}$

The display will read the value 3.

> Because the numerator has more than one term, it must be enclosed in parentheses.

✔ CHECK YOURSELF 2

Using a scientific calculator, evaluate each of the following.

(a) $\dfrac{5x}{3x - 2}$ for $x = 4$ (b) $\dfrac{2x + 9}{3x - 4}$ for $x = 3$

Simplifying Rational Expressions

Generally, we want to write rational expressions in the simplest possible form. To begin our discussion of simplifying rational expressions, let's review for a moment. As we pointed out previously, there are many parallels to your work with arithmetic fractions. Recall that

$$\frac{3}{5} = \frac{3 \cdot 2}{5 \cdot 2} = \frac{6}{10}$$

so $\dfrac{3}{5}$ and $\dfrac{6}{10}$

name equivalent fractions. In a similar fashion,

$$\frac{10}{15} = \frac{5 \cdot 2}{5 \cdot 3} = \frac{2}{3}$$

so $\dfrac{10}{15}$ and $\dfrac{2}{3}$

name equivalent fractions.

We can always multiply or divide the numerator and denominator of a fraction by the same nonzero number. The same pattern is true in algebra.

Fundamental Principle of Rational Expressions

For polynomials P, Q, and R,

$$\frac{P}{Q} = \frac{PR}{QR} \qquad \text{where } Q \neq 0 \text{ and } R \neq 0$$

This principle can be used in two ways. We can multiply or divide the numerator and denominator of a rational expression by the same nonzero polynomial. The result will always be an expression that is equivalent to the original one.

In simplifying arithmetic fractions, we used this principle to divide the numerator and denominator by all common factors. With arithmetic fractions, those common factors are generally easy to recognize. Given rational expressions where the numerator and denominator are polynomials, we must determine those factors as our first step. The most important tools for simplifying expressions are the factoring techniques in Chapter 6.

In fact, you will see that most of the methods in this chapter depend on factoring polynomials.

| **Example 3** | # Simplifying Rational Expressions |

We find the common factors 4, x, and y in the numerator and denominator. We divide the numerator and denominator by the common factor 4xy. Note that

$$\frac{4xy}{4xy} = 1$$

Simplify each rational expression. Assume the denominators are not 0.

(a) $\dfrac{4x^2y}{12xy^2} = \dfrac{4xy \cdot x}{4xy \cdot 3y}$

$= \dfrac{x}{3y}$

(b) $\dfrac{3x - 6}{x^2 - 4} = \dfrac{3(x - 2)}{(x + 2)(x - 2)}$ Factor the numerator and the denominator.

We can now divide the numerator and denominator by the common factor $x - 2$:

$$\frac{3(x - 2)}{(x + 2)(x - 2)} = \frac{3}{x + 2}$$

and the rational expression is in simplest form.

Be careful! Given the expression

$$\frac{x + 2}{x + 3}$$

We have divided the numerator and denominator by the common factor x − 2. Again note that

$$\frac{x - 2}{x - 2} = 1$$

Caution

Pick any value other than 0 for the variable *x* and substitute. You will quickly see that

$$\frac{x + 2}{x + 3} \neq \frac{2}{3}$$

students are often tempted to divide by variable *x*, as in

$$\frac{x + 2}{x + 3} \overset{?}{=} \frac{2}{3}$$

This is not a valid operation. We can only divide by common *factors*, and in the expression above the variable *x* is a *term* in both the numerator and the denominator. The numerator and denominator of a rational expression must be factored *before* common factors are divided out. Therefore,

$$\frac{x + 2}{x + 3}$$

is in its simplest possible form.

✔ CHECK YOURSELF 3

Simplify each expression.

(a) $\dfrac{36a^3b}{9ab^2}$ (b) $\dfrac{x^2 - 25}{4x + 20}$

The same techniques are used when trinomials need to be factored.

| Example 4 | **Simplifying Rational Expressions** |

Divide by the common factor $x + 1$, using the fact that

$$\dfrac{x + 1}{x + 1} = 1$$

where $x \neq -1$.

Simplify each rational expression.

(a) $\dfrac{5x^2 - 5}{x^2 - 4x - 5} = \dfrac{5(x^2 - 1)}{x^2 - 4x - 5} = \dfrac{5(x - 1)(x + 1)}{(x - 5)(x + 1)}$

$$= \dfrac{5(x - 1)}{x - 5}$$

(b) $\dfrac{2x^2 + x - 6}{2x^2 - x - 3} = \dfrac{(x + 2)(2x - 3)}{(x + 1)(2x - 3)}$

$$= \dfrac{x + 2}{x + 1}$$

In part (c) we factor by grouping in the numerator and use the sum of cubes in the denominator.

(c) $\dfrac{x^3 + 2x^2 - 3x - 6}{x^3 + 8} = \dfrac{x^2(x + 2) - 3(x + 2)}{(x + 2)(x^2 - 2x + 4)} = \dfrac{(x + 2)(x^2 - 3)}{(x + 2)(x^2 - 2x + 4)}$

$$= \dfrac{x^2 - 3}{x^2 - 2x + 4}$$

✔ CHECK YOURSELF 4

Simplify each rational expression.

(a) $\dfrac{x^2 - 5x + 6}{3x^2 - 6x}$ (b) $\dfrac{3x^2 + 14x - 5}{3x^2 + 2x - 1}$

Simplifying certain algebraic expressions involves recognizing a particular pattern. Verify for yourself that

$$3 - 9 = -(9 - 3)$$

In general, it is true that

$$a - b = -(-a + b) = -(b - a) = -1(b - a)$$

or, by dividing both sides of the equation by $b - a$,

Note that

$$\frac{a - b}{a - b} = 1$$

but

$$\frac{a - b}{b - a} = -1$$

$$\frac{a - b}{b - a} = \frac{-(b - a)}{b - a} = -1 \qquad a \neq b$$

Example 5 makes use of this result.

Example 5   Simplifying Rational Expressions

Simplify each rational expression.

Note that

$$\frac{x - 2}{2 - x} = -1$$

(a) $\dfrac{2x - 4}{4 - x^2} = \dfrac{2(x - 2)}{(2 + x)(2 - x)}$

$$= \frac{2(-1)}{2 + x} = \frac{-2}{2 + x}$$

(b) $\dfrac{9 - x^2}{x^2 + 2x - 15} = \dfrac{(3 + x)(3 - x)}{(x + 5)(x - 3)}$

$$= \frac{(3 + x)(-1)}{x + 5} = \frac{-x - 3}{x + 5}$$

Simplifying Rational Expressions

1. Completely factor both the numerator and the denominator of the expression.

2. Divide the numerator and denominator by *all* common factors.

3. The resulting expression will be in simplest form (or in lowest terms).

✔ *CHECK YOURSELF 5*

Simplify each rational expression.

(a) $\dfrac{5x - 20}{16 - x^2}$ (b) $\dfrac{x^2 - 6x - 27}{81 - x^2}$

A **rational function** is a function that is defined by a rational expression. It can be written as

$$f(x) = \frac{P}{Q}$$

where P and Q are polynomials. The function is *not* defined for any value of x for which $Q(x) = 0$.

| Example 6 | Identifying Rational Functions |

Which of the following are rational functions?

(a) $f(x) = 3x^3 - 2x + 5$ This is a rational function; it could be written over the denominator 1, and 1 is a polynomial.

(b) $f(x) = \dfrac{3x^2 - 5x + 2}{2x - 1}$ This is a rational function; it is the ratio of two polynomials.

Recall from Chapter 4 that there are no square roots of variables in a polynomial.

(c) $f(x) = 3x^3 + 3\sqrt{x}$ This is not a rational function; it is not the ratio of two polynomials.

✔ *CHECK YOURSELF 6*

Which of the following are rational functions?

(a) $f(x) = x^5 - 2x^4 - 1$ (b) $f(x) = \dfrac{x^2 - x + 7}{\sqrt{x} - 1}$ (c) $f(x) = \dfrac{3x^3 + 3x}{2x + 1}$

Simplifying Rational Functions

When we simplify a rational function, it is important that we note the x values that need to be excluded from the domain of the original function, particularly when we are trying to draw the graph of a function. The set of ordered pairs of the simplified function will be exactly the same as the set of ordered pairs of the original function. If we plug the excluded value(s) for x into the simplified expression, we get a set of ordered pairs that represent "holes" in the graph. These holes are breaks in the curve. We use an open circle to designate them on a graph.

| Example 7 | Simplifying a Rational Function |

Given the function

Note: The original function is not defined for $x = -1$.

$$f(x) = \frac{x^2 + 2x + 1}{x + 1}$$

(a) Rewrite the function in simplified form.

$$f(x) = \frac{x^2 + 2x + 1}{x + 1} = \frac{(x + 1)(x + 1)}{(x + 1)}$$

$$= x + 1 \qquad x \neq -1$$

(b) Find the ordered pair associated with the hole in the graph of the original function.

Replacing x with -1 in the simplified function yields the ordered pair $(-1, 0)$. This represents the hole in the graph of the function $f(x)$.

 **CHECK YOURSELF 7**

Given the function

$$f(x) = \frac{5x^2 - 10x}{5x}$$

(a) Rewrite the function in simplified form.

(b) Find the ordered pair associated with the hole in the graph of the original function.

Example 8   **Graphing a Rational Function**

Graph the following function.

$$f(x) = \frac{x^2 + 2x + 1}{x + 1}$$

From Example 7, we know that

$$\frac{x^2 + 2x + 1}{x + 1} = x + 1 \qquad x \neq -1$$

Therefore,

$$f(x) = x + 1 \qquad x \neq -1$$

The graph will be the graph of the line $f(x) = x + 1$, with an open circle at the point $(-1, 0)$.

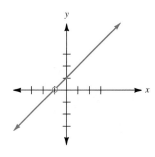

✔ *CHECK YOURSELF 8*

Graph the function $f(x) = \dfrac{5x^2 - 10x}{5x}$.

✔ *CHECK YOURSELF ANSWERS*

1. (a) $r = -7$; (b) $x = \dfrac{9}{2}$ **2.** (a) 2; (b) 3   **3.** (a) $\dfrac{4a^2}{b}$; (b) $\dfrac{x-5}{4}$

4. (a) $\dfrac{x-3}{3x}$; (b) $\dfrac{x+5}{x+1}$ **5.** (a) $\dfrac{-5}{x+4}$; (b) $\dfrac{-x-3}{x+9}$

6. (a) A rational function; (b) not a rational function; (c) a rational function

7. (a) $f(x) = x - 2$, $x \neq 0$; (b) $(0, -2)$

8.

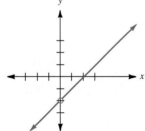

$\bigcirc$xercises · 7.1

For what values of the variable is each rational expression undefined?

1. $\dfrac{x}{x-3}$

2. $\dfrac{y}{y+7}$

3. $\dfrac{x+5}{3}$

4. $\dfrac{x-6}{4}$

5. $\dfrac{2x-3}{2x-1}$

6. $\dfrac{4x-5}{5x+2}$

7. $\dfrac{2x+5}{x}$

8. $\dfrac{3x-7}{x}$

9. $\dfrac{x(x+1)}{x+2}$

10. $\dfrac{x+2}{3x-7}$

11. $\dfrac{5-3x}{2x}$

12. $\dfrac{2x+7}{3x+\frac{1}{3}}$

Evaluate each expression, using a calculator.

13. $\dfrac{3x}{2x-1}$ for $x=2$

14. $\dfrac{4x}{5x-6}$ for $x=2$

15. $\dfrac{3x+10}{x+2}$ for $x=-6$

16. $\dfrac{4x-7}{2x-1}$ for $x=-2$

Simplify each expression. Assume the denominators are not 0.

17. $\dfrac{14}{21}$

18. $\dfrac{45}{75}$

19. $\dfrac{4x^5}{6x^2}$

20. $\dfrac{30x^8}{25x^3}$

21. $\dfrac{10x^2y^5}{25xy^2}$

22. $\dfrac{18a^2b^3}{24a^4b^3}$

23. $\dfrac{-36x^5y^3}{21x^2y^5}$

24. $\dfrac{-15x^3y^3}{-20xy^2}$

25. $\dfrac{28a^5b^3c^2}{84a^2bc^4}$

26. $\dfrac{-52p^5q^3r^2}{39p^3q^5r^2}$

27. $\dfrac{6x-24}{x^2-16}$

28. $\dfrac{x^2-25}{3x-15}$

29. $\dfrac{x^2+2x+1}{6x+6}$

30. $\dfrac{5y^2-10y}{y^2+y-6}$

31. $\dfrac{x^2-13x+36}{x^2-81}$

32. $\dfrac{2m^2+11m-21}{4m^2-9}$

33. $\dfrac{3b^2-7b-6}{b-3}$

34. $\dfrac{a^2-9b^2}{a^2+8ab+15b^2}$

35. $\dfrac{2y^2 + 3yz - 5z^2}{2y^2 + 11yz + 15z^2}$

36. $\dfrac{6x^2 - x - 2}{3x^2 - 5x + 2}$

37. $\dfrac{x^3 - 64}{x^2 - 16}$

38. $\dfrac{r^2 - rs - 6s^2}{r^3 + 8s^3}$

39. $\dfrac{a^4 - 81}{a^2 + 5a + 6}$

40. $\dfrac{x^4 - 625}{x^2 - 2x - 15}$

41. $\dfrac{xy - 2x + 3y - 6}{x^2 + 8x + 15}$

42. $\dfrac{cd - 3c + 5d - 15}{d^2 - 7d + 12}$

43. $\dfrac{x^2 + 3x - 18}{x^3 - 3x^2 - 2x + 6}$

44. $\dfrac{y^2 + 2y - 35}{y^2 - 8y + 15}$

45. $\dfrac{2m - 10}{25 - m^2}$

46. $\dfrac{5x - 20}{16 - x^2}$

47. $\dfrac{121 - x^2}{2x^2 - 21x - 11}$

48. $\dfrac{2x^2 - 7x + 3}{9 - x^2}$

In Exercises 49 to 54, identify which functions are rational functions.

49. $f(x) = -7x^2 + 2x - 5$

50. $f(x) = \dfrac{x^3 - 2x^2 + 7}{\sqrt{x} + 2}$

51. $f(x) = \dfrac{x^2 - x - 1}{x + 2}$

52. $f(x) = \dfrac{\sqrt{x} - x + 3}{x - 2}$

53. $f(x) = 5x^2 - \sqrt[3]{x}$

54. $f(x) = \dfrac{x^2 - x + 5}{x}$

For the given functions in Exercises 55 to 60, (a) rewrite the function in simplified form and (b) find the ordered pair associated with the hole in the graph of the original function.

55. $f(x) = \dfrac{x^2 - x - 2}{x + 1}$

56. $f(x) = \dfrac{x^2 + x - 12}{x + 4}$

57. $f(x) = \dfrac{3x^2 + 5x - 2}{x + 2}$

58. $f(x) = \dfrac{2x^2 - 7x + 5}{2x - 5}$

59. $f(x) = \dfrac{x^2 + 4x + 4}{5(x + 2)}$

60. $f(x) = \dfrac{x^2 - 6x + 9}{7(x - 3)}$

In Exercises 61 to 66, graph the rational functions. Indicate the coordinates of the hole in the graph.

61. $f(x) = \dfrac{x^2 - 2x - 8}{x + 2}$

62. $f(x) = \dfrac{x^2 + 4x - 5}{x + 5}$

63. $f(x) = \dfrac{x^2 + 4x + 3}{x + 1}$

64. $f(x) = \dfrac{x^2 + 7x + 10}{x + 2}$

65. $f(x) = \dfrac{x^2 - 4x + 3}{x - 1}$

66. $f(x) = \dfrac{x^2 - 6x + 8}{x - 4}$

67. Explain why the following statement is false.

$$\frac{6m^2 + 2m}{2m} = 6m^2 + 1$$

68. State and explain the fundamental principle of rational expressions.

69. The rational expression $\dfrac{x^2 - 4}{x + 2}$ can be simplified to $x - 2$. Is this reduction true for all values of x? Explain.

70. What is meant by a rational expression in lowest terms?

In Exercises 71 to 76, simplify.

71. $\dfrac{2(x + h) - 2x}{(x + h) - x}$

72. $\dfrac{-3(x + h) - (-3x)}{(x - h) - x}$

73. $\dfrac{3(x + h) - 3 - (3x - 3)}{(x + h) - x}$

74. $\dfrac{2(x + h) + 5 - (2x + 5)}{(x + h) - x}$

75. $\dfrac{(x + h)^2 - x^2}{(x + h) - x}$

76. $\dfrac{(x + h)^3 - x^3}{(x + h) - x}$

Given $f(x) = \dfrac{P(x)}{Q(x)}$, if the graphs of $P(x)$ and $Q(x)$ intersect at $(a, 0)$, then $x - a$ is a factor of both $P(x)$ and $Q(x)$. Use a graphing calculator to find the common factor for the expressions in Exercises 77 and 78.

77. $f(x) = \dfrac{x^2 + 4x - 5}{x^2 + 3x - 10}$

78. $f(x) = \dfrac{2x^2 + 11x - 21}{2x^2 + 15x + 7}$

79. Cost. A company has a setup cost of $3,500 for the production of a new product. The cost to produce a single unit is $8.75.

  (a) Write a rational function that gives the average cost per unit when x units are produced.

  (b) Find the average cost when 50 units are produced.

80. Revenue. The total revenue from the sale of a popular video is approximated by the rational function

$$R(x) = \frac{300x^2}{x^2 + 9}$$

where x is the number of months since the video has been released and $R(x)$ is the total revenue in hundreds of dollars.

(a) Find the total revenue generated by the end of the first month.

(b) Find the total revenue generated by the end of the second month.

(c) Find the total revenue generated by the end of the third month.

(d) Find the revenue in the second month only.

81. Besides holes, we sometimes encounter a different sort of "break" in the graph of a rational function. Consider the rational function

$$f(x) = \frac{1}{x - 3}$$

(a) For what value(s) of x is the function undefined?

(b) Complete the following table.

x	$f(x)$
4	
3.1	
3.01	
3.001	
3.0001	

(c) What do you observe concerning $f(x)$ as x is chosen close to 3 (but slightly larger than 3)?

(d) Complete the table.

x	$f(x)$
2	
2.9	
2.99	
2.999	
2.9999	

(e) What do you observe concerning $f(x)$ as x is chosen close to 3 (but slightly smaller than 3)?

(f) Graph the function on your graphing calculator. Describe the behavior of the graph of f near $x = 3$.

SECTION 7.2

Multiplication and Division of Rational Expressions

 OBJECTIVES

1. *Multiply and divide rational expressions*
2. *Multiply and divide two rational functions*

Once again, let's turn to an example from arithmetic to begin our discussion of multiplying rational expressions. Recall that to multiply two fractions, we multiply the numerators and multiply the denominators. For instance,

$$\frac{2}{5} \cdot \frac{3}{7} = \frac{2 \cdot 3}{5 \cdot 7} = \frac{6}{35}$$

In algebra, the pattern is exactly the same.

Multiplying Rational Expressions

For polynomials P, Q, R, and S,

$$\frac{P}{Q} \cdot \frac{R}{S} = \frac{PR}{QS} \qquad \text{where } Q \neq 0 \text{ and } S \neq 0$$

For all problems with rational expressions, assume the denominators are not 0.

| Example 1 | **Multiplying Rational Expressions** |

Multiply.

$$\frac{2x^3}{5y^2} \cdot \frac{10y}{3x^2} = \frac{20x^3y}{15x^2y^2}$$

$$= \frac{5x^2y \cdot 4x}{5x^2y \cdot 3y} \qquad \text{Divide by the common factor } 5x^2y \text{ to simplify.}$$

$$= \frac{4x}{3y}$$

 CHECK YOURSELF 1

Multiply.

$$\frac{9a^2b^3}{5ab^4} \cdot \frac{20ab^2}{27ab^3}$$

The factoring methods in Chapter 6 are used to simplify rational expressions.

Generally, you will find it best to divide by any common factors before you multiply, as Example 2 illustrates.

Example 2

Multiplying Rational Expressions

Multiply as indicated.

(a) $\dfrac{x}{x^2 - 3x} \cdot \dfrac{6x - 18}{9x}$ Factor.

$$= \dfrac{\overset{1}{\cancel{x}}}{\underset{1}{\cancel{x}}(\underset{1}{\cancel{x - 3}})} \cdot \dfrac{\overset{2}{\cancel{6}}(\overset{1}{\cancel{x - 3}})}{\underset{3}{\cancel{9}}x}$$ Divide by the common factors of 3, x, and $x - 3$.

$$= \dfrac{2}{3x}$$

(b) $\dfrac{x^2 - y^2}{5x^2 - 5xy} \cdot \dfrac{10xy}{x^2 + 2xy + y^2}$ Factor and divide by the common factors of 5, x, $x - y$, and $x + y$.

$$= \dfrac{\overset{1}{(\cancel{x + y})}\overset{1}{(\cancel{x - y})}}{\underset{1}{\cancel{5}}\underset{1}{\cancel{x}}(\cancel{x - y})} \cdot \dfrac{\overset{2}{\cancel{10}}xy}{(\underset{1}{\cancel{x + y}})(x + y)}$$

$$= \dfrac{2y}{x + y}$$

Note that

$$\dfrac{2 - x}{x - 2} = -1$$

(c) $\dfrac{4}{x^2 - 2x} \cdot \dfrac{10x - 5x^2}{8x + 24}$

$$= \dfrac{\overset{1}{\cancel{4}}}{\underset{1}{\cancel{x}}(\underset{1}{\cancel{x - 2}})} \cdot \dfrac{\overset{-1}{5\cancel{x}}(\cancel{2 - x})}{\underset{2}{\cancel{8}}(x + 3)}$$

$$= \dfrac{-5}{2(x + 3)}$$

✔ *CHECK YOURSELF 2*

Multiply as indicated.

(a) $\dfrac{x^2 - 5x - 14}{4x^2} \cdot \dfrac{8x + 56}{x^2 - 49}$ (b) $\dfrac{x}{2x - 6} \cdot \dfrac{3x - x^2}{2}$

The following algorithm summarizes our work in multiplying rational expressions.

Multiplying Rational Expressions

Step 1.   Write each numerator and denominator in completely factored form.

Step 2.   Divide by any common factors appearing in both the numerator and the denominator.

Step 3.   Multiply as needed to form the desired product.

Dividing Rational Expressions

In dividing rational expressions, you can again use your experience from arithmetic. Recall that

To divide fractions, we multiply by the reciprocal of the divisor. That is, invert the divisor (the second fraction) and multiply.

$$\frac{3}{5} \div \frac{2}{3} = \frac{3}{5} \cdot \frac{3}{2} = \frac{9}{10}$$

Once more, the pattern in algebra is identical.

Dividing Rational Expressions

For polynomials P, Q, R, and S,

$$\frac{P}{Q} \div \frac{R}{S} = \frac{P}{Q} \cdot \frac{S}{R} = \frac{PS}{QR}$$

where $Q \neq 0$, $R \neq 0$, and $S \neq 0$.

To divide rational expressions, invert the divisor and multiply as before, as Example 3 illustrates.

| **Example 3** | ## Dividing Rational Expressions |

Divide as indicated.

Invert the divisor and multiply.

(a) $\dfrac{3x^2}{8x^3y} \div \dfrac{9x^2y^2}{4y^4} = \dfrac{3x^2}{8x^3y} \cdot \dfrac{4y^4}{9x^2y^2} = \dfrac{y}{6x^3}$

Caution

Be careful! Invert the divisor, then factor.

(b) $\dfrac{2x^2 + 4xy}{9x - 18y} \div \dfrac{4x + 8y}{3x - 6y} = \dfrac{2x^2 + 4xy}{9x - 18y} \cdot \dfrac{3x - 6y}{4x + 8y}$

$$= \frac{\overset{1}{2x(x + 2y)}}{\underset{3}{9(x - 2y)}} \cdot \frac{\overset{1}{3(x - 2y)}}{\underset{2}{4(x + 2y)}} = \frac{x}{6}$$

(c) $\dfrac{2x^2 - x - 6}{4x^2 + 6x} \div \dfrac{x^2 - 4}{4x} = \dfrac{2x^2 - x - 6}{4x^2 + 6x} \cdot \dfrac{4x}{x^2 - 4}$

$$= \dfrac{(2x+3)(x-2)}{2x(2x+3)} \cdot \dfrac{4x}{(x + 2)(x-2)}$$

$$= \dfrac{2}{x + 2}$$

✔ *CHECK YOURSELF 3*

Divide and simplify.

(a) $\dfrac{5xy}{7x^3} \div \dfrac{10y^2}{14x^3}$

(b) $\dfrac{3x - 9y}{2x + 10y} \div \dfrac{x^2 - 3xy}{4x^2 + 20xy}$

(c) $\dfrac{x^2 - 9}{x^3 - 27} \div \dfrac{x^2 - 2x - 15}{2x^2 - 10x}$

We summarize our work in dividing fractions with the following algorithm.

Dividing Rational Expressions

Step 1.   Invert the divisor (the *second* rational expression) to write the problem as one of multiplication.

Step 2.   Proceed as in the algorithm for the multiplication of rational expressions.

Multiplying Rational Functions

The product of two rational functions is always a rational function. Given two rational functions $f(x)$ and $g(x)$, we can rename the product, so

$$h(x) = f(x) \cdot g(x)$$

This will always be true for values of x for which both f and g are defined. So, for example, $h(1) = f(1) \cdot g(1)$ as long as both $f(1)$ and $g(1)$ exist. Example 4 illustrates this concept.

Example 4   Multiplying Rational Functions

Given the rational functions

$$f(x) = \frac{x^2 - 3x - 10}{x + 1} \quad \text{and} \quad g(x) = \frac{x^2 - 4x - 5}{x - 5}$$

find the following.

(a) $f(0) \cdot g(0)$

If $f(0) = -10$ and $g(0) = 1$, then $f(0) \cdot g(0) = (-10)(1) = -10$.

(b) $f(5) \cdot g(5)$

Although we can find $f(5)$, $g(5)$ is undefined. The number 5 is excluded from the domain of the function. Therefore, $f(5) \cdot g(5)$ is undefined.

(c) $h(x) = f(x) \cdot g(x)$

$$h(x) = f(x) \cdot g(x)$$

$$= \frac{x^2 - 3x - 10}{x + 1} \cdot \frac{x^2 - 4x - 5}{x - 5}$$

$$= \frac{(x - 5)(x + 2)}{\overset{}{\underset{1}{\cancel{(x + 1)}}}} \cdot \frac{\overset{1}{\cancel{(x + 1)}} \overset{1}{\cancel{(x - 5)}}}{\underset{1}{\cancel{(x - 5)}}}$$

$$= (x - 5)(x + 2) \qquad x \neq -1, x \neq 5$$

(d) $h(0)$

$$h(0) = (0 - 5)(0 + 2) = -10$$

(e) $h(5)$

Although the temptation is to substitute 5 for x in part (c), notice that the function is undefined when x is -1 or 5. As was true in part (b), the function is undefined at that point.

 CHECK YOURSELF 4

Given the rational functions

$$f(x) = \frac{x^2 - 2x - 8}{x + 2} \quad \text{and} \quad g(x) = \frac{x^2 - 3x - 10}{x - 4}$$

find the following.

(a) $f(0)g(0)$ (b) $f(4)g(4)$ (c) $h(x) = f(x)g(x)$ (d) $h(0)$ (e) $h(4)$

Dividing Rational Functions

When we divide two rational functions to create a third rational function, we must be certain to exclude values for which the polynomial in the denominator is equal to zero, as Example 5 illustrates.

Example 5   ## Dividing Rational Functions

Given the rational functions

$$f(x) = \frac{x^3 - 2x^2}{x + 2} \quad \text{and} \quad g(x) = \frac{x^2 - 3x + 2}{x - 4}$$

complete the following.

(a) Find $\dfrac{f(0)}{g(0)}$.

If $f(0) = 0$ and $g(0) = -\dfrac{1}{2}$, then

$$\frac{f(0)}{g(0)} = \frac{0}{-\dfrac{1}{2}} = 0$$

(b) Find $\dfrac{f(1)}{g(1)}$.

Although we can find both $f(1)$ and $g(1)$, $g(1) = 0$, so division is undefined. The value 1 is excluded from the domain of the quotient.

(c) Find $h(x) = \dfrac{f(x)}{g(x)}$.

$$h(x) = \frac{f(x)}{g(x)}$$

Note that −2 is excluded from the domain of f and 4 is excluded from the domain of g.

$$= \frac{\dfrac{x^3 - 2x^2}{x + 2}}{\dfrac{x^2 - 3x + 2}{x - 4}}$$

Invert and multiply.

$$= \frac{x^3 - 2x^2}{x + 2} \cdot \frac{x - 4}{x^2 - 3x + 2}$$

$$= \frac{x^2\cancel{(x - 2)}^{\,1}}{x + 2} \cdot \frac{x - 4}{(x - 1)\cancel{(x - 2)}_{\,1}}$$

Because $(x - 1)(x - 2)$ is part of the denominator, 1 and 2 are excluded from the domain of h.

$$= \frac{x^2(x - 4)}{(x + 2)(x - 1)} \qquad x \neq -2, 1, 2, 4$$

(d) For which values of x is $h(x)$ undefined?

The function $h(x)$ will be undefined for any value of x that would cause division by zero. So $h(x)$ is undefined for the values $-2, 1, 2,$ and 4.

✔ *CHECK YOURSELF 5*

Given the rational functions

$$f(x) = \frac{x^2 - 2x + 1}{x + 3} \qquad \text{and} \qquad g(x) = \frac{x^2 - 5x + 4}{x - 2}$$

(a) Find $\dfrac{f(0)}{g(0)}$. (b) Find $\dfrac{f(1)}{g(1)}$. (c) Find $h(x) = \dfrac{f(x)}{g(x)}$.

(d) For which values of x is $h(x)$ undefined?

✔ *CHECK YOURSELF ANSWERS*

1. $\dfrac{4a}{3b^2}$ **2.** (a) $\dfrac{2(x + 2)}{x^2}$; (b) $\dfrac{-x^2}{4}$ **3.** (a) $\dfrac{x}{y}$; (b) 6; (c) $\dfrac{2x}{x^2 + 3x + 9}$

4. (a) -10; (b) undefined; (c) $h(x) = (x - 5)(x + 2), x \neq -2, x \neq 4$; (d) -10;
(e) undefined

5. (a) $-\dfrac{1}{6}$; (b) undefined; (c) $h(x) = \dfrac{(x - 1)(x - 2)}{(x + 3)(x - 4)}$; and (d) $x \neq -3, 1, 2, 4$

Calculating Probability.   **Probability,** or the study of chance, measures the relative likelihood of events. Suppose a situation has N equally likely possible outcomes, and an event includes E of these. Let P be the probability that the event will occur. Then

$$P = \frac{E}{N}$$

From this definition, it is clear that any probability will always be between 0 and 1: $0 \leq P \leq 1$. The probability of an event not happening is $1 - P$.

The probability of two separate, independent events, event 1 and event 2, *both* happening is the *product* of the probability of the first event and the probability of the second event: $P(E_1 \text{ and } E_2) = P(E_1) \cdot P(E_2)$. The events must be independent, which means the first event cannot have any effect on the outcome of the second.

Work with a partner to complete the following.

What is the probability that two people in your class have the same birthday? Surprisingly, it is pretty common in a class of 30 people to find at least two people who have the same birthday. You can use the probability that two people will *not* have the same birthday to figure this out.

Begin with one person from the class. This person can have a birthday on any day. What is the probability that the second person will *not* have the same birthday? (How many days are in the year?) Let us say that the second person does not have the same birthday as the first person. Now there are two days in the year when no one else in the class can have a birthday. Next we find the probability that the third person does not have the same birthday. If these two probabilities are multiplied, we get the probability that three people do not have the same birthday. Continue until all 30 people have been accounted for. What is the probability that none of them have the same birthday?

$P(\text{no two birthdays are the same among } n \text{ people}) = 1 - (P_2)(P_3)(P_4) \cdots (P_n)$

where

$$P_i = \frac{365 - (i - 1)}{365}$$

is the probability that the ith person does not have the same birthday as anyone who came before.

Exercises ▪ 7.2

Multiply or divide as indicated. Express your result in simplest form.

1. $\dfrac{x^2}{3} \cdot \dfrac{6x}{x^4}$

2. $\dfrac{-y^3}{10} \cdot \dfrac{15y}{y^6}$

3. $\dfrac{x}{8x^4} \div \dfrac{x^5}{24}$

4. $\dfrac{p^5}{8} \div \dfrac{-p^2}{12p}$

5. $\dfrac{4xy^2}{15x^3} \cdot \dfrac{25xy}{16y^3}$

6. $\dfrac{3x^3y}{10xy^3} \cdot \dfrac{5xy^2}{-9xy^3}$

7. $\dfrac{8b^3}{15ab} \div \dfrac{2ab^2}{20ab^3}$

8. $\dfrac{5x^3y^3}{8x^5} \div \dfrac{15x^3y}{32x^3y^2}$

9. $\dfrac{m^3n}{2mn} \cdot \dfrac{6mn^2}{m^3n} \div \dfrac{3mn}{5m^2n}$

10. $\dfrac{4cd^2}{5cd} \cdot \dfrac{3c^3d}{2c^2d} \div \dfrac{9cd}{20cd^3}$

11. $\dfrac{6x + 18}{4x} \cdot \dfrac{16x^3}{3x + 9}$

12. $\dfrac{a^2 - 3a}{5a} \cdot \dfrac{20a^2}{3a - 9}$

13. $\dfrac{3b - 15}{6b} \div \dfrac{4b - 20}{9b^2}$

14. $\dfrac{7m^2 + 28m}{4m} \div \dfrac{5m + 20}{12m^2}$

15. $\dfrac{x^2 - 3x - 10}{5x} \cdot \dfrac{15x^2}{3x - 15}$

16. $\dfrac{y^2 - 8y}{4y} \cdot \dfrac{12y^2}{y^2 - 64}$

17. $\dfrac{c^2 + 2c - 8}{6c} \div \dfrac{5c + 20}{18c}$

18. $\dfrac{m^2 - 64}{6m^2} \div \dfrac{2m - 16}{24m^5}$

19. $\dfrac{x^2 - x - 12}{3x - 12} \cdot \dfrac{15x^3}{x^2 - 9}$

20. $\dfrac{y^2 + 7y + 10}{y^2 + 5y} \cdot \dfrac{2y}{y^2 - 4}$

21. $\dfrac{d^2 - 3d - 18}{16d - 96} \div \dfrac{d^2 - 9}{20d}$

22. $\dfrac{b^2 + 2b - 8}{b^2 - 2b} \div \dfrac{b^2 - 16}{4b}$

23. $\dfrac{2x^2 - x - 3}{3x^2 + 7x + 4} \cdot \dfrac{3x^2 - 11x - 20}{4x^2 - 9}$

24. $\dfrac{4p^2 - 1}{2p^2 - 9p - 5} \cdot \dfrac{3p^2 - 13p - 10}{9p^2 - 4}$

25. $\dfrac{a^2 - 9}{2a^2 - 6a} \div \dfrac{2a^2 + 5a - 3}{4a^2 - 1}$

26. $\dfrac{2x^2 - 5x - 7}{4x^2 - 9} \div \dfrac{5x^2 + 5x}{2x^2 + 3x}$

27. $\dfrac{2w - 6}{w^2 + 2w} \cdot \dfrac{3w}{3 - w}$

28. $\dfrac{3y - 15}{y^2 + 3y} \cdot \dfrac{4y}{5 - y}$

29. $\dfrac{a - 7}{2a + 6} \div \dfrac{21 - 3a}{a^2 + 3a}$

30. $\dfrac{x - 5}{x^2 + 3x} \div \dfrac{25 - 5x}{2x + 6}$

31. $\dfrac{x^2 - 9y^2}{2x^2 - xy - 15y^2} \cdot \dfrac{4x + 10y}{x^2 + 3xy}$

32. $\dfrac{2a^2 - 7ab - 15b^2}{2ab - 10b^2} \cdot \dfrac{2a^2 - 3ab}{4a^2 - 9b^2}$

33. $\dfrac{3m^2 - 5mn + 2n^2}{9m^2 - 4n^2} \div \dfrac{m^3 - m^2n}{9m^2 + 6mn}$

34. $\dfrac{2x^2y - 5xy^2}{4x^2 - 25y^2} \div \dfrac{4x^2 + 20xy}{2x^2 + 15xy + 25y^2}$

35. $\dfrac{x^3 + 8}{x^2 - 4} \cdot \dfrac{5x - 10}{x^3 - 2x^2 + 4x}$

36. $\dfrac{a^3 - 27}{a^2 - 9} \div \dfrac{a^3 + 3a^2 + 9a}{3a^3 + 9a^2}$

37. Let $f(x) = \dfrac{x^2 - 3x - 4}{x + 2}$ and $g(x) = \dfrac{x^2 - 2x - 8}{x - 4}$. Find (a) $f(0) \cdot g(0)$; (b) $f(4) \cdot g(4)$; (c) $h(x) = f(x) \cdot g(x)$; (d) $h(0)$; and (e) $h(4)$.

38. Let $f(x) = \dfrac{x^2 - 4x + 3}{x + 5}$ and $g(x) = \dfrac{x^2 + 7x + 10}{x - 3}$. Find (a) $f(1) \cdot g(1)$; (b) $f(3) \cdot g(3)$; (c) $h(x) = f(x) \cdot g(x)$; (d) $h(1)$; and (e) $h(3)$.

39. Let $f(x) = \dfrac{2x^2 - 3x - 5}{x + 2}$ and $g(x) = \dfrac{3x^2 + 5x - 2}{x + 1}$. Find (a) $f(1) \cdot g(1)$; (b) $f(-2) \cdot g(-2)$; (c) $h(x) = f(x) \cdot g(x)$; (d) $h(1)$; and (e) $h(-2)$.

40. Let $f(x) = \dfrac{x^2 - 1}{x - 3}$ and $g(x) = \dfrac{x^2 - 9}{x - 1}$. Find (a) $f(2) \cdot g(2)$; (b) $f(3) \cdot g(3)$; (c) $h(x) = f(x) \cdot g(x)$; (d) $h(2)$; and (e) $h(3)$.

41. Let $f(x) = \dfrac{3x^2 + x - 2}{x - 2}$ and $g(x) = \dfrac{x^2 - 4x - 5}{x + 4}$. Find (a) $\dfrac{f(0)}{g(0)}$; (b) $\dfrac{f(1)}{g(1)}$; (c) $h(x) = \dfrac{f(x)}{g(x)}$ and (d) the values of x for which $h(x)$ is undefined.

42. Let $f(x) = \dfrac{x^2 + x}{x - 5}$ and $g(x) = \dfrac{x^2 - x - 6}{x - 5}$. Find (a) $\dfrac{f(0)}{g(0)}$; (b) $\dfrac{f(2)}{g(2)}$; (c) $h(x) = \dfrac{f(x)}{g(x)}$; and (d) the values of x for which $h(x)$ is undefined.

The results from multiplying and dividing rational expressions can be checked by using a graphing calculator. To do this, define one expression in Y_1 and the other in Y_2. Then define the operation in Y_3 as $Y_1 \cdot Y_2$ or $Y_1 \div Y_2$. Put your simplified result in Y_4 (sorry, you still must simplify algebraically). Deselect the graphs for Y_1 and Y_2. If you have correctly simplified the expression, the graphs of Y_3 and Y_4 will appear to be identical. Use this technique to check your multiplication and division in Exercises 43 to 46.

43. $\dfrac{x^3 - 3x^2 + 2x - 6}{x^2 - 9} \cdot \dfrac{5x^2 + 15x}{20x}$

44. $\dfrac{3a^3 + a^2 - 9a - 3}{15a^2 + 5a} \cdot \dfrac{3a^2 + 9}{a^4 - 9}$

45. $\dfrac{x^4 - 16}{x^2 + x - 6} \div (x^3 + 4x)$

46. $\dfrac{w^3 + 27}{w^2 + 2w - 3} \div (w^3 - 3w^2 + 9w)$

Addition and Subtraction of Rational Expressions

OBJECTIVES

1. *Add and subtract rational expressions*
2. *Add and subtract rational functions*

Adding and Subtracting Rational Expressions

Recall that adding or subtracting two arithmetic fractions with the same denominator is straightforward. The same is true in algebra. To add or subtract two rational expressions with the same denominator, we add or subtract their numerators and then write that sum or difference over the common denominator.

Adding or Subtracting Rational Expressions

$$\frac{P}{R} + \frac{Q}{R} = \frac{P+Q}{R}$$

and

$$\frac{P}{R} - \frac{Q}{R} = \frac{P-Q}{R}$$

where $R \neq 0$.

Example 1

Adding and Subtracting Rational Expressions

Perform the indicated operations.

Since we have common denominators, we simply perform the indicated operations on the numerators.

$$\frac{3}{2a^2} - \frac{1}{2a^2} + \frac{5}{2a^2} = \frac{3-1+5}{2a^2}$$

$$= \frac{7}{2a^2}$$

 CHECK YOURSELF 1

Perform the indicated operations.

$$\frac{5}{3y^2} + \frac{4}{3y^2} - \frac{7}{3y^2}$$

The sum or difference of rational expressions should always be expressed in simplest form. Consider Example 2.

| Example 2 | **Adding and Subtracting Rational Expressions** |

Add or subtract as indicated.

(a) $\dfrac{5x}{x^2-9} + \dfrac{15}{x^2-9}$ Add the numerators.

$= \dfrac{5x+15}{x^2-9}$

$= \dfrac{5(x+3)}{(x-3)(x+3)} = \dfrac{5}{x-3}$ Factor and divide by the common factor.

(b) $\dfrac{3x+y}{2x} - \dfrac{x-3y}{2x} = \dfrac{(3x+y)-(x-3y)}{2x}$ Be sure to *enclose the second numerator* in parentheses.

$= \dfrac{3x+y-x+3y}{2x}$ Remove the parentheses by *changing each sign.*

$= \dfrac{2x+4y}{2x} = \dfrac{2(x+2y)}{2x}$ Factor and divide by the common factor of 2.

$= \dfrac{x+2y}{x}$

✔ CHECK YOURSELF 2

Perform the indicated operations.

(a) $\dfrac{6a}{a^2-2a-8} + \dfrac{12}{a^2-2a-8}$ (b) $\dfrac{5x-y}{3y} - \dfrac{2x-4y}{3y}$

Now, what if our rational expressions *do not* have common denominators? In that case, we must use the least common denominator (LCD). The **least common denominator** is the simplest polynomial that is divisible by each of the individual denominators. Each expression in the desired sum or difference is then "built up" to an equivalent expression having that LCD as a denominator. We can then add or subtract as before.

By **inspection,** we mean you look at the denominators and find that the LCD is obvious (as in Example 2).

Although in many cases we can find the LCD by inspection, we can state an algorithm for finding the LCD that is similar to the one used in arithmetic.

Again, we see the key role that factoring plays in the process of working with rational expressions.

Finding the Least Common Denominator

Step 1. Write each of the denominators in completely factored form.

Step 2. Write the LCD as the product of each prime factor to the highest power to which it appears in the factored form of any individual denominators.

Example 3 illustrates the procedure.

Example 3   Finding the LCD for Two Rational Expressions

Find the LCD for each of the following pairs of rational expressions.

(a) $\dfrac{3}{4x^2}$ and $\dfrac{5}{6xy}$

Factor the denominators.

You may very well be able to find this LCD by inspecting the numerical coefficients and the variable factors.

$$4x^2 = 2^2 \cdot x^2$$
$$6xy = 2 \cdot 3 \cdot x \cdot y$$

The LCD must have the factors

$$2^2 \cdot 3 \cdot x^2 \cdot y$$

and so $12x^2y$ is the desired LCD.

(b) $\dfrac{7}{x-3}$ and $\dfrac{2}{x+5}$

Here, neither denominator can be factored. The LCD must have the factors $x - 3$ and $x + 5$. So the LCD is

It is generally best to leave the LCD in this factored form.

$$(x-3)(x+5)$$

 CHECK YOURSELF 3

Find the LCD for the following pairs of rational expressions.

(a) $\dfrac{3}{8a^3}$ and $\dfrac{5}{6a^2}$ (b) $\dfrac{4}{x+7}$ and $\dfrac{3}{x-5}$

Let's see how factoring techniques are applied in Example 4.

Example 4   Finding the LCD for Two Rational Expressions

Find the LCD for the following pairs of rational expressions.

(a) $\dfrac{2}{x^2-x-6}$ and $\dfrac{1}{x^2-9}$

Factoring, we have

$$x^2 - x - 6 = (x + 2)(x - 3)$$

and

$$x^2 - 9 = (x + 3)(x - 3)$$

The LCD of the given expressions is then

$$(x + 2)(x - 3)(x + 3)$$

> The LCD must contain *each* of the factors appearing in the original denominators.

(b) $\dfrac{5}{x^2 - 4x + 4}$ and $\dfrac{3}{x^2 + 2x - 8}$

Again, we factor:

$$x^2 - 4x + 4 = (x - 2)^2$$
$$x^2 + 2x - 8 = (x - 2)(x + 4)$$

> The LCD must contain $(x - 2)^2$ as a factor since $x - 2$ appears *twice* as a factor in the first denominator.

The LCD is then

$$(x - 2)^2(x + 4)$$

✔ *CHECK YOURSELF 4*

Find the LCD for the following pairs of rational expressions.

(a) $\dfrac{3}{x^2 - 2x - 15}$ and $\dfrac{5}{x^2 - 25}$

(b) $\dfrac{5}{y^2 + 6y + 9}$ and $\dfrac{3}{y^2 - y - 12}$

Let's look at Example 5, in which the concept of the LCD is applied in adding or subtracting rational expressions.

Example 5   Adding and Subtracting Rational Expressions

Add or subtract as indicated.

(a) $\dfrac{5}{4xy} + \dfrac{3}{2x^2}$

The LCD for $2x^2$ and $4xy$ is $4x^2y$. We rewrite each of the rational expressions with the LCD as a denominator.

> Note that in each case we are multiplying by 1 ($\frac{x}{x}$ in the first fraction and $\frac{2y}{2y}$ in the second fraction) which is why the resulting fractions are equivalent to the original ones.

$$\frac{5}{4xy} + \frac{3}{2x^2} = \frac{5 \cdot x}{4xy \cdot x} + \frac{3 \cdot 2y}{2x^2 \cdot 2y}$$

$$= \frac{5x}{4x^2y} + \frac{6y}{4x^2y} = \frac{5x + 6y}{4x^2y}$$

Multiply the first rational expression by $\frac{x}{x}$ and the second by $\frac{2y}{2y}$ to form the LCD of $4x^2y$.

(b) $\dfrac{3}{a-3} - \dfrac{2}{a}$

The LCD for a and $a - 3$ is $a(a - 3)$. We rewrite each of the rational expressions with that LCD as a denominator.

$$\dfrac{3}{a-3} - \dfrac{2}{a}$$

$$= \dfrac{3a}{a(a-3)} - \dfrac{2(a-3)}{a(a-3)} \qquad \text{Subtract the numerators.}$$

$$= \dfrac{3a - 2(a-3)}{a(a-3)} \qquad \text{Remove the parentheses, and combine like terms.}$$

$$= \dfrac{3a - 2a + 6}{a(a-3)} = \dfrac{a+6}{a(a-3)}$$

✔ CHECK YOURSELF 5

Perform the indicated operations.

(a) $\dfrac{3}{2ab} + \dfrac{4}{5b^2}$ (b) $\dfrac{5}{y+2} - \dfrac{3}{y}$

Let's proceed to Example 6, in which factoring will be required in forming the LCD.

Example 6

Adding and Subtracting Rational Expressions

Add or subtract as indicated.

(a) $\dfrac{-5}{x^2 - 3x - 4} + \dfrac{8}{x^2 - 16}$

We first factor the two denominators.

$$x^2 - 3x - 4 = (x + 1)(x - 4)$$
$$x^2 - 16 = (x + 4)(x - 4)$$

We see that the LCD must be

$$(x + 1)(x + 4)(x - 4)$$

We use the facts that

$$\dfrac{x+4}{x+4} = 1$$

and $\dfrac{x+1}{x+1} = 1$

Again, rewriting the original expressions with factored denominators gives

$$\dfrac{-5}{(x+1)(x-4)} + \dfrac{8}{(x-4)(x+4)}$$

$$= \dfrac{-5(x+4)}{(x+1)(x-4)(x+4)} + \dfrac{8(x+1)}{(x-4)(x+4)(x+1)}$$

$$= \frac{-5(x + 4) + 8(x + 1)}{(x + 1)(x - 4)(x + 4)}$$ Now add the numerators.

$$= \frac{-5x - 20 + 8x + 8}{(x + 1)(x - 4)(x + 4)}$$ Combine like terms in the numerator.

$$= \frac{3x - 12}{(x + 1)(x - 4)(x + 4)}$$ Factor.

$$= \frac{3(x - 4)}{(x + 1)(x - 4)(x + 4)}$$ Divide by the common factor $x - 4$.

$$= \frac{3}{(x + 1)(x + 4)}$$

(b) $\dfrac{5}{x^2 - 5x + 6} - \dfrac{3}{4x - 12}$

Again, factor the denominators.

$$x^2 - 5x + 6 = (x - 2)(x - 3)$$

$$4x - 12 = 4(x - 3)$$

The LCD is $4(x - 2)(x - 3)$, and proceeding as before, we have

$$\frac{5}{(x - 2)(x - 3)} - \frac{3}{4(x - 3)}$$

$$= \frac{5 \cdot 4}{4(x - 2)(x - 3)} - \frac{3(x - 2)}{4(x - 2)(x - 3)}$$

$$= \frac{20 - 3(x - 2)}{4(x - 2)(x - 3)}$$

$$= \frac{20 - 3x + 6}{4(x - 2)(x - 3)} = \frac{-3x + 26}{4(x - 2)(x - 3)}$$ Simplify the numerator by combining like terms.

✔ **CHECK YOURSELF 6**

Add or subtract as indicated.

(a) $\dfrac{-4}{x^2 - 4} + \dfrac{7}{x^2 - 3x - 10}$ (b) $\dfrac{5}{3x - 9} - \dfrac{2}{x^2 - 9}$

Example 7 looks slightly different from those you have seen thus far, but the reasoning involved in performing the subtraction is exactly the same.

Example 7	## Subtracting Rational Expressions

Subtract.

$$3 - \frac{5}{2x - 1}$$

To perform the subtraction, remember that 3 is equivalent to the fraction $\frac{3}{1}$, so

$$3 - \frac{5}{2x - 1} = \frac{3}{1} - \frac{5}{2x - 1}$$

The LCD for 1 and $2x - 1$ is just $2x - 1$. We now rewrite the first expression with that denominator.

$$3 - \frac{5}{2x - 1} = \frac{3(2x - 1)}{2x - 1} - \frac{5}{2x - 1}$$

$$= \frac{3(2x - 1) - 5}{2x - 1}$$

$$= \frac{6x - 8}{2x - 1}$$

✔ *CHECK YOURSELF 7*

Subtract.

$$\frac{4}{3x + 1} - 3$$

Example 8 uses an observation from Section 7.1. Recall that

$$a - b = -(b - a)$$

$$= -1(b - a)$$

Example 8	## Adding and Subtracting Rational Expressions

Add.

$$\frac{x^2}{x - 5} + \frac{3x + 10}{5 - x}$$

Your first thought might be to use a denominator of $(x - 5)(5 - x)$. However, we can simplify our work considerably if we multiply the numerator and denominator of the

Use

$$\frac{-1}{-1} = 1$$

Note that

$$(-1)(5 - x) = x - 5$$

The fractions now have a common denominator, and we can add as before.

second fraction by -1 to find a common denominator.

$$\frac{x^2}{x - 5} + \frac{3x + 10}{5 - x}$$

$$= \frac{x^2}{x - 5} + \frac{(-1)(3x + 10)}{(-1)(5 - x)}$$

$$= \frac{x^2}{x - 5} + \frac{-3x - 10}{x - 5}$$

$$= \frac{x^2 - 3x - 10}{x - 5}$$

$$= \frac{(x + 2)(x - 5)}{x - 5}$$

$$= x + 2$$

✔ CHECK YOURSELF 8

Add.

$$\frac{x^2}{x - 7} + \frac{10x - 21}{7 - x}$$

Adding Rational Functions

The sum of two rational functions is always a rational function. Given two rational functions $f(x)$ and $g(x)$, we can rename the sum, so $h(x) = f(x) + g(x)$. This will always be true for values of x for which both f and g are defined. So, for example, $h(-2) = f(-2) + g(-2)$ as long as both $f(-2)$ and $g(-2)$ exist.

Example 9   ## Adding Two Rational Functions

Given

$$f(x) = \frac{3x}{x + 5} \quad \text{and} \quad g(x) = \frac{x}{x - 4}$$

complete the following.

(a) Find $f(1) + g(1)$.

If $f(1) = \dfrac{1}{2}$ and $g(1) = -\dfrac{1}{3}$, then

$$f(1) + g(1) = \frac{1}{2} + \left(-\frac{1}{3}\right)$$

$$= \frac{3}{6} + \left(-\frac{2}{6}\right) = \frac{1}{6}$$

(b) Find $h(x) = f(x) + g(x)$.

$$h(x) = f(x) + g(x)$$

$$= \frac{3x}{x + 5} + \frac{x}{x - 4}$$

$$= \frac{3x(x - 4) + x(x + 5)}{(x + 5)(x - 4)} = \frac{3x^2 - 12x + x^2 + 5x}{(x + 5)(x - 4)}$$

$$= \frac{4x^2 - 7x}{(x + 5)(x - 4)} \qquad x \neq -5, 4$$

(c) Find the ordered pair $(1, h(1))$.

$$h(1) = \frac{-3}{-18} = \frac{1}{6}$$

The ordered pair is $\left(1, \dfrac{1}{6}\right)$.

✔ *CHECK YOURSELF 9*

Given

$$f(x) = \frac{x}{2x - 5} \qquad \text{and} \qquad g(x) = \frac{2x}{3x - 1}$$

complete the following.

(a) Find $f(1) + g(1)$.

(b) Find $h(x) = f(x) + g(x)$.

(c) Find the ordered pair $(1, h(1))$.

Subtracting Rational Functions

When subtracting rational functions, one must take particular care with the signs in the numerator of the expression being subtracted.

Example 10   ## Subtracting Rational Functions

Given

$$f(x) = \frac{3x}{x + 5} \qquad \text{and} \qquad g(x) = \frac{x - 2}{x - 4}$$

complete the following.

(a) Find $f(1) - g(1)$.

Because $f(1) = \dfrac{1}{2}$ and $g(1) = \dfrac{1}{3}$,

$$f(1) - g(1) = \frac{1}{2} - \frac{1}{3}$$

$$= \frac{3}{6} - \frac{2}{6}$$

$$= \frac{3 - 2}{6}$$

$$= \frac{1}{6}$$

(b) Find $h(x) = f(x) - g(x)$.

$$h(x) = \frac{3x}{x + 5} - \frac{x - 2}{x - 4}$$

$$= \frac{3x(x - 4)}{(x + 5)(x - 4)} - \frac{(x - 2)(x + 5)}{(x - 4)(x + 5)}$$

$$= \frac{3x(x - 4) - (x - 2)(x + 5)}{(x + 5)(x - 4)} \qquad \text{Subtract numerators.}$$

$$= \frac{(3x^2 - 12x) - (x^2 + 3x - 10)}{(x + 5)(x - 4)} \qquad \text{Combine like terms.}$$

$$= \frac{2x^2 - 15x + 10}{(x + 5)(x - 4)} \qquad x \neq -5, 4$$

(c) Find the ordered pair $(1, h(1))$.

$$h(1) = \frac{-3}{-18} = \frac{1}{6}$$

The ordered pair is $\left(1, \dfrac{1}{6}\right)$.

✔ *CHECK YOURSELF 10*

Given

$$f(x) = \frac{x}{2x - 5} \qquad \text{and} \qquad g(x) = \frac{2x - 1}{3x - 1}$$

complete the following.

(a) Find $f(1) - g(1)$.   (b) Find $h(x) = f(x) - g(x)$.

(c) Find the ordered pair $(1, h(1))$.

✔ *CHECK YOURSELF ANSWERS*

1. $\dfrac{2}{3y^2}$ **2.** (a) $\dfrac{6}{a - 4}$; (b) $\dfrac{x + y}{y}$ **3.** (a) $24a^3$; (b) $(x + 7)(x - 5)$

4. (a) $(x - 5)(x + 5)(x + 3)$; (b) $(y + 3)^2(y - 4)$ **5.** (a) $\dfrac{8a + 15b}{10ab^2}$; (b) $\dfrac{2y - 6}{y(y + 2)}$

6. (a) $\dfrac{3}{(x - 2)(x - 5)}$; (b) $\dfrac{5x + 9}{3(x + 3)(x - 3)}$ **7.** $\dfrac{-9x + 1}{3x + 1}$

8. $x - 3$ **9.** (a) $\dfrac{2}{3}$; (b) $h(x) = \dfrac{7x^2 - 11x}{(2x - 5)(3x - 1)}$, $x \neq \dfrac{5}{2}, \dfrac{1}{3}$; (c) $\left(1, \dfrac{2}{3}\right)$

10. (a) $-\dfrac{5}{6}$; (b) $h(x) = \dfrac{-x^2 + 11x - 5}{(2x - 5)(3x - 1)}$, $x \neq \dfrac{5}{2}, \dfrac{1}{3}$; (c) $\left(1, -\dfrac{5}{6}\right)$

Probability and Pari-Mutuel Betting.   In most gambling games, payoffs are determined by the **odds.** At horse and dog tracks, the odds (D) are a ratio that is calculated by taking into account the total amount wagered (A), the amount wagered on a particular animal (a), and the government share, called the takeout (f). The ratio is then rounded down to a comparison of integers such as 99 to 1, 3 to 1, or 5 to 2. Below is the formula that tracks use to find odds.

$$D = \frac{A(1 - f)}{a} - 1$$

Work with a partner to complete the following:

1. Assume that the government takes 10% and simplify the expression for D. Use this formula to compute the odds on each horse if a total of $10,000 was bet on all the horses and the amounts were distributed as shown in the table.

Horse	Total Amount Wagered on This Horse to Win	Odds: Amount Paid on Each Dollar Bet if Horse Wins
1	$5,000	
2	1,000	
3	2,000	
4	1,500	
5	500	

2. Odds can be used as a guide in determining the chance that a given horse will win. The probability of a horse winning is related to many variables, such as track condition, how the horse is feeling, and weather. However, the odds do reflect the consensus opinion of racing fans and can be used to give some idea of the probability.

    The relationship between odds and probability is given by the equations

$$P(\text{win}) = \frac{1}{D + 1}$$

and

$$P(\text{loss}) = 1 - P(\text{win})$$

or

$$P(\text{loss}) = 1 - \frac{1}{D + 1}$$

Solve this equation for D, the odds against the horse winning. Do the probabilities for each horse winning all add up to 1? Should they add to 1?

Perform the indicated operations. Express your results in simplest form.

1. $\dfrac{9}{4x^3} + \dfrac{3}{4x^3}$

2. $\dfrac{11}{3b^3} - \dfrac{2}{3b^3}$

3. $\dfrac{5}{3a + 7} + \dfrac{2}{3a + 7}$

4. $\dfrac{6}{5x + 3} - \dfrac{3}{5x + 3}$

5. $\dfrac{2x}{x - 3} - \dfrac{6}{x - 3}$

6. $\dfrac{6w}{w + 4} + \dfrac{24}{w + 4}$

7. $\dfrac{y^2}{2y + 8} + \dfrac{3y - 4}{2y + 8}$

8. $\dfrac{x^2}{4x - 12} - \dfrac{9}{4x - 12}$

9. $\dfrac{5m - 2}{m - 6} - \dfrac{3m + 10}{m - 6}$

10. $\dfrac{3b - 8}{b - 6} + \dfrac{b - 16}{b - 6}$

11. $\dfrac{x - 7}{x^2 - x - 6} + \dfrac{2x - 2}{x^2 - x - 6}$

12. $\dfrac{5x - 12}{x^2 - 8x + 15} - \dfrac{3x - 2}{x^2 - 8x + 15}$

13. $\dfrac{3}{2x} + \dfrac{4}{5x}$

14. $\dfrac{4}{5w} - \dfrac{3}{4w}$

15. $\dfrac{6}{a} + \dfrac{3}{a^2}$

16. $\dfrac{3}{p} - \dfrac{7}{p^2}$

17. $\dfrac{2}{m} - \dfrac{2}{n}$

18. $\dfrac{5}{x} + \dfrac{10}{y}$

19. $\dfrac{3}{4b^2} - \dfrac{5}{3b^3}$

20. $\dfrac{4}{5x^3} - \dfrac{3}{2x^2}$

21. $\dfrac{3}{b} - \dfrac{1}{b - 3}$

22. $\dfrac{4}{c} + \dfrac{3}{c + 1}$

23. $\dfrac{2}{x + 1} + \dfrac{3}{x + 2}$

24. $\dfrac{4}{y - 1} + \dfrac{2}{y + 3}$

25. $\dfrac{5}{y-3} - \dfrac{1}{y+1}$

26. $\dfrac{4}{x+5} - \dfrac{3}{x-1}$

27. $\dfrac{3w}{w-6} + \dfrac{4w}{w-2}$

28. $\dfrac{3n}{n+5} + \dfrac{n}{n-4}$

29. $\dfrac{3x}{3x-2} - \dfrac{2x}{2x+1}$

30. $\dfrac{5c}{5c-1} + \dfrac{2c}{2c-3}$

31. $\dfrac{5}{x-9} + \dfrac{4}{9-x}$

32. $\dfrac{5}{a-5} - \dfrac{3}{5-a}$

33. $\dfrac{3}{x^2-16} + \dfrac{2}{x-4}$

34. $\dfrac{5}{y^2+5y+6} + \dfrac{2}{y+2}$

35. $\dfrac{4m}{m^2-3m+2} - \dfrac{1}{m-2}$

36. $\dfrac{x}{x^2-1} - \dfrac{2}{x-1}$

As we saw in Section 7.2 exercises, the graphing calculator can be used to check our work. In Exercises 37 to 42, enter the first rational expression in Y_1 and the second in Y_2. In Y_3, you will enter either $Y_1 + Y_2$ or $Y_1 - Y_2$. Enter your algebraically simplified rational expression in Y_4. The graphs of Y_3 and Y_4 will appear to be identical if you have correctly simplified the expression.

37. $\dfrac{6y}{y^2-8y+15} + \dfrac{9}{y-3}$

38. $\dfrac{8a}{a^2-8a+12} + \dfrac{4}{a-2}$

39. $\dfrac{6x}{x^2-10x+24} - \dfrac{18}{x-6}$

40. $\dfrac{21p}{p^2-3p-10} - \dfrac{15}{p-5}$

41. $\dfrac{2}{z^2-4} + \dfrac{3}{z^2+2z-8}$

42. $\dfrac{5}{x^2-3x-10} + \dfrac{2}{x^2-25}$

Find (a) $f(1) + g(1)$; (b) $h(x) = f(x) + g(x)$; and (c) the ordered pair $(1, h(1))$.

43. $f(x) = \dfrac{3x}{x+1}$ and $g(x) = \dfrac{2x}{x-3}$

44. $f(x) = \dfrac{4x}{x-4}$ and $g(x) = \dfrac{x+4}{x+1}$

45. $f(x) = \dfrac{x}{x+1}$ and $g(x) = \dfrac{1}{x^2+2x+1}$

46. $f(x) = \dfrac{x+2}{x-4}$ and $g(x) = \dfrac{x+3}{x+4}$

Find (a) $f(1) - g(1)$; (b) $h(x) = f(x) - g(x)$; and (c) the ordered pair $(1, h(1))$.

47. $f(x) = \dfrac{x + 5}{x - 5}$ and $g(x) = \dfrac{x - 5}{x + 5}$

48. $f(x) = \dfrac{2x}{x - 4}$ and $g(x) = \dfrac{3x}{x + 7}$

49. $f(x) = \dfrac{x + 9}{4x - 36}$ and $g(x) = \dfrac{x - 9}{x^2 - 18x + 81}$

50. $f(x) = \dfrac{4x + 1}{x + 5}$ and $g(x) = -\dfrac{2}{x}$

In Exercises 51 to 60, evaluate each expression at the given variable value(s).

51. $\dfrac{5x + 5}{x^2 + 3x + 2} - \dfrac{x - 3}{x^2 + 5x - 6}, x = -4$

52. $\dfrac{y - 3}{y^2 - 6y + 8} + \dfrac{2y - 6}{y^2 - 4}, y = 3$

53. $\dfrac{2m + 2n}{m^2 - n^2} + \dfrac{m - 2n}{m^2 + 2mn + n^2}, m = 3, n = 2$

54. $\dfrac{w - 3z}{w^2 - 2wz + z^2} - \dfrac{w + 2z}{w^2 - z^2}, w = 2, z = 1$

55. $\dfrac{1}{a - 3} - \dfrac{1}{a + 3} + \dfrac{2a}{a^2 - 9}, a = 4$

56. $\dfrac{1}{m + 1} + \dfrac{1}{m - 3} - \dfrac{4}{m^2 - 2m - 3}, m = -2$

57. $\dfrac{3w^2 + 16w - 8}{w^2 + 2w - 8} + \dfrac{w}{w + 4} - \dfrac{w - 1}{w - 2}, w = 3$

58. $\dfrac{4x^2 - 7x - 45}{x^2 - 6x + 5} - \dfrac{x + 2}{x - 1} - \dfrac{x}{x - 5}, x = -3$

59. $\dfrac{a^2 - 9}{2a^2 - 5a - 3} \cdot \left(\dfrac{1}{a - 2} + \dfrac{1}{a + 3} \right), a = -3$

60. $\dfrac{m^2 - 2mn + n^2}{m^2 + 2mn - 3n^2} \cdot \left(\dfrac{2}{m - n} - \dfrac{1}{m + n} \right), m = 4, n = -3$

7.4 Complex Fractions

1. *Use the fundamental principle to simplify complex fractions*
2. *Use division to simplify complex fractions*

Our work in this section deals with two methods for simplifying complex fractions. We begin with a definition. A **complex fraction** is a fraction that has a fraction in its numerator or denominator (or both). Some examples are

$$\frac{\dfrac{5}{6}}{\dfrac{3}{4}} \qquad \frac{\dfrac{4}{x}}{\dfrac{3}{x+1}} \qquad \frac{1+\dfrac{1}{x}}{1-\dfrac{1}{x}}$$

Two methods can be used to simplify complex fractions. Method 1 involves the fundamental principle, and method 2 involves inverting and multiplying.

Method 1 for Simplifying Complex Fractions

Fundamental principle:

$$\frac{P}{Q} = \frac{PR}{QR}$$

where $Q \neq 0$ and $R \neq 0$.

Recall that by the *fundamental principle* we can always multiply the numerator and denominator of a fraction by the same nonzero quantity. In simplifying a complex fraction, we multiply the numerator and denominator by the LCD of all fractions that appear within the complex fraction.

Here the denominators are 5 and 10, so we can write

Again, we are multiplying by $\dfrac{10}{10}$ or 1.

$$\frac{\dfrac{3}{5}}{\dfrac{7}{10}} = \frac{\dfrac{3}{5} \cdot 10}{\dfrac{7}{10} \cdot 10} = \frac{6}{7}$$

Method 2 for Simplifying Complex Fractions

Our second approach interprets the complex fraction as indicating division and applies our earlier work in dividing fractions in which we *invert and multiply*.

$$\frac{\dfrac{3}{5}}{\dfrac{7}{10}} = \frac{3}{5} \div \frac{7}{10} = \frac{3}{5} \cdot \frac{10}{7} = \frac{6}{7} \qquad \text{Invert and multiply.}$$

Which method is better? The answer depends on the expression you are trying to simplify. Both approaches are effective, and you should be familiar with both. With practice you will be able to tell which method may be easier to use in a particular situation.

Let's look at the same two methods applied to the simplification of an algebraic complex fraction.

Example 1

Simplifying Complex Fractions

Simplify.

$$\frac{1 + \dfrac{2x}{y}}{2 - \dfrac{x}{y}}$$

Method 1   The LCD of 1, $\dfrac{2x}{y}$, 2, and $\dfrac{x}{y}$ is y. So we multiply the numerator and denominator by y.

$$\frac{1 + \dfrac{2x}{y}}{2 - \dfrac{x}{y}} = \frac{\left(1 + \dfrac{2x}{y}\right) \cdot y}{\left(2 - \dfrac{x}{y}\right) \cdot y}$$ Distribute y over the numerator and denominator.

$$= \frac{1 \cdot y + \dfrac{2x}{y} \cdot y}{2 \cdot y - \dfrac{x}{y} \cdot y}$$ Simplify.

$$= \frac{y + 2x}{2y - x}$$

Method 2   In this approach, we must *first work separately* in the numerator and denominator to form single fractions.

Make sure you understand the steps in forming a single fraction in the numerator and denominator.

$$\frac{1 + \dfrac{2x}{y}}{2 - \dfrac{x}{y}} = \frac{\dfrac{y}{y} + \dfrac{2x}{y}}{\dfrac{2y}{y} - \dfrac{x}{y}} = \frac{\dfrac{y + 2x}{y}}{\dfrac{2y - x}{y}}$$

$$= \frac{y + 2x}{y} \cdot \frac{y}{2y - x}$$ Invert the divisor and multiply.

$$= \frac{y + 2x}{2y - x}$$

✔ CHECK YOURSELF 1

Simplify.

$$\frac{\dfrac{x}{y} - 1}{\dfrac{2x}{y} + 2}$$

Again, simplifying a complex fraction means writing an equivalent simple fraction in lowest terms, as Example 2 illustrates.

Example 2

Simplifying Complex Fractions

Simplify.

$$\frac{1 - \dfrac{2y}{x} + \dfrac{y^2}{x^2}}{1 - \dfrac{y^2}{x^2}}$$

We choose the first method of simplification in this case. The LCD of all the fractions that appear is x^2. So we multiply the numerator and denominator by x^2.

$$\frac{1 - \dfrac{2y}{x} + \dfrac{y^2}{x^2}}{1 - \dfrac{y^2}{x^2}} = \frac{\left(1 - \dfrac{2y}{x} + \dfrac{y^2}{x^2}\right) \cdot x^2}{\left(1 - \dfrac{y^2}{x^2}\right) \cdot x^2}$$

Distribute x^2 over the numerator and denominator, and simplify.

$$= \frac{x^2 - 2xy + y^2}{x^2 - y^2}$$

Factor the numerator and denominator.

$$= \frac{(x - y)(x - y)}{(x + y)(x - y)} = \frac{x - y}{x + y}$$

Divide by the common factor $x - y$.

✔ CHECK YOURSELF 2

Simplify.

$$\frac{1 + \dfrac{5}{x} + \dfrac{6}{x^2}}{1 - \dfrac{9}{x^2}}$$

In Example 3, we will illustrate the second method of simplification for purposes of comparison.

| **Example 3** | ## Simplifying Complex Fractions |

Simplify.

$$\frac{1 - \dfrac{1}{x + 2}}{x - \dfrac{2}{x - 1}}$$

Again, take time to make sure you understand how the numerator and denominator are rewritten as single fractions.

Note: Method 2 is probably the more efficient in this case. The LCD of the denominators would be $(x + 2)(x - 1)$, leading to a somewhat more complicated process if method 1 were used.

$$\frac{1 - \dfrac{1}{x + 2}}{x - \dfrac{2}{x - 1}} = \frac{\dfrac{x + 2}{x + 2} - \dfrac{1}{x + 2}}{\dfrac{x(x - 1)}{x - 1} - \dfrac{2}{x - 1}} = \frac{\dfrac{x + 2 - 1}{x + 2}}{\dfrac{x(x - 1) - 2}{x - 1}} = \frac{\dfrac{x + 1}{x + 2}}{\dfrac{x^2 - x - 2}{x - 1}}$$

$$= \frac{x + 1}{x + 2} \cdot \frac{x - 1}{x^2 - x - 2} = \frac{\overset{1}{\cancel{x + 1}}}{x + 2} \cdot \frac{x - 1}{(x - 2)\underset{1}{\cancel{(x + 1)}}} = \frac{x - 1}{(x + 2)(x - 2)}$$

✔ *CHECK YOURSELF 3*

Simplify.

$$\frac{2 + \dfrac{5}{x - 3}}{x - \dfrac{1}{2x + 1}}$$

The following algorithm summarizes our work with complex fractions.

Simplifying Complex Fractions

Method 1

1. Multiply the numerator and denominator of the complex fraction by the LCD of all the fractions that appear within the numerator and denominator.

2. Simplify the resulting rational expression, writing the expression in lowest terms.

Method 2

1. Write the numerator and denominator of the complex fraction as single fractions, if necessary.

2. Invert the denominator and multiply as before, writing the result in lowest terms.

✔ *CHECK YOURSELF ANSWERS*

1. $\dfrac{x - y}{2(x + y)}$ **2.** $\dfrac{x + 2}{x - 3}$ **3.** $\dfrac{2x + 1}{(x - 3)(x + 1)}$

Exercises · 7.4

In Exercises 1 to 39, simplify each complex fraction.

1. $\dfrac{\dfrac{2}{3}}{\dfrac{6}{8}}$

2. $\dfrac{\dfrac{5}{6}}{\dfrac{10}{15}}$

3. $\dfrac{\dfrac{3}{4}+\dfrac{1}{3}}{\dfrac{1}{6}+\dfrac{3}{4}}$

4. $\dfrac{\dfrac{3}{4}+\dfrac{1}{2}}{\dfrac{7}{8}-\dfrac{1}{4}}$

5. $\dfrac{2+\dfrac{1}{3}}{3-\dfrac{1}{5}}$

6. $\dfrac{1+\dfrac{3}{4}}{2-\dfrac{1}{8}}$

7. $\dfrac{\dfrac{x}{8}}{\dfrac{x^2}{4}}$

8. $\dfrac{\dfrac{x^2}{12}}{\dfrac{x^5}{18}}$

9. $\dfrac{\dfrac{3}{m}}{\dfrac{6}{m^2}}$

10. $\dfrac{\dfrac{15}{x^2}}{\dfrac{20}{x^3}}$

11. $\dfrac{\dfrac{y+1}{y}}{\dfrac{y-1}{2y}}$

12. $\dfrac{\dfrac{x+3}{4x}}{\dfrac{x-3}{2x}}$

13. $\dfrac{\dfrac{a+2b}{3a}}{\dfrac{a^2+2ab}{9b}}$

14. $\dfrac{\dfrac{m-3n}{4m}}{\dfrac{m^2-3mn}{8n}}$

15. $\dfrac{\dfrac{x-3}{x^2-25}}{\dfrac{x^2+x-12}{x^2+5x}}$

16. $\dfrac{\dfrac{x+5}{x^2-6x}}{\dfrac{x^2-25}{x^2-36}}$

17. $\dfrac{2-\dfrac{1}{x}}{2+\dfrac{1}{x}}$

18. $\dfrac{3+\dfrac{1}{b}}{3-\dfrac{1}{b}}$

19. $\dfrac{\dfrac{1}{x}-\dfrac{1}{y}}{\dfrac{1}{xy}}$

20. $\dfrac{\dfrac{4}{xy}}{\dfrac{1}{y}-\dfrac{1}{x}}$

21. $\dfrac{\dfrac{x^2}{y^2}-1}{\dfrac{x}{y}+1}$

22. $\dfrac{\dfrac{m}{n}+2}{\dfrac{m^2}{n^2}-4}$

23. $\dfrac{1+\dfrac{3}{a}-\dfrac{4}{a^2}}{1+\dfrac{2}{a}-\dfrac{3}{a^2}}$

24. $\dfrac{1-\dfrac{2}{x}-\dfrac{8}{x^2}}{1-\dfrac{1}{x}-\dfrac{6}{x^2}}$

25. $\dfrac{\dfrac{x^2}{y} + 2x + y}{\dfrac{1}{y^2} - \dfrac{1}{x^2}}$

26. $\dfrac{\dfrac{a}{b} + 1 - \dfrac{2b}{a}}{\dfrac{1}{b^2} - \dfrac{4}{a^2}}$

27. $\dfrac{2 - \dfrac{2}{x+1}}{2 + \dfrac{2}{x+1}}$

28. $\dfrac{3 - \dfrac{4}{m+2}}{3 + \dfrac{4}{m+2}}$

29. $\dfrac{1 - \dfrac{1}{y-1}}{y - \dfrac{8}{y+2}}$

30. $\dfrac{1 + \dfrac{1}{x+2}}{x - \dfrac{18}{x-3}}$

31. $\dfrac{\dfrac{1}{x-3} + \dfrac{1}{x+3}}{\dfrac{1}{x-3} - \dfrac{1}{x+3}}$

32. $\dfrac{\dfrac{2}{m-2} + \dfrac{1}{m-3}}{\dfrac{2}{m-2} - \dfrac{1}{m-3}}$

33. $\dfrac{\dfrac{x}{x+1} + \dfrac{1}{x-1}}{\dfrac{x}{x-1} - \dfrac{1}{x+1}}$

34. $\dfrac{\dfrac{y}{y-4} + \dfrac{1}{y+2}}{\dfrac{4}{y-4} - \dfrac{1}{y+2}}$

35. $\dfrac{\dfrac{a+1}{a-1} - \dfrac{a-1}{a+1}}{\dfrac{a+1}{a-1} + \dfrac{a-1}{a+1}}$

36. $\dfrac{\dfrac{x+2}{x-2} - \dfrac{x-2}{x+2}}{\dfrac{x+2}{x-2} + \dfrac{x-2}{x+2}}$

37. $1 + \dfrac{1}{1 + \dfrac{1}{x}}$

38. $1 + \dfrac{1}{1 - \dfrac{1}{y}}$

39. $1 + \dfrac{1}{1 + \dfrac{1}{1 + \dfrac{1}{x}}}$

40. Extend the "continued fraction" patterns in Exercises 37 and 39 to write the next complex fraction.

41. Simplify the complex fraction in Exercise 40.

42. Outline the two different methods used to simplify a complex fraction. What are the advantages of each method?

43. Can the expression $\dfrac{x^2 + y^2}{x + y}$ be written as $\dfrac{x^2}{x} + \dfrac{y^2}{y}$? If not, is there a correct simplified form?

44. Write and simplify a complex fraction that is the reciprocal of $x + \dfrac{6}{x-1}$.

45. Let $f(x) = \dfrac{3}{x}$. Write and simplify a complex fraction whose numerator is $f(3 + h) - f(3)$ and whose denominator is h.

46. Write and simplify a complex fraction that is the arithmetic mean of $\dfrac{1}{x}$ and $\dfrac{1}{x - 1}$.

47. Use a reference work to find the first six Fibonnacci numbers. Compare this set of numbers to your results in Exercises 37, 39, and 41.

Suppose you drive at 40 mi/h from city A to city B. You then return along the same route from city B to city A at 50 mi/h. What is your average rate for the round trip? Your obvious guess would be 45 mi/h, but you are in for a surprise.

Suppose that the cities are 200 mi apart. Your time from city A to city B is the distance divided by the rate, or

$$\frac{200 \text{ mi}}{40 \text{ mi/h}} = 5 \text{ h}$$

Similarly, your time from city B to city A is

$$\frac{200 \text{ mi}}{50 \text{ mi/h}} = 4 \text{ h}$$

The total time is then 9 h, and now using *rate equals distance divided by time,* we have

$$\frac{400 \text{ mi}}{9 \text{ h}} = \frac{400}{9} \text{ mi/h} = 44\frac{4}{9} \text{ mi/h}$$

Note that the rate for the round trip is independent of the distance involved. For instance, try the same computations above if cities A and B are 400 mi apart.

The answer to the problem above is the complex fraction

$$R = \frac{2}{\dfrac{1}{R_1} + \dfrac{1}{R_2}}$$

where

$$R_1 = \text{rate going}$$
$$R_2 = \text{rate returning}$$
$$R = \text{rate for round trip}$$

Use this information to solve Exercises 48 to 51.

48. Verify that if $R_1 = 40$ mi/h and $R_2 = 50$ mi/h, then $R = 44\dfrac{4}{9}$ mi/h, by simplifying the complex fraction *after* substituting those values.

49. Simplify the given complex fraction first. *Then* substitute 40 for R_1 and 50 for R_2 to calculate R.

50. Repeat Exercise 48, where $R_1 = 50$ mi/h and $R_2 = 60$ mi/h.

51. Use the procedure in Exercise 49 with the above values for R_1 and R_2.

52. Mathematicians have shown that there are situations in which the method used to determine the number of U.S. representatives each state gets may not be fair, and a state may not get its basic quota of representatives. They give the table below of a hypothetical seven states and their populations as an example.

State	Population	Exact Quota	Rounded Quota	Actual Number of Reps.
A	325	1.625	2	2
B	788	3.940	4	4
C	548	2.740	3	3
D	562	2.810	3	3
E	4,263	21.315	21	21
F	3,219	16.095	16	15
G	295	1.475	1	2
Total	10,000	50		50

In this case, the total population of all states is 10,000, and there are 50 representatives in all, so there should be no more than 10,000/50, or 200, people per representative. The quotas are found by dividing the population by 200. Whether a state A should get an additional representative before another state E should get one is decided in this method by using the simplified inequality below. For each state, we find the ratio of the state's population to the square root of the product of the possible number of representatives and one more than this integer. We then compare these ratios for each two states. If

$$\frac{A}{\sqrt{a(a+1)}} > \frac{E}{\sqrt{e(e+1)}}$$

is true, then A gets an extra representative before E does.

(a) If you go through the process of comparing the inequality above for each pair of states, state F loses a representative to state G. Do you see how this happens? Will state F complain?

(b) Alexander Hamilton, one of the signers of the Constitution, proposed that the extra representative positions be given one at a time to states with the largest remainder until all the "extra" positions were filled. How would this affect the table? Do you agree or disagree?

53. In Italy in the 1500s, Pietro Antonio Cataldi expressed square roots as infinite, continued fractions. It is not a difficult process to follow. For instance, if you want the square root of 5, then let

$$x + 1 = \sqrt{5}$$

Squaring both sides gives

$$(x + 1)^2 = 5 \quad \text{or} \quad x^2 + 2x + 1 = 5$$

which can be written

$$x(x + 2) = 4$$

$$x = \frac{4}{x + 2}$$

One can continue replacing x with $\dfrac{4}{2+x}$:

$$x = \cfrac{4}{2 + \cfrac{4}{2 + \cfrac{4}{2 + \cfrac{4}{2 + \cdots}}}}$$

to obtain

$$\sqrt{5} - 1$$

(a) Evaluate the complex fraction above (ignore the three dots) and then add 1, and see how close it is to the square root of 5. What should you put where the ellipsis (. . .) is? Try a number you feel is close to $\sqrt{5} - 1$. How far would you have to go to get the square root correct to the nearest hundredth?

(b) Develop an infinite complex fraction for $\sqrt{10} - 1$.

In Exercises 54 and 55, use the table utility on your graphing calculator to complete the table. Comment on the equivalence of the two expressions.

54.

x	-3	-2	-1	0	1	2	3
$\dfrac{1 - \dfrac{2}{x}}{1 - \dfrac{4}{x^2}}$							
$\dfrac{x}{x + 2}$							

55.

x	-3	-2	-1	0	1	2	3
$\dfrac{-8 + \dfrac{20}{x}}{4 - \dfrac{25}{x^2}}$							
$\dfrac{-4x}{2x + 5}$							

56. Here is yet another method for simplifying a complex fraction. Suppose we want to simplify

$$\cfrac{\dfrac{3}{5}}{\dfrac{7}{10}}$$

Multiply the numerator and denominator of the complex fraction by $\dfrac{10}{7}$.

(a) What principle allows you to do this?

(b) Why was $\dfrac{10}{7}$ chosen?

(c) When learning to divide fractions, you may have heard the saying "Yours is not to reason why . . . just invert and multiply." How does this method serve to explain the "reason why" we invert and multiply?

7.5 Solving Rational Equations

7.5 OBJECTIVES

1. Solve rational equations in one variable algebraically
2. Solve literal equations involving a rational expression
3. Solve applications involving rational expressions

"One person's constant is another person's variable."

—Susan Gerhart

Rational Equations

Applications of our work in algebra will often result in equations involving rational expressions. Our objective in this section is to develop methods to find solutions for such equations.

The usual technique for solving such equations is to multiply both sides of the equation by the lowest common denominator (LCD) of all the rational expressions appearing in the equation. The resulting equation will be cleared of fractions, and we can then proceed to solve the equation as before. Example 1 illustrates the process.

Example 1

Clearing Equations of Fractions

Solve.

$$\frac{2x}{3} + \frac{x}{5} = 13$$

The LCD for 3 and 5 is 15. Multiplying both sides of the equation by 15, we have

$$15\left(\frac{2x}{3} + \frac{x}{5}\right) = 15 \cdot 13 \qquad \text{Distribute 15 on the left.}$$

$$15 \cdot \frac{2x}{3} + 15 \cdot \frac{x}{5} = 15 \cdot 13$$

$$\frac{\overset{5}{\cancel{15}} \cdot 2x}{\underset{1}{\cancel{3}}} + \frac{\overset{3}{\cancel{15}} \cdot x}{\underset{1}{\cancel{5}}} = 15 \cdot 13$$

$$10x + 3x = 195$$

Simplify. The equation is now cleared of fractions.

$$13x = 195$$

$$x = 15$$

The solution set is $\{15\}$.

To check, substitute 15 in the original equation.

$$\frac{2 \cdot 15}{3} + \frac{15}{5} \overset{?}{=} 13$$

$$10 + 3 \overset{?}{=} 13$$

$$13 = 13 \qquad \text{A true statement.}$$

So 15 is the solution for the equation.

Let's compare.

Caution

Be careful! A common mistake is to confuse an *equation* such as

$$\frac{2x}{3} + \frac{x}{5} = 13$$

and an *expression* such as

$$\frac{2x}{3} + \frac{x}{5}$$

Equation: $\dfrac{2x}{3} + \dfrac{x}{5} = 13$

Here we want to *solve the equation for x*, as in Example 1. We multiply both sides by the LCD to clear fractions and proceed as before.

Expression: $\dfrac{2x}{3} + \dfrac{x}{5}$

Here we want to find *a third fraction* that is equivalent to the given expression. We write each fraction as an equivalent fraction with the LCD as a common denominator.

$$\frac{2x}{3} + \frac{x}{5} = \frac{2x \cdot 5}{3 \cdot 5} + \frac{x \cdot 3}{5 \cdot 3}$$

$$= \frac{10x}{15} + \frac{3x}{15} = \frac{10x + 3x}{15}$$

$$= \frac{13x}{15}$$

 CHECK YOURSELF 1

Solve.

$$\frac{3x}{2} - \frac{x}{3} = 7$$

The process is similar when variables are in the denominators. Consider Example 2.

| Example 2 | **Solving an Equation Involving Rational Expressions** |

Solve.

We assume that x cannot have the value 0. Do you see why?

$$\frac{7}{4x} - \frac{3}{x^2} = \frac{1}{2x^2}$$

The LCD of $4x$, x^2, and $2x^2$ is $4x^2$. So, multiplying both sides by $4x^2$, we have

$$4x^2\left(\frac{7}{4x} - \frac{3}{x^2}\right) = 4x^2 \cdot \frac{1}{2x^2} \qquad \text{Distribute } 4x^2 \text{ on the left side.}$$

$$4x^2 \cdot \frac{7}{4x} - 4x^2 \cdot \frac{3}{x^2} = 4x^2 \cdot \frac{1}{2x^2} \qquad \text{Simplify.}$$

$$\frac{\overset{x}{\cancel{4x^2}} \cdot 7}{\underset{1}{\cancel{4x}}} - \frac{\overset{4}{\cancel{4x^2}} \cdot 3}{\underset{1}{\cancel{x^2}}} = \frac{\overset{2}{\cancel{4x^2}} \cdot 1}{\underset{1}{\cancel{2x^2}}}$$

$$7x - 12 = 2$$

$$7x = 14$$

$$x = 2$$

The solution set is $\{2\}$.

We leave the check of the solution $x = 2$ to you. Be sure to return to the original equation and substitute 2 for x.

 CHECK YOURSELF 2

Solve.

$$\frac{5}{2x} - \frac{4}{x^2} = \frac{7}{2x^2}$$

Example 3 illustrates the same solution process when there are binomials in the denominators.

| Example 3 | **Solving an Equation Involving Rational Expressions** |

Solve.

Here we assume that x cannot have the value -2 or 3.

$$\frac{4}{x + 2} + 3 = \frac{3x}{x - 3}$$

Note that multiplying *each term* by the LCD is the same as multiplying both sides of the equation by the LCD.

The LCD is $(x + 2)(x - 3)$. Multiplying by that LCD, we have

$$(x + 2)(x - 3)\left(\frac{4}{x + 2}\right) + (x + 2)(x - 3)(3) = (x + 2)(x - 3)\left(\frac{3x}{x - 3}\right)$$

Or, simplifying each term, we have

$$4(x - 3) + 3(x + 2)(x - 3) = 3x(x + 2)$$

We now clear the parentheses and proceed as before.

$$4x - 12 + 3x^2 - 3x - 18 = 3x^2 + 6x$$
$$3x^2 + x - 30 = 3x^2 + 6x$$
$$x - 30 = 6x$$
$$-5x = 30$$
$$x = -6$$

The solution set is $\{-6\}$.

Again, we leave the check of this solution to you.

✔ CHECK YOURSELF 3

Solve.

$$\frac{5}{x - 4} + 2 = \frac{2x}{x - 3}$$

Factoring plays an important role in solving equations containing rational expressions.

Example 4

Solving an Equation Involving Rational Expressions

Solve.

$$\frac{3}{x - 3} - \frac{7}{x + 3} = \frac{2}{x^2 - 9}$$

In factored form, the denominator on the right side is $(x - 3)(x + 3)$, which forms the LCD, and we multiply each term by that LCD.

$$(x - 3)(x + 3)\left(\frac{3}{x - 3}\right) - (x - 3)(x + 3)\left(\frac{7}{x + 3}\right) = (x - 3)(x + 3)\left[\frac{2}{(x - 3)(x + 3)}\right]$$

Again, simplifying each term on the right and left sides, we have

$$3(x + 3) - 7(x - 3) = 2$$
$$3x + 9 - 7x + 21 = 2$$
$$-4x = -28$$
$$x = 7$$

The solution set is $\{7\}$.

Be sure to check this result by substitution in the original equation.

 CHECK YOURSELF 4

Solve $\dfrac{4}{x-4} - \dfrac{3}{x+1} = \dfrac{5}{x^2-3x-4}$.

Whenever we multiply both sides of an equation by an expression containing a variable, there is the possibility that a proposed solution may make that multiplier 0. As we pointed out earlier, multiplying by 0 does not give an equivalent equation, and therefore verifying solutions by substitution serves not only as a check of our work but also as a check for extraneous solutions. Consider Example 5.

Example 5 | Solving an Equation Involving Rational Expressions

Solve.

Note that we must assume that $x \neq 2$.

$$\frac{x}{x-2} - 7 = \frac{2}{x-2}$$

The LCD is $x - 2$, and multiplying, we have

Note that each of the three terms gets multiplied by $x - 2$.

$$\left(\frac{x}{x-2}\right)(x-2) - 7(x-2) = \left(\frac{2}{x-2}\right)(x-2)$$

Simplifying yields

$$x - 7(x-2) = 2$$
$$x - 7x + 14 = 2$$
$$-6x = -12$$
$$x = 2$$

Caution

Because division by 0 is undefined, we conclude that 2 is *not a solution* for the original equation. It is an extraneous solution. The original equation has no solution.

To check this result, by substituting 2 for x, we have

$$\frac{2}{2-2} - 7 \overset{?}{=} \frac{2}{2-2}$$
$$\frac{2}{0} - 7 \overset{?}{=} \frac{2}{0}$$

The solution set is empty. The set is written { } or ∅.

 CHECK YOURSELF 5

Solve $\dfrac{x-3}{x-4} = 4 + \dfrac{1}{x-4}$.

Equations involving rational expressions may also lead to quadratic equations, as illustrated in Example 6.

| **Example 6** | **Solving an Equation Involving Rational Expressions** |

Solve.

Assume $x \neq 3$ and $x \neq 4$.

$$\frac{x}{x-4} = \frac{15}{x-3} - \frac{2x}{x^2-7x+12}$$

After we factor the trinomial denominator on the right, the LCD of $x - 3$, $x - 4$, and $x^2 - 7x + 12$ is $(x-3)(x-4)$. Multiplying by that LCD, we have

$$(x-3)(x-4)\left(\frac{x}{x-4}\right) = (x-3)(x-4)\left(\frac{15}{x-3}\right) - (x-3)(x-4)\left[\frac{2x}{(x-3)(x-4)}\right]$$

Simplifying yields

$$x(x-3) = 15(x-4) - 2x \qquad \text{Remove the parentheses.}$$

$$x^2 - 3x = 15x - 60 - 2x \qquad \text{Write in standard form and factor.}$$

$$x^2 - 16x + 60 = 0$$

$$(x-6)(x-10) = 0$$

So $x = 6$ or $x = 10$

Verify that 6 and 10 are both solutions for the original equation. The solution set is $\{6, 10\}$.

✔ *CHECK YOURSELF 6*

Solve $\dfrac{3x}{x+2} - \dfrac{2}{x+3} = \dfrac{36}{x^2+5x+6}$.

The following algorithm summarizes our work in solving equations containing rational expressions.

Solving Equations Containing Rational Expressions

Step 1. Clear the equation of fractions by multiplying both sides of the equation by the LCD of all the fractions that appear.

Step 2. Solve the equation resulting from step 1.

Step 3. Check all solutions by substitution in the original equation.

Rational equations are needed when we are asked to find the missing sides of a triangle. Two triangles are said to be **similar** if they have the same shape. They may or may not be the same size.

When two triangles are similar, the ratios of the related sides are equal. In the triangles below, note that side a is related to side d and side b is related to side e.

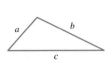 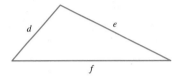

Because these are similar triangles, we can say that $\dfrac{a}{b} = \dfrac{d}{e}$. This idea is used in Example 7.

| Example 7 | **Finding the Lengths of the Sides of Similar Triangles** |

Given that the two triangles below are similar triangles, find the lengths of the indicated sides.

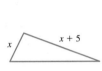 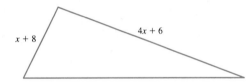

From the equation above, we know that $\dfrac{a}{b} = \dfrac{d}{e}$; in this case, we have

$$\frac{x}{x+5} = \frac{x+8}{4x+6}$$

Multiplying by the common denominator, we get the equation

$$x(4x + 6) = (x + 8)(x + 5)$$

$$4x^2 + 6x = x^2 + 13x + 40$$

$$3x^2 - 7x - 40 = 0$$

$$(3x + 8)(x - 5) = 0$$

$$x = -\frac{8}{3} \quad \text{or} \quad x = 5$$

We know that x represents the length of one side of a triangle, so it must be a positive number. We can now find the four indicated lengths.

$$x = 5$$

$$x + 5 = 10$$

$$x + 8 = 13$$

$$4x + 6 = 26$$

✔ CHECK YOURSELF 7

Given that the two triangles below are similar triangles, find the lengths of the indicated sides.

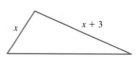

The method in this section may also be used to solve certain literal equations for a specified variable. Consider Example 8.

Example 8   Solving a Literal Equation

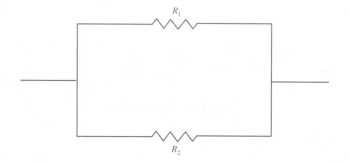

A parallel electric circuit. The symbol for a resistor is

Recall that the numbers 1 and 2 are *subscripts*. We read R_1 as "R sub 1" and R_2 as "R sub 2."

If two resistors with resistances R_1 and R_2 are connected in parallel, the combined resistance R can be found from

$$\frac{1}{R} = \frac{1}{R_1} + \frac{1}{R_2}$$

Solve the formula for R.

First, the LCD is RR_1R_2, and we multiply:

$$RR_1R_2 \cdot \frac{1}{R} = RR_1R_2 \cdot \frac{1}{R_1} + RR_1R_2 \cdot \frac{1}{R_2}$$

Simplifying yields

$$R_1R_2 = RR_2 + RR_1 \qquad \text{Factor out } R \text{ on the right.}$$

$$R_1R_2 = R(R_2 + R_1) \qquad \text{Divide by } R_2 + R_1 \text{ to isolate } R.$$

$$\frac{R_1R_2}{R_2 + R_1} = R \qquad \text{or} \qquad R = \frac{R_1R_2}{R_1 + R_2}$$

Note: This formula involves the focal length of a convex lens.

✔ *CHECK YOURSELF 8*

Solve for D_1.

$$\frac{1}{F} = \frac{1}{D_1} + \frac{1}{D_2}$$

The Distance Formula

$$d = r \cdot t \qquad \text{Distance equals rate times time.}$$

Sometimes use of this formula leads to a rational equation, as in Example 9.

Example 9

Solving a Motion Problem

A boat can travel 16 mi/h in still water. If the boat can travel 5 mi downstream in the same time it takes to travel 3 mi upstream, what is the rate of the river's current?

Step 1   We want to find the rate of the current.

Step 2   Let c be the rate of the current. Then $16 + c$ is the rate of the boat going downstream and $16 - c$ is the rate going upstream.

Step 3   We'll use a chart to help us set up an appropriate equation.

	d	r	t
Downstream	5 mi	$16 + c$	$\dfrac{5}{16 + c}$
Upstream	3 mi	$16 - c$	$\dfrac{3}{16 - c}$

We found the time by dividing distance by rate. If $d = r \cdot t$, then $t = \dfrac{d}{r}$. The key to finding our equation is noting that the time is the same upstream and downstream, so

$$\frac{5}{16 + c} = \frac{3}{16 - c}$$

Step 4   Multiplying both sides by the LCD $(16 + c)(16 - c)$ yields

$$5(16 - c) = 3(16 + c)$$
$$80 - 5c = 48 + 3c$$
$$32 = 8c$$
$$c = 4$$

The current is moving at 4 mi/h.

Step 5   To check, verify that $\dfrac{5}{16 + 4} = \dfrac{3}{16 - 4}$.

 CHECK YOURSELF 9

A plane flew 540 mi into a steady 30 mi/h wind. The pilot then returned along the same route with a tailwind. If the entire trip took $7\dfrac{1}{2}$ h, what would his speed have been in still air?

Another application that frequently results in rational equations is the work problem. Example 10 illustrates.

Example 10

Caution

Error 1: Students sometimes try adding the two times. If we add 40 min and 80 min, we get 120 min. Is this a reasonable answer? Certainly not! If one printer does the job in 40 min, why would it take 120 min for two printers to do it? It wouldn't.

Caution

Error 2: A more reasonable approach would be to give one-half of the job to each printer. The first printer would finish its half of the job in 20 min. The second would finish its half in 40 min. The first printer would be idle for the final 20 min, so we know the job could have been finished faster.

Solving a Work Problem

One computer printer can print a company's paychecks in 40 minutes (min). A second printer can print them in 80 min. If both printers are working, how long will it take to print the paychecks?

Before we learn how to solve work problems, we should look at a couple of common errors made in attempting to solve such problems.

Although the method in the margin that is labeled *error 2* does not solve the problem, it does give us guidelines for a reasonable answer. When the printers work together, it will take them somewhere between 20 and 40 min to finish the job. To find the exact amount of time, we use the work principle.

> **Work Principle**
>
> Given an object A which completes a task in time a and an object B which completes the same task in time b, we find the amount of time it takes them to complete the task together by solving the following equation for variable t.
>
> $$\frac{t}{a} + \frac{t}{b} = 1$$

Now we can solve the problem.

Step 1   We are looking for the time it takes to print the paychecks.

Step 2   Our variable t is the time it will take the printers to complete the task.

Step 3   We have the equation $\dfrac{t}{40} + \dfrac{t}{80} = 1$.

Step 4   Multiply by LCD of 80.

$$2t + t = 80$$

$$3t = 80$$

$$t = \frac{80}{3} \text{ min}$$

$$= 26\frac{2}{3} \text{ min}$$

$$= 26 \text{ min } 40 \text{ s}$$

Step 5   To verify this answer, we find what fraction of the job each printer does in this time. The first printer does $\dfrac{\frac{80}{3}}{40} = \dfrac{2}{3}$ of the job. The second printer does

$\dfrac{\frac{80}{3}}{80} = \dfrac{1}{3}$ of the job. Together, they do the entire job $\left(\dfrac{2}{3} + \dfrac{1}{3} = 1\right)$.

✔ *CHECK YOURSELF 10*

It would take Sasha 48 days to paint the house. Natasha could do it in 36 days. How long would it take them to paint the house if they worked together?

| Example 11 | **Number Analysis** |

The numerator of a fraction exceeds the denominator by 5. When the numerator is increased by 4 and the denominator is decreased by 8, the fraction is equivalent to 2. Find the fraction.

Step 1   We want to find the original fraction.

Step 2   Let x represent the denominator of the original fraction.

The original numerator is $x + 5$.

The original fraction is $\dfrac{x + 5}{x}$.

The new numerator is $(x + 5) + 4$, or $x + 9$.

The new denominator is $x - 8$.

The new fraction is $\dfrac{x + 9}{x - 8}$.

Step 3   Since the new fraction must have the value of 2, the equation is

$$\frac{x + 9}{x - 8} = 2$$

Step 4   Multiplying both sides of the equation by the LCD of $x - 8$ yields

$$x + 9 = 2(x - 8)$$

$$x + 9 = 2x - 16$$

$$9 + 16 = 2x - x$$

$$25 = x$$

The original denominator is 25.

The original numerator is $x + 5$, or 30.

The original fraction is $\dfrac{30}{25}$.

Step 5   You should check by returning to the original statement of the problem to see that the answer gives a new fraction equivalent to 2.

✔ *CHECK YOURSELF 11*

Five added to twice the numerator of a fraction results in the denominator of the fraction. When 1 is added to both the numerator and denominator, the resulting fraction is $\dfrac{1}{3}$. Find the original fraction.

✔ *CHECK YOURSELF ANSWERS*

1. {6}   **2.** {3}   **3.** {9}   **4.** {−11}   **5.** { } or ∅   **6.** $\left\{-5, \dfrac{8}{3}\right\}$

7. The lengths are 6 and 9 on the first triangle and 8 and 12 on the second.

8. $\dfrac{FD_2}{D_2 - F}$ **9.** 150 mi/h   **10.** $20\dfrac{4}{7}$ days   **11.** $\dfrac{3}{11}$

Exercises ▪ 7.5

In Exercises 1 to 8, decide whether each of the following is an expression or an equation. If it is an equation, find a solution. If it is an expression, write it as a single fraction.

1. $\dfrac{x}{4} - \dfrac{x}{5} = 2$

2. $\dfrac{x}{4} - \dfrac{x}{7} = 3$

3. $\dfrac{x}{2} - \dfrac{x}{5}$

4. $\dfrac{x}{7} - \dfrac{x}{14}$

5. $\dfrac{3x + 1}{4} = x - 1$

6. $\dfrac{3x - 1}{2} - \dfrac{x}{5} - \dfrac{x + 3}{4}$

7. $\dfrac{x}{4} = \dfrac{x}{12} + \dfrac{1}{2}$

8. $\dfrac{2x - 1}{3} + \dfrac{x}{2}$

In Exercises 9 to 50, solve each equation.

9. $\dfrac{x}{3} + \dfrac{3}{2} = \dfrac{x}{6} + \dfrac{7}{3}$

10. $\dfrac{x}{12} - \dfrac{2}{3} = \dfrac{x}{6} + \dfrac{3}{4}$

11. $\dfrac{4}{x} + \dfrac{3}{4} = \dfrac{10}{x}$

12. $\dfrac{3}{x} = \dfrac{5}{3} - \dfrac{7}{x}$

13. $\dfrac{5}{4x} - \dfrac{1}{2} = \dfrac{1}{2x}$

14. $\dfrac{7}{6x} - \dfrac{1}{3} = \dfrac{1}{2x}$

15. $\dfrac{4}{x + 5} = \dfrac{3}{x + 3}$

16. $\dfrac{5}{x - 2} = \dfrac{4}{x + 1}$

17. $\dfrac{9}{x} + 2 = \dfrac{2x}{x + 3}$

18. $\dfrac{6}{x} + 3 = \dfrac{3x}{x + 1}$

19. $\dfrac{3}{x + 2} - \dfrac{5}{x} = \dfrac{13}{x + 2}$

20. $\dfrac{7}{x} - \dfrac{2}{x - 3} = \dfrac{6}{x}$

21. $\dfrac{3}{2} + \dfrac{2}{2x - 4} = \dfrac{1}{x - 2}$

22. $\dfrac{10}{2x + 6} + \dfrac{2}{x + 3} = \dfrac{1}{2}$

23. $\dfrac{x}{3x + 12} + \dfrac{x - 1}{x + 4} = \dfrac{5}{3}$

24. $\dfrac{x}{4x - 12} - \dfrac{x - 4}{x - 3} = \dfrac{1}{8}$

25. $\dfrac{16 - 3x}{x + 6} + \dfrac{x + 3}{5} = \dfrac{3x - 2}{15}$

26. $\dfrac{x + 1}{x - 2} - \dfrac{x + 3}{x} = \dfrac{6}{x^2 - 2x}$

27. $\dfrac{1}{x - 2} - \dfrac{2}{x + 2} = \dfrac{2}{x^2 - 4}$

28. $\dfrac{1}{x + 4} + \dfrac{1}{x - 4} = \dfrac{12}{x^2 - 16}$

29. $\dfrac{7}{x + 5} - \dfrac{1}{x - 5} = \dfrac{x}{x^2 - 25}$

30. $\dfrac{2}{x - 2} = \dfrac{3}{x + 2} + \dfrac{x}{x^2 - 4}$

31. $\dfrac{11}{x+2} - \dfrac{5}{x^2-x-6} = \dfrac{1}{x-3}$

32. $\dfrac{3}{x-4} - \dfrac{4}{x^2-3x-4} = \dfrac{1}{x+1}$

33. $\dfrac{5}{x-2} - \dfrac{3}{x+3} = \dfrac{24}{x^2+x-6}$

34. $\dfrac{3}{x+1} - \dfrac{5}{x+6} = \dfrac{2}{x^2+7x+6}$

35. $\dfrac{x}{x-3} - 2 = \dfrac{3}{x-3}$

36. $\dfrac{x}{x-5} + 2 = \dfrac{5}{x-5}$

37. $\dfrac{2}{x^2-3x} - \dfrac{1}{x^2+2x} = \dfrac{2}{x^2-x-6}$

38. $\dfrac{2}{x^2-x} - \dfrac{4}{x^2+5x-6} = \dfrac{3}{x^2+6x}$

39. $\dfrac{2}{x^2-4x+3} - \dfrac{3}{x^2-9} = \dfrac{2}{x^2+2x-3}$

40. $\dfrac{2}{x^2-4} - \dfrac{1}{x^2+x-2} = \dfrac{3}{x^2-3x+2}$

41. $\dfrac{7}{x-5} - \dfrac{3}{x+5} = \dfrac{40}{x^2-25}$

42. $\dfrac{3}{x-3} - \dfrac{18}{x^2-9} = \dfrac{5}{x+3}$

43. $\dfrac{2x}{x-3} + \dfrac{2}{x-5} = \dfrac{3x}{x^2-8x+15}$

44. $\dfrac{x}{x-4} = \dfrac{5x}{x^2-x-12} - \dfrac{3}{x+3}$

45. $\dfrac{2x}{x+2} = \dfrac{5}{x^2-x-6} - \dfrac{1}{x-3}$

46. $\dfrac{3x}{x-1} = \dfrac{2}{x-2} - \dfrac{2}{x^2-3x+2}$

47. $\dfrac{7}{x-2} + \dfrac{16}{x+3} = 3$

48. $\dfrac{5}{x-2} + \dfrac{6}{x+2} = 2$

49. $\dfrac{11}{x-3} - 1 = \dfrac{10}{x+3}$

50. $\dfrac{17}{x-4} - 2 = \dfrac{10}{x+2}$

In Exercises 51 and 52, two similar triangles are given. Find the indicated sides.

51.

52.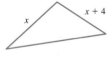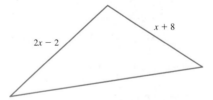

In Exercises 53 to 58, solve each equation for the indicated variable.

53. $\dfrac{1}{x} = \dfrac{1}{a} - \dfrac{1}{b}$ for x

54. $\dfrac{1}{x} = \dfrac{1}{a} + \dfrac{1}{b}$ for a

55. $\dfrac{1}{R} = \dfrac{1}{R_1} + \dfrac{1}{R_2}$ for R_1

56. $\dfrac{1}{F} = \dfrac{1}{D_1} + \dfrac{1}{D_2}$ for D_2

57. $y = \dfrac{x+1}{x-1}$ for x

58. $y = \dfrac{x-3}{x-2}$ for x

Solve the following problems.

59. Number analysis. One number is 4 times another number. The sum of the reciprocals of the numbers is $\frac{5}{24}$. Find the two numbers.

60. Number analysis. The sum of the reciprocals of two consecutive even integers is equal to 10 times the reciprocal of the product of those integers. Find the two integers.

61. Number analysis. If the same number is subtracted from the numerator and denominator of $\frac{11}{15}$, the result is $\frac{1}{3}$. Find that number.

62. Number analysis. If the numerator of $\frac{8}{9}$ is multiplied by a number and that same number is subtracted from the denominator, the result is 10. What is that number?

63. Motion. A motorboat can travel 20 mi/h in still water. If the boat can travel 3 mi downstream on a river in the same time it takes to travel 2 mi upstream, what is the rate of the river's current?

64. Motion. Janet and Michael took a canoeing trip, traveling 6 mi upstream along a river, against a 2 mi/h current. They then returned downstream to the starting point of their trip. If their entire trip took 4 h, what was their rate in still water?

65. Motion. A plane flew 720 mi with a steady 30 mi/h tailwind. The pilot then returned to the starting point, flying against the same wind. If the round-trip flight took 10 h, what was the plane's airspeed?

66. Motion. A small jet has an airspeed (the rate in still air) of 300 mi/h. During one day's flights, the pilot noted that the plane could fly 85 mi with a tailwind in the same time it took to fly 65 mi against the same wind. What was the rate of the wind?

67. Work. One computer printer can print a company's weekly payroll checks in 60 min. A second printer would take 90 min to complete the job. How long would it take the two printers, operating together, to print the checks?

68. Work. An electrician can wire a house in 20 h. If she works with an apprentice, the same job can be completed in 12 h. How long would it take the apprentice, working alone, to wire the house?

69. What special considerations must be made when an equation contains rational expressions with variables in the denominator?

7.6 Solving Rational Inequalities

7.6 OBJECTIVES

1. Determine the zeros of a rational function
2. Determine for what values of a variable a rational expression is positive and for what values it is negative
3. Solve a rational inequality

In Section 7.5, we solved rational equations. This section will focus on solutions for rational inequalities. When we work with inequalities, it is usually helpful to first solve the related equation. We will begin the discussion in this section by finding the zeros for some rational functions. Recall that a zero of a function is a value for x such that $f(x) = 0$.

Example 1

Finding the Zeros for a Rational Function

Find the zeros for each function.

(a) $f(x) = \dfrac{x - 6}{x + 5}$

To find the zeros of a function, we set the expression on the left equal to zero and then find the values for the variable.

$$\frac{x - 6}{x + 5} = 0$$

For a fraction to equal zero, its numerator must equal zero.

$$x - 6 = 0$$

$$x = 6$$

This is the zero of the function; we say

$$f(6) = 0$$

(b) $f(x) = \dfrac{x^2 - 4x - 5}{x + 2}$

$$\frac{x^2 - 4x - 5}{x + 2} = 0$$

$x^2 - 4x - 5 = 0$ The numerator of a fraction must be 0 for the fraction to be 0.

$(x - 5)(x + 1) = 0$ Since the product is 0, one or both of the factors must equal 0.

$x = 5$ or $x = -1$

The zeros of the function are -1 and 5.

✔ CHECK YOURSELF 1

Find the zeros for each function.

(a) $f(x) = \dfrac{2x - 5}{x - 5}$

(b) $f(x) = \dfrac{x^2 + 2x - 24}{x - 3}$

In Example 1, we found the values of x for which the function was equal to zero. In Example 2, we will find the values of x for which a function is positive.

Example 2   **Finding the Positive Values for a Rational Function**

Find the values of x for which each function is positive.

(a) $f(x) = \dfrac{x - 6}{x + 5}$

In order for a fraction to be positive, either the numerator and denominator are both positive or they are both negative. We will first find the values for which both numerator and denominator are positive.

$x - 6$ is positive when x is greater than 6

$x + 5$ is positive when x is greater than -5

Let's visualize these facts on a number line.

We can see that both are positive only when x is greater than 6. We write $x > 6$.

When are both negative?

$x - 6$ is negative when x is less than 6

$x + 5$ is negative when x is less than -5

In interval notation this would be written as $(6, \infty)$.

In interval notation this would be written as $(-\infty, -5)$.

Using the number line on the previous page, we see that both are negative only when x is less than -5. We write $x < -5$.

Combining these two cases, we find that the function is positive when $x > 6$ or $x < -5$.

(b) $f(x) = \dfrac{4x - 5}{x + 2}$

Again, we will first look for values of x for which both the numerator and the denominator are positive.

When is the numerator positive?

$$4x - 5 > 0$$

$$4x > 5$$

This is written as $\left(\dfrac{5}{4}, \infty\right)$ in interval notation.

$$x > \frac{5}{4}$$

When is the denominator positive?

$$x + 2 > 0$$

This is written as $(-2, \infty)$ in interval notation.

so

$$x > -2$$

Using a number line to visualize these facts, we see that both the numerator and the denominator are positive when $x > \dfrac{5}{4}$.

Now we will find when both the numerator and the denominator are negative, since the fraction will then be positive.

When is the numerator negative?

$$4x - 5 < 0$$

$$4x < 5$$

This is written as $\left(-\infty, \dfrac{5}{4}\right)$ in interval notation.

$$x < \frac{5}{4}$$

When is the denominator negative?

$$x + 2 < 0$$

This is written as $(-\infty, -2)$ in interval notation.

so

$$x < -2$$

Using the number line on the previous page, we find that both the numerator and the denominator are negative when $x < -2$.

The function is positive when

$$x > \frac{5}{4} \quad \text{or} \quad x < -2$$

✔ CHECK YOURSELF 2

Find the values of x for which each function is positive.

(a) $f(x) = \dfrac{2x - 5}{x - 5}$

(b) $f(x) = \dfrac{2x - 5}{x + 3}$

Rational Inequalities

To solve inequalities involving rational expressions, we use the same properties of division over the real numbers demonstrated in Example 2.

1. The quotient of two positive numbers is always positive.

2. The quotient of two negative numbers is always positive.

3. The quotient of a positive number and a negative number is always negative.

We solve rational inequalities by using sign graphs, as Example 3 illustrates.

Example 3   ## Solving a Rational Inequality

Solve.

$$\frac{x - 3}{x + 2} < 0$$

The inequality states that the quotient of $x - 3$ and $x + 2$ must be negative (less than 0). This means that the numerator and denominator must have opposite signs.

We start by finding the critical points. These are points where either the numerator or the denominator is 0. In this case, the critical points are 3 and -2.

The solution depends on determining whether the numerator and denominator are positive or negative. To visualize the process, start with a number line and label it as shown below.

Examining the sign of the numerator and denominator, we see that

For any x less than -2, the quotient is positive (quotient of two negatives).

For any x between -2 and 3, the quotient is negative (quotient of a negative and a positive).

For any x greater than 3, the quotient is positive (quotient of two positives).

We return to the original inequality

$$\frac{x - 3}{x + 2} < 0$$

This inequality is true only when the quotient is negative, that is, when x is between -2 and 3. This solution set can be written as $\{x \mid -2 < x < 3\}$ and represented on a graph as follows.

✔ CHECK YOURSELF 3

Solve and graph the solution set.

$$\frac{x - 4}{x + 2} > 0$$

The solution process illustrated in Example 3 is valid only when the rational expression is isolated on one side of the inequality and is related to 0. If this is not the case, we must write an equivalent inequality as the first step, as Example 4 illustrates.

Example 4   **Solving a Rational Inequality**

Solve.

$$\frac{2x - 3}{x + 1} \geq 1$$

Since the rational expression is not related to 0, we use the following procedure.

We have subtracted 1 from both sides.

$$\frac{2x - 3}{x + 1} - 1 \geq 0 \qquad \text{Form a common denominator on the left side.}$$

$$\frac{2x - 3}{x + 1} - \frac{x + 1}{x + 1} \geq 0 \qquad \text{Combine the expressions on the left side.}$$

$$\frac{2x - 3 - (x + 1)}{x + 1} \geq 0 \qquad \text{Simplify.}$$

$$\frac{x - 4}{x + 1} \geq 0$$

We can now proceed as before since the rational expression is related to 0. The critical points are 4 and -1, and the sign graph is formed as shown below.

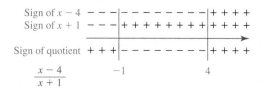

We want a *positive* quotient.

From the sign graph the solution is

$$\{x \mid x < -1 \text{ or } x \geq 4\}$$

The graph is shown below.

Note that 4 is included, but -1 cannot be included in the solution set. Why?

✔ CHECK YOURSELF 4

Solve and graph the solution set.

$$\frac{2x - 3}{x - 2} \leq 1$$

✔ CHECK YOURSELF ANSWERS

1. (a) $\dfrac{5}{2}$; (b) -6 and 4   **2.** (a) $x < \dfrac{5}{2}$ or $x > 5$; (b) $x < -3$ or $x > \dfrac{5}{2}$

3. $\{x \mid x < -2 \text{ or } x > 4\}$

4. $\{x \mid 1 \leq x < 2\}$

In Exercises 1 to 10, find the zeros of each function.

1. $f(x) = \dfrac{x}{10} - \dfrac{12}{5}$

2. $f(x) = \dfrac{4x}{3} - \dfrac{x}{6}$

3. $f(x) = \dfrac{12}{x+5} - \dfrac{5}{x}$

4. $f(x) = \dfrac{1}{x-2} - \dfrac{3}{x}$

5. $f(x) = \dfrac{1}{x-3} + \dfrac{2}{x} - \dfrac{5}{3x}$

6. $f(x) = \dfrac{2}{x} - \dfrac{1}{x+1} - \dfrac{3}{x^2+x}$

7. $f(x) = 1 + \dfrac{39}{x^2} - \dfrac{16}{x}$

8. $f(x) = x - \dfrac{72}{x} + 1$

9. $f(x) = \dfrac{x^2 + 3x - 10}{x+1}$

10. $f(x) = \dfrac{2x^2 - x - 6}{x+4}$

In Exercises 11 to 14, find the values of x for which the function is positive.

11. $f(x) = \dfrac{x-3}{x+2}$

12. $f(x) = \dfrac{x+5}{x-1}$

13. $f(x) = \dfrac{3x-5}{x+3}$

14. $f(x) = \dfrac{2x+3}{x-4}$

In Exercises 15 to 28, solve each inequality, and graph the solution set on a number line.

15. $\dfrac{x-2}{x+1} < 0$

16. $\dfrac{x+3}{x-2} > 0$

17. $\dfrac{x-4}{x-2} > 0$

18. $\dfrac{x+6}{x-3} < 0$

19. $\dfrac{x-5}{x+3} \le 0$

20. $\dfrac{x+3}{x-2} \le 0$

21. $\dfrac{2x-1}{x+3} \geq 0$

22. $\dfrac{3x-2}{x-4} \leq 0$

23. $\dfrac{x}{x-3} + \dfrac{2}{x-3} \leq 0$

24. $\dfrac{x}{x+5} - \dfrac{3}{x+5} > 0$

25. $\dfrac{x}{x+3} \leq \dfrac{4}{x+3}$

26. $\dfrac{x}{x-5} \geq \dfrac{2}{x-5}$

27. $\dfrac{2x-5}{x-2} > 1$

28. $\dfrac{2x+3}{x+4} \geq 1$

29. In solving the inequality $\dfrac{x}{x-1} > 5$, is it incorrect to find the solution by multiplying both sides by $x - 1$? Why? What technique should be used?

Examining Rational Functions

  OBJECTIVES

1. *Evaluate a rational function*
2. *Find the domain for a rational function*
3. *Find the zeros for a rational function*
4. *Use a graphing calculator to graph a rational function*

Recall that a rational expression is the ratio of two polynomial expressions.

$$R = \frac{P}{Q}$$

Similarly, a rational function is the ratio of two polynomial functions.

$$f(x) = \frac{P(x)}{Q(x)}$$

In Example 1, we will evaluate a rational function.

Example 1 Evaluating a Rational Function

Find $f(-2)$ and $f(3)$.

$$f(x) = \frac{x^2 - 3x - 4}{x + 5}$$

$$f(-2) = \frac{(-2)^2 - 3(-2) - 4}{(-2) + 5} = \frac{4 + 6 - 4}{3} = 2 \qquad \text{Substitute } -2 \text{ for } x \text{ in the original function.}$$

$$f(3) = \frac{(3)^2 - 3(3) - 4}{3 + 5} = \frac{9 - 9 - 4}{8} = -\frac{1}{2} \qquad \text{Substitute } 3 \text{ for } x \text{ in the original function.}$$

  CHECK YOURSELF 1

Find $f(-3)$ and $f(2)$.

$$f(x) = \frac{x^2 + 2x - 15}{x - 3}$$

When we work with a rational function, as with rational expressions, it is important to remember that division by zero is undefined.

Example 2

Determining the Domain of a Rational Function

Find the domain for the function

$$f(x) = \frac{x^2 - 3x - 4}{x + 5}$$

Recall that the domain of a function is the set of all variable values that produce a valid result. In the case of a rational function, the only real values that are excluded from the domain are values that would result in division by zero.

In this case, we want to make certain that

$$x + 5 \neq 0$$

That means that x cannot be equal to -5. All other real numbers are part of the domain. We write this, using set-builder notation, as follows:

$$D = \{x \,|\, x \in \mathbb{R}, x \neq -5\}$$

 CHECK YOURSELF 2

Find the domain of the function.

$$f(x) = \frac{x^2 + 2x - 15}{x - 3}$$

As with polynomials, it is sometimes useful to know the zeros of a rational function.

Example 3

Determining the Zeros of a Rational Function

Find the zeros of the function

$$f(x) = \frac{x^2 - 3x - 4}{x + 5}$$

Recall that the zeros of a function are the variable values that make the function equal zero. In this case, we want to determine when

$$x^2 - 3x - 4 = 0$$

Factoring, we find

$$(x + 1)(x - 4) = 0$$

The zeros are the x values -1 and 4.

✔ CHECK YOURSELF 3

Find the zeros of the function.

$$f(x) = \frac{x^2 + 2x - 15}{x - 3}$$

To this point, we have

1. Evaluated a rational function $f(x)$ for given values of x

2. Found excluded values in the domain of a rational function

3. Found the zeros of a rational function

All these items are of interest when we look at a graph of the function. In subsequent math classes, you will be expected to find such a graph by analysis. In this text, we expect that you will create the graph with a graphing calculator so that you can see the significance of items 1, 2, and 3 above.

Example 4   Graphing a Rational Function

Using a graphing calculator, graph the function

$$f(x) = \frac{x^2 - 3x - 4}{x + 5}$$

Graph this function in your graphing calculator. Change the window to see what happens in the third quadrant.

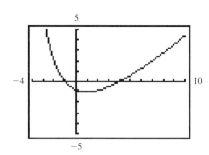

Confirm that the following points are on the graph:

$(-2, 2)$ and $\left(3, -\dfrac{1}{2}\right)$ These are the points found in Example 1.

$(-1, 0)$ and $(4, 0)$ These are the zeros of the function found in Example 3.

Also note that there is no y value associated with an x value of -5. This is the excluded value we found in Example 2.

✔ *CHECK YOURSELF 4*

Using a graphing calculator, graph the function.

$$f(x) = \frac{x^2 - 2x - 15}{x - 3}$$

In Example 5, we combine several of the previous ideas.

Example 5

Graphing a Rational Function

Let $f(x) = \dfrac{x^2 + 5x + 4}{x - 1}$.

(a) Find the domain of $f(x)$.

First, we find the values of x to be excluded from the domain of $f(x)$. This is done by determining the values of x for which the denominator is equal to 0. We want to make sure that

$$x - 1 \neq 0$$

or $$x \neq 1$$

Thus the domain of $f(x)$ is $\{x \mid x \in \mathbb{R}, x \neq 1\}$.

(b) Find the zeros of $f(x)$.

We find the zeros of $f(x)$ by setting the numerator equal to 0.

$$x^2 + 5x + 4 = 0$$

$$(x + 4)(x + 1) = 0$$

$$x + 4 = 0 \quad \text{or} \quad x + 1 = 0$$

$$x = -4 \quad \text{or} \quad x = -1$$

So the zeros of $f(x)$ are -4 and -1.

(c) Using a graphing calculator, graph $f(x)$.

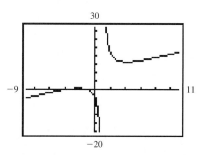

Note that (1) $(-4, 0)$ and $(-1, 0)$ are points on the graph and (2) there is no point on the graph that intersects the line $x = 1$.

✔ *CHECK YOURSELF 5*

Let $f(x) = \dfrac{x^2 - 7x + 10}{x + 5}$.

(a) Find the domain of $f(x)$.

(b) Find the zeros of $f(x)$.

(c) Using a graphing calculator, graph $f(x)$.

✔ *CHECK YOURSELF ANSWERS*

1. $f(-3) = 2$, $f(2) = 7$

2. $\{x \mid x \in \mathbb{R}, x \neq 3\}$ **3.** -5

4.

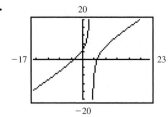

5. (a) $\{x \mid x \in \mathbb{R}, x \neq -5\}$; (b) 2, 5

(c)

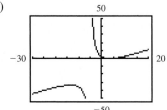

Exercises · 7.7

In Exercises 1 to 10, (a) evaluate the given rational function for the values given; (b) find the domain of the function; (c) find the zeros of the function; and (d) graph the function, using a graphing calculator.

1. $f(x) = \dfrac{x^2 - 2x - 8}{x - 1}$; $f(-1)$ and $f(2)$

2. $f(x) = \dfrac{x^2 - x - 6}{x + 4}$; $f(-3)$ and $f(10)$

3. $f(x) = \dfrac{2x^2 - x - 3}{x + 5}$; $f(1)$ and $f(3)$

4. $f(x) = \dfrac{x^2 + x - 6}{x + 4}$; $f(-3)$ and $f(2)$

5. $f(x) = \dfrac{x^2 - 6x + 8}{x + 5}$; $f(-2)$ and $f(2)$

6. $f(x) = \dfrac{6x^2 + x - 2}{x + 6}$; $f(-2)$ and $f(0)$

7. $f(x) = \dfrac{3x^2 + 5x - 2}{x - 1}$; $f(-3)$ and $f(-2)$

8. $f(x) = \dfrac{x^2 - 4x + 4}{x + 2}$; $f(2)$ and $f(3)$

9. $f(x) = \dfrac{x^2 + x - 2}{x - 3}$; $f(-2)$ and $f(4)$

10. $f(x) = \dfrac{4x^2 - 1}{x - 3}$; $f(-2)$ and $f(0)$

Summary for Chapter 7

Example	Topic	Reference
	Simplifying Rational Expressions	7.1
$\dfrac{x^2 - 5x}{x - 3}$ is a rational expression. The variable x cannot have the value 3.	**Rational expressions** have the form $\dfrac{P}{Q}$ in which P and Q are polynomials and $Q(x) \neq 0$ for all x.	**p. 568**
This uses the fact that $\dfrac{R}{R} = 1$ when $R \neq 0$.	**Fundamental Principle of Rational Expressions** For polynomials P, Q, and R, $$\frac{P}{Q} = \frac{PR}{QR} \qquad \text{when } Q \neq 0 \text{ and } R \neq 0$$ This principle can be used in two ways. We can multiply or divide the numerator and denominator of a rational expression by the same nonzero polynomial.	**p. 570**
$\dfrac{x^2 - 4}{x^2 - 2x - 8}$ $= \dfrac{(x-2)(x+2)}{(x-4)(x+2)}$ $= \dfrac{x-2}{x-4}$	**Simplifying Rational Expressions** To simplify a rational expression, use the following algorithm. 1. Completely factor both the numerator and the denominator of the expression. 2. Divide the numerator and denominator by *all* common factors. 3. The resulting expression will be in simplest form (or in lowest terms).	**p. 573**
	Identifying Rational Functions A rational function is a function that is defined by a rational expression. It can be written as $$f(x) = \frac{P}{Q} \qquad \text{in which } P \text{ and } Q \text{ are polynomials, } Q \neq 0.$$	**p. 574**
	Simplifying Rational Functions When we simplify a rational function, it is important that we note the x values that need to be excluded, particularly when we are trying to draw the graph of a function. The set of ordered pairs of the simplified function will be exactly the same as the set of ordered pairs of the original function. If we plug the excluded value(s) for x into a simplified expression, we get a set of ordered pairs that represent "holes" in the graph. These holes are breaks in the curve. We use an open circle to designate them on a graph.	**p. 574**

	Multiplication and Division of Rational Expressions	**7.2**
$\dfrac{2x-6}{x^2-9} \cdot \dfrac{x^2+3x}{6x+24}$ $= \dfrac{2(x-3)}{(x-3)(x+3)} \cdot \dfrac{x(x+3)}{6(x+4)}$ $= \dfrac{x}{3(x+4)}$	**Multiplying Rational Expressions** For polynomials P, Q, R, and S, $$\frac{P}{Q} \cdot \frac{R}{S} = \frac{PR}{QS} \qquad \text{when } Q \neq 0,\, S \neq 0$$ In practice, we apply the following algorithm to multiply two rational expressions. 1. Write each numerator and denominator in completely factored form. 2. Divide by any common factors appearing in both the numerator and the denominator. 3. Multiply as needed to form the desired product.	**p. 581** **and** **p. 583**
$\dfrac{5y}{2y-8} \div \dfrac{10y^2}{y^2-y-12}$ $= \dfrac{5y}{2y-8} \cdot \dfrac{y^2-y-12}{10y^2}$ $= \dfrac{5y}{2(y-4)} \cdot \dfrac{(y-4)(y+3)}{10y^2}$ $= \dfrac{y+3}{4y}$	**Dividing Rational Expressions** For polynomials P, Q, R, and S, $$\frac{P}{Q} \div \frac{R}{S} = \frac{P}{Q} \cdot \frac{S}{R} = \frac{PS}{QR} \qquad \text{when } Q \neq 0,\, R \neq 0,\, S \neq 0$$ To divide two rational expressions, apply the following algorithm. 1. Invert the divisor (the *second* rational expression) to write the problem as one of multiplication. 2. Proceed as in the algorithm for the multiplication of rational expressions.	**p. 583** **and** **p. 584**
	Addition and Subtraction of Rational Expressions	**7.3**
$\dfrac{5w}{w^2-16} - \dfrac{20}{w^2-16}$ $= \dfrac{5w-20}{w^2-16}$ $= \dfrac{5(w-4)}{(w+4)(w-4)}$ $= \dfrac{5}{w+4}$	**Adding and Subtracting Rational Expressions** To add or subtract rational expressions with the same denominator, add or subtract their numerators and then write that sum over the common denominator. The result should be written in lowest terms. In symbols, $$\frac{P}{R} + \frac{Q}{R} = \frac{P+Q}{R}$$ and $$\frac{P}{R} - \frac{Q}{R} = \frac{P-Q}{R}$$ when $R \neq 0$.	**p. 591**

To find the LCD for

$$\frac{2}{x^2 + 2x + 1} \quad \text{and} \quad \frac{3}{x^2 + x}$$

write

$$x^2 + 2x + 1 = (x + 1)(x + 1)$$
$$x^2 + x = x(x + 1)$$

The LCD is

$$x(x + 1)(x + 1)$$

Least Common Denominator The **least common denominator (LCD)** of a group of rational expressions is the simplest polynomial that is divisible by each of the individual denominators of the rational expressions. To find the LCD, use the following algorithm.

1. Write each of the denominators in completely factored form.

2. Write the LCD as the product of each prime factor, to the highest power to which it appears in the factored form of any individual denominators.

p. 592

$$\frac{2}{(x + 1)^2} - \frac{3}{x(x + 1)}$$

$$= \frac{2 \cdot x}{(x + 1)^2 x} - \frac{3(x + 1)}{x(x + 1)(x + 1)}$$

$$= \frac{2x - 3(x + 1)}{x(x + 1)(x + 1)}$$

$$= \frac{-x - 3}{x(x + 1)(x + 1)}$$

To add or subtract rational expressions with different denominators, we first find the LCD by the procedure outlined above. We then rewrite each of the rational expressions with that LCD as a common denominator. Then we can add or subtract as before.

p. 592

Complex Fractions

7.4

Simplify $\dfrac{1 - \dfrac{2}{x}}{1 - \dfrac{4}{x^2}}$.

Method 1:

$$\frac{\left(1 - \dfrac{2}{x}\right)x^2}{\left(1 - \dfrac{4}{x^2}\right)x^2}$$

$$= \frac{x^2 - 2x}{x^2 - 4} = \frac{x(x - 2)}{(x + 2)(x - 2)}$$

$$= \frac{x}{x + 2}$$

Method 2:

$$\frac{\dfrac{x - 2}{x}}{\dfrac{x^2 - 4}{x^2}}$$

$$= \frac{x - 2}{x} \cdot \frac{x^2}{x^2 - 4}$$

$$= \frac{x - 2}{x} \cdot \frac{x^2}{(x + 2)(x - 2)}$$

$$= \frac{x}{x + 2}$$

Complex fractions are fractions that have a fraction in their numerator or denominator (or both).

There are two commonly used methods for simplifying complex fractions: Methods 1 and 2.

p. 606

Method 1

1. Multiply the numerator and denominator of the complex fraction by the LCD of all the fractions that appear within the numerator and denominator.

2. Simplify the resulting rational expression, writing the result in lowest terms.

Method 2

1. Write the numerator and denominator of the complex fraction as single fractions, if necessary.

2. Invert the denominator and multiply as before, writing the result in lowest terms.

Solving Rational Equations

7.5

p. 615

Solve.

$$\frac{3}{x-3} - \frac{2}{x+2} = \frac{19}{x^2 - x - 6}$$

Multiply by the LCD

$(x-3)(x+2)$:

$$3(x+2) - 2(x-3) = 19$$
$$3x + 6 - 2x + 6 = 19$$
$$x = 7$$

Check:

$$\frac{3}{4} - \frac{2}{9} \stackrel{?}{=} \frac{19}{36}$$

$$\frac{19}{36} = \frac{19}{36} \qquad \text{True}$$

To solve an equation involving rational expressions, apply the following algorithm.

1. Clear the equation of fractions by multiplying both sides of the equation by the LCD of all the fractions that appear.

2. Solve the equation resulting from step 1.

3. Check all solutions by substitution in the original equation.

Solving Rational Inequalities

7.6

p. 633

Solve.

$$\frac{x+2}{x-1} \geq 0$$

Critical points are -2 and 1.

Sign of $x + 2$ $\quad - - -|+ + + + +|+ + +$

Sign of $x - 1$ $\quad - - -|- - - - -|+ + +$

Sign of quotient $(x+2)(x-1)$ $\quad + + +|- - - - -|+ + +$
$\quad -2 \qquad 1$

The solution set is

$$\{x \mid x \leq -2 \text{ or } x > 1\}$$

To solve a rational inequality, we must first find the critical points. These are points where either the numerator or the denominator is 0. A sign graph is then used to determine where the quotient is either negative or positive. The solution set is then determined.

Examining Rational Functions | 7.7

p. 640

$$f(x) = \frac{x^2 + 4x - 21}{x + 3}$$

1. $f(1) = \dfrac{1 + 4 - 21}{4} = -\dfrac{16}{4}$

   $= -4$

2. The excluded values are found by setting $x + 3 = 0$ and solving. Thus the excluded value of x is -3, and the domain of $f(x)$ is $\{x \mid x \in \mathbb{R}, x \neq -3\}$.

3. The zeros are found by solving

   $$x^2 + 4x - 21 = 0$$

   $$(x + 7)(x - 3) = 0$$

   $x + 7 = 0 \quad$ or $\quad x - 3 = 0$

   $\quad x = -7 \quad$ or $\quad x = 3$

4.

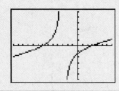

1. Evaluate the function for certain specified values of x.

2. Find the domain of a rational function by determining the excluded values. These are found by setting the denominator equal to 0 and solving.

3. Find the zeros of a rational function by setting the numerator equal to 0 and solving.

4. Use this information to verify the graph obtained by using a graphing calculator.

Summary Exercises ▪ 7

This summary exercise set is provided to give you practice with each of the objectives in the chapter. Each exercise is keyed to the appropriate chapter section.

[7.1] For what value(s) of the variable will each of the following rational expressions be defined?

1. $\dfrac{x}{2}$

2. $\dfrac{3}{y}$

3. $\dfrac{2}{x - 5}$

4. $\dfrac{4x}{3x - 2}$

[7.1] Simplify each of the following rational expressions.

5. $\dfrac{18x^5}{24x^3}$

6. $\dfrac{15m^3n}{-5mn^2}$

7. $\dfrac{7y - 49}{y - 7}$

8. $\dfrac{5x - 20}{x^2 - 16}$

9. $\dfrac{9 - x^2}{x^2 + 2x - 15}$

10. $\dfrac{3w^2 + 8w - 35}{2w^2 + 13w + 15}$

11. $\dfrac{6a^2 - ab - b^2}{9a^2 - b^2}$

12. $\dfrac{6w - 3z}{8w^3 - z^3}$

[7.1] Graph the following rational functions. Indicate the coordinates of the hole in the graph.

13. $f(x) = \dfrac{x^2 - 3x - 4}{x + 1}$

14. $f(x) = \dfrac{x^2 + x - 6}{x - 2}$

[7.2] Multiply or divide as indicated. Express your results in simplest form.

15. $\dfrac{x^7}{36} \cdot \dfrac{24}{x^4}$

16. $\dfrac{a^3b}{4ab^2} \div \dfrac{ab}{12ab^2}$

17. $\dfrac{6y - 18}{9y} \cdot \dfrac{10}{5y - 15}$

18. $\dfrac{m^2 - 3m}{m^2 - 5m + 6} \cdot \dfrac{m^2 - 4}{m^2 + 7m + 10}$

19. $\dfrac{a^2 - 2a}{a^2 - 4} \div \dfrac{2a^2}{3a + 6}$

20. $\dfrac{r^2 + 2rs}{r^3 - r^2s} \div \dfrac{5r + 10s}{r^2 - 2rs + s^2}$

21. $\dfrac{x^2 - 2xy - 3y^2}{x^2 - xy - 2y^2} \cdot \dfrac{x^2 - 4y^2}{x^2 - 8xy + 15y^2}$

22. $\dfrac{w^3 + 3w^2 + 2w + 6}{w^4 - 4} \div (w^3 + 27)$

23. Let $f(x) = \dfrac{x^2 - 16}{x - 5}$ and $g(x) = \dfrac{x^2 - 25}{x + 4}$. Find (a) $f(3) \cdot g(3)$; (b) $h(x) = f(x) \cdot g(x)$; and (c) $h(3)$.

24. Let $f(x) = \dfrac{2x^2 - 5x - 3}{x - 4}$ and $g(x) = \dfrac{x^2 - 3x - 4}{2x^2 + 5x + 2}$. Find (a) $f(3) \cdot g(3)$; (b) $h(x) = f(x) \cdot g(x)$; and (c) $h(3)$.

[7.3] Perform the indicated operations. Express your results in simplified form.

25. $\dfrac{5x + 7}{x + 4} - \dfrac{2x - 5}{x - 4}$

26. $\dfrac{5}{6x^2} + \dfrac{3}{4x}$

27. $\dfrac{2}{x - 5} - \dfrac{1}{x}$

28. $\dfrac{2}{y + 5} + \dfrac{3}{y + 4}$

29. $\dfrac{2}{3m - 3} - \dfrac{5}{2m - 2}$

30. $\dfrac{7}{x - 3} - \dfrac{5}{3 - x}$

31. $\dfrac{6}{3x + 3} - \dfrac{6}{5x - 5}$

32. $\dfrac{2a}{a^2 - 9a + 20} + \dfrac{8}{a - 4}$

33. $\dfrac{2}{s - 1} - \dfrac{6s}{s^2 + s - 2}$

34. $\dfrac{4}{x^2 - 9} - \dfrac{3}{x^2 - 4x + 3}$

35. $\dfrac{x^2 - 14x - 8}{x^2 - 2x - 8} + \dfrac{2x}{x - 4} - \dfrac{3}{x + 2}$

36. $\dfrac{w^2 + 2wz + z^2}{w^2 - wz - 2z^2} \cdot \left(\dfrac{3}{w + z} - \dfrac{1}{w - z} \right)$

37. Let $f(x) = \dfrac{2x}{x - 2}$ and $g(x) = \dfrac{x}{x - 3}$. Find (a) $f(4) + g(4)$; (b) $h(x) = f(x) + g(x)$; and (c) the ordered

pair $(4, h(4))$.

38. Let $f(x) = \dfrac{x + 2}{x - 2}$ and $g(x) = \dfrac{x + 1}{x - 7}$. Find (a) $f(3) + g(3)$; (b) $h(x) = f(x) + g(x)$; and (c) the ordered

pair $(3, h(3))$.

[7.4] Simplify each of the following complex fractions.

39. $\dfrac{\dfrac{x^3}{15}}{\dfrac{x^5}{10}}$

40. $\dfrac{\dfrac{y - 1}{y^2 - 4}}{\dfrac{y^2 - 1}{y^2 - y - 2}}$

41. $\dfrac{1 + \dfrac{a}{b}}{1 - \dfrac{a}{b}}$

42. $\dfrac{2 - \dfrac{x}{y}}{4 - \dfrac{x^2}{y^2}}$

43. $\dfrac{\dfrac{1}{r^2} - \dfrac{1}{s^2}}{\dfrac{1}{r} - \dfrac{1}{s}}$

44. $\dfrac{1 - \dfrac{1}{x + 2}}{1 + \dfrac{1}{x + 2}}$

45. $\dfrac{1 - \dfrac{2}{x - 1}}{x + \dfrac{3}{x - 4}}$

46. $\dfrac{\dfrac{w}{w + 1} - \dfrac{1}{w - 1}}{\dfrac{w}{w - 1} + \dfrac{1}{w + 1}}$

47. $\dfrac{1}{1 - \dfrac{1}{1 - \dfrac{1}{y - 1}}}$

48. $1 - \dfrac{1}{1 + \dfrac{1}{1 - \dfrac{1}{x}}}$

49. $\dfrac{1 - \dfrac{1}{x - 1}}{x - \dfrac{8}{x + 2}}$

50. $\dfrac{1}{1 - \dfrac{1}{1 + \dfrac{1}{y + 1}}}$

[7.5] Solve each of the following equations.

51. $\dfrac{1}{2x} + \dfrac{1}{3x} = \dfrac{1}{6}$

52. $\dfrac{5}{2x^2} - \dfrac{1}{4x} = \dfrac{1}{x}$

53. $\dfrac{x}{x-2} + 1 = \dfrac{x+4}{x-2}$

54. $\dfrac{2x-1}{x-3} - \dfrac{5}{x-3} = 1$

55. $\dfrac{2}{3x+1} = \dfrac{1}{x+2}$

56. $\dfrac{5}{x+1} + \dfrac{1}{x-2} = \dfrac{7}{x+1}$

57. $\dfrac{4}{x-1} - \dfrac{5}{3x-7} = \dfrac{3}{x-1}$

58. $\dfrac{7}{x} - \dfrac{1}{x-3} = \dfrac{9}{x^2-3x}$

59. $\dfrac{2}{x-3} - \dfrac{11}{x^2-9} = \dfrac{3}{x+3}$

60. $\dfrac{5}{x+3} + \dfrac{1}{x-5} = \dfrac{1}{x+3}$

61. $\dfrac{2}{x-4} = \dfrac{x}{x-2} - \dfrac{x+4}{x^2-6x+8}$

62. $\dfrac{x}{x-5} = \dfrac{3x}{x^2-7x+10} + \dfrac{8}{x-2}$

[7.5] In Exercises 63 and 64, two similar triangles are given. Find the integer lengths of the indicated sides.

63.

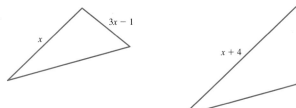

64.

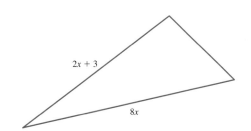

[7.5] Solve the following.

65. Number analysis. The sum of the reciprocals of two consecutive integers is equal to 11 times the reciprocal of the product of those two integers. What are the two integers?

66. Motion. Karl drove 224 mi on the expressway for a business meeting. On his return, he decided to use a shorter route of 200 mi, but road construction slowed his speed by 6 mi/h. If the trip took the same time each way, what was his average speed in each direction?

67. Work. An electrician can wire a certain model home in 20 h while it would take her apprentice 30 h to wire the same model. How long would it take the two of them, working together, to wire the house?

68. Motion. A light plane took 1 h longer to fly 540 mi on the first portion of a trip than to fly 360 mi on the second. If the rate was the same for each portion, what was the flying time for each leg of the trip?

[7.6] In Exercises 69 to 72, find the zeros for each function.

69. $f(x) = \dfrac{3x}{5} - \dfrac{x}{10}$

70. $f(x) = \dfrac{x - 2}{3} - \dfrac{2x + 1}{5}$

71. $f(x) = \dfrac{2}{x} - \dfrac{3}{x - 1}$

72. $f(x) = 1 - \dfrac{14}{x^2} - \dfrac{5}{x}$

[7.6] In Exercises 73 and 74, find the values of x for which the function is positive.

73. $f(x) = \dfrac{x + 2}{x - 4}$

74. $f(x) = \dfrac{4x - 1}{x + 2}$

[7.6] Solve each of the following inequalities.

75. $\dfrac{x - 2}{x + 1} < 0$

76. $\dfrac{x + 3}{x - 2} > 0$

77. $\dfrac{x - 4}{x - 2} > 0$

78. $\dfrac{x + 6}{x + 3} < 0$

79. $\dfrac{x - 5}{x + 3} \le 0$

80. $\dfrac{x + 3}{x - 2} \le 0$

81. $\dfrac{2x - 1}{x + 3} \ge 0$

82. $\dfrac{3x - 2}{x - 4} \le 0$

83. $\dfrac{x}{x - 3} + \dfrac{2}{x - 3} \le 0$

84. $\dfrac{x}{x + 5} - \dfrac{3}{x + 5} > 0$

85. $\dfrac{x}{x + 3} \le \dfrac{4}{x + 3}$

86. $\dfrac{x}{x - 5} \ge \dfrac{2}{x - 5}$

87. $\dfrac{2x - 5}{x - 2} > 1$

88. $\dfrac{2x + 3}{x + 4} \ge 1$

[7.7] In Exercises 89 and 90, (a) evaluate the function for the values given; (b) find the domain of the function; (c) find the zeros of the function; and (d) graph the function.

89. $f(x) = \dfrac{x^2 + 2x - 8}{x + 1}; f(-2)$ and $f(8)$

90. $f(x) = \dfrac{x^2 - 8x + 15}{x - 2}; f(1)$ and $f(3)$

The purpose of this self-test is to help you check your progress and to review for a chapter test in class. Allow yourself about 1 hour to take the test. When you are done, check your answers in the back of the book. If you missed any answers, be sure to go back and review the appropriate sections in the chapter and their exercises.

Simplify each of the following rational expressions.

1. $\dfrac{-21x^5y^3}{28xy^5}$

2. $\dfrac{3w^2 + w - 2}{3w^2 - 8w + 4}$

3. $\dfrac{x^3 + 2x^2 - 3x}{x^3 - 3x^2 + 2x}$

Graph the following.

4. $f(x) = \dfrac{x^2 - 5x + 4}{x - 4}$

Multiply or divide as indicated.

5. $\dfrac{3ab^2}{5ab^3} \cdot \dfrac{20a^2b}{21b}$

6. $\dfrac{m^2 - 3m}{m^2 - 9} \div \dfrac{4m}{m^2 - m - 12}$

7. $\dfrac{x^2 - 3x}{5x^2} \cdot \dfrac{10x}{x^2 - 4x + 3}$

8. $\dfrac{x^2 + 3xy}{2x^3 - x^2y} \div \dfrac{x^2 + 6xy + 9y^2}{4x^2 - y^2}$

9. $\dfrac{9x^2 - 9x - 4}{6x^2 - 11x + 3} \cdot \dfrac{15 - 10x}{3x - 4}$

Add or subtract as indicated.

10. $\dfrac{5}{x - 2} - \dfrac{1}{x}$

11. $\dfrac{2}{x + 3} + \dfrac{12}{x^2 - 9}$

12. $\dfrac{6x}{x^2 - x - 2} - \dfrac{2}{x + 1}$

13. $\dfrac{3}{x^2 - 3x - 4} + \dfrac{5}{x^2 - 16}$

Simplify each of the following complex fractions.

14. $\dfrac{3 - \dfrac{x}{y}}{9 - \dfrac{x^2}{y^2}}$

15. $\dfrac{1 - \dfrac{10}{z + 3}}{2 - \dfrac{12}{z - 1}}$

Solve the following equation.

16. $\dfrac{5}{x} - \dfrac{x - 3}{x + 2} = \dfrac{22}{x^2 + 2x}$

Solve the following and graph the solution set.

17. $\dfrac{x + 4}{x - 3} \le 0$

18. $\dfrac{x + 7}{x + 3} \ge 3$

19. If the numerator of $\dfrac{4}{7}$ is multiplied by a number and that same number is added to the denominator, the result is $\dfrac{6}{5}$. What was that number?

20. (a) Find the zeros of the function $f(x) = \dfrac{4x - 1}{3x + 2}$.

(b) Find the values of x for which the function defined in (a) is positive.

Cumulative Test ▪ 0–7

This test is provided to help you in the process of reviewing previous chapters. Answers are provided in the back of the book. If you miss any answers, be sure to go back and review the appropriate chapter sections.

1. Solve the equation $5x - 3(2x + 6) = 4 - (3x - 2)$.

2. If $f(x) = 5x^4 - 3x^2 + 7x - 9$, find $f(-1)$.

3. Find the equation of the line that is parallel to the line $6x + 7y = 42$ and has a y intercept of $(0, -3)$.

4. Find the x and y intercepts of the equation $7x - 6y = -42$.

Simplify each of the following polynomial functions.

5. $f(x) = 3x - 2[x - (3x - 1)] + 6x(x - 2)$

6. $f(x) = x(2x - 1)(x + 3)$

7. Find the domain of the function $f(x) = \dfrac{x^2 + x}{x}$.

8. Evaluate the expression $6^2 - (16 \div 8 \cdot 2) - 4^2$.

Factor each of the following completely.

9. $6x^3 + 7x^2 - 3x$

10. $16x^{16} - 9y^8$

Simplify each of the following rational expressions.

11. $\dfrac{5}{x - 1} - \dfrac{2x + 6}{x^2 + 2x - 3}$

12. $\dfrac{x + 1}{x^2 - 5x - 6} \div \dfrac{x^2 - 1}{x - 6}$

13. $\dfrac{1 - \dfrac{3}{x + 3}}{\dfrac{1}{x^2 - 9}}$

14. If $f(x) = \dfrac{1}{x - 5}$ and $g(x) = 8x + 6$, find (a) $f(x) + g(x)$; (b) $\dfrac{f(x)}{g(x)}$; and (c) domain of $\dfrac{f(x)}{g(x)}$.

Solve the following equations.

15. $7x + (x - 10) = -12(x - 5)$ **16.** $|-9x - 6| = 2$

17. $-4(7x + 6) = 8(5x + 12)$

Solve the following inequalities.

18. $-4(-2x - 7) > -6x$ **19.** $|5x - 4| < 3$

20. $-6|2x + 6| \le -12$

21. Solve the equation $\dfrac{5}{x} = \dfrac{2}{x + 3}$ for x.

22. Find the zeros of the function $f(x) = \dfrac{3}{x} - \dfrac{5}{x + 2}$.

23. Simplify the expression $\left(\dfrac{a^{-2}b}{a^3b^{-2}}\right)^2$.

Solve the following applications.

24. When each works alone, Barry can mow a lawn in 3 h less time than Don. When they work together, it takes 2 h. How long does it take each to do the job by himself?

25. The length of a rectangle is 2 cm less than twice the width. The area of the rectangle is 180 cm^2. Find the length and width of the rectangle.

8

Systems of Linear Equations and Inequalities

Successful businesses juggle many factors, including workers' schedules, available machine time, and costs of raw material and storage. Getting the most efficient mix of these factors is crucial if the business is to make money. Systems of linear equations can be used to find the most efficient ways to combine these costs.

For example, the owner of a small bakery must decide how much of several kinds of bread to make based on the time it takes to produce each kind and the profit to be made on each one. The owner knows the bakery requires 0.75 h of oven time and 1.25 h of preparation time for every 8 loaves of bread, and that a coffee cake takes 1 h of oven time and 1.25 h of preparation time for every 6 coffee cakes. In 1 day the bakery has 12 h of oven time and 16 h of preparation time available. The owner knows that she clears a profit of $0.50 for every loaf of bread sold and $1.75 for every coffee cake sold. But she has found that on a regular day she sells no more than 12 coffee cakes. Given these constraints, how many of each type of product can be made in a day? Which combination of products would give the highest profit if all the products were sold?

These questions can be answered by graphing a system of inequalities that model the constraints, where b is the number of loaves of bread and c is the number of coffee cakes.

8.1 Solving Systems of Linear Equations by Graphing

1. Recognize the solution of a linear system of equations
2. Solve a system by graphing
3. Use slopes to identify consistent systems

In Section 2.1, we defined a solution set as "the set of all values for the variable that makes an equation a true statement." If we have the equation

$$2x - 3x + 5 = x - 7$$

the solution set is {6}. This tells us that 6 is the only value for the variable x that makes the equation a true statement.

When we studied equations in two variables in Chapter 3, we found that a solution to a two-variable equation was an ordered pair. Given the equation

$$y = 2x - 5$$

one possible solution to the equation is the ordered pair $(1, -3)$. There are an infinite number of other possible solutions, so we don't list them in a solution set.

In this section, we are introducing a topic that has many applications in chemistry, business, economics, and physics. Each of these areas has occasion to solve a system of equations.

A **system of equations** is a set of two or more related equations.

Our goal in this chapter is to solve linear systems of equations.

A **solution** for a linear system of equations in two variables is an ordered pair of real numbers (x, y) that satisfies all the equations in the system.

Over the next couple of sections, we will look at different ways in which a linear system of equations can be solved. Our first look will be at a graphical method of solving a system.

Example 1

Solving a System by Graphing

Solve the system by graphing.

$$2x + y = 4$$
$$x - y = 5$$

Solve each equation for y and then graph.

$$y_1 = -2x + 4$$

and

$$y_2 = x - 5$$

We can *approximate* the solution by tracing the curves near their intersection.

We graph the lines corresponding to the two equations of the system.

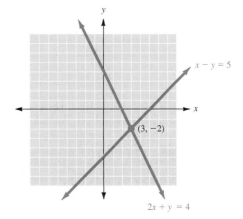

Because there are two variables in the equations, we conclude that we are searching for ordered pairs here. We want all ordered pairs that make *both* statements true.

Each equation has an infinite number of solutions (ordered pairs) corresponding to points on a line. The point of intersection, here $(3, -2)$, is the *only* point lying on both lines, and so $(3, -2)$ is the only ordered pair satisfying both equations and $(3, -2)$ is the solution for the system. The solution set is $\{(3, -2)\}$.

✔ CHECK YOURSELF 1

Solve the system by graphing.

$$3x - y = 2$$
$$x + y = 6$$

In Example 1, the two lines are nonparallel and intersect at only one point. The system has a unique solution corresponding to that point. Such a system is called a **consistent system.** In Example 2, we examine a system representing two lines that have no point of intersection.

Example 2

Solving a System by Graphing

Solve the system by graphing.

$$2x - y = 4$$
$$6x - 3y = 18$$

The lines corresponding to the two equations are graphed below.

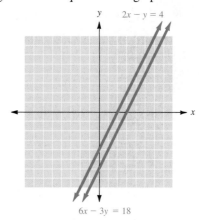

The lines are distinct and parallel. There is no point at which they intersect, so the system has no solution. We call such a system an **inconsistent system.**

   CHECK YOURSELF 2

Solve the system, if possible.

$$3x - \ y = 1$$
$$6x - 2y = 3$$

Sometimes the equations in a system have the same graph.

Example 3   ## Solving a System by Graphing

Solve the system by graphing.

$$2x - \ y = 2$$
$$4x - 2y = 4$$

The equations are graphed, as follows.

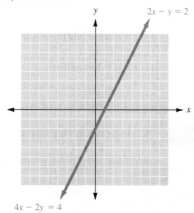

The lines have the same graph, so they have an infinite number of solutions in common. We call such a system a **dependent system.**

✔ CHECK YOURSELF 3

Solve the system by graphing.

$$6x - 3y = 12$$
$$y = 2x - 4$$

We've now seen the three possible types of solutions to a system of two linear equations. There will be a single solution (a consistent system), an infinite number of solutions (a dependent system), or no solution (an inconsistent system).

Note that, for both the dependent system and the inconsistent system, the slopes of the two lines in the system must be the same. (Do you see why that is true?) Given any two lines with different slopes, they will intersect at exactly one point. This idea is used in Example 4.

Example 4

Identifying the Type of a System

For each system, determine the number of solutions, and identify the type of system.

(a) $y = 2x - 5$

$y = 2x + 9$

Both lines have a slope of 2, but different y intercepts. We have two distinct parallel lines, and therefore there are no solutions. The system is inconsistent.

(b) $y = 3x + 7$

$y = -\dfrac{1}{3}x + 2$

These lines are perpendicular. There is one solution. The system is consistent.

The slopes are $\dfrac{2}{3}$ and $-\dfrac{3}{5}$.

(c) $2x - 3y = 7$

$3x + 5y = 2$

The lines have different slopes. There is a single solution. The system is consistent.

Solving $2x - 3y = 12$ for y, we have $y = \dfrac{2}{3}x - 4$

(d) $y = \dfrac{2}{3}x - 6$

$2x - 3y = 12$

Both lines have a slope of $\dfrac{2}{3}$, but different y intercepts. There are no solutions. The system is inconsistent.

✔ CHECK YOURSELF 4

For each system, determine the number of solutions, and identify the type of system.

(a) $y = 2x - 1$ (b) $y = -3x - 2$

       $y = 3x + 7$ $y = -\dfrac{1}{3}x + 4$

(c) $6x - 3y = 4$ (d) $y = \dfrac{1}{2}x - 4$

     $-2x + \ y = 9$ $x - y = 6$

A graphing calculator can be a very useful tool for solving systems. To employ this tool, we must first solve each equation for y, although we do not necessarily need to write them in slope-intercept form. Then we may have to explore a bit to obtain a viewing window that shows the point of intersection. Finally, we should be able to use an "intersection" utility to estimate the desired coordinates. Consider the following example.

Example 5

Using a Graphing Calculator to Solve a System

Solve the given system on a graphing calculator. Estimate coordinates to the nearest hundredth.

$$37x + 15y = 2{,}531$$
$$45x + 29y = 3{,}946$$

First, we solve each equation for y.

$$15y = 2{,}531 - 37x$$ There is no need to put this in slope-intercept form.

$$y = \frac{2{,}531 - 37x}{15}$$

We are referring in this example to a TI graphing calculator, but if you have a different type, you will surely find similar notation and utilities.

In the calculator, we enter this as

$$Y_1 = (2{,}531 - 37X)/15$$

Similarly,

$$29y = 3{,}946 - 45x$$

$$y = \frac{3{,}946 - 45x}{29}$$

So we enter

$$Y_2 = (3{,}946 - 45X)/29$$

Let us view the graphs of Y_1 and Y_2 in the standard viewing window:

Such a window usually shows
x values from −10 to 10, and
y values from −10 to 10.

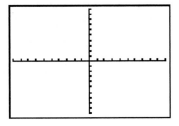

Using this window, we do not see any parts of the graphs. So we explore function values in the TABLE utility:

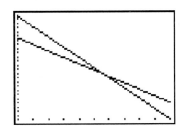

We chose 70 and 95 to be
sure to include the four y
values highlighted in the table.

Note that when x is 30, Y_1 is higher than Y_2. But when x is 40, Y_1 is lower than Y_2. This means that the lines must cross when x is somewhere between 30 and 40. If we define a window where x values run from to 30 to 40 and y values run from, say, 70 to 95, we should see the intersection point when we graph.

Finally, we may use the calculator's INTERSECT utility to find the coordinates of the intersection point:

$$x = 35.701005 \quad \text{and} \quad y = 80.670854$$

The solution set is
{(35.70, 80.67)}.

Rounding to the nearest hundredth, we see that the solution for the system is (35.70, 80.67).

✔ *CHECK YOURSELF 5*

Solve the given system on a graphing calculator. Estimate coordinates to the nearest hundredth.

$$19x + 83y = 4,587$$

$$36x + 51y = 4,229$$

✔ *CHECK YOURSELF ANSWERS*

1.

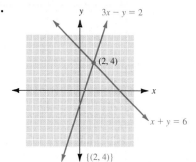

2.

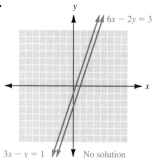

3.

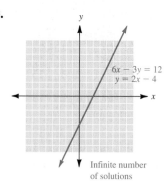

4. (a) one solution, consistent;
    (b) one solution, consistent;
    (c) no solutions, inconsistent;
    (d) one solution, consistent

5. $\{(57.98, 41.99)\}$

Exercises · 8.1

In Exercises 1 to 24, graph the system of equations and then solve the system.

1. $x + y = 6$
$x - y = 4$

2. $x - y = 8$
$x + y = 2$

3. $x + y = 5$
$-x + y = 7$

4. $x + y = 7$
$-x + y = 3$

5. $x + 2y = 4$
$x - y = 1$

6. $3x + y = 6$
$x + y = 4$

7. $3x - y = 21$
$3x + y = 15$

8. $x - 2y = -2$
$x + 2y = 6$

9. $x + 3y = 12$
$2x - 3y = 6$

10. $2x - y = 4$
$2x - y = 6$

11. $3x + 2y = 12$
$y = 3$

12. $5x - y = 11$
$2x - y = 8$

13. $x - y = 4$
$2x - 2y = 8$

14. $2x - y = 8$
$x = 2$

15. $x - 4y = -4$
$x + 2y = 8$

16. $4x + y = -7$
$-2x + y = 5$

17. $3x - 2y = 6$
$2x - y = 5$

18. $4x + 3y = 12$
$x + y = 2$

19. $3x - y = 3$
$3x - y = 6$

20. $3x - 6y = 9$
$x - 2y = 3$

21. $2y = 3$
$\quad x - 2y = -3$

22. $x + y = -6$
$\quad -x + 2y = 6$

23. $x = -5$
$\quad\; y = 3$

24. $x = -3$
$\quad\; y = 5$

Use a graphing calculator to solve each of the following exercises. Estimate your answer to the nearest hundredth. You may need to adjust the viewing window to see the point of intersection.

25. $88x + 57y = 1{,}909$
$\quad\; 95x + 48y = 1{,}674$

26. $32x + 45y = 2{,}303$
$\quad\; 29x - 38y = 1{,}509$

27. $25x - 65y = 5{,}312$
$\quad\; -21x + 32y = 1{,}256$

28. $-27x + 76y = 1{,}676$
$\quad\; 56x - 2y = -678$

29. $15x + 20y = 79$
$\quad\; 7x + 5y = 115$

30. $23x - 31y = 1{,}915$
$\quad\; 15x + 42y = 1{,}107$

For each system, determine whether it is consistent, inconsistent, or dependent.

31. $y = 3x + 7$
$\quad y = 7x - 2$

32. $y = 2x - 5$
$\quad y = -2x + 9$

33. $y = 7x - 1$
$\quad y = 7x + 8$

34. $y = -5x + 9$
$\quad y = -5x - 11$

35. $3x + 4y = 12$
$\quad 9x - 5y = 10$

36. $2x - 4y = 11$
$\quad -8x + 16y = 15$

37. $7x - 2y = 5$
$\quad 14x - 4y = 10$

38. $3x + 2y = 8$
$\quad 6x - 4y = -12$

39. Find values for m and b in the following system so that the solution to the system is $(1, 2)$.

$$mx + 3y = 8$$
$$-3x + 4y = b$$

40. Find values for m and b in the following system so that the solution to the system is $(-3, 4)$.

$$5x + 7y = b$$
$$mx + y = 22$$

41. Complete the following statements in your own words:

"To solve an equation means to . . . ."

"To solve a system of equations means to . . . ."

42. A system of equations such as the one below is sometimes called a *2-by-2* system of linear equations.

$$3x + 4y = 1$$
$$x - 2y = 6$$

Explain this term.

43. Complete this statement in your own words: "All the points on the graph of the equation $2x + 3y = 6$ . . . ." Exchange statements with other students. Do you agree with other students' statements?

44. Does a system of linear equations always have a solution? How can you tell without graphing that a system of two equations will be graphed as two parallel lines? Give some examples to explain your reasoning.

45. Suppose we have the following linear system:

$$Ax + By = C$$
$$Dx + Ey = F$$

    (a) Write the slope of the line determined by the first equation.

    (b) Write the slope of the line determined by the second equation.

    (c) What must be true about the given coefficients in order to guarantee that the system is consistent?

Systems of Equations in Two Variables with Applications

OBJECTIVES

1. Solve a system by the addition method
2. Solve a system by the substitution method
3. Use a system of equations to solve an application

Graphical solutions to linear systems are excellent for seeing and estimating solutions. The drawback comes in precision. No matter how carefully one graphs the lines, the displayed solution rarely leads to one that is exact. This problem is exaggerated when the solution includes fractional values. In this section, we will look at a couple of methods that result in exact solutions.

One method for solving systems of linear equations in two variables is the **addition method.** This method of solving systems is based on finding equivalent systems. Two systems are equivalent if they have the same solution set.

An equivalent system is formed whenever

1. One of the equations is multiplied by a nonzero number.

2. One of the equations is replaced by the sum of a constant multiple of another equation and that equation.

Example 1 illustrates the addition method for solving systems.

| Example 1 | Solving a System by the Addition Method |

Solve the system by the addition method.

$$5x - 2y = 12 \qquad (1)$$

$$3x + 2y = 12 \qquad (2)$$

The addition method is sometimes called **solution by elimination** for this reason.

In this case, adding the equations will eliminate variable y, and we have

$$8x = 24$$

$$x = 3 \qquad (3)$$

Now equation (3) can be paired with either of the original equations to form an equivalent system. We let $x = 3$ in equation (1):

The solution should be checked by substituting these values into equation (2). Here

$$3(3) + 2\left(\frac{3}{2}\right) \stackrel{?}{=} 12$$

$$9 + 3 \stackrel{?}{=} 12$$

$$12 = 12$$

is a true statement.

$$5(3) - 2y = 12$$

$$15 - 2y = 12$$

$$-2y = -3$$

$$y = \frac{3}{2}$$

and $\left(3, \frac{3}{2}\right)$ is the solution for our system. The solution set, then, is $\left\{\left(3, \frac{3}{2}\right)\right\}$.

✔ CHECK YOURSELF 1

Solve the system by the addition method.

$$4x - 3y = 19$$

$$-4x + 5y = -25$$

Remember that multiplying one or both of the equations by a nonzero constant produces an equivalent system.

Example 1 and Check Yourself 1 were straightforward in that adding the equations of the system immediately eliminated one of the variables. Example 2 illustrates a common situation in which we must multiply one or both of the equations by a nonzero constant before the addition method is applied.

Example 2

Solving a System by the Addition Method

Solve the system by the addition method.

All these solutions can be approximated by graphing the lines and locating the intersection point. This is particularly useful when the coordinates are not integers (the technical term for such solutions is *ugly*).

$$3x - 5y = 19 \qquad \textbf{(4)}$$

$$5x + 2y = 11 \qquad \textbf{(5)}$$

It is clear that adding the equations of the given system will *not* eliminate one of the variables. Therefore, we must use multiplication to form an equivalent system. The choice of multipliers depends on which variable we decide to eliminate. Here we have decided to eliminate y. We multiply equation (4) by 2 and equation (5) by 5. We then have a system that is equivalent to the original system, but easier to solve.

Note that the coefficients of y are now *opposites* of each other.

$$6x - 10y = 38$$

$$25x + 10y = 55$$

Adding now eliminates y and yields

$$31x = 93$$

$$x = 3 \qquad \textbf{(6)}$$

Pairing equation (6) with equation (4) gives an equivalent system, and we can substitute 3 for x in equation (4):

$$3 \cdot (3) - 5y = 19$$
$$9 - 5y = 19$$
$$-5y = 10$$
$$y = -2$$

Again, the solution should be checked in both equations (4) and (5).

The solution set for the system is $\{(3, -2)\}$.

✔ *CHECK YOURSELF 2*

Solve the system by the addition method.

$$2x + 3y = -18$$
$$3x - 5y = 11$$

The following algorithm summarizes the addition method of solving linear systems of two equations in two variables.

Solving by the Addition Method

Step 1.   If necessary, multiply one or both of the equations by a constant so that one of the variables can be eliminated by addition.

Step 2.   Add the equations of the equivalent system formed in step 1.

Step 3.   Solve the equation found in step 2.

Step 4.   Substitute the value found in step 3 into either of the equations of the original system to find the corresponding value of the remaining variable. The ordered pair formed is the solution to the system.

Step 5.   Check the solution by substituting the pair of values found in step 4 into both of the original equations.

Example 3 illustrates two special situations you may encounter while applying the addition method.

Example 3   **Solving a System by the Addition Method**

Solve each system by the addition method.

(a) $4x + 5y = 20$ **(7)**

$\ 8x + 10y = 19$ **(8)**

Multiply equation (7) by -2. Then

If these equations are graphed, we have two parallel lines.

$$-8x - 10y = -40$$

We add the two left sides to get 0 and the two right sides to get -21.

$$\underline{8x + 10y = 19}$$

$$0 = -21$$

The result $0 = -21$ is a *false* statement, which means that there is no point of intersection. Therefore, the system is inconsistent, and there is no solution.

(b) $5x - 7y = 9$ **(9)**

$15x - 21y = 27$ **(10)**

Multiply equation (9) by -3. We then have

$$-15x + 21y = -27$$

We add the two equations.

$$\underline{15x - 21y = 27}$$

$$0 = 0$$

The solution set could be written $\{(x, y) \mid 5x - 7y = 9\}$. This means the set of all ordered pairs (x, y) that make $5x - 7y = 9$ a true statement.

Both variables have been eliminated, and the result is a *true* statement. If the two original equations were graphed, we would see that the two lines coincide. Thus there are an infinite number of solutions, one for each point on that line. Recall that this a *dependent system.*

✔ **CHECK YOURSELF 3**

Solve each system by the addition method, if possible.

(a) $3x + 2y = 8$ (b) $x - 2y = 8$

      $9x + 6y = 11$ $3x - 6y = 24$

The results of Example 3 can be summarized as follows.

When a system of two linear equations is solved:

1. If a false statement such as $3 = 4$ is obtained, then the system is inconsistent and has no solution.

2. If a true statement such as $8 = 8$ is obtained, then the system is dependent and has an infinite number of solutions.

The Substitution Method

A third method for finding the solutions of linear systems in two variables is called the **substitution method.** You may very well find the substitution method more difficult to apply in solving certain systems than the addition method, particularly when the equations involved in the substitution lead to fractions. However, the substitution method does have important extensions to systems involving higher-degree equations, as you will see in later mathematics classes.

To outline the technique, we solve one of the equations from the original system for one of the variables. That expression is then substituted into the *other* equation of the system to provide an equation in a single variable. That equation is solved, and the corresponding value for the other variable is found as before, as Example 4 illustrates.

Example 4   Solving a System by the Substitution Method

(a) Solve the system by the substitution method.

$$2x - 3y = -3 \tag{11}$$

$$y = 2x - 1 \tag{12}$$

Since equation (12) is already solved for y, we substitute $2x - 1$ for y in equation (11).

$$2x - 3(2x - 1) = -3$$

We now have an equation in the single variable x.

Solving for x gives

$$2x - 6x + 3 = -3$$

$$-4x = -6$$

$$x = \frac{3}{2}$$

As a partial check, we substitute these values in equation (11) and have

$$2\left(\frac{3}{2}\right) - 3 \cdot 2 \overset{?}{=} -3$$

$$3 - 6 \overset{?}{=} -3$$

$$-3 = -3$$

a true statement!

We now substitute $\frac{3}{2}$ for x in equation (12).

$$y = 2\left(\frac{3}{2}\right) - 1$$

$$= 3 - 1 = 2$$

The solution set for our system is $\left\{\left(\frac{3}{2}, 2\right)\right\}$.

(b) Solve the system by the substitution method.

$$2x + 3y = 16 \tag{13}$$

$$3x - y = 2 \tag{14}$$

We start by solving equation (14) for y.

Why did we choose to solve for y in equation (14)? We could have solved for x, so that

$$x = \frac{y + 2}{3}$$

We simply chose the easier case to avoid fractions.

$$3x - y = 2$$

$$-y = -3x + 2$$

$$y = 3x - 2 \tag{15}$$

Substituting in equation (13) yields

$$2x + 3(3x - 2) = 16$$
$$2x + 9x - 6 = 16$$
$$11x = 22$$
$$x = 2$$

We now substitute 2 for x in equation (15).

$$y = 3 \cdot (2) - 2$$
$$= 6 - 2 = 4$$

The solution set for the system is $\{(2, 4)\}$. We leave the check of this result to you.

The solution should be checked in *both* equations of the original system.

✔ *CHECK YOURSELF 4*

Solve each system by the substitution method.

(a) $2x + 3y = 6$ (b) $3x + 4y = -3$

  $x = 3y + 6$ $x + 4y = -1$

The following algorithm summarizes the substitution method for solving linear systems of two equations in two variables.

Solving by the Substitution Method

Step 1.   If necessary, solve one of the equations of the original system for one of the variables.

Step 2.   Substitute the expression obtained in step 1 into the *other* equation of the system to write an equation in a single variable.

Step 3.   Solve the equation found in step 2.

Step 4.   Substitute the value found in step 3 into the equation derived in step 1 to find the corresponding value of the remaining variable. The ordered pair formed is the solution for the system.

Step 5.   Check the solution by substituting the pair of values found in step 4 into *both* equations of the original system.

A natural question at this point is, How do you decide which solution method to use? First, the graphical method can generally provide only approximate solutions. When exact solutions are necessary, one of the algebraic methods must be applied. Which method to use depends totally on the given system.

If you can easily solve for a variable in one of the equations, the substitution method should work well. However, if solving for a variable in either equation of the system leads to fractions, you may find the addition approach more efficient.

Solving Applications

We are now ready to apply our equation-solving skills to solving various applications or word problems. Being able to extend these skills to problem solving is an important goal, and the procedures developed here are used throughout the rest of the book.

Although we consider applications from a variety of areas in this section, all are approached with the same five-step strategy presented here to begin the discussion.

Solving Applications

Step 1.   Read the problem carefully to determine the unknown quantities.

Step 2.   Choose variables to represent the unknown quantities.

Step 3.   Translate the problem to the language of algebra to form a system of equations.

Step 4.   Solve the system of equations, and answer the question of the original problem.

Step 5.   Verify your solution by returning to the original problem.

Example 5

Solving a Mixture Problem

A coffee merchant has two types of coffee beans, one selling for $8 per pound (lb) and the other for $10 per pound. The beans are to be mixed to provide 100 lb of a mixture selling for $9.50 per pound. How much of each type of coffee bean should be used to form 100 lb of the mixture?

Step 1   The unknowns are the amounts of the two types of beans.

Step 2   We use two variables to represent the two unknowns. Let x be the amount of the $8 beans and y the amount of the $10 beans.

Step 3   We now want to establish a system of two equations. One equation will be based on the *total amount* of the mixture, the other on the mixture's *value*.

Since we use *two* variables, we must form *two* equations.

$$x + y = 100 \quad \text{The mixture must weigh 100 lb.} \tag{16}$$

The total value of the mixture comes from $100(9.50) = 950$

$$8x + 10y = 950 \tag{17}$$

Value of $8 beans   Value of $10 beans   Total value

Step 4   An easy approach to the solution of the system is to multiply equation (16) by -8 and add to eliminate x.

$$
\begin{aligned}
-8x - 8y &= -800 \\
8x + 10y &= 950 \\
\hline
2y &= 150 \\
y &= 75 \text{ lb}
\end{aligned}
$$

By substitution in equation (16), we have

$$x = 25 \text{ lb}$$

We should use 25 lb of $8 beans and 75 lb of $10 beans.

Note that

8(25) + 10(75) = 200 + 750
                = 950

and 9.50(100) = 950 also.

Step 5   To check the result, show that the value of the $8 beans, added to the value of the $10 beans, equals the desired value of the mixture.

✔ *CHECK YOURSELF 5*

Peanuts, which sell for $2.40 per pound, and cashews, which sell for $6 per pound, are to be mixed to form a 60-lb mixture selling for $3 per pound. How much of each type of nut should be used?

A related problem is illustrated in Example 6.

| Example 6 | # Solving a Mixture Problem |

A chemist has a 25% and a 50% acid solution. How much of each solution should be used to form 200 milliliters (ml) of a 35% acid solution?

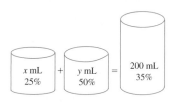

Drawing a sketch of a problem is often a valuable part of the problem-solving strategy.

Step 1   The unknowns in this case are the amounts of the 25% and 50% solutions to be used in forming the mixture.

Step 2   Again we use two variables to represent the two unknowns. Let x be the amount of the 25% solution and y the amount of the 50% solution. Let's draw a picture before proceeding to form a system of equations.

Step 3   Now, to form our two equations, we want to consider two relationships: the *total amounts* combined and the *amounts of acid* combined.

Total amounts combined.

Amounts of acid combined.

$$x + \quad y = 200 \qquad \textbf{(18)}$$
$$0.25x + 0.50y = 0.35(200) \qquad \textbf{(19)}$$

Step 4   Now, clear equation (19) of decimals by multiplying equation (19) by 100. The solution then proceeds as before, with the result

$$x = 120 \text{ ml} \qquad (25\% \text{ solution})$$
$$y = 80 \text{ ml} \qquad (50\% \text{ solution})$$

We need 120 ml of the 25% solution and 80 ml of the 50% solution.

Step 5   To check, show that the amount of acid in the 25% solution, (0.25)(120), added to the amount in the 50% solution, (0.50)(80), equals the correct amount in the mixture, (0.35)(200). We leave that to you.

✔ CHECK YOURSELF 6

A pharmacist wants to prepare 300 ml of a 20% alcohol solution. How much of a 30% solution and a 15% solution should be used to form the desired mixture?

Applications that involve a constant rate of travel, or speed, require the use of the distance formula

$$d = rt$$

where

$$d = \text{distance traveled}$$

$$r = \text{rate or speed}$$

$$t = \text{time}$$

Example 7 illustrates this approach.

Example 7   Solving a Distance-Rate-Time Problem

A boat can travel 36 mi downstream in 2 h. Coming back upstream, the boat takes 3 h. What is the rate of the boat in still water? What is the rate of the current?

Step 1   We want to find the two rates.

Step 2   Let x be the rate of the boat in still water and y the rate of the current.

Downstream the rate is then

$$x + y$$

Upstream, the rate is

$$x - y$$

Step 3   To form a system, think about the following. Downstream, the rate of the boat is *increased* by the effect of the current. Upstream, the rate is *decreased*.

    In many applications, it helps to lay out the information in tabular form. Let's try that strategy here.

	d	r	t
Downstream	36	$x + y$	2
Upstream	36	$x - y$	3

Since $d = rt$, from the table we can easily form two equations:

$$36 = (x + y)(2) \tag{20}$$

$$36 = (x - y)(3) \tag{21}$$

Step 4   We clear equations (20) and (21) of parentheses and simplify, to write the equivalent system

$$x + y = 18$$

$$x - y = 12$$

Solving, we have

$$x = 15 \text{ mi/h}$$

$$y = 3 \text{ mi/h}$$

The rate of the current is 3 mi/h, and the rate of the boat in still water is 15 mi/h.

Step 5   To check, verify the $d = rt$ equation in *both* the upstream and the downstream cases. We leave that to you.

✔ *CHECK YOURSELF 7*

A plane flies 480 mi in an easterly direction, with the wind, in 4 h. Returning westerly along the same route, against the wind, the plane takes 6 h. What is the rate of the plane in still air? What is the rate of the wind?

Systems of equations in problem solving have many applications in a business setting. Example 8 illustrates one such application.

Example 8

Solving a Business-Based Application

A manufacturer produces a standard model and a deluxe model of a 13-inch (in.) television set. The standard model requires 12 h of labor to produce, while the deluxe model requires 18 h. The company has 360 h of labor available per week. The plant's capacity is a total of 25 sets per week. If all the available time and capacity are to be used, how many of each type of set should be produced?

Step 1   The unknowns in this case are the number of standard and deluxe models that can be produced.

The choices for x and y could have been reversed.

Step 2   Let x be the number of standard models and y the number of deluxe models.

Step 3   Our system will come from the two given conditions that fix the total number of sets that can be produced and the total labor hours available.

$$x + y = 25 \longleftarrow \text{Total number of sets}$$

$$12x + 18y = 360 \longleftarrow \text{Total labor hours available}$$

Labor hours, standard sets

Labor hours, deluxe sets

Step 4   Solving the system in step 3, we have

$$x = 15 \quad \text{and} \quad y = 10$$

which tells us that to use all the available capacity, the plant should produce 15 standard sets and 10 deluxe sets per week.

Step 5   We leave the check of this result to the reader.

✔ CHECK YOURSELF 8

A manufacturer produces standard cassette players and compact disk players. Each cassette player requires 2 h of electronic assembly, and each CD requires 3 h. The cassette players require 4 h of case assembly and the CDs 2 h. The company has 120 h of electronic assembly time available per week and 160 h of case assembly time. How many of each type of unit can be produced each week if all available assembly time is to be used?

Let's look at one final application that leads to a system of two equations.

| **Example 9** | ## Solving a Business-Based Application |

Two car rental agencies have the following rate structures for a subcompact car. Company A charges $20 per day plus 15¢ per mile. Company B charges $18 per day plus 16¢ per mile. If you rent a car for 1 day, for what number of miles will the two companies have the same total charge?

Letting c represent the total a company will charge and m the number of miles driven, we calculate the following.

For company A:

You first saw this type of linear model in exercises in Section 5.2.

$$c(m) = 20 + 0.15m \qquad \textbf{(22)}$$

For company B:

$$c(m) = 18 + 0.16m \qquad \textbf{(23)}$$

The system can be solved most easily by substitution. Substituting $18 + 0.16m$ for $c(m)$ in equation (22) gives

$$18 + 0.16m = 20 + 0.15m$$
$$0.01m = 2$$
$$m = 200 \text{ mi}$$

The graph of the system is shown below.

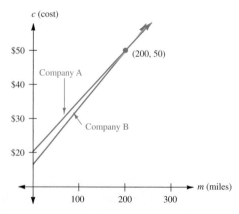

From the graph, how would you make a decision about which agency to use?

✔ CHECK YOURSELF 9

For a compact car, the same two companies charge $27 per day plus 20¢ per mile and $24 per day plus 22¢ per mile. For a 2-day rental, for what mileage will the charges be the same? What is the total charge?

✔ CHECK YOURSELF ANSWERS

1. $\left\{\left(\dfrac{5}{2}, -3\right)\right\}$ 2. $\{(-3, -4)\}$

3. (a) Inconsistent system, no solution; (b) dependent system, an infinite number of solutions

4. (a) $\left\{\left(4, -\dfrac{2}{3}\right)\right\}$; (b) $\left\{\left(-2, \dfrac{3}{4}\right)\right\}$

5. 50 lb of peanuts and 10 lb of cashews

6. 100 ml of the 30% and 200 ml of the 15%   7. 100 mi/h plane and 20 mi/h wind

8. 30 cassette players and 20 CD players   9. At 300 mi, $114 charge

In Exercises 1 to 14, solve each system by the addition method. If a unique solution does not exist, state whether the system is inconsistent or dependent.

1. $2x - y = 1$
   $-2x + 3y = 5$

2. $x + 3y = 12$
   $2x - 3y = 6$

3. $4x + 3y = -5$
   $5x + 3y = -8$

4. $2x + 3y = 1$
   $5x + 3y = 16$

5. $x + y = 3$
   $3x - 2y = 4$

6. $x - y = -2$
   $2x + 3y = 21$

7. $2x + y = 8$
   $-4x - 2y = -16$

8. $2x + 3y = 11$
   $x - y = 3$

9. $5x - 2y = 31$
   $4x + 3y = 11$

10. $2x - y = 4$
    $6x - 3y = 10$

11. $3x - 2y = 7$
    $-6x + 4y = -15$

12. $3x + 4y = 0$
    $5x - 3y = -29$

13. $-2x + 7y = 2$
    $3x - 5y = -14$

14. $5x - 2y = 3$
    $10x - 4y = 6$

In Exercises 15 to 26, solve each system by the substitution method. If a unique solution does not exist, state whether the system is inconsistent or dependent.

15. $x - y = 7$
    $y = 2x - 12$

16. $x - y = 4$
    $x = 2y - 2$

17. $3x + 2y = -18$
    $x = 3y + 5$

18. $3x - 18y = 4$
    $x = 6y + 2$

19. $10x - 2y = 4$
    $y = 5x - 2$

20. $5x - 2y = 8$
    $y = -x + 3$

21. $3x + 4y = 9$
    $y = 3x + 1$

22. $6x - 5y = 27$
    $x = 5y + 2$

23. $x - 7y = 3$
    $2x - 5y = 15$

24. $4x + 3y = -11$
    $5x + y = -11$

25. $3x - 9y = 7$
    $-x + 3y = 2$

26. $5x - 6y = 21$
    $x - 2y = 5$

In Exercises 27 to 32, solve each system by any method discussed in this section.

27. $2x - 3y = 4$
    $x = 3y + 6$

28. $5x + y = 2$
    $5x - 3y = 6$

29. $4x - 3y = 0$
    $5x + 2y = 23$

30. $7x - 2y = -17$
    $x + 4y = 4$

31. $3x - y = 17$
    $5x + 3y = 5$

32. $7x + 3y = -51$
    $y = 2x + 9$

680

In Exercises 33 to 36, solve each system by any method discussed in this section. (*Hint*: You should multiply to clear fractions as your first step.)

33. $\frac{1}{4}x - \frac{1}{6}y = 2$

$\frac{3}{4}x + y = 3$

34. $\frac{1}{5}x - \frac{1}{2}y = 0$

$x - \frac{3}{2}y = 4$

35. $\frac{2}{3}x + \frac{3}{5}y = -3$

$\frac{1}{3}x + \frac{2}{5}y = -3$

36. $\frac{3}{8}x - \frac{1}{2}y = -5$

$\frac{1}{4}x + \frac{3}{2}y = 4$

In Exercises 37 to 44, each application can be solved by the use of a system of linear equations. Match the application with the appropriate system below.

(a) $12x + 5y = 116$
     $8x + 12y = 112$

(b) $x + y = 4{,}000$
     $0.03x + 0.05y = 180$

(c) $x + y = 200$
     $0.20x + 0.60y = 90$

(d) $x + y = 36$
     $y = 3x - 4$

(e) $2(x + y) = 36$
     $3(x - y) = 36$

(f) $x + y = 200$
     $5.50x + 4y = 980$

(g) $L = 2W + 3$
     $2L + 2W = 36$

(h) $x + y = 120$
     $2.20x + 5.40y = 360$

37. **Number problem.** One number is 4 less than 3 times another. If the sum of the numbers is 36, what are the two numbers?

38. **Recreation.** Suppose a movie theater sold 200 adult and student tickets for a showing with a revenue of $980. If the adult tickets cost $5.50 and the student tickets cost $4, how many of each type of ticket were sold?

39. **Geometry.** The length of a rectangle is 3 cm more than twice its width. If the perimeter of the rectangle is 36 cm, find the dimensions of the rectangle.

40. **Business.** An order of 12 dozen roller-ball pens and 5 dozen ballpoint pens costs $116. A later order for 8 dozen roller-ball pens and 12 dozen ballpoint pens costs $112. What was the cost of 1 dozen of each type of pen?

41. **Mixture problem.** A candy merchant wants to mix peanuts selling at $2.20 per pound with cashews selling at $5.40 per pound to form 120 lb of a mixed-nut blend that will sell for $3 per pound. What amount of each type of nut should be used?

42. **Investment.** Donald has investments totaling $4,000 in two accounts—one a savings account paying 3% interest and the other a bond paying 5%. If the annual interest from the two investments was $180, how much did he have invested at each rate?

43. **Mixture problem.** A chemist wants to combine a 20% alcohol solution with a 60% solution to form 200 ml of a 45% solution. How much of each solution should be used to form the mixture?

44. **Motion problem.** Xian was able to make a downstream trip of 36 mi in 2 h. Returning upstream, he took 3 h to make the trip. How fast can his boat travel in still water? What was the rate of the river's current?

In Exercises 45 to 66, solve by choosing a variable to represent each unknown quantity and writing a system of equations.

45. Mixture problem. Suppose 750 tickets were sold for a concert with a total revenue of $5,300. If adult tickets were $8 and students tickets were $4.50, how many of each type of ticket were sold?

46. Mixture problem. Theater tickets sold for $7.50 on the main floor and $5 in the balcony. The total revenue was $3,250, and there were 100 more main-floor tickets sold than balcony tickets. Find the number of each type of ticket sold.

47. Geometry. The length of a rectangle is 3 in. less than twice its width. If the perimeter of the rectangle is 84 in., find the dimensions of the rectangle.

48. Geometry. The length of a rectangle is 5 cm more than 3 times its width. If the perimeter of the rectangle is 74 cm, find the dimensions of the rectangle.

49. Mixture problem. A garden store sold 8 bags of mulch and 3 bags of fertilizer for $24. The next purchase was for 5 bags of mulch and 5 bags of fertilizer. The cost of that purchase was $25. Find the cost of a single bag of mulch and a single bag of fertilizer.

50. Mixture problem. The cost of an order for 10 computer disks and 3 packages of paper was $22.50. The next order was for 30 disks and 5 packages of paper, and its cost was $53.50. Find the price of a single disk and a single package of paper.

51. Mixture problem. A coffee retailer has two grades of decaffeinated beans, one selling for $4 per pound and the other for $6.50 per pound. She wishes to blend the beans to form a 150-lb mixture that will sell for $4.75 per pound. How many pounds of each grade of bean should be used in the mixture?

52. Mixture problem. A candy merchant sells jelly beans at $3.50 per pound and gumdrops at $4.70 per pound. To form a 200-lb mixture that will sell for $4.40 per pound, how many pounds of each type of candy should be used?

53. **Investment.** Cheryl decided to divide $12,000 into two investments—one a time deposit that pays 4% annual interest and the other a bond that pays 6%. If her annual interest was $640, how much did she invest at each rate?

54. **Investment.** Miguel has $1,500 more invested in a mutual fund paying 5% interest than in a savings account paying 3%. If he received $155 in interest for 1 year, how much did he have invested in the two accounts?

55. **Science.** A chemist mixes a 10% acid solution with a 50% acid solution to form 400 ml of a 40% solution. How much of each solution should be used in the mixture?

56. **Science.** A laboratory technician wishes to mix a 70% saline solution and a 20% saline solution to prepare 500 ml of a 40% solution. What amount of each solution should be used?

57. **Motion.** A boat traveled 36 mi up a river in 3 h. Returning downstream, the boat took 2 h. What is the boat's rate in still water, and what is the rate of the river's current?

58. **Motion.** A jet flew east a distance of 1,800 mi with the jetstream in 3 h. Returning west, against the jetstream, the jet took 4 h. Find the jet's speed in still air and the rate of the jetstream.

59. **Number problem.** The sum of the digits of a two-digit number is 8. If the digits are reversed, the new number is 36 more than the original number. Find the original number. (*Hint:* If u represents the units digit of the number and t the tens digit, the original number can be represented by $10t + u$.)

60. **Number problem.** The sum of the digits of a two-digit number is 10. If the digits are reversed, the new number is 54 less than the original number. What was the original number?

61. **Business.** A manufacturer produces a battery-powered calculator and a solar model. The battery-powered model requires 10 min of electronic assembly and the solar model 15 min. There is 450 min of assembly time available per day. Both models require 8 min for packaging, and 280 min of packaging time is available per day. If the manufacturer wants to use all the available time, how many of each unit should be produced per day?

62. **Business.** A small tool manufacturer produces a standard model and a cordless model power drill. The standard model takes 2 h of labor to assemble and the cordless model 3 h. There is 72 h of labor available per week for the drills. Material costs for the standard drill are $10, and for the cordless drill they are $20. The company wishes to limit material costs to $420 per week. How many of each model drill should be produced in order to use all the available resources?

63. Economics. In economics, a demand equation gives the quantity D that will be demanded by consumers at a given price p, in dollars. Suppose that $D = 210 - 4p$ for a particular product.

A supply equation gives the supply S that will be available from producers at price p. Suppose also that for the same product $S = 10p$.

The equilibrium point is that point where the supply equals the demand (here, where $S = D$). Use the given equations to find the equilibrium point.

64. Economics. Suppose the demand equation for a product is $D = 150 - 3p$ and the supply equation is $S = 12p$. Find the equilibrium point for the product.

65. Consumer affairs. Two car rental agencies have the following rate structure for compact cars.

Company A: $30/day and 22¢/mi

Company B: $28/day and 26¢/mi

For a 2-day rental, at what number of miles will the charges be the same?

66. Construction. Two construction companies submit the following bids.

Company A: $5,000 plus $15 per square foot of building

Company B: $7,000 plus $12.50 per square foot of building

For what number of square feet of building will the bids of the two companies be the same?

Certain systems that are not linear can be solved with the methods of this section if we first substitute to change variables. For instance, the system

$$\frac{1}{x} + \frac{1}{y} = 4$$

$$\frac{1}{x} - \frac{3}{y} = -6$$

can be solved by the substitutions $u = \dfrac{1}{x}$ and $v = \dfrac{1}{y}$. That gives the system $u + v = 4$ and $u - 3v = -6$. The system is then solved for u and v, and the corresponding values for x and y are found. Use this method to solve the systems in Exercises 67 to 70.

67. $\dfrac{1}{x} + \dfrac{1}{y} = 4$

$\dfrac{1}{x} - \dfrac{3}{y} = -6$

68. $\dfrac{1}{x} + \dfrac{3}{y} = 1$

$\dfrac{4}{x} + \dfrac{3}{y} = 3$

69. $\dfrac{2}{x} + \dfrac{3}{y} = 4$

$\dfrac{2}{x} - \dfrac{6}{y} = 10$

70. $\dfrac{4}{x} - \dfrac{3}{y} = -1$

$\dfrac{12}{x} - \dfrac{1}{y} = 1$

Writing the equation of a line through two points can be done by the following method. Given the coordinates of two points, substitute each pair of values into the equation $y = mx + b$. This gives a system of two equations in variables m and b, which can be solved as before.

In Exercises 71 and 72, write the equation of the line through each of the following pairs of points, using the method outlined above.

71. $(2, 1)$ and $(4, 4)$

72. $(-3, 7)$ and $(6, 1)$

In Exercises 73 and 74, use your calculator to approximate the solution to each system. Express each coordinate to the nearest tenth.

73. $y = 2x - 3$

$2x + 3y = 1$

74. $3x - 4y = -7$

$2x + 3y = -1$

For Exercises 75 and 76, adjust the viewing window on your calculator so that you can see the point of intersection for the two lines representing the equations in the system. Then approximate the solution, expressing each coordinate to the nearest tenth.

75. $5x - 12y = 8$

$7x + 2y = 44$

76. $9x - 3y = 10$

$x + 5y = 58$

77. We have discussed three different methods of solving a system of two linear equations in two unknowns: the graphical method, the addition method, and the substitution method. Discuss the strengths and weaknesses of each method.

78. Determine a system of two linear equations for which the solution is $(3, 4)$. Are there other systems that have the same solution? If so, determine at least one more and explain why this can be true.

79. Suppose we have the following linear system:

$$Ax + By = C \qquad\qquad (1)$$

$$Dx + Ey = F \qquad\qquad (2)$$

(a) Multiply equation (1) by $-D$, multiply equation (2) by A, and add. This will allow you to eliminate x. Solve for y and indicate what must be true about the coefficients in order for a unique value for y to exist.

(b) Now return to the original system and eliminate y instead of x. [*Hint*: Try multiplying equation (1) by E and equation (2) by $-B$.] Solve for x and again indicate what must be true about the coefficients for a unique value for x to exist.

8.3 Systems of Linear Equations in Three Variables

OBJECTIVES

1. Find ordered triples associated with each equation
2. Solve by the addition method
3. Look at a graphical interpretation of a solution
4. Solve an application of a system with three variables

Suppose an application involves three quantities that we want to label x, y, and z. A typical equation used for the solution might be

$$2x + 4y - z = 8$$

This is called a **linear equation in three variables.** A solution for such an equation is an **ordered triple** (x, y, z) of real numbers that satisfies the equation. For example, the ordered triple $(2, 1, 0)$ is a solution for the equation above since substituting 2 for x, 1 for y, and 0 for z results in the following true statement.

$$2 \cdot 2 + 4 \stackrel{?}{=} 8$$

$$4 + 4 \stackrel{?}{=} 8$$

$$8 = 8 \quad \text{True}$$

Of course, other solutions, in fact infinitely many, exist. You might want to verify that $(1, 1, -2)$ and $(3, 1, 2)$ are also solutions. To extend the concepts of Section 8.2, we want to consider systems of three linear equations in three variables, such as

$$x + y + z = 5$$
$$2x - y + z = 9$$
$$x - 2y + 3z = 16$$

For a unique solution to exist, when *three variables* are involved, we must have *three equations.*

The choice of which variable to eliminate is yours. Generally, you should pick the variable that allows the easiest computation.

The solution for such a system is the set of all ordered triples that satisfy each equation of the system. In this case, you should verify that $(2, -1, 4)$ is a solution for the system since that ordered triple makes each equation a true statement.

Let's turn now to the solution process itself. In this section, we will consider the addition method. We will then apply what we have learned to solving applications.

The Addition Method

The central idea is to choose *two pairs* of equations from the system and, by the addition method, to eliminate the *same variable* from each of those pairs. The method is best illustrated by example. So let's proceed to see how the solution for the previous system was determined.

686

Example 1	### Solving a Linear System in Three Variables

Solve the system.

$$x + y + z = 5 \qquad \textbf{(1)}$$
$$2x - y + z = 9 \qquad \textbf{(2)}$$
$$x - 2y + 3z = 16 \qquad \textbf{(3)}$$

First, we choose two of the equations and the variable to eliminate. Variable y seems convenient in this case. Pairing equations (1) and (2) and then adding, we have

Any pair of equations could have been selected.

$$
\begin{aligned}
x + y + z &= 5 \\
2x - y + z &= 9 \\
\hline
3x \qquad + 2z &= 14 \qquad \textbf{(4)}
\end{aligned}
$$

We now want to choose a different pair of equations to eliminate y. Using equations (1) and (3) this time, we multiply equation (1) by 2 and then add the result to equation (3):

$$
\begin{aligned}
2x + 2y + 2z &= 10 \\
x - 2y + 3z &= 16 \\
\hline
3x \qquad + 5z &= 26 \qquad \textbf{(5)}
\end{aligned}
$$

We now have equations (4) and (5) in variables x and z.

$$3x + 2z = 14$$
$$3x + 5z = 26$$

Since we are now dealing with a system of two equations in two variables, any of the methods of Section 8.2 apply. We have chosen to multiply equation (4) by -1 and then add that result to equation (5). This yields

$$3z = 12$$
$$z = 4$$

Substituting $z = 4$ in equation (4) gives

$$3x + 2 \cdot (4) = 14$$
$$3x + 8 = 14$$
$$3x = 6$$
$$x = 2$$

Any of the original equations could have been used.

Finally, letting $x = 2$ and $z = 4$ in equation (1) gives

$$(2) + y + (4) = 5$$
$$y = -1$$

To check, substitute these values into the other equations of the original system.

and $(2, -1, 4)$ is shown to be the solution for the system. The solution set is $\{(2, -1, 4)\}$.

CHECK YOURSELF 1

Solve the system.

$$x - 2y + z = 0$$
$$2x + 3y - z = 16$$
$$3x - y - 3z = 23$$

One or more of the equations of a system may already have a missing variable. The elimination process is simplified in that case, as Example 2 illustrates.

Example 2

Solving a Linear System in Three Variables

Solve the system.

$$2x + y - z = -3 \qquad \textbf{(6)}$$
$$y + z = 2 \qquad \textbf{(7)}$$
$$4x - y + z = 12 \qquad \textbf{(8)}$$

Noting that equation (7) involves only y and z, we must simply find another equation in those same two variables. Multiply equation (6) by -2 and add the result to equation (8) to eliminate x.

We now have a *second* equation in y and z.

$$-4x - 2y + 2z = 6$$
$$\underline{4x - y + z = 12}$$
$$-3y + 3z = 18$$
$$y - z = -6 \qquad \textbf{(9)}$$

We now form a system consisting of equations (7) and (9) and solve as before.

$$y + z = 2$$
$$\underline{y - z = -6} \qquad \text{Adding eliminates } z.$$
$$2y = -4$$
$$y = -2$$

From equation (7), if $y = -2$,

$$(-2) + z = 2$$
$$z = 4$$

and from equation (6), if $y = -2$ and $z = 4$,

$$2x + (-2) - (4) = -3$$
$$2x = 3$$
$$x = \frac{3}{2}$$

The solution set for the system is

$$\left\{ \left(\frac{3}{2}, -2, 4 \right) \right\}$$

✔ CHECK YOURSELF 2

Solve the system.

$$x + 2y - z = -3$$
$$x - y + z = 2$$
$$x - z = 3$$

The following algorithm summarizes the procedure for finding the solutions for a linear system of three equations in three variables.

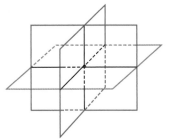

Three planes intersecting at a point

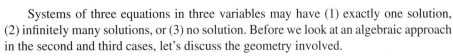

Solving a System of Three Equations in Three Unknowns

Step 1.   Choose a pair of equations from the system and use the addition method to eliminate one of the variables.

Step 2.   Choose a *different* pair of equations and eliminate the *same* variable.

Step 3.   Solve the system of two equations in two variables determined in steps 1 and 2.

Step 4.   Substitute the values found above into one of the original equations and solve for the remaining variable.

Step 5.   The solution is the ordered triple of values found in steps 3 and 4. It can be checked by substituting into the other equations of the original system.

Systems of three equations in three variables may have (1) exactly one solution, (2) infinitely many solutions, or (3) no solution. Before we look at an algebraic approach in the second and third cases, let's discuss the geometry involved.

The graph of a linear equation in three variables is a plane (a flat surface) in three dimensions. Two distinct planes either will be parallel or will intersect in a line.

If three distinct planes intersect, that intersection will be either a single point (as in our first example) or a line (think of three pages in an open book—they intersect along the binding of the book).

Let's look at examples involving dependent systems (infinitely many solutions) and inconsistent systems (no solutions).

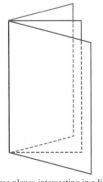

Three planes intersecting in a line

| Example 3 | Solving a Dependent Linear System in Three Variables |

Solve the system.

$$x + 2y - z = 5 \tag{10}$$

$$x - y + z = -2 \tag{11}$$

$$-5x - 4y + z = -11 \tag{12}$$

We begin as before by choosing two pairs of equations from the system and eliminating the same variable from each of the pairs. Adding equations (10) and (11) gives

$$2x + y = 3 \tag{13}$$

Adding equations (10) and (12) gives

$$-4x - 2y = -6 \tag{14}$$

Now consider the system formed by equations (13) and (14). We multiply equation (13) by 2 and add again:

$$
\begin{array}{r}
4x + 2y = 6 \\
-4x - 2y = -6 \\
\hline
0 = 0
\end{array}
$$

There are ways of representing the solutions, as you will see in later courses.

This true statement tells us that the system has an infinite number of solutions (lying along a straight line). Again, such a system is dependent.

 CHECK YOURSELF 3

Solve the system.

$$2x - y + 3z = 3$$

$$-x + y - 2z = 1$$

$$y - z = 5$$

We have seen that a linear system in three variables can have exactly one solution (a consistent system) or, as in Example 3, infinitely many solutions (a dependent system). Consider now a third possibility: no solutions (an inconsistent system). It is illustrated in Example 4.

| Example 4 | Solving an Inconsistent Linear System in Three Variables |

Solve the system.

$$3x + y - 3z = 1 \qquad \textbf{(15)}$$

$$-2x - y + 2z = 1 \qquad \textbf{(16)}$$

$$-x - y + \ z = 2 \qquad \textbf{(17)}$$

This time we eliminate variable y. Adding equations (15) and (16), we have

$$x - z = 2 \qquad \textbf{(18)}$$

Adding equations (15) and (17) gives

$$2x - 2z = 3 \qquad \textbf{(19)}$$

Now, multiply equation (18) by -2 and add the result to equation (19).

$$
\begin{aligned}
-2x + 2z &= -4 \\
2x - 2z &= 3 \\
\hline
0 &= -1
\end{aligned}
$$

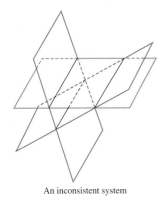

An inconsistent system

All the variables have been eliminated, and we have arrived at a contradiction, $0 = -1$. This means that the system is *inconsistent* and has no solutions. There is *no* point common to all three planes. The solution set is the empty set $\varnothing$.

✔ *CHECK YOURSELF 4*

Solve the system.

$$
\begin{aligned}
x - \ y - z &= 0 \\
-3x + 2y + z &= 1 \\
3x - \ y + z &= -1
\end{aligned}
$$

As a closing note, we have by no means illustrated all possible types of inconsistent and dependent systems. Other possibilities involve either distinct parallel planes or planes that coincide. The solution techniques in these additional cases are, however, similar to those illustrated above.

Solving Applications

In many instances, if an application involves three unknown quantities, you will find it useful to assign three variables to those quantities and then build a system of three equations from the given relationships in the problem. The extension of our problem-solving strategy is natural, as Example 5 illustrates.

| Example 5 | ## Solving a Number Problem |

The sum of the digits of a three-digit number is 12. The tens digit is 2 less than the hundreds digit, and the units digit is 4 less than the sum of the other two digits. What is the number?

Step 1   The three unknowns are, of course, the three digits of the number.

Sometimes it helps to choose variable letters that relate to the words, as is done here.

Step 2   We now want to assign variables to each of three digits. Let u be the units digit, t the tens digit, and h the hundreds digit.

Take a moment now to go back to the original problem and pick out those conditions. That skill is a crucial part of the problem-solving strategy.

Step 3   There are three conditions given in the problem that allow us to write the necessary three equations. From those conditions

$$h + t + u = 12$$
$$t = h - 2$$
$$u = h + t - 4$$

Step 4   There are various ways to approach the solution. To use addition, write the system in the equivalent form

$$h + t + u = 12$$
$$-h + t = -2$$
$$-h - t + u = -4$$

and solve by our earlier methods. The solution, which you can verify, is $h = 5$, $t = 3$, and $u = 4$. The desired number is 534.

Step 5   To check, you should show that the digits of 534 meet each of the conditions of the original problem.

   CHECK YOURSELF 5

The sum of the measures of the angles of a triangle is 180°. In a given triangle, the measure of the second angle is twice the measure of the first. The measure of the third angle is 30° less than the sum of the measures of the first two. Find the measure of each angle.

Let's continue with a slightly different application that will lead to a system of three equations.

| Example 6 | **Solving an Investment Application** |

Monica decided to divide a total of $42,000 into three investments: a savings account paying 5% interest, a time deposit paying 7%, and a bond paying 9%. Her total annual interest from the three investments was $2,600, and the interest from the savings account was $200 less than the total interest from the other two investments. How much did she invest at each rate?

Step 1   The three amounts are the unknowns.

Again, we choose letters that suggest the unknown quantities—s for savings, t for time deposit, and b for bond.

Step 2   We let s be the amount invested at 5%, t the amount at 7%, and b the amount at 9%. Note that the interest from the savings account is then $0.05s$, and so on.

A table will help with the next step.

For 1 year, the interest formula is

$$I = Pr$$

(interest equals principal times rate).

	5%	7%	9%
Principal	s	t	b
Interest	$0.05s$	$0.07t$	$0.09b$

Step 3   Again there are three conditions in the given problem. By using the table above, they lead to the following equations.

Total invested.

Total interest.

The savings interest was $200 less than that from the other two investments.

$$s + t + b = 42,000$$
$$0.05s + 0.07t + 0.09b = 2,600$$
$$0.05s = 0.07t + 0.09b - 200$$

Step 4   We clear the equation of decimals and solve as before, with the result

$$s = \$24,000 \qquad t = \$11,000 \qquad b = \$7,000$$

Step 5   We leave the check of this solution to you.

Find the interest earned from each investment, and verify that the conditions of the problem are satisfied.

✔ *CHECK YOURSELF 6*

Glenn has a total of $11,600 invested in three accounts: a savings account paying 6% interest, a stock paying 8%, and a mutual fund paying 10%. The annual interest from the stock and mutual fund is twice that from the savings account, and the mutual fund returned $120 more than the stock. How much did Glenn invest in each account?

✔ *CHECK YOURSELF ANSWERS*

1. $\{(5, 1, -3)\}$ **2.** $\{(1, -3, -2)\}$

3. The system is dependent (there are an infinite number of solutions).

4. The system is inconsistent (there are no solutions).

5. The three angles are 35°, 70°, and 75°.

6. $5,000 in savings, $3,000 in stocks, and $3,600 in the mutual fund

Exercises · 8.3

In Exercises 1 to 20, solve each system of equations. If a unique solution does not exist, state whether the system is inconsistent or dependent.

1.
$$x - y + z = 3$$
$$2x + y + z = 8$$
$$3x + y - z = 1$$

2.
$$x - y - z = 2$$
$$2x + y + z = 8$$
$$x + y + z = 6$$

3.
$$x + y + z = 1$$
$$2x - y + 2z = -1$$
$$-x - 3y + z = 1$$

4.
$$x - y - z = 6$$
$$-x + 3y + 2z = -11$$
$$3x + 2y + z = 1$$

5.
$$x + y + z = 1$$
$$-2x + 2y + 3z = 20$$
$$2x - 2y - z = -16$$

6.
$$x + y + z = -3$$
$$3x + y - z = 13$$
$$3x + y - 2z = 18$$

7.
$$2x + y - z = 2$$
$$-x - 3y + z = -1$$
$$-4x + 3y + z = -4$$

8.
$$x + 4y - 6z = 8$$
$$2x - y + 3z = -10$$
$$3x - 2y + 3z = -18$$

9.
$$3x - y + z = 5$$
$$x + 3y + 3z = -6$$
$$x + 4y - 2z = 12$$

10.
$$2x - y + 3z = 2$$
$$x - 2y + 3z = 1$$
$$4x - y + 5z = 5$$

11.
$$x + 2y + z = 2$$
$$2x + 3y + 3z = -3$$
$$2x + 3y + 2z = 2$$

12.
$$x - 4y - z = -3$$
$$x + 2y + z = 5$$
$$3x - 7y - 2z = -6$$

13.
$$x + 3y - 2z = 8$$
$$3x + 2y - 3z = 15$$
$$4x + 2y + 3z = -1$$

14.
$$x + y - z = 2$$
$$3x + 5y - 2z = -5$$
$$5x + 4y - 7z = -7$$

15.
$$x + y - z = 2$$
$$x - 2z = 1$$
$$2x - 3y - z = 8$$

16.
$$x + y + z = 6$$
$$x - 2y = -7$$
$$4x + 3y + z = 7$$

17.
$$x - 3y + 2z = 1$$
$$16y - 9z = 5$$
$$4x + 4y - z = 8$$

18.
$$x - 4y + 4z = -1$$
$$y - 3z = 5$$
$$3x - 4y + 6z = 1$$

19.
$$x + 2y - 4z = 13$$
$$3x + 4y - 2z = 19$$
$$3x + 2z = 3$$

20.
$$x + 2y - z = 6$$
$$-3x - 2y + 5z = -12$$
$$x - 2z = 3$$

Solve Exercises 21 to 32 by choosing a variable to represent each unknown quantity and writing a system of equations.

21. Number problem. The sum of three numbers is 16. The largest number is equal to the sum of the other two, and 3 times the smallest number is 1 more than the largest. Find the three numbers.

694

22. **Number problem.** The sum of three numbers is 24. Twice the smallest number is 2 less than the largest number, and the largest number is equal to the sum of the other two. What are the three numbers?

23. **Coin problem.** A cashier has 25 coins consisting of nickels, dimes, and quarters with a value of $4.90. If the number of dimes is 1 less than twice the number of nickels, how many of each type of coin does she have?

24. **Recreation.** A theater has tickets at $6 for adults, $3.50 for students, and $2.50 for children under 12 years old. A total of 278 tickets were sold for one showing with a total revenue of $1,300. If the number of adult tickets sold was 10 less than twice the number of student tickets, how many of each type of ticket were sold for the showing?

25. **Geometry.** The perimeter of a triangle is 19 cm. If the length of the longest side is twice that of the shortest side and 3 cm less than the sum of the lengths of the other two sides, find the lengths of the three sides.

26. **Geometry.** The measure of the largest angle of a triangle is 10° more than the sum of the measures of the other two angles and 10° less than 3 times the measure of the smallest angle. Find the measures of the three angles of the triangle.

27. **Investments.** Jovita divides $12,000 into three investments: a savings account paying 4% annual interest, a bond paying 6%, and a money market fund paying 9%. The annual interest from the three accounts is $860, and she has 2 times as much invested in the bond as in the savings account. What amount does she have invested in each account?

28. **Investments.** Adrienne has $10,000 invested in a savings account paying 5%, a time deposit paying 7%, and a bond paying 10%. She has $1,000 less invested in the bond than in her savings account, and she earned $700 in annual interest. What has she invested in each account?

29. **Number problem.** The sum of the digits of a three-digit number is 9, and the tens digit of the number is twice the hundreds digit. If the digits are reversed in order, the new number is 99 more than the original number. What is the original number?

30. **Number problem.** The sum of the digits of a three-digit number is 9. The tens digit is 3 times the hundreds digit. If the digits are reversed in order, the new number is 99 less than the original number. Find the original three-digit number.

31. **Motion.** Roy, Sally, and Jeff drive a total of 50 mi to work each day. Sally drives twice as far as Roy, and Jeff drives 10 mi farther than Sally. Use a system of three equations in three unknowns to find how far each person drives each day.

32. Consumer affairs. A parking lot has spaces reserved for motorcycles, cars, and vans. There are 5 more spaces reserved for vans than for motorcycles. There are 3 times as many car spaces as van and motorcycle spaces combined. If the parking lot has 180 total reserved spaces, how many of each type are there?

The solution process illustrated in this section can be extended to solving systems of more than three variables in a natural fashion. For instance, if four variables are involved, eliminate one variable in the system and then solve the resulting system in three variables as before. Substituting those three values into one of the original equations will provide the value for the remaining variable and the solution for the system.

In Exercises 33 and 34, use this procedure to solve the system.

33.
$$x + 2y + 3z + w = 0$$
$$-x - y - 3z + w = -2$$
$$x - 3y + 2z + 2w = -11$$
$$-x + y - 2z + w = 1$$

34.
$$x + y - 2z - w = 4$$
$$x - y + z + 2w = 3$$
$$2x + y - z - w = 7$$
$$x - y + 2z + w = 2$$

In some systems of equations there are more equations than variables. We can illustrate this situation with a system of three equations in two variables. To solve this type of system, pick any two of the equations and solve this system. Then substitute the solution obtained into the third equation. If a true statement results, the solution used is the solution to the entire system. If a false statement occurs, the system has no solution.

In Exercises 35 and 36, use this procedure to solve each system.

35.
$$x - y = 5$$
$$2x + 3y = 20$$
$$4x + 5y = 38$$

36.
$$3x + 2y = 6$$
$$5x + 7y = 35$$
$$7x + 9y = 8$$

37. Experiments have shown that cars (C), trucks (T), and buses (B) emit different amounts of air pollutants. In one such experiment, a truck emitted 1.5 pounds (lb) of carbon dioxide (CO_2) per passenger-mile and 2 grams (g) of nitrogen oxide (NO) per passenger-mile. A car emitted 1.1 lb of CO_2 per passenger-mile and 1.5 g of NO per passenger-mile. A bus emitted 0.4 lb of CO_2 per passenger-mile and 1.8 g of NO per passenger-mile. A total of 85 mi was driven by the three vehicles, and 73.5 lb of CO_2 and 149.5 g of NO were collected. Use the following system of equations to determine the miles driven by each vehicle.

$$T + C + B = 85.0$$
$$1.5T + 1.1C + 0.4B = 73.5$$
$$2T + 1.5C + 1.8B = 149.5$$

38. Experiments have shown that cars (C), trucks (T), and trains (R) emit different amounts of air pollutants. In one such experiment, a truck emitted 0.8 lb of carbon dioxide per passenger-mile and 1 g of nitrogen oxide per passenger-mile. A car emitted 0.7 lb of CO_2 per passenger-mile and 0.9 g of NO per passenger-mile. A train emitted 0.5 lb of CO_2 per passenger-mile and 4 g of NO per passenger-mile. A total of 141 mi was driven by the three vehicles, and 82.7 lb of CO_2 and 424.4 g of NO were collected. Use the following system of equations to determine the miles driven by each vehicle.

$$T + \quad C + \quad R = 141.0$$
$$0.8T + 0.7C + 0.5R = \quad 82.7$$
$$T + 0.9C + \quad 4R = 424.4$$

39. In Chapter 11 you will learn about quadratic functions and their graphs. A quadratic function has the form $y = ax^2 + bx + c$, where a, b, and c are specific numbers and $a \neq 0$. Three distinct points on the graph are enough to determine the equation.

(a) Suppose that (1, 5), (2, 10), and (3, 19) are on the graph of $y = ax^2 + bx + c$. Substituting the pair (1, 5) into this equation (that is, let $x = 1$ and $y = 5$) yields $5 = a + b + c$. Substituting each of the other ordered pairs yields $10 = 4a + 2b + c$ and $19 = 9a + 3b + c$. Solve the resulting system of equations to determine the values of a, b, and c. Then write the equation of the function.

(b) Repeat the work of part (a), using the three points (1, 2), (2, 9), and (3, 22).

8.4 Matrices (Optional)

8.4 OBJECTIVES

1. Write the augmented matrix for a system of equations
2. Solve a system of equations by using matrices

In Section 8.3 we learned that we needed three independent equations to solve a system of equations with three variables. It is always the case that the number of equations needed is equal to the number of variables in the system. If we had 10 variables, we would need to have 10 equations to solve the system.

Such systems are usually solved with the help of computers. The technique is always some variation of the matrix method introduced in this section. One variation is introduced in Section A.1 of Appendix A on determinants and Cramer's rule. We will begin with some definitions.

A **matrix** is a rectangular array of numbers. Each number in the matrix is called an **element** of the matrix. For example,

$$\begin{bmatrix} 2 & 3 & 1 \\ -1 & 3 & 2 \end{bmatrix}$$

is a matrix with elements 2, 3, 1, -1, 3, and 2. The matrix above has 2 rows and 3 columns. It is called a 2×3 (read "2 by 3") matrix. We always give the number of rows and *then* the number of columns.

Now we will see how a matrix of numbers is associated with a system of linear equations. For the system of equations

$$x - 2y = -1$$
$$2x + y = 8$$

the **coefficient matrix** is

$$\begin{bmatrix} 1 & -2 \\ 2 & 1 \end{bmatrix}$$

and the **constant matrix** is

$$\begin{bmatrix} -1 \\ 8 \end{bmatrix}$$

The matrix

$$\begin{bmatrix} 1 & -2 & \vdots & -1 \\ 2 & 1 & \vdots & 8 \end{bmatrix}$$

Note: The credit for inventing matrices is generally given to the English mathematician Arthur Cayley (1821–1895).

Note: The plural of *matrix* is *matrices*.

Note: *Brackets* [ ] are used to enclose the elements of a matrix. You may also encounter the use of large parentheses to designate a matrix. In that case, the matrix would look like this:

$$\begin{pmatrix} 2 & 3 & 1 \\ -1 & 3 & 2 \end{pmatrix}$$

Note: The dotted vertical line is just for convenience, to separate the coefficients and the constants.

is called the **augmented matrix** of the system, and it is formed by adjoining the constant matrix to the coefficient matrix so that the matrix contains both the coefficients of the variables and the constants from the system.

Example 1	**Writing an Augmented Matrix**

Write the augmented matrix for the given systems.

(a) $5x - 2y = 7$

$3x + 4y = 8$

The augmented matrix for the system is

$$\begin{bmatrix} 5 & -2 & \vdots & 7 \\ 3 & 4 & \vdots & 8 \end{bmatrix}$$

(b) $2x - 3y = 9$

$y = 6$

The augmented matrix for the system is

Note: Zero is used for the *x* coefficient in the second equation.

$$\begin{bmatrix} 2 & -3 & \vdots & 9 \\ 0 & 1 & \vdots & 6 \end{bmatrix}$$

✔ CHECK YOURSELF 1

Write the augmented matrices for each of the given systems.

(a) $3x - 5y = -2$

$2x + 3y = 5$

(b) $x - 3y = 3$

$y = 2$

Your work in the Check Yourself Exercise 1(b) led to a particular form of augmented matrix. The matrix

Note: Note the form of the coefficient matrix. We have 1s along the *main diagonal* (from the upper left to the lower right) and 0s below that diagonal.

$$\begin{bmatrix} 1 & -3 & \vdots & 3 \\ 0 & 1 & \vdots & 2 \end{bmatrix}$$

is associated with the system

$$x - 3y = 3$$

$$y = 2$$

Given a system in this form, it is particularly easy to solve for the desired variables by a process called **back substitution.**

From the second equation we know that $y = 2$. Substituting that value back into the first equation, we have

$$x - 3(2) = 3$$
$$x - 6 = 3$$
$$x = 9$$

and $\{(9, 2)\}$ is the solution set for our system.

Our goal in this section will be to transform the augmented matrix of a given system to the form just illustrated, so that the solutions can be easily found by back substitution.

How do we go about transforming the augmented matrix? The rules are based on our earlier work with linear systems.

An equivalent system is formed when

Note: Recall that an equivalent system has the same solutions as the original system.

1. Two equations are interchanged.

2. An equation is multiplied by a nonzero constant.

3. An equation is replaced by adding a constant multiple of another equation to that equation.

Now each property above produces a corresponding property that can be applied to the rows of the augmented matrix rather than to the actual equations of the system. These are called the **elementary row operations,** and they will always produce the augmented matrix of an equivalent system of equations.

Rules and Properties: Elementary Row Operations

1. Two rows can be interchanged.

2. A row can be multiplied by a nonzero constant.

3. A row can be replaced by adding a nonzero multiple of another row to that row.

We will now apply these elementary row operations to the solution of a linear system.

Example 2   Solving a System of Two Equations

Solve the system, using elementary row operations.

$$2x - 3y = 2$$
$$x + 2y = 8$$

First, we write the augmented matrix of the given system.

$$\begin{bmatrix} 2 & -3 & | & 2 \\ 1 & 2 & | & 8 \end{bmatrix}$$

Think about our objective. We want to use back substitution to solve an equivalent system whose augmented matrix has the form

$$\left[\begin{array}{cc|c} 1 & a & b \\ 0 & 1 & c \end{array}\right]$$

Note: Compare this form to the augmented matrix in our discussion of back substitution. In the coefficient matrix we want 1s along the main diagonal and 0s below.

To arrive at this form, we start with the first column. Interchanging the rows of the augmented matrix will give a 1 in the top position, as desired:

$$\left[\begin{array}{cc|c} 2 & -3 & 2 \\ 1 & 2 & 8 \end{array}\right] \xrightarrow{R_1 \leftrightarrow R_2} \left[\begin{array}{cc|c} 1 & 2 & 8 \\ 2 & -3 & 2 \end{array}\right]$$

Note: The notation $R_1 \leftrightarrow R_2$ is an abbreviation indicating that we have interchanged those rows.

Now multiply row 1 by -2 and add the result to row 2. This will give a 0 in the lower portion of column 1.

$$\left[\begin{array}{cc|c} 1 & 2 & 8 \\ 2 & -3 & 2 \end{array}\right] \xrightarrow{-2R_1 + R_2} \left[\begin{array}{cc|c} 1 & 2 & 8 \\ 0 & -7 & -14 \end{array}\right]$$

Note: The notation $-2R_1 + R_2$ means R_2 has been replaced by the sum of -2 times row 1 and row 2.

We now multiply row 2 by the constant $-\dfrac{1}{7}$, to produce a 1 in the second position of the second row:

$$\left[\begin{array}{cc|c} 1 & 2 & 8 \\ 0 & -7 & -14 \end{array}\right] \xrightarrow{-\frac{1}{7}R_2} \left[\begin{array}{cc|c} 1 & 2 & 8 \\ 0 & 1 & 2 \end{array}\right]$$

Note: The notation $-\dfrac{1}{7}R_2$ means we have multiplied row 2 by $-\dfrac{1}{7}$.

The final matrix is in the desired form with 1s along the main diagonal and 0s below and represents the equivalent system

$$x + 2y = 8$$
$$y = 2$$

and we can apply back substitution for our solution. Since $y = 2$, in the first equation, we have

$$x + 2(2) = 8$$
$$x = 4$$

The solution set for our system is $\{(4, 2)\}$.

✔ CHECK YOURSELF 2

Use elementary row operations to solve the system.

$$-3x - 2y = -1$$
$$x + 3y = -9$$

We will extend our matrix approach to consider a system of three linear equations in three unknowns. Again our procedure is to find the augmented matrix of an equivalent system with 1s along the main diagonal of the coefficient matrix and 0s below that diagonal.

| Example 3 | ## Solving a System of Three Equations |

Use elementary row operations to solve the system.

$$x - 2y + z = 10$$

$$-3x + y + 2z = 5$$

$$2x + 3y - z = -9$$

The augmented matrix for the system is

$$\left[\begin{array}{ccc|c} 1 & -2 & 1 & 10 \\ -3 & 1 & 2 & 5 \\ 2 & 3 & -1 & -9 \end{array}\right]$$

Again we start with the first column. The top entry is already 1, and we want 0s below that entry.

$$\xrightarrow{3R_1 + R_2} \left[\begin{array}{ccc|c} 1 & -2 & 1 & 10 \\ 0 & -5 & 5 & 35 \\ 2 & 3 & -1 & -9 \end{array}\right]$$

$$\xrightarrow{-2R_1 + R_3} \left[\begin{array}{ccc|c} 1 & -2 & 1 & 10 \\ 0 & -5 & 5 & 35 \\ 0 & 7 & -3 & -29 \end{array}\right]$$

Note: The order of obtaining the 1s and 0s should be followed carefully.

We now want a 1 in the center position of column 2 and a 0 below that 1:

$$\xrightarrow{-\frac{1}{5}R_2} \left[\begin{array}{ccc|c} 1 & -2 & 1 & 10 \\ 0 & 1 & -1 & -7 \\ 0 & 7 & -3 & -29 \end{array}\right]$$

$$\xrightarrow{-7R_2 + R_3} \left[\begin{array}{ccc|c} 1 & -2 & 1 & 10 \\ 0 & 1 & -1 & -7 \\ 0 & 0 & 4 & 20 \end{array}\right]$$

Finally, we want a 1 in the third position of the third column.

$$\xrightarrow{\frac{1}{4}R_3} \left[\begin{array}{ccc|c} 1 & -2 & 1 & 10 \\ 0 & 1 & -1 & -7 \\ 0 & 0 & 1 & 5 \end{array}\right]$$

The augmented matrix is now in the desired form and corresponds to the system

$$x - 2y + z = 10$$
$$y - z = -7$$
$$z = 5$$

Substituting $z = 5$ in the second equation produces

$$y - 5 = -7$$
$$y = -2$$

Substituting $y = -2$ and $z = 5$ in the first equation gives

$$x - 2(-2) + 5 = 10$$
$$x + 4 + 5 = 10$$
$$x = 1$$

and $\{(1, -2, 5)\}$ is the solution set for our system.

✔ CHECK YOURSELF 3

Use elementary row operations to solve the system.

$$x + 2y - 4z = -9$$
$$2x + 5y - 10z = -21$$
$$-3x - 5y + 11z = 28$$

Will our work with matrices provide solutions for all linear systems? From your previous work with inconsistent and dependent systems, you should realize that the answer is no. Examples 4 and 5 illustrate.

Example 4

Solving a System by Using Elementary Row Operations

Use elementary row operations to solve the system, if possible.

$$x + 2y = 2$$
$$-3x - 6y = -5$$

The augmented matrix for the system is

$$\begin{bmatrix} 1 & 2 & \vdots & 2 \\ -3 & -6 & \vdots & -5 \end{bmatrix}$$

Note: In general, if we have a row with all 0s as coefficients and a nonzero constant on the right, the system is inconsistent.

Adding 3 times row 1 to row 2, we have

$$\xrightarrow{3R_1 + R_2} \begin{bmatrix} 1 & 2 & \vdots & 2 \\ 0 & 0 & \vdots & 1 \end{bmatrix}$$

Note that row 2 now gives $0x + 0y = 1$, which implies that $0 = 1$, a contradiction. This means that the given system is inconsistent. There are *no* solutions.

✔ CHECK YOURSELF 4

Use elementary row operations to solve the system, if possible.

$$2x - 4y = 6$$
$$-x + 2y = -2$$

Example 5 considers one final case.

Example 5 | Using Elementary Row Operations

Use elementary row operations to solve the system, if possible.

$$x - 4y = 4$$
$$2x - 8y = 8$$

The augmented matrix for this system is

$$\begin{bmatrix} 1 & -4 & \vdots & 4 \\ 2 & -8 & \vdots & 8 \end{bmatrix}$$

In this case we multiply row 1 by -2 and add that result to row 2:

$$\xrightarrow{-2R_1 + R_2} \begin{bmatrix} 1 & -4 & \vdots & 4 \\ 0 & 0 & \vdots & 0 \end{bmatrix}$$

The bottom row of 0s represents the statement

$$0x + 0y = 0$$

which is, of course, true for any values of x and y. This means that the original system was dependent and has an infinite number of solutions, in this case all values (x, y) that satisfy the equation $x - 4y = 4$.

✔ CHECK YOURSELF 5

Use elementary row operations to solve the system, if possible.

$$3x - 6y = -9$$
$$-x + 2y = 3$$

━━━

✔ CHECK YOURSELF ANSWERS

1. (a) $\begin{bmatrix} 3 & -5 & | & -2 \\ 2 & 3 & | & 5 \end{bmatrix}$; (b) $\begin{bmatrix} 1 & -3 & | & 3 \\ 0 & 1 & | & 2 \end{bmatrix}$

2. $\{(3, -4)\}$ **3.** $\{(-3, 5, 4)\}$ **4.** Inconsistent system, no solutions

5. Dependent system, infinite number of solutions

Exercises · 8.4

Write the augmented matrix for each of the systems of equations.

1. $2x - 3y = 5$
$x + 4y = 2$

2. $x + 5y = 3$
$-2x + y = -1$

3. $x - 5y = 6$
$y = 2$

4. $x + 3y = 6$
$y = 3$

5. $x + 2y - z = 3$
$x + 3z = 1$
$y - 2z = 4$

6. $x - 2y + 5z = 3$
$3y - 2z = 1$
$z = 4$

Write the systems of equations corresponding to each of the augmented matrices.

7. $\begin{bmatrix} 1 & 2 & | & 3 \\ 1 & 5 & | & -6 \end{bmatrix}$

8. $\begin{bmatrix} 1 & 2 & | & 3 \\ -3 & 1 & | & -3 \end{bmatrix}$

9. $\begin{bmatrix} 1 & 3 & | & 5 \\ 0 & 1 & | & 2 \end{bmatrix}$

10. $\begin{bmatrix} 1 & -2 & | & 4 \\ 0 & 1 & | & 3 \end{bmatrix}$

11. $\begin{bmatrix} 1 & 2 & 0 & | & 4 \\ 0 & 1 & 5 & | & 3 \\ 1 & 1 & 1 & | & 1 \end{bmatrix}$

12. $\begin{bmatrix} 1 & 2 & -1 & | & 3 \\ 0 & 1 & 4 & | & 1 \\ 0 & 0 & 1 & | & -5 \end{bmatrix}$

Solve each of the systems, using elementary row operations.

13. $x + 3y = -5$
$2x + y = 0$

14. $x - 3y = -9$
$-3x + y = 11$

15. $x - 5y = 20$
$-4x + 3y = -29$

16. $x + 2y = -10$
$5x - y = 16$

17. $3x - 2y = 14$
$x + 5y = -18$

18. $4x + 3y = 0$
$x + 5y = -17$

19. $5x + 3y = 11$
$-x - 2y = 2$

20. $6x - y = 24$
$-x + 3y = 13$

21. $x - 3y = 5$
$-4x + 12y = -18$

22. $-3x + 6y = -15$
$x - 2y = 5$

706

Solve each of the systems, using elementary row operations.

23. $x + y - z = -1$
$\qquad -3y + 2z = 6$
$\quad 2x \qquad + 3z = 2$

24. $x + 2y \qquad = -3$
$\qquad x + 3y - 6z = -9$
$\quad 2x \qquad - 3z = 0$

25. $x \qquad - z = 2$
$\quad 2x - y + z = 3$
$\qquad y - 2z = -2$

26. $x - 2y + 5z = 10$
$\quad -3x + 4y \qquad = -12$
$\qquad x \qquad + z = 3$

27. $x \qquad - z = 6$
$\quad x + 2y + z = -7$
$\quad 2x - y + 2z = 1$

28. $x + 2y + 3z = 1$
$\quad -2x - 3y - 4z = -1$
$\qquad y + 3z = 3$

29. $x + 2y + 3z = 1$
$\quad x + 3y + z = 2$
$\quad 3x + 2y + z = -1$

30. $2x + 2y + z = 3$
$\quad 3x + 2y + 2z = 7$
$\quad x + y + z = 2$

Graphing Linear Inequalities in Two Variables

1. *Graph linear inequalities in two variables*
2. *Graph a region defined by linear inequalities*

What does the solution set look like when we are faced with an inequality in two variables? We will see that it is a set of ordered pairs best represented by a shaded region. The general form for a linear inequality in two variables is

$$Ax + By < C$$

where A and B cannot both be 0. The symbol $<$ can be replaced with $>$, $\leq$, or $\geq$. Some examples are

$$y < -2x + 6 \qquad x - 2y \leq 4 \qquad \text{and} \qquad 2x - 3y \geq x + 5y$$

As was the case with an equation, the solution set of a linear inequality is a set of ordered pairs of real numbers. However, in the case of the linear inequalities, we will find that the solution sets will be all the points in an entire region of the plane, called a **half plane.**

To determine such a solution set, let's start with the first inequality listed above. To graph the solution set of

$$y < -2x + 6$$

we begin by writing the corresponding linear equation

$$y = -2x + 6$$

First, note that the graph of $y = -2x + 6$ is simply a straight line.

Now, to graph the solution set of $y < -2x + 6$, we must include all ordered pairs that satisfy that inequality. For instance if $x = 1$, we have

$$y < -2 \cdot 1 + 6$$

$$y < 4$$

So we want to include all points of the form $(1, y)$, where $y < 4$. Of course, since $(1, 4)$ is *on* the corresponding line, this means that we want all points *below* the line along the vertical line $x = 1$. The result will be similar for any choice of x, and our solution set will then contain all points below the line $y = -2x + 6$. We can then graph the solution set as the shaded region shown. We have the following definition.

$y < -2x + 6$

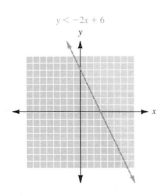

The line is dashed to indicate that points on the line are *not* included.

We call the graph of the equation

$$Ax + By = C$$

the **boundary line** of the half planes.

In general, the solution set of an inequality of the form

$$Ax + By < C \quad \text{or} \quad Ax + By > C$$

will be a half plane either above or below the corresponding line determined by

$$Ax + By = C$$

How do we decide which half plane represents the desired solution set? The use of a **test point** provides an easy answer. Choose any point *not* on the line. Then substitute the coordinates of that point into the given inequality. If the coordinates satisfy the inequality (result in a true statement), then shade the region or half plane that includes the test point; if not, shade the opposite half plane. Example 1 illustrates the process.

Example 1   Graphing a Linear Inequality

Graph the linear inequality

$$x - 2y < 4$$

First, we graph the corresponding equation

$$x - 2y = 4$$

to find the boundary line. Now to decide on the appropriate half plane, we need a test point *not* on the line. As long as the line *does not pass through the origin*, we can always use $(0, 0)$ as a test point. It provides the easiest computation.

Here letting $x = 0$ and $y = 0$, we have

$$(0) - 2 \cdot (0) \overset{?}{<} 4$$

$$0 < 4$$

Because this a true statement, we proceed to shade the half plane including the origin (the test point), as shown.

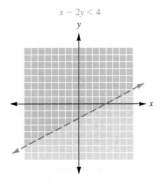

$$x - 2y < 4$$

✔ CHECK YOURSELF 1

Graph the solution set of $3x + 4y > 12$.

The graphs of some linear inequalities will include the boundary line. That will be the case whenever equality is included with the inequality statement, as illustrated in Example 2.

Example 2   Graphing a Linear Inequality

Graph the inequality

$$2x + 3y \geq 6$$

First, we graph the boundary line, here corresponding to $2x + 3y = 6$. Note that we use a solid line in this case since equality is included in the original statement.

Again, we choose a convenient test point not on the line. As before, the origin will provide the simplest computation.

Substituting $x = 0$ and $y = 0$, we have

$$2 \cdot (0) + 3 \cdot (0) \overset{?}{\geq} 6$$

$$0 \geq 6$$

A solid boundary line means that points on the line are solutions.

This is a *false* statement. Hence the graph will consist of all points on the *opposite* side from the origin. The graph will then be the upper half plane, as shown.

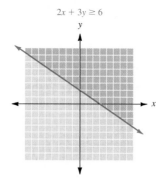

$2x + 3y \geq 6$

✔ CHECK YOURSELF 2

Graph the solution set of $x - 3y \leq 6$.

Example 3

Graphing a Linear Inequality

Graph the solution set of

$$y \leq 2x$$

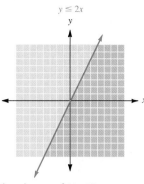

$y \leq 2x$

We proceed as before by graphing the boundary line (it is solid since equality is included). The only difference between this and previous examples is that we *cannot use the origin* as a test point. Do you see why?

Choosing (1, 1) as our test point gives the statement

$$(1) \overset{?}{\leq} 2 \cdot (1)$$

$$1 \leq 2$$

Because the statement is *true,* we shade the half plane *including* the test point (1, 1).

The choice of (1, 1) is arbitrary. We simply want *any* point *not* on the line.

✔ *CHECK YOURSELF 3*

Graph the solution set of $3x + y > 0$.

Let's consider a special case of graphing linear inequalities in the rectangular coordinate system.

Example 4

Graphing a Linear Inequality

Graph the solution set of $x > 3$.

Here we specify the rectangular coordinate system to indicate we want a two-dimensional graph.

First, we draw the boundary line (a dashed line because equality is not included) corresponding to

$$x = 3$$

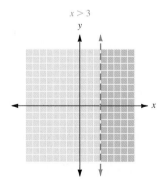

$x > 3$

We can choose the origin as a test point in this case, and that results in the false statement

$$0 > 3$$

We then shade the half plane *not* including the origin. In this case, the solution set is represented by the half plane to the right of the vertical boundary line.

As you may have observed, in this special case choosing a test point is not really necessary. Because we want values of x that are *greater than* 3, we want those ordered pairs that are to the *right* of the boundary line.

✔ *CHECK YOURSELF 4*

Graph the solution set of

$$y \leq 2$$

in the rectangular coordinate system.

Applications of linear inequalities will often involve more than one inequality condition. Consider Example 5.

Example 5	## Graphing a Region Defined by Linear Inequalities

Graph the region satisfying the following conditions.

$$3x + 4y \leq 12$$

$$x \geq 0$$

$$y \geq 0$$

$3x + 4y \leq 12$
$x \geq 0$
$y \geq 0$

The solution set in this case must satisfy *all three conditions*. As before, the solution set of the first inequality is graphed as the half plane *below* the boundary line. The second and third inequalities mean that x and y must also be nonnegative. Therefore, our solution set is restricted to the first quadrant (and the appropriate segments of the x and y axes), as shown.

✔ CHECK YOURSELF 5

Graph the region satisfying the following conditions.

$$3x + 4y < 12$$

$$x \geq 0$$

$$y \geq 0$$

The following algorithm summarizes our work in graphing linear inequalities in two variables.

To Graph a Linear Inequality

1. Replace the inequality symbol with an equality symbol to form the equation of the boundary line of the solution set.

2. Graph the boundary line. Use a dashed line if equality is not included ($<$ or $>$). Use a solid line if equality is included ($\leq$ or $\geq$).

3. Choose any convenient test point *not* on the boundary line.

4. If the inequality is *true* for the test point, shade the half plane *including* the test point. If the inequality is *false* for the test point, shade the half plane *not including* the test point.

✔ *CHECK YOURSELF ANSWERS*

1.

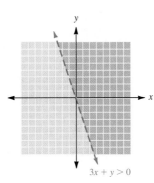

$3x + 4y > 12$

2.

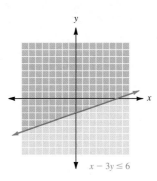

$x - 3y \leq 6$

3.

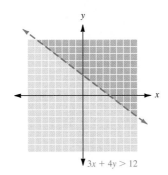

$3x + y > 0$

4.

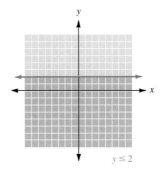

$y \leq 2$

5.

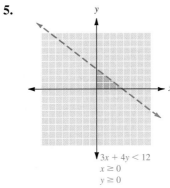

$3x + 4y < 12$
$x \geq 0$
$y \geq 0$

Exercises · 8.5

In Exercises 1 to 24, graph the solution sets of the linear inequalities.

1. $x + y < 4$

2. $x + y \geq 6$

3. $x - y \geq 3$

4. $x - y < 5$

5. $y \geq 2x + 1$

6. $y < 3x - 4$

7. $2x + 3y < 6$

8. $3x - 4y \geq 12$

9. $x - 4y > 8$

10. $2x + 5y \leq 10$

11. $y \geq 3x$

12. $y \leq -2x$

13. $x - 2y > 0$

14. $x + 4y \leq 0$

15. $x < 3$

16. $y < -2$

17. $y > 3$

18. $x \leq -4$

19. $3x - 6 \leq 0$

20. $-2y > 6$

21. $0 < x < 1$

22. $-2 \leq y \leq 1$

23. $1 \leq x \leq 3$

24. $1 < y < 5$

In Exercises 25 to 28, graph the region satisfying each set of conditions.

25. $0 \le x \le 3$
$2 \le y \le 4$

26. $1 \le x \le 5$
$0 \le y \le 3$

27. $x + 2y \le 4$
$x \ge 0$
$y \ge 0$

28. $2x + 3y \le 6$
$x \ge 0$
$y \ge 0$

29. Assume that you are working only with the variable x. Describe the solution to the statement $x > -1$.

30. Now, assume that you are working in two variables x and y. Describe the solution to the statement $x > -1$.

31. Manufacturing. A manufacturer produces a standard model and a deluxe model of a 13-in. television set. The standard model requires 12 h to produce, while the deluxe model requires 18 h. The labor available is limited to 360 h per week.

   If x represents the number of standard model sets produced per week and y represents the number of deluxe models, draw a graph of the region representing the feasible values for x and y. Keep in mind that the values for x and y must be nonnegative since they represent a quantity of items. (This will be the solution set for the system of inequalities.)

32. Manufacturing. A manufacturer produces standard record turntables and CD players. The turntables require 10 h of labor to produce while CD players require 20 h. Let x represent the number of turntables produced and y the number of CD players.

   If the labor hours available are limited to 300 h per week, graph the region representing the feasible values for x and y.

33. Serving capacity. A hospital food service can serve at most 1,000 meals per day. Patients on a normal diet receive 3 meals per day, and patients on a special diet receive 4 meals per day. Write a linear inequality that describes the number of patients that can be served per day and draw its graph.

34. Time on job. The movie and TV critic for the local radio station spends 3 to 7 h daily reviewing movies and fewer than 4 h reviewing TV shows. Let x represent the hours watching movies and y represent the time spent watching TV. Write two inequalities that model the situation, and graph their intersection.

In Exercises 35 to 38, write an inequality for the shaded region shown in the figure.

35.

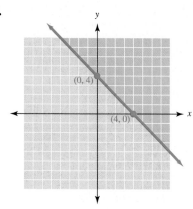

36.

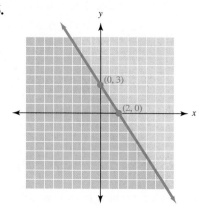

37.

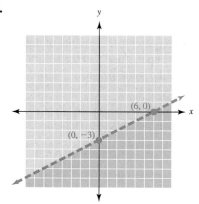

38.

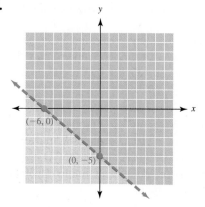

8.6 Systems of Linear Inequalities in Two Variables

1. Graph systems of linear inequalities
2. Solve an application of a system of linear inequalities

"When I'm working on a problem, I never think about beauty. I think only how to solve the problem. But when I have finished, if the solution is not beautiful, I know it is wrong."

—Richard Buckminster Fuller

Our previous work in this chapter dealt with finding the solution set of a system of linear equations. That solution set represented the points of intersection of the graphs of the equations in the system. In this section, we extend that idea to include systems of linear inequalities.

In this case, the solution set is all ordered pairs that satisfy each inequality. **The graph of the solution set of a system of linear inequalities** is then the intersection of the graphs of the individual inequalities. Let's look at an example.

| Example 1 | ## Solving a System of Linear Inequalities |

Solve the following system of linear inequalities by graphing.

$$x + y > 4$$
$$x - y < 2$$

We start by graphing each inequality separately. The boundary line is drawn, and using $(0, 0)$ as a test point, we see that we should shade the half plane above the line in both graphs.

Note that the boundary line is dashed to indicate that points on the line do *not* represent solutions.

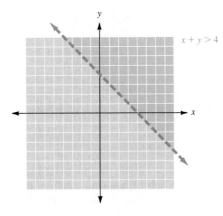

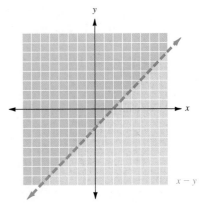

In practice, the graphs of the two inequalities are combined on the same set of axes, as shown below. The graph of the solution set of the original system is the intersection of the graphs drawn on the previous page.

Points on the lines are not included in the solution.

We want to show all ordered pairs that satisfy *both* statements.

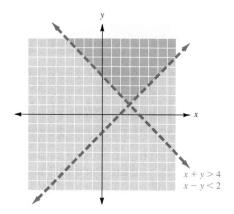

$x + y > 4$
$x - y < 2$

✔ CHECK YOURSELF 1

Solve the following system of linear inequalities by graphing.

$$2x - y < 4$$

$$x + y < 3$$

Most applications of systems of linear inequalities lead to **bounded regions.** This requires a system of three or more inequalities, as shown in Example 2.

Example 2   Solving a System of Linear Inequalities

Solve the following system of linear inequalities by graphing.

$$x + 2y \leq 6$$

$$x + y \leq 5$$

$$x \geq 2$$

$$y \geq 0$$

On the same set of axes, we graph the boundary line of each of the inequalities. We then choose the appropriate half planes (indicated by the arrow that is perpendicular to the line) in each case, and we locate the intersection of those regions for our graph.

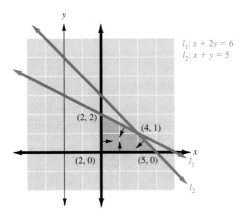

The vertices of the shaded region are given because they have particular significance in later applications of this concept. Can you see how the coordinates of the vertices were determined?

✔ *CHECK YOURSELF 2*

Solve the following system of linear inequalities by graphing.

$$2x - y \le 8 \qquad x \ge 0$$

$$x + y \le 7 \qquad y \ge 0$$

Let's expand on Example 8 in Section 8.2 to see an application of our work with systems of linear inequalities. Consider Example 3.

| Example 3 | **Solving a Business-Based Application** |

A manufacturer produces a standard model and a deluxe model of a 13-in. television set. The standard model requires 12 h of labor to produce, and the deluxe model requires 18 h. The labor available is limited to 360 h per week. Also, the plant capacity is limited to producing a total of 25 sets per week. Draw a graph of the region representing the number of sets that can be produced, given these conditions.

As suggested earlier, we let x represent the number of standard model sets produced and y the number of deluxe model sets. Since the labor is limited to 360 h, we have

The total labor is limited to (or less than or equal to) 360 h.

$$12x \quad + \quad 18y \quad \le \quad 360 \qquad\qquad \textbf{(1)}$$

    ↑   ↑
12 h per   18 h per
standard set   deluxe set

The total production, here $x + y$ sets, is limited to 25, so we can write

$$x + y \le 25 \qquad\qquad \textbf{(2)}$$

We have $x \geq 0$ and $y \geq 0$ since the number of sets produced cannot be negative.

For convenience in graphing, we divide both members of inequality (1) by 6, to write the equivalent system

$$2x + 3y \leq 60$$
$$x + y \leq 25$$
$$x \geq 0$$
$$y \geq 0$$

We now graph the system of inequalities as before. The shaded area represents all possibilities in terms of the number of sets that can be produced.

The shaded area is called the **feasible region.** All points in the region meet the given conditions of the problem and represent possible production options.

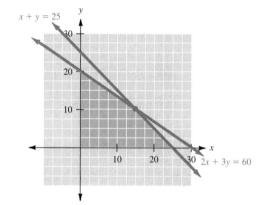

✔ *CHECK YOURSELF 3*

A manufacturer produces TVs and CD players. The TVs require 10 h of labor to produce and the CD players require 20 h. The labor hours available are limited to 300 h per week. Existing orders require that at least 10 TVs and at least 5 CD players be produced per week. Draw a graph of the region representing the possible production options.

✔ *CHECK YOURSELF ANSWERS*

1. $2x - y < 4$

$x + y < 3$

2. $2x - y \leq 8$

$x + y \leq 7$

$x \geq 0$

$y \geq 0$

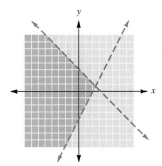

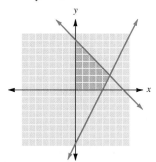

3. Let x be the number of TVs and y be the number of CD players. The system is

$10x + 20y \leq 300$

$x \geq 10$

$y \geq 5$

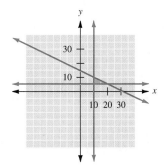

Solve each system of linear inequalities graphically.

1. $x + 2y \leq 4$
$x - y \geq 1$

2. $3x - y > 6$
$x + y < 6$

3. $3x + y < 6$
$x + y > 4$

4. $2x + y \geq 8$
$x + y \geq 4$

5. $x + 3y \leq 12$
$2x - 3y \leq 6$

6. $x - 2y > 8$
$3x - 2y > 12$

7. $3x + 2y \leq 12$
$x \geq 2$

8. $2x + y \leq 6$
$y \geq 1$

9. $2x + y \leq 8$
$x > 1$
$y > 2$

10. $3x - y \leq 6$
$x \geq 1$
$y \leq 3$

11. $x + 2y \leq 8$
$2 \leq x \leq 6$
$y \geq 0$

12. $x + y < 6$
$0 \leq y \leq 3$
$x \geq 1$

13. $3x + y \leq 6$

     $x + y \leq 4$

        $x \geq 0$

        $y \geq 0$

14. $x - 2y \geq -2$

     $x + 2y \leq 6$

        $x \geq 0$

        $y \geq 0$

15. $4x + 3y \leq 12$

     $x + 4y \leq 8$

        $x \geq 0$

        $y \geq 0$

16. $2x + y \leq 8$

     $x + y \geq 3$

        $x \geq 0$

        $y \geq 0$

17. $x - 4y \leq -4$

     $x + 2y \leq 8$

        $x \geq 2$

18. $x - 3y \geq -6$

     $x + 2y \geq 4$

        $x \leq 4$

In Exercises 19 and 20, draw the appropriate graph.

19. Manufacturing. A manufacturer produces both two-slice and four-slice toasters. The two-slice toaster takes 6 h of labor to produce, and the four-slice toaster takes 10 h. The labor available is limited to 300 h per week, and the total production capacity is 40 toasters per week. Draw a graph of the feasible region, given these conditions, where x is the number of two-slice toasters and y is the number of four-slice toasters.

20. Production. A small firm produces both AM and AM/FM car radios. The AM radios take 15 h to produce, and the AM/FM radios take 20 h. The number of production hours is limited to 300 h per week. The plant's capacity is limited to a total of 18 radios per week, and existing orders require that at least 4 AM radios and at least 3 AM/FM radios be produced per week. Draw a graph of the feasible region, given these conditions, where x is the number of AM radios and y the number of AM/FM radios.

21. When you solve a system of linear inequalities, it is often easier to shade the region that is not part of the solution, rather than the region that is. Try this method, then describe its benefits.

22. Describe a system of linear inequalities for which there is no solution.

23. Write the system of inequalities whose graph is the shaded region.

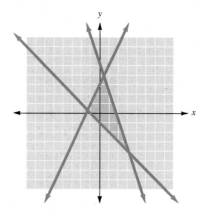

24. Write the system of inequalities whose graph is the shaded region.

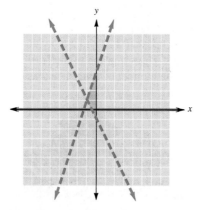

Summary for Chapter 8

Example	Topic	Reference
	Solving Systems of Linear Equations by Graphing	8.1
The solution for the system $$2x - y = 7$$ $$x + y = 2$$ is $(3, -1)$. It is the only ordered pair that will satisfy each equation.	A **system of linear equations** is two or more linear equations considered together. A solution for a linear system in two variables is an ordered pair of real numbers (x, y) that satisfies both equations in the system. There are three solution techniques: the graphing method, the addition method, and the substitution method.	**p. 658**
To solve $$2x - y = 7$$ $$x + y = 2$$ by graphing One solution—a consistent system No solutions—an inconsistent system An infinite number of solutions—a dependent system.	**Solving by the Graphing Method** Graph each equation of the system on the same set of coordinate axes. If a solution exists, it will correspond to the point of intersection of the two lines. Such a system is called a **consistent system.** If a solution does not exist, there is no point at which the two lines intersect. Such lines are parallel, and the system is called an **inconsistent system.** If there are an infinite number of solutions, the lines coincide. Such a system is called a **dependent system.** You may or may not be able to determine exact solutions for the system of equations with this method. Consistent system No solutions—an inconsistent system An infinite number of solutions—a dependent system	**p. 659**

Systems of Linear Equations in Two Variables with Applications

8.2

To solve	**Solving by the Addition Method**	**p. 670**

To solve

$$5x - 2y = 11 \quad (1)$$
$$2x + 3y = 12 \quad (2)$$

Multiply equation (1) by 3 and equation (2) by 2. Then add to eliminate y.

$$19x = 57$$
$$x = 3$$

Substituting 3 for x in equation (1), we have

$$15 - 2y = 11$$
$$y = 2$$

So (3, 2) is the solution.

Solving by the Addition Method

1. If necessary, multiply one or both of the equations by a constant so that one of the variables can be eliminated by addition.

2. Add the equations of the equivalent system formed in step 1.

3. Solve the equation found in step 2.

4. Substitute the value found in step 3 into either of the equations of the original system to find the corresponding value of the remaining variable. The ordered pair formed is the solution to the system.

5. Check the solution by substituting the pair of values found in step 4 into the other equation of the original system.

p. 670

To solve

$$3x - 2y = 6 \quad (3)$$
$$6x + y = 2 \quad (4)$$

by substitution, solve (4) for y.

$$y = -6x + 2 \quad (5)$$

Substituting in (3) gives

$$3x - 2(-6x + 2) = 6$$

and

$$x = \frac{2}{3}$$

Substituting $\frac{2}{3}$ for x in (5) gives

$$y = (-6)\left(\frac{2}{3}\right) + 2$$

$$= -4 + 2 = -2$$

The solution is

$$\left(\frac{2}{3}, -2\right)$$

Solving by the Substitution Method

1. If necessary, solve one of the equations of the original system for one of the variables.

2. Substitute the expression obtained in step 1 into the other equation of the system to write an equation in a single variable.

3. Solve the equation found in step 2.

4. Substitute the value found in step 3 into the equation derived in step 1 to find the corresponding value of the remaining variable. The ordered pair formed is the solution for the system.

5. Check the solution by substituting the pair of values found in step 4 in *both* equations of the original system.

p. 673

Applications of Systems of Linear Equations

Also determine the condition that relates the unknown quantities.

Use a different letter for each variable.

A table or a sketch often helps in writing the equations of the system.

1. Read the problem carefully to determine the unknown quantities.
2. Choose a variable to represent any unknown.
3. Translate the problem to the language of algebra to form a system of equations.
4. Solve the system of equations by any of the methods discussed, and answer the question in the original problem.
5. Verify your solution by returning to the original problem.

p. 674

Systems of Linear Equations in Three Variables

8.3

Solve.
$$x + y - z = 6 \quad (6)$$
$$2x - 3y + z = -9 \quad (7)$$
$$3x + y + 2z = 2 \quad (8)$$

Adding (6) and (7) gives
$$3x - 2y = -3 \quad (9)$$

Multiplying (6) by 2 and adding the result to (8) give
$$5x + 3y = 14 \quad (10)$$

The system consisting of (9) and (10) is solved as before and
$$x = 1 \qquad y = 3$$

Substituting these values into (6) gives
$$z = -2$$

The solution is $(1, 3, -2)$.

A solution for a linear system of three equations in three variables is an ordered triple of numbers (x, y, z) that satisfies each equation in the system.

Solving a System of Three Equations in Three Unknowns

1. Choose a pair of equations from the system and use the addition method to eliminate one of the variables.
2. Choose a different pair of equations and eliminate the same variable.
3. Solve the system of two equations in two variables determined in steps 1 and 2.
4. Substitute the values found above into one of the original equations and solve for the remaining variable.
5. The solution is the ordered triple of values found in steps 3 and 4. It can be checked by substituting into the other equations of the original system.

p. 689

To graph		8.5

Graphing Linear Inequalities in Two Variables

To graph

$$x - 2y < 4$$

In general, the solution set of an inequality of the form

$$ax + by < c \qquad \text{or} \qquad ax + by > c$$

will be a **half plane** either above or below the **boundary line** determined by

$$ax + by = c$$

The boundary line is included in the graph if equality is included in the statement of the original inequality. Such a line is solid. The boundary line is dashed if it is not included in the graph.

To graph a linear inequality:

1. Replace the inequality symbol with an equality symbol to form the equation of the boundary line of the solution set.

2. Graph the boundary line. Use a dashed line if equality is not included ($<$ or $>$). Use a solid line if equality is included ($\leq$ or $\geq$).

3. Choose any convenient test point *not* on the boundary line.

4. If the inequality is *true* for the test point, shade the half plane *including* the test point. If the inequality is *false* for the test point, shade the half plane *not including* the test point.

p. 712

Systems of Linear Inequalities in Two Variables

8.6

To solve

$$x + 2y \leq 8$$
$$x + \ y \leq 6$$
$$x \geq 0$$
$$y \geq 0$$

by graphing

A **system of linear inequalities** is two or more linear inequalities considered together. The **graph of the solution set** of a system of linear inequalities is the intersection of the graphs of the individual inequalities.

Solving Systems of Linear Inequalities by Graphing

1. Graph each inequality, shading the appropriate half plane on the same set of coordinate axes.

2. The graph of the system is the intersection of the regions shaded in step 1.

p. 717

Summary Exercises · 8

This summary exercise set will give you practice with each of the objectives in the chapter.

[8.1] Graph the system of equations and then solve the system.

1. $x + y = 8$
$x - y = 4$

2. $x + 2y = 8$
$x - y = 5$

3. $2x + 3y = 12$
$2x + y = 8$

4. $x + 4y = 8$
$y = 1$

In Exercises 5 and 6, use a graphing calculator to solve. Estimate your answer to the nearest hundredth. You may need to adjust the viewing window to see the point of intersection.

5. $44x + 35y = 1{,}115$
$11x - 27y = 850$

6. $15x - 48y = 935$
$25x + 51y = 1{,}051$

[8.2] Solve each of the following systems by the addition method. If a unique solution does not exist, state whether the given system is inconsistent or dependent.

7. $x + 2y = 7$
$x - y = 1$

8. $x + 3y = 14$
$4x + 3y = 29$

9. $3x - 5y = 5$
$-x + y = -1$

10. $x - 4y = 12$
$2x - 8y = 24$

11. $6x + 5y = -9$
$-5x + 4y = 32$

12. $3x + y = 8$
$-6x - 2y = -10$

13. $5x - y = -17$
$4x + 3y = -6$

14. $4x - 3y = 1$
$6x + 5y = 30$

15. $x - \dfrac{1}{2}y = 8$
$\dfrac{2}{3}x + \dfrac{3}{2}y = -2$

16. $\dfrac{1}{5}x - 2y = 4$
$\dfrac{3}{5}x + \dfrac{2}{3}y = -8$

[8.2] Solve each of the following systems by the substitution method. If a unique solution does not exist, state whether the given system is inconsistent or dependent.

17. $2x + y = 23$
$x = y + 4$

18. $x - 5y = 26$
$y = x - 10$

19. $3x + y = 7$
$y = -3x + 5$

20. $2x - 3y = 13$
$x = 3y + 9$

21. $5x - 3y = 13$
$x - y = 3$

22. $4x - 3y = 6$
$x + y = 12$

23. $3x - 2y = -12$
$6x + y = 1$

24. $x - 4y = 8$
$-2x + 8y = -16$

[8.2] Solve each of the following problems by choosing a variable to represent each unknown quantity. Then write a system of equations that will allow you to solve for each variable.

25. Number problem. One number is 2 more than 3 times another. If the sum of the two numbers is 30, find the two numbers.

26. Money value. Suppose that a cashier has 78 $5 and $10 dollar bills with a value of $640. How many of each type of bill does she have?

27. Ticket sales. Tickets for a basketball game sold at $7 for an adult ticket and $4.50 for a student ticket. If the revenue from 1,200 tickets was $7,400, how many of each type of ticket were sold?

28. Purchase price. A purchase of 8 blank cassette tapes and 4 blank videotapes costs $36. A second purchase of 4 cassette tapes and 5 videotapes costs $30. What is the price of a single cassette tape and of a single videotape?

29. Rectangles. The length of a rectangle is 4 cm less than twice its width. If the perimeter of the rectangle is 64 cm, find the dimensions of the rectangle.

30. Mixture. A grocer in charge of bulk foods wishes to combine peanuts selling for $2.25 per pound and cashews selling for $6 per pound. What amount of each nut should be used to form a 120-lb mixture selling for $3 per pound?

31. **Investments.** Reggie has two investments totaling $17,000—one a savings account paying 6%, the other a time deposit paying 8%. If his annual interest is $1,200, what does he have invested in each account?

32. **Mixtures.** A pharmacist mixes a 20% alcohol solution and a 50% alcohol solution to form 600 ml of a 40% solution. How much of each solution should he use in forming the mixture?

33. **Motion.** A jet flying east, with the wind, makes a trip of 2,200 mi in 4 h. Returning against the wind, the jet can travel only 1,800 mi in 4 h. What is the plane's rate in still air? What is the rate of the wind?

34. **Number problem.** The sum of the digits of a two-digit number is 9. If the digits are reversed, the new number is 45 more than the original number. What was the original number?

35. **Work.** A manufacturer produces $5\frac{1}{4}$ in. computer disk drives and $3\frac{1}{2}$ in. drives. The $5\frac{1}{4}$ in. drives require 20 min of component assembly time; the $3\frac{1}{2}$ in. drives, 25 min. The manufacturer has 500 min of component assembly time available per day. Each drive requires 30 min for packaging and testing, and 690 min of that time is available per day. How many of each of the drives should be produced daily to use all the available time?

36. **Equilibrium price.** If the demand equation for a product is $D = 270 - 5p$ and the supply equation is $S = 13p$, find the equilibrium point.

37. **Rental charges.** Two car rental agencies have the following rates for the rental of a compact automobile:

    Company A: $18 per day plus 12¢ per mile
    Company B: $20 per day plus 10¢ per mile

    For a 3-day rental, at what number of miles will the charges from the two companies be the same?

[8.3] Solve each of the following systems by the addition method. If a unique solution does not exist, state whether the given system is inconsistent or dependent.

38. $x - y + z = 0$
    $x + 4y - z = 14$
    $x + y - z = 6$

39. $x - y + z = 3$
    $3x + y + 2z = 15$
    $2x - y + 2z = 7$

40. $x - y - z = 2$
    $-2x + 2y + z = -5$
    $-3x + 3y + z = -10$

41. $x - y = 3$
    $2y + z = 5$
    $x + 2z = 7$

42. $x + y + z = 2$
$x + 3y - 2z = 13$
$y - 2z = 7$

43. $x + y - z = -1$
$x - y - 2z = 2$
$-5x - y - z = -1$

44. $2x + 3y + z = 7$
$-2x - 9y + 2z = 1$
$4x - 6y + 3z = 10$

[8.3] Solve each of the following problems by choosing a variable to represent each unknown quantity.

45. Number problem. The sum of three numbers is 15. The largest number is 4 times the smallest number, and it is also 1 more than the sum of the other two numbers. Find the three numbers.

46. Number problem. The sum of the digits of a three-digit number is 16. The tens digit is 3 times the hundreds digit, and the units digit is 1 more than the hundreds digit. What is the number?

47. Tickets sold. A theater has orchestra tickets at $10, box seat tickets at $7, and balcony tickets at $5. For one performance, a total of 360 tickets was sold, and the total revenue was $3,040. If the number of orchestra tickets sold was 40 more than that of the other two types combined, how many of each type of ticket were sold for the performance?

48. Triangles. The measure of the largest angle of a triangle is 15° less than 4 times the measure of the smallest angle and 30° more than the sum of the measures of the other two angles. Find the measures of the three angles of the triangle.

49. Investments. Rachel divided $12,000 into three investments: a savings account paying 5%, a stock paying 7%, and a mutual fund paying 9%. Her annual interest from the investments was $800, and the amount that she had invested at 5% was equal to the sum of the amounts invested in the other accounts. How much did she have invested in each type of account?

[8.4] Find the augmented matrix for each system of equations.

50. $x - 3y = -9$
$-3x + y = 11$

51. $5x + 3y = 11$
$-x - 2y = 2$

Solve each system by using elementary row operations.

52. $x - 3y = -9$
$-3x + y = 11$

53. $5x + 3y = 11$
$-x - 2y = 2$

[8.5] Graph the solution set for each of the following linear inequalities.

54. $y < 2x + 1$ **55.** $y \geq -2x + 3$ **56.** $3x + 2y \geq 6$ **57.** $3x - 5y < 15$

58. $y < -2x$ **59.** $4x - y \geq 0$ **60.** $y \geq -3$ **61.** $x < 4$

[8.6] Solve each of the following systems of linear inequalities.

62. $x - y < 7$ **63.** $x - 2y \leq -2$ **64.** $x - 6y < 6$ **65.** $2x + y \leq 8$
   $x + y > 3$ $x + 2y \leq \ \ 6$ $-x + \ \ y < 4$ $x \geq 1$
                                                                                    $y \geq 0$

66. $2x + y \leq 6$ **67.** $4x + y \leq 8$ **68.** $4x + 2y \leq 8$ **69.** $3x + y \leq 6$
      $x \geq 1$ $x \geq 0$ $x + \ \ y \leq 3$ $x + y \leq 4$
      $y \geq 0$ $y \geq 2$ $x \geq 0$ $x \geq 0$
                                                                 $y \geq 0$ $y \geq 0$

Self-Test 8

The purpose of this self-test is to help you check your progress and to review for a chapter test in class. Allow yourself about 1 hour to take the test. When you are done, check your answers in the back of the book. If you missed any answers, be sure to go back and review the appropriate sections in the chapter and the exercises that are provided.

Solve each of the following systems. If a unique solution does not exist, state whether the given system is inconsistent or dependent.

1. $3x + y = -5$
$5x - 2y = -23$

2. $4x - 2y = -10$
$y = 2x + 5$

3. $9x - 3y = 4$
$-3x + y = -1$

4. $5x - 3y = 5$
$3x + 2y = -16$

5. $x - 2y = 5$
$2x + 5y = 10$

6. $5x - 3y = 20$
$4x + 9y = -3$

Solve each of the following systems.

7. $x - y + z = 1$
$-2x + y + z = 8$
$x + 5z = 19$

8. $x + 3y - 2z = -6$
$3x - y + 2z = 8$
$-2x + 3y - 4z = -11$

Solve each of the following by choosing a variable to represent each unknown quantity. Then write a system of equations that will allow you to solve for each variable.

9. An order for 30 computer disks and 12 printer ribbons totaled $147. A second order for 12 more disks and 6 additional ribbons cost $66. What was the cost per individual disk and ribbon?

10. A candy dealer wants to combine jawbreakers selling for $2.40 per pound and licorice selling for $3.90 per pound to form a 100-lb mixture that will sell for $3 per pound. What amount of each type of candy should be used?

11. A small electronics firm assembles 5-in. portable television sets and 12-in. models. The 5-in. set requires 9 h of assembly time; the 12-in. set, 6 h. Each unit requires 5 h for packaging and testing. If 72 h of assembly time and 50 h of packaging and testing time are available per week, how many of each type of set should be finished if the firm wishes to use all its available capacity?

12. Hans decided to divide $16,000 into three investments: a savings account paying 3% annual interest, a bond paying 5%, and a mutual fund paying 7%. His annual interest from the three investments was $900, and he has as much in the mutual fund as in the savings account and bond combined. What amount did he invest in each type?

13. The fence around a rectangular yard requires 260 ft of fencing. The length is 20 ft less than twice the width. Find the dimensions of the yard.

Graph the solution set in each of the following.

14. $5x + 6y \leq 30$

15. $x + 3y > 6$

16. $4x - 8 \leq 0$

17. $2y + 4 > 0$

Solve each of the following systems of linear inequalities.

18. $x - 2y < 6$
$x + y < 3$

19. $3x + 4y \geq 12$
$x \geq 1$

20. $x + 2y \leq 8$
$x + y \leq 6$
$x \geq 0$
$y \geq 0$

Cumulative Test ▪ 0–8

This test is provided to help you in the process of reviewing the previous chapters. Answers are provided in the back of the book. If you missed any answers, be sure to go back and review the appropriate sections.

Solve each of the following.

1. $3x - 2(x + 5) = 12 - 3x$

2. $2x - 7 < 3x - 5$

3. $|2x - 3| = 5$

4. $|3x + 5| \leq 7$

5. $|5x - 4| > 21$

6. $2x + 3(x - 2) = -4(x + 1) + 16$

Graph each of the following.

7. $5x + 7y = 35$

8. $2x + 3y < 6$

9. Solve the equation $P = P_0 + IRT$ for R.

10. Find the slope of the line connecting $(4, 6)$ and $(3, -1)$.

11. Write an equation of the line that passes through the points $(-1, 4)$ and $(5, -2)$.

Simplify the following expressions.

12. $(2x + 1)(x - 3)$

13. $(3x - 2)^2$

14. Completely factor the expression $x^3 - 3x^2 - 5x + 15$.

15. Write an equation of the line passing through the point $(3, 2)$ and parallel to the line $4x - 5y = 20$.

16. Find $f(-5)$ if $f(x) = 3x^2 - 4x - 5$.

Solve each of the following systems of equations.

17. $2x + 3y = 6$
$5x + 3y = -24$

18. $x + y + z = 3$
$2x - y + 2z = 0$
$-x - 3y + z = -9$

Solve each of the following applications.

19. The length of a rectangle is 3 cm more than twice its width. If the perimeter of the rectangle is 54 cm, find the dimensions of the rectangle.

20. The sum of the digits of a two-digit number is 10. If the digits are reversed, the number is 36 less than the original number. What was the original number?

21. Simplify the expression.

$$\left(\frac{x^{-3}y^{-4}}{x^2y^{-7}}\right)^3$$

Simplify each of the following.

22. $\dfrac{6}{x-2} + \dfrac{2}{x}$

23. $\dfrac{x^2 - x - 42}{x + 6} \cdot \dfrac{x^2 + 7x}{x^2 - 49}$

24. $\dfrac{y^2 - 6y}{4y + 12} \div \dfrac{2y - 12}{y^2 - 9}$

25. $\dfrac{\dfrac{3}{xy} - 1}{\dfrac{4}{x^2y}}$

9 Graphical Solutions

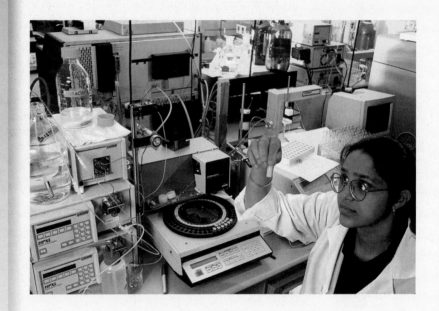

Quality control is exercised in nearly all manufacturing processes. In the pharmaceutical-making process, great caution must be exercised to ensure that the medicines and drugs are pure and contain precisely what is indicated on the label. Guaranteeing such purity is a task that the quality control division of the pharmaceutical company assumes.

A lab technician working in quality control must run a series of tests on samples of every ingredient, even simple ingredients such as salt (NaCl). One such test is a measure of how much weight is lost as a sample is dried. The technician must set up a 3-h procedure that involves cleaning and drying bottles and stoppers and then weighing them while they are empty and again when they contain samples of the substance to be heated and dried. At the end of the procedure, to compute the percentage of weight loss from drying, the technician uses the formula

$$L = \frac{W_g - W_f}{W_g - T} \cdot 100$$

where

L = percentage loss in drying

W_g = weight of container and sample

W_f = weight of container and sample after drying process completed

T = weight of empty container

The pharmaceutical company may have a standard of acceptability for this substance. For instance, the substance may not be acceptable if the loss of weight from drying is greater than 10%. The technician would then use the following inequality to calculate acceptable weight loss:

$$10 \geq \frac{W_g - W_f}{W_g - T} \cdot 100$$

Such inequalities are more useful when solved for one of the variables, here W_f or T. In Chapter 2, you learned how to solve such an inequality. In this chapter you will begin to analyze the graph of such an equation.

9.1 Solving Equations in One Variable Graphically

1. Rewrite a linear equation in one variable as $f(x) = g(x)$
2. Find the point of intersection of $f(x)$ and $g(x)$
3. Interpret the point of intersection of $f(x)$ and $g(x)$
4. Solve a linear equation in one variable by writing it as the functional equality $f(x) = g(x)$

"Descartes commanded the future from his study more than Napolean from the throne."

—Oliver Wendell Holmes

In Chapter 2, we solved linear equations in one variable algebraically. In this section, we will look at graphical methods for the same type of equations.

While you are studying the sections in this chapter, keep in mind the following: The techniques presented here are not meant to replace algebraic methods. They should be seen as a powerful, alternative approach to solving a variety of statements. You will find that they are particularly useful in later sections, when we solve absolute value statements and quadratic inequalities.

In our first example, we will solve a simple linear equation. The graphical method may seem cumbersome, but once you master it, you will find it quite helpful, particularly if you are a visual learner.

| Example 1 | **Solving a Linear Equation Graphically** |

Graphically solve the following equation.

$$2x - 6 = 0$$

We note that this is a one-variable equation. We are interested in values of x that make this true.

Step 1 Let each side of the equation represent a function of x.

$$f(x) = 2x - 6$$

$$g(x) = 0$$

Step 2   Graph the two functions on the same set of axes.

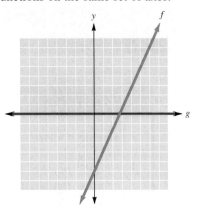

Note: We ask the question, When is the graph of *f* equal to the graph of *g*? Specifically, for what values of *x* does this occur?

Step 3   Find the point of intersection of the two graphs. The *x* coordinate of this point represents the solution to the original equation.

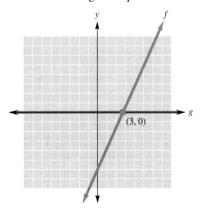

The two lines intersect on the *x* axis at the point $(3, 0)$. Again, because we are solving an equation in one variable (x), we are only interested in *x* values. Thus, the solution is $x = 3$, and the solution set is $\{3\}$.

✔   *CHECK YOURSELF 1*

Graphically solve the following equation.

$$-3x + 6 = 0$$

The same three-step process is used for solving any equation. In Example 2, we look for a point of intersection that is *not* on the *x* axis.

Example 2	**Solving a Linear Equation Graphically**

Graphically solve the following equation.

$$2x - 6 = -3x + 4$$

Step 1   Let each side of the equation represent a function of x.

$$f(x) = 2x - 6$$
$$g(x) = -3x + 4$$

Step 2   Graph the two functions on the same set of axes.

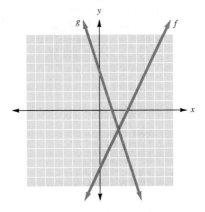

Step 3   Find the point of intersection of the two graphs. Because we want the x coordinate of this point, we suggest the following: Draw a vertical line from the point of intersection $(2, -2)$ to the x axis, marking a point there. This is done to emphasize that we are interested only in the x value: 2. The solution set for the original equation is $\{2\}$.

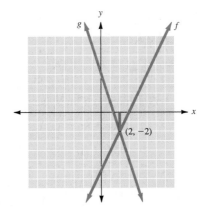

✔ CHECK YOURSELF 2

Graphically solve the following equation.

$$-2x - 5 = x - 2$$

The following algorithm summarizes our work in graphically solving a linear equation.

Solving a Linear Equation Graphically

Step 1.   Let each side of the equation represent a function of x.

Step 2.   Graph the two functions on the same set of axes.

Step 3.   Find the point of intersection of the two graphs. Draw a vertical line from the point of intersection to the x axis, marking a point there. The x value at the indicated point represents the solution to the original equation.

We often apply some algebra even when we are taking a graphical approach. Consider Example 3.

Example 3   Solving a Linear Equation Graphically

Solve the following equation graphically.

$$2(x + 3) = -3x - 4$$

Use the distributive property to rid the left side of parentheses.

$$2x + 6 = -3x - 4$$

Now let

$$f(x) = 2x + 6$$

$$g(x) = -3x - 4$$

Graphing both lines, we get

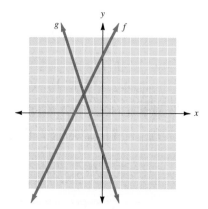

The point of intersection is $(-2, 2)$. Draw a vertical line to the x axis and mark a point. The desired x value is -2. The solution set for the original equation is $\{-2\}$.

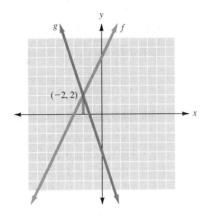

✔ *CHECK YOURSELF 3*

Graphically solve the following equation.

$$3(x - 2) = -4x + 1$$

A graphing calculator can certainly be used to solve equations in the manner just described. Using such a tool, we do not need to apply algebraic ideas such as the distributive property. Let's demonstrate this on the same equation seen in Example 3.

Example 4

Solving a Linear Equation with a Graphing Calculator

Solve the following equation, using a graphing calculator.

$$2(x + 3) = -3x - 4$$

As before, let each side define a function.

$$Y_1 = 2(x + 3)$$
$$Y_2 = -3x - 4$$

When we graph these in the "standard viewing," we see the following:

This window typically shows x values from -10 to 10 and y values from -10 to 10.

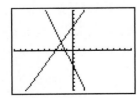
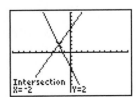

Using the INTERSECT utility, we then see the view shown to the right.

Note that the calculator reports the intersection point as $(-2, 2)$. Since we are interested only in the x value, the solution is $x = -2$. The solution set is $\{-2\}$.

✔ CHECK YOURSELF 4

Solve the following equation, using a graphing calculator.

$$-2x - 5 = x - 2$$

The graphing calculator is particularly effective when we are solving equations with "messy" coefficients. With the INTERSECT utility, we can obtain a solution to any desired level of accuracy. Consider Example 5.

Example 5

Solving a Linear Equation with a Graphing Calculator

Solve the following equation, using a graphing calculator. Give the solution accurate to the nearest hundredth.

$$2.05(x - 4.83) = -3.17(x + 0.29)$$

 In the calculator we define

$$Y_1 = 2.05(x - 4.83)$$

$$Y_2 = -3.17(x + 0.29)$$

In the standard viewing window, we see this:

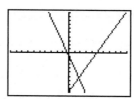 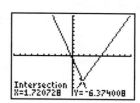

With the INTERSECT utility, we find the intersection point to be $(1.720728, -6.374008)$. Because we want only the x value, the solution (rounded to the nearest hundredth) is 1.72. The solution set is $\{1.72\}$.

 CHECK YOURSELF 5

Solve the following equation, using a graphing calculator. Give the solution accurate to the nearest hundredth.

$$-0.87x + 1.14 = -2.69(x + 4.05)$$

In Example 6, we turn to a business application.

| Example 6 | # Solving a Business Application |

A manufacturer can produce and sell x items per week at the following cost in dollars.

$$C(x) = 30x + 800$$

The revenue from selling those items is given by

$$R(x) = 110x$$

Use a graphical approach to find the break-even point, which is the number of units at which the revenue equals the cost. That is, we wish to graphically solve the equation

$$110x = 30x + 800$$

Graphing the two functions, we have

Try graphing these functions on your graphing calculator.

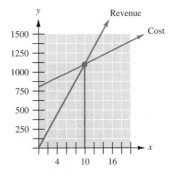

Drawing vertically from the intersection point to the x axis, we see that the desired x value (the break-even point) is 10 items per week. Note that if the company sells more than 10 units, it makes a profit since the revenue exceeds the cost.

✔ CHECK YOURSELF 6

A manufacturer can produce and sell x items per week at a cost of

$$C(x) = 30x + 1,800$$

The revenue from selling those items is given by

$$R(x) = 120x$$

Use a graphical approach to find the break-even point.

✔ *CHECK YOURSELF ANSWERS*

1. $f(x) = -3x + 6$
$g(x) = 0$

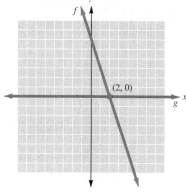

Solution set: {2}

2. $f(x) = -2x - 5$
$g(x) = x - 2$

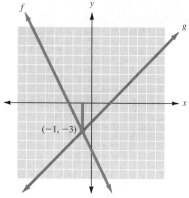

Solution set: {−1}

3. $f(x) = 3(x - 2) = 3x - 6$
$g(x) = -4x + 1$

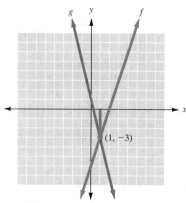

Solution set: {1}

4. $Y_1 = -2x - 5$
$Y_2 = x - 2$

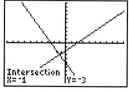

Solution set: {−1}

5. $Y_1 = -0.87x + 1.14$
$Y_2 = -2.69(x + 4.05)$

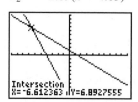

Solution set: {−6.61}

6. 20 items

Exercises · 9.1

Solve the following equations graphically. Do not use a calculator.

1. $2x - 8 = 0$ **2.** $4x + 12 = 0$ **3.** $7x - 7 = 0$ **4.** $2x - 6 = 0$

5. $5x - 8 = 2$ **6.** $4x + 5 = -3$ **7.** $2x - 3 = 7$ **8.** $5x + 9 = 4$

9. $4x - 2 = 3x + 1$ **10.** $6x + 1 = x + 6$ **11.** $\dfrac{7}{5}x - 3 = \dfrac{3}{10}x + \dfrac{5}{2}$ **12.** $2x - 3 = 3x - 2$

13. $3(x - 1) = 4x - 5$ **14.** $2(x + 1) = 5x - 7$ **15.** $7\left(\dfrac{1}{5}x - \dfrac{1}{7}\right) = x + 1$ **16.** $2(3x - 1) = 12x + 4$

Solve the following equations with a graphing calculator. Give your solution to the nearest hundredth.

17. $3.10(x - 2.57) = -4.15(x + 0.28)$

18. $4.17(x + 3.56) = 2.89(x + 0.35)$

19. $5.67(x - 2.13) = 1.14(x - 1.23)$

20. $3.61(x + 4.13) = 2.31(x - 2.59)$

21. The following graph represents the rates that two different car rental agencies charge. The x axis represents the number of miles driven (in hundreds of miles), and the y axis represents the total charge. How would you use this graph to decide which agency to use?

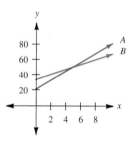

22. Business. A firm producing flashlights finds that its fixed cost is $2,400 per week, and its variable cost is $4.50 per flashlight. The revenue is $7.50 per flashlight, so the cost and revenue equations are, respectively,

$$C(x) = 4.50x + 2,400 \qquad \text{and} \qquad R(x) = 7.50x$$

Find the break-even point for the firm (the point at which the revenue equals the cost). Use a graphical approach.

23. Business. A company that produces portable television sets determines that its fixed cost is $8,750 per month. The variable cost is $70 per set, and the revenue is $105 per set. The cost and revenue equations, respectively, are

$$C(x) = 70x + 8,750 \qquad \text{and} \qquad R(x) = 105x$$

Find the number of sets the company must produce and sell in order to break even. Use a graphical approach.

24. Graphs can be used to solve distance, time, and rate problems because graphs make pictures of the action.

(a) Consider this earlier exercise: "Robert left on a trip, traveling at 45 mi/h. One-half hour later, Laura discovered that Robert forgot his luggage, and so she left along the same route, traveling at 54 mi/h, to catch up with him. When did Laura catch up with Robert?" How could drawing a graph help solve this problem? If you graph Robert's distance as a function of time and Laura's distance as a function of time, what does the slope of each line correspond to in the problem?

(b) Use a graph to solve this problem: Marybeth and Sam left her mother's house to drive home to Minneapolis along the interstate. They drove an average of 60 mi/h. After they had been gone for $\frac{1}{2}$ h, Marybeth's mother realized they had left their laptop computer. She grabbed it, jumped into her car, and pursued the two at 70 mi/h. Marybeth and Sam also noticed the missing computer, but not until 1 h after they had left. When they noticed that it was missing, they slowed to 45 mi/h while they considered what to do. After driving for another $\frac{1}{2}$ h, they turned around and drove back toward the home of Marybeth's mother at 65 mi/h. Where did they pass each other? How long had Marybeth's mother been driving when they met?

(c) Now that you have become experts at this, try solving this problem by drawing a graph. It will require that you think about the slope and perhaps make several guesses when drawing the graphs. If you ride your new bicycle to class, it takes you 1.2 h. If you drive, it takes you 40 min. If you drive in traffic an average of 15 mi/h faster than you can bike, how far away from school do you live? Write an explanation of how you solved this problem by using a graph.

25. Graphing. The family next door to you is trying to decide which health maintenance organization (HMO) to join. One parent has a job with health benefits for the employee only, but the rest of the family can be covered if the employee agrees to a payroll deduction. The choice is between The Empire Group, which would cost the family $185 per month for coverage and $25.50 for each office visit, and Group Vitality, which costs $235 per month and $4.00 for each office visit.

(a) Write an equation showing total yearly costs for each HMO. Graph the cost per year as a function of the number of visits, and put both graphs on the same axes.

(b) Write a note to the family explaining when The Empire Group would be better and when Group Vitality would be better. Explain how they can use your data and graph to help make a good decision. What other issues might be of concern to them?

Solving Linear Inequalities in One Variable Graphically

OBJECTIVE

1. *Solve linear inequalities in one variable graphically*

In Section 9.1, we looked at the graphical approach to solving a linear equation. In this section, we will use the graphs of linear functions to determine the solutions of a linear inequality.

Linear inequalities in one variable x are obtained from linear equations by replacing the symbol for equality ($=$) with one of the inequality symbols ($<$, $>$, $\leq$, $\geq$).

The general form for a linear inequality in one variable is

$$x < a$$

where the symbol $<$ can be replaced with $>$, $\leq$, or $\geq$. Examples of linear inequalities in one variable include

$$x \geq -3 \qquad 2x + 5 > 7 \qquad 2x - 3 \leq 5x + 6$$

Recall that the solution set for an equation is the set of all values for the variable (or ordered pair) that make the equation a true statement. Similarly, the solution set for an inequality is the set of all values that make the inequality a true statement. Example 1 looks at the graphical approach to solving an inequality.

Example 1

Solving a Linear Inequality Graphically

Use a graph to find the solution set to the inequality

$$2x + 5 > 7$$

First, rewrite the inequality as a comparison of two functions. Here $f(x) > g(x)$, in which $f(x) = 2x + 5$ and $g(x) = 7$.

Now graph the two functions on a single set of axes.

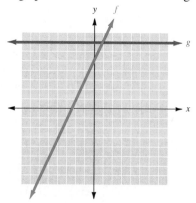

Here we ask the question, For what values of x is the graph of f above the graph of g?

Next, draw a vertical dotted line through the point of intersection of the two functions. In this case, there will be a vertical line through the point $(1, 7)$.

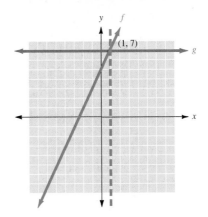

A dotted line is used to indicate that the x value of 1 is not included.

The solution set is every x value that results in $f(x)$ being greater than $g(x)$, which is every x value to the right of the dotted line.

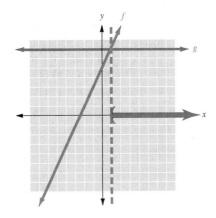

Note: The solution set will be all the x values that make the original statement $2x + 5 > 7$ true.

Finally, we express the solution set in set-builder notation as $\{x \mid x > 1\}$.

  CHECK YOURSELF 1

Solve the inequality $3x - 2 < 4$ graphically.

In Example 1, the function $g(x) = 7$ resulted in a horizontal line. In Example 2, we see that the same method works for comparing any two functions.

Example 2   Solving an Inequality Graphically

Solve the inequality graphically.

$$2x - 3 \geq 5x$$

First, rewrite the inequality as a comparison of two functions. Here, $f(x) \geq g(x)$, $f(x) = 2x - 3$, and $g(x) = 5x$. Now graph the two functions on a single set of axes.

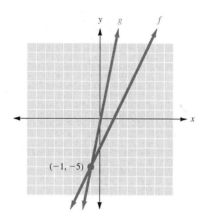

As in Example 1, draw a vertical line through the point of intersection of the two functions. The vertical line will go through the point $(-1, -5)$. In this case, the line is included (greater than or *equal to*), so the line is solid, not dotted.

Again, we need to mark every x value that makes the statement true. In this case, that is every x for which the line representing $f(x)$ is above or intersects the line representing $g(x)$. That is the region in which $f(x)$ is greater than or equal to $g(x)$. We mark the x values to the left of the line, but we also want to include the x value on the line, so we make it a bracket rather than a parenthesis.

A solid line is used to indicate that the x value of -1 is included.

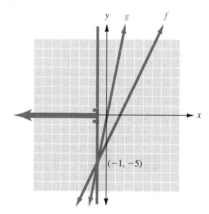

Finally, we express the solutions in set notation. We see that the solution set is every x value less than or equal to -1, so we write

$$\{x \mid x \leq -1\}$$

✔ CHECK YOURSELF 2

Solve the inequality graphically.

$$3x + 2 \geq -2x - 8$$

The following algorithm summarizes our work in this section.

Solving an Inequality in One Variable Graphically

Step 1.   Rewrite the inequality as a comparison of two functions.

$$f(x) < g(x) \qquad f(x) > g(x) \qquad f(x) \leq g(x) \qquad f(x) \geq g(x)$$

Step 2.   Graph the two functions on a single set of axes.

Step 3.   Draw a vertical line through the point of intersection of the two graphs. Use a dotted line if equality is not included (< or >). Use a solid line if equality is included (≤ or ≥).

Step 4.   Mark the x values that make the inequality a true statement.

Step 5.   Write the solutions in set-builder notation.

It is possible for a linear inequality to have no solutions, or to be true for all real numbers. Using graphical methods makes these situations clear.

Example 3   Solving a Linear Inequality Graphically

Solve the following inequalities graphically.

(a) $5 + \dfrac{1}{2}x > \dfrac{x+4}{2}$

Let

$$f(x) = 5 + \frac{1}{2}x = \frac{1}{2}x + 5$$

and

$$g(x) = \frac{x+4}{2} = \frac{1}{2}x + 2$$

We note that the graphs of f and g have the same slope of $\dfrac{1}{2}$ and therefore are parallel. The graphs are shown below.

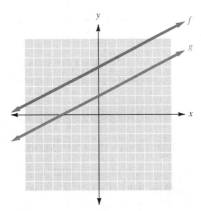

We ask, For what values of x is the graph of f above the graph of g? Clearly, f is *always* above g. So the original statement is true for *all* real numbers, and the solution set is $\mathbb{R}$.

(b) $\dfrac{9 - x}{3} \leq \dfrac{-x - 6}{3}$

Let $$f(x) = \frac{9 - x}{3} = 3 - \frac{x}{3} = -\frac{1}{3}x + 3$$

and $$g(x) = \frac{-x - 6}{3} = -\frac{1}{3}x - 2$$

Again, we note that the graphs of f and g have the same slope.

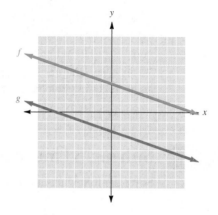

Now we ask, For what values of x is the graph of f below (or equal to) the graph of g? The answer here is "Never!" The original statement is never true, and the solution set is empty, or $\varnothing$.

✔ CHECK YOURSELF 3

Graphically solve the following inequalities.

(a) $2(3 - x) < \dfrac{2 - 4x}{2}$

(b) $2 + \dfrac{2}{5}x \geq \dfrac{2x - 15}{5}$

Solving a business application graphically can be handled effectively with a graphing calculator. This is illustrated in Example 4.

| Example 4 | Solving a Business Application Graphically |

A company's cost per item is $14.29, and its fixed cost per week is $1,735. If the company charges $25.98 per item, what number of items must be manufactured and sold per week for the company to be making a profit?

The company's cost function is

$$C(x) = 14.29x + 1,735$$

and the revenue function is

$$R(x) = 25.98x$$

where x is the number of items made and sold per week.

We want to know when revenue exceeds cost, so we need to solve the following:

$$R(x) > C(x)$$

$$25.98x > 14.29x + 1,735$$

Using a graphical approach, we define functions in the calculator:

$$Y_1 = 25.98x$$

$$Y_2 = 14.29x + 1,735$$

To learn approximately where the two graphs cross, we explore using the TABLE utility:

X	Y1	Y2
0	0	1735
100	2598	3164
200	5196	4593

X=

Note that when $x = 100$, Y_1 (revenue) is lower than Y_2 (cost). But when $x = 200$, Y_1 is higher than Y_2. So the graphs must intersect between 100 and 200 on the x axis, and a suitable viewing window could be $100 \leq x \leq 200$, $2,600 \leq y \leq 5,200$. Using this, we see (figure, below left):

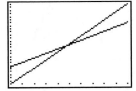 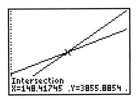

and we find the intersection (figure, above right) at (148.41745, 3,855.8854). Since x needs to be a whole number, we conclude that the company makes a profit when x is at least 149, or when $x \geq 149$.

✔ CHECK YOURSELF 4

A company's cost per item is $8.62, and its fixed cost per week is $1,270. If the company charges $12.95 per item, what number of items must be manufactured and sold per week to make a profit?

✔ *CHECK YOURSELF ANSWERS*

1. $f(x) = 3x - 2$
$g(x) = 4$

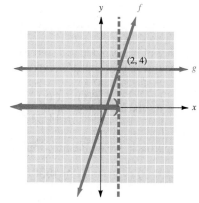

Solution set: $\{x \mid x < 2\}$

2. $f(x) = 3x + 2$
$g(x) = -2x - 8$

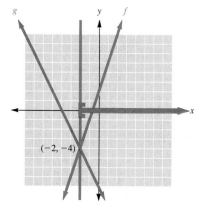

Solution set: $\{x \mid x \geq -2\}$

3. (a) $f(x) = -2x + 6$
$g(x) = -2x + 1$

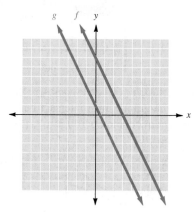

Solution set: $\varnothing$

(b) $f(x) = \dfrac{2}{5}x + 2$

$g(x) = \dfrac{2}{5}x - 3$

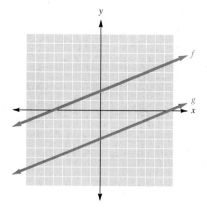

Solution set: $\mathbb{R}$

4. $x \geq 294$ items

Exercises · 9.2

In Exercises 1 to 16, solve the linear inequalities graphically.

1. $2x < 8$

2. $-x < 4$

3. $\dfrac{x+3}{2} < -1$

4. $\dfrac{-3x+3}{4} > -3$

5. $6x \geq 6$

6. $-3x \leq 6$

7. $7x - 7 < -2x + 2$

8. $7x + 2 > x - 4$

9. $\dfrac{14x+4}{3} > 2(4x - 1)$

10. $2(3x + 1) < 4(x + 1)$

11. $6(1 + x) \geq 2(3x - 5)$

12. $2(x - 5) \geq 2x - 1$

13. $7x > \dfrac{9x-5}{2}$

14. $-4x - 12 < x + 8$

15. $4x - 6 \leq 2x - 2(5x - 12)$

16. $5x + 3 > 2(4 - x) + 7x$

In Exercises 17 to 22, solve the applications.

17. Business. The cost to produce x units of wire is $C(x) = 50x + 5{,}000$, and the revenue generated is $R(x) = 60x$. Find all values of x for which the product will at least break even.

18. Business. Find the values of x for which a product will at least break even if the cost is $C(x) = 85x + 900$ and the revenue is given by $R(x) = 105x$.

19. Car rental. Tom and Jean went to Salem, Massachusetts, for 1 week. They needed to rent a car for a week, so they checked out two rental firms. Wheels, Inc., wanted $28 per day with no mileage fee. Downtown Edsel wanted $98 per week and 14¢ per mile. Set up equations to express the rates of the two firms, and then decide when each deal should be taken.

20. Mileage. A fuel company has a fleet of trucks. The annual operating cost per truck is $C(x) = 0.58x + 7,800$, where x is the number of miles traveled by a truck per year. What number of miles will yield an operating cost that is less than $25,000?

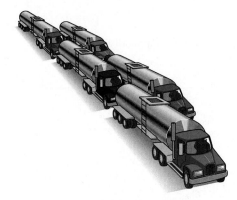

21. Wedding. Amanda and Joe are having their wedding reception at the Richland Fire Hall. They can spend at the most $3,000 for the reception. If the hall charges a $250 cleanup fee plus $25 per person, find the greatest number of people they can invite.

22. Tuition. A nearby college charges annual tuition of $6,440. Meg makes no more than $1,610 per year in her summer job. What is the fewest number of summers that she must work in order to make enough for 1 year's tuition?

23. Graphing. Explain to a relative how a graph is helpful in solving each inequality below. Be sure to include the significance of the point at which the lines meet (or what happens if the lines do not meet).

  (a) $3x - 2 < 5$ (b) $3x - 2 \le 4 - x$ (c) $4(x - 1) \ge 2 + 4x$

24. College. Look at the data here about enrollment in college. Assume that the changes occurred at a constant rate over the years. Make one linear graph for men and one for women, but on the same set of axes. What conclusions could you draw from reading the graph?

Year	Number (in Millions) of Men in the United States Enrolled in College	Number (in Millions) of Women in the United States Enrolled in College
1960	2.3	1.2
1998	6.4	7.8

9.3 Solving Absolute Value Equations Graphically

1. Graph an absolute value function
2. Solve absolute value equations in one variable graphically

In Section 2.7 we learned to solve absolute value equations algebraically. In this section, we will examine a graphical method for solving similar equations.

To demonstrate the graphical method, we will first look at the graph of an absolute value function. We will start by looking at the graph of the function $f(x) = |x|$. All other graphs of absolute value functions are variations of this graph.

The graph can be found using a graphing calculator (most graphing calculators use $\boxed{\text{abs}}$ to represent the absolute value). We will develop the graph from a table of values.

Graph the function

$$y = |x|$$

as $Y_1 = \text{abs}(x)$

x	$f(x) = |x|$
-3	3
-2	2
-1	1
0	0
1	1
2	2

Plotting these ordered pairs, we see a pattern emerge. The graph is like a large V that has its vertex at the origin. The slope of the line to the right of 0 is 1, and the slope of the line to the left of 0 is -1.

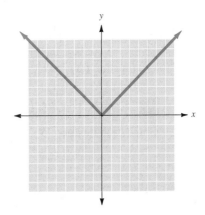

Let us now see what happens to the graph when we add or subtract some constant inside the absolute value bars.

Example 1

Graphing an Absolute Value Function

Graph each function.

The equation

$$f(x) = |x - 3|$$

would be entered as

$$Y_1 = abs(x - 3)$$

(a) $f(x) = |x - 3|$

Again, we start with a table of values.

x	$f(x)$
-2	5
-1	4
0	3
1	2
2	1
3	0
4	1
5	2

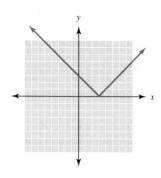

Then we plot the points associated with the set of ordered pairs. The graph is shown to the left.

    The graph of the function $f(x) = |x - 3|$ is the same shape as the graph of the function $f(x) = |x|$; it has just shifted to the right 3 units.

(b) $f(x) = |x + 1|$

We begin with a table of values.

x	$f(x)$
-2	1
-1	0
0	1
1	2
2	3
3	4

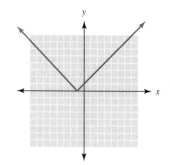

Again, you will find the graph in the margin.

    Note that the graph of $f(x) = |x + 1|$ is the same shape as the graph of the function $f(x) = |x|$, except that it has shifted 1 unit to the left.

✔ CHECK YOURSELF 1

Graph each function.

(a) $f(x) = |x - 2|$ (b) $f(x) = |x + 3|$

We can summarize what we have discovered about the horizontal shift of the graph of an absolute value function.

Note: If a is negative, $x - a$ will be x plus some positive number. For example, if $a = -2$, $x - a = x - (-2) = x + 2$.

The graph of the function $f(x) = |x - a|$ will be the same shape as the graph of $f(x) = |x|$ except that the graph will be shifted a units

    To the right if a is positive

    To the left if a is negative

We will now use these methods to solve equations that contain an absolute value expression.

Example 2	**Solving an Absolute Value Equation Graphically**

Graphically find the solution set for the equation

$$|x - 3| = 4$$

We graph the function associated with each side of the equation.

$$f(x) = |x - 3| \qquad \text{and} \qquad g(x) = 4$$

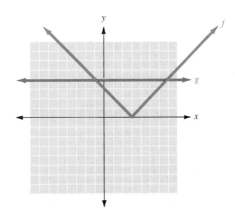

Then we draw a vertical line through each of the intersection points.

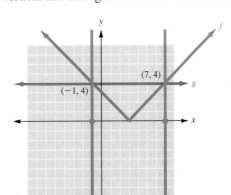

We ask the question, For what values of x do f and g coincide?

Looking at the x values of the two vertical lines, we find the solutions to the original equation. There are two x values that make the statement true: -1 and 7. The solution set is $\{-1, 7\}$.

✔ CHECK YOURSELF 2

Graphically find the solution set for the equation.

$$|x - 2| = 3$$

Example 3 illustrates a case involving an absolute value expression and a linear (but not simply constant) expression.

Example 3

Solving an Absolute Value Equation Graphically

Solve the following equation graphically.

$$|x - 2| = 4 - \frac{1}{2}x$$

Let

$$f(x) = |x - 2|$$

and

$$g(x) = 4 - \frac{1}{2}x = -\frac{1}{2}x + 4$$

We graph each function.

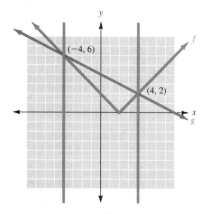

Draw a vertical line through each point of intersection. The vertical lines hit the x axis when $x = -4$ and when $x = 4$. There are therefore two solutions, and the solution set is $\{-4, 4\}$.

✔ CHECK YOURSELF 3

Solve the following equation graphically.

$$|x + 3| = \frac{1}{3}x + 5$$

In each of the examples of this section, we have found two solutions. It is also quite possible for an absolute value equation to have one solution, no solution, or even an infinite number of solutions. Fortunately, the situation will be clear if we are employing graphical methods.

Example 4

Solving an Absolute Value Equation Graphically

Solve the following equation graphically.

$$|x + 2| = 2x + 7$$

Let $f(x) = |x + 2|$

and $g(x) = 2x + 7$

Graph the functions.

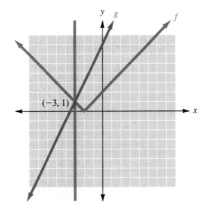

Since the slope of g is 2, we are convinced that the graph of g will meet the graph of f exactly once. The point of intersection is $(-3, 1)$. Drawing a vertical through this point, we note the x value of -3. The solution set is $\{-3\}$.

✔ CHECK YOURSELF 4

Solve the following equation graphically.

$$|x - 4| = 6 - 2x$$

If we are creating graphs by hand, it can be very difficult to locate the point(s) of intersection. But as you gain experience with a graphing calculator (changing the viewing window and finding intersection points), you will find that you can apply the graphical methods shown here to solve "messy" equations with great accuracy.

✔ CHECK YOURSELF ANSWERS

1. (a)

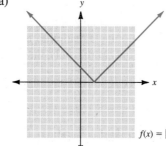

$f(x) = |x - 2|$

(b)

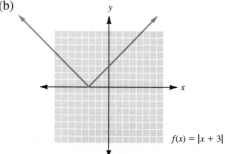

$f(x) = |x + 3|$

2.

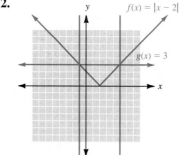

$f(x) = |x - 2|$

$g(x) = 3$

Solution set: $\{-1, 5\}$

3.

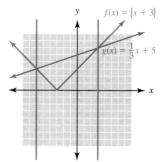

$f(x) = |x + 3|$

$g(x) = \frac{1}{3}x + 5$

Solution set: $\{-6, 3\}$

4.

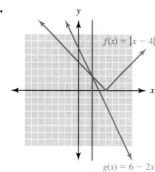

$f(x) = |x - 4|$

$g(x) = 6 - 2x$

Solution set: $\{2\}$

In Exercises 1 to 6, graph each function.

1. $f(x) = |x - 3|$

2. $f(x) = |x + 2|$

3. $f(x) = |x + 3|$

4. $f(x) = |x - 4|$

5. $f(x) = |x - (-3)|$

6. $f(x) = |x - (-5)|$

In Exercises 7 to 12, solve the equations graphically.

7. $|x| = 3$

8. $|x| = 5$

9. $|x - 2| = 5$

10. $|x - 5| = 3$

11. $|x + 2| = 4$

12. $|x + 4| = 2$

13. $|x - 3| = 5 - \dfrac{1}{3}x$

14. $|x + 1| = 4 - 2x$

15. $|x + 3| = \dfrac{1}{3}x + \dfrac{14}{3}$

16. $|x + 1| = \dfrac{1}{3}x + 5$

Determine the function represented by each graph.

17.

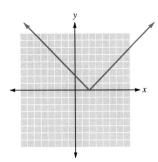

18.

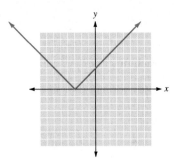

19.

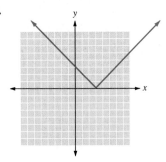

20.

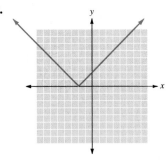

Graph each function. (*Hint:* It may be useful to make a table of values.)

21. $f(x) = -|x|$ **22.** $f(x) = -|x - 2|$ **23.** $f(x) = -|x + 3|$

24. What observations can you make concerning the graphs in Exercises 21 to 23? Based on these observations, make up a new function similar to those, predict the shape and position of the graph, and then create the graph to check your predictions.

Graph each function. (*Hint:* It may be useful to make a table of values.)

25. $f(x) = |x| - 3$ **26.** $f(x) = |x| + 2$ **27.** $f(x) = |x + 1| + 4$

28. $f(x) = |x + 2| - 3$

29. What observations can you make concerning the graphs in Exercises 25 to 28? Based on these observations, make up a new function similar to those, predict the shape and position of the graph, and then create the graph to check your predictions.

In Exercises 30 to 34, graph each function. (*Hint:* It may be useful to make a table of values.)

30. $f(x) = 2|x|$ **31.** $f(x) = 3|x|$ **32.** $f(x) = \dfrac{1}{2}|x|$

33. $f(x) = -2|x|$ **34.** $f(x) = -\dfrac{1}{2}|x|$

35. What observations can you make concerning the graphs in Exercises 30 to 34? Based on these observations, make up a new function similar to those, predict the shape and position of the graph, and then create the graph to check your predictions.

Use graphing techniques to solve each equation.

36. $|x - 2| = |x + 4|$ **37.** $|x - 3| = |x + 5|$

38. $|x - 1| = |x + 5|$ **39.** $|x - 3| = |x + 1|$

Solving Absolute Value Inequalities Graphically

OBJECTIVES

1. *Solve absolute value inequalities in one variable graphically*
2. *Solve absolute value inequalities in one variable algebraically*

In Section 9.3, we looked at a graphical method for solving an absolute value equation. In this section, we will look at a graphical method for solving absolute value inequalities.

Absolute value inequalities in one variable x are obtained from absolute value equations by replacing the symbol for equality (=) with one of the inequality symbols ($<, >, \leq, \geq$).

The general form for an absolute value inequality in one variable is

$$|x - a| < b$$

where the symbol $<$ can be replaced with $>$, $\leq$, or $\geq$. Examples of absolute value inequalities in one variable include

$$|x| < 6 \qquad |x - 4| \geq 2 \qquad |3x - 5| \leq 8$$

Example 1

Solving an Absolute Value Inequality Graphically

Graphically solve

$$|x| < 6$$

As we did in previous sections, we begin by letting each side of the inequality represent a function. Here

$$f(x) = |x| \qquad \text{and} \qquad g(x) = 6$$

Now we graph both functions on the same set of axes.

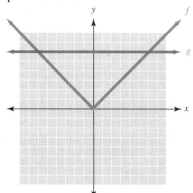

We now ask the question, For what values of x is f *below* g?

768

We next draw a vertical dotted line (equality is not included) through the points of intersection of the two graphs.

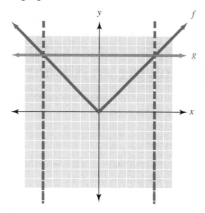

The solution set is any value of x for which the graph of $f(x)$ is below the graph of $g(x)$.

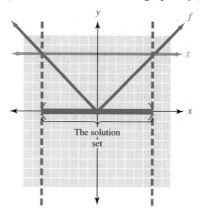

Keep in mind that we are searching for x values that make the original statement true.

In set notation, we write $\{x \mid -6 < x < 6\}$.

✔ *CHECK YOURSELF 1*

Graphically solve the inequality.

$$|x| < 3$$

The graphical method of Example 1 relates to the following general statement from Section 2.8.

Absolute Value Inequalities—Property 1

For any positive number p, if

$$|x| < p$$

then
$$-p < x < p$$

Before we continue with a graphical approach, let's review the use of this property in solving an absolute value inequality.

| **Example 2** | **Solving an Absolute Value Inequality Algebraically** |

Solve the following inequality algebraically and graph the solution set on a number line.

$$|x - 3| < 5$$

With property 1 we can *translate* an absolute value inequality to an inequality *not* containing an absolute value, which can be solved by our earlier methods.

From property 1, we know that the given absolute value inequality is equivalent to the compound inequality

$$-5 < x - 3 < 5$$

Solve as before.

$$-5 < x - 3 < 5 \qquad \text{Add 3 to all three parts.}$$
$$-2 < x < 8$$

The solution set is

$$\{x \mid -2 < x < 8\}$$

Note that the solution set is an open interval on the number line.

The graph of the solution set is shown below.

✔ *CHECK YOURSELF 2*

Solve the following inequality algebraically and graph the solution set on a number line.

$$|x - 4| \leq 8$$

In Example 3, we will look at a graphical method for solving the same inequality.

| **Example 3** | **Solving an Absolute Value Inequality Graphically** |

Solve the following inequality graphically, and graph the solution set on a number line.

$$|x - 3| < 5$$

Let $f(x) = |x - 3|$ and $g(x) = 5$, and graph both functions on the same set of axes.

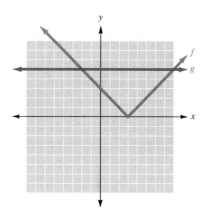

Here again, we ask, For what values of x is f below g?

Drawing a vertical dotted line through the intersection points, we find the set of x values for which $f(x) < g(x)$.

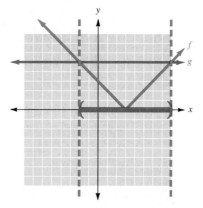

We see that the desired x values are those which lie between -2 and 8. The solution set is $\{x \mid -2 < x < 8\}$. The graph of the solution set is

Note that the graph of the solution set shown here is precisely the portion of the x axis that has been marked in the two-dimensional graph above.

✔ *CHECK YOURSELF 3*

Solve the following inequality graphically, and graph the solution set on a number line.

$$|x - 4| \leq 8$$

We have seen that the solution set for the statement $|x| < 6$ is the set of all numbers between -6 and 6. Now, how does the result change for the statement $|x| > 6$? Solving graphically will make this clear.

Example 4	Solving an Absolute Value Inequality Graphically

Solve the following inequality graphically, and graph the solution set on a number line.

$$|x| > 6$$

As before, we define

$$f(x) = |x| \quad \text{and} \quad g(x) = 6$$

We graph both functions on the same set of axes, and we draw vertical dotted lines through the points of intersection

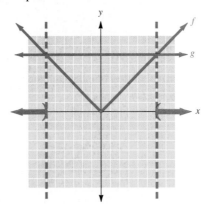

Now we ask, For what values of x is the graph of f *above* the graph of g?

We see that f is *above* g when $x < -6$ or when $x > 6$. The solution set is $\{x \mid x < -6$ or $x > 6\}$. The graph of the solution set is

✔ *CHECK YOURSELF 4*

Solve the following inequality graphically, and graph the solution set on a number line.

$$|x| > 3$$

The solution set for the statement $|x| > 6$ is the set of all numbers that are greater than 6 or less than -6. We can describe these numbers with the compound inequality

$$x < -6 \quad \text{or} \quad x > 6$$

This relates to the following general statement from Section 2.8:

Absolute Value Inequalities—Property 2

For any positive number p, if

$$|x| > p$$

then $x < -p \quad$ or $\quad x > p$

We review now the use of this property in solving an absolute value inequality.

<table>
<tr><td>**Example 5**</td><td></td></tr>
</table>

Solving an Absolute Value Inequality Algebraically

Solve the following inequality algebraically, and graph the solution set on a number line.

$$|2 - x| > 8$$

From property 2, we know that the given absolute inequality is equivalent to the compound inequality

$$2 - x < -8 \qquad \text{or} \qquad 2 - x > 8$$

Again we translate the absolute value inequality to the compound inequality not containing an absolute value.

Solving as before, we have

$$
\begin{array}{ccc}
2 - x < -8 & \text{or} & 2 - x > 8 \\
-x < -10 & & -x > 6 \\
x > 10 & & x < -6
\end{array}
$$

When we divide by a negative number, we reverse the inequality.

The solution set is $\{x \mid x < -6 \text{ or } x > 10\}$, and the graph of the solution set is shown below.

✔ *CHECK YOURSELF 5*

Solve the following inequality algebraically, and graph the solution set on a number line.

$$|3 - x| \geq 4$$

A property that can be useful in working with absolute values is given below.

Absolute Value Expressions—Property 3

For any real numbers a and b

$$|a - b| = |b - a|$$

Our final example looks at a graphical approach to solving the same inequality studied in Example 5.

Example 6

Solving an Absolute Value Inequality Graphically

Solve the following inequality graphically, and graph the solution set on a number line.

$$|2 - x| > 8$$

Let $f(x) = |2 - x|$ and $g(x) = 8$. Using property 3, we know that $|2 - x| = |x - 2|$, so let's define f as

$$f(x) = |x - 2|$$

Graphing f and g on the same set of axes, we have

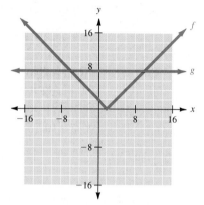

Drawing a vertical dotted line through each intersection point, we mark (on the x axis) all x values for which $f(x) > g(x)$.

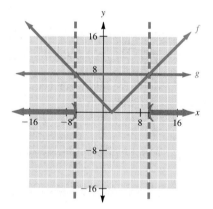

The solution set is therefore $\{x \mid x < -6 \text{ or } x > 10\}$, and the graph of the solution set is

✔ *CHECK YOURSELF 6*

Graphically solve the inequality, and graph the solution set on a number line.

$$|3 - x| \geq 4$$

✔ *CHECK YOURSELF ANSWERS*

1. $\{x \mid -3 < x < 3\}$

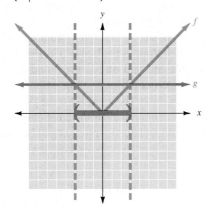

2. $\{x \mid -4 \le x \le 12\}$

3. $\{x \mid -4 \le x \le 12\}$

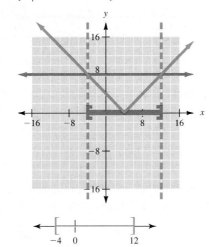

4. $\{x \mid x < -3 \text{ or } x > 3\}$

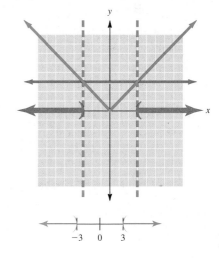

5. $\{x \mid x \le -1 \text{ or } x \ge 7\}$

6. $\{x \mid x \le -1 \text{ or } x \ge 7\}$

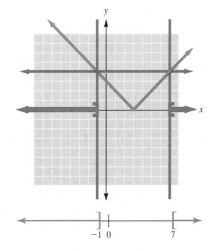

Determining Heating Costs.   Ed and Sharon's heating system has just broken down. They can replace their current electric heat pump system with another electric heat pump system or replace it with an oil heating system or a gas system. Ed knows that in order to install an oil heating system, he will need a new air conditioner and extensive duct remodeling in his house. He also knows that a gas heating system will require that gas lines be laid on his street and then connected to the house. Both Ed and Sharon are concerned about installing another heat pump system because it does not seem to work well in their cold Northeast climate. They both want to install a system that will be economical in the long run. Ed did some research and collected installation costs and yearly operational costs. These data appear in the following chart.

Type of Heating	Installation Costs	Operational Costs (per Year)
Reinstall an electric system	$ 4,200	$900
Convert to oil system	8,500	550
Convert to gas system	13,000	425

1. Use 5-year intervals to construct a table that shows the total cost of heating over a 35-year period for the three heating systems.

2. Determine the average yearly heating costs for each of the three systems after (a) 5 years, (b) 10 years, (c) 15 years, (d) 20 years, (e) 25 years, (f) 30 years, and (g) 35 years.

3. The total cost of each system is a function composed of the operational costs multiplied by the number of years of operation added to the installation costs. For each of the three systems, develop a function that represents the total cost.

4. Graph the three straight lines developed in step 3 on the same coordinate axis.

5. Using the graphs in step 4, estimate the following:
    (a) When does the oil system become cheaper than the electric system?
    (b) When does the gas system become cheaper than the electric system?
    (c) When does the gas system become cheaper than the oil system?

6. Using the functions developed in step 3, determine the following:
    (a) When does the oil system become cheaper than the electric system?
    (b) When does the gas system become cheaper than the electric system?
    (c) When does the gas system become cheaper than the oil system?

7. Do the following:
    (a) Contact your local electric and gas utility and obtain the current rate schedule as well as the cost of installing a new heating system that would be adequate for a 2,000 ft² home with 8-ft ceilings.
    (b) Contact a local oil company and obtain the current prices as well as installation costs of a new heating system that would be adequate for a 2,000 ft² home with 8-ft ceilings.
    (c) Use the information from (a) and (b) and construct a table similar to that developed in step 1.
    (d) Develop a function that represents the total heating costs for each of the three systems based on the information you obtained in (a) and (b).
    (e) Based solely on the costs associated with the heating systems, which one would you choose to install in your house?

8. In addition to cost, what other factors would you consider in choosing a new heating system for your home?

9. Considering all the factors involved, decide what type of heating system you would pick for a new home you are building, and write a paragraph explaining your decision.

Exercises · 9.4

In Exercises 1 to 16, solve each inequality algebraically. Graph the solution set on a number line.

1. $|x| < 5$

2. $|x| > 3$

3. $|x| \geq 7$

4. $|x| \leq 4$

5. $|x - 4| > 2$

6. $|x + 5| < 3$

7. $|x + 6| \leq 4$

8. $|x - 7| > 5$

9. $|3 - x| > 5$

10. $|5 - x| < 3$

11. $|x - 7| < 0$

12. $|x + 5| \geq 0$

13. $|2x - 5| < 3$

14. $|3x - 1| > 8$

15. $|3x + 4| \geq 5$

16. $|2x + 3| \leq 9$

In Exercises 17 to 28, solve each inequality graphically.

17. $|x| < 4$

18. $|x| < 6$

19. $|x| \geq 5$

20. $|x| \geq 2$

21. $|x - 3| < 4$　　　　**22.** $|x - 1| < 5$　　　　**23.** $|x - 2| \geq 5$　　　　**24.** $|x + 2| > 4$

25. $|x + 1| \leq 5$　　　　**26.** $|x + 4| > 1$　　　　**27.** $|x + 2| \geq -2$　　　　**28.** $|x - 4| > -1$

29. Manufacturing.　Combustion engines get their power from the force exerted by burning fuel on a piston inside a chamber. The piston is forced down out of the cylinder by the force of a small explosion caused by burning fuel mixed with air. The piston in turn moves a piston rod, which transfers the motion to the work of the engine. The rod is attached to a flywheel, which pushes the piston back into the cylinder to begin the process all over. Cars usually have four to eight of these cylinders and pistons. It is crucial that the piston and the cylinder fit well together, with just a thin film of oil separating the sides of the piston and the sides of the cylinder. When these are manufactured, the measurements for each part must be accurate. But there is always some error. How much error is a matter for the engineers to set and for the quality control department to check.

Suppose the diameter of the cylinder is meant to be 7.6 cm, and the engineer specifies that this part must be manufactured to within 0.1 mm of that measurement. This figure is called the **tolerance.** As parts come off the assembly line, someone in quality control takes samples and measures the cylinders and the pistons. Given this information, complete the following.

(a) Write an absolute value statement about the diameter d_c of the cylinder.

(b) If the diameter of the piston is to be 7.59 cm with a tolerance of 0.1 mm, write an absolute value statement about the diameter d_p of the piston.

(c) Investigate all the possible ways these two parts will fit together. If the two parts have to be within 0.1 mm of each other for the engine to run well, is there a problem with the way the parts may be paired together? Write your answer and use a graph to explain.

(d) Accuracy in machining the parts is expensive, so the tolerance should be close enough to make sure the engine runs correctly, but not so close that the cost is prohibitive. If you think a tolerance of 0.1 mm is too large, find another that you think would work better. If it is too small, how much can it be enlarged and still have the engine run according to design (that is, so $|d_c - d_p| \leq 0.1$ mm)? Write the tolerance, using absolute value signs. Explain your reasoning if you think a tolerance of 0.1 mm is not workable.

(e) After you have decided on the appropriate tolerance for these parts, think about the quality control engineer's job. Hazard a few educated opinions to answer these questions: How many parts should be pulled off the line and measured? How often? How many parts can reasonably be expected to be outside the expected tolerance before the whole line is shut down and the tools are corrected?

Summary for Chapter 9

Example	Topic	Reference
	Solving Equations in One Variable Graphically	9.1
To solve $2x - 6 = 8x$, let $f(x) = 2x - 6$ $\quad g(x) = 8x$ then graph both lines. The intersection occurs when $x = -1$.	**Finding a Graphical Solution for an Equation** **Step 1** Let each side of the equation represent a function of x. **Step 2** Graph the two functions on the same set of axes. **Step 3** Find the point of intersection of the two graphs. The x value at this indicated point represents the solution to the original equation.	p. 742
	Solving Linear Inequalities in One Variable Graphically	9.2
To solve $2x - 6 > 8x$, let $f(x) = 2x - 6$ $\quad g(x) = 8x$ then graph both lines.	**Finding the Solution for an Inequality in One Variable** **Step 1** Rewrite the inequality as a comparison of two functions. $f(x) < g(x) \quad f(x) > g(x) \quad f(x) \leq g(x) \quad f(x) \geq g(x)$ **Step 2** Graph the two functions on a single set of axes. **Step 3** Draw a vertical line through each point of intersection of the two functions. Use a dotted line if equality is not included ($<$ or $>$). Use a solid line if equality is included ($\leq$ or $\geq$).	p. 753

Draw a dotted line where they intersect.

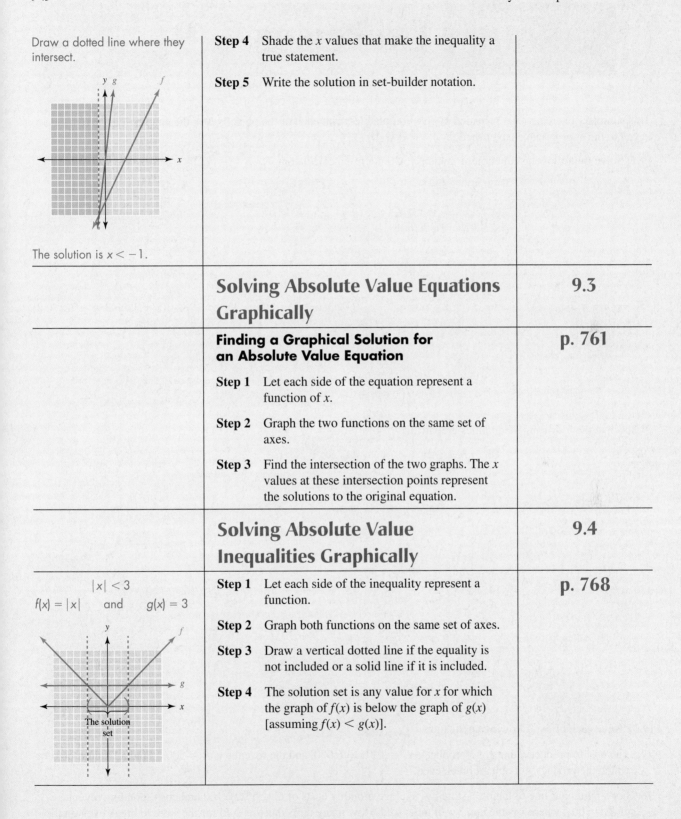

The solution is $x < -1$.

Step 4 Shade the x values that make the inequality a true statement.

Step 5 Write the solution in set-builder notation.

	Solving Absolute Value Equations Graphically	**9.3**
	Finding a Graphical Solution for an Absolute Value Equation	**p. 761**
	Step 1 Let each side of the equation represent a function of x.	
	Step 2 Graph the two functions on the same set of axes.	
	Step 3 Find the intersection of the two graphs. The x values at these intersection points represent the solutions to the original equation.	
	Solving Absolute Value Inequalities Graphically	**9.4**
	Step 1 Let each side of the inequality represent a function.	**p. 768**
	Step 2 Graph both functions on the same set of axes.	
	Step 3 Draw a vertical dotted line if the equality is not included or a solid line if it is included.	
	Step 4 The solution set is any value for x for which the graph of $f(x)$ is below the graph of $g(x)$ [assuming $f(x) < g(x)$].	

$|x| < 3$

$f(x) = |x|$ and $g(x) = 3$

Summary Exercises · 9

This summary exercise set is provided to give you practice with each of the objectives in the chapter. Each exercise is keyed to the appropriate chapter section.

[9.1] Solve the following equations graphically. Do not use a calculator.

1. $3x - 6 = 0$

2. $4x + 3 = 7$

3. $3x + 5 = x + 7$

4. $4x - 3 = x - 6$

5. $\dfrac{6x - 1}{2} = 2(x - 1)$

6. $3x + 2 = 2x - 1$

7. $3(x - 2) = 2(x - 1)$

8. $3(x + 1) + 3 = -7(x + 2)$

[9.2] Solve the following inequalities graphically.

9. $5 < 2x + 1$

10. $-3 \geq -2x + 3$

11. $3x + 2 \geq 6 - x$

12. $3x - 5 < 15 - 2x$

13. $x + 6 < -2x - 3$

14. $4x - 3 \geq 7 - x$

15. $x \geq -3 - 5x$

16. $x < 4 + 2x$

[9.2] Solve each of the following applications.

17. The cost to produce x units of a product is $C = 100x + 6{,}000$, and the revenue is $R = 500x$. Find all values of x for which the product will at least break even.

18. For a particular line of lamps, a store has average monthly costs of $C = 55x + 180$ and corresponding revenue of $R = 100x$, where x is the number of units sold. How many units must be sold for the store to break even on the lamps for the month?

782

[9.3] Solve the following equations graphically.

19. $|x + 3| = 5$

20. $|x - 2| = 7$

21. $|2 - x| = 3$

22. $|4 - x| = 2$

23. $|x + 2| = 4 + \dfrac{1}{2}x$

24. $|x - 3| = 5 - \dfrac{1}{3}x$

Solve the following inequalities graphically.

25. $|x| < 7$

26. $|x| \le 6$

27. $|x - 1| > 6$

28. $|x + 3| \le 4$

29. $|x + 5| \ge 2$

30. $|x - 1| < 8$

[9.4] In Exercises 31 to 35, solve each inequality algebraically. Graph the solution on a number line.

31. $|x| \le 4$

32. $|x| > 3$

33. $|x - 1| \le 2$

34. $|x + 2| \ge 1$

35. $|x - 1| < 0$

The purpose of this self-test is to help you check your progress and to review for a chapter test in class. Allow yourself about 1 hour to take the test. When you are done, check your answers in the back of the book. If you missed any answers, be sure to go back and review the appropriate sections in the chapter and the exercises that are provided.

Solve the following equations graphically.

1. $4x - 7 = 5$

2. $6 - x = 4(x - 1)$

3. $-8x + 11 = 2x - 9$

4. $6(x - 1) = -3(x - 4)$

Solve the following inequalities graphically.

5. $5x - 3 < 7$

6. $2x - 1 \leq 3(x - 1)$

Solve the following equations graphically.

7. $|x + 3| = 7$

8. $|x - 5| = 9$

Solve the following inequalities graphically.

9. $|x| \leq 3$

10. $|x + 1| \geq 4$

11. $|x - 1| > 5$

12. $|x + 2| < 4$

13. $|x - 1| < 4$

14. $|3 + x| \geq 2$

15. $|3 - x| \leq 5$

16. $|2 - x| > 5$

Solve the following applications.

17. By traveling 30 mi/h for one period of time and 40 mi/h for another, Maria traveled 230 mi. Had she gone 10 mi/h faster throughout, she would have traveled 300 mi. How many miles did she travel at each rate?

18. A company that manufactures boxed mints sells each box for $3. The company incurs a cost of $1.50 per box with a total fixed cost of $30,000. How many boxes of mints must be sold for the company to break even?

Solve the following inequalities for x.

19. $|2x + 1| < 8$

20. $|3x - 2| \geq 9$

Solve the following equations algebraically.

1. $4x - 2(x + 1) = -5 - (x + 1)$

2. $|x - 3| = 5$

3. If $f(x) = 4x^3 - 3x^2 + 5x - 9$, find $f(-1)$.

4. Find the equation of the line that is perpendicular to $3x + 2y = 6$ and has a y intercept of -2.

5. Simplify $4x - 3(x + 1) + 6x(x - 1) - 2x^2$.

Factor completely.

6. $x^3 - 9x$

7. $2x^2 - 5x - 3$

8. $x^2 - 2x - 10 + 5x$

Simplify.

9. $\dfrac{4}{x + 1} - \dfrac{3}{x + 1}$

10. $\dfrac{x + 1}{x^2 + x - 6} - \dfrac{x + 2}{x + 3}$

11. $\dfrac{2x - 3}{x^2 - 16} \div \dfrac{2x^2 - x - 3}{x + 4}$

12. Solve the following system.

$$3x + 2y = 2$$
$$x + 3y = -4$$

13. Find the x and y intercepts of the equation $4x + 2y = 12$.

Solve the following inequalities algebraically.

14. $x - 3 \le 2x + 5$

15. $|x - 3| \le 6$

16. $-3\,|x + 1| < -6$

If $f(x) = \dfrac{1}{x - 5}$ and $g(x) = x + 6$, find the following.

17. $f + g$

18. $f \div g$

Graph the following functions.

19. $f(x) = -\dfrac{2}{3}x + 5$

20. $f(x) = |x + 4|$

Solve the following equations graphically.

21. $2x + 1 = 3x + 4$

22. $|x + 4| = 5$

Solve the following inequalities graphically.

23. $4x - 5 < 3$

24. $|x - 2| \le 3$

25. $|x + 1| \ge 1$

10 Radicals and Exponents

As we have all experienced firsthand, consumer goods increase in price from year to year. This increase is usually measured by the Consumer Price Index (CPI), which measures a change in the prices of such everyday goods and services as energy, food, shelter, apparel, transportation, medical care, and utilities. The percentage change in the CPI is a reflection of the purchasing power of the dollar and indicates the rate of inflation.

Since many labor contracts and government benefits programs such as Social Security increase or decrease along with the CPI, the method used to calculate this index is hotly debated by economists and statisticians. Beginning in April 1997, the Bureau of Labor Statistics began releasing an experimental CPI that uses a **geometric mean formula.** This new method may more accurately reflect the true cost-of-living increase or decrease because it takes into consideration that consumers' buying habits change as prices fluctuate. For instance, consumers may switch from romaine lettuce to iceberg lettuce or spinach if the price of romaine lettuce is too high.

The formula for a **geometric mean** involves radicals and rational exponents, topics that we will study in this chapter.

10.1 Roots and Radicals

1. Use radical notation
2. Evaluate expressions that contain radicals
3. Use a calculator to estimate or evaluate radical expressions
4. Simplify expressions that contain radicals

In Chapter 4 we discussed the properties of integer exponents. Over the next several sections, we will be working toward an extension of those properties. To achieve that objective, we must develop a notation that "reverses" the power process.

A statement such as

$$x^2 = 9$$

is read as "x squared equals 9."

In this section we are concerned with the relationship between the base x and the number 9. Equivalently, we can say that "x is the square root of 9."

We know from experience that x must be 3 (because $3^2 = 9$) or -3 [because $(-3)^2 = 9$]. We see that 9 has the two square roots, 3 and -3. In fact, every positive number has *two* square roots, one positive and one negative. In general,

Note: We will see later that a negative number has *no* real square roots.

If $x^2 = a$, we say x is a *square root* of a.

We also know that

$$3^3 = 27$$

and similarly we call 3 a *cube root* of 27. Here 3 is the *only* real number with that property. Every real number (positive or negative) has *one* real cube root.

Roots

In general, we can state that if

$$x^n = a$$

then x is an *nth root* of a.

Note: The symbol $\sqrt{}$ first appeared in print in 1525. In Latin, "radix" means root, and this was contracted to a small r. The present symbol may have been used because it resembled the manuscript form of that small r.

We are now ready for new notation. The symbol $\sqrt{}$ is called a *radical sign*. We saw above that 3 was the positive square root of 9.

We call 3 the *principal square root* of 9, and we write

$$\sqrt{9} = 3$$

In some applications we will want to indicate the negative square root; to do so, we must write

$$-\sqrt{9} = -3$$

to indicate the negative root.

Note: You will see this used later in our work with quadratic equations in Chapter 11.

If both square roots need to be indicated, we can write

$$\pm\sqrt{9} = \pm 3$$

Every radical expression contains three parts, as shown below. The principal *n*th root of *a* is written as

Note: The index of 2 for square roots is generally not written. We understand that

$$\sqrt{a}$$

is the principal square root of *a*.

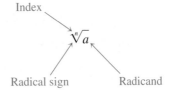

Index
$\sqrt[n]{a}$
Radical sign   Radicand

| Example 1 | **Evaluating Radical Expressions** |

Evaluate, if possible.

(a) $\sqrt{49} = 7$

(b) $-\sqrt{49} = -7$

(c) $\pm\sqrt{49} = \pm 7$

Note: Notice that neither 7^2 nor $(-7)^2$ equals -49.

(d) $\sqrt{-49}$ is not a real number.

Let's examine part (d) more carefully. Suppose that for some real number *x*,

$$x = \sqrt{-49}$$

By our earlier definition, this means that

$$x^2 = -49$$

Note: We consider imaginary numbers in detail in Section 10.6.

which is impossible. There is no real square root for -49. We call $\sqrt{-49}$ an *imaginary number.*

✔ CHECK YOURSELF 1

Evaluate, if possible.

(a) $\sqrt{64}$ (b) $-\sqrt{64}$ (c) $\pm\sqrt{64}$ (d) $\sqrt{-64}$

Our next example considers cube roots.

| Example 2 | **Evaluating Radical Expressions** |

Evaluate, if possible.

(a) $\sqrt[3]{64} = 4$ Because $4^3 = 64$

Note: Notice that the cube root of a *negative* number is *negative*.

(b) $-\sqrt[3]{64} = -4$

(c) $\sqrt[3]{-64} = -4$ Because $(-4)^3 = -64$

✔ CHECK YOURSELF 2

Evaluate.

(a) $\sqrt[3]{125}$ (b) $-\sqrt[3]{125}$ (c) $\sqrt[3]{-125}$

Note: The word *indices* is the plural of *index*.

Let's consider radicals with other indices in Example 3.

| Example 3 | **Evaluating Radical Expressions** |

Evaluate, if possible.

(a) $\sqrt[4]{81} = 3$ Because $3^4 = 81$

Note: In general, an *even* root of a *negative* number is *not real*; it is *imaginary*.

(b) $\sqrt[4]{-81}$ is not a real number.

(c) $\sqrt[5]{32} = 2$ Because $2^5 = 32$

(d) $\sqrt[5]{-32} = -2$ Because $(-2)^5 = -32$

✔ CHECK YOURSELF 3

Evaluate, if possible.

(a) $\sqrt[4]{16}$ (b) $\sqrt[5]{243}$ (c) $\sqrt[4]{-16}$ (d) $\sqrt[5]{-243}$

Note: All the numbers of our previous examples and exercises were chosen so that the results would be *rational numbers.* That is, our radicands were

Perfect squares: 1, 4, 9, 16, 25, . . .

Perfect cubes: 1, 8, 27, 64, 125, . . .

and so on.

The square root of a number that is *not* a perfect square (or the cube root of a number that is *not* a perfect cube) is not a rational number.

Expressions such as $\sqrt{2}$, $\sqrt{3}$, and $\sqrt{5}$ are *irrational numbers.* A calculator with a square root key $\boxed{\sqrt{}}$ will give decimal approximations for such numbers.

| Example 4 | **Estimating Radical Expressions** |

Note: On some calculators, the square root is shown as the "2nd function" or "inverse" of x^2. If that is the case, press the $\boxed{\text{2nd function}}$ key and then the $\boxed{x^2}$ key. On graphing calculators, you press the $\boxed{\sqrt{}}$ key, then enclose the radicand in parentheses.

Using a calculator, find the decimal approximation for each of the following. Round all answers to three decimal places.

(a) $\sqrt{17}$

Enter 17 in your calculator and press the $\boxed{\sqrt{}}$ key. The display will read 4.123105626 (if your calculator displays 10 digits). If this is rounded to three decimal places, the result is 4.123.

(b) $\sqrt{28}$

The display should read 5.291502622. Rounded to three decimal places, the result is 5.292.

(c) $\sqrt{-11}$

Enter -11 by first entering 11 and then pressing the $\boxed{+/-}$ key. Take the square root by pressing the $\boxed{\sqrt{}}$ key. The display will read ERROR. This indicates that -11 does not have a real square root.

✔ *CHECK YOURSELF 4*

Use a calculator to find the decimal approximation for each of the following. Round each answer to three decimal places.

(a) $\sqrt{13}$ (b) $\sqrt{38}$ (c) $\sqrt{-21}$

Note: Not all scientific calculators have this key.

To evaluate roots other than square roots by using scientific calculators, the key marked $\boxed{y^x}$ can be used together with the $\boxed{\text{INV}}$ key. (On some calculators, the $\boxed{\text{INV}}$ key is $\boxed{\text{2nd F}}$.)

Example 5	**Estimating Radical Expressions**

Using a calculator, find a decimal approximation for each of the following. Round each answer to three decimal places.

Note: Again, depending upon your calculator, you may have only an eight-digit display.

(a) $\sqrt[4]{12}$

Enter 12 and press $\boxed{\text{INV}}$ $\boxed{y^x}$. Then enter 4 and press $\boxed{=}$. The display will read 1.861209718. Rounded to three decimal places, the result is 1.861.

(b) $\sqrt[5]{27}$

Enter 27 and press $\boxed{\text{INV}}$ $\boxed{y^x}$. Then enter 5 and press $\boxed{=}$. The display will read 1.933182045. Rounded to three decimal places, the result is 1.933.

✔ *CHECK YOURSELF 5*

Use a calculator to find the decimal approximation for each of the following. Round each answer to three decimal places.

(a) $\sqrt[4]{35}$ (b) $\sqrt[5]{29}$

A certain amount of caution should be exercised in dealing with principal even roots. For example, consider the statement

$$\sqrt{x^2} = x \tag{1}$$

Note: Because $x = 2$, $\sqrt{x^2} = x$.

First, let $x = 2$ in equation (1).

$$\sqrt{2^2} = \sqrt{4} = 2 \tag{2}$$

Note: Because here $x = -2$, $\sqrt{x^2} \neq x$.

Now let $x = -2$.

$$\sqrt{(-2)^2} = \sqrt{4} = 2 \tag{3}$$

We see that statement (1) is not true when x is negative, but we can write

$$\sqrt{x^2} = \begin{cases} x & \text{when } x \geq 0 \\ -x & \text{when } x < 0 \end{cases}$$

From your earlier work with absolute values you will remember that

$$|x| = \begin{cases} x & \text{when } x \geq 0 \\ -x & \text{when } x < 0 \end{cases}$$

Note: Statement (4) can be extended to

$$\sqrt[n]{x^n} = |x|$$

when n is even.

and we can summarize the discussion by writing

$$\sqrt{x^2} = |x| \tag{4}$$

| **Example 6** | **Evaluating Radical Expressions** |

Evaluate.

(a) $\sqrt{5^2} = 5$

Note: Alternatively we could write

$$\sqrt{(-4)^2} = \sqrt{16} = 4$$

(b) $\sqrt{(-4)^2} = |-4| = 4$

(c) $\sqrt[4]{2^4} = 2$

(d) $\sqrt[4]{(-3)^4} = |-3| = 3$

✔ CHECK YOURSELF 6

Evaluate.

(a) $\sqrt{6^2}$ (b) $\sqrt{(-6)^2}$ (c) $\sqrt[4]{3^4}$ (d) $\sqrt[4]{(-3)^4}$

Note: The case for roots with indices that are odd does *not* require the use of absolute value, as illustrated in Example 6. For instance,

$$\sqrt[3]{3^3} = \sqrt[3]{27} = 3$$

$$\sqrt[3]{(-3)^3} = \sqrt[3]{-27} = -3$$

and we see that

$$\sqrt[n]{x^n} = x \qquad \text{when } n \text{ is odd}$$

To summarize, we can write

$$\sqrt[n]{x^n} = \begin{cases} |x| & \text{when } n \text{ is even} \\ x & \text{when } n \text{ is odd} \end{cases}$$

Let's turn now to a final example in which variables are involved in the radicand.

| **Example 7** | **Simplifying Radical Expressions** |

Simplify the following.

(a) $\sqrt[3]{a^3} = a$

(b) $\sqrt{16m^2} = 4|m|$

Note: Notice that we can determine the power of the variable in our root by dividing the power in the radicand by the index. In Example 7(d), $8 \div 4 = 2$.

(c) $\sqrt[5]{32x^5} = 2x$

(d) $\sqrt[4]{x^8} = x^2$ Because $(x^2)^4 = x^8$

(e) $\sqrt[3]{27y^6} = 3y^2$ Do you see why?

✔ CHECK YOURSELF 7

Simplify.

(a) $\sqrt[4]{x^4}$ (b) $\sqrt{49w^2}$ (c) $\sqrt[5]{a^{10}}$ (d) $\sqrt[3]{8y^9}$

✔ CHECK YOURSELF ANSWERS

1. (a) 8; (b) -8; (c) ± 8; (d) not a real number

2. (a) 5; (b) -5; (c) -5

3. (a) 2; (b) 3; (c) not a real number; (d) -3

4. (a) 3.606; (b) 6.164; (c) not a real number

5. (a) 2.432; (b) 1.961   **6.** (a) 6; (b) 6; (c) 3; (d) 3

7. (a) $|x|$; (b) $7|w|$; (c) a^2; (d) $2y^3$

Exercises · 10.1

Evaluate each of the following roots where possible.

1. $\sqrt{49}$ **2.** $\sqrt{36}$ **3.** $-\sqrt{36}$ **4.** $-\sqrt{81}$

5. $\pm\sqrt{81}$ **6.** $\pm\sqrt{49}$ **7.** $\sqrt{-49}$ **8.** $\sqrt{-25}$

9. $\sqrt[3]{27}$ **10.** $\sqrt[3]{64}$ **11.** $\sqrt[3]{-64}$ **12.** $-\sqrt[3]{125}$

13. $-\sqrt[3]{216}$ **14.** $\sqrt[3]{-27}$ **15.** $\sqrt[4]{81}$ **16.** $\sqrt[5]{32}$

17. $\sqrt[5]{-32}$ **18.** $\sqrt[4]{-81}$ **19.** $-\sqrt[4]{16}$ **20.** $\sqrt[5]{-243}$

21. $\sqrt[4]{-16}$ **22.** $-\sqrt[5]{32}$ **23.** $-\sqrt[5]{243}$ **24.** $-\sqrt[4]{625}$

25. $\sqrt{\dfrac{4}{9}}$ **26.** $\sqrt{\dfrac{9}{25}}$ **27.** $\sqrt[3]{\dfrac{8}{27}}$ **28.** $\sqrt[3]{-\dfrac{27}{64}}$

29. $\sqrt{6^2}$ **30.** $\sqrt{9^2}$ **31.** $\sqrt{(-3)^2}$ **32.** $\sqrt{(-5)^2}$

33. $\sqrt[3]{4^3}$ **34.** $\sqrt[3]{(-5)^3}$ **35.** $\sqrt[4]{3^4}$ **36.** $\sqrt[4]{(-2)^4}$

Simplify each of the following roots.

37. $\sqrt{x^2}$ **38.** $\sqrt[3]{w^3}$ **39.** $\sqrt[5]{y^5}$ **40.** $\sqrt[7]{z^7}$

41. $\sqrt{9x^2}$ **42.** $\sqrt{81y^2}$ **43.** $\sqrt{a^4b^6}$ **44.** $\sqrt{w^6z^{10}}$

45. $\sqrt{16x^4}$ **46.** $\sqrt{49y^6}$ **47.** $\sqrt[4]{y^{20}}$ **48.** $\sqrt[3]{a^{18}}$

49. $\sqrt[4]{m^8n^{12}}$ **50.** $\sqrt[3]{a^6b^9}$ **51.** $\sqrt[3]{125a^3}$ **52.** $\sqrt[3]{-27x^3}$

53. $\sqrt[5]{32x^5y^{15}}$ **54.** $\sqrt[5]{-32m^{10}n^5}$

Using a calculator, evaluate the following. Round each answer to three decimal places.

55. $\sqrt{15}$ **56.** $\sqrt{29}$ **57.** $\sqrt{156}$ **58.** $\sqrt{213}$

59. $\sqrt{-15}$ **60.** $\sqrt{-79}$ **61.** $\sqrt[3]{83}$ **62.** $\sqrt[3]{97}$

63. $\sqrt[5]{123}$ **64.** $\sqrt[5]{283}$ **65.** $\sqrt[3]{-15}$ **66.** $\sqrt[5]{-29}$

Label each of the following statements as true or false.

67. $\sqrt{16x^{16}} = 4x^4$

68. $\sqrt{36c^2} = 6c$

69. $\sqrt[3]{(4x^6y^9)^3} = 4x^6y^9$

70. $\sqrt[4]{(x-4)^4} = x - 4$

71. $\sqrt{x^4 + 16} = x^2 + 4$

72. $\sqrt[3]{x^8 + 27} = x^2 + 3$

73. $\sqrt{16x^{-4}y^{-4}}$ is not a real number.

74. $\sqrt[3]{-8x^6y^6}$ is not a real number.

75. Is there any prime number whose square root is an integer? Explain your answer.

76. Determine two consecutive integers whose square roots are also consecutive integers.

77. Try the following, using your calculator.

   (a) Choose a number greater than 1 and find its square root. Then find the square root of the result and continue in this manner, observing the successive square roots. Do these numbers seem to be approaching a certain value? If so, what?

   (b) Choose a number greater than 0 but less than 1 and find its square root. Then find the square root of the result, and continue in this manner, observing successive square roots. Do these numbers seem to be approaching a certain value? If so, what?

78. (a) Can a number be equal to its own square root?

   (b) Other than the number(s) found in part (a), is a number always greater than its square root? Investigate.

79. Let a and b be positive numbers. If a is greater than b, is it always true that the square root of a is greater than the square root of b? Investigate.

80. Suppose that a weight is attached to a string of length L, and the other end of the string is held fixed. If we pull the weight and then release it, allowing the weight to swing back and forth, we can observe the behavior of a simple pendulum. The period T is the time required for the weight to complete a full cycle, swinging forward and then back. The following formula may be used to describe the relationship between T and L.

$$T = 2\pi \sqrt{\frac{L}{g}}$$

If L is expressed in centimeters, then $g = 980$ cm/s^2. For each of the following string lengths, calculate the corresponding period. Round to the nearest tenth of a second.

   (a) 30 cm   (b) 50 cm   (c) 70 cm   (d) 90 cm   (e) 110 cm

81. In parts (a) through (f), evaluate when possible.

   (a) $\sqrt{4 \cdot 9}$ (b) $\sqrt{4} \cdot \sqrt{9}$ (c) $\sqrt{9 \cdot 16}$

   (d) $\sqrt{9} \cdot \sqrt{16}$ (e) $\sqrt{(-4)(-25)}$ (f) $\sqrt{-4} \cdot \sqrt{-25}$

   (g) Based on parts (a) through (f), make a general conjecture concerning $\sqrt{ab}$. Be careful to specify any restrictions on possible values for a and b.

82. In parts (a) through (d), evaluate when possible.

   (a) $\sqrt{9 + 16}$ (b) $\sqrt{9} + \sqrt{16}$ (c) $\sqrt{36 + 64}$ (d) $\sqrt{36} + \sqrt{64}$

   (e) Based on parts (a) through (d), what can you say about $\sqrt{a + b}$ and $\sqrt{a} + \sqrt{b}$?

10.2 Simplifying Radical Expressions

 OBJECTIVES

1. Simplify a radical expression by using the product property
2. Simplify a radical expression by using the quotient property

Note: A precise set of conditions for a radical to be in simplified form will follow in this section.

In the last section, we introduced the radical notation. For some applications, we will want to make sure that all radical expressions are written in *simplified form*. To accomplish this objective, we will need two basic properties. In stating these properties, and in our subsequent examples, we will assume that all variables represent positive real numbers whenever the index of a radical is even. To develop our first property, consider an expression such as

$$\sqrt{25 \cdot 4}$$

One approach to simplify the expression would be

$$\sqrt{25 \cdot 4} = \sqrt{100} = 10$$

Now what happens if we separate the original radical as follows?

$$\sqrt{25 \cdot 4} = \sqrt{25} \cdot \sqrt{4}$$
$$= 5 \cdot 2 = 10$$

The result in either case is the same, and this suggests our first property for radicals.

Note: As we stated in the first paragraph, a and b are assumed to be positive real numbers when n is an even integer.

Rules and Properties: Product Property for Radicals
$$\sqrt[n]{ab} = \sqrt[n]{a} \cdot \sqrt[n]{b}$$

In words, the radical of a product is equal to the product of the radicals.

The second property we will need is similar.

Note: To convince yourself that this must be the case, at least for square roots, let $a = 100$ and $b = 4$ and evaluate both sides of the equation.

Rules and Properties: Quotient Property for Radicals
$$\sqrt[n]{\frac{a}{b}} = \frac{\sqrt[n]{a}}{\sqrt[n]{b}}$$

In words, the radical of a quotient is the quotient of the radicals.

Caution

Be careful! Students sometimes assume that because

$$\sqrt{ab} = \sqrt{a} \cdot \sqrt{b}$$

it should also be true that

$$\sqrt{a + b} = \sqrt{a} + \sqrt{b}$$

This is *not* true.

Note: You can easily see that this is *not* true. Let $a = 9$ and $b = 16$ in the statement.

With these two properties, we are now ready to define the simplified form for a radical expression. A radical is in simplified form if the following three conditions are satisfied.

Simplified Form for a Radical Expression

1. The radicand has no factor raised to a power greater than or equal to the index.

2. No fraction appears in the radical.

3. No radical appears in a denominator.

Our initial example deals with satisfying the first of the above conditions. Essentially, we want to find the largest perfect-square factor (in the case of a square root) in the radicand and then apply the product property to simplify the expression.

| **Example 1** | **Simplifying Radical Expressions** |

Write each expression in simplified form.

Note: The largest perfect-square factor of 18 is 9.

(a) $\sqrt{18} = \sqrt{9 \cdot 2}$

$\qquad\quad = \sqrt{9} \cdot \sqrt{2}$ Apply the product property.

$\qquad\quad = 3\sqrt{2}$

Note: The largest perfect-square factor of 75 is 25.

(b) $\sqrt{75} = \sqrt{25 \cdot 3}$

$\qquad\quad = \sqrt{25} \cdot \sqrt{3}$

$\qquad\quad = 5\sqrt{3}$

Note: The largest perfect-square factor of $27x^3$ is $9x^2$. Note that the exponent must be *even* in a perfect square.

(c) $\sqrt{27x^3} = \sqrt{9x^2 \cdot 3x}$

$\qquad\quad = \sqrt{9x^2} \cdot \sqrt{3x} = 3x\sqrt{3x}$

(d) $\sqrt{72a^3b^4} = \sqrt{36a^2b^4 \cdot 2a}$

$\qquad\quad\quad = \sqrt{36a^2b^4} \cdot \sqrt{2a}$

$\qquad\quad\quad = 6ab^2\sqrt{2a}$

✔ CHECK YOURSELF 1

Write each expression in simplified form.

(a) $\sqrt{45}$ (b) $\sqrt{200}$ (c) $\sqrt{75p^5}$ (d) $\sqrt{98m^3n^4}$

Writing a cube root in simplest form involves finding factors of the radicand that are perfect cubes, as illustrated in Example 2. The process illustrated in this example is extended in an identical fashion to simplify radical expressions with any index.

Example 2   Simplifying Radical Expressions

Write each expression in simplified form.

(a) $\sqrt[3]{48} = \sqrt[3]{8 \cdot 6}$

$\phantom{(a) \sqrt[3]{48}} = \sqrt[3]{8} \cdot \sqrt[3]{6} = 2\sqrt[3]{6}$

Note: In a perfect cube, the exponent must be a *multiple of* 3.

(b) $\sqrt[3]{24x^4} = \sqrt[3]{8x^3 \cdot 3x}$

$\phantom{(b) \sqrt[3]{24x^4}} = \sqrt[3]{8x^3} \cdot \sqrt[3]{3x} = 2x\sqrt[3]{3x}$

(c) $\sqrt[3]{54a^7b^4} = \sqrt[3]{27a^6b^3 \cdot 2ab}$

$\phantom{(c) \sqrt[3]{54a^7b^4}} = \sqrt[3]{27a^6b^3} \cdot \sqrt[3]{2ab} = 3a^2b\sqrt[3]{2ab}$

✔ CHECK YOURSELF 2

Write each expression in simplified form.

(a) $\sqrt[3]{128w^4}$ (b) $\sqrt[3]{40x^5y^7}$ (c) $\sqrt[4]{48a^8b^5}$

Satisfying our second condition for a radical to be in simplified form (no fractions should appear inside the radical) requires the second property for radicals. Consider the following example.

Example 3   Simplifying Radical Expressions

Write each expression in simplified form.

Note: Apply the quotient property.

(a) $\sqrt{\dfrac{5}{9}} = \dfrac{\sqrt{5}}{\sqrt{9}}$

$\phantom{(a) \sqrt{\dfrac{5}{9}}} = \dfrac{\sqrt{5}}{3}$

(b) $\sqrt{\dfrac{a^4}{25}} = \dfrac{\sqrt{a^4}}{\sqrt{25}} = \dfrac{a^2}{5}$

(c) $\sqrt[3]{\dfrac{5x^2}{8}} = \dfrac{\sqrt[3]{5x^2}}{\sqrt[3]{8}} = \dfrac{\sqrt[3]{5x^2}}{2}$

✔ *CHECK YOURSELF 3*

Write each expression in simplified form.

(a) $\sqrt{\dfrac{7}{16}}$ (b) $\sqrt{\dfrac{3}{25a^2}}$ (c) $\sqrt[3]{\dfrac{5x}{27}}$

Our next example also begins with the application of the quotient property for radicals. However, an additional step is required because, as we will see, the third condition (no radicals can appear in a denominator) must also be satisfied during the process.

Example 4

Rationalizing the Denominator

Write $\sqrt{\dfrac{3}{5}}$ in simplified form.

$$\sqrt{\dfrac{3}{5}} = \dfrac{\sqrt{3}}{\sqrt{5}}$$

The application of the quotient property satisfies the second condition—there are now no fractions *inside* a radical. However, we now have a radical in the denominator, violating the third condition. The expression will not be simplified until that radical is removed.

To remove the radical in the denominator, we multiply the numerator and denominator by the *same* expression, here $\sqrt{5}$. This is called *rationalizing the denominator*.

Note: The value of the expression is *not* changed as we multiply by $\dfrac{\sqrt{5}}{\sqrt{5}}$, or 1.

Note: The point here is to arrive at a perfect square inside the radical in the denominator. This is done by multiplying the numerator and denominator by $\sqrt{5}$ because

$$\sqrt{5} \cdot \sqrt{5} = \sqrt{5^2} = \sqrt{25}$$

$$\dfrac{\sqrt{3}}{\sqrt{5}} = \dfrac{\sqrt{3} \cdot \sqrt{5}}{\sqrt{5} \cdot \sqrt{5}}$$

$$= \dfrac{\sqrt{15}}{\sqrt{25}} = \dfrac{\sqrt{15}}{5}$$

✔ CHECK YOURSELF 4

Simplify $\sqrt{\dfrac{3}{7}}$.

Let's look at some further examples that involve rationalizing the denominator of an expression.

| Example 5 | **Rationalizing the Denominator** |

Note: We multiply numerator and denominator by $\sqrt{2}$. Why did we choose $\sqrt{2}$? Note that

$$\sqrt{8} = \sqrt{2^3}$$

so

$$\sqrt{8}\sqrt{2} = \sqrt{2^3}\sqrt{2}$$
$$= \sqrt{2^4}$$
$$= 2^2$$
$$= 4$$

Write each expression in simplified form.

(a) $\dfrac{3}{\sqrt{8}} = \dfrac{3 \cdot \sqrt{2}}{\sqrt{8} \cdot \sqrt{2}}$

$\qquad\quad = \dfrac{3\sqrt{2}}{\sqrt{16}} = \dfrac{3\sqrt{2}}{4}$

(b) $\sqrt[3]{\dfrac{5}{4}} = \dfrac{\sqrt[3]{5}}{\sqrt[3]{4}}$

Now note that

$$\sqrt[3]{4} \cdot \sqrt[3]{2} = \sqrt[3]{8} = 2$$

Note: Why did we use $\sqrt[3]{2}$? Note that

$$\sqrt[3]{4} \cdot \sqrt[3]{2} = \sqrt[3]{2^2} \cdot \sqrt[3]{2}$$
$$= \sqrt[3]{2^3}$$

and the exponent is a multiple of 3.

so multiplying the numerator and denominator by $\sqrt[3]{2}$ will produce a perfect cube inside the radical in the denominator. Continuing, we have

$$\dfrac{\sqrt[3]{5}}{\sqrt[3]{4}} = \dfrac{\sqrt[3]{5} \cdot \sqrt[3]{2}}{\sqrt[3]{4} \cdot \sqrt[3]{2}}$$

$$= \dfrac{\sqrt[3]{10}}{\sqrt[3]{8}} = \dfrac{\sqrt[3]{10}}{2}$$

✔ CHECK YOURSELF 5

Simplify each expression.

(a) $\dfrac{5}{\sqrt{12}}$ (b) $\sqrt[3]{\dfrac{2}{9}}$

As our final example, we illustrate the process of rationalizing a denominator when variables are involved in a rational expression.

Example 6   Rationalizing Variable Denominators

Simplify each expression.

(a) $\sqrt{\dfrac{8x^3}{3y}}$

By the quotient property we have

$$\sqrt{\frac{8x^3}{3y}} = \frac{\sqrt{8x^3}}{\sqrt{3y}}$$

Because the numerator can be simplified in this case, let's start with that procedure.

$$\frac{\sqrt{8x^3}}{\sqrt{3y}} = \frac{\sqrt{4x^2} \cdot \sqrt{2x}}{\sqrt{3y}} = \frac{2x\sqrt{2x}}{\sqrt{3y}}$$

Multiplying the numerator and denominator by $\sqrt{3y}$ will rationalize the denominator.

$$\frac{2x\sqrt{2x} \cdot \sqrt{3y}}{\sqrt{3y} \cdot \sqrt{3y}} = \frac{2x\sqrt{6xy}}{\sqrt{9y^2}} = \frac{2x\sqrt{6xy}}{3y}$$

(b) $\dfrac{2}{\sqrt[3]{3x}}$

To satisfy the third condition, we must remove the radical from the denominator. For this we need a perfect cube inside the radical in the denominator. Multiplying the numerator and denominator by $\sqrt[3]{9x^2}$ will provide the perfect cube. So

Note:

$$\sqrt[3]{9x^2} = \sqrt[3]{3^2x^2}$$

so

$$\sqrt[3]{3x} \cdot \sqrt[3]{9x^2} = \sqrt[3]{3^3x^3}$$

and each exponent is a multiple of 3.

$$\frac{2\sqrt[3]{9x^2}}{\sqrt[3]{3x} \cdot \sqrt[3]{9x^2}} = \frac{2\sqrt[3]{9x^2}}{\sqrt[3]{27x^3}}$$

$$= \frac{2\sqrt[3]{9x^2}}{3x}$$

✔ CHECK YOURSELF 6

Simplify each expression.

(a) $\sqrt{\dfrac{12a^3}{5b}}$ (b) $\dfrac{3}{\sqrt[3]{2w^2}}$

Example 7 demonstrates using the distance formula.

| **Example 7** | **Finding the Distance Between Two Points** |

Find the distance between each pair of points.

(a) $(3, -5)$ and $(-5, -5)$

Let $(x_1, y_1) = (3, -5)$ and $(x_2, y_2) = (-5, -5)$. Plugging those values into the distance formula, we find

$$d = \sqrt{(-5 - 3)^2 + [-5 - (-5)]^2} = \sqrt{(-8)^2 + 0^2} = \sqrt{64} = 8$$

The distance between the two points is 8 units.

(b) $(-4, 7)$ and $(1, 5)$

Let $(x_1, y_1) = (-4, 7)$ and $(x_2, y_2) = (1, 5)$. Plugging those values into the distance formula, we find

$$d = \sqrt{[1 - (-4)]^2 + (5 - 7)^2} = \sqrt{5^2 + (-2)^2} = \sqrt{29}$$

The distance between the two points is $\sqrt{29}$ units.

(c) $(2, -1)$ and $(-4, 5)$

Let $(x_1, y_1) = (2, -1)$ and $(x_2, y_2) = (-4, 5)$. Plugging those values into the distance formula, we find

$$d = \sqrt{(-4 - 2)^2 + [5 - (-1)]^2} = \sqrt{(-6)^2 + 6^2} = \sqrt{72} = \sqrt{36 \cdot 2} = 6\sqrt{2}$$

The distance between the two points is $6\sqrt{2}$ units.

✔ *CHECK YOURSELF 7*

Find the distance between each pair of points.

(a) $(-2, 7)$ and $(-5, 7)$

(b) $(3, -5)$ and $(7, -4)$

(c) $(-1, 8)$ and $(2, 2)$

The following algorithm summarizes our work in simplifying radical expressions.

Simplifying Radical Expressions

Step 1.   To satisfy the first condition: Determine the largest perfect-square factor of the radicand. Apply the product property to "remove" that factor from inside the radical.

Note: In the case of a cube root, steps 1 and 2 would refer to perfect cubes, etc.

Step 2.   To satisfy the second condition: Use the quotient property to write the expression in the form

$$\frac{\sqrt{a}}{\sqrt{b}}$$

If b is a perfect square, remove the radical in the denominator. If that is not the case, proceed to step 3.

Step 3.   Multiply the numerator and denominator of the radical expression by an appropriate radical to remove the radical in the denominator. Simplify the resulting expression when necessary.

✔ CHECK YOURSELF ANSWERS

1. (a) $3\sqrt{5}$; (b) $10\sqrt{2}$; (c) $5p^2\sqrt{3p}$; (d) $7mn^2\sqrt{2m}$

2. (a) $4w\sqrt[3]{2w}$; (b) $2xy^2\sqrt[3]{5x^2y}$; (c) $2a^2b\sqrt[4]{3b}$

3. (a) $\dfrac{\sqrt{7}}{4}$; (b) $\dfrac{\sqrt{3}}{5a}$; (c) $\dfrac{\sqrt[3]{5x}}{3}$

4. $\dfrac{\sqrt{21}}{7}$ 5. (a) $\dfrac{5\sqrt{3}}{6}$; (b) $\dfrac{\sqrt[3]{6}}{3}$

6. (a) $\dfrac{2a\sqrt{15ab}}{5b}$; (b) $\dfrac{3\sqrt[3]{4w}}{2w}$

7. (a) 3 units; (b) $\sqrt{17}$ units; (c) $3\sqrt{5}$ units

Exercises ▪ 10.2

Use the product property to write each expression in simplified form.

1. $\sqrt{12}$ **2.** $\sqrt{24}$ **3.** $\sqrt{50}$ **4.** $\sqrt{28}$ **5.** $-\sqrt{108}$

6. $\sqrt{32}$ **7.** $\sqrt{52}$ **8.** $-\sqrt{96}$ **9.** $\sqrt{60}$ **10.** $\sqrt{150}$

11. $-\sqrt{125}$ **12.** $\sqrt{128}$ **13.** $\sqrt[3]{16}$ **14.** $\sqrt[3]{-54}$ **15.** $\sqrt[3]{-48}$

16. $\sqrt[3]{250}$ **17.** $\sqrt[3]{135}$ **18.** $\sqrt[3]{-160}$ **19.** $\sqrt[4]{32}$ **20.** $\sqrt[4]{96}$

21. $\sqrt[4]{243}$

Use the product property to write each expression in simplified form. Assume that all variables represent positive real numbers.

22. $\sqrt{18z^2}$ **23.** $\sqrt{45a^2}$ **24.** $\sqrt{63x^4}$ **25.** $\sqrt{54w^4}$

26. $\sqrt{98m^3}$ **27.** $\sqrt{75a^5}$ **28.** $\sqrt{80x^2y^3}$ **29.** $\sqrt{108p^5q^2}$

30. $\sqrt[3]{40b^3}$ **31.** $\sqrt[3]{16x^3}$ **32.** $\sqrt[3]{48p^9}$ **33.** $\sqrt[3]{-80a^6}$

34. $\sqrt[3]{54m^7}$ **35.** $\sqrt[3]{250x^{13}}$ **36.** $\sqrt[3]{56x^6y^5z^4}$ **37.** $-\sqrt[3]{250a^4b^{15}c^9}$

38. $\sqrt[4]{32x^8}$ **39.** $\sqrt[4]{96w^5z^{13}}$ **40.** $\sqrt[4]{128a^{12}b^{17}}$ **41.** $\sqrt[5]{64w^{10}}$

42. $\sqrt[5]{96a^5b^{12}}$

Use the quotient property to write each expression in simplified form. Assume that all variables represent positive real numbers.

43. $\sqrt{\dfrac{5}{16}}$ **44.** $\sqrt{\dfrac{a^6}{49}}$ **45.** $\sqrt{\dfrac{5}{9y^4}}$ **46.** $\sqrt{\dfrac{7}{25x^2}}$

47. $\sqrt[3]{\dfrac{5}{8}}$ **48.** $\sqrt[3]{\dfrac{4x^2}{27}}$

Find the distance between each pair of points.

49. $(2, -6)$ and $(2, -9)$ **50.** $(-3, 7)$ and $(4, 7)$

51. $(-7, 1)$ and $(-4, 0)$ **52.** $(-17, -5)$ and $(-12, -3)$

53. $(22, -13)$ and $(18, -9)$ **54.** $(-12, -17)$ and $(-9, -11)$

Write each expression in simplified form. Assume that all variables represent positive real numbers.

55. $\dfrac{5}{\sqrt{7}}$ **56.** $\sqrt{\dfrac{5}{8}}$ **57.** $\dfrac{7}{\sqrt{12}}$ **58.** $\dfrac{\sqrt{5}}{\sqrt{11}}$

59. $\dfrac{2\sqrt{3}}{\sqrt{10}}$ **60.** $\dfrac{3\sqrt{5}}{\sqrt{3}}$ **61.** $\sqrt[3]{\dfrac{7}{4}}$ **62.** $\sqrt[3]{\dfrac{5}{9}}$

63. $\dfrac{5}{\sqrt[3]{16}}$ **64.** $\sqrt{\dfrac{3}{x}}$ **65.** $\sqrt{\dfrac{12}{w}}$ **66.** $\dfrac{\sqrt{18}}{\sqrt{a}}$

67. $\dfrac{\sqrt{8m^3}}{\sqrt{5n}}$ **68.** $\sqrt{\dfrac{24x^5}{7y}}$ **69.** $\sqrt[3]{\dfrac{5}{y}}$ **70.** $\sqrt[3]{\dfrac{7}{x^2}}$

71. $\dfrac{3}{\sqrt[3]{2x}}$ **72.** $\dfrac{5}{\sqrt[3]{3a}}$ **73.** $\sqrt[3]{\dfrac{2}{5x^2}}$ **74.** $\sqrt[3]{\dfrac{5}{7w^2}}$

75. $\dfrac{\sqrt[3]{5}}{\sqrt[3]{4a^2}}$ **76.** $\dfrac{\sqrt[3]{2}}{\sqrt[3]{9m^2}}$ **77.** $\sqrt[3]{\dfrac{a^5}{b^7}}$ **78.** $\sqrt[3]{\dfrac{w^7}{z^{10}}}$

Label each of the following statements as true or false.

79. $\sqrt{16x^{16}} = 4x^8$ **80.** $\sqrt{x^2 + y^2} = x + y$ **81.** $\dfrac{\sqrt{x^2 - 25}}{\sqrt{x - 5}} = \sqrt{x + 5}$

82. $\sqrt[3]{x^6} \cdot \sqrt[3]{x^3 - 1} = x^2\sqrt[3]{x - 1}$ **83.** $\sqrt[3]{(8b^6)^2} = \left(\sqrt[3]{8b^6}\right)^2$ **84.** $\dfrac{\sqrt[3]{8x^3}}{\sqrt[3]{2x}} = \sqrt[3]{4x^2}$

Simplify.

85. $\dfrac{7\sqrt{x^2y^4} \cdot \sqrt{36xy}}{6\sqrt{x^{-6}y^{-2}} \cdot \sqrt{49x^{-1}y^{-3}}}$ **86.** $\dfrac{3\sqrt[3]{32c^{12}d^2} \cdot \sqrt[3]{2c^5d^4}}{4\sqrt[3]{9c^8d^{-2}} \cdot \sqrt[3]{3c^{-3}d^{-4}}}$

87. Explain the difference between a pair of binomials in which the middle sign is changed and the opposite of a binomial. To illustrate, use $4 - \sqrt{7}$.

88. Determine the missing binomial in the following:

$$(\sqrt{3} - 2)(\qquad) = -1$$

89. Use a calculator to evaluate the following expressions in parts (a) through (d). Round your answer to the nearest hundredth.

(a) $3\sqrt{5} + 4\sqrt{5}$ (b) $7\sqrt{5}$ (c) $2\sqrt{6} + 3\sqrt{6}$ (d) $5\sqrt{6}$

(e) Based on parts (a) through (d), make a conjecture concerning $a\sqrt{m} + b\sqrt{m}$. Check your conjecture on an example of your own that is similar to parts (a) through (d).

10.3 Operations on Radical Expressions

10.3 OBJECTIVES

1. Add two radical expressions
2. Subtract two radical expressions
3. Multiply two radical expressions
4. Divide two radical expressions

The addition and subtraction of radical expressions exactly parallel our earlier work with polynomials containing like terms. Let's review for a moment.

To add $3x^2 + 4x^2$, we have

Note: This uses the distributive property.

$$3x^2 + 4x^2 = (3 + 4)x^2$$
$$= 7x^2$$

Keep in mind that we were able to simplify or combine the above expressions because of like terms in x^2. (Recall that like terms have the same variable factor raised to the same power.)

We *cannot* combine terms such as

$$4a^3 + 3a^2 \qquad \text{or} \qquad 3x - 5y$$

By extending these ideas, radical expressions can be combined *only* if they are *similar,* that is, if the expressions contain the same radicand with the same index. This is illustrated in Example 1.

| Example 1 | **Adding or Subtracting Radical Expressions** |

Add or subtract as indicated.

Note: Apply the distributive property again.

(a) $3\sqrt{7} + 2\sqrt{7} = (3 + 2)\sqrt{7}$
$$= 5\sqrt{7}$$

(b) $7\sqrt{3} - 4\sqrt{3} = (7 - 4)\sqrt{3} = 3\sqrt{3}$

(c) $5\sqrt{10} - 3\sqrt{10} + 2\sqrt{10} = (5 - 3 + 2)\sqrt{10}$
$$= 4\sqrt{10}$$

Note: The expressions have different radicands, 5 and 3.

(d) $2\sqrt{5} + 3\sqrt{3}$ cannot be combined or simplified.

Note: The expressions have different indices, 2 and 3.

(e) $\sqrt{7} + \sqrt[3]{7}$ cannot be simplified.

(f) $5\sqrt{x} + 2\sqrt{x} = (5 + 2)\sqrt{x}$
$$= 7\sqrt{x}$$

(g) $5\sqrt{3ab} - 2\sqrt{3ab} + 3\sqrt{3ab} = (5 - 2 + 3)\sqrt{3ab} = 6\sqrt{3ab}$

Note: The radicands are *not* the same.

(h) $\sqrt[3]{3x^2} + \sqrt[3]{3x}$ cannot be simplified.

✔ **CHECK YOURSELF 1**

Add or subtract as indicated.

(a) $5\sqrt{3} + 2\sqrt{3}$ (b) $7\sqrt{5} - 2\sqrt{5} + 3\sqrt{5}$ (c) $2\sqrt{3} + 3\sqrt{2}$

(d) $\sqrt{2y} + 5\sqrt{2y} - 3\sqrt{2y}$ (e) $2\sqrt[3]{3m} - 5\sqrt[3]{3m}$ (f) $\sqrt{5x} - \sqrt[3]{5x}$

Often it is necessary to simplify radical expressions by the methods of Section 10.2 before they can be combined. Example 2 illustrates how the product property is applied.

Example 2

Adding or Subtracting Radical Expressions

Add or subtract as indicated.

(a) $\sqrt{48} + 2\sqrt{3}$

In this form, the radicals cannot be combined. However, note that the first radical can be simplified by our earlier methods because 48 has the perfect-square factor 16.

$$\sqrt{48} = \sqrt{16 \cdot 3} = 4\sqrt{3}$$

With this result we can proceed as before.

$$\sqrt{48} + 2\sqrt{3} = 4\sqrt{3} + 2\sqrt{3}$$
$$= (4 + 2)\sqrt{3} = 6\sqrt{3}$$

Note: Notice that each of the radicands has a perfect-square factor. The reader should provide the details for the simplification of each radical.

(b) $\sqrt{50} - \sqrt{32} + \sqrt{98} = 5\sqrt{2} - 4\sqrt{2} + 7\sqrt{2}$
$$= (5 - 4 + 7)\sqrt{2} = 8\sqrt{2}$$

(c) $x\sqrt{2x} + 3\sqrt{8x^3}$

Note that

$$3\sqrt{8x^3} = 3\sqrt{4x^2 \cdot 2x}$$
$$= 3\sqrt{4x^2} \cdot \sqrt{2x}$$
$$= 3 \cdot 2x\sqrt{2x} = 6x\sqrt{2x}$$

So

Note: We can now combine the similar radicals.

$$x\sqrt{2x} + 3\sqrt{8x^3} = x\sqrt{2x} + 6x\sqrt{2x}$$
$$= (x + 6x)\sqrt{2x} = 7x\sqrt{2x}$$

(d) $\sqrt[3]{2a} - \sqrt[3]{16a} + \sqrt[3]{54a} = \sqrt[3]{2a} - 2\sqrt[3]{2a} + 3\sqrt[3]{2a}$

$$= (1 - 2 + 3)\sqrt[3]{2a}$$

$$= 2\sqrt[3]{2a}$$

✔ CHECK YOURSELF 2

Add or subtract as indicated.

(a) $\sqrt{125} + 3\sqrt{5}$ (b) $\sqrt{75} - \sqrt{27} + \sqrt{48}$

(c) $5\sqrt{24y^3} - y\sqrt{6y}$ (d) $\sqrt[3]{81x} - \sqrt[3]{3x} + \sqrt[3]{24x}$

It may also be necessary to apply the quotient property before combining rational expressions. Consider the following example.

| Example 3 | **Adding or Subtracting Radical Expressions** |

Add or subtract as indicated.

(a) $2\sqrt{6} + \sqrt{\dfrac{2}{3}}$

We apply the quotient property to the *second term* and rationalize the denominator.

Note: Multiply by $\dfrac{\sqrt{3}}{\sqrt{3}}$, or 1.

$$\sqrt{\frac{2}{3}} = \frac{\sqrt{2}}{\sqrt{3}} = \frac{\sqrt{2} \cdot \sqrt{3}}{\sqrt{3} \cdot \sqrt{3}} = \frac{\sqrt{6}}{3}$$

So

$$2\sqrt{6} + \sqrt{\frac{2}{3}} = 2\sqrt{6} + \frac{\sqrt{6}}{3}$$

Note: Note that $\dfrac{\sqrt{6}}{3}$ and $\dfrac{1}{3}\sqrt{6}$ are equivalent.

$$= \left(2 + \frac{1}{3}\right)\sqrt{6} = \frac{7}{3}\sqrt{6}$$

(b) $\sqrt{20x} - \sqrt{\dfrac{x}{5}}$

Again we first simplify the two expressions. So

$$\sqrt{20x} - \sqrt{\frac{x}{5}} = 2\sqrt{5x} - \frac{\sqrt{x} \cdot \sqrt{5}}{\sqrt{5} \cdot \sqrt{5}}$$

$$= 2\sqrt{5x} - \frac{\sqrt{5x}}{5}$$

$$= \left(2 - \frac{1}{5}\right)\sqrt{5x} = \frac{9}{5}\sqrt{5x}$$

✔ *CHECK YOURSELF 3*

Add or subtract as indicated.

(a) $3\sqrt{7} + \sqrt{\dfrac{1}{7}}$ (b) $\sqrt{40x} - \sqrt{\dfrac{2x}{5}}$

Our next example illustrates how our earlier methods for adding fractions may have to be applied in working with radical expressions.

Example 4

Adding Radical Expressions

Add $\dfrac{\sqrt{5}}{3} + \dfrac{2}{\sqrt{5}}$.

Our first step is to rationalize the denominator of the second fraction, to write the sum as

$$\frac{\sqrt{5}}{3} + \frac{2\sqrt{5}}{\sqrt{5} \cdot \sqrt{5}}$$

or

$$\frac{\sqrt{5}}{3} + \frac{2\sqrt{5}}{5}$$

The LCD of the fractions is 15, and rewriting each fraction with that denominator, we have

$$\frac{\sqrt{5} \cdot 5}{3 \cdot 5} + \frac{2\sqrt{5} \cdot 3}{5 \cdot 3} = \frac{5\sqrt{5} + 6\sqrt{5}}{15}$$

$$= \frac{11\sqrt{5}}{15}$$

✔ *CHECK YOURSELF 4*

Subtract $\dfrac{3}{\sqrt{10}} - \dfrac{\sqrt{10}}{5}$.

In Section 10.2 we introduced the product and quotient properties for radical expressions. At that time they were used for simplifying radicals.

If we turn those properties around, we have our rules for the multiplication and division of radical expressions. For multiplication:

Rules and Properties: Multiplying Radical Expressions

$$\sqrt[n]{a} \cdot \sqrt[n]{b} = \sqrt[n]{a \cdot b}$$

In words, the product of two roots is the root of the product of the radicands.

The use of this multiplication rule is illustrated in Example 5. Again we assume that all variables represent positive real numbers.

| **Example 5** | ## Multiplying Radical Expressions |

Multiply.

Note: Just multiply the radicands.

(a) $\sqrt{7} \cdot \sqrt{5} = \sqrt{7 \cdot 5} = \sqrt{35}$

(b) $\sqrt{3x} \cdot \sqrt{10y} = \sqrt{3x \cdot 10y}$
$$= \sqrt{30xy}$$

(c) $\sqrt[3]{4x} \cdot \sqrt[3]{7x} = \sqrt[3]{4x \cdot 7x}$
$$= \sqrt[3]{28x^2}$$

✔ *CHECK YOURSELF 5*

Multiply.

(a) $\sqrt{6} \cdot \sqrt{7}$ (b) $\sqrt{5a} \cdot \sqrt{11b}$ (c) $\sqrt[3]{3y} \cdot \sqrt[3]{5y}$

Keep in mind that all radical expressions should be written in simplified form. Often we have to apply the methods of Section 10.2 to simplify a product once it has been formed.

| **Example 6** | ## Multiplying Radical Expressions |

Multiply and simplify.

Note: $\sqrt{18}$ is *not* in simplified form. 9 is a perfect-square factor of 18.

(a) $\sqrt{3} \cdot \sqrt{6} = \sqrt{18}$
$$= \sqrt{9 \cdot 2} = \sqrt{9}\sqrt{2}$$
$$= 3\sqrt{2}$$

(b) $\sqrt{5x} \cdot \sqrt{15x} = \sqrt{75x^2}$
$$= \sqrt{25x^2 \cdot 3} = \sqrt{25x^2} \cdot \sqrt{3}$$
$$= 5x\sqrt{3}$$

Note: Now we want a factor that is a *perfect cube*.

(c) $\sqrt[3]{4a^2b} \cdot \sqrt[3]{10a^2b^2} = \sqrt[3]{40a^4b^3} = \sqrt[3]{8a^3b^3 \cdot 5a}$
$$= \sqrt[3]{8a^3b^3} \cdot \sqrt[3]{5a} = 2ab\sqrt[3]{5a}$$

✔ *CHECK YOURSELF 6*

Multiply and simplify.

(a) $\sqrt{10} \cdot \sqrt{20}$ (b) $\sqrt{6x} \cdot \sqrt{15x}$

(c) $\sqrt[3]{9p^2q^2} \cdot \sqrt[3]{6pq^2}$

We are now ready to combine multiplication with the techniques for the addition and subtraction of radicals. This will allow us to multiply radical expressions with more than one term. Consider the following examples.

Example 7   Using the Distributive Property

Multiply and simplify.

Note: We distribute $\sqrt{2}$ over the sum $\sqrt{5} + \sqrt{7}$ to multiply.

(a) $\sqrt{2}(\sqrt{5} + \sqrt{7})$

Distributing $\sqrt{2}$, we have

$$\sqrt{2} \cdot \sqrt{5} + \sqrt{2} \cdot \sqrt{7} = \sqrt{10} + \sqrt{14}$$

The expression cannot be simplified further.

Note: Distribute $\sqrt{3}$.

(b) $\sqrt{3}(\sqrt{6} + 2\sqrt{15}) = \sqrt{3} \cdot \sqrt{6} + \sqrt{3} \cdot 2\sqrt{15}$

$$= \sqrt{18} + 2\sqrt{45}$$

$$= 3\sqrt{2} + 6\sqrt{5}$$

Note: Alternately we could choose to simplify $\sqrt{8x}$ in the original expression as our first step. We leave it to the reader to verify that the result would be the same.

(c) $\sqrt{x}(\sqrt{2x} + \sqrt{8x}) = \sqrt{x} \cdot \sqrt{2x} + \sqrt{x} \cdot \sqrt{8x}$

$$= \sqrt{2x^2} + \sqrt{8x^2}$$

$$= x\sqrt{2} + 2x\sqrt{2} = 3x\sqrt{2}$$

✔ *CHECK YOURSELF 7*

Multiply and simplify.

(a) $\sqrt{3}(\sqrt{10} + \sqrt{2})$ (b) $\sqrt{2}(3 + 2\sqrt{6})$

(c) $\sqrt{a}(\sqrt{3a} + \sqrt{12a})$

If both of the radical expressions involved in a multiplication statement have two terms, we must apply the patterns for multiplying polynomials developed in Chapter 4. The following example illustrates.

| **Example 8** | **Multiplying Radical Binomials** |

Multiply and simplify.

(a) $(\sqrt{3} + 1)(\sqrt{3} + 5)$

To write the desired product, we use the FOIL pattern for multiplying binomials.

$$(\sqrt{3} + 1)(\sqrt{3} + 5)$$

$$= \overset{\text{First}}{\overbrace{\sqrt{3} \cdot \sqrt{3}}} + \overset{\text{Outer}}{\overbrace{5 \cdot \sqrt{3}}} + \overset{\text{Inner}}{\overbrace{1 \cdot \sqrt{3}}} + \overset{\text{Last}}{\overbrace{1 \cdot 5}}$$

Note: Combine the outer and inner products.

$$= 3 + 6\sqrt{3} + 5$$

Note: Combine the first and last products.

$$= 8 + 6\sqrt{3}$$

(b) $(\sqrt{6} + \sqrt{2})(\sqrt{6} - \sqrt{2})$

Multiplying as before, we have

$$\sqrt{6} \cdot \sqrt{6} - \sqrt{6} \cdot \sqrt{2} + \sqrt{6} \cdot \sqrt{2} - \sqrt{2} \cdot \sqrt{2} = 6 - 2 = 4$$

Note: Notice that sum of the outer and inner products $-\sqrt{12} + \sqrt{12}$ is 0.

Note: Two binomial radical expressions that differ *only* in the sign of the second term are called *conjugates* of each other. So

$$\sqrt{6} + \sqrt{2} \quad \text{and} \quad \sqrt{6} - \sqrt{2}$$

Note: Notice the form of the product

$$(a + b)(a - b)$$

gives

$$a^2 - b^2$$

when *a* and *b* are square roots. The product will be rational.

are conjugates, and their product does *not* contain a radical—the product is a rational number. That will always be the case with two conjugates. This will have particular significance later in this section.

(c) $(\sqrt{2} + \sqrt{5})^2 = (\sqrt{2} + \sqrt{5})(\sqrt{2} + \sqrt{5})$

Multiplying as before, we have

$$(\sqrt{2} + \sqrt{5})^2 = \sqrt{2} \cdot \sqrt{2} + \sqrt{2} \cdot \sqrt{5} + \sqrt{2} \cdot \sqrt{5} + \sqrt{5} \cdot \sqrt{5}$$

Note: We write the square as a product and apply the multiplication pattern for binomials.

$$= 2 + \sqrt{10} + \sqrt{10} + 5$$

$$= 7 + 2\sqrt{10}$$

Note: This square can also be evaluated by using our earlier formula for the square of a binomial

$$(a + b)^2 = a^2 + 2ab + b^2$$

in which $a = \sqrt{2}$ and $b = \sqrt{5}$.

✔ *CHECK YOURSELF 8*

Multiply and simplify.

(a) $(\sqrt{2} + 3)(\sqrt{2} + 5)$ (b) $(\sqrt{5} - \sqrt{3})(\sqrt{5} + \sqrt{3})$ (c) $(\sqrt{7} - \sqrt{3})^2$

We are now ready to state our basic rule for the division of radical expressions. Again, it is simply a restatement of our earlier quotient property.

> **Rules and Properties: Dividing Radical Expressions**
>
> $$\frac{\sqrt[n]{a}}{\sqrt[n]{b}} = \sqrt[n]{\frac{a}{b}}$$
>
> In words, the quotient of two roots is the root of the quotient of the radicands.

Although we illustrate the use of this property in one of the examples that follow, the division of rational expressions is most often carried out by rationalizing the denominator. This process can be divided into two types of problems, those with a monomial divisor and those with binomial divisors. The following series of examples will illustrate.

Example 9

Note: Use

$$\frac{\sqrt{5}}{\sqrt{5}} = 1$$

Note: Notice that

$$\sqrt[3]{2} \cdot \sqrt[3]{4} = \sqrt[3]{2} \cdot \sqrt[3]{2^2}$$
$$= \sqrt[3]{2^3}$$
$$= 2$$

Dividing Radical Expressions

Simplify each expression. Again assume that all variables represent positive real numbers.

(a) $\dfrac{3}{\sqrt{5}} = \dfrac{3 \cdot \sqrt{5}}{\sqrt{5} \cdot \sqrt{5}} = \dfrac{3\sqrt{5}}{5}$ We multiply numerator and denominator by $\sqrt{5}$ to rationalize the denominator.

(b) $\dfrac{\sqrt{7x}}{\sqrt{10y}} = \dfrac{\sqrt{7x} \cdot \sqrt{10y}}{\sqrt{10y} \cdot \sqrt{10y}}$

$= \dfrac{\sqrt{70xy}}{10y}$

(c) $\dfrac{3}{\sqrt[3]{2}} = \dfrac{3\sqrt[3]{4}}{\sqrt[3]{2} \cdot \sqrt[3]{4}}$ In this case we want a perfect cube in the denominator, and so we multiply numerator and denominator by $\sqrt[3]{4}$.

$= \dfrac{3\sqrt[3]{4}}{2}$

Note that these division problems are identical to those we saw earlier in Section 10.2 when we were simplifying radical expressions. They are shown here to illustrate this case of division with radicals.

✔ CHECK YOURSELF 9

Simplify each expression.

(a) $\dfrac{5}{\sqrt{7}}$ (b) $\dfrac{\sqrt{3a}}{\sqrt{5b}}$ (c) $\dfrac{5}{\sqrt[3]{9}}$

Our division rule can be particularly useful if the radicands in the numerator and denominator have common factors. Consider the following example.

| **Example 10** | **Dividing Radical Expressions** |

Simplify

$$\frac{\sqrt{10}}{\sqrt{15a}}$$

Note: Notice the common factor of 5 in the radicand of the numerator and denominator.

We apply the division rule so that the radicand can be reduced as a fraction:

$$\frac{\sqrt{10}}{\sqrt{15a}} = \sqrt{\frac{10}{15a}} = \sqrt{\frac{2}{3a}}$$

Now we use the quotient property and rationalize the denominator:

$$\sqrt{\frac{2}{3a}} = \frac{\sqrt{2}}{\sqrt{3a}} = \frac{\sqrt{2} \cdot \sqrt{3a}}{\sqrt{3a} \cdot \sqrt{3a}}$$

$$= \frac{\sqrt{6a}}{3a}$$

✔ CHECK YOURSELF 10

Simplify $\dfrac{\sqrt{15}}{\sqrt{18x}}$.

We now turn our attention to a second type of division problem involving radical expressions. Here the divisors (the denominators) are binomials. This will use the idea of conjugates that we saw in Example 8, earlier in this section. The following example illustrates.

| **Example 11** | **Rationalizing Radical Denominators** |

Rationalize each denominator.

Note: If a radical expression has a sum or difference in the denominator, multiply the numerator and denominator by the *conjugate* of the denominator to rationalize.

(a) $\dfrac{6}{\sqrt{6} + \sqrt{2}}$

Recall that $\sqrt{6} - \sqrt{2}$ is the conjugate of $\sqrt{6} + \sqrt{2}$, and the product of conjugates is *always a rational number.* Therefore, to rationalize the denominator, we multiply by $\sqrt{6} - \sqrt{2}$.

Note: See Example 8(b) for the details of the multiplication in the denominator.

$$\frac{6}{\sqrt{6} + \sqrt{2}} = \frac{6(\sqrt{6} - \sqrt{2})}{(\sqrt{6} + \sqrt{2})(\sqrt{6} - \sqrt{2})}$$

$$= \frac{6(\sqrt{6} - \sqrt{2})}{4}$$

$$= \frac{3(\sqrt{6} - \sqrt{2})}{2}$$

(b) $\dfrac{\sqrt{5} + \sqrt{3}}{\sqrt{5} - \sqrt{3}}$

Multiply numerator and denominator by $\sqrt{5} + \sqrt{3}$.

Note: Combine like terms, factor, and divide the numerator and denominator by 2 to simplify.

$$\frac{(\sqrt{5} + \sqrt{3})(\sqrt{5} + \sqrt{3})}{(\sqrt{5} - \sqrt{3})(\sqrt{5} + \sqrt{3})} = \frac{5 + \sqrt{15} + \sqrt{15} + 3}{5 - 3}$$

$$= \frac{8 + 2\sqrt{15}}{2} = \frac{2(4 + \sqrt{15})}{2}$$

$$= 4 + \sqrt{15}$$

✔ CHECK YOURSELF 11

Rationalize the denominator.

(a) $\dfrac{4}{\sqrt{3} - \sqrt{2}}$ (b) $\dfrac{\sqrt{6} + \sqrt{3}}{\sqrt{6} - \sqrt{3}}$

✔ CHECK YOURSELF ANSWERS

1. (a) $7\sqrt{3}$; (b) $8\sqrt{5}$; (c) Cannot be simplified; (d) $3\sqrt{2y}$; (e) $-3\sqrt[3]{3m}$; (f) Cannot be simplified

2. (a) $8\sqrt{5}$; (b) $6\sqrt{3}$; (c) $9y\sqrt{6y}$; (d) $4\sqrt[3]{3x}$

3. (a) $\dfrac{22}{7}\sqrt{7}$; (b) $\dfrac{9}{5}\sqrt{10x}$ 4. $\dfrac{\sqrt{10}}{10}$ 5. (a) $\sqrt{42}$; (b) $\sqrt{55ab}$; (c) $\sqrt[3]{15y^2}$

6. (a) $10\sqrt{2}$; (b) $3x\sqrt{10}$; (c) $3pq\sqrt[3]{2q}$

7. (a) $\sqrt{30} + \sqrt{6}$; (b) $3\sqrt{2} + 4\sqrt{3}$; (c) $3a\sqrt{3}$

8. (a) $17 + 8\sqrt{2}$; (b) 2; (c) $10 - 2\sqrt{21}$ 9. (a) $\dfrac{5\sqrt{7}}{7}$; (b) $\dfrac{\sqrt{15ab}}{5b}$; (c) $\dfrac{5\sqrt[3]{3}}{3}$

10. $\dfrac{\sqrt{30x}}{6x}$ 11. (a) $4(\sqrt{3} + \sqrt{2})$; (b) $3 + 2\sqrt{2}$

Exercises · 10.3

Add or subtract as indicated. Assume that all variables represent positive real numbers.

1. $3\sqrt{5} + 4\sqrt{5}$

2. $5\sqrt{6} + 3\sqrt{6}$

3. $11\sqrt{3a} - 8\sqrt{3a}$

4. $2\sqrt{5w} + 3\sqrt{5w}$

5. $7\sqrt{m} + 6\sqrt{n}$

6. $8\sqrt{a} - 6\sqrt{b}$

7. $2\sqrt[3]{2} + 7\sqrt[3]{2}$

8. $5\sqrt[4]{3} - 2\sqrt[4]{3}$

9. $8\sqrt{6} - 2\sqrt{6} + 3\sqrt{6}$

10. $8\sqrt{3} + 2\sqrt{3} - 7\sqrt{3}$

Simplify the radical expressions when necessary. Then add or subtract as indicated. Again assume that all variables represent positive real numbers.

11. $\sqrt{20} + \sqrt{5}$

12. $\sqrt{27} + \sqrt{3}$

13. $4\sqrt{28} - \sqrt{63}$

14. $2\sqrt{40} + \sqrt{90}$

15. $\sqrt{98} - \sqrt{18} + \sqrt{8}$

16. $\sqrt{108} - \sqrt{27} + \sqrt{75}$

17. $\sqrt[3]{81} + \sqrt[3]{3}$

18. $\sqrt[3]{16} - \sqrt[3]{2}$

19. $\sqrt{54w} - \sqrt{24w}$

20. $\sqrt{27p} + \sqrt{75p}$

21. $\sqrt{18x^3} + \sqrt{8x^3}$

22. $\sqrt{125y^3} - \sqrt{20y^3}$

23. $\sqrt[3]{16w^5} + 2w\sqrt[3]{2w^2} - \sqrt[3]{2w^5}$ **24.** $\sqrt[4]{2z^7} - z\sqrt[4]{32z^3} + \sqrt[4]{162z^7}$

25. $\sqrt{3} + \sqrt{\dfrac{1}{3}}$

26. $\sqrt{6} - \sqrt{\dfrac{1}{6}}$

27. $\sqrt[3]{48} - \sqrt[3]{\dfrac{3}{4}}$

28. $\sqrt[3]{96} + \sqrt[3]{\dfrac{4}{9}}$

29. $\dfrac{\sqrt{6}}{2} + \dfrac{1}{\sqrt{6}}$

30. $\dfrac{\sqrt{10}}{2} - \dfrac{1}{\sqrt{10}}$

31. $\dfrac{\sqrt{12}}{3} - \dfrac{1}{\sqrt{3}}$

32. $\dfrac{\sqrt{20}}{5} + \dfrac{2}{\sqrt{5}}$

Multiply each of the following expressions.

33. $\sqrt{a} \cdot \sqrt{11}$

34. $\sqrt{10} \cdot \sqrt{w}$

35. $\sqrt{3} \cdot \sqrt{7} \cdot \sqrt{2}$

36. $\sqrt{5} \cdot \sqrt{7} \cdot \sqrt{3}$

37. $\sqrt[3]{4} \cdot \sqrt[3]{9}$

38. $\sqrt[3]{5} \cdot \sqrt[3]{7}$

Multiply and simplify each of the following expressions.

39. $\sqrt{3} \cdot \sqrt{12}$

40. $\sqrt{5} \cdot \sqrt{20}$

41. $\sqrt[3]{9p^2} \cdot \sqrt[3]{6p}$

42. $\sqrt[3]{25x^2} \cdot \sqrt[3]{10x^2}$

43. $\sqrt[3]{4x^2y} \cdot \sqrt[3]{10xy^3}$

44. $\sqrt[3]{18r^2s^2} \cdot \sqrt[3]{9r^2s}$

45. $\sqrt{2}(\sqrt{3} + 5)$

46. $\sqrt{3}(\sqrt{5} - 7)$

47. $\sqrt{3}(5\sqrt{2} - \sqrt{18})$

48. $\sqrt{2}(2\sqrt{10} + \sqrt{40})$

49. $\sqrt{x}(\sqrt{3x} + \sqrt{27x})$

50. $\sqrt{y}(\sqrt{8y} - \sqrt{2y})$

51. $\sqrt[3]{4}(\sqrt[3]{4} + \sqrt[3]{32})$

52. $\sqrt[3]{6}(\sqrt[3]{32} - \sqrt[3]{4})$

53. $(\sqrt{2} + 3)(\sqrt{2} - 4)$

54. $(\sqrt{3} - 1)(\sqrt{3} + 5)$

55. $(\sqrt{2} + 3\sqrt{5})(\sqrt{2} - 2\sqrt{5})$

56. $(\sqrt{6} - 2\sqrt{3})(\sqrt{6} - 3\sqrt{3})$

57. $(\sqrt{5} + 3)(\sqrt{5} - 3)$

58. $(\sqrt{10} + 2)(\sqrt{10} - 2)$

59. $(\sqrt{a} + \sqrt{3})(\sqrt{a} - \sqrt{3})$

60. $(\sqrt{m} + \sqrt{7})(\sqrt{m} - \sqrt{7})$

61. $(\sqrt{3} - 5)^2$

62. $(\sqrt{5} + \sqrt{2})^2$

63. $(\sqrt{a} + 3)^2$

64. $(\sqrt{x} - 4)^2$

65. $(\sqrt{x} + \sqrt{y})^2$

66. $(\sqrt{r} - \sqrt{s})^2$

Rationalize the denominator in each of the following expressions. Simplify when necessary.

67. $\dfrac{\sqrt{3}}{\sqrt{7}}$

68. $\dfrac{\sqrt{5}}{\sqrt{3}}$

69. $\dfrac{\sqrt{2a}}{\sqrt{3b}}$

70. $\dfrac{\sqrt{5x}}{\sqrt{6y}}$

71. $\dfrac{3}{\sqrt[3]{4}}$

72. $\dfrac{2}{\sqrt[3]{9}}$

73. $\dfrac{1}{2 + \sqrt{3}}$

74. $\dfrac{2}{3 - \sqrt{2}}$

75. $\dfrac{8}{3 - \sqrt{5}}$

76. $\dfrac{20}{4 + \sqrt{6}}$

77. $\dfrac{\sqrt{6} + \sqrt{3}}{\sqrt{6} - \sqrt{3}}$

78. $\dfrac{\sqrt{7} - \sqrt{5}}{\sqrt{7} + \sqrt{5}}$

79. $\dfrac{\sqrt{w} + 3}{\sqrt{w} - 3}$

80. $\dfrac{\sqrt{x} - 5}{\sqrt{x} + 5}$

81. $\dfrac{\sqrt{x} - \sqrt{y}}{\sqrt{x} + \sqrt{y}}$

82. $\dfrac{\sqrt{m} + \sqrt{n}}{\sqrt{m} - \sqrt{n}}$

Simplify each of the following radical expressions.

83. $x\sqrt[3]{8x^4} + 4\sqrt[3]{27x^7}$

84. $\sqrt[3]{8x^2} - \sqrt[3]{27x^2}$

85. $\dfrac{\sqrt{2x^2 + 3x}}{\sqrt{x}}$

86. $\dfrac{\sqrt{x^2 - 9}}{\sqrt{x + 3}}$

10.4 Solving Radical Equations

1. Solve an equation containing a radical expression
2. Solve an equation containing two radical expressions

In this section, we wish to establish procedures for solving equations involving radicals. The basic technique we will use involves raising both sides of an equation to some power. However, doing so requires some caution.

For example, let's begin with the equation $x = 1$. Squaring both sides gives us

$$x^2 = 1$$

$$x^2 - 1 = 0$$

$$(x + 1)(x - 1) = 0$$

so the solutions appear to be 1 and -1.

Clearly -1 is not a solution to the original equation. We refer to -1 as an *extraneous solution.*

We must be aware of the possibility of extraneous solutions anytime we raise both sides of an equation to any *even power.* Having said that, we are now prepared to introduce the power property of equality.

Rules and Properties: The Power Property of Equality

Given any two expressions a and b and any positive integer n,

If $a = b$ then $a^n = b^n$

Note that although in applying the power property you will never lose a solution, you will often find an extraneous one as a result of raising both sides of an equation to some power. Because of this, it is very important that you *check all solutions.*

| Example 1 | ### Solving a Radical Equation |

Note: Notice that

$$(\sqrt{x + 2})^2 = x + 2$$

That is why squaring both sides of the equation removes the radical.

Solve $\sqrt{x + 2} = 3$.

Squaring each side, we have

$$(\sqrt{x + 2})^2 = 3^2$$
$$x + 2 = 9$$
$$x = 7$$

Substituting 7 into the original equation, we find

$$\sqrt{7 + 2} \overset{?}{=} 3$$
$$\sqrt{9} \overset{?}{=} 3$$
$$3 = 3$$

Because this is a true statement, we have found the solution for the equation, $x = 7$.

✔ *CHECK YOURSELF 1*

Solve the equation $\sqrt{x - 5} = 4$.

| Example 2 | ### Solving a Radical Equation |

Note: Applying the power property will only remove the radical if that radical is isolated on one side of the equation.

Solve $\sqrt{4x + 5} + 1 = 0$.

We must *first isolate the radical* on the left side:

$$\sqrt{4x + 5} = -1$$

Note: Notice that on the right $(-1)^2 = 1$.

Then, squaring both sides, we have

$$(\sqrt{4x + 5})^2 = (-1)^2$$
$$4x + 5 = 1$$

and solving for x, we find that

$$x = -1$$

Now we will check the solution by substituting -1 for x in the original equation:

Note: $\sqrt{1} = 1$, the principal root.

$$\sqrt{4(-1) + 5} + 1 \overset{?}{=} 0$$
$$\sqrt{1} + 1 \overset{?}{=} 0$$

Note: This is clearly a false statement, so -1 is *not* a solution for the original equation.

and $2 \neq 0$

Because -1 is an extraneous solution, there are *no solutions* to the original equation.

CHECK YOURSELF 2

Solve $\sqrt{3x - 2} + 2 = 0$.

Let's consider an example in which the procedure we have described will involve squaring a binomial.

Example 3

Note: These problems can also be solved graphically. With a graphing utility, plot the two graphs $Y_1 = \sqrt{x + 3}$ and $Y_2 = x + 1$. Note that the graphs have one point of intersection, where $x = 1$.

Note: We solved similar equations in Section 6.4.

Note: Verify this for yourself by substituting 1 and then -2 for x in the original equation.

Caution

Be careful! Sometimes (as in this example), one side of the equation contains a binomial. In that case, we must remember the middle term when we square the binomial. The square of a binomial is *always a trinomial.*

Solving a Radical Equation

Solve $\sqrt{x + 3} = x + 1$.

We can square each side, as before.

$$(\sqrt{x + 3})^2 = (x + 1)^2$$

$$x + 3 = x^2 + 2x + 1$$

Simplifying this gives us the quadratic equation

$$x^2 + x - 2 = 0$$

Factoring, we have

$$(x - 1)(x + 2) = 0$$

which gives us the possible solutions

$$x = 1 \qquad \text{or} \qquad x = -2$$

Now we check for extraneous solutions and find that $x = 1$ is a valid solution, but that $x = -2$ does not yield a true statement.

 CHECK YOURSELF 3

Solve $\sqrt{x - 5} = x - 7$.

It is not always the case that one of the solutions is extraneous. We may have zero, one, or two valid solutions when we generate a quadratic from a radical equation.

In Example 4 we see a case in which both of the solutions derived will satisfy the equation.

Example 4	## Solving a Radical Equation

Solve $\sqrt{7x + 1} - 1 = 2x$.

First, *we must isolate the term involving the radical.*

Note: Again, with a graphing utility plot $Y_1 = \sqrt{7x + 1}$ and $Y_2 = 2x + 1$. Where do they intersect?

$$\sqrt{7x + 1} = 2x + 1$$

We can now square both sides of the equation.

$$7x + 1 = 4x^2 + 4x + 1$$

Now we write the quadratic equation in standard form.

$$4x^2 - 3x = 0$$

Factoring, we have

$$x(4x - 3) = 0$$

which yields two possible solutions

$$x = 0 \qquad \text{or} \qquad x = \frac{3}{4}$$

Checking the solutions by substitution, we find that both values for x give true statements, as follows.

Letting x be 0, we have

$$\sqrt{7(0) + 1} - 1 \overset{?}{=} 2(0)$$
$$\sqrt{1} - 1 \overset{?}{=} 0$$

or

$$0 = 0 \qquad \text{A true statement.}$$

Letting x be $\dfrac{3}{4}$, we have

$$\sqrt{7\left(\frac{3}{4}\right) + 1} - 1 \overset{?}{=} 2\left(\frac{3}{4}\right)$$

$$\sqrt{\frac{25}{4}} - 1 \overset{?}{=} \frac{3}{2}$$

$$\frac{5}{2} - 1 \overset{?}{=} \frac{3}{2}$$

$$\frac{3}{2} = \frac{3}{2} \qquad \text{Again a true statement.}$$

✔ *CHECK YOURSELF 4*

Solve $\sqrt{5x + 1} - 1 = 3x$.

Sometimes when an equation involves more than one radical, we must apply the power property more than once. In such a case, it is generally best to avoid having to work with two radicals on the same side of the equation. Example 5 illustrates one approach to the solution of such equations.

| **Example 5** | **Solving an Equation Containing Two Radicals** |

Solve $\sqrt{x-2} - \sqrt{2x-6} = 1$.

First we isolate $\sqrt{x-2}$ by adding $\sqrt{2x-6}$ to both sides of the equation. This gives

$$\sqrt{x-2} = 1 + \sqrt{2x-6}$$

Note: $1 + \sqrt{2x-6}$ is a binomial of the form $a + b$, in which a is 1 and b is $\sqrt{2x-6}$. The square on the right then has the form $a^2 + 2ab + b^2$.

Then squaring each side, we have

$$x - 2 = 1 + 2\sqrt{2x-6} + 2x - 6$$

We now isolate the radical that remains on the right side.

$$-x + 3 = 2\sqrt{2x-6}$$

We must square again to remove that radical.

$$x^2 - 6x + 9 = 4(2x - 6)$$

Now we solve the quadratic equation that results.

$$x^2 - 14x + 33 = 0$$
$$(x - 3)(x - 11) = 0$$

So

$$x = 3 \quad \text{or} \quad x = 11$$

is the possible solution. Checking the possible solutions, you will find that $x = 3$ yields the only valid solution. You should verify that for yourself.

✔ *CHECK YOURSELF 5*

Solve $\sqrt{x+3} - \sqrt{2x+4} + 1 = 0$.

Earlier in this section, we noted that extraneous roots were possible whenever we raised both sides of the equation to an *even power*. In Example 6, we will raise both sides of the equation to an odd power. We will still check the solutions, but in this case it will simply be a check of our work and not a search for extraneous solutions.

Example 6	Solving a Radical Equation

Note: Because a *cube root* is involved, we *cube* both sides to remove the radical.

Solve $\sqrt[3]{x^2 + 23} = 3$.

Cubing each side, we have

$$x^2 + 23 = 27$$

which results in the quadratic equation

$$x^2 - 4 = 0$$

This has two solutions

$$x = 2 \qquad \text{or} \qquad x = -2$$

Checking the solutions, we find that both result in true statements. Again you should verify this result.

✔ CHECK YOURSELF 6

Solve $\sqrt[3]{x^2 - 8} - 2 = 0$.

We summarize our work in this section in the following algorithm for solving equations involving radicals.

Solving Equations Involving Radicals

Step 1.   Isolate a radical on one side of the equation.

Step 2.   Raise each side of the equation to the smallest power that will eliminate the isolated radical.

Step 3.   If any radicals remain in the equation derived in step 2, return to step 1 and continue the solution process.

Step 4.   Solve the resulting equation to determine any possible solutions.

Step 5.   Check all solutions to determine whether extraneous solutions may have resulted from step 2.

✔ CHECK YOURSELF ANSWERS

1. $\{21\}$ **2.** No solution   **3.** $\{9\}$ **4.** $\left\{0, -\dfrac{1}{9}\right\}$ **5.** $\{6\}$ **6.** $\{4, -4\}$

Exercises · 10.4

Solve each of the following equations. Be sure to check your solutions.

1. $\sqrt{x} = 2$

2. $\sqrt{x} - 3 = 0$

3. $2\sqrt{y} - 1 = 0$

4. $3\sqrt{2z} = 9$

5. $\sqrt{m + 5} = 3$

6. $\sqrt{y + 7} = 5$

7. $\sqrt{2x + 4} - 4 = 0$

8. $\sqrt{3x + 3} - 6 = 0$

9. $\sqrt{3x - 2} + 2 = 0$

10. $\sqrt{4x + 1} + 3 = 0$

11. $\sqrt{x - 1} = \sqrt{1 - x}$

12. $\sqrt{x + 1} = \sqrt{1 + x}$

13. $\sqrt{w + 3} = \sqrt{3 + w}$

14. $\sqrt{w - 3} = \sqrt{3 - w}$

15. $\sqrt{2x - 3} + 1 = 3$

16. $\sqrt{3x + 1} - 2 = -1$

17. $2\sqrt{3z + 2} - 1 = 5$

18. $3\sqrt{4q - 1} - 2 = 7$

19. $\sqrt{15 - 2x} = x$

20. $\sqrt{48 - 2y} = y$

21. $\sqrt{x + 5} = x - 1$

22. $\sqrt{2x - 1} = x - 8$

23. $\sqrt{3m - 2} + m = 10$

24. $\sqrt{2x + 1} + x = 7$

25. $\sqrt{t + 9} + 3 = t$

26. $\sqrt{2y + 7} + 4 = y$

27. $\sqrt{6x + 1} - 1 = 2x$

28. $\sqrt{7x + 1} - 1 = 3x$

29. $\sqrt[3]{x - 5} = 3$

30. $\sqrt[3]{x + 6} = 2$

31. $\sqrt[3]{x^2 - 1} = 2$

32. $\sqrt[3]{x^2 + 11} = 3$

Solve each of the following equations. Be sure to check your solutions.

33. $\sqrt{2x} = \sqrt{x + 1}$

34. $\sqrt{3x} = \sqrt{5x - 1}$

35. $2\sqrt{3r} = \sqrt{r + 11}$

36. $5\sqrt{2q - 7} = \sqrt{15q}$

37. $\sqrt{x + 2} + 1 = \sqrt{x + 4}$

38. $\sqrt{x + 5} - 1 = \sqrt{x + 3}$

39. $\sqrt{4m - 3} - 2 = \sqrt{2m - 5}$

40. $\sqrt{2c - 1} = \sqrt{3c + 1} - 1$

41. $\sqrt{x + 1} + \sqrt{x} = 1$

42. $\sqrt{z - 1} - \sqrt{6 - z} = 1$

43. $\sqrt{5x + 6} - \sqrt{x + 3} = 3$

44. $\sqrt{5y + 6} - \sqrt{3y + 4} = 2$

45. $\sqrt{y^2 + 12y} - 3\sqrt{5} = 0$

46. $\sqrt{x^2 + 2x} - 2\sqrt{6} = 0$

47. $\sqrt{\dfrac{x - 3}{x + 2}} = \dfrac{2}{3}$

48. $\dfrac{\sqrt{x - 2}}{x - 2} = \dfrac{x - 5}{\sqrt{x - 2}}$

49. $\sqrt{\sqrt{t} + 5} = 3$

50. $\sqrt{\sqrt{s} - 1} = \sqrt{s - 7}$

51. For what values of x is $\sqrt{(x-1)^2} = x - 1$ a true statement?

52. For what values of x is $\sqrt[3]{(x-1)^3} = x - 1$ a true statement?

Solve for the indicated variable.

53. $h = \sqrt{pq}$ for q

54. $c = \sqrt{a^2 + b^2}$ for a

55. $v = \sqrt{2gR}$ for R

56. $v = \sqrt{2gR}$ for g

57. $r = \sqrt{\dfrac{S}{2\pi}}$ for S

58. $r = \sqrt{\dfrac{3V}{4\pi}}$ for V

59. $r = \sqrt{\dfrac{2V}{\pi h}}$ for V

60. $r = \sqrt{\dfrac{2V}{\pi h}}$ for h

61. $d = \sqrt{(x-1)^2 + (y-2)^2}$ for x

62. $d = \sqrt{(x-1)^2 + (y-2)^2}$ for y

A weight suspended on the end of a string is a *pendulum*. The most common example of a pendulum (this side of Edgar Allen Poe) is the kind found in many clocks.

The regular back-and-forth motion of the pendulum is *periodic,* and one such cycle of motion is called a *period.* The time, in seconds, that it takes for one period is given by the radical equation

$$t = 2\pi \sqrt{\dfrac{l}{g}}$$

in which g is the force of gravity (10 m/s^2) and l is the length of the pendulum.

63. Find the period (to the nearest hundredth of a second) if the pendulum is 0.9 m long.

64. Find the period if the pendulum is 0.049 m long.

65. Solve the equation for length l.

66. How long would the pendulum be if the period were exactly 1 s?

Solve each of the following applications.

67. The sum of an integer and its square root is 12. Find the integer.

68. The difference between an integer and its square root is 12. What is the integer?

69. The sum of an integer and twice its square root is 24. What is the integer?

70. The sum of an integer and 3 times its square root is 40. Find the integer.

The function $d = \sqrt{2h}$ can be used to estimate the distance d to the horizon (in miles) from a given height (in feet).

71. If a plane flies at 30,000 ft, how far away is the horizon?

72. Janine was looking out across the ocean from her hotel room on the beach. Her eyes were 250 ft above the ground. She saw a ship on the horizon. Approximately how far was the ship from her?

73. Given a distance d to the horizon, what altitude would allow you to see that far?

When a car comes to a sudden stop, you can determine the skidding distance (in feet) for a given speed (in miles per hour) by using the formula $s = 2\sqrt{5x}$, in which s is skidding distance and x is speed. Calculate the skidding distance for the following speeds.

74. 55 mi/h   **75.** 65 mi/h   **76.** 75 mi/h   **77.** 40 mi/h

78. Given the skidding distance s, what formula would allow you to calculate the speed in miles per hour?

79. Use the formula obtained in Exercise 78 to determine the speed of a car in miles per hour if the skid marks were 35 ft long.

For each given equation, use a graphing calculator to solve. Express solutions to the nearest hundredth. (*Hint:* Define Y_1 by the expression on the left side of the equation, and define Y_2 by the expression on the right side. Graph these functions and locate any intersection points. For each such point, the x value represents a solution.)

80. $\sqrt{x + 4} = x - 3$ **81.** $\sqrt{2 - x} = x + 4$ **82.** $3 - 2\sqrt{x + 4} = 2x - 5$ **83.** $5 - 3\sqrt{2 - x} = 3 - 4x$

10.5 Rational Exponents

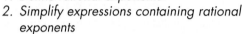

10.5 OBJECTIVES

1. Define rational exponents
2. Simplify expressions containing rational exponents
3. Use a calculator to estimate the value of an expression containing rational exponents
4. Write an expression in radical or exponential form

In Section 10.1, we discussed the radical notation, along with the concept of roots. In this section, we use that concept to develop a new notation, using exponents that provide an alternate way of writing these roots.

That new notation involves **rational numbers as exponents.** To start the development, we extend all the previous properties of exponents to include rational exponents.

Given that extension, suppose that

$$a = 4^{1/2} \tag{1}$$

Squaring both sides of the equation yields

$$a^2 = (4^{1/2})^2$$

or

Note: We will see later in this chapter that the property $(x^m)^n = x^{mn}$ holds for rational numbers m and n.

$$a^2 = 4^{(1/2)(2)}$$
$$a^2 = 4^1$$
$$a^2 = 4 \tag{2}$$

From equation (2) we see that a is the number whose square is 4; that is, a is the principal square root of 4. Using our earlier notation, we can write

$$a = \sqrt{4}$$

But from (1)

$$a = 4^{1/2}$$

and to be consistent, we must have

Note: $4^{1/2}$ indicates the *principal square root* of 4.

$$4^{1/2} = \sqrt{4}$$

This argument can be repeated for any exponent of the form $\dfrac{1}{n}$, so it seems reasonable to make the following definition.

Rational Exponents

If a is any real number and n is a positive integer $(n > 1)$, then

$$a^{1/n} = \sqrt[n]{a}$$

We restrict a so that a is nonnegative when n is even. In words, $a^{1/n}$ indicates the principal nth root of a.

Example 1 illustrates the use of rational exponents to represent roots.

Example 1

Writing Expressions in Radical Form

Write each expression in radical form and then simplify.

(a) $25^{1/2} = \sqrt{25} = 5$

Note: $27^{1/3}$ is the *cube root* of 27.

(b) $27^{1/3} = \sqrt[3]{27} = 3$

(c) $-36^{1/2} = -\sqrt{36} = -6$

(d) $(-36)^{1/2} = \sqrt{-36}$ is not a real number.

Note: $32^{1/5}$ is the *fifth root* of 32.

(e) $32^{1/5} = \sqrt[5]{32} = 2$

✔ CHECK YOURSELF 1

Write each expression in radical form and simplify.

(a) $8^{1/3}$ (b) $-64^{1/2}$ (c) $81^{1/4}$

We are now ready to extend our exponent notation to allow *any* rational exponent, again assuming that our previous exponent properties must still be valid. Note that

$$a^{m/n} = (a^{1/n})^m = (a^m)^{1/n}$$

Note: This is because

$$\frac{m}{n} = (m)\left(\frac{1}{n}\right) = \left(\frac{1}{n}\right)(m)$$

From our earlier work, we know that $a^{1/n} = \sqrt[n]{a}$, and combining this with the above observation, we offer the following definition for $a^{m/n}$.

Note: The two radical forms for $a^{m/n}$ are equivalent, and the choice of which form to use generally depends on whether we are evaluating numerical expressions or rewriting expressions containing variables in radical form.

For any real number a and positive integers m and n with $n > 1$,

$$a^{m/n} = (\sqrt[n]{a})^m = \sqrt[n]{a^m}$$

This new extension of our rational exponent notation is applied in Example 2.

| Example 2 | **Simplifying Expressions with Rational Exponents** |

Simplify each expression.

(a) $9^{3/2} = (9^{1/2})^3 = (\sqrt{9})^3$

$\qquad\quad = 3^3 = 27$

(b) $\left(\dfrac{16}{81}\right)^{3/4} = \left(\left(\dfrac{16}{81}\right)^{1/4}\right)^3 = \left(\sqrt[4]{\dfrac{16}{81}}\right)^3$

$\qquad\qquad\quad = \left(\dfrac{2}{3}\right)^3 = \dfrac{8}{27}$

(c) $(-8)^{2/3} = ((-8)^{1/3})^2 = (\sqrt[3]{-8})^2$

$\qquad\qquad\quad = (-2)^2 = 4$

Note: This illustrates why we use $(\sqrt[n]{a})^m$ for $a^{m/n}$ when evaluating numerical expressions. The numbers involved will be smaller and easier to work with.

In (a) we could also have evaluated the expression as

$$9^{3/2} = \sqrt{9^3} = \sqrt{729}$$

$$= 27$$

✔ *CHECK YOURSELF 2*

Simplify each expression.

(a) $16^{3/4}$ (b) $\left(\dfrac{8}{27}\right)^{2/3}$ (c) $(-32)^{3/5}$

Now we want to extend our rational exponent notation. Using the definition of negative exponents, we can write

$$a^{-m/n} = \dfrac{1}{a^{m/n}}$$

Example 3 illustrates the use of negative rational exponents.

| Example 3 | **Simplifying Expressions with Rational Exponents** |

Simplify each expression.

(a) $16^{-1/2} = \dfrac{1}{16^{1/2}} = \dfrac{1}{4}$

(b) $27^{-2/3} = \dfrac{1}{27^{2/3}} = \dfrac{1}{(\sqrt[3]{27})^2} = \dfrac{1}{3^2} = \dfrac{1}{9}$

✔ CHECK YOURSELF 3

Simplify each expression.

(a) $16^{-1/4}$ (b) $81^{-3/4}$

Graphing calculators can be used to evaluate expressions that contain rational exponents by using the $\boxed{\wedge}$ key and the parentheses keys.

| Example 4 | Estimating Powers by Using a Calculator |

Using a graphing calculator, evaluate each of the following. Round all answers to three decimal places.

(a) $45^{2/5}$

Enter 45 and press the $\boxed{\wedge}$ key. Then use the following keystrokes:

$\boxed{(}\ 2\ \boxed{\div}\ 5\ \boxed{)}$

Press $\boxed{\text{ENTER}}$, and the display will read 4.584426407. Rounded to three decimal places, the result is 4.584.

Note: If you are using a scientific calculator, try using the $\boxed{y^x}$ key in place of the $\boxed{\wedge}$ key.

(b) $38^{-2/3}$

Enter 38 and press the $\boxed{\wedge}$ key. Then use the following keystrokes:

$\boxed{(}\ \boxed{(-)}\ 2\ \boxed{\div}\ 3\ \boxed{)}$

Press $\boxed{\text{ENTER}}$, and the display will read 0.088473037. Rounded to three decimal places, the result is 0.088.

Note: The $\boxed{(-)}$ key changes the sign of the exponent to minus.

✔ CHECK YOURSELF 4

Evaluate each of the following by using a calculator. Round each answer to three decimal places.

(a) $23^{3/5}$ (b) $18^{-4/7}$

As we mentioned earlier in this section, we assume that all our previous exponent properties will continue to hold for rational exponents. Those properties are restated here.

Rules and Properties: Properties of Exponents

For any nonzero real numbers a and b and rational numbers m and n,

1. Product rule $a^m \cdot a^n = a^{m+n}$

2. Quotient rule $\dfrac{a^m}{a^n} = a^{m-n}$

3. Power rule $(a^m)^n = a^{mn}$

4. Product-power rule $(ab)^m = a^m b^m$

5. Quotient-power rule $\left(\dfrac{a}{b}\right)^m = \dfrac{a^m}{b^m}$

We restrict a and b to being nonnegative real numbers when m or n indicates an even root.

Example 5 illustrates the use of our extended properties to simplify expressions involving rational exponents. Here, we assume that all variables represent positive real numbers.

Example 5

Simplifying Expressions

Simplify each expression.

Note: Product rule—add the exponents.

(a) $x^{2/3} \cdot x^{1/2} = x^{2/3+1/2}$

$$= x^{4/6+3/6} = x^{7/6}$$

Note: Quotient rule—subtract the exponents.

(b) $\dfrac{w^{3/4}}{w^{1/2}} = w^{3/4-1/2}$

$$= w^{3/4-2/4} = w^{1/4}$$

Note: Power rule—multiply the exponents.

(c) $(a^{2/3})^{3/4} = a^{(2/3)(3/4)}$

$$= a^{1/2}$$

✔ CHECK YOURSELF 5

Simplify each expression.

(a) $z^{3/4} \cdot z^{1/2}$ (b) $\dfrac{x^{5/6}}{x^{1/3}}$ (c) $(b^{5/6})^{2/5}$

As you would expect from your previous experience with exponents, simplifying expressions often involves using several exponent properties.

| Example 6 | **Simplifying Expressions** |

Simplify each expression.

(a) $(x^{2/3} \cdot y^{5/6})^{3/2}$

$\qquad = (x^{2/3})^{3/2} \cdot (y^{5/6})^{3/2}$ Product-power rule

$\qquad = x^{(2/3)(3/2)} \cdot y^{(5/6)(3/2)} = xy^{5/4}$ Power rule

(b) $\left(\dfrac{r^{-1/2}}{s^{1/3}}\right)^6 = \dfrac{(r^{-1/2})^6}{(s^{1/3})^6}$ Quotient-power rule

$\qquad\qquad = \dfrac{r^{-3}}{s^2} = \dfrac{1}{r^3 s^2}$ Power rule

(c) $\left(\dfrac{4a^{-2/3} \cdot b^2}{a^{1/3} \cdot b^{-4}}\right)^{1/2} = \left(\dfrac{4b^2 \cdot b^4}{a^{1/3} \cdot a^{2/3}}\right)^{1/2} = \left(\dfrac{4b^6}{a}\right)^{1/2}$ We simplify inside the parentheses as the first step.

$\qquad\qquad\qquad = \dfrac{(4b^6)^{1/2}}{a^{1/2}} = \dfrac{4^{1/2}(b^6)^{1/2}}{a^{1/2}}$

$\qquad\qquad\qquad = \dfrac{2b^3}{a^{1/2}}$

✔ *CHECK YOURSELF 6*

Simplify each expression.

(a) $(a^{3/4} \cdot b^{1/2})^{2/3}$ (b) $\left(\dfrac{w^{1/2}}{z^{-1/4}}\right)^4$ (c) $\left(\dfrac{8x^{-3/4}y}{x^{1/4} \cdot y^{-5}}\right)^{1/3}$

We can also use the relationships between rational exponents and radicals to write expressions involving rational exponents as radicals and vice versa.

| Example 7 | **Writing Expressions in Radical Form** |

Write each expression in radical form.

Note: Here we use $a^{m/n} = \sqrt[n]{a^m}$, which is generally the preferred form in this situation.

(a) $a^{3/5} = \sqrt[5]{a^3}$

(b) $(mn)^{3/4} = \sqrt[4]{(mn)^3}$

$\qquad\qquad\quad = \sqrt[4]{m^3 n^3}$

Note: Notice that the exponent applies *only* to the variable y.

(c) $2y^{5/6} = 2\sqrt[6]{y^5}$

Note: Now the exponent applies to $2y$ because of the parentheses.

(d) $(2y)^{5/6} = \sqrt[6]{(2y)^5}$

$\qquad\qquad = \sqrt[6]{32y^5}$

✔ *CHECK YOURSELF 7*

Write each expression in radical form.

(a) $(ab)^{2/3}$ (b) $3x^{3/4}$ (c) $(3x)^{3/4}$

Example 8 **Writing Expressions in Exponential Form**

Using rational exponents, write each expression and simplify.

(a) $\sqrt[3]{5x} = (5x)^{1/3}$

(b) $\sqrt{9a^2b^4} = (9a^2b^4)^{1/2}$
$$= 9^{1/2}(a^2)^{1/2}(b^4)^{1/2} = 3ab^2$$

(c) $\sqrt[4]{16w^{12}z^8} = (16w^{12}z^8)^{1/4}$
$$= 16^{1/4}(w^{12})^{1/4}(z^8)^{1/4} = 2w^3z^2$$

✔ *CHECK YOURSELF 8*

Using rational exponents, write each expression and simplify.

(a) $\sqrt{7a}$ (b) $\sqrt[3]{27p^6q^9}$ (c) $\sqrt[4]{81x^8y^{16}}$

✔ *CHECK YOURSELF ANSWERS*

1. (a) $\sqrt[3]{8} = 2$; (b) $-\sqrt{64} = -8$; (c) $\sqrt[4]{81} = 3$ **2.** (a) 8; (b) $\frac{4}{9}$; (c) -8

3. (a) $\frac{1}{2}$; (b) $\frac{1}{27}$ **4.** (a) 6.562; (b) 0.192   **5.** (a) $z^{5/4}$; (b) $x^{1/2}$; (c) $b^{1/3}$

6. (a) $a^{1/2}b^{1/3}$; (b) w^2z; (c) $\dfrac{2y^2}{x^{1/3}}$ **7.** (a) $\sqrt[3]{a^2b^2}$; (b) $3\sqrt[4]{x^3}$; (c) $\sqrt[4]{27x^3}$

8. (a) $(7a)^{1/2}$; (b) $(27p^6q^9)^{1/3} = 3p^2q^3$; (c) $(81x^8y^{16})^{1/4} = 3x^2y^4$

Exercises · 10.5

In Exercises 1 to 12, use the definition of $a^{1/n}$ to evaluate each expression.

1. $49^{1/2}$

2. $100^{1/2}$

3. $-25^{1/2}$

4. $(-81)^{1/2}$

5. $(-49)^{1/2}$

6. $-49^{1/2}$

7. $27^{1/3}$

8. $(-64)^{1/3}$

9. $81^{1/4}$

10. $-81^{1/4}$

11. $\left(\dfrac{4}{9}\right)^{1/2}$

12. $\left(\dfrac{27}{8}\right)^{1/3}$

In Exercises 13 to 22, use the definition of $a^{m/n}$ to evaluate each expression.

13. $27^{2/3}$

14. $16^{3/2}$

15. $(-8)^{4/3}$

16. $64^{2/3}$

17. $32^{2/5}$

18. $-81^{3/4}$

19. $81^{3/2}$

20. $(-243)^{3/5}$

21. $\left(\dfrac{8}{27}\right)^{2/3}$

22. $\left(\dfrac{16}{9}\right)^{3/2}$

In Exercises 23 to 32, use the definition of $a^{-m/n}$ to evaluate the following expression. Use your calculator to check each answer.

23. $49^{-1/2}$

24. $27^{-1/3}$

25. $81^{-1/4}$

26. $121^{-1/2}$

27. $9^{-3/2}$

28. $16^{-3/4}$

29. $64^{-5/6}$

30. $16^{-3/2}$

31. $\left(\dfrac{4}{25}\right)^{-1/2}$

32. $\left(\dfrac{27}{8}\right)^{-2/3}$

In Exercises 33 to 76, use the properties of exponents to simplify each expression. Assume all variables represent positive real numbers.

33. $x^{1/2} \cdot x^{1/2}$

34. $a^{2/3} \cdot a^{1/3}$

35. $y^{5/7} \cdot y^{1/7}$

36. $m^{1/4} \cdot m^{5/4}$

37. $b^{2/3} \cdot b^{3/2}$

38. $p^{5/6} \cdot p^{2/3}$

39. $\dfrac{x^{2/3}}{x^{1/3}}$

40. $\dfrac{a^{5/6}}{a^{1/6}}$

41. $\dfrac{s^{7/5}}{s^{2/5}}$

42. $\dfrac{z^{9/2}}{z^{3/2}}$

43. $\dfrac{w^{7/6}}{w^{1/2}}$

44. $\dfrac{b^{7/6}}{b^{2/3}}$

45. $(x^{3/4})^{4/3}$

46. $(y^{4/3})^{3/4}$

47. $(a^{2/5})^{3/2}$

48. $(p^{3/4})^{2/3}$

49. $(y^{-2/3})^{6}$

50. $(w^{-2/3})^{6}$

51. $(a^{2/3} \cdot b^{3/2})^{6}$

52. $(p^{3/4} \cdot q^{5/2})^{4}$

53. $(x^{2/7} \cdot y^{3/7})^{7}$

54. $(3m^{3/4} \cdot n^{5/4})^{4}$

55. $(s^{3/4} \cdot t^{1/4})^{4/3}$

56. $(x^{5/2} \cdot y^{5/7})^{2/5}$

10-49

835

57. $(8p^{3/2} \cdot q^{5/2})^{2/3}$

58. $(16a^{1/3} \cdot b^{2/3})^{3/4}$

59. $(x^{3/5} \cdot y^{3/4} \cdot z^{3/2})^{2/3}$

60. $(p^{5/6} \cdot q^{2/3} \cdot r^{5/3})^{3/5}$

61. $\dfrac{a^{5/6} \cdot b^{3/4}}{a^{1/3} \cdot b^{1/2}}$

62. $\dfrac{x^{2/3} \cdot y^{3/4}}{x^{1/2} \cdot y^{1/2}}$

63. $\dfrac{(r^{-1} \cdot s^{1/2})^3}{r \cdot s^{-1/2}}$

64. $\dfrac{(w^{-2} \cdot z^{-1/4})^6}{w^{-8}z^{1/2}}$

65. $\left(\dfrac{x^{12}}{y^8}\right)^{1/4}$

66. $\left(\dfrac{q^{12}}{p^{28}}\right)^{1/4}$

67. $\left(\dfrac{m^{-1/4}}{n^{1/2}}\right)^4$

68. $\left(\dfrac{r^{1/5}}{s^{-1/2}}\right)^{10}$

69. $\left(\dfrac{r^{-1/2} \cdot s^{3/4}}{t^{1/4}}\right)^4$

70. $\left(\dfrac{a^{1/3} \cdot b^{-1/6}}{c^{-1/6}}\right)^6$

71. $\left(\dfrac{8x^3 \cdot y^{-6}}{z^{-9}}\right)^{1/3}$

72. $\left(\dfrac{16p^{-4} \cdot q^6}{r^2}\right)^{-1/2}$

73. $\left(\dfrac{16m^{-3/5} \cdot n^2}{m^{1/5} \cdot n^{-2}}\right)^{1/4}$

74. $\left(\dfrac{27x^{5/6} \cdot y^{-4/3}}{x^{-7/6} \cdot y^{5/3}}\right)^{1/3}$

75. $\left(\dfrac{x^{3/2} \cdot y^{1/2}}{z^2}\right)^{1/2}\left(\dfrac{x^{3/4} \cdot y^{3/2}}{z^{-3}}\right)^{1/3}$

76. $\left(\dfrac{p^{1/2} \cdot q^{4/3}}{r^{-4}}\right)^{3/4}\left(\dfrac{p^{15/8} \cdot q^{-3}}{r^6}\right)^{1/3}$

In Exercises 77 to 84, write each expression in radical form. Do not simplify.

77. $a^{3/4}$

78. $m^{5/6}$

79. $2x^{2/3}$

80. $3m^{-2/5}$

81. $3x^{2/5}$

82. $2y^{-3/4}$

83. $(3x)^{2/5}$

84. $(2y)^{-3/4}$

In Exercises 85 to 88, write each expression using rational exponents, and simplify where necessary.

85. $\sqrt{7a}$

86. $\sqrt{25w^4}$

87. $\sqrt[3]{8m^6n^9}$

88. $\sqrt[5]{32r^{10}s^{15}}$

In Exercises 89 to 92, evaluate each expression, using a calculator. Round each answer to three decimal places.

89. $46^{3/5}$

90. $23^{2/7}$

91. $12^{-2/5}$

92. $36^{-3/4}$

93. Describe the difference between x^{-2} and $x^{1/2}$.

94. Some rational exponents, such as $\dfrac{1}{2}$, can easily be rewritten as terminating decimals (0.5). Others, such as $\dfrac{1}{3}$, cannot. What is it that determines which rational numbers can be rewritten as terminating decimals?

In Exercises 95 to 104, apply the appropriate multiplication patterns. Then simplify your result.

95. $a^{1/2}(a^{3/2} + a^{3/4})$

96. $2x^{1/4}(3x^{3/4} - 5x^{-1/4})$

97. $(a^{1/2} + 2)(a^{1/2} - 2)$

98. $(w^{1/3} - 3)(w^{1/3} + 3)$

99. $(m^{1/2} + n^{1/2})(m^{1/2} - n^{1/2})$

100. $(x^{1/3} + y^{1/3})(x^{1/3} - y^{1/3})$

101. $(x^{1/2} + 2)^2$

102. $(a^{1/3} - 3)^2$

103. $(r^{1/2} + s^{1/2})^2$

104. $(p^{1/2} - q^{1/2})^2$

As is suggested by several of the preceding exercises, certain expressions containing rational exponents are factorable. For instance, to factor $x^{2/3} - x^{1/3} - 6$, let $u = x^{1/3}$. Note that $x^{2/3} = (x^{1/3})^2 = u^2$.
    Substituting, we have $u^2 - u - 6$, and factoring yields $(u - 3)(u + 2)$ or $(x^{1/3} - 3)(x^{1/3} + 2)$.

In Exercises 105 to 110, use this technique to factor each expression.

105. $x^{2/3} + 4x^{1/3} + 3$ **106.** $y^{2/5} - 2y^{1/5} - 8$

107. $a^{4/5} - 7a^{2/5} + 12$ **108.** $w^{4/3} + 3w^{2/3} - 10$

109. $x^{4/3} - 4$ **110.** $x^{2/5} - 16$

In Exercises 111 to 120, perform the indicated operations. Assume that n represents a positive integer and that the denominators are not zero.

111. $x^{3n} \cdot x^{2n}$ **112.** $p^{1-n} \cdot p^{n+3}$ **113.** $(y^2)^{2n}$ **114.** $(a^{3n})^3$

115. $\dfrac{r^{n+2}}{r^n}$ **116.** $\dfrac{w^n}{w^{n-3}}$ **117.** $(a^3 \cdot b^2)^{2n}$ **118.** $(c^4 \cdot d^2)^{3m}$

119. $\left(\dfrac{x^{n+2}}{x^n}\right)^{1/2}$ **120.** $\left(\dfrac{b^n}{b^{n-3}}\right)^{1/3}$

In Exercises 121 to 124, write each expression in exponent form, simplify, and give the result as a single radical.

121. $\sqrt{\sqrt{x}}$ **122.** $\sqrt[3]{\sqrt{a}}$ **123.** $\sqrt[4]{\sqrt{y}}$ **124.** $\sqrt{\sqrt[3]{w}}$

In Exercises 125 to 130, simplify each expression. Write your answer in scientific notation.

125. $(4 \times 10^8)^{1/2}$ **126.** $(8 \times 10^6)^{1/3}$ **127.** $(16 \times 10^{-12})^{1/4}$

128. $(9 \times 10^{-4})^{1/2}$ **129.** $(16 \times 10^{-8})^{1/2}$ **130.** $(16 \times 10^{-8})^{3/4}$

131. While investigating rainfall runoff in a region of semiarid farmland, a researcher encounters the following expression:

$$t = C\left(\frac{L}{xy^2}\right)^{1/3}$$

Evaluate t when $C = 20$, $L = 600$, $x = 3$, and $y = 5$.

132. The average velocity of water in an open irrigation ditch is given by the formula

$$V = \frac{1.5x^{2/3}y^{1/2}}{z}$$

Evaluate V when $x = 27$, $y = 16$, and $z = 12$.

133. Use the properties of exponents to decide what x should be to make each statement true. Explain your choices regarding which properties of exponents you decide to use.

(a) $(a^{2/3})^x = a$ (b) $(a^{5/6})^x = \dfrac{1}{a}$ (c) $a^{2x} \cdot a^{3/2} = 1$ (d) $(\sqrt{a^{2/3}})^x = a$

134. The geometric mean is used to measure average inflation rates or interest rates. If prices increased by 15% over 5 years, then the average *annual* rate of inflation is obtained by taking the fifth root of 1.15:

$$(1.15)^{1/5} = 1.0283 \quad \text{or} \quad {\sim}2.8\%$$

The 1 is added to 0.15 because we are taking the original price and adding 15% of that price. We could write that as

$$P + 0.15P$$

Factoring, we get

$$P + 0.15P = P(1 + 0.15)$$

$$= P(1.15)$$

From December 1990 through February 1997, the Bureau of Labor Statistics computed an inflation rate of 16.2%, which is equivalent to an annual growth rate of 2.46%. From December 1990 through February 1997 is 75 months. To what exponent was 1.162 raised to obtain this average annual growth rate?

135. On your calculator, try evaluating $(-9)^{4/2}$ in the following two ways:

(a) $[(-9)^4]^{1/2}$ (b) $[(-9)^{1/2}]^4$

Discuss the results.

SECTION 10.6 Complex Numbers

10.6 OBJECTIVES

1. Define a complex number
2. Add and subtract complex numbers
3. Multiply and divide complex numbers

Radicals such as

$$\sqrt{-4} \quad \text{and} \quad \sqrt{-49}$$

are *not* real numbers because no real number squared produces a negative number. Our work in this section will extend our number system to include these **imaginary numbers,** which will allow us to consider radicals such as $\sqrt{-4}$.

First we offer a definition.

The Imaginary Number *i*

The number *i* is defined as

$$i = \sqrt{-1}$$

Note that this means that

$$i^2 = -1$$

This definition of the number *i* gives us an alternate means of indicating the square root of a negative number.

Rules and Procedures: Writing an Imaginary Number

When *a* is a positive real number,

$$\sqrt{-a} = \sqrt{a}i \quad \text{or} \quad i\sqrt{a}$$

Example 1 | **Using the Number *i***

Write each expression as a multiple of *i*.

(a) $\sqrt{-4} = \sqrt{4}i = 2i$

(b) $-\sqrt{-9} = -\sqrt{9}i = -3i$

Note: We simplify $\sqrt{8}$ as $2\sqrt{2}$. Note that we write the *i in front of* the radical to make it clear that *i* is *not part* of the radicand.

(c) $\sqrt{-8} = \sqrt{8}i = 2\sqrt{2}i$ or $2i\sqrt{2}$

(d) $\sqrt{-7} = \sqrt{7}i$ or $i\sqrt{7}$

✔ *CHECK YOURSELF 1*

Write each radical as a multiple of *i*.

(a) $\sqrt{-25}$ (b) $\sqrt{-24}$

We are now ready to define complex numbers in terms of the number *i*.

Note: The term *imaginary number* was introduced by René Descartes in 1637. Euler used *i* to indicate $\sqrt{-1}$ in 1748, but it was not until 1832 that Gauss used the term *complex number*.

Complex Number

A **complex number** is any number that can be written in the form

$$a + bi$$

in which *a* and *b* are real numbers and

$$i = \sqrt{-1}$$

Note: The first application of these numbers was made by Charles Steinmetz (1865–1923) in explaining the behavior of electric circuits.

Note: Also, 5*i* is called a **pure imaginary** number.

Note: The real numbers can be considered a subset of the set of complex numbers.

The form $a + bi$ is called the **standard form** of a complex number. We call *a* the **real part** of the complex number and *b* the **imaginary part.** Some examples follow.

$3 + 7i$ is an example of a complex number with real part 3 and imaginary part 7.

$5i$ is also a complex number because it can be written as $0 + 5i$.

-3 is a complex number because it can be written as $-3 + 0i$.

The basic operations of addition and subtraction on complex numbers are defined here.

Rules and Properties: Adding and Subtracting Complex Numbers

For the complex numbers $a + bi$ and $c + di$,

$$(a + bi) + (c + di) = (a + c) + (b + d)i$$
$$(a + bi) - (c + di) = (a - c) + (b - d)i$$

In words, we add or subtract the real parts and the imaginary parts of the complex numbers.

Example 2 illustrates the use of these definitions.

| Example 2 | **Adding and Subtracting Complex Numbers** |

Perform the indicated operations.

Note: The regrouping is essentially a matter of combining like terms.

(a) $(5 + 3i) + (6 - 7i) = (5 + 6) + (3 - 7)i$
$$= 11 - 4i$$

(b) $5 + (7 - 5i) = (5 + 7) + (-5i)$
$$= 12 - 5i$$

(c) $(8 - 2i) - (3 - 4i) = (8 - 3) + [-2 - (-4)]i$
$$= 5 + 2i$$

✔ *CHECK YOURSELF 2*

Perform the indicated operations.

(a) $(4 - 7i) + (3 - 2i)$ (b) $-7 + (-2 + 3i)$ (c) $(-4 + 3i) - (-2 - i)$

Because complex numbers are binomial, the product of two complex numbers is found by applying our earlier multiplication pattern for binomials, as Example 3 illustrates.

| Example 3 | **Multiplying Complex Numbers** |

Multiply.

(a) $(2 + 3i)(3 - 4i)$
$$= 2 \cdot 3 + 2(-4i) + (3i)3 + (3i)(-4i)$$

Note: We can replace i^2 with -1 because of the definition of i, and we usually do so because of the resulting simplification.

$$= 6 + (-8i) + 9i + (-12i^2)$$
$$= 6 - 8i + 9i + (-12)(-1)$$
$$= 6 + i + 12$$
$$= 18 + i$$

(b) $(1 - 2i)(3 - 4i)$
$$= 1 \cdot 3 + 1(-4i) + (-2i)3 + (-2i)(-4i)$$
$$= 3 + (-4i) + (-6i) + 8i^2$$
$$= 3 - 10i + 8(-1)$$
$$= 3 - 10i - 8$$
$$= -5 - 10i$$

✔ *CHECK YOURSELF 3*

Multiply $(2 - 5i)(3 - 2i)$.

Example 3 suggests the following pattern for multiplication on complex numbers.

> **Rules and Properties: Multiplying Complex Numbers**
>
> For the complex numbers $a + bi$ and $c + di$,
>
> $$(a + bi)(c + di) = ac + adi + bci + bdi^2$$
> $$= ac + adi + bci - bd$$
> $$= (ac - bd) + (ad + bc)i$$

This formula for the general product of two complex numbers can be memorized. However, you will find it much easier to get used to the multiplication pattern as it is applied to complex numbers than to memorize this formula.

There is one particular product form that will seem very familiar. We call $a + bi$ and $a - bi$ **complex conjugates.** For instance,

$$3 + 2i \qquad \text{and} \qquad 3 - 2i$$

are complex conjugates.

Consider the product

$$(3 + 2i)(3 - 2i) = 3^2 - (2i)^2$$
$$= 9 - 4i^2 = 9 - 4(-1)$$
$$= 9 + 4 = 13$$

The product of $3 + 2i$ and $3 - 2i$ is a real number. In general, we can write the product of two complex conjugates as

$$(a + bi)(a - bi) = a^2 + b^2$$

The fact that this product is always a real number will be very useful when we consider the division of complex numbers later in this section.

| Example 4 | **Multiplying Complex Numbers** |

Multiply.

Note: We could get the same result by applying the formula above with $a = 7$ and $b = 4$.

$$(7 - 4i)(7 + 4i) = 7^2 - (4i)^2$$
$$= 7^2 - 4^2(-1)$$
$$= 7^2 + 4^2$$
$$= 49 + 16 = 65$$

✔ *CHECK YOURSELF 4*

Multiply $(5 + 3i)(5 - 3i)$.

We are now ready to discuss the division of complex numbers. Generally, we find the quotient by multiplying the numerator and denominator by the conjugate of the denominator, as Example 5 illustrates.

| Example 5 | **Dividing Complex Numbers** |

Divide.

Note: Think of $3i$ as $0 + 3i$ and of its conjugate as $0 - 3i$, or $-3i$.

(a) $\dfrac{6 + 9i}{3i}$

$$\frac{6 + 9i}{3i} = \frac{(6 + 9i)(-3i)}{(3i)(-3i)}$$
The conjugate of $3i$ is $-3i$, and so we multiply the numerator and denominator by $-3i$.

Note: Multiplying the numerator and denominator in the original expression by i would yield the same result. Try it yourself.

$$= \frac{-18i - 27i^2}{-9i^2}$$

$$= \frac{-18i - 27(-1)}{(-9)(-1)}$$

$$= \frac{27 - 18i}{9} = 3 - 2i$$

Note: We multiply by $\dfrac{3 - 2i}{3 - 2i}$, which equals 1.

(b) $\dfrac{3 - i}{3 + 2i} = \dfrac{(3 - i)(3 - 2i)}{(3 + 2i)(3 - 2i)}$

$$= \frac{9 - 6i - 3i + 2i^2}{9 - 4i^2}$$

$$= \frac{9 - 9i - 2}{9 + 4}$$

Note: To write a complex number in standard form, we separate the real component from the imaginary.

$$= \frac{7 - 9i}{13} = \frac{7}{13} - \frac{9}{13}i$$

(c) $\dfrac{2 + i}{4 - 5i} = \dfrac{(2 + i)(4 + 5i)}{(4 - 5i)(4 + 5i)}$

$$= \frac{8 + 4i + 10i + 5i^2}{16 - 25i^2}$$

$$= \frac{8 + 14i - 5}{16 + 25}$$

$$= \frac{3 + 14i}{41} = \frac{3}{41} + \frac{14}{41}i$$

✔ *CHECK YOURSELF 5*

Divide.

(a) $\dfrac{5 + i}{5 - 3i}$ (b) $\dfrac{4 + 10i}{2i}$

We have seen that $i = \sqrt{-1}$ and $i^2 = -1$. We can use these two values to develop a table for the powers of i.

$$i = i$$

$$i^2 = -1$$

$$i^3 = i^2 \cdot i = -1 \cdot i = -i$$

$$i^4 = i^2 \cdot i^2 = -1 \cdot (-1) = 1$$

$$i^5 = i^4 \cdot i = 1 \cdot i = i$$

$$i^6 = i^4 \cdot i^2 = 1 \cdot (-1) = -1$$

$$i^7 = i^4 \cdot i^3 = 1 \cdot (-i) = -i$$

$$i^8 = i^4 \cdot i^4 = 1 \cdot 1 = 1$$

This pattern i, -1, $-i$, 1 repeats forever. You will see it again in the exercise set (and then in many subsequent math classes!).

We conclude this section with the following diagram, which summarizes the structure of the system of complex numbers.

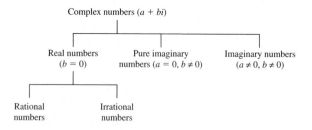

✔ *CHECK YOURSELF ANSWERS*

1. (a) $5i$; (b) $2i\sqrt{6}$ **2.** (a) $7 - 9i$; (b) $-9 + 3i$; (c) $-2 + 4i$

3. $-4 - 19i$ **4.** 34 **5.** (a) $\dfrac{11}{17} + \dfrac{10}{17}i$; (b) $5 - 2i$

Exercises · 10.6

In Exercises 1 to 10, write each root as a multiple of i. Simplify your results where necessary.

1. $\sqrt{-16}$ **2.** $\sqrt{-36}$ **3.** $-\sqrt{-121}$ **4.** $-\sqrt{-25}$

5. $\sqrt{-21}$ **6.** $\sqrt{-23}$ **7.** $\sqrt{-12}$ **8.** $\sqrt{-24}$

9. $-\sqrt{-108}$ **10.** $-\sqrt{-192}$

In Exercises 11 to 26, perform the indicated operations.

11. $(5 + 4i) - (6 + 5i)$

12. $(2 + 3i) + (4 + 5i)$

13. $(-3 - 2i) + (2 + 3i)$

14. $(-5 - 3i) + (-2 + 7i)$

15. $(5 + 4i) - (3 + 2i)$

16. $(7 + 6i) - (3 + 5i)$

17. $(8 - 5i) - (3 + 2i)$

18. $(7 - 3i) - (-2 - 5i)$

19. $(3 + i) + (4 + 5i) - 7i$

20. $(3 - 2i) + (2 + 3i) + 7i$

21. $(2 + 3i) - (3 - 5i) + (4 + 3i)$

22. $(5 - 7i) + (7 + 3i) - (2 - 7i)$

23. $(7 + 3i) - [(3 + i) - (2 - 5i)]$

24. $(8 - 2i) - [(4 + 3i) - (-2 + i)]$

25. $(5 + 3i) + (-5 - 3i)$

26. $(9 - 11i) + (-9 + 11i)$

In Exercises 27 to 42, find each product. Write your answer in standard form.

27. $3i(3 + 5i)$ **28.** $2i(7 + 3i)$ **29.** $6i(2 - 5i)$

30. $2i(6 + 3i)$ **31.** $-2i(4 - 3i)$ **32.** $-5i(2 - 7i)$

33. $6i\left(\dfrac{2}{3} + \dfrac{5}{6}i\right)$ **34.** $4i\left(\dfrac{1}{2} + \dfrac{3}{4}i\right)$ **35.** $(4 + 3i)(4 - 3i)$

36. $(5 - 2i)(3 - i)$ **37.** $(4 - 3i)(2 + 5i)$ **38.** $(7 + 2i)(3 - 2i)$

39. $(-2 - 3i)(-3 + 4i)$ **40.** $(-5 - i)(-3 - 4i)$ **41.** $(7 - 3i)^2$

42. $(3 + 7i)^2$

In Exercises 43 to 50, write the conjugate of each complex number. Then find the product of the given number and the conjugate.

43. $3 - 2i$ **44.** $5 + 2i$ **45.** $3 + 2i$ **46.** $7 - i$

47. $-3 - 2i$ **48.** $-5 - 7i$ **49.** $5i$ **50.** $-3i$

In Exercises 51 to 62, find each quotient, and write your answer in standard form.

51. $\dfrac{3 + 2i}{i}$

52. $\dfrac{5 - 3i}{-i}$

53. $\dfrac{5 - 2i}{3i}$

54. $\dfrac{8 + 12i}{-4i}$

55. $\dfrac{3}{2 + 5i}$

56. $\dfrac{5}{2 - 3i}$

57. $\dfrac{13}{2 + 3i}$

58. $\dfrac{-17}{3 + 5i}$

59. $\dfrac{2 + 3i}{4 + 3i}$

60. $\dfrac{4 - 2i}{5 - 3i}$

61. $\dfrac{3 - 4i}{3 + 4i}$

62. $\dfrac{7 + 2i}{7 - 2i}$

63. The first application of complex numbers was suggested by the Norwegian surveyor Caspar Wessel in 1797. He found that complex numbers could be used to represent distance and direction on a two-dimensional grid. Why would a surveyor care about such a thing?

64. To what sets of numbers does 1 belong?

We defined $\sqrt{-4} = \sqrt{4}\,i = 2i$ in the process of expressing the square root of a negative number as a multiple of i. Particular care must be taken with products where two negative radicands are involved. For instance,

$$\sqrt{-3} \cdot \sqrt{-12} = (i\sqrt{3})(i\sqrt{12})$$

$$= i^2\sqrt{36} = (-1)\sqrt{36} = -6$$

is correct. However, if we try to apply the product property for radicals, we have

$$\sqrt{-3} \cdot \sqrt{-12} \overset{?}{=} \sqrt{(-3)(-12)} = \sqrt{36} = 6$$

which is *not* correct. The property $\sqrt{a} \cdot \sqrt{b} = \sqrt{ab}$ is not applicable in the case where a and b are both negative. Radicals such as $\sqrt{-a}$ must be written in the standard form $i\sqrt{a}$ *before* multiplying to use the rules for real-valued radicals.

In Exercises 65 to 72, find each product.

65. $\sqrt{-5} \cdot \sqrt{-7}$

66. $\sqrt{-3} \cdot \sqrt{-10}$

67. $\sqrt{-2} \cdot \sqrt{-18}$

68. $\sqrt{-4} \cdot \sqrt{-25}$

69. $\sqrt{-6} \cdot \sqrt{-15}$

70. $\sqrt{-5} \cdot \sqrt{-30}$

71. $\sqrt{-10} \cdot \sqrt{-10}$

72. $\sqrt{-11} \cdot \sqrt{-11}$

In Exercises 73 to 80, simplify each power of i.

73. i^{10}

74. i^9

75. i^{20}

76. i^{15}

77. i^{38}

78. i^{40}

79. i^{51}

80. i^{61}

81. Show that a square root of i is $\dfrac{\sqrt{2}}{2} + \dfrac{\sqrt{2}}{2}i$. That is, $\left(\dfrac{\sqrt{2}}{2} + \dfrac{\sqrt{2}}{2}i\right)^2 = i$.

Summary for Chapter 10

Example	Topic	Reference
	Roots and Radicals	**10.1**
$\sqrt{25} = 5$ 5 is the principal square root of 25 because $5^2 = 25$. $-\sqrt{49} = -7$ because $(-7)^2 = 49$.	**Square Roots** Every positive number has two square roots. The positive or principal square root of a number a is denoted $$\sqrt{a}$$ The negative square root is written as $$-\sqrt{a}$$	**p. 788**
$\sqrt[3]{27} = 3$ $\sqrt[3]{-64} = -4$ $\sqrt[4]{81} = 3$ $\sqrt{4^2} = 4$ $\sqrt{(-5)^2} = 5$ $\sqrt[3]{(-3)^3} = -3$ $\sqrt{m^2} = |m|$ $\sqrt[3]{27x^3} = 3x$	**Higher Roots** Cube roots, fourth roots, and so on are denoted by using an index and a radical. The principal nth root of a is written as Index $\rightarrow$ $\sqrt[n]{a}$ $\leftarrow$ Radicand Radical sign **Radicals Containing Variables** In general, $$\sqrt[n]{x^n} = \begin{cases} |x| & \text{if } n \text{ is even} \\ x & \text{if } n \text{ is odd} \end{cases}$$	**p. 790** **p. 792**
	Simplifying Radical Expressions	**10.2**
	Simplifying radical expressions entails applying two properties for radicals.	
$\sqrt{35} = \sqrt{5 \cdot 7}$ $\phantom{\sqrt{35}} = \sqrt{5} \cdot \sqrt{7}$	**Product Property** $$\sqrt[n]{ab} = \sqrt[n]{a} \cdot \sqrt[n]{b}$$	**p. 797**
$\sqrt{\dfrac{2}{5}} = \dfrac{\sqrt{2}}{\sqrt{5}}$	**Quotient Property** $$\sqrt[n]{\dfrac{a}{b}} = \dfrac{\sqrt[n]{a}}{\sqrt[n]{b}} \qquad b \neq 0$$	**p. 797**
$\sqrt{18x^3} = \sqrt{9x^2 \cdot 2x}$ $\phantom{\sqrt{18x^3}} = \sqrt{9x^2} \cdot \sqrt{2x}$ $\phantom{\sqrt{18x^3}} = 3x\sqrt{2x}$ $\sqrt{\dfrac{5}{9}} = \dfrac{\sqrt{5}}{\sqrt{9}} = \dfrac{\sqrt{5}}{3}$	**Simplified Form for Radicals** A radical is in *simplified form* if the following three conditions are satisfied. 1. The radicand has no factor raised to a power greater than or equal to the index. 2. No fraction appears in the radical. 3. No radical appears in a denominator.	**p. 798**

$$\sqrt{\frac{3}{7x}} = \frac{\sqrt{3}}{\sqrt{7x}} = \frac{\sqrt{3} \cdot \sqrt{7x}}{\sqrt{7x} \cdot \sqrt{7x}}$$

$$= \frac{\sqrt{21x}}{\sqrt{49x^2}} = \frac{\sqrt{21x}}{7x}$$

Note: Satisfying the third condition may require *rationalizing the denominator.*

Operations on Radical Expressions | 10.3

$$8\sqrt{5} + 3\sqrt{5} = (8+3)\sqrt{5}$$
$$= 11\sqrt{5}$$

$$2\sqrt{18} - 4\sqrt{2}$$
$$= 2\sqrt{9 \cdot 2} - 4\sqrt{2}$$
$$= 2\sqrt{9} \cdot \sqrt{2} - 4\sqrt{2}$$
$$= 2 \cdot 3\sqrt{2} - 4\sqrt{2}$$
$$= 6\sqrt{2} - 4\sqrt{2} = (6-4)\sqrt{2}$$
$$= 2\sqrt{2}$$

Radical expressions may be combined by using addition or subtraction only if they are *similar,* that is, if they have the same radicand with the same index.

Similar radicals are combined by application of the distributive property.

p. 807

$$\sqrt{3x} \cdot \sqrt{6x^2} = \sqrt{18x^3}$$
$$= \sqrt{9x^2 \cdot 2x}$$
$$= \sqrt{9x^2} \cdot \sqrt{2x}$$
$$= 3x\sqrt{2x}$$

$$\sqrt{2}(5 + \sqrt{8}) = \sqrt{2} \cdot 5$$
$$+ \sqrt{2} \cdot \sqrt{8}$$
$$= 5\sqrt{2} + 4$$

$$(3 + \sqrt{2})(5 - \sqrt{2})$$
$$= 15 - 3\sqrt{2} + 5\sqrt{2} - 2$$
$$= 13 + 2\sqrt{2}$$

Multiplication   To multiply two radical expressions, we use

$$\sqrt[n]{a} \cdot \sqrt[n]{b} = \sqrt[n]{ab}$$

and simplify the product.

If binomial expressions are involved, we use the distributive property or the FOIL method.

p. 810

$$\frac{5}{\sqrt{8}} = \frac{5 \cdot \sqrt{2}}{\sqrt{8} \cdot \sqrt{2}} = \frac{5\sqrt{2}}{\sqrt{16}}$$
$$= \frac{5\sqrt{2}}{4}$$

Note: $3 + \sqrt{5}$ is the conjugate of $3 - \sqrt{5}$.

$$\frac{2}{3 - \sqrt{5}} = \frac{2(3 + \sqrt{5})}{(3 - \sqrt{5})(3 + \sqrt{5})}$$
$$= \frac{2(3 + \sqrt{5})}{4}$$
$$= \frac{3 + \sqrt{5}}{2}$$

Division   To divide two radical expressions, rationalize the denominator by multiplying the numerator and denominator by the appropriate radical.

If the divisor (the denominator) is a binomial, multiply the numerator and denominator by the conjugate of the denominator.

p. 814

	Solving Radical Equations	10.4
If $\sqrt{x+1} = 5$ then $(\sqrt{x+1})^2 = 5^2$ $x+1 = 25$ $x = 24$	**Power Property of Equality** If $a = b$ then $a^n = b^n$	**p. 819**

Given $\sqrt{x} + \sqrt{x+7} = 7$ $\sqrt{x} = 7 - \sqrt{x+7}$ $x = 49 - 14\sqrt{x+7}$ $\quad + (x+7)$ $x = 56 + x - 14\sqrt{x+7}$ $-56 = -14\sqrt{x+7}$ $4 = \sqrt{x+7}$ $16 = x + 7$ $x = 9$	**Solving Equations Involving Radicals** **Step 1** Isolate a radical on one side of the equation. **Step 2** Raise each side of the equation to the smallest power that will eliminate the isolated radical. **Step 3** If any radicals remain in the equation derived in step 2, return to step 1 and continue the solution process. **Step 4** Solve the resulting equation to determine any possible solutions. **Step 5** Check all solutions to determine whether extraneous solutions may have resulted from step 2.	**p. 824**

	Rational Exponents	10.5
$36^{1/2} = \sqrt{36} = 6$ $-27^{1/3} = -\sqrt[3]{27} = -3$ $243^{1/5} = \sqrt[5]{243} = 3$ $25^{-1/2} = \dfrac{1}{\sqrt{25}} = \dfrac{1}{5}$ $27^{2/3} = (\sqrt[3]{27})^2$ $\quad = 3^2 = 9$ $(a^4 b^8)^{3/4} = \sqrt[4]{(a^4 b^8)^3}$ $\quad = \sqrt[4]{a^{12} b^{24}}$ $\quad = a^3 b^6$	Rational exponents are an alternate way of indicating roots. We use the following definition. If a is any real number and n is a positive integer $(n > 1)$, $$a^{1/n} = \sqrt[n]{a}$$ We restrict a so that a is nonnegative when n is even. We also define the following. For any real number a and positive integers m and n, with $n > 1$, then $$a^{m/n} = (\sqrt[n]{a})^m = \sqrt[n]{a^m}$$	**p. 829**

	Properties of Exponents The following five properties for exponents continue to hold for rational exponents. **Product Rule**	**p. 832**
$x^{1/2} \cdot x^{1/3} = x^{1/2+1/3} = x^{5/6}$	$$a^m \cdot a^n = a^{m+n}$$	
	Quotient Rule	**p. 832**
$\dfrac{x^{3/2}}{x^{1/2}} = x^{3/2-1/2} = x^{2/2} = x$	$$\dfrac{a^m}{a^n} = a^{m-n}$$	
	Power Rule	**p. 832**
$(x^{1/3})^5 = x^{1/3 \cdot 5} = x^{5/3}$	$$(a^m)^n = a^{m \cdot n}$$	

$(2xy)^{1/2} = 2^{1/2}x^{1/2}y^{1/2}$	**Product-Power Rule** $$(ab)^m = a^m b^m$$	**p. 832**
$\left(\dfrac{x^{1/3}}{3}\right)^2 = \dfrac{(x^{1/3})^2}{3^2}$ $= \dfrac{x^{2/3}}{9}$	**Quotient-Power Rule** $$\left(\frac{a}{b}\right)^m = \frac{a^m}{b^m}$$	**p. 832**

	Complex Numbers	**10.6**
$\sqrt{-16} = 4i$ $\sqrt{-8} = 2i\sqrt{2}$	The number i is defined as $$i = \sqrt{-1}$$ Note that this means that $$i^2 = -1$$ A **complex number** is any number that can be written in the form $$a + bi$$ in which a and b are real numbers and $$i = \sqrt{-1}$$	**p. 840**
$(2 + 3i) + (-3 - 5i)$ $= (2 - 3) + (3 - 5)i$ $= -1 - 2i$ $(5 - 2i) - (3 - 4i)$ $= (5 - 3) + [-2 - (-4)]i$ $= 2 + 2i$	**Addition and Subtraction** For the complex numbers $a + bi$ and $c + di$, $$(a + bi) + (c + di) = (a + c) + (b + d)i$$ and $\quad (a + bi) - (c + di) = (a - c) + (b - d)i$	**p. 840**
$(2 + 5i)(3 - 4i)$ $= 6 - 8i + 15i - 20i^2$ $= 6 + 7i - 20(-1)$ $= 26 + 7i$	**Multiplication** For the complex numbers $a + bi$ and $c + di$, $$(a + bi)(c + di) = (ac - bd) + (ad + bc)i$$ **Note:** It is generally easier to use the FOIL multiplication pattern and the definition of i than to apply the above formula.	**p. 842**
$\dfrac{3 + 2i}{3 - 2i} = \dfrac{(3 + 2i)(3 + 2i)}{(3 - 2i)(3 + 2i)}$ $= \dfrac{9 + 6i + 6i + 4i^2}{9 - 4i^2}$ $= \dfrac{9 + 12i + 4(-1)}{9 - 4(-1)}$ $= \dfrac{5 + 12i}{13}$ $= \dfrac{5}{13} + \dfrac{12}{13}i$	**Division** To divide two complex numbers, we multiply the numerator and denominator by the complex conjugate of the denominator and write the result in standard form.	**p. 843**

Summary Exercises ▪ 10

This summary exercise set is provided to give you practice with each of the objectives of the chapter. Each exercise is keyed to the appropriate chapter section.

[10.1] Evaluate each of the following roots over the set of real numbers.

1. $\sqrt{121}$

2. $-\sqrt{64}$

3. $\sqrt{-81}$

4. $\sqrt[3]{64}$

5. $\sqrt[3]{-64}$

6. $\sqrt[4]{81}$

7. $\sqrt{\dfrac{9}{16}}$

8. $\sqrt[3]{-\dfrac{8}{27}}$

9. $\sqrt{8^2}$

Simplify each of the following expressions. Assume that all variables represent positive real numbers for all subsequent exercises in this exercise set.

10. $\sqrt{4x^2}$

11. $\sqrt{a^4}$

12. $\sqrt{36y^2}$

13. $\sqrt{49w^4z^6}$

14. $\sqrt[3]{x^9}$

15. $\sqrt[3]{-27b^6}$

16. $\sqrt[3]{8r^3s^9}$

17. $\sqrt[4]{16x^4y^8}$

18. $\sqrt[5]{32p^5q^{15}}$

[10.2] Use the product property to write each of the following expressions in simplified form.

19. $\sqrt{45}$

20. $-\sqrt{75}$

21. $\sqrt{60x^2}$

22. $\sqrt{108a^3}$

23. $\sqrt[3]{32}$

24. $\sqrt[3]{-80w^4z^3}$

Use the quotient property to write each of the following expressions in simplified form.

25. $\sqrt{\dfrac{9}{16}}$

26. $\sqrt{\dfrac{7}{36}}$

27. $\sqrt{\dfrac{y^4}{49}}$

28. $\sqrt{\dfrac{2x}{9}}$

29. $\sqrt{\dfrac{5}{16x^2}}$

30. $\sqrt[3]{\dfrac{5a^2}{27}}$

[10.3] Simplify each of the following expressions if necessary. Then add or subtract as indicated.

31. $7\sqrt{10} + 4\sqrt{10}$

32. $5\sqrt{3x} - 2\sqrt{3x}$

33. $7\sqrt[3]{2x} + 3\sqrt[3]{2x}$

34. $8\sqrt{10} - 3\sqrt{10} + 2\sqrt{10}$

35. $\sqrt{72} + \sqrt{50}$

36. $\sqrt{54} - \sqrt{24}$

37. $9\sqrt{7} - 2\sqrt{63}$

38. $\sqrt{20} - \sqrt{45} + 2\sqrt{125}$

39. $2\sqrt[3]{16} + 3\sqrt[3]{54}$

40. $\sqrt{27w^3} - w\sqrt{12w}$

41. $\sqrt[3]{128a^5} + 6a\sqrt[3]{2a^2}$

42. $\sqrt{20} + \dfrac{3}{\sqrt{5}}$

43. $\sqrt{72x} - \sqrt{\dfrac{x}{2}}$

44. $\sqrt[3]{81a^4} - a\sqrt[3]{\dfrac{a}{9}}$

45. $\dfrac{\sqrt{15}}{3} - \dfrac{1}{\sqrt{15}}$

[10.3] Multiply and simplify each of the following expressions.

46. $\sqrt{3x} \cdot \sqrt{7y}$

47. $\sqrt{6x^2} \cdot \sqrt{18}$

48. $\sqrt[3]{4a^2b} \cdot \sqrt[3]{ab^2}$

49. $\sqrt{5}(\sqrt{3} + 2)$

50. $\sqrt{6}(\sqrt{8} - \sqrt{2})$

51. $\sqrt{a}(\sqrt{5a} + \sqrt{125a})$

52. $(\sqrt{3} + 5)(\sqrt{3} - 7)$

53. $(\sqrt{7} - \sqrt{2})(\sqrt{7} + \sqrt{3})$

54. $(\sqrt{5} - 2)(\sqrt{5} + 2)$

55. $(\sqrt{7} - \sqrt{3})(\sqrt{7} + \sqrt{3})$

56. $(2 + \sqrt{3})^2$

57. $(\sqrt{5} - \sqrt{2})^2$

Rationalize the denominator, and write each of the following expressions in simplified form.

58. $\sqrt{\dfrac{3}{7}}$

59. $\dfrac{\sqrt{12}}{\sqrt{x}}$

60. $\dfrac{\sqrt{10a}}{\sqrt{5b}}$

61. $\sqrt[3]{\dfrac{3}{a^2}}$

62. $\dfrac{2}{\sqrt[3]{3x}}$

63. $\dfrac{\sqrt[3]{x^2}}{\sqrt[3]{y^5}}$

Divide and simplify each of the following expressions.

64. $\dfrac{1}{3 + \sqrt{2}}$

65. $\dfrac{11}{5 - \sqrt{3}}$

66. $\dfrac{\sqrt{5} - 2}{\sqrt{5} + 2}$

67. $\dfrac{\sqrt{x} - 3}{\sqrt{x} + 3}$

[10.4] Solve each of the following equations. Be sure to check your solutions.

68. $\sqrt{x - 5} = 4$

69. $\sqrt{3x - 2} + 2 = 5$

70. $\sqrt{y + 7} = y - 5$

71. $\sqrt{2x - 1} + x = 8$

72. $\sqrt[3]{5x + 2} = 3$

73. $\sqrt[3]{x^2 + 2} - 3 = 0$

74. $\sqrt{z + 7} = 1 + \sqrt{z}$

75. $\sqrt{4x + 5} - \sqrt{x - 1} = 3$

[10.5] Evaluate each of the following expressions.

76. $49^{1/2}$

77. $-100^{1/2}$

78. $(-27)^{1/3}$

79. $16^{1/4}$

80. $64^{2/3}$

81. $25^{3/2}$

82. $\left(\dfrac{4}{9}\right)^{3/2}$

83. $49^{-1/2}$

84. $81^{-3/4}$

Use the properties of exponents to simplify each of the following expressions.

85. $x^{3/2} \cdot x^{5/2}$ **86.** $b^{2/3} \cdot b^{3/2}$ **87.** $\dfrac{r^{8/5}}{r^{3/5}}$ **88.** $\dfrac{a^{5/4}}{a^{1/2}}$

89. $(x^{3/5})^{2/3}$ **90.** $(y^{-4/3})^6$ **91.** $(x^{4/5}y^{3/2})^{10}$ **92.** $(16x^{1/3} \cdot y^{2/3})^{3/4}$

93. $\left(\dfrac{x^{-2}y^{-1/6}}{x^{-4}y}\right)^3$ **94.** $\left(\dfrac{27y^3z^{-6}}{x^{-3}}\right)^{1/3}$

Write each of the following expressions in radical form.

95. $x^{3/4}$ **96.** $(w^2z)^{2/5}$ **97.** $3a^{2/3}$ **98.** $(3a)^{2/3}$

Write each of the following expressions, using rational exponents, and simplify when necessary.

99. $\sqrt[5]{7x}$ **100.** $\sqrt{16w^4}$ **101.** $\sqrt[3]{27p^3q^9}$ **102.** $\sqrt[4]{16a^8b^{16}}$

[10.6] Write each of the following roots as a multiple of i. Simplify your result.

103. $\sqrt{-49}$ **104.** $\sqrt{-13}$ **105.** $-\sqrt{-60}$

Perform the indicated operations.

106. $(2 + 3i) + (3 - 5i)$ **107.** $(7 - 3i) + (-3 - 2i)$

108. $(5 - 3i) - (2 + 5i)$ **109.** $(-4 + 2i) - (-1 - 3i)$

Find each of the following products.

110. $4i(7 - 2i)$ **111.** $(5 - 2i)(3 + 4i)$

112. $(3 - 4i)^2$ **113.** $(2 - 3i)(2 + 3i)$

Find each of the following quotients, and write your answer in standard form.

114. $\dfrac{5 - 15i}{5i}$ **115.** $\dfrac{10}{3 - 4i}$

116. $\dfrac{3 - 2i}{3 + 2i}$ **117.** $\dfrac{5 + 10i}{2 + i}$

Self-Test • 10

The purpose of this self-test is to help you check your progress and to review for a chapter test in class. Allow yourself about 1 hour to take the test. When you are done, check your answers in the back of the book. If you missed any problems, be sure to go back and review the appropriate sections in the chapter and the exercises that are provided.

Simplify each expression. Assume that all variables represent positive real numbers in all subsequent problems.

1. $\sqrt{49a^4}$

2. $\sqrt[3]{-27w^6z^9}$

Use the product or quotient properties to write each expression in simplified form.

3. $\sqrt[3]{9p^7q^5}$

4. $\dfrac{7x}{\sqrt{64y^2}}$

Rationalize the denominator, and write each expression in simplified form.

5. $\sqrt{\dfrac{5x}{8y}}$

6. $\dfrac{3}{\sqrt[3]{9x}}$

Simplify each expression if necessary. Then add or subtract as indicated.

7. $\sqrt{3x^3} + x\sqrt{75x} - \sqrt{27x^3}$

8. $\sqrt[3]{54m^4} + m\sqrt[3]{16m}$

Multiply or divide as indicated. Then simplify your result.

9. $\sqrt{6x}(\sqrt{18x} - \sqrt{2x})$

10. $\dfrac{\sqrt{6} - \sqrt{3}}{\sqrt{6} + \sqrt{3}}$

11. $\sqrt{7x^3}\ \sqrt{2x^4}$

12. $\sqrt[3]{4x^5}\ \sqrt[3]{8x^6}$

Solve the following equations. Be sure to check your solutions.

13. $\sqrt{x - 7} - 2 = 0$

14. $\sqrt{3w + 4} + w = 8$

Use the properties of exponents to simplify each expression.

15. $(16x^4)^{3/2}$

16. $(27m^{3/2}n^{-6})^{2/3}$

17. $\left(\dfrac{16r^{-1/3}s^{5/3}}{rs^{-7/3}}\right)^{3/4}$

18. $\left(\dfrac{4x^{1/5}y^{1/5}}{x^{-7/5}y^{3/5}}\right)^{5/2}$

Use a calculator to evaluate the following. Round each answer to the nearest tenth.

19. $\sqrt{43}$

20. $\sqrt[3]{\dfrac{73}{27}}$

Write the expression in radical form and simplify.

21. $(a^7b^3)^{2/5}$

Write the expression, using rational exponents. Then simplify.

22. $\sqrt[3]{125p^9q^6}$

Perform the indicated operations.

23. $(-2 + 3i) - (-5 - 7i)$

24. $(5 - 3i)(-4 + 2i)$

25. $\dfrac{10 - 20i}{3 - i}$

854

Cumulative Test ■ 0–10

This test is provided to help you in the process of reviewing the previous chapters. Answers are provided in the back of the book. If you missed any answers, be sure to go back and review the appropriate chapter section.

1. Solve the equation $7x - 6(x - 1) = 2(5 + x) + 11$.

2. If $f(x) = 3x^6 - 4x^3 + 9x^2 - 11$, find $f(-1)$.

3. Find the equation of the line that has a y intercept of $(0, -6)$ and is parallel to the line $6x - 4y = 18$.

4. Solve the equation $|3x - 5| = 4$.

Simplify each of the following expressions.

5. $5x^2 - 8x + 11 - (-3x^2 - 2x + 8) - (-2x^2 - 4x + 3)$

6. $(5x + 3)(2x - 9)$

Factor each of the following completely.

7. $2x^3 + x^2 - 3x$

8. $9x^4 - 36y^4$

9. $4x^2 + 8xy - 5x - 10y$

Simplify each of the following rational expressions.

10. $\dfrac{2x^2 + 13x + 15}{6x^2 + 7x - 3}$

11. $\dfrac{3}{x - 5} - \dfrac{2}{x - 1}$

12. $\dfrac{a^2 - 4a}{a^2 - 6a + 8} \cdot \dfrac{a^2 - 4}{2a^2}$

13. $\dfrac{a^2 - 9}{a^2 - a - 12} \div \dfrac{a^2 - a - 6}{a^2 - 2a - 8}$

Simplify each of the following radical expressions.

14. $\sqrt{3x^3y}\,\sqrt{4x^5y^6}$

15. $(\sqrt{3}-5)(\sqrt{2}+3)$

Graph each equation.

16. $y = 3x - 5$

17. $x = -5$

18. $2x - 3y = 12$

19. Solve the system of equations.

$$4x - 3y = 15$$
$$x + y = 2$$

20. Solve the following inequalities algebraically.

(a) $5x - (2 - 3x) \geq 6 + 10x$ (b) $|x - 2| < 8$

Simplify the following.

21. $\left(\dfrac{x^2y^{-3}}{x^5y^4}\right)^{-2}$

22. $(-12x^3y^2)(-18xy^3)$

23. $\left(\dfrac{64x^3y^9}{27}\right)^{1/3}$

Solve the following equations.

24. $\sqrt{x + 3} = 2$

25. $|x + 1| = |2x|$

11 Quadratic Functions

Running power lines from a power plant, wind farm, or hydroelectric plant to a city is a very costly enterprise. Land must be cleared, towers built, and conducting wires strung from tower to tower across miles of countryside. Typical construction designs run from about 300- to 1200-ft spans, with towers about 75 to 200 ft high. Of course, if a lot of towers are needed, and many of them must be tall, the construction costs skyrocket.

Power line construction carries a unique set of problems. Towers must be built tall enough and close enough to keep the conducting lines well above the ground. The sag of these wires (how much they droop from the towers) is a function of the weight of the conductor, the span length, and the tension in the wires. The amount of this sag, measured in feet, is approximated by the following formula:

$$\text{Sag} = \frac{wS^2}{8T} \qquad \begin{aligned} \text{where } w &= \text{weight of wires in pounds/foot} \\ S &= \text{span length in feet} \\ T &= \text{tension in wires measured in pounds} \end{aligned}$$

The actual curve of the power lines is called a **catenary curve.** The curve that we use to approximate the sag is a **parabola.**

857

SECTION 11.1

Solving Quadratic Equations by Completing the Square

11.1 OBJECTIVES

1. Solve quadratic equations by using the square root method
2. Solve quadratic equations by completing the square

Recall that a quadratic equation is an equation of the form $ax^2 + bx + c = 0$, where a is not equal to zero.

In Section 6.4, we solved quadratic equations by factoring and using the zero-product principle. However, not all equations are solvable by that method. In this section, we will learn another technique that can be used to solve some quadratic equations. This technique is called the **square root method.** Let's begin by reviewing the factoring method.

| Example 1 | Solving Equations by Factoring |

Solve the quadratic equation $2x^2 + x - 10 = 0$ by factoring.

Write the equation in factored form:

$$(x - 2)(2x + 5) = 0$$

Using the zero-product principle, we set each factor equal to 0.

$$x - 2 = 0 \quad \text{and} \quad 2x + 5 = 0$$

Solving, we have

$$x = 2 \qquad x = -\frac{5}{2} \quad \text{or} \quad \left\{ -\frac{5}{2}, 2 \right\}$$

 CHECK YOURSELF 1

Solve each of the following equations by factoring.

(a) $x^2 + x - 12 = 0$ (b) $3x^2 - x - 10 = 0$

858

| Example 2 | ## Solving Equations by Factoring |

Solve the quadratic equation $x^2 = 16$ by factoring.

We write the equation in standard form:

$$x^2 - 16 = 0$$

Here, we factor the quadratic member of the equation as a difference of squares.

Factoring, we have

$$(x + 4)(x - 4) = 0$$

Finally, the solutions are

$$x = -4 \quad \text{or} \quad x = 4 \quad \text{or} \quad \{\pm 4\}$$

✔ *CHECK YOURSELF 2*

Solve each of the following quadratic equations.

(a) $x^2 = 25$ (b) $5x^2 = 180$

The Square Root Method

The equation in Example 2 could have been solved in an alternative fashion. We could have used what is called the **square root method.** Again, given the equation

$$x^2 = 16$$

we can write the equivalent statement

Note: Be sure to include *both* the positive and the negative square roots when you use the square root method.

$$x = \sqrt{16} \quad \text{or} \quad x = -\sqrt{16}$$

This yields the solutions

$$x = 4 \quad \text{or} \quad x = -4 \quad \text{or} \quad \{\pm 4\}$$

This discussion leads us to the following general result.

Square Root Property

If $x^2 = k$, where k is any real number, then

$$x = \sqrt{k} \quad \text{or} \quad x = -\sqrt{k}$$

Example 3 further illustrates the use of this property.

Example 3   Using the Square Root Method

Solve each equation by using the square root method.

(a) $x^2 = 9$

By the square root property,

$$x = \sqrt{9} \quad \text{or} \quad x = -\sqrt{9}$$
$$= 3 \qquad\qquad = -3 \quad \text{or} \quad \{\pm 3\}$$

(b) $x^2 - 17 = 0$

Add 17 to both sides of the equation.

$$x^2 = 17$$

If a calculator were used, $\sqrt{17} = 4.123$ (rounded to three decimal places).

$$x = \sqrt{17} \quad \text{or} \quad x = -\sqrt{17} \quad \text{or} \quad \{\pm\sqrt{17}\} \quad \text{or} \quad \{-\sqrt{17}, \sqrt{17}\}$$

(c) $4x^2 - 3 = 0$

$$4x^2 = 3$$

$$x^2 = \frac{3}{4}$$

$$x = \pm\sqrt{\frac{3}{4}} = \pm\frac{\sqrt{3}}{\sqrt{4}}$$

$$x = \pm\frac{\sqrt{3}}{2} \quad \text{or} \quad \left\{\pm\frac{\sqrt{3}}{2}\right\}$$

In Example 3(d), we see that complex number solutions may result.

(d) $x^2 + 1 = 0$

$$x^2 = -1$$

$$x = \pm\sqrt{-1}$$

$$x = \pm i \quad \text{or} \quad \{\pm i\}$$

✔ CHECK YOURSELF 3

Solve each equation.

(a) $x^2 = 5$ (b) $x^2 - 2 = 0$ (c) $9x^2 - 8 = 0$ (d) $x^2 + 9 = 0$

We can also use the approach in Example 3 to solve an equation of the form

$$(x + 3)^2 = 16$$

As before, by the square root property we have

$$x + 3 = \pm 4 \qquad \text{Add } -3 \text{ to both sides of the equation.}$$

Solving for x yields

$$x = -3 \pm 4$$

which means that there are two solutions:

$$x = -3 + 4 \qquad \text{or} \qquad x = -3 - 4$$
$$= 1 \qquad\qquad\qquad = -7 \qquad \text{or} \qquad \{-7, 1\}$$

Example 4   Using the Square Root Method

Use the square root method to solve each equation.

(a) $(x - 5)^2 - 5 = 0$

$$(x - 5)^2 = 5$$

$$x - 5 = \pm\sqrt{5}$$

The two solutions $5 + \sqrt{5}$ and $5 - \sqrt{5}$ are abbreviated as $5 \pm \sqrt{5}$.

$$x = 5 \pm \sqrt{5} \qquad \text{or} \qquad \{5 \pm \sqrt{5}\}$$

(b) $9(y + 1)^2 - 2 = 0$

$$9(y + 1)^2 = 2$$

$$(y + 1)^2 = \frac{2}{9}$$

$$y + 1 = \pm\sqrt{\frac{2}{9}}$$

$$y = -1 \pm \frac{\sqrt{2}}{3}$$

$$= \frac{-3}{3} \pm \frac{\sqrt{2}}{3}$$

$$= \frac{-3 \pm \sqrt{2}}{3} \qquad \text{or} \qquad \left\{\frac{-3 \pm \sqrt{2}}{3}\right\}$$

✔ *CHECK YOURSELF 4*

Using the square root method, solve each equation.

(a) $(x - 2)^2 - 3 = 0$ (b) $4(x - 1)^2 = 3$

Completing the Square

Not all quadratic equations can be solved directly by factoring or using the square root method. We must extend our techniques.

The square root method is useful in this process because any quadratic equation can be written in the form

$$(x + h)^2 = k$$

which yields the solution

$$x = -h \pm \sqrt{k}$$

If $(x + h)^2 = k$, then

$$x + h = \pm\sqrt{k}$$

and

$$x = -h \pm \sqrt{k}$$

The process of changing an equation in standard form

$$ax^2 + bx + c = 0$$

to the form

$$(x + h)^2 = k$$

is called the method of **completing the square,** and it is based on the relationship between the middle term and the last term of any perfect-square trinomial.

Let's look at three perfect-square trinomials to see whether we can detect a pattern:

$$x^2 + 4x + \ \ 4 = (x + 2)^2 \tag{1}$$

$$x^2 - 6x + \ \ 9 = (x - 3)^2 \tag{2}$$

$$x^2 + 8x + 16 = (x + 4)^2 \tag{3}$$

Note that this relationship is true *only* if the leading, or x^2, coefficient is 1. That will be important later.

Note that in each case the last (or constant) term is the square of one-half of the coefficient of x in the middle (or linear) term. For example, in equation (2),

$$x^2 - 6x + 9 = (x - 3)^2$$

$$\frac{1}{2} \text{ of this coefficient is } -3,$$

$$\text{and } (-3)^2 = 9, \text{ the constant.}$$

Verify this relationship for yourself in equation (3). To summarize, in perfect-square trinomials, the constant is always the square of one-half the coefficient of x.

We are now ready to use the above observation in the solution of quadratic equations by completing the square. Consider Example 5.

Example 5   Completing the Square to Solve an Equation

Solve $x^2 + 8x - 7 = 0$ by completing the square.

First, we rewrite the equation with the constant on the *right-hand side:*

$$x^2 + 8x = 7$$

When you graph the related function $y = x^2 + 8x - 7$, you will note that the x values for the x intercepts are just below 1 and just above -9. Be certain that you see how these points relate to the exact solutions

$$-4 + \sqrt{23} \quad \text{and} \quad -4 - \sqrt{23}$$

Remember that if $(x + h)^2 = k$, then

$$x + h = \pm\sqrt{k}$$

so $x = -h \pm \sqrt{k}$

Our objective is to have a perfect-square trinomial on the left-hand side. We know that we must add the square of one-half of the x coefficient to complete the square. In this case, that value is 16, so now we add 16 to each side of the equation.

$$x^2 + 8x + 16 = 7 + 16 \qquad \frac{1}{2} \cdot 8 = 4 \text{ and } 4^2 = 16$$

Factor the perfect-square trinomial on the left, and combine like terms on the right to yield

$$(x + 4)^2 = 23$$

Now the square root property yields

$$x + 4 = \pm\sqrt{23}$$

Subtracting 4 from both sides of the equation gives

$$x = -4 \pm \sqrt{23} \qquad \text{or} \qquad \{-4 \pm \sqrt{23}\}$$

 CHECK YOURSELF 5

Solve $x^2 - 6x - 2 = 0$ by completing the square.

Example 6   Completing the Square to Solve an Equation

Solve $x^2 + 5x - 3 = 0$ by completing the square.

$$x^2 + 5x - 3 = 0 \qquad \text{Add 3 to both sides.}$$

$$x^2 + 5x = 3 \qquad \text{Make the left-hand side a perfect square.}$$

$$x^2 + 5x + \left(\frac{5}{2}\right)^2 = 3 + \left(\frac{5}{2}\right)^2$$

$$\left(x + \frac{5}{2}\right)^2 = \frac{37}{4} \qquad \text{Take the square root of both sides.}$$

$$x + \frac{5}{2} = \pm\frac{\sqrt{37}}{2} \qquad \text{Solve for } x.$$

$$x = \frac{-5 \pm \sqrt{37}}{2} \qquad \text{or} \qquad \left\{\frac{-5 \pm \sqrt{37}}{2}\right\}$$

Add the square of one-half of the x coefficient to both sides of the equation. Note that

$$\frac{1}{2} \cdot 5 = \frac{5}{2}$$

 CHECK YOURSELF 6

Solve $x^2 + 3x - 7 = 0$ by completing the square.

Some equations have nonreal complex solutions, as Example 7 illustrates.

| Example 7 | **Completing the Square to Solve an Equation** |

Solve $x^2 + 4x + 13 = 0$ by completing the square.

Note that the graph of $y = x^2 + 4x + 13$ does not intercept the x axis.

$$x^2 + 4x + 13 = 0 \qquad \text{Subtract 13 from both sides.}$$

$$x^2 + 4x = -13 \qquad \text{Add } \left[\frac{1}{2}(4)\right]^2 \text{ to both sides.}$$

$$x^2 + 4x + 4 = -13 + 4 \qquad \text{Factor the left-hand side.}$$

$$(x + 2)^2 = -9 \qquad \text{Take the square root of both sides.}$$

$$x + 2 = \pm\sqrt{-9} \qquad \text{Simplify the radical.}$$

$$x + 2 = \pm i\sqrt{9}$$

$$x + 2 = \pm 3i$$

$$x = -2 \pm 3i \qquad \text{or} \qquad \{-2 \pm 3i\}$$

 CHECK YOURSELF 7

Solve $x^2 + 10x + 41 = 0$.

Example 8 illustrates a situation in which the leading coefficient of the quadratic member is not equal to 1. As you will see, an extra step will be required.

| Example 8 | **Completing the Square to Solve an Equation** |

Solve $4x^2 + 8x - 7 = 0$ by completing the square.

Caution

Before you can complete the square on the left, the coefficient of x^2 must be equal to 1. Otherwise, we must *divide* both sides of the equation by that coefficient.

$$4x^2 + 8x - 7 = 0 \qquad \text{Add 7 to both sides.}$$

$$4x^2 + 8x = 7 \qquad \text{Divide both sides by 4.}$$

$$x^2 + 2x = \frac{7}{4} \qquad \text{Now, complete the square on the left.}$$

$$x^2 + 2x + 1 = \frac{7}{4} + 1 \qquad \text{The left side is now a perfect square.}$$

$$(x + 1)^2 = \frac{11}{4}$$

$$x + 1 = \pm\sqrt{\frac{11}{4}}$$

$$x = -1 \pm \sqrt{\frac{11}{4}}$$

$$= -1 \pm \frac{\sqrt{11}}{2}$$

$$= \frac{-2 \pm \sqrt{11}}{2} \qquad \text{or} \qquad \left\{\frac{-2 \pm \sqrt{11}}{2}\right\}$$

✔ *CHECK YOURSELF 8*

Solve $4x^2 - 8x + 3 = 0$ by completing the square.

The following algorithm summarizes our work in this section with solving quadratic equations by completing the square.

Completing the Square

Step 1. Isolate the constant on the right side of the equation.

Step 2. Divide both sides of the equation by the coefficient of the x^2 term if that coefficient is not equal to 1.

Step 3. Add the square of one-half of the coefficient of the linear term to both sides of the equation. This will give a perfect-square trinomial on the left side of the equation.

Step 4. Write the left side of the equation as the square of a binomial, and simplify on the right side.

Step 5. Use the square root property, and then solve the resulting linear equations.

✔ *CHECK YOURSELF ANSWERS*

1. (a) $\{-4, 3\}$; (b) $\left\{-\dfrac{5}{3}, 2\right\}$ **2.** (a) $\{-5, 5\}$; (b) $\{-6, 6\}$

3. (a) $\{\sqrt{5}, -\sqrt{5}\}$; (b) $\{\sqrt{2}, -\sqrt{2}\}$; (c) $\left\{\dfrac{2\sqrt{2}}{3}, -\dfrac{2\sqrt{2}}{3}\right\}$; (d) $\{3i, -3i\}$

4. (a) $\{2 \pm \sqrt{3}\}$; (b) $\left\{\dfrac{2 \pm \sqrt{3}}{2}\right\}$ **5.** $\{3 \pm \sqrt{11}\}$ **6.** $\left\{\dfrac{-3 \pm \sqrt{37}}{2}\right\}$

7. $\{-5 \pm 4i\}$ **8.** $\left\{\dfrac{1}{2}, \dfrac{3}{2}\right\}$

Exercises ▪ 11.1

In Exercises 1 to 8, solve by factoring or completing the square.

1. $x^2 + 9x + 14 = 0$

2. $x^2 + 5x + 6 = 0$

3. $z^2 - 2z - 35 = 0$

4. $q^2 - 5q - 24 = 0$

5. $2x^2 - 5x - 3 = 0$

6. $3x^2 + 10x - 8 = 0$

7. $6y^2 - y - 2 = 0$

8. $21z^2 + z - 2 = 0$

In Exercises 9 to 20, use the square root method to find solutions for the equations.

9. $x^2 = 121$

10. $x^2 = 144$

11. $y^2 = 7$

12. $p^2 = 18$

13. $2x^2 - 12 = 0$

14. $5x^2 = 65$

15. $2t^2 + 12 = 4$

16. $3u^2 - 5 = -32$

17. $(x + 1)^2 = 12$

18. $(2x - 3)^2 = 5$

19. $(3z + 1)^2 - 5 = 0$

20. $(3p - 4)^2 + 9 = 0$

In Exercises 21 to 32, find the constant that must be added to each binomial expression to form a perfect-square trinomial.

21. $x^2 + 18x$

22. $r^2 - 14r$

23. $y^2 - 8y$

24. $w^2 + 16w$

25. $x^2 - 3x$

26. $z^2 + 7z$

27. $n^2 + n$

28. $x^2 - x$

29. $x^2 + \dfrac{1}{5}x$

30. $x^2 - \dfrac{1}{3}x$

31. $x^2 - \dfrac{1}{6}x$

32. $y^2 - \dfrac{1}{4}y$

In Exercises 33 to 54, solve each equation by completing the square.

33. $x^2 + 10x = 4$

34. $x^2 - 14x - 7 = 0$

35. $y^2 - 2y = 8$

36. $z^2 + 4z - 72 = 0$

37. $x^2 - 2x - 5 = 0$

38. $x^2 - 3x = 10$

39. $x^2 + 10x + 13 = 0$

40. $x^2 + 3x - 17 = 0$

41. $z^2 - 5z - 7 = 0$

42. $q^2 - 8q + 20 = 0$

43. $m^2 - 3m - 5 = 0$

44. $y^2 + y - 5 = 0$

866

45. $x^2 + \dfrac{1}{2}x = 1$

46. $x^2 - \dfrac{1}{3}x = 2$

47. $2x^2 + 2x - 1 = 0$

48. $5x^2 - 6x = 3$

49. $3x^2 - 8x = 2$

50. $4x^2 + 8x - 1 = 0$

51. $3x^2 - 2x + 12 = 0$

52. $7y^2 - 2y + 3 = 0$

53. $x^2 + 10x + 28 = 0$

54. $x^2 - 2x + 10 = 0$

55. Why must the leading coefficient of the quadratic member be set equal to 1 before you can use the technique of completing the square?

56. What relationship exists between the solution(s) of a quadratic equation and the graph of a quadratic function?

In Exercises 57 to 62, find the constant that must be added to each binomial to form a perfect-square trinomial. Let x be the variable; other letters represent constants.

57. $x^2 + 2ax$

58. $x^2 + 2abx$

59. $x^2 + 3ax$

60. $x^2 + abx$

61. $a^2x^2 + 2ax$

62. $a^2x^2 + 4abx$

In Exercises 63 and 64, solve each equation by completing the square.

63. $x^2 + 2ax = 4$

64. $x^2 + 2ax - 8 = 0$

In Exercises 65 to 68, use your graphing utility to find the graph. Approximate the x intercepts for each graph. (You may have to adjust the viewing window to see both intercepts.) Round your answer to the nearest tenth.

65. $y = x^2 + 12x - 2$

66. $y = x^2 - 14x - 7$

67. $y = x^2 - 2x - 8$

68. $y = x^2 + 4x - 72$

69. On your graphing calculator, view the graph of $f(x) = x^2 + 1$.
   (a) What can you say about the x intercepts of the graph?
   (b) Determine the zeros of the function, using the square root method.
   (c) How does your answer to part (a) relate to your answer to part (b)?

70. Consider the following representation of "completing the square": Suppose we wish to complete the square for $x^2 + 10x$. A square with dimensions x by x has area equal to x^2.

We divide the quantity $10x$ by 2 and get $5x$. If we extend the base x by 5 units and draw the rectangle attached to the square, the rectangle's dimensions are 5 by x with an area of $5x$.

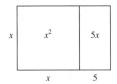

Now we extend the height by 5 units, and we draw another rectangle whose area is $5x$.

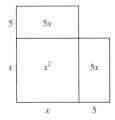

(a) What is the total area represented in the figure so far?

(b) How much area must be added to the figure to "complete the square"?

(c) Write the area of the completed square as a binomial squared.

71. Repeat the process described in Exercise 70 with $x^2 + 16x$.

11.2 The Quadratic Formula

OBJECTIVES

1. *Solve quadratic equations by using the quadratic formula*
2. *Determine the nature of the solutions of a quadratic equation by using the discriminant*
3. *Use the Pythagorean theorem to solve a geometric application*

Every quadratic equation can be solved by using the quadratic formula. In this section, first we will describe how the quadratic formula is derived, and then we will examine its use. Recall that a quadratic equation is any equation that can be written in the form

$$ax^2 + bx + c = 0 \qquad \text{where } a \neq 0$$

Deriving the Quadratic Formula

Step 1.   Isolate the constant on the right side of the equation.

$$ax^2 + bx = -c$$

Step 2.   Divide both sides by the coefficient of the x^2 term.

$$x^2 + \frac{b}{a}x = -\frac{c}{a}$$

Step 3.   Add the square of one-half the x coefficient to both sides.

$$x^2 + \frac{b}{a}x + \frac{b^2}{4a^2} = -\frac{c}{a} + \frac{b^2}{4a^2}$$

Step 4.   Factor the left side to write it as the square of a binomial. Then apply the square root property.

$$\left(x + \frac{b}{2a}\right)^2 = \frac{-4ac + b^2}{4a^2}$$

$$x + \frac{b}{2a} = \pm\sqrt{\frac{b^2 - 4ac}{4a^2}}$$

Step 5.   Solve the resulting linear equations.

$$x = -\frac{b}{2a} \pm \frac{\sqrt{b^2 - 4ac}}{2a}$$

Step 6.   Simplify.

$$= \frac{-b \pm \sqrt{b^2 - 4ac}}{2a}$$

We now use the result derived above to state the **quadratic formula,** a formula that allows us to find the solutions for any quadratic equation.

The Quadratic Formula

Given any quadratic equation in the form

$$ax^2 + bx + c = 0 \qquad \text{where } a \neq 0$$

the two solutions to the equation are found by using the formula

$$x = \frac{-b \pm \sqrt{b^2 - 4ac}}{2a}$$

Our first example uses an equation in standard form.

Example 1 | Using the Quadratic Formula

Note that the equation is in standard form.

Solve, using the quadratic formula.

$$6x^2 - 7x - 3 = 0$$

First, we determine the values for a, b, and c. Here,

$$a = 6 \qquad b = -7 \qquad c = -3$$

Since $b^2 - 4ac = 121$ is a perfect square, the two solutions in this case are rational numbers.

Substituting those values into the quadratic formula, we have

$$x = \frac{-(-7) \pm \sqrt{(-7)^2 - 4(6)(-3)}}{2(6)}$$

Simplifying inside the radical gives us

$$x = \frac{7 \pm \sqrt{121}}{12}$$

$$= \frac{7 \pm 11}{12}$$

Compare these solutions to the x intercepts of the graph of

$$y = 6x^2 - 7x - 3$$

This gives us the solutions

$$x = \frac{3}{2} \qquad \text{or} \qquad x = -\frac{1}{3} \qquad \text{or} \qquad \left\{\frac{3}{2}, -\frac{1}{3}\right\}$$

Note that since the solutions for the equation of this example are rational, the original equation could have been solved by our earlier method of factoring.

 CHECK YOURSELF 1

Solve, using the quadratic formula.

$$3x^2 + 2x - 8 = 0$$

To use the quadratic formula, we often must write the equation in standard form. Example 2 illustrates this approach.

| **Example 2** | ## Using the Quadratic Formula |

Solve by using the quadratic formula.

$$9x^2 = 12x - 4$$

The equation *must be in standard form* to determine a, b, and c.

First, we must write the equation in standard form.

$$9x^2 - 12x + 4 = 0$$

Second, we find the values of a, b, and c. Here,

$$a = 9 \qquad b = -12 \qquad c = 4$$

Substituting these values into the quadratic formula, we find

$$x = \frac{-(-12) \pm \sqrt{(-12)^2 - 4(9)(4)}}{2(9)}$$

$$= \frac{12 \pm \sqrt{0}}{18}$$

The graph of
$$y = 9x^2 - 12x + 4$$
intercepts the x axis only at
the point $\left(\dfrac{2}{3}, 0\right)$.

and simplifying yields

$$x = \frac{2}{3} \qquad \text{or} \qquad \left\{\frac{2}{3}\right\}$$

  CHECK YOURSELF 2

Use the quadratic formula to solve the equation.

$$4x^2 - 4x = -1$$

Thus far in this section our examples and exercises have led to solutions that are rational numbers. That is not always the case, as Example 3 illustrates.

| **Example 3** | ## Using the Quadratic Formula |

Using the quadratic formula, solve

$$x^2 - 3x = 5$$

Once again, to use the quadratic formula, we write the equation in standard form.

$$x^2 - 3x - 5 = 0$$

We now determine values for a, b, and c and substitute.

$$x = \frac{-(-3) \pm \sqrt{(-3)^2 - 4(1)(-5)}}{2(1)}$$

Decimal approximations for the solutions can be found with a calculator. To the nearest hundredth, we have -1.19 and 4.19.

Simplifying as before, we have

$$x = \frac{3 \pm \sqrt{29}}{2} \qquad \text{or} \qquad \left\{ \frac{3 \pm \sqrt{29}}{2} \right\}$$

✔ **CHECK YOURSELF 3**

Using the quadratic formula, solve $2x^2 = x + 7$.

Example 4 requires some special care in simplifying the solution.

Example 4

Using the Quadratic Formula

Using the quadratic formula, solve

Caution

Students are sometimes tempted to reduce this result to

$$\frac{6 \pm 2\sqrt{3}}{6} \stackrel{?}{=} 1 \pm 2\sqrt{3}$$

This is *not a valid step*. We must divide *each of the terms* in the numerator by 2 when simplifying the expression.

$$3x^2 - 6x + 2 = 0$$

Here, we have $a = 3$, $b = -6$, and $c = 2$. Substituting gives

$$x = \frac{-(-6) \pm \sqrt{(-6)^2 - 4(3)(2)}}{2(3)}$$

$$= \frac{6 \pm \sqrt{12}}{6} \qquad \text{We now look for the largest perfect-square factor of 12, the radicand.}$$

Simplifying, we note that $\sqrt{12}$ is equal to $\sqrt{4 \cdot 3}$, or $2\sqrt{3}$. We can then write the solutions as

$$x = \frac{6 \pm 2\sqrt{3}}{6} = \frac{2(3 \pm \sqrt{3})}{6} = \frac{3 \pm \sqrt{3}}{3}$$

✔ **CHECK YOURSELF 4**

Solve by using the quadratic formula.

$$x^2 - 4x = 6$$

Let's examine a case in which the solutions are nonreal complex numbers.

Example 5	**Using the Quadratic Formula**

Solve by using the quadratic formula.

$$x^2 - 2x + 2 = 0$$

Labeling the coefficients, we find that

$$a = 1 \qquad b = -2 \qquad c = 2$$

The solutions will be nonreal any time $b^2 - 4ac$ is negative.

Applying the quadratic formula, we have

$$x = \frac{2 \pm \sqrt{-4}}{2}$$

and noting that $\sqrt{-4}$ is $2i$, we can simplify to

The graph of $y = x^2 - 2x + 2$ does not intercept the x axis, so there are no real solutions.

$$x = 1 \pm i \qquad \text{or} \qquad \{1 \pm i\}$$

 CHECK YOURSELF 5

Solve by using the quadratic formula.

$$x^2 - 4x + 6 = 0$$

In attempting to solve a quadratic equation, you should first try the factoring method. If this method does not work, you can apply the quadratic formula or the square root method to find the solution. The following algorithm outlines the steps.

> **Solving a Quadratic Equation by Using the Quadratic Formula**
>
> Step 1.   Write the equation in standard form (one side is equal to 0).
>
> $$ax^2 + bx + c = 0$$
>
> Step 2.   Determine the values for a, b, and c.
>
> Step 3.   Substitute those values into the quadratic formula.
>
> $$x = \frac{-b \pm \sqrt{b^2 - 4ac}}{2a}$$
>
> Step 4.   Simplify.

Although the solutions are not necessarily distinct or real, every second-degree equation has two solutions.

Given a quadratic equation, the radicand $b^2 - 4ac$ determines the number of real solutions. For example, if $b^2 - 4ac$ is a negative number, we would be taking the square root of that negative number, and therefore we would obtain two nonreal solutions. Because of the information that this quantity $b^2 - 4ac$ gives us, it has a name: it is called the **discriminant.**

Graphically, we can see the number of real solutions as the number of times the related quadratic function intercepts the x axis.

The Discriminant

Given the equation $ax^2 + bx + c = 0$, the quantity $b^2 - 4ac$ is called the **discriminant.**

$$\text{If } b^2 - 4ac \begin{cases} < 0 & \text{there are } \textit{no real solutions,} \text{ but two nonreal solutions} \\ = 0 & \text{there is } \textit{one real solution} \text{ (a double solution)} \\ > 0 & \text{there are } \textit{two distinct real solutions} \end{cases}$$

| Example 6 | ## Analyzing the Discriminant |

How many real solutions are there for each of the following quadratic equations?

(a) $x^2 + 7x - 15 = 0$

The discriminant $[49 - 4(1)(-15)]$ is 109. This indicates that there are two real solutions.

We could find the two nonreal solutions by using the quadratic formula.

(b) $3x^2 - 5x + 7 = 0$

The discriminant $b^2 - 4ac = -59$ is negative. There are no real solutions.

(c) $9x^2 - 12x + 4 = 0$

The discriminant is 0. There is exactly one real solution (a double solution).

    CHECK YOURSELF 6

How many real solutions are there for each of the following quadratic equations?

(a) $2x^2 - 3x + 2 = 0$ (b) $3x^2 + x - 11 = 0$

(c) $4x^2 - 4x + 1 = 0$ (d) $x^2 = -5x - 7$

Frequently, as in Examples 3 and 4, the solutions of a quadratic equation involve square roots. When we are solving algebraic equations, it is generally best to leave solutions in this form. However, if an equation resulting from an application has been solved by the use of the quadratic formula, we will often estimate the root and sometimes accept only positive solutions. Consider the following two applications involving thrown balls that can be solved by using the quadratic formula.

| Example 7 | Solving a Thrown-Ball Application |

If a ball is thrown upward from the ground, the equation to find the height h of such a ball thrown with an initial velocity of 80 ft/s is

Here h measures the height above the ground, in feet, t seconds (s) after the ball is thrown upward.

$$h(t) = 80t - 16t^2$$

Find the time it takes the ball to reach a height of 48 ft.

First we substitute 48 for $h(t)$, and then we rewrite the equation in standard form.

Note that the result of dividing by 16

$$16t^2 - 80t + 48 = 0$$

$$\frac{0}{16} = 0$$

To simplify the computation, we divide both sides of the equation by the common factor 16. This yields

is 0 on the right.

$$t^2 - 5t + 3 = 0$$

We solve for t as before, using the quadratic equation, with the result

$$t = \frac{5 \pm \sqrt{13}}{2}$$

There are two solutions because the ball reaches the height *twice*, once on the way up and once on the way down.

This gives us two solutions, $\dfrac{5 + \sqrt{13}}{2}$ and $\dfrac{5 - \sqrt{13}}{2}$. But, because we have specified units of time, we generally estimate the answer to the nearest tenth or hundredth of a second.

In this case, estimating to the nearest tenth of a second gives solutions of 0.7 and 4.3 s.

✔ *CHECK YOURSELF 7*

The equation to find the height h of a ball thrown with an initial velocity of 64 ft/s is

$$h(t) = 64t - 16t^2$$

Find the time it takes the ball to reach a height of 32 ft, estimating to the nearest tenth of a second.

| Example 8 | Solving a Thrown-Ball Application |

The height h of a ball thrown downward from the top of a 240-ft building with an initial velocity of 64 ft/s is given by

$$h(t) = 240 - 64t - 16t^2$$

At what time will the ball reach a height of 176 ft?

Let $h(t) = 176$, and write the equation in standard form.

$$176 = 240 - 64t - 16t^2$$
$$0 = 64 - 64t - 16t^2$$
$$16t^2 + 64t - 64 = 0$$

The graph of

$$h(t) = 240 - 64t - 16t^2$$

shows the height h at any time t.

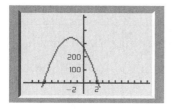

The ball has a height of 176 ft at approximately 0.8 s.

Divide both sides of the equation by 16 to simplify the computation.

$$t^2 + 4t - 4 = 0$$

Applying the quadratic formula with $a = 1$, $b = 4$, and $c = -4$ yields

$$t = -2 \pm 2\sqrt{2}$$

Estimating these solutions, we have $t = -4.8$ and $t = 0.8$ s, but of these two values only the *positive value* makes any sense. (To accept the negative solution would be to say that the ball reached the specified height before it was thrown.)

 CHECK YOURSELF 8

The height h of a ball thrown upward from the top of a 96-ft building with an initial velocity of 16 ft/s is given by

$$h(t) = 96 + 16t - 16t^2$$

When will the ball have a height of 32 ft? (Estimate your answer to the nearest tenth of a second.)

Another geometric result that generates quadratic equations in applications is the **Pythagorean theorem.** You may recall from earlier algebra courses that the theorem gives an important relationship between the lengths of the sides of a right triangle (a triangle with a 90° angle).

The Pythagorean Theorem

In any right triangle, the square of the longest side (the hypotenuse) is equal to the sum of the squares of the two shorter sides (the legs).

$$c^2 = a^2 + b^2$$

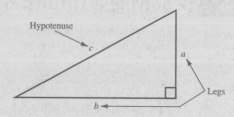

In Example 9, the solution of the quadratic equation again contains radicals. When working with applications, we are usually interested in decimal approximations rather than exact radicals. Checking the "reasonableness" of our answer is very important, but we should not expect a decimal approximation to check *exactly*.

Example 9

A Triangular Application

One leg of a right triangle is 4 cm longer than the other leg. The length of the hypotenuse of the triangle is 12 cm. Find the length of the two legs, accurate to the nearest hundredth.

As in any geometric problem, a sketch of the information will help us visualize.

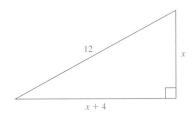

We assign variable x to the shorter leg and $x + 4$ to the other leg.

Remember: The sum of the squares of the legs of the triangle is equal to the square of the hypotenuse.

Now we apply the Pythagorean theorem to write an equation for the solution.

$$x^2 + (x + 4)^2 = (12)^2$$

$$x^2 + x^2 + 8x + 16 = 144$$

or

$$2x^2 + 8x - 128 = 0$$

Dividing both sides by 2, we have the equivalent equation

$$x^2 + 4x - 64 = 0$$

Dividing both sides of a quadratic equation by a common factor is always a prudent step. It simplifies your work with the quadratic formula.

Using the quadratic formula, we get

$$x = \frac{-4 \pm \sqrt{272}}{2}$$

Since we are interested in lengths rounded to the nearest hundredth, there is no need to simplify the radical expression. Using a calculator, we can reject one solution (do you see why?), and we find $x = 6.25$ cm for the other.

This means that the two legs have lengths of 6.25 and 10.25 cm. How do we check? The sides must "reasonably" satisfy the Pythagorean theorem:

Do you see why we should not expect an "exact" check here?

$$6.25^2 + 10.25^2 \approx 12^2$$

$$144.125 \approx 144$$

This indicates that our answer is reasonable.

✔ *CHECK YOURSELF 9*

One leg of a right triangle is 2 cm longer than the other. The hypotenuse is 1 cm less than twice the length of the shorter leg. Find the length of each side of the triangle, accurate to the nearest tenth.

✔ *CHECK YOURSELF ANSWERS*

1. $\left\{-2, \dfrac{4}{3}\right\}$ **2.** $\left\{\dfrac{1}{2}\right\}$ **3.** $\left\{\dfrac{1 \pm \sqrt{57}}{4}\right\}$ **4.** $\{2 \pm \sqrt{10}\}$

5. $\{2 \pm i\sqrt{2}\}$ **6.** (a) None; (b) two; (c) one; (d) none   **7.** 0.6 and 3.4 s

8. 2.6 s   **9.** Approximately 4.3, 6.3, and 7.7 cm

Exercises · 11.2

In Exercises 1 to 8, solve each quadratic equation first by factoring and then by using the quadratic formula.

1. $x^2 - 5x - 14 = 0$

2. $x^2 - 2x - 35 = 0$

3. $t^2 + 8t - 65 = 0$

4. $q^2 + 3q - 130 = 0$

5. $3x^2 + x - 10 = 0$

6. $3x^2 + 2x - 1 = 0$

7. $16t^2 - 24t + 9 = 0$

8. $6m^2 - 23m + 10 = 0$

In Exercises 9 to 20, solve each quadratic equation by (a) completing the square and (b) using the quadratic formula.

9. $x^2 - 4x - 7 = 0$

10. $x^2 + 6x - 1 = 0$

11. $x^2 + 3x - 27 = 0$

12. $t^2 + 4t - 7 = 0$

13. $3x^2 - 5x + 1 = 0$

14. $2x^2 - 6x + 1 = 0$

15. $2q^2 - 2q - 1 = 0$

16. $3r^2 - 2r + 4 = 0$

17. $3x^2 - x - 2 = 0$

18. $2x^2 - 5x + 3 = 0$

19. $2y^2 - y - 5 = 0$

20. $3m^2 + 2m - 1 = 0$

In Exercises 21 to 42, solve each equation by using the quadratic formula.

21. $x^2 - 4x + 3 = 0$

22. $x^2 - 7x + 3 = 0$

23. $p^2 - 8p + 16 = 0$

24. $u^2 + 7u - 30 = 0$

25. $-x^2 + 3x + 5 = 0$

26. $2x^2 - 3x - 7 = 0$

27. $-3s^2 + 2s - 1 = 0$

28. $5t^2 - 2t - 2 = 0$

(*Hint:* Clear each of the following equations of fractions first, or remove grouping symbols, as needed.)

29. $2x^2 - \dfrac{1}{2}x - 5 = 0$

30. $3x^2 + \dfrac{1}{3}x - 3 = 0$

31. $5t^2 - 2t - \dfrac{2}{3} = 0$

32. $3y^2 + 2y + \dfrac{3}{4} = 0$

33. $(x - 2)(x + 3) = 4$

34. $(x + 1)(x - 8) = 3$

35. $(t + 1)(2t - 4) - 7 = 0$

36. $(2w + 1)(3w - 2) = 1$

37. $2x - 7 = \dfrac{3}{x}$

38. $x + 3 = \dfrac{1}{x}$

39. $2t - \dfrac{3}{t} = 3$

40. $4p - \dfrac{1}{p} = 6$

41. $\dfrac{5}{y^2} + \dfrac{2}{y} - 1 = 0$

42. $\dfrac{6}{x^2} - \dfrac{2}{x} = 1$

In Exercises 43 to 50, for each quadratic equation, find the value of the discriminant and give the number of real solutions.

43. $2x^2 - 5x = 0$

44. $3x^2 + 8x = 0$

45. $m^2 - 18m + 81 = 0$

46. $4p^2 + 12p + 9 = 0$

47. $3x^2 - 7x + 1 = 0$

48. $2x^2 - x + 5 = 0$

49. $2w^2 - 5w + 11 = 0$

50. $6q^2 - 5q + 2 = 0$

In Exercises 51 to 62, find all the solutions of each quadratic equation. Use any applicable method.

51. $x^2 - 8x + 16 = 0$

52. $4x^2 + 12x + 9 = 0$

53. $3t^2 - 7t + 1 = 0$

54. $2z^2 - z + 5 = 0$

55. $5y^2 - 2y = 0$

56. $7z^2 - 6z - 2 = 0$

57. $(x - 1)(2x + 7) = -6$

58. $4x^2 - 3 = 0$

59. $x^2 + 9 = 0$

60. $(4x - 5)(x + 2) = 1$

61. $x - 3 - \dfrac{10}{x} = 0$

62. $1 + \dfrac{2}{x} + \dfrac{2}{x^2} = 0$

63. Free fall. The equation

$$h(t) = 112t - 16t^2$$

is the equation for the height of an arrow, shot upward from the ground with an initial velocity of 112 ft/s, where t is the time, in seconds, after the arrow leaves the ground.

(a) Find the time it takes for the arrow to reach a height of 112 ft.

(b) Find the time it takes for the arrow to reach a height of 144 ft.

Express your answers to the nearest tenth of a second.

64. Free fall. The equation

$$h(t) = 320 - 32t - 16t^2$$

is the equation for the height of a ball, thrown downward from the top of a 320-ft building with an initial velocity of 32 ft/s, where t is the time after the ball is thrown down from the top of the building.

(a) Find the time it takes for the ball to reach a height of 240 ft.

(b) Find the time it takes for the ball to reach a height of 96 ft.

Express your answer to the nearest tenth of a second.

65. Number problem. The product of two consecutive integers is 72. What are the two integers?

66. Number problem. The sum of the squares of two consecutive whole numbers is 61. Find the two whole numbers.

67. Rectangles. The width of a rectangle is 3 ft less than its length. If the area of the rectangle is 70 ft², what are the dimensions of the rectangle?

68. Rectangles. The length of a rectangle is 5 cm more than its width. If the area of the rectangle is 84 cm², find the dimensions.

69. Triangles. One leg of a right triangle is twice the length of the other. The hypotenuse is 6 m long. Find the length of each leg.

70. Triangles. One leg of a right triangle is 2 ft longer than the shorter side. If the length of the hypotenuse is 14 ft, how long is each leg?

71. Thrown ball. If a ball is thrown vertically upward from the ground, with an initial velocity of 64 ft/s, its height h after t s is given by $h(t) = 64t - 16t^2$.

   (a) How long does it take the ball to return to the ground? [*Hint:* Let $h(t) = 0$]

   (b) How long does it take the ball to reach a height of 48 ft on the way up?

72. Thrown ball. If a ball is thrown vertically upward from the ground, with an initial velocity of 96 ft/s, its height h after t s is given by $h(t) = 96t - 16t^2$.

   (a) How long does it take the ball to return to the ground?

   (b) How long does it take the ball to pass through a height of 128 ft on the way back down to the ground?

73. Cost. Suppose that the cost $C(x)$, in dollars, of producing x chairs is given by

$$C(x) = 2{,}400 - 40x + 2x^2$$

How many chairs can be produced for $5,400?

74. Profit. Suppose that the profit $T(x)$, in dollars, of producing and selling x appliances is given by

$$T(x) = -3x^2 + 240x - 1{,}800$$

How many appliances must be produced and sold to achieve a profit of $3,000?

75. Triangles. One leg of a right triangle is 1 in. shorter than the other leg. The hypotenuse is 3 in. longer than the shorter side. Find the length of each side.

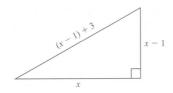

76. Triangles. The hypotenuse of a given right triangle is 5 cm longer than the shorter leg. The length of the shorter leg is 2 cm less than the length of the longer leg. Find the lengths of the three sides.

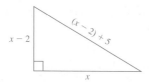

77. Profit. A small manufacturer's weekly profit, in dollars, is given by

$$P(x) = -3x^2 + 270x$$

Find the number of items x that must be produced to realize a profit of $5,100.

78. Profit. Suppose the profit, in dollars, is given by

$$P(x) = -2x^2 + 240x$$

Now how many items must be sold to realize a profit of $5,100?

79. Equilibrium price. The demand equation for a certain computer chip is given by

$$D = -2p + 14$$

The supply equation is predicted to be

$$S = -p^2 + 16p - 2$$

Find the equilibrium price.

80. Equilibrium price. The demand equation for a certain type of print is predicted to be

$$D = -200p + 36,000$$

The supply equation is predicted to be

$$S = -p^2 + 400p - 24,000$$

Find the equilibrium price.

81. Thrown ball. If a ball is thrown upward from the roof of a building 70 m tall with an initial velocity of 15 m/s, its approximate height h after t s is given by

$$h(t) = 70 + 15t - 5t^2$$

Note: The difference between this equation and the one we used in Example 8 has to do with the units used. When we used feet, the t^2 coefficient was -16 (from the fact that the acceleration due to gravity is approximately 32 ft/s^2). When we use meters as the height, the t^2 coefficient is -5 (that same acceleration becomes approximately 10 m/s^2). Use this information to solve the following.

(a) How long does it take the ball to fall back to the ground?

(b) When will the ball reach a height of 80 m?

82. **Thrown ball.** Changing the initial velocity to 25 m/s will only change the t coefficient. Our new equation becomes

$$h(t) = 70 + 25t - 5t^2$$

   (a) How long will it take the ball to return to the thrower?

   (b) When will the ball reach a height of 85 m?

   The only part of the height equation that we have not discussed is the constant. You have probably noticed that the constant is always equal to the initial height of the ball (70 m in our previous exercises). Now, let's have *you* develop an equation.

83. **Thrown ball.** A ball is thrown upward from the roof of a 100-m building with an initial velocity of 20 m/s. Use this information to solve the following.

   (a) Find the equation for the height h of the ball after t s.

   (b) How long will it take the ball to fall back to the ground?

   (c) When will the ball reach a height of 75 m?

   (d) Will the ball ever reach a height of 125 m? (*Hint:* Check the discriminant.)

84. **Thrown ball.** A ball is thrown upward from the roof of a 100-ft building with an initial velocity of 20 ft/s. Use this information to solve the following.

   (a) Find the height h of the ball after t s.

   (b) How long will it take the ball to fall back to the ground?

   (c) When will the ball reach a height of 80 ft?

   (d) Will the ball ever reach a height of 120 ft? Explain.

85. Can the solution of a quadratic equation with integer coefficients include one real and one imaginary number? Justify your answer.

86. Explain how the discriminant is used to predict the nature of the solutions of a quadratic equation.

In Exercises 87 to 94, solve each equation for x.

87. $x^2 + y^2 = z^2$

88. $2x^2y^2z^2 = 1$

89. $x^2 - 36a^2 = 0$

90. $ax^2 - 9b^2 = 0$

91. $2x^2 + 5ax - 3a^2 = 0$

92. $3x^2 - 16bx + 5b^2 = 0$

93. $2x^2 + ax - 2a^2 = 0$

94. $3x^2 - 2bx - 2b^2 = 0$

95. Given that the polynomial $x^3 - 3x^2 - 15x + 25 = 0$ has as one of its solutions $x = 5$, find the other two solutions. (*Hint:* If you divide the given polynomial by $x - 5$, the quotient will be a quadratic equation. The remaining solutions will be the solutions for *that* equation.)

96. Given that $2x^3 + 2x^2 - 5x - 2 = 0$ has as one of its solutions $x = -2$, find the other two solutions. (*Hint:* In this case, divide the original polynomial by $x + 2$.)

97. Find all the zeros of the function $f(x) = x^3 + 1$.

98. Find the zeros of the function $f(x) = x^2 + x + 1$.

99. Find all six solutions to the equation $x^6 - 1 = 0$. (*Hint:* Factor the left-hand side of the equation first as the difference of squares, then as the sum and difference of cubes.)

100. Find all six solutions to $x^6 = 64$.

101. (a) Use the quadratic formula to solve $x^2 - 3x - 5 = 0$. For each solution give a decimal approximation to the nearest tenth.

  (b) Graph the function $f(x) = x^2 - 3x - 5$ on your graphing calculator. Use a zoom utility and estimate the x intercepts to the nearest tenth.

  (c) Describe the connection between parts (a) and (b).

102. (a) Solve the following equation, using any appropriate method:

$$x^2 - 2x = 3$$

  (b) Graph the following functions on your graphing calculator:

$$f(x) = x^2 - 2x \quad \text{and} \quad g(x) = 3$$

  Estimate the points of intersection of the graphs of f and g. In particular, note the x coordinates of these points.

  (c) Describe the connection between parts (a) and (b).

Solving Quadratic Equations by Graphing

 11.3 OBJECTIVES

1. Find an axis of symmetry
2. Find a vertex
3. Graph a parabola
4. Solve quadratic equations by graphing
5. Solve an application involving a quadratic equation

In Section 3.3, we learned to graph a linear equation. We discovered that the graph of every linear equation in two variables was a straight line. In this section, we will consider the graph of a quadratic equation in two variables. Then we'll see how such a graph may be used to solve a quadratic equation (and later, a quadratic inequality) in one variable.

Consider an equation of the form

$$y = ax^2 + bx + c \qquad a \neq 0$$

This equation is quadratic in x and linear in y. Its graph will always be a curve called a parabola.

In an equation of the form
$$y = ax^2 + bx + c \qquad a \neq 0$$
the parabola opens upward or downward, as follows:

1. If $a > 0$, the parabola opens *upward*.

2. If $a < 0$, the parabola opens *downward*.
$$y = ax^2 + bx + c$$

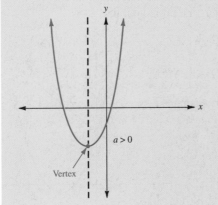

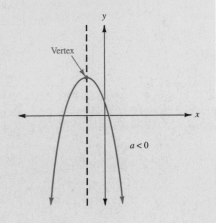

Two concepts regarding the parabola can be made by observation. Consider the illustrations on the previous page.

1. There is always a **minimum** (or lowest) point on the parabola if it opens upward. There is always a **maximum** (or highest) point on the parabola if it opens downward. In either case, that maximum or minimum value occurs at the **vertex** of the parabola.

2. Every parabola has an **axis of symmetry.** In the case of parabolas that open upward or downward, that axis of symmetry is a vertical line midway between any pair of symmetric points on the parabola. Also, the point where this axis of symmetry intersects the parabola is the vertex of the parabola.

The following figure summarizes these observations.

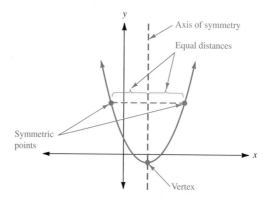

Our objective is to be able to quickly sketch a parabola. This can be done with *as few as three points* if those points are carefully chosen. For this purpose you will want to find the vertex and two symmetric points.

First, let's see how the coordinates of the vertex can be determined from the standard equation

$$y = ax^2 + bx + c \tag{1}$$

In equation (1), if $x = 0$, then $y = c$, and so $(0, c)$ gives the point where the parabola intersects the y axis (the y intercept).

Look at the sketch. To determine the coordinates of the symmetric point (x_1, c), note that it lies along the horizontal line $y = c$.

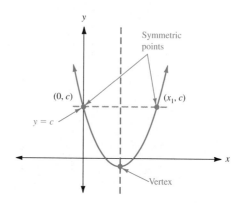

Therefore, let $y = c$ in equation (1):

$$c = ax^2 + bx + c$$

$$0 = ax^2 + bx$$

$$0 = x(ax + b)$$

and

$$x = 0 \qquad \text{or} \qquad x = -\frac{b}{a}$$

We now know that

$$(0, c) \qquad \text{and} \qquad \left(-\frac{b}{a}, c\right)$$

Note: To ensure that we remember the negative, we sometimes write

$$x = -\frac{b}{2a}$$

are the coordinates of the symmetric points shown. Since the axis of symmetry must be midway between these points, the x value along that axis is given by

$$x = \frac{0 + (-b/a)}{2} = -\frac{b}{2a} \qquad \textbf{(2)}$$

Since the vertex for any parabola lies on the axis of symmetry, we can now state the following general result.

Note: The y coordinate of the vertex can be found most easily by substituting the value found for x into the original equation.

Vertex of a Parabola

If

$$y = ax^2 + bx + c \qquad a \neq 0$$

then the x coordinate of the vertex of the corresponding parabola is

$$x = -\frac{b}{2a}$$

We now know how to find the vertex of a parabola, and if two symmetric points can be determined, we are well on our way to the desired graph. Perhaps the simplest case occurs when the quadratic member of the given equation is factorable. In most cases, the two x intercepts will then give two symmetric points that are very easily found. Example 1 illustrates such a case.

Example 1   Graphing a Parabola

Graph the equation

$$y = x^2 + 2x - 8$$

Sketch the information to help you solve the problem. Begin by drawing—as a dashed line—the axis of symmetry.

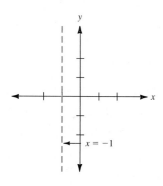

At this point you can plot the vertex along the axis of symmetry.

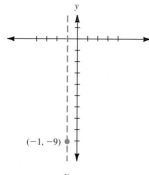

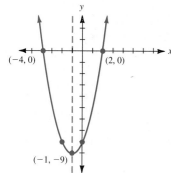

First, find the axis of symmetry. In this equation, $a = 1$, $b = 2$, and $c = -8$. We then have

$$x = -\frac{b}{2a} = -\frac{2}{2 \cdot 1} = -\frac{2}{2} = -1$$

Thus, $x = -1$ is the axis of symmetry.

Second, find the vertex. Since the vertex of the parabola lies on the axis of symmetry, let $x = -1$ in the original equation. If $x = -1$,

$$y = (-1)^2 + 2(-1) - 8 = -9$$

and $(-1, -9)$ is the vertex of the parabola.

Third, find two symmetric points. Note that the quadratic member in this case is factorable, and so setting $y = 0$ in the original equation will quickly give two symmetric points (the x intercepts):

$$0 = x^2 + 2x - 8$$
$$= (x + 4)(x - 2)$$

So when $y = 0$,

$$x + 4 = 0 \qquad \text{or} \qquad x - 2 = 0$$
$$x = -4 \qquad\qquad x = 2$$

and our x intercepts are $(-4, 0)$ and $(2, 0)$.

Fourth, draw a smooth curve connecting the points found above, to form the parabola.

Note: You can choose to find additional pairs of symmetric points at this time if necessary. For instance, the symmetric points $(0, -8)$ and $(-2, -8)$ are easily located.

✔ CHECK YOURSELF 1

Graph the equation

$$y = -x^2 - 2x + 3$$

(*Hint:* Since the coefficient of x^2 is negative, the parabola opens downward.)

A similar process will work if the quadratic member of the given equation is *not* factorable. In that case, one of two things happens:

1. The x intercepts are irrational and therefore not particularly helpful in the graphing process.

2. The x intercepts do not exist.

Consider Example 2.

Example 2

Graphing a Parabola

Graph the function

$$f(x) = x^2 - 6x + 3$$

First, find the axis of symmetry. Here $a = 1$, $b = -6$, and $c = 3$. So

$$x = \frac{-b}{2a} = \frac{-(-6)}{2(1)} = \frac{6}{2} = 3$$

Thus, $x = 3$ is the axis of symmetry.

Second, find the vertex. If $x = 3$,

$$f(3) = 3^2 - 6 \cdot 3 + 3 = -6$$

and $(3, -6)$ is the vertex of the desired parabola.

Third, find two symmetric points. Here the quadratic member is not factorable, so we need to find another pair of symmetric points.

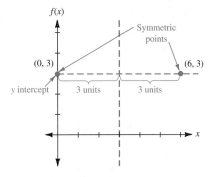

Note that $(0, 3)$ is the y intercept of the parabola. We found the axis of symmetry at $x = 3$ in step 1. Note that the symmetric point to $(0, 3)$ lies along the horizontal line through the y intercept at the same distance (3 units) from the axis of symmetry. Hence, $(6, 3)$ is our symmetric point.

Fourth, draw a smooth curve connecting the points found above to form the parabola.

Note: An alternate method is available in step 3. Observing that $(0, 3)$ is the y intercept and that the symmetric point lies along the line $y = 3$, set $f(x) = 3$ in the original equation:

$$3 = x^2 - 6x + 3$$
$$0 = x^2 - 6x$$
$$0 = x(x - 6)$$

so

$$x = 0 \quad \text{or} \quad x - 6 = 0$$
$$x = 6$$

and $(0, 3)$ and $(6, 3)$ are the desired symmetric points.

✔ CHECK YOURSELF 2

Graph the function.

$$f(x) = x^2 + 4x + 5$$

Thus far the coefficient of x^2 has been 1 or -1. Example 3 shows the effect of different coefficients on the shape of the graph of a quadratic equation.

Example 3 | **Graphing a Parabola**

Graph the equation

$$y = 3x^2 - 6x + 5$$

First, find the axis of symmetry.

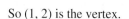

$$x = \frac{-b}{2a} = \frac{-(-6)}{2(3)} = \frac{6}{6} = 1$$

Second, find the vertex. If $x = 1$,

$$y = 3(1)^2 - 6 \cdot 1 + 5 = 2$$

So $(1, 2)$ is the vertex.

Third, find symmetric points. Again the quadratic member is not factorable, and we use the y intercept $(0, 5)$ and its symmetric point $(2, 5)$.

Fourth, connect the points with a smooth curve to form the parabola. Compare this curve to those in previous examples. Note that the parabola is "tighter" about the axis of symmetry. That is the effect of the larger x^2 coefficient.

✔ CHECK YOURSELF 3

Graph the equation.

$$y = \frac{1}{2}x^2 - 3x - 1$$

The following algorithm summarizes our work thus far in this section.

To Graph a Parabola

Step 1. Find the axis of symmetry.

Step 2. Find the vertex.

Step 3. Determine two symmetric points.

Note: You can use the x intercepts if the quadratic member of the given equation is factorable. Otherwise use the y intercept and its symmetric point.

Step 4. Draw a smooth curve connecting the points found above to form the parabola. You may choose to find additional pairs of symmetric points at this time.

We have seen that quadratic equations can be solved in three different ways: by factoring (Section 6.4), by completing the square (Section 11.1), or by using the quadratic formula (Section 11.2). We will now look at a fourth technique for solving quadratic equations, a graphical method. Unlike the other methods, the graphical technique may yield only an approximation of the solution(s). This, however, may be perfectly adequate in applications; in such situations, we are generally more interested in the decimal form of a number than in the "exact radical" form. And, using current technology, we are able to approximate solutions with great accuracy.

To solve the equation

$$ax^2 + bx + c = 0$$

recall, from our previous work, that we define functions f and g as follows:

$$f(x) = ax^2 + bx + c$$

$$g(x) = 0$$

and we ask, For what values of x do the two graphs intersect? Now, the graph of f is a parabola, and the graph of g is simply the x axis. So solutions for the original equation are simply the x values where the parabola intersects the x axis. We need only look at the x intercepts!

Example 4

Solving a Quadratic Equation Graphically

Use a graphing calculator to solve the following equation. Give solutions accurate to the nearest thousandth.

$$0.4x^2 - x - 2.5 = 0$$

In the calculator, we define $Y_1 = 0.4x^2 - x - 2.5$, and we view the graph in the standard viewing window:

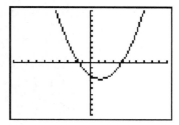

We are interested in the x intercepts. Your calculator will have a "ZERO" or "ROOT" utility which allows you to locate the x intercepts. Using this, we find

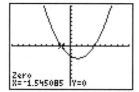 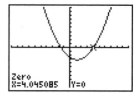

So, to the nearest thousandth, the solution set is $\{-1.545, 4.045\}$.

 CHECK YOURSELF 4

Use a graphing calculator to solve the following equation. Give solutions accurate to the nearest thousandth.

$$-0.3x^2 - 0.4x + 2.75 = 0$$

We often need to apply our knowledge of parabolas when solving graphically on a calculator. Consider the next example.

Example 5 ## Solving a Quadratic Equation Graphically

Use a graphing calculator to solve the following equation. Give solutions accurate to the nearest thousandth.

$$-x^2 + 16x + 160 = 0$$

When we define $Y_1 = -x^2 + 16x + 160$ and view the graph in the standard viewing window, we see

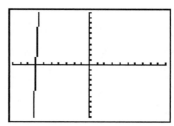

We expect to see a parabola, and we expect that it opens down (do you see why?). Therefore we must be looking at the left portion of the parabola. Further, we know that the graph climbs for a while, turns, and comes back down to cross the x axis somewhere to the right. While we have this view, we find the x intercept (to the nearest thousandth) to be $(-6.967, 0)$.

Using the TABLE utility, we explore:

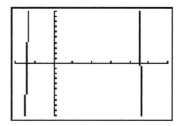

We see that the graph must cross the x axis somewhere between $x = 20$ and $x = 30$. If we simply change the window to be $-10 \leq x \leq 30$, with x scale of 5, we see this:

We find the right-hand x intercept to be $(22.967, 0)$. So the solution set is $\{-6.967, 22.967\}$.

✔ *CHECK YOURSELF 5*

Use a graphing calculator to solve the following equation. Give solutions accurate to the nearest thousandth.

$$x^2 + 24x - 265 = 0$$

From graphs of equations of the form $y = ax^2 + bx + c$, we know that if $a > 0$, then the vertex is the lowest point on the graph (the minimum value). Also, if $a < 0$, then the vertex is the highest point on the graph (the maximum value). We can use this result to solve a variety of problems in which we want to find the maximum or minimum value of a variable. The following are just two of many typical examples.

| Example 6 | **An Application Involving a Quadratic Function** |

A software company sells a word processing program for personal computers. It has found that the monthly profit P, in dollars, from selling x copies of the program is approximated by

$$P(x) = -0.3x^2 + 90x - 1,500$$

Find the number of copies of the program that should be sold in order to maximize the profit.

Since the relating equation is quadratic, the graph must be a parabola. Also since the coefficient of x^2 is negative, the parabola must open downward, and thus the vertex will give the maximum value for the profit P. To find the vertex,

View a graph of $P(x)$ using this window: $0 \leq x \leq 300$ and $-1{,}500 \leq y \leq 6{,}000$. Then determine how to use the MAXIMUM utility in your calculator to find the vertex.

$$x = \frac{-b}{2a} = \frac{-90}{2(-0.3)} = \frac{-90}{-0.6} = 150$$

The maximum profit must then occur when $x = 150$, and we substitute that value into the original equation:

$$P(150) = -0.3(150)^2 + (90)(150) - 1{,}500$$

$$= \$5{,}250$$

The maximum profit will occur when 150 copies are sold per month, and that profit will be $5,250.

✔   CHECK YOURSELF 6

A company that sells portable radios finds that its weekly profit P, in dollars, and the number of radios sold x are related by

$$P(x) = -0.2x^2 + 40x - 100$$

Find the number of radios that should be sold to have the largest weekly profit. Also find the amount of that profit.

Example 7   An Application Involving a Quadratic Function

A farmer has 3,600 ft of fence and wishes to enclose the largest possible rectangular area with that fencing. Find the largest possible area that can be enclosed.

As usual, when dealing with geometric figures, we start by drawing a sketch of the problem.

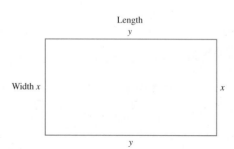

First, we can write the area A as

Area = length × width

$$A = xy \tag{3}$$

Also, since 3,600 ft of fence is to be used, we know that

The perimeter of the region is
$$2x + 2y$$

$$2x + 2y = 3,600 \tag{4}$$

$$2y = 3,600 - 2x$$

$$y = 1,800 - x$$

Substituting for y in equation (3), we have

Note that $A(x)$ is a quadratic function.

$$A(x) = x(1,800 - x) \tag{5}$$

$$= 1,800x - x^2$$

$$= -x^2 + 1,800x$$

Again, the graph for A is a parabola opening downward, and the largest possible area will occur at the vertex. As before,

The width x is 900 ft. Since from equation (4)
$$y = 1,800 - 900$$
$$= 900 \text{ ft}$$

$$x = \frac{-1,800}{2(-1)} = \frac{-1,800}{-2} = 900$$

the length is also 900 ft. The desired region is a square.

and the largest possible area is

$$A(x) = -(900)^2 + 1,800(900) = 810,000 \text{ ft}^2$$

✔ CHECK YOURSELF 7

We want to enclose three sides of the largest possible rectangular area by using 900 ft of fence. Assume that an existing wall makes the fourth side. What will be the dimensions of the rectangle?

✔ CHECK YOURSELF ANSWERS

1.

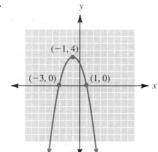

2.

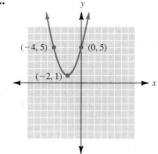

3.

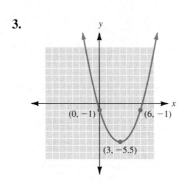

4. $\{-3.767, 2.434\}$ **5.** $\{-32.224, 8.224\}$

6. 100 radios, \$1,900   **7.** Width 225 ft, length 450 ft

Electric Power Costs.   You and a partner in a small engineering firm have been asked to help calculate the costs involved in running electric power a distance of 3 mi from a public utility company to a nearby community. You have decided to consider three tower heights: 75 ft, 100 ft, and 120 ft. The 75-ft towers cost \$850 each to build and install; the 100-ft towers cost \$1,110 each; and the 120-ft towers cost \$1,305 each. Spans between towers typically run 300 to 1,200 ft. The conducting wires weigh 1,900 pounds (lb) per 1,000 ft, and the tension on the wires is 6,000 lb. To be safe, the conducting wires should never be closer than 45 ft to the ground.

Work with a partner to complete the following.

1. Develop a plan for the community showing all three scenarios and their costs. Remember that hot weather will expand the wires and cause them to sag more than normally. Therefore, allow for about a 10% margin of error when you calculate the sag amount.

2. Write an accompanying letter to the town council, explaining your recommendation. Be sure to include any equations you used to help you make your decision. Remember to look at the formula at the beginning of the chapter, which gives the amount of sag for wires given a certain weight per foot, span length, and tension on the conductors.

Exercises · 11.3

In Exercises 1 to 8, match each graph with one of the equations at the left.

(a) $y = x^2 + 2$
(b) $y = 2x^2 - 1$
(c) $y = 2x + 1$
(d) $y = x^2 - 3x$
(e) $y = -x^2 - 4x$
(f) $y = -2x + 1$
(g) $y = x^2 + 2x - 3$
(h) $y = -x^2 + 6x - 8$

1.

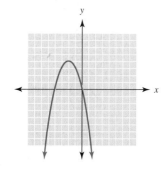

2.

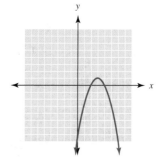

3.

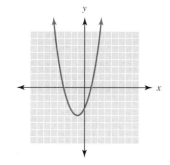

4.

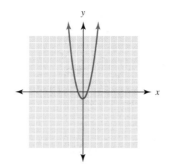

5.

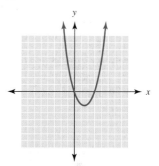

6.

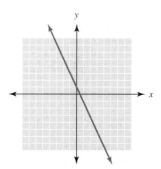

7.

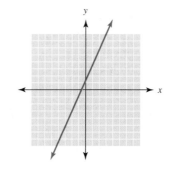

8.

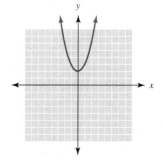

In Exercises 9 to 14, which of the given conditions apply to the graphs of the following equations? Note that more than one condition may apply.

(a) The parabola opens upward.   (b) The parabola opens downward.

(c) The parabola has two x intercepts.   (d) The parabola has one x intercept.

(e) The parabola has no x intercept.

9. $y = x^2 - 3$ **10.** $y = -x^2 + 4x$ **11.** $y = x^2 - 3x - 4$

12. $y = x^2 - 2x + 2$ **13.** $y = -x^2 - 3x + 10$ **14.** $y = x^2 - 8x + 16$

In Exercises 15 to 28, find the equation of the axis of symmetry, the coordinates of the vertex, and the x intercepts. Sketch the graph of each equation.

15. $y = x^2 - 4x$ **16.** $y = x^2 - 1$

17. $y = -x^2 + 4$ **18.** $y = x^2 + 2x$

19. $y = -x^2 - 2x$ **20.** $y = -x^2 - 3x$

21. $y = x^2 - 6x + 5$ **22.** $y = x^2 + x - 6$

23. $y = x^2 - 5x + 6$ **24.** $y = x^2 + 6x + 5$

25. $y = x^2 - 6x + 8$ **26.** $y = -x^2 - 3x + 4$

27. $y = -x^2 - 6x - 5$ **28.** $y = -x^2 + 6x - 8$

In Exercises 29 to 40, find the equation of the axis of symmetry, the coordinates of the vertex, and at least two symmetric points. Sketch the graph of each equation. (*Note:* A sample answer is provided for the two symmetric points.)

29. $y = x^2 - 2x - 1$ **30.** $y = x^2 + 4x + 6$

31. $y = x^2 - 4x - 1$ **32.** $y = -x^2 + 6x - 5$

33. $y = -x^2 + 3x - 3$ **34.** $y = x^2 + 5x + 3$

35. $y = 2x^2 + 4x - 1$ **36.** $y = \dfrac{1}{2}x^2 - x - 1$

37. $y = -\dfrac{1}{3}x^2 + x - 3$ **38.** $y = -2x^2 - 4x - 1$

39. $y = 3x^2 + 12x + 5$ **40.** $y = -3x^2 + 6x + 1$

In Exercises 41 to 48, use the graph of the related parabola to estimate the solutions to each equation. Round answers to the nearest thousandth.

41. $0 = x^2 + x - 12$ **42.** $0 = x^2 + 3x + 2$ **43.** $0 = 6x^2 - 19x$ **44.** $0 = 7x^2 - 15x$

45. $0 = 9x^2 + 12x - 7$ **46.** $0 = 3x^2 + 9x + 5$ **47.** $0 = x^2 + 2x - 7$ **48.** $0 = x^2 - 8x + 11$

49. Profit. A company's weekly profit P is related to the number of items sold by $P(x) = -0.3x^2 + 60x - 400$. Find the number of items that should be sold each week in order to maximize the profit. Then find the amount of that weekly profit.

50. Profit. A company's monthly profit P is related to the number of items sold by $P(x) = -0.2x^2 + 50x - 800$. How many items should be sold each month to obtain the largest possible profit? What is the amount of that profit?

51. Area. A builder wants to enclose the largest possible rectangular area with 2,000 ft of fencing. What should be the dimensions of the rectangle, and what will the area of the rectangle be?

52. Area. A farmer wants to enclose a rectangular area along a river on three sides. If 1,600 ft of fencing is to be used, what dimensions will give the maximum enclosed area? Find that maximum area.

53. Motion. A ball is thrown upward into the air with an initial velocity of 96 ft/s. If h gives the height of the ball at time t, then the equation relating h and t is

$$h(t) = -16t^2 + 96t$$

Find the maximum height the ball will attain.

54. Motion. A ball is thrown upward into the air with an initial velocity of 64 ft/s. If h gives the height of the ball at time t, then the equation relating h and t is

$$h(t) = -16t^2 + 64t$$

Find the maximum height the ball will attain.

55. Motion. A ball is thrown upward with an initial velocity v. After 3 s, it attains a height of 120 ft. Find the initial velocity, using the equation

$$h(t) = -16t^2 + v \cdot t$$

For each of the following quadratic functions, use your graphing calculator to determine (a) the vertex of the parabola and (b) the range of the function.

56. $f(x) = 2(x - 3)^2 + 1$ **57.** $g(x) = 3(x + 4)^2 + 2$ **58.** $f(x) = -(x - 1)^2 + 2$

59. $g(x) = -(x + 2)^2 - 1$ **60.** $f(x) = 3(x + 1)^2 - 2$ **61.** $g(x) = -2(x - 4)^2$

62. Explain how to determine the domain and range of the function $f(x) = a(x - h)^2 + k$.

In Exercises 63 to 66, describe a viewing window that would include the vertex and all intercepts for the graph of each function.

63. $f(x) = 3x^2 - 25$ **64.** $f(x) = 9x^2 - 5x - 7$

65. $f(x) = -2x^2 + 5x - 7$ **66.** $f(x) = -5x^2 + 2x + 7$

11.4 Solving Quadratic Inequalities

1. Solve a quadratic inequality graphically
2. Solve a quadratic inequality algebraically

A **quadratic inequality** is an inequality that can be written in the form

$$ax^2 + bx + c < 0 \qquad \text{where } a \neq 0$$

Note that the inequality symbol $<$ can be replaced by the symbol $>$, $\leq$, or $\geq$ in the above definition.

In Chapter 8, solutions to linear inequalities such as $4x + 2 < 0$ were analyzed graphically. Recall that, given the graph of the function $f(x) = 4x + 2$,

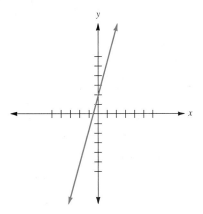

the solution to the inequality is the set of all x values associated with points on the line that are *below* the x axis.

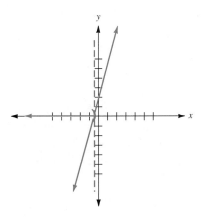

By using interval notation, the solution could be written as

$\left(-\infty, -\dfrac{1}{2}\right)$

In this case, each x to the left of $-\dfrac{1}{2}$ is associated with a point on the line that is below the x axis. The solution set for the inequality $4x + 2 < 0$ is the set

$\left\{ x \mid x < -\dfrac{1}{2} \right\}.$

The same principle can be applied to solving quadratic inequalities. Example 1 illustrates this concept.

Example 1

Solving a Quadratic Inequality Graphically

Solve the inequality graphically, and graph the solution set on a number line.

$$x^2 - x - 12 \leq 0$$

First, use the techniques in Section 11.3 to graph the function $f(x) = x^2 - x - 12$.

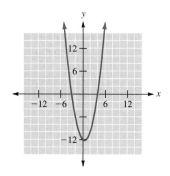

Next, looking at the graph, determine the values of x that make $x^2 - x - 12 \leq 0$ a true statement.

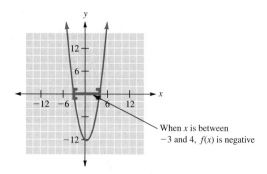

When x is between -3 and 4, $f(x)$ is negative

Notice that the graph is below the x axis for values of x between -3 and 4. The graph intercepts the x axis when x is -3 or 4. The solution set for the inequality is $\{x \mid -3 \leq x \leq 4\}$.

In interval notation this would be $[-3, 4]$.

The graph of the solution set is

✔ *CHECK YOURSELF 1*

Graphically solve the inequality, and graph the solution set on a number line.

$$x^2 - 3x - 10 \geq 0$$

Algebraic methods can also be used to find the solutions to a quadratic inequality. Subsequent examples in this section will discuss algebraic methods. When solving an equation or inequality algebraically, it is always a good idea to compare the graph to the algebraic solutions.

Example 2

Note: If we expand the binomial product, we get
$$x^2 - 2x - 3 < 0$$
Looking at the graph of
$$y = x^2 - 2x - 3$$
where is y less than 0 on the graph?

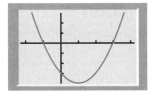

Note: $x + 1$ is negative if x is less than -1; $x + 1$ is positive if x is greater than -1.

Note: $x - 3$ is negative if x is less than 3; $x - 3$ is positive if x is greater than 3.

Solving a Quadratic Inequality Algebraically

Solve $(x - 3)(x + 1) < 0$, and graph the solution set on a number line.

We start by finding the solutions of the corresponding quadratic equation. So

$$(x - 3)(x + 1) = 0$$

has solutions 3 and -1, called the *critical values* of x.

Our process depends on determining where each factor is positive or negative. To help visualize that process, we start with a number line and label it as shown below. We begin with our first critical value of -1.

We now continue in the same manner with the second critical value of 3.

In practice, we combine the two steps above for the following result.

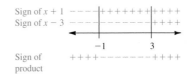

Examining the signs of the factors, we see that in this case

For any x less than -1, the product is positive because both factors are negative.

For any x between -1 and 3, the product is negative because the factors have opposite signs.

For any x greater than 3, the product is again positive because both factors are positive.

We return to the original inequality:

$$(x - 3)(x + 1) < 0$$

Note: The product of the two binomials must be negative.

We can see that this is true only between -1 and 3. In set notation, the solution set can be written as

$$\{x \mid -1 < x < 3\}$$

In interval notation, this would be $(-1, 3)$.

On a number line, the graph of the solution set is

✔ *CHECK YOURSELF 2*

Solve algebraically and graph the solution set.

$$(x - 2)(x + 4) < 0$$

We now consider an example in which the quadratic member of the inequality must be factored.

Example 3

Solving a Quadratic Inequality Algebraically

Solve $x^2 - 5x + 4 > 0$, and graph the solution set on a number line.

Factoring the quadratic member, we have

$$(x - 1)(x - 4) > 0$$

Note: Examine the graph of $y = x^2 - 5x + 4$. For what values of x is y (the graph) greater than zero?

The critical values are 1 and 4, and we form the sign graph as before.

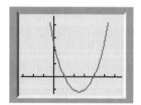

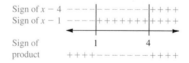

In this case, we want those values of x for which the product is *positive,* and we can see from the sign graph above that the solution set is

$$\{x \mid x < 1 \text{ or } x > 4\}$$

The graph of the solution set is shown below.

CHECK YOURSELF 3

Solve algebraically and graph the solution set.

$$2x^2 - x - 3 > 0$$

The method used in the previous examples works *only* when one side of the inequality is factorable and the other is 0. It is sometimes necessary to rewrite the inequality in an equivalent form, as Example 4 illustrates.

Example 4

Note: Use a calculator to graph both $f(x) = x^2 - 3x - 4$ and $g(x) = 6$. Where is $f(x)$ above $g(x)$? Compare this to the algebraic solution.

Note: Both factors are negative if x is less than -2. Both factors are positive if x is greater than 5.

Solving a Quadratic Inequality Algebraically

Solve $(x + 1)(x - 4) \geq 6$.

First, we multiply to clear the parentheses.

$$x^2 - 3x - 4 \geq 6$$

Now we subtract 6 from both sides so that the inequality is *related to 0*:

$$x^2 - 3x - 10 \geq 0$$

Factoring the quadratic member, we have

$$(x - 5)(x + 2) \geq 0$$

We can now proceed with the sign graph method as before.

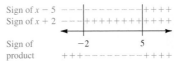

From the sign graph we see that the solution set is

$$\{x \mid x \leq -2 \text{ or } x \geq 5\}$$

The graph of the solution set is shown below.

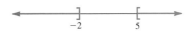

 CHECK YOURSELF 4

Solve algebraically and graph the solution set.

$$(x - 5)(x + 7) \leq -11$$

The following algorithm summarizes our work to this point in solving a quadratic inequality.

Solving a Quadratic Inequality Algebraically

Step 1.   Clear the inequality of parentheses and fractions.

Step 2.   Rewrite the inequality such that the expression is related to zero.

Step 3.   Factor the quadratic member.

Step 4.   Use a sign graph to find the values for x that make the inequality a true statement.

Step 5.   Use a number line to graph the solution set.

✔ CHECK YOURSELF ANSWERS

1. $\{x \mid x \le -2 \text{ or } x \ge 5\}$

2. $\{x \mid -4 < x < 2\}$

3. $\left\{x \mid x < -1 \text{ or } x > \dfrac{3}{2}\right\}$

4. $\{x \mid -6 \le x \le 4\}$

Trajectory and Height.   So far in this chapter you have done many exercises involving balls that have been thrown upward with varying velocities. How does trajectory—the angle at which the ball is thrown—affect its height and time in the air?

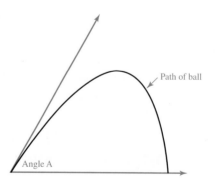

Path of ball

Angle A

If you throw a ball from ground level with an initial upward velocity of 70 ft/s, the equation

$$h = -16t^2 + \partial(70)t$$

gives you the height h in feet t s after the ball has been thrown at a certain angle A. The value of ∂ in the equation depends on the angle A and is given in the accompanying table.

Measure of Angle A in Degrees	Value of ∂
0	0.000
5	0.087
10	0.174
15	0.259
20	0.342
25	0.423
30	0.500
35	0.574
40	0.643
45	0.707
50	0.766
55	0.819
60	0.866
65	0.906
70	0.940
75	0.966
80	0.985
85	0.996
90 (straight up)	1.000

Investigate the following questions and write your conclusions to each one in complete sentences, showing all charts and graphs. Indicate what initial velocity you are using in each case.

1. Suppose an object is thrown from the ground level with an initial upward velocity of 70 ft/s and at an angle of 45°. What will be the height of the ball in 1 s (nearest tenth of a foot)? How long is the ball in the air?

2. Does the ball stay in the air longer if the angle of the throw is greater?

3. If you double the angle of the throw, will the ball stay in the air double the length of time?

4. If you double the angle of the throw, will the ball go twice as high?

5. Is the height of the ball directly related to the angle at which you throw it? That is, does the ball go higher if the angle of the throw is larger?

6. Repeat this exercise, using another initial upward velocity.

Exercises · 11.4

In Exercises 1 to 8, solve each inequality graphically and graph the solution set on a number line.

1. $(x - 3)(x + 4) < 0$

2. $(x - 2)(x + 5) > 0$

3. $(x - 3)(x + 4) > 0$

4. $(x - 2)(x + 5) < 0$

5. $(x - 3)(x + 4) \leq 0$

6. $(x - 2)(x + 5) \geq 0$

7. $(x - 3)(x + 4) \geq 0$

8. $(x - 2)(x + 5) \leq 0$

In Exercises 9 to 38, solve each inequality algebraically and graph the solution set on a number line.

9. $x^2 - 3x - 4 > 0$

10. $x^2 - 2x - 8 < 0$

11. $x^2 + x - 12 \leq 0$

12. $x^2 - 2x - 15 \geq 0$

13. $x^2 - 5x + 6 \geq 0$

14. $x^2 + 7x + 10 \leq 0$

15. $x^2 + 2x \leq 24$

16. $x^2 - 3x > 18$

17. $x^2 > 27 - 6x$

18. $x^2 \leq 7x - 12$

19. $2x^2 + x - 6 \leq 0$

20. $3x^2 - 10x - 8 < 0$

21. $4x^2 + x < 3$

22. $5x^2 - 13x \geq 6$

23. $x^2 - 16 \leq 0$

24. $x^2 - 9 > 0$

25. $x^2 \geq 25$

26. $x^2 < 49$

27. $4 - x^2 < 0$

28. $36 - x^2 \geq 0$

29. $x^2 - 4x \leq 0$

30. $x^2 + 5x > 0$

31. $x^2 \geq 6x$

32. $x^2 < 3x$

33. $4x > x^2$

34. $6x \leq x^2$

35. $x^2 - 4x + 4 \leq 0$

36. $x^2 + 6x + 9 \geq 0$

37. $(x + 3)(x - 6) \leq 10$

38. $(x + 4)(x - 5) > 22$

39. Can a quadratic inequality be solved if the quadratic member of the inequality is not factorable? If so, explain how the solution can be found. If not, explain why not.

40. Is it necessary to relate a quadratic inequality to 0 in order to solve it? Why or why not?

An inequality of the form

$$(x - a)(x - b)(x - c) < 0$$

can be solved by using a sign graph to consider the signs of *all three factors*. In Exercises 41 to 46, use this suggestion to solve each inequality. Then graph the solution set.

41. $x(x - 2)(x + 1) < 0$

42. $x(x + 3)(x - 2) \geq 0$

43. $(x - 3)(x + 2)(x - 1) \geq 0$

44. $(x - 5)(x + 1)(x - 4) < 0$

45. $x^3 - 2x^2 - 15x \leq 0$

46. $x^3 + 2x^2 - 24x > 0$

47. Profit. A small manufacturer's weekly profit is given by

$$P(x) = -2x^2 + 220x$$

where x is the number of items manufactured and sold. Find the number of items that must be manufactured and sold if the profit is to be greater than or equal to $6,000.

48. Profit. Suppose that a company's profit is given by

$$P(x) = -2x^2 + 360x$$

How many items must be produced and sold so that the profit will be at least $16,000?

49. Free fall. If a ball is thrown vertically upward from the ground with an initial velocity of 80 ft/s, its approximate height is given by

$$h(t) = -16t^2 + 80t$$

where t is the time (in seconds) after the ball was released. When will the ball have a height of at least 96 ft?

50. Free fall. Suppose a ball's height (in meters) is given by

$$h(t) = -5t^2 + 20t$$

When will the ball have a height of at least 15 m?

In Exercises 51 to 54, use your calculator to find the graph for each equation. Use that graph to estimate the solution of the related inequality.

51. $y = x^2 + 6x; \ x^2 + 6x < 0$

52. $y = x^2 - 49; \ x^2 - 49 \geq 0$

53. $y = x^2 - 5x - 6; \ x^2 - 5x - 6 \leq 0$

54. $y = 2x^2 + 7x - 15; \ 2x^2 + 7x - 15 > 0$

Summary for Chapter 11

Example	Topic	Reference
	Solving Quadratic Equations by Completing the Square	11.1
To solve: $(x-3)^2 = 5$ $x - 3 = \pm\sqrt{5}$ $x = 3 \pm \sqrt{5}$	**Square Root Property** If $x^2 = k$, when k is any real number, then $x = \sqrt{k}$ or $x = -\sqrt{k}$.	**p. 859**
To solve: $$x^2 + x = \frac{1}{2}$$ $$x^2 + x + \left(\frac{1}{2}\right)^2 = \frac{1}{2} + \left(\frac{1}{2}\right)^2$$ $$\left(x + \frac{1}{2}\right)^2 = \frac{3}{4}$$ $$x + \frac{1}{2} = \pm\sqrt{\frac{3}{4}}$$ $$x = \frac{-1 \pm \sqrt{3}}{2}$$	**Completing the Square** 1. Isolate the constant on the right side of the equation. 2. Divide both sides of the equation by the coefficient of the x^2 term if that coefficient is not equal to 1. 3. Add the square of one-half of the coefficient of the linear term to both sides of the equation. This will give a perfect-square trinomial on the left side of the equation. 4. Write the left side of the equation as the square of a binomial, and simplify on the right side. 5. Use the square root property, and then solve the resulting linear equations.	**p. 865**
	The Quadratic Formula	11.2
To solve: $$x^2 - 2x = 4$$ write the equation as $$x^2 - 2x - 4 = 0$$ $a = 1 \quad b = -2 \quad c = -4$ $$x = \frac{-(-2) \pm \sqrt{(-2)^2 - 4(1)(-4)}}{2 \cdot 1}$$ $$= \frac{2 \pm \sqrt{20}}{2}$$ $$= \frac{2 \pm 2\sqrt{5}}{2}$$ $$= 1 \pm \sqrt{5}$$	Any quadratic equation can be solved by using the following algorithm. 1. Write the equation in standard form (set it equal to 0). $$ax^2 + bx + c = 0$$ 2. Determine the values for a, b, and c. 3. Substitute those values into the quadratic formula $$x = \frac{-b \pm \sqrt{b^2 - 4ac}}{2a}$$ 4. Write the solutions in simplest form.	**p. 873**

The Quadratic Formula (continued)	11.2

Given $$2x^2 - 5x + 3 = 0$$ $a = 2 \quad b = -5 \quad c = 3$ $b^2 - 4ac = 25 - 4(2)(3)$ $\quad\quad\quad = 25 - 24$ $\quad\quad\quad = 1$ There are two distinct solutions.	**The Discriminant** The expression $b^2 - 4ac$ is called the **discriminant** for a quadratic equation. There are three possibilities: 1. If $b^2 - 4ac < 0$, there are no real solutions (but two imaginary solutions). 2. If $b^2 - 4ac = 0$, there is one real solution (a double solution). 3. If $b^2 - 4ac > 0$, there are two distinct real solutions.	**p. 874**

Solving Quadratic Equations by Graphing	11.3

Graph the equation $$y = x^2 - 4x - 12$$ 1. Find the axis of symmetry. $$x = \frac{-b}{2a} = \frac{-(-4)}{2(1)}$$ $$\quad = \frac{4}{2} = 2$$ so $x = 2$ is the axis of symmetry. 2. Find the vertex. Let $x = 2$ in the original equation. $$y = (2)^2 - 4(2) - 12$$ $$y = 4 - 8 - 12$$ $$y = -16$$ The vertex is $(2, -16)$. 3. Find two symmetric points. $$0 = x^2 - 4x - 12$$ $$0 = (x - 6)(x + 2)$$ $$x - 6 = 0 \quad x + 2 = 0$$ $$x = 6 \quad\quad x = -2$$ Two symmetric points are $(6, 0)$ and $(-2, 0)$.	**Axis of Symmetry** The axis of symmetry is a vertical line midway between any pair of symmetric points on a parabola. The axis of symmetry passes through the vertex of the parabola. **Vertex of a Parabola** If $$y = ax^2 + bx + c \quad\quad a \neq 0$$ then the x coordinate of the vertex of the corresponding parabola is $$x = \frac{-b}{2a}$$ **To Graph a Parabola:** 1. Find the axis of symmetry. 2. Find the vertex. 3. Determine two symmetric points. **Note:** You can use the x intercepts if the equation is factorable. Otherwise use the y intercept and its symmetric point.	**p. 886** **p. 887** **p. 890**

4. Draw a smooth curve connecting the points found.

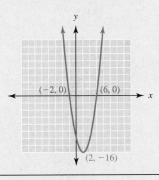

4. Draw a smooth curve connecting the points found in step 3 to form the parabola. You may choose to find additional pairs of symmetric points at this time.

Solve the following equation graphically.

$$Y = 0.5x^2 + 3x - 2$$

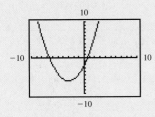

$$\{-6.606, 0.606\}$$

To solve the equation

$$ax^2 + bx + c = 0$$

1. Graph the function

$$Y = ax^2 + bx + c$$

2. Use the ZERO or ROOT utility to determine the x intercepts of the graph. These values are the solutions to the original equation.

p. 892

Solving Quadratic Inequalities

11.4

Solve $x^2 - 3x - 18 < 0$.
Graph $f(x) = x^2 - 3x - 18$.

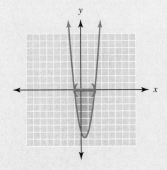

The graph is above the x axis for all x in the interval $-3 < x < 6$. So the solution set is

$$\{x \mid -3 < x < 6\}$$

Solving a Quadratic Inequality Graphically To solve a quadratic inequality graphically

1. Graph $f(x) = ax^2 + bx + c$.

2. The x values of the points on the graph that are **above** the x axis form the solution set of

$$ax^2 + bx + c > 0$$

and the x values of the points on the graph that are **below** the x axis form the solution set of

$$ax^2 + bx + c < 0$$

p. 902

To solve

$$x^2 - 3x < 18$$

$$x^2 - 3x - 18 < 0$$

$$(x - 6)(x + 3) < 0$$

Critical values of x are

$$-3 \quad \text{and} \quad 6$$

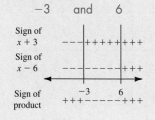

The solution set is

$$\{x \mid -3 < x < 6\}$$

Solving a Quadratic Inequality Algebraically   p. 906

1. Clear the inequality of parentheses and fractions.

2. Rewrite the inequality such that the expression is related to zero.

3. Factor the quadratic member and find the critical values of x.

4. Use a sign graph to find the values for x that make the inequality a true statement.

5. Use a number line to graph the solution set.

Summary Exercises ▪ 11

This summary exercise set is provided to give you practice with each of the objectives in the chapter. Each exercise is keyed to the appropriate chapter section.

[11.1] Solve each of the following equations, using the square root method.

1. $x^2 - 12 = 0$

2. $3y^2 - 15 = 0$

3. $(x - 2)^2 = 20$

4. $(3x - 2)^2 - 15 = 0$

[11.1] Find the constant that must be added to each of the following binomials to form a perfect-square trinomial.

5. $x^2 - 14x$

6. $y^2 + 3y$

[11.1] Solve the following equations by completing the square.

7. $x^2 - 4x - 5 = 0$

8. $x^2 - 8x - 9 = 0$

9. $w^2 - 10w - 3 = 0$

10. $y^2 + 3y - 1 = 0$

11. $2x^2 - 6x + 1 = 0$

12. $3x^2 + 4x - 1 = 0$

[11.2] Solve each of the following equations by using the quadratic formula.

13. $x^2 - 5x - 24 = 0$

14. $w^2 + 10w + 25 = 0$

15. $x^2 = 5x - 2$

16. $2y^2 - 5y + 2 = 0$

17. $3y^2 + 4y = 1$

18. $3y^2 + 4y + 7 = 0$

19. $(x - 5)(x + 3) = 13$

20. $\dfrac{1}{x^2} - \dfrac{4}{x} + 1 = 0$

21. $3x^2 + 2x + 5 = 0$

22. $(x - 1)(2x + 3) = -5$

[11.2] For each of the following quadratic equations, use the discriminant to determine the number of real solutions.

23. $x^2 - 3x + 3 = 0$

24. $x^2 + 4x = 2$

25. $4x^2 - 12x + 9 = 0$

26. $2x^2 + 3 = 3x$

27. Number problem. The sum of two integers is 12, and their product is 32. Find the two integers.

28. Number problem. The product of two consecutive, positive, even integers is 80. What are the two integers?

29. Number problem. Twice the square of a positive integer is 10 more than 8 times that integer. Find the integer.

30. Rectangles. The length of a rectangle is 2 ft more than its width. If the area of the rectangle is 80 ft^2, what are the dimensions of the rectangle?

31. **Rectangles.** The length of a rectangle is 3 cm less than twice its width. The area of the rectangle is 35 cm². Find the length and width of the rectangle.

32. **Rectangles.** An open box is formed by cutting 3-in. squares from each corner of a rectangular piece of cardboard which is 3 in. longer than it is wide. If the box is to have a volume of 120 in.³, what must be the size of the original piece of cardboard?

33. **Profit.** Suppose that a manufacturer's weekly profit P is given by

$$P(x) = -3x^2 + 240x$$

where x is the number of items manufactured and sold. Find the number of items that must be manufactured and sold if the profit is to be at least \$4,500.

34. **Thrown ball.** If a ball is thrown vertically upward from the ground with an initial velocity of 64 ft/s, its approximate height is given by

$$h(t) = -16t^2 + 64t$$

When will the ball reach a height of at least 48 ft?

35. **Rectangle.** The length of a rectangle is 1 cm more than twice its width. If the length is doubled, the area of the new rectangle is 36 cm² more than that of the old. Find the dimensions of the original rectangle.

36. **Triangle.** One leg of a right triangle is 4 in. longer than the other. The hypotenuse of the triangle is 8 in. longer than the shorter leg. What are the lengths of the three sides of the triangle?

37. **Rectangle.** The diagonal of a rectangle is 9 ft longer than the width of the rectangle, and the length is 7 ft more than its width. Find the dimensions of the rectangle.

38. **Thrown ball.** If a ball is thrown vertically upward from the ground, the height h after t s is given by

$$h(t) = 128t - 16t^2$$

(a) How long does it take the ball to return to the ground?

(b) How long does it take the ball to reach a height of 240 ft on the way up?

39. **Triangle.** One leg of a right triangle is 2 m longer than the other. If the length of the hypotenuse is 8 m, find the length of the other two legs.

40. **Thrown ball.** Suppose that the height (in meters) of a golf ball, hit off a raised tee, is approximated by

$$h(t) = -5t^2 + 10t + 10$$

t s after the ball is hit. When will the ball hit the ground?

[11.1–11.2] Find the zeros of the following functions.

41. $f(x) = x^2 - x - 2$ 42. $f(x) = 6x^2 + 7x + 2$

43. $f(x) = -2x^2 - 7x - 6$ 44. $f(x) = -x^2 - 1$

[11.3] Find the equation of the axis of symmetry and the coordinates for the vertex of each of the following.

45. $f(x) = x^2$

46. $f(x) = x^2 + 2$

47. $f(x) = x^2 - 5$

48. $f(x) = (x - 3)^2$

49. $f(x) = (x + 2)^2$

50. $f(x) = -(x - 3)^2$

51. $f(x) = (x + 3)^2 + 1$

52. $f(x) = -(x + 2)^2 - 3$

53. $f(x) = -(x - 5)^2 - 2$

54. $f(x) = 2(x - 2)^2 - 5$

55. $f(x) = -x^2 + 2x$

56. $f(x) = x^2 - 4x + 3$

57. $f(x) = -x^2 - x + 6$

58. $f(x) = x^2 + 4x + 5$

59. $f(x) = -x^2 - 6x + 4$

[11.3] Use a graphing calculator to estimate the solutions to each equation. Round answers to the nearest thousandth.

60. $0 = x^2 - 2x - 5$

61. $0 = 2x^2 + 5x - 9$

62. $0 = x^2 - 5x + 5$

63. $0 = x^2 - 7x + 5$

64. $0 = 2x^2 - 4x - 5$

[11.3] Graph each of the following functions.

65. $f(x) = x^2$

66. $f(x) = x^2 + 2$

67. $f(x) = x^2 - 5$

68. $f(x) = (x - 3)^2$

69. $f(x) = (x + 2)^2$

70. $f(x) = -(x - 3)^2$

71. $f(x) = (x + 3)^2 + 1$

72. $f(x) = -(x + 2)^2 - 3$

73. $f(x) = x^2 - 4x$

74. $f(x) = -x^2 + 2x$

75. $f(x) = x^2 + 2x - 3$

76. $f(x) = x^2 - 4x + 3$

77. $f(x) = -x^2 - x + 6$ **78.** $f(x) = -x^2 + 3x + 4$ **79.** $f(x) = x^2 + 4x + 5$ **80.** $f(x) = x^2 - 6x + 4$

81. $f(x) = x^2 - 2x + 4$ **82.** $f(x) = -x^2 + 2x - 2$ **83.** $f(x) = 2x^2 - 4x + 1$ **84.** $f(x) = \dfrac{1}{2}x^2 - 4x$

[11.4] In Exercises 85 to 88, solve the inequalities graphically and graph the solution set.

85. $(x - 2)(x + 5) > 0$ **86.** $(x - 1)(x - 6) < 0$

87. $(x + 1)(x + 3) \le 0$ **88.** $(x + 4)(x - 5) \ge 0$

In Exercises 89 to 94, solve the inequalities algebraically and graph the solution set.

89. $x^2 - 5x - 24 \le 0$ **90.** $x^2 + 4x \ge 21$

91. $x^2 \ge 64$ **92.** $x^2 + 5x \ge 0$

93. $(x + 2)(x - 6) < 9$ **94.** $(x - 1)(x + 2) \ge 4$

95. Motion. If a ball is thrown vertically upward from the ground with an initial velocity of 64 ft/s, its approximate height is given by $h(t) = -16t^2 + 64t$. When will the ball reach a height of at least 48 ft?

96. Cost. Suppose that the cost, in dollars, of producing x stereo systems is given by the equation $C(x) = 3{,}000 - 60x + 3x^2$. How many systems can be produced if the cost cannot exceed \$7,500?

Self-Test 11

The purpose of this self-test is to help you check your progress and to review for a chapter test in class. Allow yourself about 1 hour to take the test. When you are done, check your answers in the back of the book. If you missed any answers, be sure to go back and review the appropriate sections in the chapter and the exercises that are provided.

Solve each of the following equations by factoring.

1. $2x^2 + 7x + 3 = 0$
2. $6x^2 = 10 - 11x$
3. $4x^3 - 9x = 0$

Solve each of the following equations, using the square root method.

4. $4w^2 - 20 = 0$
5. $(x - 1)^2 = 10$
6. $4(x - 1)^2 = 23$

Solve each of the following equations by completing the square.

7. $m^2 + 3m - 1 = 0$
8. $2x^2 - 10x + 3 = 0$

Solve each of the following equations, using the quadratic formula.

9. $x^2 - 5x - 3 = 0$
10. $x^2 + 4x = 7$

11. Find the zeros of the function $f(x) = 3x^2 - 10x - 8$.

Solve.

12. The product of two consecutive, positive, odd integers is 63. Find the two integers.

13. Suppose that the height (in feet) of a ball thrown upward from a raised platform is approximated by

$$h(t) = -16t^2 + 32t + 32$$

  t seconds after the ball has been released. How long will it take the ball to hit the ground?

Find the equation of the axis of symmetry and the coordinates of the vertex of each of the following.

14. $y = -3(x + 2)^2 + 1$
15. $y = x^2 - 4x - 5$
16. $y = -2x^2 + 6x - 3$

17. $y = (x - 3)^2 - 2$
18. $y = x^2 - 6x + 2$

Graph each of the following functions.

19. $f(x) = (x - 5)^2$
20. $f(x) = (x + 2)^2 - 3$
21. $f(x) = -2(x - 3)^2 - 1$
22. $f(x) = 3x^2 + 9x + 2$

Use a graphing calculator to estimate the solutions to each equation. Round your answers to the nearest thousandth.

23. $0 = x^2 + 3x - 7$
24. $0 = 4x^2 + 2x - 5$

Solve the following inequalities and graph the solution set.

25. $(x + 3)(x - 1) > 0$
26. $x^2 + 5x - 14 < 0$
27. $x^2 - 3x \geq 18$

28. $3x^2 + x - 10 \leq 0$
29. $x^2 - 3x \leq 0$
30. $x^2 + 5x > 0$

Cumulative Test • 0–11

This test is provided to help you in the process of reviewing the previous chapters. Answers are provided in the back of the book. If you missed any answers, be sure to go back and review the appropriate chapter section.

Graph each of the following equations.

1. $2x - 3y = 6$

2. $y = -\dfrac{1}{3}x - 2$

3. $y = 4$

Find the slope of the line determined by each set of points.

4. $(-4, 7)$ and $(-3, 4)$

5. $(-2, 3)$ and $(-5, -1)$

6. Let $f(x) = 6x^2 - 5x + 1$. Evaluate $f(-2)$.

7. Simplify the function $f(x) = (x^2 - 1)(x + 3)$.

8. Completely factor the expression $x^3 + x^2 - 6x$.

9. Simplify the expression $\dfrac{2}{x + 2} - \dfrac{3x - 2}{x^2 - x - 6}$.

10. Simplify the expression $\sqrt{72x^3y^5}$.

Solve each equation.

11. $2x - 7 = 0$

12. $3x - 5 = 5x + 3$

13. $0 = (x - 3)(x + 5)$

14. $x^2 - 3x + 2 = 0$

15. $x^2 + 7x - 30 = 0$

16. $x^2 - 3x - 3 = 0$

17. $(x - 3)^2 = 5$

18. $|x - 2| = 4$

19. $\dfrac{x}{3} - \dfrac{4}{9} = \dfrac{5}{18}$

Solve each inequality.

20. $x - 2 \le 7$

21. $|x - 1| \ge 8$

22. $x^2 - x - 6 \le 0$

Solve the following word problems. Show the equation used for the solution.

23. Five times a number decreased by 7 is -72. Find the number.

24. One leg of a right triangle is 4 ft longer than the shorter leg. If the hypotenuse is 28 ft, how long is each leg?

25. Suppose that a manufacturer's weekly profit P is given by

$$P(x) = -4x^2 + 320x$$

where x is the number of units manufactured and sold. Find the number of items that must be manufactured and sold to guarantee a profit of at least $4,956.

920

12 Conic Sections

Large cities often commission fireworks artists to choreograph elaborate displays on holidays. Such displays look like beautiful paintings in the sky, in which the fireworks seem to dance to well-known popular and classical music. The displays are feats of engineering and very accurate timing. Suppose the designer wants a second set of rockets of a certain color and shape to be released after the first set of a different color and shape reaches a specific height and explodes. He must know the strength of the initial liftoff and use a quadratic equation to determine the proper time for setting off the second round.

The equation $h = -16t^2 + 100t$ gives the height in feet t seconds after the rockets are shot into the air if the initial velocity is 100 ft/s. Using this equation, the designer knows how high the rocket will ascend and when it will begin to fall. He can time the next round to achieve the effect he wishes. Displays that involve large banks of fireworks in shows that last up to an hour are programmed using computers, but quadratic equations are at the heart of the mechanism that creates the beautiful effects.

SECTION 12.1 More on the Parabola

 OBJECTIVES

1. *Recognize the general form for a conic section*
2. *Graph a horizontal parabola*

In Chapter 11, we discussed the graph of a function with the general form

$$y = ax^2 + bx + c \qquad \textbf{(1)}$$

Such a graph is called a **parabola.** The parabola is one example from a family of curves that are called conic sections. A **conic section** is a curve formed when a plane cuts through, or forms a section of, a cone. The conic sections comprise four curves—the **parabola, circle, ellipse,** and **hyperbola.** Examples of these curves are shown in the figure.

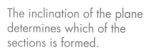

The inclination of the plane determines which of the sections is formed.

The names *ellipse, parabola,* and *hyperbola* are attributed to Apollonius, a third-century B.C.E. Greek mathematician and astronomer.

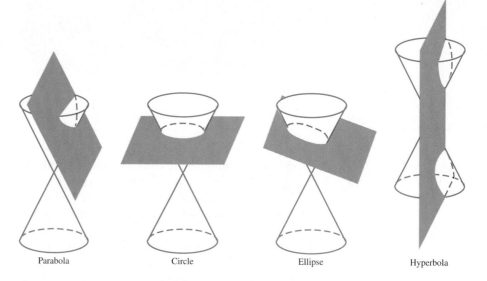

| Parabola | Circle | Ellipse | Hyperbola |

Algebraically, a conic section is formed from any second-degree equation from the general form

$$ax^2 + by^2 + cxy + dx + ey + f = 0 \qquad \textbf{(2)}$$

In the case of the parabola we examined in Section 11.3, the values of b and c are each zero. This leaves the form

$$ax^2 + dx + ey + f = 0 \qquad \textbf{(3)}$$

922

Note that if a, d, e, and f are each given nonzero values, an equation in the form of equation (3) can be rewritten into the form of equation (1). Example 1 illustrates this.

| Example 1 | ## Rewriting Quadratic Equations |

Given the standard form for a conic section

$$ax^2 + by^2 + cxy + dx + ey + f = 0$$

and the values $a = -3$, $b = 0$, $c = 0$, $d = 6$, $e = -2$, and $f = -8$, rewrite the equation in standard quadratic form.

Substituting the coefficient values, we have the equation

$$-3x^2 + 6x - 2y - 8 = 0$$

or

$$-2y = 3x^2 - 6x + 8$$

so, dividing by the coefficient of the y term, we have

$$y = -\frac{3}{2}x^2 + 3x - 4$$

✔ *CHECK YOURSELF 1*

Given the standard form for a conic section

$$ax^2 + by^2 + cxy + dx + ey + f = 0$$

and the values $a = 5$, $b = 0$, $c = 0$, $d = 6$, $e = -3$, and $f = -9$, rewrite the equation in standard quadratic form.

So far we have dealt with equations of the form

$$y = ax^2 + bx + c$$

Suppose we reverse the roles of x and y. We then have

$$x = ay^2 + by + c$$

which is quadratic in y but not in x. The graph of such an equation is once again a parabola, but this time the parabola is horizontally oriented.

In an equation of the form

$$x = ay^2 + by + c \qquad a \neq 0$$

the parabola opens leftward or rightward, as follows:

1. If $a > 0$, the parabola opens *rightward*.
2. If $a < 0$, the parabola opens *leftward*.

$$x = ay^2 + by + c$$

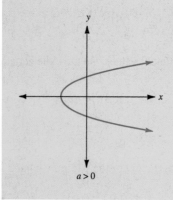

$a > 0$

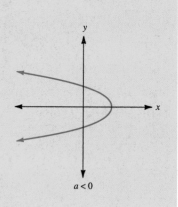

$a < 0$

Much of what we did earlier is easily extended to a horizontally oriented parabola. Example 2 illustrates the changes in the process.

Example 2

Graphing a Parabola

Graph the equation

$$x = y^2 + 4y - 5$$

First, find the axis of symmetry. Now the axis of symmetry is horizontal with the equation

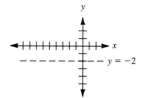

$$y = \frac{-b}{2a}$$

So
$$y = \frac{-b}{2a} = \frac{-(4)}{2(1)} = -2$$

Second, find the vertex. If $y = -2$,

$$x = (-2)^2 + 4(-2) - 5 = -9$$

So $(-9, -2)$ is the vertex.

Third, find two symmetric points. Here the quadratic member is factorable, so set $x = 0$ in the original equation. That gives the y intercepts:

$$0 = y^2 + 4y - 5$$
$$= (y + 5)(y - 1)$$
$$y + 5 = 0 \quad \text{or} \quad y - 1 = 0$$
$$y = -5 \qquad\qquad y = 1$$

The y intercepts then are at $(0, -5)$ and $(0, 1)$.

Fourth, draw a smooth curve through the points already found to form the parabola.

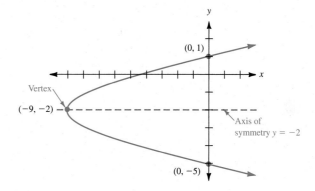

✔ CHECK YOURSELF 2

Graph the equation $x = -y^2 - 2y + 3$.

The parabola is known to have a very interesting reflective property. For any given parabola, there is a special point, known as the **focus,** that lies on the axis of symmetry and is "inside" the parabola. Imagine a parabolic bowl that is aimed at a specific source, such as the sun or a satellite. All rays or signals that enter the bowl parallel to the axis of symmetry will be reflected to that single point, the focus. This is the principle behind solar ovens and satellite dishes.

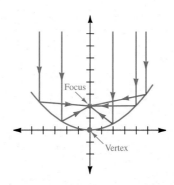

Similarly, if a source of light is placed at the focus, the light rays from this source will bounce off the walls of the parabolic surface, and the light rays emerge parallel to each other (and, of course, to the axis of symmetry). This is essentially how search-lights, flashlights, and automobile headlights work.

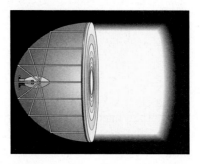

The location of the focus is actually quite easy to determine if we know the equation of the parabola. Let the distance between the vertex and the focus be p. Whether the axis of symmetry is vertical ($y = ax^2 + bx + c$) or horizontal ($x = ay^2 + by + c$), the value of p can be found from the following:

$$p = \frac{1}{4|a|}$$

The focus can now be located by first finding the vertex and then moving the appropriate distance p.

✔ CHECK YOURSELF ANSWERS

1. $y = \dfrac{5}{3}x^2 + 2x - 3$ **2.** $x = -y^2 - 2y + 3$

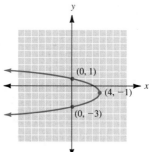

Exercises · 12.1

In Exercises 1 to 6, use the given values of a, b, c, d, e, and f to rewrite the standard form for a conic section

$$ax^2 + by^2 + cxy + dx + ey + f = 0$$

into standard quadratic form.

1. $a = 3, b = 0, c = 0, d = 2, e = -1, f = -5$

2. $a = -2, b = 0, c = 0, d = -3, e = 1, f = 4$

3. $a = -1, b = 0, c = 0, d = 3, e = 2, f = 5$

4. $a = 4, b = 0, c = 0, d = -5, e = 3, f = 7$

5. $a = -2, b = 0, c = 0, d = 4, e = -2, f = 3$

6. $a = -3, b = 0, c = 0, d = -2, e = -4, f = 5$

In Exercises 7 to 16, sketch the graph of each equation by finding the axis of symmetry, the coordinates of the vertex, and at least two symmetric points. Label these key characteristics on the graph.

7. $x = y^2 + 4y$

8. $x = y^2 - 4y$

9. $x = y^2 + 8y + 12$

10. $x = y^2 - 2y + 3$

11. $x = -y^2 + 6y - 5$

12. $x = -y^2 + 2y + 3$

13. $x = -2y^2 - 4y - 5$

14. $x = -3y^2 - 6y - 5$

15. $x = 3y^2 - 6y - 1$

16. $x = 2y^2 - 16y + 37$

17. Under what conditions will the graph of $x = a(y - k)^2 + h$ have no y intercepts?

18. Discuss similarities and differences between the graphs of $y = x^2 + 3x + 4$ and $x = y^2 + 3y + 4$. Use both graphs in your discussion.

To graph $x = y^2 + 3y + 4$ using a graphing utility, rewrite the equation as a quadratic equation in y.

$$y^2 + 3y + (-x + 4) = 0$$

Then use the quadratic formula to solve for y and enter the resulting equations.

$$Y_1 = \frac{-3 + \sqrt{9 - 4(-x + 4)}}{2(1)}$$

$$Y_2 = \frac{-3 - \sqrt{9 - 4(-x + 4)}}{2(1)}$$

In Exercises 19 to 24 use the technique shown in Exercise 18 to graph each parabola.

19. $x = y^2 - 4$ **20.** $x = y^2 - 9$ **21.** $x = -y^2 - 4y + 5$

22. $x = -y^2 + 6y - 5$ **23.** $x = 2y^2 + 3y - 2$ **24.** $x = 3y^2 + 2y - 5$

Each equation below defines a relation (x, y). Write the domain of each relation. (*Hint:* Determine the vertex and whether the parabola opens to the left or to the right.)

25. $x = y^2 + 6y$ **26.** $x = y^2 - 8y$ **27.** $x = -y^2 + 6y - 7$

28. $x = -y^2 - 4y + 3$

12.2 Circles

1. Identify the graph of an equation as a line, a parabola, or a circle
2. Write the equation of a circle in standard form and graph the circle

The second conic section we look at is the circle. The circle can be described by using the standard form for a conic section

$$ax^2 + by^2 + cxy + dx + ey + f = 0$$

but we will develop the standard form for a circle through the definition of a circle.

> A **circle** is the set of all points in the plane equidistant from a fixed point, called the **center** of the circle. The distance between the center of the circle and any point on the circle is called the **radius** of the circle.

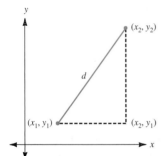

The distance formula is central to any discussion of conic sections.

The Distance Formula

The distance d between two points (x_1, y_1) and (x_2, y_2) is given by

$$d = \sqrt{(x_2 - x_1)^2 + (y_2 - y_1)^2}$$

We can use the distance formula to derive the algebraic equation of a circle, given its center and its radius.

Suppose a circle has its center at a point with coordinates (h, k) and radius r. If (x, y) represents any point on the circle, then, by its definition, the distance from (h, k) to (x, y) is r. Applying the distance formula, we have

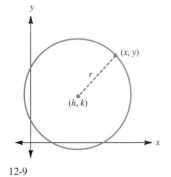

$$r = \sqrt{(x - h)^2 + (y - k)^2}$$

Squaring both sides of the equation gives the equation of the circle

$$r^2 = (x - h)^2 + (y - k)^2$$

In general, we can write the following equation of a circle.

A special case is the circle centered at the origin with radius r. Then $(h, k) = (0, 0)$, and its equation is

$$x^2 + y^2 = r^2$$

Equation of a Circle

The equation of a circle with center (h, k) and radius r is

$$(x - h)^2 + (y - k)^2 = r^2 \qquad \textbf{(1)}$$

Equation (1) can be used in two ways. Given the center and radius of the circle, we can write its equation; or given its equation, we can find the center and radius of a circle.

Example 1

Finding the Equation of a Circle

Find the equation of a circle with center at $(2, -1)$ and radius 3. Sketch the circle.

$(x - 2)^2 + (y + 1)^2 = 9$

Let $(h, k) = (2, -1)$ and $r = 3$. Applying equation (1) yields

$$(x - 2)^2 + [y - (-1)]^2 = 3^2$$
$$(x - 2)^2 + (y + 1)^2 = 9$$

To sketch the circle, we locate the center of the circle. Then we determine four points 3 units to the right and left and up and down from the center of the circle. Drawing a smooth curve through those four points completes the graph.

✔ *CHECK YOURSELF 1*

Find the equation of the circle with center at $(-2, 1)$ and radius 5. Sketch the circle.

Now, given an equation for a circle, we can also find the radius and center and then sketch the circle. We start with an equation in the special form of equation (1).

Example 2

Finding the Center and Radius of a Circle

Find the center and radius of the circle with equation

$$(x - 1)^2 + (y + 2)^2 = 9$$

Remember, the general form is

$$(x - h)^2 + (y - k)^2 = r^2$$

The circle can be graphed on the calculator by solving for y, then graphing both the upper half and lower half of the circle. In this case,

$$(x - 1)^2 + (y + 2)^2 = 9$$

$$(y + 2)^2 = 9 - (x - 1)^2$$

$$y + 2 = \pm \sqrt{9 - (x - 1)^2}$$

$$y = -2 \pm \sqrt{9 - (x - 1)^2}$$

Now graph the two functions

$$y = -2 + \sqrt{9 - (x - 1)^2}$$

and

$$y = -2 - \sqrt{9 - (x - 1)^2}$$

on your calculator. (The display screen may need to be squared to obtain the shape of a circle.)

Our equation "fits" this form when it is written as

Note: $y + 2 = y - (-2)$

$$(x - 1)^2 + [y - (-2)]^2 = 3^2$$

So the center is at $(1, -2)$, and the radius is 3. The graph is shown.

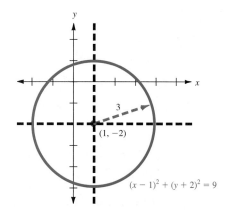

$(1, -2)$

$(x - 1)^2 + (y + 2)^2 = 9$

✔ CHECK YOURSELF 2

Find the center and radius of the circle with equation

$$(x + 3)^2 + (y - 2)^2 = 16$$

Sketch the circle.

To graph the equation of a circle that is not in standard form, we *complete the square*. Let's see how completing the square can be used in graphing the equation of a circle.

Example 3

To recognize the equation as having the form of a circle, note that the coefficients of x^2 and y^2 are equal.

The linear terms in x and y show a translation of the center away from the origin.

Finding the Center and Radius of a Circle

Find the center and radius of the circle with equation

$$x^2 + 2x + y^2 - 6y = -1$$

Then sketch the circle.

We could, of course, simply substitute values of x and try to find the corresponding values for y. A much better approach is to rewrite the original equation so that it matches the standard form.

First, add 1 to both sides to complete the square in x.

$$x^2 + 2x + 1 + y^2 - 6y = -1 + 1$$

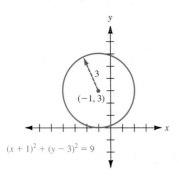

$(x + 1)^2 + (y - 3)^2 = 9$

Then add 9 to both sides to complete the square in y.

$$x^2 + 2x + 1 + y^2 - 6y + 9 = -1 + 1 + 9$$

We can factor the two trinomials on the left (they are both perfect squares) and simplify on the right.

$$(x + 1)^2 + (y - 3)^2 = 9$$

The equation is now in standard form, and we can see that the center is at $(-1, 3)$ and the radius is 3. The sketch of the circle is shown. Note the "translation" of the center to $(-1, 3)$.

✔ **CHECK YOURSELF 3**

Find the center and radius of the circle with equation

$$x^2 - 4x + y^2 + 2y = -1$$

Sketch the circle.

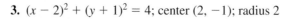

✔ **CHECK YOURSELF ANSWERS**

1. $(x + 2)^2 + (y - 1)^2 = 25$ **2.** Center $(-3, 2)$; radius 4

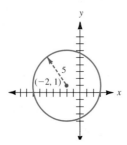
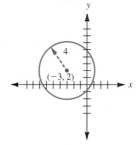

3. $(x - 2)^2 + (y + 1)^2 = 4$; center $(2, -1)$; radius 2

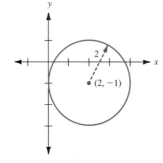

Decide whether the graph of each equation is a line, a parabola, a circle, or none of these.

1. $y = x^2 - 2x + 5$

2. $y^2 + x^2 = 81$

3. $x = 3y + 5$

4. $2y - 3x = 12$

5. $(x - 3)^2 + (y + 2)^2 = 10$

6. $y + 2(x - 3)^2 = 5$

7. $x^2 + 4x + y^2 - 6y = 3$

8. $5x - 3y = 15$

9. $y^2 - 4x^2 = 36$

10. $x^2 + (y - 3)^2 = 9$

11. $y = -3x^2 + 7x - 5$

12. $2x^2 - 3y^2 + 6y = 13$

Find the center and the radius for each circle.

13. $x^2 + y^2 = 36$

14. $x^2 + y^2 = 72$

15. $(x - 3)^2 + (y + 1)^2 = 16$

16. $(x - 4)^2 + y^2 = 49$

17. $x^2 + 4x + y^2 = 21$

18. $x^2 + y^2 - 6y = 72$

19. $x^2 - 6x + y^2 + 8y = 16$

20. $x^2 - 5x + y^2 - 3y = 8$

Find the center and radius of the circle. Then graph it.

21. $x^2 + y^2 = 4$

22. $x^2 + y^2 = 25$

23. $4x^2 + 4y^2 = 36$

24. $9x^2 + 9y^2 = 144$

25. $(x - 1)^2 + y^2 = 9$

26. $x^2 + (y + 2)^2 = 16$

27. $(x - 4)^2 + (y + 1)^2 = 16$ **28.** $(x + 3)^2 + (y + 2)^2 = 25$ **29.** $x^2 + y^2 - 4y = 12$

30. $x^2 - 6x + y^2 = 0$ **31.** $x^2 - 4x + y^2 + 2y = -1$ **32.** $x^2 - 2x + y^2 - 6y = 6$

In Exercises 33 to 36, write the equation of the circle pictured.

33.

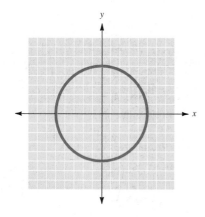

34.

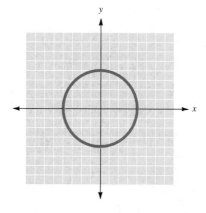

35.

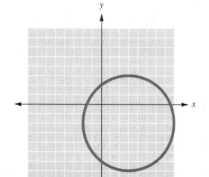

36.

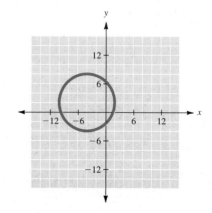

37. Describe the graph of $x^2 + y^2 - 2x - 4y + 5 = 0$.

38. Describe how completing the square is used in graphing circles.

39. A solar oven is constructed in the shape of a hemisphere. If the equation

$$x^2 + y^2 = 500$$

describes the circumference of the oven in centimeters, what is its radius?

40. A solar oven in the shape of a hemisphere is to have a diameter of 80 cm. Write the equation that describes the circumference of this oven.

41. A solar water heater is constructed in the shape of a cylinder, with the water supply pipe at its center. If the water heater has a diameter of $\dfrac{4}{3}$ m, what is the equation that describes its circumference?

42. A solar water heater is constructed in the shape of a cylinder having a circumference described by the equation

$$9x^2 + 9y^2 - 16 = 0$$

What is its diameter if the units for the equation are meters?

In Exercises 43 to 46, use the technique shown below to graph each circle.

A circle can be graphed on a calculator by plotting the upper and lower semicircles on the same axes. For example, to graph $x^2 + y^2 = 16$, we solve for y:

$$y = \pm\sqrt{16 - x^2}$$

This is then graphed as two separate functions,

$$Y_1 = \sqrt{16 - x^2} \qquad \text{and} \qquad Y_2 = -\sqrt{16 - x^2}$$

43. $x^2 + y^2 = 36$

44. $(x - 3)^2 + y^2 = 9$

45. $(x + 5)^2 + y^2 = 36$

46. $(x - 2)^2 + (y + 1)^2 = 25$

Each of the following equations defines a relation (x, y). Write the domain and the range of each relation.

47. $(x + 3)^2 + (y - 2)^2 = 16$

48. $(x - 1)^2 + (y - 5)^2 = 9$

49. $x^2 + (y - 3)^2 = 25$

50. $(x + 2)^2 + y^2 = 36$

Ellipses

> 1. *Identify an equation whose graph is an ellipse*
> 2. *Sketch the graph of an ellipse*

Let's turn now to the third conic section, the ellipse. It can be described as an "oval-shaped" curve and has the following geometric description.

Ellipses occur frequently in nature. The planets have elliptical orbits with the sun at one focus.

The reflecting properties of the ellipse are also interesting. Rays from one focus are reflected by the ellipse in such a way that they always pass through the other focus.

> An **ellipse** is the set of all points (x, y) such that the sum of the distances from (x, y) to two fixed points, called the *foci* of the ellipse, is constant.

The following sketch illustrates the definition in two particular cases:

1. When the foci are located on the x axis and are symmetric about the origin

2. When the foci are located on the y axis and are symmetric about the origin

In either case, $d_1 + d_2$ is constant.

An Ellipse

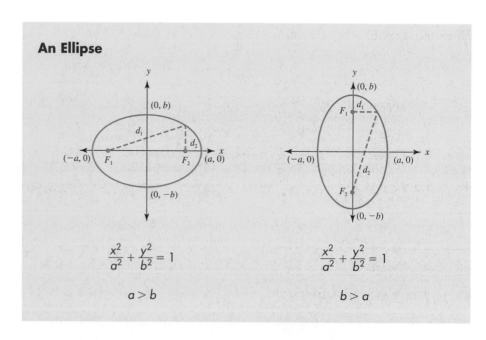

$$\frac{x^2}{a^2} + \frac{y^2}{b^2} = 1$$

$$a > b$$

$$\frac{x^2}{a^2} + \frac{y^2}{b^2} = 1$$

$$b > a$$

To quickly sketch ellipses of the preceding forms, we need to determine only *four points*—the points where the ellipse intercepts the coordinate axes.

936

Fortunately those points are easily found when the ellipse is written in standard form:

$$\frac{x^2}{a^2} + \frac{y^2}{b^2} = 1 \qquad \textbf{(1)}$$

The x intercepts are $(a, 0)$ and $(-a, 0)$. The y intercepts are $(0, b)$ and $(0, -b)$. Let's use this information to sketch an ellipse.

Example 1   Graphing an Ellipse

Sketch the ellipse

$$\frac{x^2}{9} + \frac{y^2}{4} = 1$$

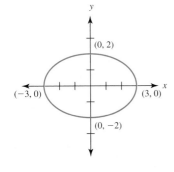

1. The equation is in standard form.

2. Find the x intercepts. From equation (1), $a^2 = 9$, and the x intercepts are $(3, 0)$ and $(-3, 0)$.

3. Find the y intercepts. From equation (1), $b^2 = 4$, and the y intercepts are $(0, 2)$ and $(0, -2)$.

4. Plot the intercepts found above, and draw a smooth curve to form the desired ellipse.

✔ CHECK YOURSELF 1

Sketch the ellipse.

$$\frac{x^2}{16} + \frac{y^2}{9} = 1$$

Example 2   Graphing an Ellipse

Sketch the ellipse with equation

$$9x^2 + 4y^2 = 36$$

Step 1   Since this equation is *not* in standard form (the right side is *not* 1), we divide both sides of the equation by the constant 36:

$$\frac{9x^2}{36} + \frac{4y^2}{36} = \frac{36}{36}$$

$$\frac{x^2}{4} + \frac{y^2}{9} = 1$$

We can now proceed as before. Comparing the derived equation with that in standard form, we deduce steps 2 and 3.

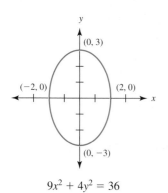

$9x^2 + 4y^2 = 36$

Step 2   The x intercepts are $(2, 0)$ and $(-2, 0)$.

Step 3   The y intercepts are $(0, 3)$ and $(0, -3)$.

Step 4   We connect the intercepts with a smooth curve to complete the sketch of the ellipse.

✔   *CHECK YOURSELF 2*

Sketch the ellipse.

$$25x^2 + 4y^2 = 100$$

(*Hint:* First write the equation in standard form by dividing both sides of the equation by 100.)

The following algorithm summarizes our work with graphing ellipses.

Graphing the Ellipse

1. Write the given equation in standard form.

2. From that standard form, determine the x intercepts.

3. Also determine the y intercepts.

4. Plot the four intercepts and connect the points with a smooth curve, to complete the sketch.

✔   *CHECK YOURSELF ANSWERS*

1. $\dfrac{x^2}{16} + \dfrac{y^2}{9} = 1$ **2.** $\dfrac{x^2}{4} + \dfrac{y^2}{25} = 1$

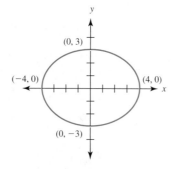   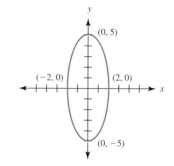

Sketch the graph of each equation. If necessary, rewrite the equation in standard form.

1. $\dfrac{x^2}{4} + \dfrac{y^2}{9} = 1$

2. $\dfrac{x^2}{16} + \dfrac{y^2}{9} = 1$

3. $\dfrac{x^2}{9} + \dfrac{y^2}{25} = 1$

4. $\dfrac{x^2}{36} + \dfrac{y^2}{25} = 1$

5. $x^2 + 9y^2 = 36$

6. $4x^2 + y^2 = 16$

7. $4x^2 + 9y^2 = 36$

8. $25x^2 + 4y^2 = 100$

9. $4x^2 + 25y^2 = 100$

10. $9x^2 + 16y^2 = 144$

11. $25x^2 + 9y^2 = 225$

12. $16x^2 + 9y^2 = 144$

13. A semielliptical archway over a one-way road has a height of 10 ft and a width of 40 ft (see the figure). Will a truck that is 10 ft wide and 9 ft high clear the opening of the highway?

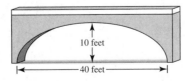

10 feet

40 feet

14. A truck that is 8 ft wide is carrying a load that reaches 7 ft above the ground. Will the truck clear a semielliptical arch that is 10 ft high and 30 ft wide?

In Exercises 15 to 18, use the technique shown below to graph each ellipse.

An ellipse can be graphed on a calculator by plotting the upper and lower halves on the same axes. For example, to graph $9x^2 + 16y^2 = 144$, we solve for y:

$$Y_1 = \frac{\sqrt{144 - 9x^2}}{4} \quad \text{and} \quad Y_2 = -\frac{\sqrt{144 - 9x^2}}{4}$$

Then graph each equation as a separate function.

15. $4x^2 + 16y^2 = 64$ **16.** $9x^2 + 36y^2 = 324$ **17.** $25x^2 + 9y^2 = 225$ **18.** $4x^2 + 9y^2 = 36$

Each of the following equations defines a relation (x, y). Write the domain and range of each relation.

19. $\dfrac{x^2}{9} + \dfrac{y^2}{16} = 1$ **20.** $\dfrac{x^2}{25} + \dfrac{y^2}{36} = 1$

In Exercises 21 and 22, write the equation of the ellipse pictured.

21.

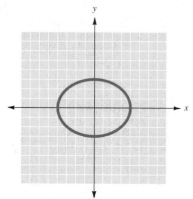

22.

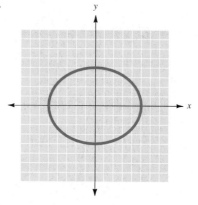

SECTION

12.4 Hyperbolas

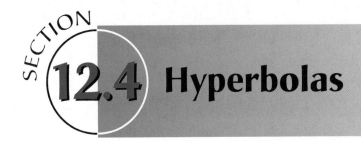

12.4 OBJECTIVES

1. *Identify an equation whose graph is a hyperbola*
2. *Sketch the graph of a hyperbola*

Our discussion now turns to the last of the conic sections, the hyperbola. As you will see, the geometric description of the hyperbola (and hence the corresponding standard form) is quite similar to that of the ellipse.

> A **hyperbola** is the set of all points (x, y) such that the absolute value of the differences from (x, y) to each of two fixed points, called the *foci* of the hyperbola, is constant.

The following sketch illustrates the definition in the case where the foci are located on the x axis and are symmetric about the origin.

This is the first of two special cases we will investigate in this section.

The Hyperbola

$$\frac{x^2}{a^2} - \frac{y^2}{b^2} = 1$$

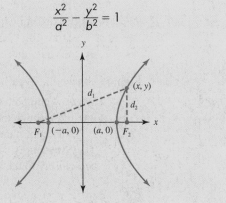

The difference $|d_1 - d_2|$ remains constant for any point on the hyperbola.

Before we try to sketch a hyperbola from its equation, let's examine the standard form more carefully. For

$$\frac{x^2}{a^2} - \frac{y^2}{b^2} = 1 \tag{1}$$

the graph is a hyperbola that opens to the right and left and is symmetric about the x axis. The points where this hyperbola intercepts the x axis are called the **vertices** of the hyperbola. The vertices of the hyperbola are located at $(a, 0)$ and at $(-a, 0)$.

As we move away from the center of the hyperbola, the *branches* of the hyperbola will approach two straight lines called the **asymptotes** of the hyperbola. The equations of the two asymptotes of the hyperbola are given by

$$y = \frac{b}{a}x \quad \text{and} \quad y = -\frac{b}{a}x$$

These asymptotes prove to be extremely useful aids in sketching the hyperbola. In fact, for most purposes, the vertices and the asymptotes are the only tools that we will need. Example 1 illustrates.

Although we show these equations, you will see an easier method for finding the asymptotes in our first example.

| **Example 1** | ## Graphing a Hyperbola |

Sketch the hyperbola

$$\frac{x^2}{9} - \frac{y^2}{4} = 1$$

The equation of the hyperbola also has both x^2 and y^2 terms. Here the coefficients of those terms have *opposite* signs. If the x^2 coefficient is *positive*, the hyperbola will open *horizontally*.

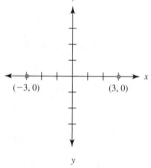

Step 1   The equation is in standard form.

Step 2   Find and plot the vertices.

From the standard form we can see that $a^2 = 9$ and $a = 3$ or -3. The vertices of the hyperbola then occur at $(3, 0)$ and $(-3, 0)$.

Step 3   Sketch the asymptotes.

Here is an easy way to sketch the asymptotes. Note again from the standard form (1) that $b^2 = 4$, so $b = 2$ or -2. Plot the points $(0, 2)$ and $(0, -2)$ on the y axis.

Draw (using dashed lines) the rectangle whose sides are parallel to the x and y axes and which pass through the points determined in steps 2 and 3.

Draw the diagonals of the rectangle (again using dashed lines), and then extend those diagonals to form the desired asymptotes.

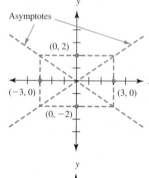

Step 4   Sketch the hyperbola.

We now complete our task by sketching the hyperbola as two smooth curves, passing through the vertices and approaching the asymptotes.

It is important to remember that the asymptotes are *not* a part of the graph. They are simply used as aids in sketching the graph as the branches get "closer and closer" to the lines.

✔ CHECK YOURSELF 1

Sketch the hyperbola.

$$\frac{x^2}{16} - \frac{y^2}{9} = 1$$

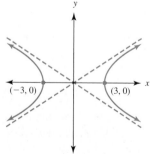

We now want to consider a second case of the hyperbola and its standard form. Suppose that the foci of the hyperbola are now on the y axis and symmetric about the origin. A sketch of such a hyperbola, and the equation that corresponds, follows.

The Hyperbola

$$\frac{y^2}{b^2} - \frac{x^2}{a^2} = 1 \qquad\qquad (2)$$

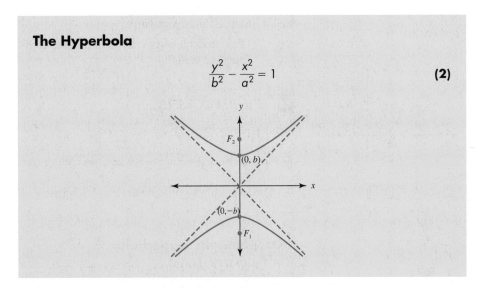

Some observations about this case are in order:

Here the vertices are on the y axis.

1. The vertices of the hyperbola are now at $(0, b)$ and $(0, -b)$.

The asymptotes are the same as before.

2. The asymptotes of the hyperbola have the equations

$$y = \frac{b}{a}x \qquad \text{and} \qquad y = -\frac{b}{a}x$$

Example 2 illustrates the sketching of a hyperbola in this case.

Example 2 — Graphing a Hyperbola

Sketch the hyperbola.

$$4y^2 - 25x^2 = 100$$

You can recognize this equation as corresponding to a hyperbola since the coefficients of the squared terms are opposite in sign. Since the y^2 coefficient is positive, the hyperbola will open vertically.

Step 1   Write the equation in the standard form of equation (2) by dividing both sides by 100:

$$\frac{4y^2}{100} - \frac{25x^2}{100} = \frac{100}{100}$$

$$\frac{y^2}{25} - \frac{x^2}{4} = 1$$

Step 2   Find the vertices.

From the standard form of equation (2), we see that since $b^2 = 25$, $b = 5$, or $b = -5$, so the vertices are at $(0, 5)$ and at $(0, -5)$.

Step 3   Sketch the asymptotes.

Also from the standard form of equation (2), we see that since $a^2 = 4$, $a = 2$ or $a = -2$.

Plot $(2, 0)$ and $(-2, 0)$ on the x axis, and complete the dashed rectangle as before. The diagonals once again extend to form the asymptotes.

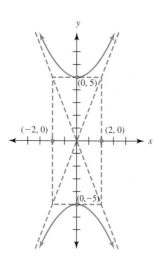

Step 4   Sketch the hyperbola.

Draw smooth curves, through the intercepts, that approach the asymptotes to complete the graph.

✔ *CHECK YOURSELF 2*

Sketch the hyperbola $9y^2 - 4x^2 = 36$.

The following algorithm summarizes our work with sketching hyperbolas.

Graphing the Hyperbola

1. Write the given equation in standard form.
2. Determine the vertices of the hyperbola.

   If the x^2 coefficient is positive, the vertices are at $(a, 0)$ and $(-a, 0)$ on the x axis.

   If the y^2 coefficient is positive, the vertices are at $(0, b)$ and $(0, -b)$ on the y axis.
3. Sketch the asymptotes of the hyperbola.

   Plot points $(a, 0)$, $(-a, 0)$, $(0, b)$, and $(0, -b)$. Form a rectangle from these points. The diagonals (extended) are the asymptotes of the hyperbola.
4. Sketch the hyperbola.

   Draw smooth curves, through the intercepts and approaching the asymptotes.

The following chart shows all the equation forms considered in this chapter.

Curve	Example	Recognizing the Curve
Straight line	$4x - 3y = 12$	The equation involves x and/or y to the first power.
Parabola	$y = x^2 - 3x$ or $x = y^2 - 2y + 3$	Only one term, in x or in y, may be squared. The other variable appears to the first power.
Circle	$x^2 + 4x + y^2 = 5$	The equation has both x^2 and y^2 terms. The coefficients of those terms are equal.
Ellipse	$4x^2 + 9y^2 = 36$	The equation has both x^2 and y^2 terms. The coefficients of those terms have the same algebraic sign but different coefficients.
Hyperbola	$4x^2 - 9y^2 = 36$ or $9y^2 - 16x^2 = 144$	The equation has both x^2 and y^2 terms. The coefficients of those terms have different algebraic signs.

✔ *CHECK YOURSELF ANSWERS*

1. $\dfrac{x^2}{16} - \dfrac{y^2}{9} = 1$ **2.** $\dfrac{y^2}{4} - \dfrac{x^2}{9} = 1$

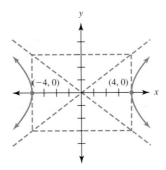   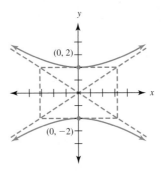

Exercises · 12.4

Identify the graph of each of the following equations as one of the conic sections (a parabola, circle, ellipse, or hyperbola).

1. $x^2 + y^2 = 16$

2. $\dfrac{x^2}{4} - \dfrac{y^2}{16} = 1$

3. $y = x^2 - 4$

4. $\dfrac{x^2}{16} + \dfrac{y^2}{9} = 1$

5. $9x^2 - 4y^2 = 36$

6. $x^2 = 4y$

7. $y^2 - 4x^2 = 4$

8. $x = y^2 - 2y + 1$

9. $x^2 - 6x + y^2 + 2x = 2$

10. $4x^2 + 25y^2 = 100$

11. $9y^2 - 16x^2 = 144$

12. $y = x^2 - 6x + 8$

Match each of the curves shown with the appropriate equation given below.

(a) $4x^2 + 25y^2 = 100$

(b) $y = x^2 - 2x - 3$

(c) $x = \dfrac{1}{2}y^2 - 2y$

(d) $\dfrac{x^2}{9} + \dfrac{y^2}{16} = 1$

(e) $\dfrac{y^2}{25} - \dfrac{x^2}{4} = 1$

(f) $16x^2 - 9y^2 = 144$

(g) $(x - 2)^2 + (y - 2)^2 = 9$

(h) $x^2 + y^2 = 16$

13.

14.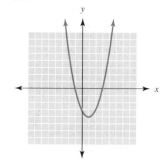

15.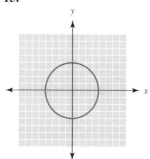

16.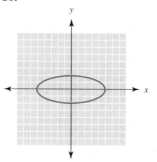

17.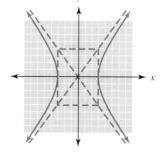

18.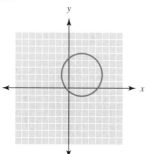

19.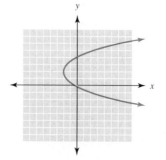

20.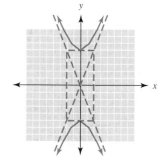

Graph the following hyperbolas by finding the vertices and asymptotes. If necessary, write the equation in standard form.

21. $\dfrac{x^2}{9} - \dfrac{y^2}{9} = 1$ **22.** $\dfrac{y^2}{9} - \dfrac{x^2}{4} = 1$ **23.** $\dfrac{y^2}{16} - \dfrac{x^2}{9} = 1$ **24.** $\dfrac{x^2}{25} - \dfrac{y^2}{16} = 1$

25. $\dfrac{x^2}{36} - \dfrac{y^2}{9} = 1$ **26.** $\dfrac{y^2}{25} - \dfrac{x^2}{9} = 1$ **27.** $x^2 - 9y^2 = 36$ **28.** $4y^2 - 9x^2 = 36$

29. $9x^2 - 4y^2 = 36$ **30.** $9y^2 - 4x^2 = 36$ **31.** $16y^2 - 9x^2 = 144$ **32.** $4x^2 - 9y^2 = 36$

33. $25y^2 - 4x^2 = 100$ **34.** $9x^2 - 25y^2 = 225$

In Exercises 35 to 38, use a graphing utility to graph each equation. *Hint:* Solve for y and define two separate functions in the calculator to graph the top half and bottom half of each hyperbola.

35. $y^2 - 16x^2 = 16$ **36.** $y^2 - 25x^2 = 25$ **37.** $16x^2 - 9y^2 = 144$ **38.** $25x^2 - 16y^2 = 400$

Example	Topic	Reference
	More on the Parabola	12.1
	The *conic sections* are the curves formed when a plane cuts through a cone. These include the *parabola, circle, ellipse,* and *hyperbola*.	p. 922 **and** p. 924

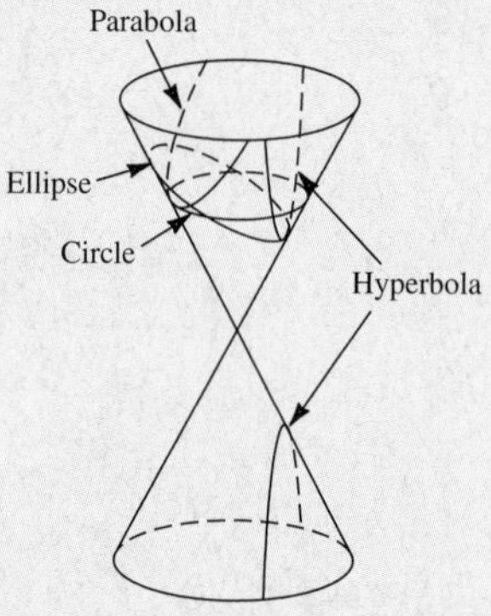

A conic section is formed from any second-degree equation with the general form

$$ax^2 + by^2 + cxy + dx + ey + f = 0$$

The graph of

$$x = ay^2 + by + c \qquad a \neq 0$$

is a parabola with horizontal axis of symmetry.

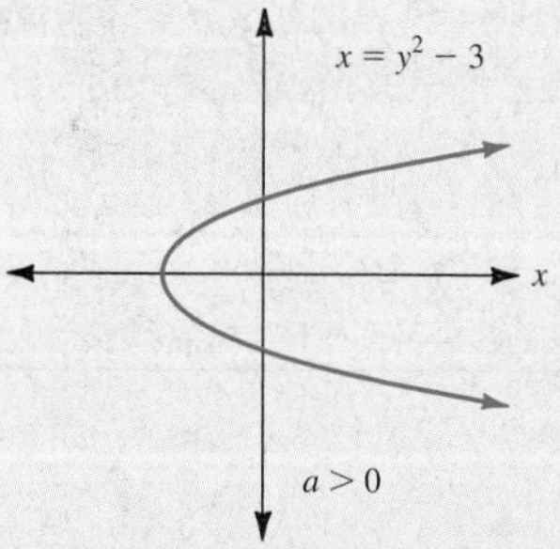

If $a > 0$, the parabola opens *rightward*.

$x = y^2 - 3$

$a > 0$

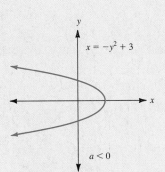

$x = -y^2 + 3$

$a < 0$

If $x = -y^2 - 2y + 3$, the axis of symmetry has the equation

$$y = -\frac{-2}{2(-1)} = -1$$

When $y = -1$,

$$x = -(-1)^2 - 2(-1) + 3$$
$$= -1 + 2 + 3$$
$$= 4$$

so the vertex is $(4, -1)$.

If $a < 0$, the parabola opens *leftward*.

The vertex of the parabola is on the axis of symmetry with the equation

$$y = -\frac{b}{2a}$$

Circles

12.2

p. 930

Given the equation

$$(x - 2)^2 + (y + 3)^2 = 4$$

we see that the center is at $(2, -3)$ and the radius is 2.

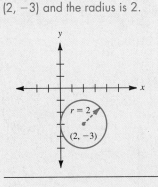

$r = 2$

$(2, -3)$

The standard form for the circle with center (h, k) and radius r is

$$(x - h)^2 + (y - k)^2 = r^2$$

Determining the center and radius of the circle from its equation allows us to easily graph the circle.
Note: Completing the square may be used to derive an equivalent equation in standard form if the original equation is not in the form.

	Ellipses	**12.3**
Graph the equation $$\frac{x^2}{9} + \frac{y^2}{4} = 1$$ 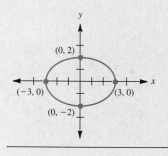	**Ellipse** The standard form for the *ellipse,* whose foci are located on either the x or y axis and are symmetric about the origin, is $$\frac{x^2}{a^2} + \frac{y^2}{b^2} = 1$$ The x intercepts for the ellipse are $(a, 0)$ and $(-a, 0)$. The y intercepts for the ellipse are $(0, b)$ and $(0, -b)$. Determining the four intercepts of an ellipse allows us to sketch its graph. **Note:** If the given equation is not in standard form, we can divide both sides of the equation by the appropriate constant to derive the standard form.	**p. 936**
	Hyperbolas	**12.4**
Sketch the hyperbola $$\frac{x^2}{4} - \frac{y^2}{9} = 1$$ The vertices are at $(2, 0)$ and $(-2, 0)$. The asymptotes are $y = \frac{3}{2}x$ and $y = -\frac{3}{2}x$. 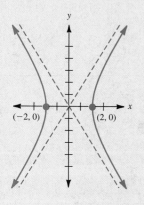	**Hyperbola** The standard form for the *hyperbola* whose foci are located on the x axis and are symmetric about the origin is $$\frac{x^2}{a^2} - \frac{y^2}{b^2} = 1$$ The vertices for this hyperbola are on the x axis, $(a, 0)$ and $(-a, 0)$. The asymptotes of the hyperbola have the equations $$y = \pm \frac{b}{a}x$$ Determining and sketching the vertices and asymptotes of a hyperbola will allow us to sketch its graph quickly. **Note:** Again divide both sides of the given equation by the appropriate constant if the equation is not in standard form. If the foci of the hyperbola are located on the y axis and are symmetric about the origin, the standard form is $$\frac{y^2}{b^2} - \frac{x^2}{a^2} = 1$$ The vertices for this hyperbola are on the y axis, $(0, b)$ and $(0, -b)$. The equations for the asymptotes of the hyperbola remain the same as before.	**p. 941**

This summary exercise set is provided to give you practice with each of the objectives in the chapter. Each chapter is keyed to the appropriate chapter section.

[12.1] In Exercises 1 and 2, use the given values of a, b, c, d, e, and f to rewrite the standard form for a conic section

$$ax^2 + by^2 + cxy + dx + ey + f = 0$$

into standard quadratic form.

1. $a = 2, b = 0, c = 0, d = 3, e = -5, f = -1$

2. $a = -3, b = 0, c = 0, d = -2, e = -3, f = 2$

[12.1] In Exercises 3 to 8, find the equation of the axis of symmetry, the coordinates of the vertex, and at least two symmetric points. Sketch the graph of each equation.

3. $x = -y^2 + 4y$

4. $x = y^2 - 4y$

5. $x = y^2 - 5y + 6$

6. $x = -y^2 + 3y - 2$

7. $x = 2y^2 - 4y + 9$

8. $x = -3y^2 + 5y - 8$

[12.2] Find the center and radius of the graph of each equation.

9. $x^2 + y^2 = 81$

10. $x^2 + y^2 = 50$

11. $4x^2 + 4y^2 = 36$

12. $5x^2 + 5y^2 = 25$

13. $(x - 3)^2 + y^2 = 36$

14. $(x - 2)^2 + y^2 = 9$

15. $(x + 2)^2 + (y - 1)^2 = 25$

16. $x^2 + 6x + y^2 + 4y = 12$

17. $x^2 + 8x + y^2 + 10y = 23$

18. $x^2 - 8x + y^2 + 2y = 32$

[12.2] Graph each of the following.

19. $x^2 + y^2 = 16$

20. $4x^2 + 4y^2 = 36$

21. $x^2 + (y + 3)^2 = 25$

22. $(x - 2)^2 + y^2 = 9$

23. $(x - 1)^2 + (y - 2)^2 = 16$

24. $(x + 3)^2 + (y + 3)^2 = 25$

25. $x^2 + y^2 - 4y - 5 = 0$

26. $x^2 - 2x + y^2 - 6y = 6$

[12.3–12.4] Sketch the graph of each of the following equations.

27. $\dfrac{x^2}{36} + \dfrac{y^2}{4} = 1$

28. $\dfrac{x^2}{49} + \dfrac{y^2}{9} = 1$

29. $\dfrac{x^2}{25} + \dfrac{y^2}{9} = 1$

30. $\dfrac{x^2}{4} + \dfrac{y^2}{16} = 1$

31. $9x^2 + 4y^2 = 36$

32. $16x^2 + 9y^2 = 144$

33. $\dfrac{x^2}{9} - \dfrac{y^2}{4} = 1$

34. $\dfrac{y^2}{16} - \dfrac{x^2}{4} = 1$

35. $4x^2 - 9y^2 = 36$

36. $16x^2 - 9y^2 = 144$

[12.1–12.4] In Exercises 37 to 54, decide whether the graph of the equation is a line, parabola, circle, ellipse, or hyperbola.

37. $x + y = -4$

38. $x + y^2 = 5$

39. $4x^2 + 4y^2 = 36$

40. $3x + 3y = 36$

41. $y = (x - 3)^2$

42. $(x + 1)^2 + y^2 = 18$

43. $y = (x - 1)^2 + 1$

44. $x = y^2 + 4y + 4$

45. $\dfrac{x^2}{4} + \dfrac{y^2}{25} = 1$

46. $x^2 - 6x + y^2 + 6y = 18$

47. $\dfrac{x^2}{16} - \dfrac{y^2}{36} = 1$

48. $9x^2 - 4y^2 = 36$

49. $16x^2 + 4y^2 = 64$

50. $x^2 = -y^2 + 18$

51. $\dfrac{x^2}{4} - \dfrac{y^2}{9} = 1$

52. $4x^2 + 4y^2 = 36$

53. $6x + 4y = 15$

54. $5x^2 + y - 5 = 0$

In Exercises 55 to 58, each of the given equations defines a relation (x, y). Write the domain and range of each relation.

55. $x^2 + y^2 = 121$

56. $(x - 1)^2 + (y - 2)^2 = 25$

57. $\dfrac{x^2}{16} + \dfrac{y^2}{9} = 1$

58. $\dfrac{x^2}{9} + \dfrac{y^2}{25} = 1$

In Exercises 59 and 60, write the equation of the curve pictured.

59.

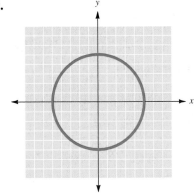

60.

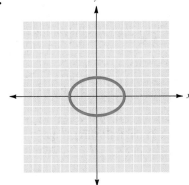

Self-Test 12

The purpose of this self-test is to help you check your progress and to review for a chapter test in class. Allow yourself about 1 hour to take the test. When you are done, check your answers with the answers in the back of the book. If you missed any answers, be sure to go back and review the appropriate sections in the chapter and their exercises.

1. Given the values $a = 2$, $b = 0$, $c = 0$, $d = -6$, $e = 2$, and $f = 3$, rewrite the standard form for a conic section $(ax^2 + by^2 + cxy + dx + ey + f = 0)$ into standard quadratic form.

Sketch the graph of each equation by finding the axis of symmetry, the coordinates of the vertex, and at least two symmetric points. Label these key characteristics on the graph.

2. $x = y^2 + 4y$

3. $x = 2y^2 - 12y + 17$

4. $x = -3y^2 + 12y - 6$

Find the coordinates for the center and the radius and then graph each equation.

5. $(x - 3)^2 + (y + 2)^2 = 36$

6. $x^2 + 2x + y^2 - 4y - 20 = 0$

7. $x^2 + 6x + y^2 + 5 = 0$

8. $(x - 2)^2 + (y + 3)^2 = 9$

Sketch the graph of each of the following equations.

9. $\dfrac{x^2}{25} + \dfrac{y^2}{9} = 1$

10. $16x^2 + 4y^2 = 64$

11. $\dfrac{x^2}{9} - \dfrac{y^2}{16} = 1$

12. $4x^2 - 25y^2 = 100$

For each of the following equations, decide whether its graph is a line, parabola, circle, ellipse, or hyperbola.

13. $5x - 2y - 3 = 0$

14. $3x^2 + 2y = 9$

15. $5x^2 + 5y^2 = 125$

16. $(x - 4)^2 + y^2 = 36$

17. $4y^2 - 7x = 15$

18. $7x^2 + 3y^2 = 21$

19. $-6x^2 + 3y^2 = 18$

20. $2x^2 + 2y^2 + 4x - 6y = 12$

954

Cumulative Test ▪ 0–12

This test is provided to help you in the process of reviewing the previous chapters. Answers are provided in the back of the book. If you missed any answers, be sure to go back and review the appropriate sections.

Solve each of the following.

1. $3x - 2(x + 5) = 12 - 3x$

2. $2x - 7 < 3x - 5$

3. $|2x - 3| = 5$

4. $|3x + 5| \leq 7$

5. $|5x - 4| > 21$

6. $x^2 - 5x - 24 = 0$

Graph each of the following.

7. $5x + 7y = 35$

8. $2x + 3y \leq 6$

9. Find the distance between the points $(-1, 2)$ and $(4, -22)$.

10. Find the slope of the line containing $(4, 6)$ and $(3, -1)$.

11. Write the function form of the equation of the line that passes through the points $(-1, 4)$ and $(5, -2)$.

Simplify the following polynomials.

12. $(2x + 1)(x - 3)$

13. $(3x - 2)^2$

14. Completely factor the function $f(x) = x^3 - 3x^2 - 5x + 15$.

Graph the following.

15. $y = x^2 - 6x + 5$

16. $(x + 1)^2 + (y - 2)^2 = 25$

17. $\dfrac{x^2}{64} + \dfrac{y^2}{9} = 1$

Solve the following system of equations.

18. $2x + 3y = \quad 6$
$5x + 3y = -24$

Solve each of the following applications.

19. Geometry. The length of a rectangle is 3 cm more than twice its width. If the perimeter of the rectangle is 54 cm, find the dimensions of the rectangle.

20. Number problem. The sum of the digits of a two-digit number is 10. If the digits are reversed, the new number is 36 less than the original number. What was the original number?

13

Exponential and Logarithmic Functions

Pharmacologists researching the effects of drugs use exponential and logarithmic functions to model drug absorption and elimination. After a drug is taken orally, it is distributed throughout the body via the circulatory system. Once in the bloodstream, the drug is carried to the body's organs, where it is first absorbed and then eliminated again into the bloodstream. For a medicine or drug to be effective, there must be enough of the substance in the body to achieve the desired effect but not enough to cause harm. This therapeutic level is maintained by taking the proper dosage at timed intervals determined by the rate at which the body absorbs or eliminates the medicine.

The rate at which the body eliminates the drug is proportional to the amount of the drug present. That is, the more drug there is, the faster the drug is eliminated. The amount of a drug dosage P still left after a number of hours t is affected by the **half-life** of the drug. In this case, the half-life is how many hours it takes for the body to use up or eliminate one-half of the drug dosage.

If P is the amount of an initial dose and H is the time it takes the body to eliminate one-half dose of a drug, then the amount of the drug still remaining in the system after t units of time is

$$A(t) = Pe^{t\left(\frac{-\ln 2}{H}\right)}$$

If the amount of an initial dose of a drug is 30 mg and if the half-life of the drug in the body is 4 h, the amount in milligrams of the drug still in the body t hours after one dose is given by

$$A(t) = 30e^{-0.173t}$$

Inverse Relations and Functions

 OBJECTIVES

1. *Find the inverse of a function given in equation form*
2. *Find the inverse of a function given in table form*
3. *Graph a relation and its inverse*
4. *Identify a one-to-one function*
5. *Determine whether the inverse of a function is also a function*
6. *Restrict the domain of a function*

Composition of functions (introduced in Chapter 4) leads us to the following question: Can we chain together (i.e., compose) two functions in such a way that one function "undoes" the other?

Suppose, for example, that $f(x) = \dfrac{x-5}{3}$ and $g(x) = 3x + 5$. Let's pick a convenient x value for f, say, $x = 8$. Now $f(8) = \dfrac{8-5}{3} = \dfrac{3}{3} = 1$. The function f turns 8 into 1. Now let g act on this result: $g(1) = 3(1) + 5 = 8$. The function g turns 1 back into 8.

When we view the composition of g and f, acting on 8, we see this effect of g "undoing" f's actions:

$$(g \circ f)(8) = g(f(8)) = g(1) = 8$$

In general, a function g that undoes the action of f is called the **inverse** of f.

Inverse Functions

Functions f and g are said to be **inverse functions** if

$$(g \circ f)(x) = x \qquad \text{for all } x \text{ in domain of } f$$

and $\qquad (f \circ g)(x) = x \qquad \text{for all } x \text{ in domain of } g$

If g is the inverse of f, we denote the function g as f^{-1}.

Note: The notation f^{-1} has a different meaning from the negative exponent, as in x^{-1} or $\dfrac{1}{x}$.

A natural question now is, given a function f, how do we find the inverse function f^{-1}? One way to find such a function f^{-1} is to (1) analyze the actions of f, noting the order of operations involved, and then (2) define f^{-1} by using the *opposite* operations *in the reverse order.*

| Example 1 | Finding the Inverse of a Function |

Given $f(x) = \dfrac{x - 5}{3}$, find its inverse function f^{-1}.

When we substitute a value for x into f, two actions occur in the following order:

1. The number 5 is subtracted.
2. Division by 3 occurs.

To design an inverse function f^{-1}, we use the *opposite* operations indicated above *in the reverse order:*

1. Multiplication by 3 occurs.
2. The number 5 is added.

So we conclude that

$$f^{-1}(x) = 3x + 5$$

To verify, we must check that $(f^{-1} \circ f)(x) = x$ and $(f \circ f^{-1})(x) = x$:

$$(f^{-1} \circ f)(x) = f^{-1}(f(x)) = f^{-1}\left(\dfrac{x - 5}{3}\right) = 3\left(\dfrac{x - 5}{3}\right) + 5 = x - 5 + 5 = x$$

$$(f \circ f^{-1})(x) = f(f^{-1}(x)) = f(3x + 5) = \dfrac{3x + 5 - 5}{3} = \dfrac{3x}{3} = x$$

✔ CHECK YOURSELF 1

Given $f(x) = \dfrac{x + 1}{4}$, find f^{-1}.

To develop another technique for finding inverses, let us revisit functions defined by tables.

| Example 2 | Finding the Inverse of a Function |

Find the inverse of the function f:

x	$f(x)$
-4	8
0	6
2	5
1	-2

The inverse of f is easily found by reversing the order of the input values and output values:

x	$f^{-1}(x)$
8	−4
6	0
5	2
−2	1

Note that while f turns −4 into 8, for example, f^{-1} turns 8 back into −4.

✔ *CHECK YOURSELF 2*

Find the inverse of the function f.

x	$f(x)$
−2	5
1	0
4	−4
8	2

In Example 1, if we write y in place of $f(x)$, we see that we are just interchanging the roles of x and y in order to create the inverse f^{-1}. This suggests the following technique for finding the inverse of a function that is given in equation form.

Finding the Inverse of a Function

Step 1.   Given a function $f(x)$ in equation form, write y in place of $f(x)$.

Step 2.   Interchange the variables x and y.

Step 3.   Solve for y.

Step 4.   Write $f^{-1}(x)$ in place of y.

Example 3	**Finding the Inverse of a Function**

Find the inverse of $f(x) = 2x - 4$.

Begin by writing y in place of $f(x)$:

$$y = 2x - 4 \qquad \text{Interchange } x \text{ and } y.$$

$$x = 2y - 4 \qquad \text{Solve for } y. \text{ Add 4 to both sides.}$$

$$x + 4 = 2y \qquad \text{Divide both sides by 2.}$$

$$\frac{x + 4}{2} = y \qquad \text{Write } f^{-1}(x) \text{ in place of } y.$$

So

$$f^{-1}(x) = \frac{x + 4}{2}$$

You should verify this result by checking that $(f^{-1} \circ f)(x) = x$ and $(f \circ f^{-1})(x) = x$.

✔ *CHECK YOURSELF 3*

Find the inverse of $f(x) = \dfrac{x - 7}{5}$.

The graphs of relations and their inverses are connected in an interesting way. First, note that the graphs of the ordered pairs (a, b) and (b, a) always have symmetry about the line $y = x$.

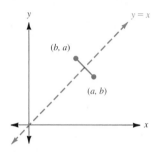

Now, with the above symmetry in mind, let's consider Example 4.

Example 4

Graphing a Relation and Its Inverse

Graph the relation f from Example 3 along with its inverse.

Recall that

$$f(x) = 2x - 4$$

and

$$f^{-1}(x) = \frac{1}{2}x + 2$$

The graphs of f and f^{-1} are shown below.

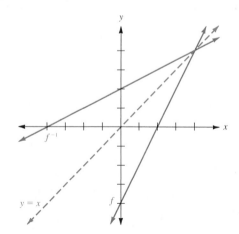

Note that the graphs of f and f^{-1} are symmetric about the line $y = x$. That symmetry follows from our earlier observation about the pairs (a, b) and (b, a) because we simply reversed the roles of x and y in forming the inverse relation.

✔ CHECK YOURSELF 4

Find the inverse of the function $f(x) = 3x + 6$ and graph both $f(x)$ and $f^{-1}(x)$ along with its inverse.

From our work thus far, we have seen techniques for finding the inverse of a function. However, it is quite possible that the inverse obtained may not be a function.

| Example 5 | **Finding the Inverse of a Function** |

Find the inverses of the following functions.

(a) $f = \{(1, 3), (2, 4), (3, 9)\}$

Its inverse is

Note: The elements of the ordered pairs have been interchanged.

$$\{(3, 1), (4, 2), (9, 3)\}$$

which is also a function.

(b) $g = \{(1, 3), (2, 6), (3, 6)\}$

Its inverse is

Note: It is not a function because 6 is mapped to both 2 and 3.

$$\{(3, 1), (6, 2), (6, 3)\}$$

which is *not* a function.

✔ CHECK YOURSELF 5

Write the inverses for each of the following relations. Which of the inverses are also functions?

(a) $\{(-1, 2), (0, 3), (1, 4)\}$ (b) $\{(2, 5), (3, 7), (4, 5)\}$

Can we predict in advance whether the inverse of a function will also be a function? The answer is yes.

We already know that for a relation to be a function, no element in its domain can be associated with more than one element in its range. Since, in creating an inverse, the *x* values and *y* values are interchanged, the inverse of a function *f* will also be a function only if *no element in the range of* f *can be associated with more than one element in its domain*. That is, no two ordered pairs of *f* can have the same *y* coordinate. This leads us to the following definition.

One-to-One Function

A function *f* is **one-to-one** if no two distinct domain elements are paired with the same range element.

We then have the following property.

Rules and Properties: Inverse of a Function

The inverse of a function *f* is also a function if *f* is one-to-one.

Note that in Example 5(a)

$$f = \{(1, 3), (2, 4), (3, 9)\}$$

is a one-to-one function and its inverse is also a function. However, the function in Example 5(b)

$$g = \{(1, 3), (2, 6), (3, 6)\}$$

is *not* a one-to-one function, and its inverse is *not* a function.

Our result regarding a one-to-one function and its inverse also has a convenient graphical interpretation. Below we graph the function *g* from Example 5:

$$g = \{(1, 3), (2, 6), (3, 6)\}$$

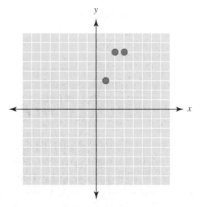

Again, *g* is *not* a one-to-one function, since two pairs, namely (2, 6) and (3, 6), have the same range element. As a result, a horizontal line may be drawn that passes through two points.

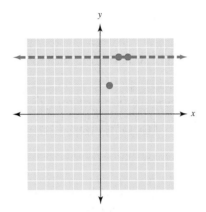

This means that when we form the inverse by reversing the coordinates, the resulting relation will *not* be a function. Points (6, 2) and (6, 3) are part of the inverse, and the resulting graph fails the vertical line test.

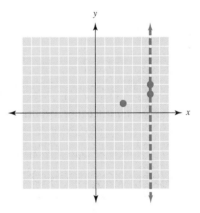

This leads us to the **horizontal line test:**

Note: In Section 5.2, we referred to the *vertical line test* to determine whether a relation was a function. The *horizontal line test* determines whether a function is one-to-one.

Horizontal Line Test

A function is one-to-one if no horizontal line can pass through two or more points on its graph.

Now we have a graphical way to determine whether the inverse of a function *f* will also be a function.

Rules and Properties: Inverse of a Function

The inverse of a function *f* is also a function if the graph of *f* passes the horizontal line test.

This is a very useful property, as Example 6 illustrates.

Example 6   ## Identifying One-to-One Functions

For each function, determine (i) whether the function is one-to-one and (ii) whether the inverse is also a function.

(a)

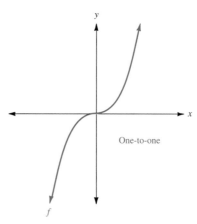

One-to-one

(i) Because no horizontal line passes through two or more points of the graph, *f* is one-to-one.

(ii) Because *f* is one-to-one, the inverse is also a function.

(b)

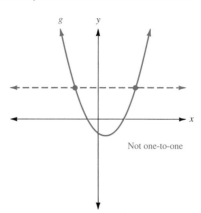

Not one-to-one

(i) Because a horizontal line can meet the graph of *g* at two points, *g* is *not* a one-to-one function.

(ii) Because *g* is not one-to-one, the inverse is not a function.

✔ *CHECK YOURSELF 6*

For each function, determine (i) whether the function is one-to-one and (ii) whether the inverse is also a function.

(a)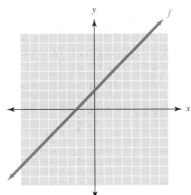

(b)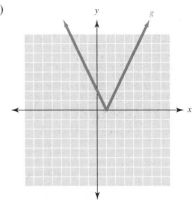

When a function is not one-to-one, we can restrict the domain of the function so that it will be one-to-one (and, as a result, the inverse will be a function). Consider the following example.

| Example 7 | # Restricting the Domain of a Function |

Given the function $f(x) = x^2 + 3$:

(a) Restrict the domain of f so that f is one-to-one.

The graph of f is a parabola, shifted vertically upward 3 units (below left):

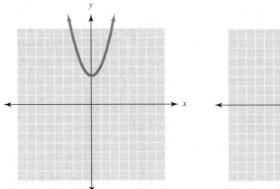

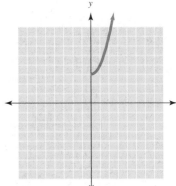

We restrict the domain of f to be $D = \{x \mid x \geq 0\}$ so the graph will pass the horizontal line test. (See graph above right.) The restricted f is stated as follows:

$$f(x) = x^2 + 3 \qquad x \geq 0$$

We could also restrict the domain of f to be

$$D = \{x \mid x \leq 0\}$$

in order to pass the horizontal line test.

(b) Using the restricted f, find f^{-1}.

Letting y replace $f(x)$, we write

$$y = x^2 + 3 \qquad \text{Interchange } x \text{ and } y.$$
$$x = y^2 + 3$$

This would produce a parabola with horizontal axis, failing the vertical line test. Solve for y.

$$x - 3 = y^2$$
$$y = \pm\sqrt{x - 3}$$

We choose $y = \sqrt{x - 3}$ (do you see why?) and write

$$f^{-1}(x) = \sqrt{x - 3}$$

(c) Graph the restricted f and f^{-1} on the same axes.

The graphs are shown:

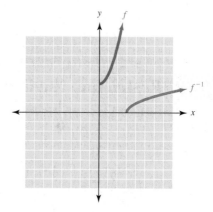

Observe that the graphs of f (restricted) and f^{-1} are symmetric with respect to the $y = x$ line.

Note that the domain of the restricted f (all real numbers greater than or equal to 0) is the same as the range of f^{-1}. Further, the range of the restricted f (all real numbers greater than or equal to 3) is the same as the domain of f^{-1}.

✔ *CHECK YOURSELF 7*

Given the function $f(x) = x^2 - 2$,

(a) Restrict the domain of f so that f is one-to-one.

(b) Using the restricted f, find f^{-1}.

(c) Graph the restricted f and f^{-1} on the same axes.

✔ *CHECK YOURSELF ANSWERS*

1. $f^{-1}(x) = 4x - 1$

4. $f^{-1}(x) = \dfrac{x - 6}{3}$

2.

x	$f^{-1}(x)$
5	-2
0	1
-4	4
2	8

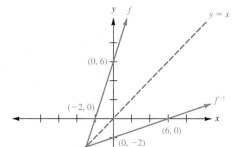

3. $f^{-1}(x) = 5x + 7$

5. (a) $\{(2, -1), (3, 0), (4, 1)\}$; a function
   (b) $\{(5, 2), (7, 3), (5, 4)\}$; not a function

6. (a) One-to-one; the inverse is a function
   (b) Not one-to-one; the inverse is not a function

7. (a) $f(x) = x^2 - 2, x \geq 0$ (b) $f^{-1}(x) = \sqrt{x + 2}$
   (c)

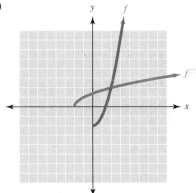

Exercises · 13.1

Find the inverse function f^{-1} for the given function f.

1. $f(x) = 3x + 5$

2. $f(x) = -3x - 7$

3. $f(x) = \dfrac{x - 1}{2}$

4. $f(x) = \dfrac{x + 1}{3}$

5. $f(x) = 2x - 3$

6. $f(x) = -5x + 3$

7. $f(x) = \dfrac{x + 4}{3}$

8. $f(x) = \dfrac{x - 5}{7}$

9. $f(x) = \dfrac{x}{3} + 5$

10. $f(x) = \dfrac{2x}{5} - 7$

Find the inverse of each function. In each case, determine whether the inverse is also a function.

11.

x	$f(x)$
-2	-5
2	6
3	4
4	3

12.

x	$f(x)$
-5	2
-3	3
1	3
5	2

13.

x	$f(x)$
-4	3
-2	7
3	5
7	-4

14.

x	$f(x)$
-5	2
-3	-4
0	2
6	-4

15. $f = \{(2, 3), (3, 4), (4, 5)\}$

16. $g = \{(1, 4), (2, 3), (3, 4)\}$

17. $f = \{(1, 5), (2, 5), (3, 5)\}$

18. $g = \{(4, 7), (2, 6), (6, 9)\}$

19. $f = \{(2, 3), (3, 5), (4, 7)\}$

20. $g = \{(-1, 0), (2, 0), (0, -1)\}$

Find the inverse function f^{-1} for the given function f. (*Hint:* Use the method demonstrated in Example 3.)

21. $f(x) = \dfrac{6 - 5x}{3}$ **22.** $f(x) = \dfrac{4x - 7}{3}$ **23.** $f(x) = \dfrac{11x}{5} + 2$ **24.** $f(x) = 5 - \dfrac{7x}{3}$

For each function f, find its inverse f^{-1}. Then graph both on the same set of axes.

25. $f(x) = 3x - 6$ **26.** $f(x) = 4x + 8$

27. $f(x) = -2x + 6$ **28.** $f(x) = -3x - 6$

Determine whether the given function is one-to-one. In each case decide whether the inverse is a function.

29. $f = \{(-3, 5), (-2, 3), (0, 2), (1, 4), (6, 5)\}$ **30.** $g = \{(-3, 7), (0, 4), (2, 5), (4, 1)\}$

31.

x	$f(x)$
-3	4
0	-3
2	1
6	2
8	0

32.

x	$g(x)$
-2	-6
-1	2
3	0
4	2

33.

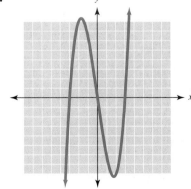

34.

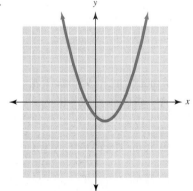

35.

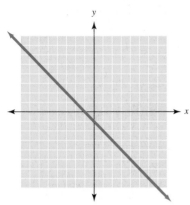

36.

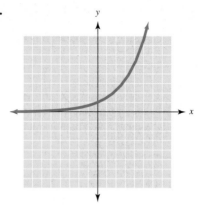

37. An inverse process is an operation that undoes a procedure. If the procedure is wrapping a present, describe in detail the inverse process.

38. If the procedure is the series of steps that take you from home to your classroom, describe the inverse process.

If $f(x) = 3x - 6$, then $f^{-1}(x) = \frac{1}{3}x + 2$. Given these two functions, in Exercises 39 to 44, find each of the following.

39. $f(6)$ **40.** $f^{-1}(6)$ **41.** $f(f^{-1}(6))$

42. $f^{-1}(f(6))$ **43.** $f(f^{-1}(x))$ **44.** $f^{-1}(f(x))$

If $g(x) = \frac{x+1}{2}$, then $g^{-1}(x) = 2x - 1$. Given these two functions, in Exercises 45 to 50, find each of the following.

45. $g(3)$ **46.** $g^{-1}(3)$ **47.** $g(g^{-1}(3))$

48. $g^{-1}(g(3))$ **49.** $g(g^{-1}(x))$ **50.** $g^{-1}(g(x))$

Given $h(x) = 2x + 8$, in Exercises 51 to 56, find each of the following.

51. $h(4)$ **52.** $h^{-1}(4)$ **53.** $h(h^{-1}(4))$

54. $h^{-1}(h(4))$ **55.** $h(h^{-1}(x))$ **56.** $h^{-1}(h(x))$

Suppose that f and g are one-to-one functions.

57. If $f(5) = 7$, find $f^{-1}(7)$.   **58.** If $g^{-1}(4) = 9$, find $g(9)$.

Let f be a linear function; that is, let $f(x) = mx + b$.

59. Find $f^{-1}(x)$.

60. Based on Exercise 59, if the slope of f is 3, what is the slope of f^{-1}?

61. Based on Exercise 59, if the slope of f is $\frac{2}{5}$, what is the slope of f^{-1}?

13.2 Exponential Functions

13.2 OBJECTIVES

1. Graph an exponential function
2. Solve an application of exponential functions
3. Solve an elementary exponential equation

Up to this point in the book, we have worked with polynomial functions and other functions in which the variable was used as a base. We now want to turn to a new classification of functions, the **exponential function.**

Exponential functions are functions whose defining equations involve the variable as an *exponent*. The introduction of these functions will allow us to consider many further applications, including population growth and radioactive decay.

Exponential Functions

An *exponential function* is a function that can be expressed in the form

$$f(x) = b^x$$

where $b > 0$ and $b \neq 1$. We call b the *base* of the exponential function.

The following are examples of exponential functions.

$$f(x) = 2^x \qquad g(x) = 3^x \qquad h(x) = \left(\frac{1}{2}\right)^x$$

As we have done with other new functions, we begin by finding some function values. We then use that information to graph the function.

| Example 1 | Graphing an Exponential Function |

Graph the exponential function

$$f(x) = 2^x$$

First, choose convenient values for x.

Note:

$$2^{-2} = \frac{1}{2^2} = \frac{1}{4}$$

$$f(0) = 2^0 = 1 \qquad f(-1) = 2^{-1} = \frac{1}{2} \qquad f(1) = 2^1 = 2$$

$$f(-2) = 2^{-2} = \frac{1}{4} \qquad f(2) = 2^2 = 4 \qquad f(-3) = 2^{-3} = \frac{1}{8}$$

Next, form a table from these values. Then plot the corresponding points and connect them with a smooth curve for the desired graph.

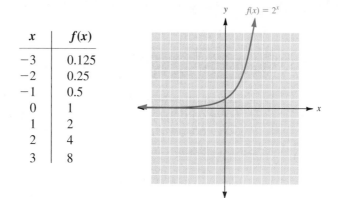

x	$f(x)$
-3	0.125
-2	0.25
-1	0.5
0	1
1	2
2	4
3	8

A vertical line will cross the graph at one point at most. The same is true for a horizontal line.

There is no value for x such that

$$2^x = 0$$

so the graph never touches the x axis.

We call $y = 0$ (or the x axis) the **horizontal asymptote.**

Let's examine some characteristics of the graph of the exponential function. First, the vertical line test shows that this is indeed the graph of a function. Also note that the horizontal line test shows that the function is one-to-one.

The graph *approaches* the x axis on the left, but it does *not intersect* the x axis. The y intercept is $(0, 1)$ (because $2^0 = 1$ by definition). To the right the functional values get larger. We say that the values *grow without bound.* This same language may be applied to linear or quadratic functions.

 CHECK YOURSELF 1

Sketch the graph of the exponential function

$$g(x) = 3^x$$

Let's look at an example in which the base of the function is less than 1.

Example 2 | Graphing an Exponential Function

Graph the exponential function

$$f(x) = \left(\frac{1}{2}\right)^x$$

Recall that

$$\left(\frac{1}{2}\right)^x = 2^{-x}$$

First, choose convenient values for x.

$$f(0) = \left(\frac{1}{2}\right)^0 = 1 \qquad f(-1) = \left(\frac{1}{2}\right)^{-1} = 2 \qquad f(1) = \left(\frac{1}{2}\right)^1 = \frac{1}{2}$$

$$f(-2) = \left(\frac{1}{2}\right)^{-2} = 4 \qquad f(2) = \left(\frac{1}{2}\right)^2 = \frac{1}{4} \qquad f(-3) = \left(\frac{1}{2}\right)^{-3} = 8$$

$$f(3) = \left(\frac{1}{2}\right)^3 = \frac{1}{8}$$

Again, form a table of values and graph the desired function.

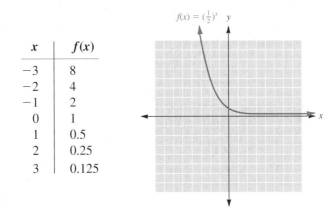

x	$f(x)$
-3	8
-2	4
-1	2
0	1
1	0.5
2	0.25
3	0.125

Let's compare this graph and that of Example 1. Clearly, this graph also represents a one-to-one function. As was true in Example 1, the graph does not intersect the x axis but approaches that axis, here on the right. The values for the function again grow without bound, but this time on the left. The y intercept for both graphs occurs at (0, 1).

The base of a *growth function* is *greater than* 1.

Note that the graph of Example 1 is *increasing* (going up) as we move from left to right. That function is an example of a **growth function.**

The base of a *decay function* is *less than* 1 but greater than 0.

The graph of Example 2 is *decreasing* (going down) as we move from left to right. It is an example of a **decay function.**

✔ CHECK YOURSELF 2

Sketch the graph of the exponential function

$$g(x) = \left(\frac{1}{3}\right)^x$$

The following algorithm summarizes our work thus far in this section.

Graphing an Exponential Function

Step 1. Establish a table of values by considering the function in the form $y = b^x$.

Step 2. Plot points from that table of values and connect them with a smooth curve to form the graph.

Step 3. If $b > 1$, the graph increases from left to right. If $0 < b < 1$, the graph decreases from left to right.

Step 4. All graphs will have the following in common:
(a) The y intercept will be (0, 1).
(b) The graphs will approach, but not touch, the x axis.
(c) The graphs will represent one-to-one functions.

The use of the letter *e* as a base originated with Leonhard Euler (1707–1783), and *e* is sometimes called *Euler's number* for that reason.

We used bases of 2 and $\frac{1}{2}$ for the exponential functions of our examples because they provided convenient computations. A far more important base for an exponential function is an irrational number named *e*. In fact, when *e* is used as a base, the function defined by

$$f(x) = e^x$$

is called *the* exponential function.

Graph $y = e^x$ on your calculator. You may find the $\boxed{e^x}$ key to be the second (or inverse) function to the ln *x* key. Note that e^1 is approximately 2.71828.

The significance of this number will be made clear in later courses, particularly calculus. For our purposes, *e* can be approximated as

$$e \approx 2.71828$$

The graph of $f(x) = e^x$ is shown below. Of course, it is very similar to the graphs seen earlier in this section.

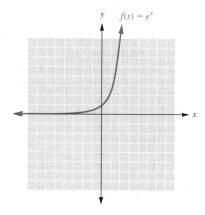

Exponential expressions involving base *e* occur frequently in real-world applications. Example 3 illustrates this approach.

Example 3   An Exponential Application

Be certain that you enclose the multiplication (0.05 × 5) in parentheses, or else the calculator will misinterpret your intended order of operation.

(a) Suppose that the population of a city is presently 20,000 and that the population is expected to grow at a rate of 5% per year. The equation

$$P(t) = 20{,}000e^{0.05t}$$

gives the town's population after *t* years. Find the population in 5 years.

Let $t = 5$ in the original equation to obtain

$$P(5) = 20{,}000e^{(0.05)(5)} \approx 25{,}681$$

which is the population expected 5 years from now.

Continuous compounding will
give the highest accumulation
of interest at any rate.
However, daily compounding
will result in an amount of
interest that is only slightly
less.

Note that in 9 years the
amount in the account is a
little more than *double* the
original principal.

(b) Suppose $1,000 is invested at an annual rate of 8%, compounded continuously. The equation

$$A(t) = 1,000e^{0.08t}$$

gives the amount in the account after t years. Find the amount after 9 years.
     Let $t = 9$ in the original equation to obtain

$$A(9) = 1,000e^{(0.08)(9)} \approx 2,054$$

which is the amount in the account after 9 years.

✔ **CHECK YOURSELF 3**

If $1,000 is invested at an annual rate of 6%, compounded continuously, then the equation for the amount in the account after t years is

$$A(t) = 1,000e^{0.06t}$$

Use your calculator to find the amount in the account after 12 years.

In some applications, we see a variation of the basic exponential function $f(x) = b^x$. What happens to the graph of such a function if we add (or subtract) a constant? We will see that the graph of the new function $f(x) = b^x + k$ is a familiar graph translated vertically k units. Consider the next example.

| Example 4 |

Graphing a Variation of an Exponential Function

Graph the function $f(x) = 2^x - 3$.

Again, we create a table of values and connect points with a smooth curve.

Note: Graph this function on
your graphing calculator. On
the same screen, also graph
the line $y = -3$.

x	$f(x)$
-3	-2.875
-2	-2.75
-1	-2.5
0	-2
1	-1
2	1
3	5

Note that the graph of f appears to be the familiar graph of $g(x) = 2^x$ shifted down 3 units. Instead of a y intercept of $(0, 1)$, we see a y intercept of $(0, -2)$. Instead of a horizontal asymptote of $y = 0$, we see a horizontal asymptote of $y = -3$.

✔ *CHECK YOURSELF 4*

Sketch the graph of the function $f(x) = \left(\dfrac{1}{2}\right)^x + 2$.

We can generalize the result seen above as follows:

Graphing a Function of the Form $y = b^x + k$

All graphs will have the following in common:

1. The y intercept will be $(0, k + 1)$.
2. There will be a horizontal asymptote with equation $y = k$.

As we observed, the exponential function is always one-to-one. This yields an important property that can be used to solve certain types of equations involving exponents.

If $b > 0$ and $b \neq 1$, then

$$b^m = b^n \qquad \text{if and only if} \qquad m = n \qquad \textbf{(1)}$$

| Example 5 | **Solving an Exponential Equation** |

(a) Solve $2^x = 8$ for x.

We recognize that 8 is a power of 2, and we can write the equation as

$$2^x = 2^3 \qquad \text{Write with equal bases.}$$

Applying property (1) above, we have

$$x = 3 \qquad \text{Set exponents equal.}$$

and 3 is the solution. The solution set is $\{3\}$.

(b) Solve $3^{2x} = 81$ for x.

Since $81 = 3^4$, we can write

$$3^{2x} = 3^4$$
$$2x = 4$$
$$x = 2$$

The answer can easily be checked by substitution. Letting $x = 2$ gives

$$3^{2(2)} = 3^4 = 81$$

We see that 2 is the solution for the equation. The solution set is $\{2\}$.

Note:

$$\frac{1}{16} = \frac{1}{2^4} = 2^{-4}$$

To verify the solution:

$$2^{-5+1} \overset{?}{=} \frac{1}{16}$$

$$2^{-4} \overset{?}{=} \frac{1}{16}$$

$$\frac{1}{16} = \frac{1}{16}$$

(c) Solve $2^{x+1} = \dfrac{1}{16}$ for x.

Again, we write $\dfrac{1}{16}$ as a power of 2, so that

$$2^{x+1} = 2^{-4}$$

Then

$$x + 1 = -4$$

$$x = -5$$

The solution set is $\{-5\}$.

✔ *CHECK YOURSELF 5*

Solve each of the following equations for x.

(a) $2^x = 16$ (b) $4^{x+1} = 64$ (c) $3^{2x} = \dfrac{1}{81}$

✔ *CHECK YOURSELF ANSWERS*

1. $y = g(x) = 3^x$

2. $y = \left(\dfrac{1}{3}\right)^x$

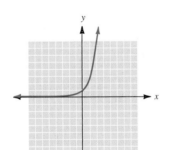

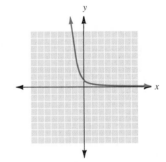

3. $2,054.43

4.

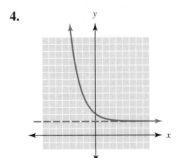

5. (a) $\{4\}$; (b) $\{2\}$; (c) $\{-2\}$

Exercises ▪ 13.2

Match the graphs in Exercises 1 to 8 with the appropriate equation.

(a) $y = \left(\dfrac{1}{2}\right)^x$ (b) $y = 2x - 1$ (c) $y = 2^x$ (d) $y = x^2$

(e) $y = 1^x$ (f) $y = 5^x$ (g) $x = 2^y$ (h) $x = y^2$

1.

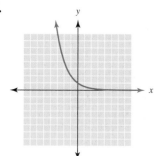

2.

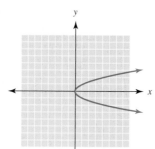

3.

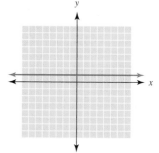

4.

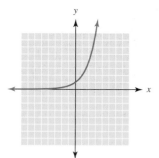

5.

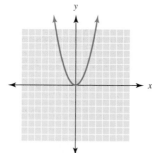

6.

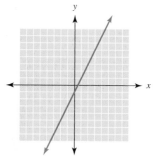

7.

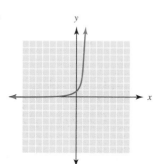

8.

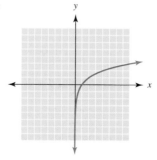

In Exercises 9 to 12, let $f(x) = 4^x$ and find each of the following.

9. $f(0)$ **10.** $f(4)$ **11.** $f(2)$ **12.** $f(-2)$

In Exercises 13 to 16, let $g(x) = 4^{x+1}$ and find each of the following.

13. $g(-1)$ **14.** $g(1)$ **15.** $g(2)$ **16.** $g(-2)$

980

In Exercises 17 to 20, let $h(x) = 4^x + 1$ and find each of the following.

17. $h(-1)$ **18.** $h(1)$ **19.** $h(2)$ **20.** $h(-2)$

In Exercises 21 to 24, let $f(x) = \left(\dfrac{1}{4}\right)^x$ and find each of the following.

21. $f(1)$ **22.** $f(-1)$ **23.** $f(0)$ **24.** $f(2)$

Graph each exponential function.

25. $y = 4^x$ **26.** $y = \left(\dfrac{1}{4}\right)^x$ **27.** $y = \left(\dfrac{2}{3}\right)^x$ **28.** $y = \left(\dfrac{3}{2}\right)^x$

29. $y = 3 \cdot 2^x$ **30.** $y = 2 \cdot 3^x$ **31.** $y = 3^x$ **32.** $y = 2^{x-1}$

33. $y = 2^{2x}$ **34.** $y = \left(\dfrac{1}{2}\right)^{2x}$ **35.** $y = e^{-x}$ **36.** $y = e^{2x}$

Graph each function. Sketch each horizontal asymptote as a dotted line.

37. $y = 3^x + 2$ **38.** $y = \left(\dfrac{1}{3}\right)^x - 4$ **39.** $y = e^x - 1$

Solve each exponential equation for x.

40. $2^x = 128$ **41.** $4^x = 64$ **42.** $10^x = 10,000$ **43.** $5^x = 625$

44. $3^x = \dfrac{1}{9}$ **45.** $2^x = \dfrac{1}{16}$ **46.** $4^{2x} = 64$ **47.** $3^{2x} = 81$

48. $3^{x-1} = 81$ **49.** $4^{x-1} = 16$ **50.** $3^{x-1} = \dfrac{1}{27}$ **51.** $2^{x-3} = \dfrac{1}{16}$

Suppose it takes 1 h for a certain bacterial culture to double by dividing in half. If there are 100 bacteria in the culture to start, then the number of bacteria in the culture after x hours is given by $N(x) = 100 \cdot 2^x$. In Exercises 52 to 55, use this function to find each of the following.

52. The number of bacteria in the culture after 2 h

53. The number of bacteria in the culture after 3 h

54. The number of bacteria in the culture after 5 h

55. Graph the relationship between the number of bacteria in the culture and the number of hours. Be sure to choose an appropriate scale for the N axis.

The half-life of radium is 1,690 years. That is, after a 1,690-year period, one-half of the original amount of radium will have decayed into another substance. If the original amount of radium was 64 grams (g), the formula relating the amount of radium left after time t is given by $R(t) = 64 \cdot 2^{-t/1,690}$. In Exercises 56 to 59, use that formula to find each of the following.

56. The amount of radium left after 1,690 years

57. The amount of radium left after 3,380 years

58. The amount of radium left after 5,070 years

59. Graph the relationship between the amount of radium remaining and time. Be sure to use appropriate scales for the R and t axes.

If \$1,000 is invested in a savings account with an interest rate of 8%, compounded annually, the amount in the account after t years is given by $A(t) = 1,000(1 + 0.08)^t$. In Exercises 60 to 63, use a calculator to find each of the following.

60. The amount in the account after 2 years

61. The amount in the account after 5 years

62. The amount in the account after 9 years

63. Graph the relationship between the amount in the account and time. Be sure to choose appropriate scales for the A and t axes.

The so-called learning curve in psychology applies to learning a skill, such as typing, in which the performance level progresses rapidly at first and then levels off with time. One can approximate N, the number of words per minute that a person can type after t weeks of training, with the equation $N = 80(1 - e^{-0.06t})$. Use a calculator to find the following.

64. (a) N after 10 weeks, (b) N after 20 weeks, (c) N after 30 weeks

65. Graph the relationship between the number of words per minute N and the number of weeks of training t.

66. Find two different calculators that have $\boxed{e^x}$ keys. Describe how to use the function on each of the calculators.

67. Are there any values of x for which e^x produces an exact answer on the calculator? Why are other answers not exact?

A possible calculator sequence for evaluating the expression

$$\left(1 + \frac{1}{n}\right)^n$$

where $n = 10$ is

$$\boxed{(}\ 1\ \boxed{+}\ 1\ \boxed{\div}\ 10\boxed{)}\ \boxed{\wedge}\ 10\ \boxed{ENTER}$$

which will yield approximately 2.5937.

In Exercises 68 to 72, use that sequence to find $\left(1 + \frac{1}{n}\right)^n$ for the following values of n. Round your answers to the nearest ten-thousandth.

68. $n = 100$ **69.** $n = 1,000$ **70.** $n = 10,000$

71. $n = 100,000$ **72.** $n = 1,000,000$

73. What did you observe from the experiment above?

74. Graph the exponential function defined by $y = 2^x$.

75. Graph the function defined by $x = 2^y$ on the same set of axes as the previous graph. What do you observe? (*Hint:* To graph $x = 2^y$, choose convenient values for y and then the corresponding values for x.)

76. Suppose you have a large piece of paper whose thickness is 0.003 in. If you tear the paper in half and stack the pieces, the height of the stack is (0.003)(2) in., or 0.006 in. If you now tear the stack in half again and then stack the pieces, the stack is $(0.003)(2)(2) = (0.003)(2^2)$ in., or 0.012 in. high.

(a) Define a function that gives the height h of the stack (in inches) after n tears.

(b) After which tear will the stack's height exceed 8 in.?

(c) Compute the height of the stack after the 15th tear. You will need to convert your answer to the appropriate units.

77. Use your graphing calculator to find all three points of intersection of the graphs of $f(x) = x^2$ and $g(x) = 2^x$. Give coordinates accurate to two decimal places.

SECTION 13.3 Logarithmic Functions

 OBJECTIVES

1. Graph a logarithmic function
2. Convert between logarithmic and exponential equations
3. Evaluate a logarithmic expression
4. Solve an elementary logarithmic equation

Given our experience with the exponential function in Section 13.2 and our earlier work with the inverse of a function in Section 13.1, we now can introduce the logarithmic function.

John Napier (1550–1617), a Scotsman, is credited with the invention of logarithms. The development of the logarithm grew out of a desire to ease the work involved in numerical computations, particularly in the field of astronomy. Today the availability of inexpensive scientific calculators has made the use of logarithms as a computational tool unnecessary.

However, the concept of the logarithm and the properties of the logarithmic function that we describe in a later section still are very important in the solutions of particular equations, in calculus, and in the applied sciences.

Again, the applications for this new function are numerous. The Richter scale for measuring the intensity of an earthquake and the decibel scale for measuring the intensity of sound both make use of logarithms.

To develop the idea of a logarithmic function, we must return to the definition of an exponential function

$$f(x) = b^x \qquad b > 0, b \neq 1 \tag{1}$$

Letting y replace $f(x)$ and interchanging the roles of x and y, we have the inverse function

$$x = b^y \tag{2}$$

Napier also coined the word *logarithm* from the Greek words "logos"(a ratio) and "arithmos" (a number).

Presently, we have no way to solve the equation $x = b^y$ for y. So, to write the inverse, equation (2), in a more useful form, we offer the following definition.

Recall that f is a one-to-one function, so its inverse is also a function.

The **logarithm of x to base b** is denoted

$$\log_b x$$

and $\qquad\qquad y = \log_b x \qquad$ if and only if $\qquad x = b^y$

We can now write an inverse function, using this new notation, as

Note that the restrictions on the base are the same as those used for the exponential function.

$$f^{-1}(x) = \log_b x \qquad b > 0, b \neq 1 \tag{3}$$

In general, any function defined in this form is called a **logarithmic function.**

At this point we should stress the meaning of this new relationship. Consider the equivalent forms illustrated here.

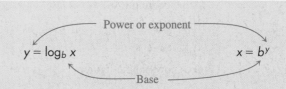

The logarithm y is the power to which we must raise b to get x. In other words, a *logarithm is simply a power or an exponent.* We return to this thought later when using the exponential and logarithmic forms of equivalent equations.

We begin our work by graphing a typical logarithmic function.

Example 1

Graphing a Logarithmic Function

Graph the logarithmic function

$$y = \log_2 x$$

Since $y = \log_2 x$ is equivalent to the exponential form

The base is 2, and the logarithm or power is y.

$$x = 2^y$$

we can find ordered pairs satisfying this equation by choosing convenient values for y and calculating the corresponding values for x.

Letting y take on values from -3 to 3 yields the table of values shown below. As before, we plot points from the ordered pairs and connect them with a smooth curve to form the graph of the function.

What do the vertical and horizontal line tests tell you about this graph?

x	y
$\dfrac{1}{8}$	-3
$\dfrac{1}{4}$	-2
$\dfrac{1}{2}$	-1
1	0
2	1
4	2
8	3

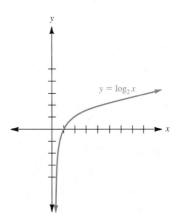

We observe that the graph represents a one-to-one function whose domain is $\{x \mid x > 0\}$ and whose range is the set of all real numbers.

For base 2 (or for any base greater than 1) the function will always be increasing over its domain.

Recall from Section 13.1 that the graphs of a function and its inverse are always reflections of each other about the line $y = x$. Since we have defined the logarithmic function as the inverse of an exponential function, we can anticipate the same relationship.

The graphs of

$$f(x) = 2^x \quad \text{and} \quad f^{-1}(x) = \log_2 x$$

are shown in the figure.

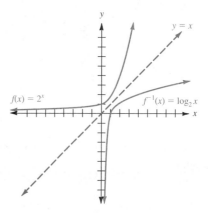

We see that the graphs of f and f^{-1} are indeed reflections of each other about the line $y = x$. In fact, this relationship provides an alternate method of sketching $y = \log_b x$. We can sketch the graph of $y = b^x$ and then reflect that graph about line $y = x$ to form the graph of the logarithmic function.

✔ CHECK YOURSELF 1

Graph the logarithmic function defined by

$$y = \log_3 x$$

(*Hint:* Consider the equivalent form $x = 3^y$.)

Because of the relationship between the graphs of $y = b^x$ and $y = \log_b x$, we can make the following observations:

1. Since the point $(0, 1)$ is on the graph of $y = b^x$ (it is the y intercept), the point $(1, 0)$ is on the graph of $y = \log_b x$ (it is the x intercept);

2. Since the line $y = 0$ is an asymptote for $y = b^x$ (it is horizontal), the line $x = 0$ is an asymptote for $y = \log_b x$ (it is vertical).

For our later work in this chapter, it will be necessary for us to be able to convert back and forth between exponential and logarithmic forms. The conversion is straightforward. You need only keep in mind the basic relationship.

Again, this tells us that a logarithm is an exponent or a power.

$$y = \log_b x \quad \text{means the same as} \quad x = b^y$$

Look at the following example.

| **Example 2** | **Writing Equations in Logarithmic Form** |

Convert to logarithmic form.

The base is 3, and the exponent or power is 4.

(a) $3^4 = 81$ is equivalent to $\log_3 81 = 4$.

(b) $10^3 = 1{,}000$ is equivalent to $\log_{10} 1{,}000 = 3$.

(c) $2^{-3} = \dfrac{1}{8}$ is equivalent to $\log_2 \dfrac{1}{8} = -3$.

(d) $9^{1/2} = 3$ is equivalent to $\log_9 3 = \dfrac{1}{2}$.

  CHECK YOURSELF 2

Convert each statement to logarithmic form.

(a) $4^3 = 64$ (b) $10^{-2} = 0.01$ (c) $3^{-3} = \dfrac{1}{27}$ (d) $27^{1/3} = 3$

Example 3 shows how to write a logarithmic expression in exponential form.

| **Example 3** | **Writing Equations in Exponential Form** |

Convert to exponential form.

Here, the base is 2; the logarithm, which is the power, is 3.

(a) $\log_2 8 = 3$ is equivalent to $2^3 = 8$.

(b) $\log_{10} 100 = 2$ is equivalent to $10^2 = 100$.

(c) $\log_3 \dfrac{1}{9} = -2$ is equivalent to $3^{-2} = \dfrac{1}{9}$.

(d) $\log_{25} 5 = \dfrac{1}{2}$ is equivalent to $25^{1/2} = 5$.

CHECK YOURSELF 3

Convert to exponential form.

(a) $\log_2 32 = 5$ (b) $\log_{10} 1{,}000 = 3$

(c) $\log_4 \dfrac{1}{16} = -2$ (d) $\log_{27} 3 = \dfrac{1}{3}$

Certain logarithms can be directly calculated by changing an expression to the equivalent exponential form, as Example 4 illustrates.

Example 4

Evaluating Logarithmic Expressions

(a) Evaluate $\log_3 27$.

If $x = \log_3 27$, in exponential form we have

Recall that $b^m = b^n$ if and only if $m = n$.

$$3^x = 27$$
$$3^x = 3^3$$
$$x = 3$$

We then have $\log_3 27 = 3$.

(b) Evaluate $\log_{10} \dfrac{1}{10}$.

If $x = \log_{10} \dfrac{1}{10}$, we can write

Rewrite each side as a power of the same base.

$$10^x = \dfrac{1}{10}$$
$$= 10^{-1}$$

We then have $x = -1$ and

$$\log_{10} \dfrac{1}{10} = -1$$

 CHECK YOURSELF 4

Evaluate each logarithm.

(a) $\log_2 64$ (b) $\log_3 \dfrac{1}{27}$

The relationship between exponents and logarithms also allows us to solve certain equations involving logarithms where two of the quantities in the equation $y = \log_b x$ are known, as Example 5 illustrates.

Example 5   **Solving Logarithmic Equations**

(a) Solve $\log_5 x = 3$ for x.

Since $\log_5 x = 3$, in exponential form we have

$$x = 5^3$$
$$= 125$$

The solution set is $\{125\}$.

(b) Solve $y = \log_4 \dfrac{1}{16}$ for y.

The original equation is equivalent to

$$4^y = \frac{1}{16}$$
$$= 4^{-2}$$

We then have $y = -2$ as the solution. The solution set is $\{-2\}$.

(c) Solve $\log_b 81 = 4$ for b.

In exponential form the equation becomes

$$b^4 = 81$$
$$b = 3$$

Keep in mind that the base must be *positive*, so we do not consider the possible solution $b = -3$.

The solution set is $\{3\}$.

✔ *CHECK YOURSELF 5*

Solve each of the following equations for the variable.

(a) $\log_4 x = 4$ (b) $\log_b \dfrac{1}{8} = -3$ (c) $y = \log_9 3$

Loudness can be measured in **bels (B),** a unit named for Alexander Graham Bell. This unit is rather large, so a more practical unit is the **decibel (dB),** a unit that is one-tenth as large.

Variable I_0 is the intensity of the minimum sound level detectable by the human ear.

The **decibel scale** is used in measuring the loudness of various sounds.

If I represents the intensity of a given sound and I_0 represents the intensity of a "threshold sound," then the decibel (dB) rating of the given sound is given by

$$L = 10 \log_{10} \frac{I}{I_0}$$

where $I_0 = 10^{-16}$ watt per square centimeter (W/cm²). Consider Example 6.

| **Example 6** | **A Decibel Application** |

(a) A whisper has intensity $I = 10^{-14}$. Its decibel rating is

$$L = 10 \log_{10} \frac{10^{-14}}{10^{-16}}$$

$$= 10 \log_{10} 10^2$$

$$= 10 \cdot 2$$

$$= 20$$

(b) A rock concert has intensity $I = 10^{-4}$. Its decibel rating is

$$L = 10 \log_{10} \frac{10^{-4}}{10^{-16}}$$

$$= 10 \log_{10} 10^{12}$$

$$= 10 \cdot 12$$

$$= 120$$

✔ CHECK YOURSELF 6

Ordinary conversation has intensity $I = 10^{-12}$. Find its rating on the decibel scale.

The scale was named after Charles Richter, a U.S. geologist.

Geologists use the **Richter scale** to convert seismographic readings, which give the intensity of the shock waves of an earthquake, to a measure of the magnitude of that earthquake.

The magnitude M of an earthquake is given by

A *zero-level* earthquake is the quake of least intensity that is measurable by a seismograph.

$$M = \log_{10} \frac{a}{a_0}$$

where a is the intensity of its shock waves and a_0 is the intensity of the shock wave of a zero-level earthquake.

| **Example 7** | **A Richter Scale Application** |

How many times stronger is an earthquake measuring 5 on the Richter scale than one measuring 4 on the Richter scale?

Suppose a_1 is the intensity of the earthquake with magnitude 5 and a_2 is the intensity of the earthquake with magnitude 4. Then

On your calculator, the [log] key is actually $\log_{10} x$.

$$5 = \log_{10} \frac{a_1}{a_0} \qquad \text{and} \qquad 4 = \log_{10} \frac{a_2}{a_0}$$

We convert these logarithmic expressions to exponential form.

$$10^5 = \frac{a_1}{a_0} \quad \text{and} \quad 10^4 = \frac{a_2}{a_0}$$

or

$$a_1 = a_0 \cdot 10^5 \quad \text{and} \quad a_2 = a_0 \cdot 10^4$$

The ratio of a_1 to a_2 is
$$\frac{a_1}{a_2}$$

We want the ratio of the intensities of the two earthquakes, so

$$\frac{a_1}{a_2} = \frac{a_0 \cdot 10^5}{a_0 \cdot 10^4} = 10^1 = 10$$

The earthquake of magnitude 5 is *10 times stronger* than the earthquake of magnitude 4.

✔ *CHECK YOURSELF 7*

How many times stronger is an earthquake of magnitude 6 than one of magnitude 4?

✔ *CHECK YOURSELF ANSWERS*

1. $y = \log_3 x$

2. (a) $\log_4 64 = 3$; (b) $\log_{10} 0.01 = -2$; (c) $\log_3 \frac{1}{27} = -3$; (d) $\log_{27} 3 = \frac{1}{3}$

3. (a) $2^5 = 32$; (b) $10^3 = 1{,}000$; (c) $4^{-2} = \frac{1}{16}$; (d) $27^{1/3} = 3$

4. (a) $\log_2 64 = 6$; (b) $\log_3 \frac{1}{27} = -3$ **5.** (a) $\{256\}$; (b) $\{2\}$; (c) $\left\{\frac{1}{2}\right\}$

6. 40 dB   **7.** 100 times

Exercises · 13.3

Sketch the graph of the function defined by each equation.

1. $y = \log_4 x$ **2.** $y = \log_{10} x$ **3.** $y = \log_2 (x - 1)$ **4.** $y = \log_3 (x + 1)$

5. $y = \log_8 x$ **6.** $y = \log_3 x + 1$

Convert each statement to logarithmic form.

7. $2^5 = 32$ **8.** $3^5 = 243$ **9.** $10^2 = 100$ **10.** $5^3 = 125$

11. $3^0 = 1$ **12.** $10^0 = 1$ **13.** $4^{-2} = \dfrac{1}{16}$ **14.** $3^{-4} = \dfrac{1}{81}$

15. $10^{-2} = \dfrac{1}{100}$ **16.** $3^{-3} = \dfrac{1}{27}$ **17.** $16^{1/2} = 4$ **18.** $125^{1/3} = 5$

19. $64^{-1/3} = \dfrac{1}{4}$ **20.** $36^{-1/2} = \dfrac{1}{6}$ **21.** $27^{2/3} = 9$ **22.** $9^{3/2} = 27$

23. $27^{-2/3} = \dfrac{1}{9}$ **24.** $16^{-3/2} = \dfrac{1}{64}$

Convert each statement to exponential form.

25. $\log_2 16 = 4$ **26.** $\log_5 5 = 1$ **27.** $\log_5 1 = 0$ **28.** $\log_3 27 = 3$

29. $\log_{10} 10 = 1$ **30.** $\log_2 32 = 5$ **31.** $\log_5 125 = 3$ **32.** $\log_{10} 1 = 0$

992

33. $\log_3 \dfrac{1}{27} = -3$ **34.** $\log_5 \dfrac{1}{25} = -2$ **35.** $\log_{10} 0.001 = -3$ **36.** $\log_{10} \dfrac{1}{1,000} = -3$

37. $\log_{16} 4 = \dfrac{1}{2}$ **38.** $\log_{125} 5 = \dfrac{1}{3}$ **39.** $\log_8 4 = \dfrac{2}{3}$ **40.** $\log_9 27 = \dfrac{3}{2}$

41. $\log_{25} \dfrac{1}{5} = -\dfrac{1}{2}$ **42.** $\log_{64} \dfrac{1}{16} = -\dfrac{2}{3}$

Evaluate each logarithm.

43. $\log_2 64$ **44.** $\log_3 81$ **45.** $\log_4 64$ **46.** $\log_{10} 10,000$

47. $\log_3 \dfrac{1}{81}$ **48.** $\log_4 \dfrac{1}{64}$ **49.** $\log_{10} \dfrac{1}{100}$ **50.** $\log_5 \dfrac{1}{25}$

51. $\log_{25} 5$ **52.** $\log_{27} 3$

Solve each equation for the unknown variable.

53. $y = \log_5 25$ **54.** $\log_2 x = 4$ **55.** $\log_b 256 = 4$ **56.** $y = \log_3 1$

57. $\log_{10} x = 2$ **58.** $\log_b 125 = 3$ **59.** $y = \log_5 5$ **60.** $y = \log_3 81$

61. $\log_{3/2} x = 3$ **62.** $\log_b \dfrac{4}{9} = 2$ **63.** $\log_b \dfrac{1}{25} = -2$ **64.** $\log_3 x = -3$

65. $\log_{10} x = -3$ **66.** $y = \log_2 \dfrac{1}{16}$ **67.** $y = \log_8 \dfrac{1}{64}$ **68.** $\log_b \dfrac{1}{100} = -2$

69. $\log_{27} x = \dfrac{1}{3}$ **70.** $y = \log_{100} 10$ **71.** $\log_b 5 = \dfrac{1}{2}$ **72.** $\log_{64} x = \dfrac{2}{3}$

73. $y = \log_{27} \dfrac{1}{9}$ **74.** $\log_b \dfrac{1}{8} = -\dfrac{3}{4}$

Solve using the decibel formula

$$L = 10 \log_{10} \dfrac{I}{I_0}$$

where $I_0 = 10^{-16}$ W/cm^2.

75. Sound. A television commercial has a volume with intensity $I = 10^{-11}$ W/cm^2. Find its rating in decibels.

76. Sound. The sound of a jet plane on takeoff has an intensity $I = 10^{-2}$ W/cm^2. Find its rating in decibels.

77. Sound. The sound of a computer printer has an intensity of $I = 10^{-9}$ W/cm^2. Find its rating in decibels.

78. Sound. The sound of a busy street has an intensity of $I = 10^{-8}$ W/cm^2. Find its rating in decibels.

The formula for the decibel rating L can be solved for the intensity of the sound as $I = I_0 \cdot 10^{L/10}$. Use this formula in Exercises 79 to 83.

79. Sound. Find the intensity of the sound in an airport waiting area if the decibel rating is 80.

80. Sound. Find the intensity of the sound of conversation in a crowded room if the decibel rating is 70.

81. Sound. What is the ratio of intensity of a sound of 80 dB to that of 70 dB?

82. Sound. What is the ratio of intensity of a sound of 60 dB to one measuring 40 dB?

83. Sound. What is the ratio of intensity of a sound of 70 dB to one measuring 40 dB?

84. Derive the formula for intensity provided above. (*Hint*: First divide both sides of the decibel formula by 10. Then write the equation in exponential form.)

Solve by using the earthquake formula

$$M = \log_{10} \frac{a}{a_0}$$

85. Earthquakes. An earthquake has an intensity a of $10^6 \cdot a_0$, where a_0 is the intensity of the zero-level earthquake. What was its magnitude?

86. Earthquakes. The great San Francisco earthquake of 1906 had an intensity of $10^{8.3} \cdot a_0$. What was its magnitude?

87. Earthquakes. An earthquake can begin causing damage to buildings with a magnitude of 5 on the Richter scale. Find its intensity in terms of a_0.

88. Earthquakes. An earthquake may cause moderate building damage with a magnitude of 6 on the Richter scale. Find its intensity in terms of a_0.

89. The **learning curve** describes the relationship between learning and time. Its graph is a logarithmic curve in the first quadrant. Describe that curve as it relates to learning.

90. In which scientific fields would you expect to again encounter a discussion of logarithms?

The *half-life* of a radioactive substance is the time it takes for one-half of the original amount of the substance to decay to a nonradioactive element. The half-life of radioactive waste is very important in figuring how long the waste must be kept isolated from the environment in some sort of storage facility. Half-lives of various radioactive waste products vary from a few seconds to millions of years. It usually takes at least 10 half-lives for a radioactive waste product to be considered safe.

The half-life of a radioactive substance can be determined by the following formula.

$$\log_e \frac{1}{2} = -\lambda x$$

where λ = radioactive decay constant

                                         x = half-life

Approximate the half-lives of the following important radioactive waste products, given the radioactive decay constant (RDC).

91. Plutonium 239, RDC = 0.000029   **92.** Strontium 90, RDC = 0.024755

93. Thorium 230, RDC = 0.000009   **94.** Cesium 135, RDC = 0.00000035

95. How many years will it be before each waste product will be considered safe?

96. (a) Evaluate $\log_2 (4 \cdot 8)$. (b) Evaluate $\log_2 4$ and $\log_2 8$. (c) Write an equation that connects the result of part (a) with the results of part (b).

97. (a) Evaluate $\log_3 (9 \cdot 81)$. (b) Evaluate $\log_3 9$ and $\log_3 81$. (c) Write an equation that connects the result of part (a) with the results of part (b).

98. Based on Exercises 96 and 97, propose a statement that connects $\log_a (mn)$ with $\log_a m$ and $\log_a n$.

13.4 Properties of Logarithms

OBJECTIVES

1. Apply the properties of logarithms
2. Evaluate logarithmic expressions with any base
3. Solve applications involving logarithms
4. Estimate the value of an antilogarithm

As mentioned earlier, logarithms were developed as aids to numerical computations. The early utility of the logarithm was due to the properties that we will discuss in this section. Even with the advent of the scientific calculator, that utility remains important today. We can apply these same properties to applications in a variety of areas that lead to exponential or logarithmic equations.

Since a logarithm is, by definition, an exponent, it seems reasonable that our knowledge of the properties of exponents should lead to useful properties for logarithms. That is, in fact, the case.

Logarithmic Properties

We start with two basic facts that follow immediately from the definition of the logarithm.

The properties follow from the facts that

$$b^1 = b \quad \text{and} \quad b^0 = 1$$

For $b > 0$ and $b \neq 1$,

1. $\log_b b = 1$
2. $\log_b 1 = 0$

We know that the logarithmic function $y = \log_b x$ and the exponential function $y = b^x$ are inverses of each other. So, for $f(x) = b^x$, we have $f^{-1}(x) = \log_b x$.

For any one-to-one function f,

The inverse has "undone" whatever f did to x.

$$f^{-1}(f(x)) = x \quad \text{for any } x \text{ in domain of } f$$

and

$$f(f^{-1}(x)) = x \quad \text{for any } x \text{ in domain of } f^{-1}$$

Since $f(x) = b^x$ is a one-to-one function, we can apply the above to the case where

$$f(x) = b^x \quad \text{and} \quad f^{-1}(x) = \log_b x$$

to derive the following.

For property 3,

$$f^{-1}(f(x)) = f^{-1}(b^x) = \log_b b^x$$

But in general, for any one-to-one function f,

$$f^{-1}(f(x)) = x$$

3. $\log_b b^x = x$
4. $b^{\log_b x} = x \quad$ for $x > 0$

Since logarithms are exponents, we can again turn to the familiar exponent rules to derive some further properties of logarithms. Consider the following.

We know that

$$\log_b M = x \qquad \text{if and only if} \qquad M = b^x$$

and

$$\log_b N = y \qquad \text{if and only if} \qquad N = b^y$$

Then

$$M \cdot N = b^x \cdot b^y = b^{x+y} \tag{1}$$

From equation (1) we see that $x + y$ is the power to which we must raise b to get the product MN. In logarithmic form, that becomes

$$\log_b MN = x + y \tag{2}$$

Now, since $x = \log_b M$ and $y = \log_b N$, we can substitute in equation (2) to write

$$\log_b MN = \log_b M + \log_b N \tag{3}$$

This is the first of the basic logarithmic properties presented here. The remaining properties may all be proved by arguments similar to those presented in equations (1) to (3).

In all cases, $M, N > 0$, $b > 0$, $b \neq 1$, and p is any real number.

Properties of Logarithms

Product Property

$$\log_b MN = \log_b M + \log_b N$$

Quotient Property

$$\log_b \frac{M}{N} = \log_b M - \log_b N$$

Power Property

$$\log_b M^p = p \log_b M$$

Many applications of logarithms require using these properties to write a single logarithmic expression as the sum or difference of simpler expressions, as Example 1 illustrates.

Example 1 | **Using the Properties of Logarithms**

Expand, using the properties of logarithms.

(a) $\log_b xy = \log_b x + \log_b y$ Product property

(b) $\log_b \dfrac{xy}{z} = \log_b xy - \log_b z$ Quotient property

$\qquad\qquad = \log_b x + \log_b y - \log_b z$ Product property

(c) $\log_{10} x^2 y^3 = \log_{10} x^2 + \log_{10} y^3$ Product property

$= 2 \log_{10} x + 3 \log_{10} y$ Power property

Recall $\sqrt{a} = a^{1/2}$.

(d) $\log_b \sqrt{\dfrac{x}{y}} = \log_b \left(\dfrac{x}{y}\right)^{1/2}$ Definition of exponent

$= \dfrac{1}{2} \log_b \dfrac{x}{y}$ Power property

$= \dfrac{1}{2} (\log_b x - \log_b y)$ Quotient property

✔ CHECK YOURSELF 1

Expand each expression, using the properties of logarithms.

(a) $\log_b x^2 y^3 z$ (b) $\log_{10} \sqrt{\dfrac{xy}{z}}$

In some cases, we will reverse the process and use the properties to write a single logarithm, given a sum or difference of logarithmic expressions.

Example 2   ## Rewriting Logarithmic Expressions

Write each expression as a single logarithm with coefficient 1.

(a) $2 \log_b x + 3 \log_b y$

$= \log_b x^2 + \log_b y^3$ Power property

$= \log_b x^2 y^3$ Product property

(b) $5 \log_{10} x + 2 \log_{10} y - \log_{10} z$

$= \log_{10} x^5 y^2 - \log_{10} z$

$= \log_{10} \dfrac{x^5 y^2}{z}$ Quotient property

(c) $\dfrac{1}{2} (\log_2 x - \log_2 y)$

$= \dfrac{1}{2} \left(\log_2 \dfrac{x}{y}\right)$

$= \log_2 \left(\dfrac{x}{y}\right)^{1/2}$ Power property

$= \log_2 \sqrt{\dfrac{x}{y}}$

✔ *CHECK YOURSELF 2*

Write each expression as a single logarithm with coefficient 1.

(a) $3 \log_b x + 2 \log_b y - 2 \log_b z$ (b) $\frac{1}{3}(2 \log_2 x - \log_2 y)$

Example 3 illustrates the basic concept of the use of logarithms as a computational aid.

Example 3

We have written the logarithms correct to three decimal places and will follow this practice throughout the remainder of this chapter.

Keep in mind, however, that this is an approximation and that $10^{0.301}$ will only approximate 2. Verify this with your calculator.

We have extended the product rule for logarithms.

Note that $\log_b 1 = 0$ for any base b.

Evaluating Logarithmic Expressions

Suppose $\log_{10} 2 = 0.301$ and $\log_{10} 3 = 0.477$. Given these values, find the following:

(a) $\log_{10} 6$

Since $6 = 2 \cdot 3$,

$$
\begin{aligned}
\log_{10} 6 &= \log_{10} (2 \cdot 3) \\
&= \log_{10} 2 + \log_{10} 3 \\
&= 0.301 + 0.477 \\
&= 0.778
\end{aligned}
$$

(b) $\log_{10} 18$

Since $18 = 2 \cdot 3 \cdot 3$,

$$
\begin{aligned}
\log_{10} 18 &= \log_{10} (2 \cdot 3 \cdot 3) \\
&= \log_{10} 2 + \log_{10} 3 + \log_{10} 3 \\
&= 1.255
\end{aligned}
$$

(c) $\log_{10} \dfrac{1}{9}$

Since $\dfrac{1}{9} = \dfrac{1}{3^2}$,

$$
\begin{aligned}
\log_{10} \frac{1}{9} &= \log_{10} \frac{1}{3^2} \\
&= \log_{10} 1 - \log_{10} 3^2 \\
&= 0 - 2 \log_{10} 3 \\
&= -0.954
\end{aligned}
$$

(d) $\log_{10} 16$

Since $16 = 2^4$,

$$
\begin{aligned}
\log_{10} 16 &= \log_{10} 2^4 = 4 \log_{10} 2 \\
&= 1.204
\end{aligned}
$$

Verify each answer with your calculator.

(e) $\log_{10} \sqrt{3}$

Since $\sqrt{3} = 3^{1/2}$,

$$\log_{10} \sqrt{3} = \log_{10} 3^{1/2} = \frac{1}{2} \log_{10} 3$$

$$= 0.239$$

✔ *CHECK YOURSELF 3*

Given the values for $\log_{10} 2$ and $\log_{10} 3$, find each of the following.

(a) $\log_{10} 12$ (b) $\log_{10} 27$ (c) $\log_{10} \sqrt[3]{2}$

Logarithms to Particular Bases

You can easily check the results in Example 3 by using the $\boxed{\log}$ key on a scientific calculator. For instance, in Example 3(d), to find $\log_{10} 16$, enter

On a graphing calculator, enter the $\boxed{\log}$ first, then 16.

$$16 \;\boxed{\log}$$

and the result (to three decimal places) will be 1.204. As you can see, the $\boxed{\log}$ key on your calculator provides logarithms to base 10, which is one of two types of logarithms used most frequently in mathematics:

Logarithms to base 10

Logarithms to base e

Of course, the use of logarithms to base 10 is convenient because our number system has base 10. We call logarithms to base 10 **common logarithms,** and it is customary to omit the base in writing a common (or base-10) logarithm. So

Note: When no base for log is written, it is assumed to be 10.

$$\log N \quad \text{means} \quad \log_{10} N$$

The following table shows the common logarithms for various powers of 10.

Exponential Form	Logarithmic Form
$10^3 = 1{,}000$	$\log 1{,}000 = 3$
$10^2 = 100$	$\log 100 = 2$
$10^1 = 10$	$\log 10 = 1$
$10^0 = 1$	$\log 1 = 0$
$10^{-1} = 0.1$	$\log 0.1 = -1$
$10^{-2} = 0.01$	$\log 0.01 = -2$
$10^{-3} = 0.001$	$\log 0.001 = -3$

| **Example 4** | **Approximating Logarithms with a Calculator** |

Verify each of the following with a calculator.

The number 4.8 lies between 1 and 10, so log 4.8 lies between 0 and 1.

(a) $\log 4.8 = 0.681$

(b) $\log 48 = 1.681$

(c) $\log 480 = 2.681$

(d) $\log 4{,}800 = 3.681$

(e) $\log 0.48 = -0.319$

Note that
$$480 = 4.8 \times 10^2$$
and
$$\begin{aligned}
&\log(4.8 \times 10^2) \\
&= \log 4.8 + \log 10^2 \\
&= \log 4.8 + 2 \\
&= 2 + \log 4.8
\end{aligned}$$

✔ *CHECK YOURSELF 4*

Use your calculator to find each of the following logarithms, rounded to three decimal places.

(a) $\log 2.3$ (b) $\log 23$ (c) $\log 230$

(d) $\log 2{,}300$ (e) $\log 0.23$ (f) $\log 0.023$

The value of log 0.48 is really $-1 + 0.681$. Your calculator will combine the signed numbers.

Note: A solution is **neutral** with pH = 7, **acidic** if the pH is less than 7, and **basic** if the pH is greater than 7.

Let's look at an application of common logarithms from chemistry. Common logarithms are used to define the pH of a solution. This is a scale that measures whether the solution is acidic or basic.

The pH of a solution is defined as

$$pH = -\log[H^+]$$

where $[H^+]$ is the hydrogen ion concentration, in moles per liter (mol/l), in the solution.

| **Example 5** | **A pH Application** |

Find the pH of each of the following. Determine whether each is a base or an acid.

(a) Rainwater: $[H^+] = 1.6 \times 10^{-7}$

From the definition,

Note the use of the product rule here.

Also, in general, $\log_b b^x = x$, so $\log 10^{-7} = -7$.

$$\begin{aligned}
pH &= -\log[H^+] \\
&= -\log(1.6 \times 10^{-7}) \\
&= -(\log 1.6 + \log 10^{-7}) \\
&= -[0.204 + (-7)] \\
&= -(-6.796) = 6.796
\end{aligned}$$

The rain is just slightly acidic.

(b) Household ammonia: $[H^+] = 2.3 \times 10^{-8}$

$$pH = -\log(2.3 \times 10^{-8})$$
$$= -(\log 2.3 + \log 10^{-8})$$
$$= -[0.362 + (-8)]$$
$$= 7.638$$

The ammonia is slightly basic.

(c) Vinegar: $[H^+] = 2.9 \times 10^{-3}$

$$pH = -\log(2.9 \times 10^{-3})$$
$$= -(\log 2.9 + \log 10^{-3})$$
$$= 2.538$$

The vinegar is very acidic.

✔ CHECK YOURSELF 5

Find the pH for the following solutions. Are they acidic or basic?

(a) Orange juice: $[H^+] = 6.8 \times 10^{-5}$ (b) Drain cleaner: $[H^+] = 5.2 \times 10^{-13}$

Many applications require reversing the process. That is, given the logarithm of a number, we must be able to find that number. The process is straightforward.

| Example 6 | **Using a Calculator to Estimate Antilogarithms** |

Because it is a one-to-one function, the logarithmic function has an inverse.

On a graphing calculator, enter

2nd log 2.1567 ⃞ ENTER

Suppose that $\log x = 2.1567$. We want to find a number x whose logarithm is 2.1567. Using a scientific calculator requires one of the following sequences:

$$2.1567 \boxed{10^x} \quad \text{or} \quad 2.1567 \boxed{INV} \boxed{\log}$$

Both give the result 143.45, often called the **antilogarithm** of 2.1567.

✔ CHECK YOURSELF 6

Approximate the value of the antilogarithm of x.

(a) $\log x = 0.828$ (b) $\log x = 1.828$

(c) $\log x = 2.828$ (d) $\log x = -0.172$

Let's return to the application from chemistry for an example requiring the use of the antilogarithm.

Example 7

A pH Application

Suppose that the pH for tomato juice is 6.2. Find the hydrogen ion concentration $[H^+]$.

Recall from our earlier formula that

$$pH = -\log [H^+]$$

In this case, we have

$$6.2 = -\log [H^+]$$

or

$$\log [H^+] = -6.2$$

The desired value for $[H^+]$ is then the antilogarithm of -6.2, and we use the following scientific calculator sequence:

$$6.2 \boxed{+/-} \boxed{INV} \boxed{\log}$$

On a graphing calculator, enter

$\boxed{2nd} \boxed{\log} \boxed{(-)} 6.2 \boxed{)} \boxed{ENTER}$

The result is 0.00000063, and we can write

$$[H^+] = 6.3 \times 10^{-7}$$

✔ *CHECK YOURSELF 7*

The pH for eggs is 7.8. Find $[H^+]$ for eggs.

Natural logarithms are also called **Napierian logarithms** after Napier. The importance of this system of logarithms was not fully understood until later developments in the calculus.

As we mentioned, there are two systems of logarithms in common use. The second type of logarithm uses the number e as a base, and we call logarithms to base e the **natural logarithms.** As with common logarithms, a convenient notation has developed, as the following definition shows.

> The **natural logarithm** is a logarithm to base e, and it is denoted ln x, where
>
> $$\ln x = \log_e x$$

The restrictions on the domain of the natural logarithmic function are the same as before. The function is defined only if $x > 0$.

By the general definition of a logarithm,

$$y = \ln x \qquad \text{means the same as} \qquad x = e^y$$

and this leads us directly to the following rules.

In general,

$$\log_b b^x = x \qquad b \neq 1$$

$$\ln 1 = 0 \qquad\qquad \text{Since } e^0 = 1$$

$$\ln e = 1 \qquad\qquad \text{Since } e^1 = e$$

$$\ln e^2 = 2 \quad \text{and} \quad \ln e^{-3} = -3$$

Example 8

Estimating Natural Logarithms

To find other natural logarithms, we can again turn to a calculator. To find the value of ln 2, use the sequence

$$2 \ \boxed{\ln}$$

or, with a graphing calculator,

$$\boxed{\ln} \ 2 \ \boxed{)} \ \boxed{\text{ENTER}}$$

The result is 0.693 (to three decimal places).

✔ CHECK YOURSELF 8

Use a calculator to find each of the following.

(a) ln 3   (b) ln 6   (c) ln 4   (d) ln $\sqrt{3}$

Of course, the properties of logarithms are applied in an identical fashion, no matter what the base.

Example 9

Evaluating Logarithms

If ln 2 = 0.693 and ln 3 = 1.099, find the following.

Recall that

$$\log_b MN = \log_b M + \log_b N$$

$$\log_b M^p = p \log_b M$$

(a) $\ln 6 = \ln (2 \cdot 3) = \ln 2 + \ln 3 = 1.792$

(b) $\ln 4 = \ln 2^2 = 2 \ln 2 = 1.386$

(c) $\ln \sqrt{3} = \ln 3^{1/2} = \dfrac{1}{2} \ln 3 = 0.550$

Again, verify these results with your calculator.

✔ CHECK YOURSELF 9

Use ln 2 = 0.693 and ln 3 = 1.099 to find the following.

(a) ln 12   (b) ln 27

The natural logarithm function plays an important role in both theoretical and applied mathematics. Example 10 illustrates just one of the many applications of this function.

| Example 10 | A Learning Curve Application |

A class of students took a final mathematics examination and received an average score of 76. In a psychological experiment, the students are retested at weekly intervals over the same material. If t is measured in weeks, then the new average score after t weeks is given by

Recall that we read $S(t)$ as "S of t" which means that S is a function of t.

$$S(t) = 76 - 5 \ln (t + 1)$$

Complete the following.

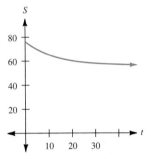

(a) Find the score after 10 weeks.

$$S(t) = 76 - 5 \ln (10 + 1)$$
$$= 76 - 5 \ln 11 \approx 64$$

(b) Find the score after 20 weeks.

$$S(t) = 76 - 5 \ln (20 + 1) \approx 61$$

This is an example of a **forgetting curve.** Note how it drops more rapidly at first. Compare this curve to the learning curve drawn in Section 13.2, Exercise 65.

(c) Find the score after 30 weeks.

$$S(t) = 76 - 5 \ln (30 + 1) \approx 59$$

✔ CHECK YOURSELF 10

The average score for a group of biology students, retested after time t (in months), is given by

$$S(t) = 83 - 9 \ln (t + 1)$$

Find the average score after

(a) 3 months   (b) 6 months

We conclude this section with one final property of logarithms. This property will allow us to quickly find the logarithm of a number to any base. Although work with logarithms with bases other than 10 or e is relatively infrequent, the relationship between logarithms of different bases is interesting in itself. Consider the following argument.
    Suppose that

$$x = \log_2 5$$

or $$2^x = 5$$ **(4)**

Taking the logarithm to base 10 of both sides of equation (4) yields

$$\log 2^x = \log 5$$

or $x \log 2 = \log 5$ Use the power property of logarithms.   **(5)**

(Note that we omit the 10 for the base and write log 2, for example.) Now, dividing both sides of equation (5) by log 2, we have

$$x = \frac{\log 5}{\log 2}$$

We can now find a value for x with the calculator. Dividing with the calculator log 5 by log 2, we get an approximate answer of 2.3219.

Since $x = \log_2 5$ and $x = \dfrac{\log 5}{\log 2}$, then

Caution

Do not cancel the logs.

$$\log_2 5 = \frac{\log 5}{\log 2}$$

Generalizing our result, we find the following.

Change-of-Base Formula

For the positive real numbers a and x,

$$\log_a x = \frac{\log_b x}{\log_b a}$$

Note that the logarithm on the left side has base a while the logarithms on the right side have base b. This allows us to calculate the logarithm to base a of any positive number, given the corresponding logarithms to base b (or any other base), as Example 11 illustrates.

Example 11

Evaluating Logarithms

Find $\log_5 15$.

From the change-of-base formula with $a = 5$ and $b = 10$,

We have written $\log_{10} 15$ rather than log 15 to emphasize the change-of-base formula.

$$\log_5 15 = \frac{\log_{10} 15}{\log_{10} 5}$$

$$= 1.683$$

Note: $\log_5 5 = 1$ and $\log_5 25 = 2$, so the result for $\log_5 15$ must be between 1 and 2.

The graphing calculator sequence for the above computation is

$$\boxed{\log}\ 15\ \boxed{)}\ \boxed{\div}\ \boxed{\log}\ 5\ \boxed{)}\ \boxed{\text{ENTER}}$$

On a scientific calculator, we type

$$15\boxed{\log}\ \boxed{\div}\ 5\ \boxed{\log}\ \boxed{=}$$

The result is 1.683, rounded to the thousandth.

✔ *CHECK YOURSELF 11*

Use the change-of-base formula to find $\log_8 32$.

Caution

A *common error* is to write

$$\frac{\log 15}{\log 5} = \log 15 - \log 5$$

This is *not* a logarithmic property. A true statement would be

$$\log \frac{15}{5} = \log 15 - \log 5$$

but

$$\log \frac{15}{5} \quad \text{and} \quad \frac{\log 15}{\log 5}$$

are *not* the same.

Note: The $\log_e x$ is called the **natural log** of x. We use $\ln x$ to designate the natural log of x. A special case of the change-of-base formula allows us to find natural logarithms in terms of common logarithms:

$$\ln x = \frac{\log x}{\log e}$$

so

$$\ln x \approx \frac{\log x}{0.434}$$

or, since

$$\frac{1}{0.434} \approx 2.304, \text{ then}$$

$$\ln x \approx 2.304 \log x$$

Of course, since all modern calculators have both the log function key and the ln function key, this conversion formula is now rarely used.

✔ *CHECK YOURSELF ANSWERS*

1. (a) $2 \log_b x + 3 \log_b y + \log_b z$; (b) $\frac{1}{2}(\log_{10} x + \log_{10} y - \log_{10} z)$

2. (a) $\log_b \frac{x^3 y^2}{z^2}$; (b) $\log_2 \sqrt[3]{\frac{x^2}{y}}$ **3.** (a) 1.079; (b) 1.431; (c) 0.100

4. (a) 0.362; (b) 1.362; (c) 2.362; (d) 3.362; (e) -0.638; (f) -1.638

5. (a) 4.167, acidic; (b) 12.284, basic   **6.** (a) 6.73; (b) 67.3; (c) 673; (d) 0.673

7. $[H^+] = 1.6 \times 10^{-8}$ **8.** (a) 1.099; (b) 1.792; (c) 1.386; (d) 0.549

9. (a) 2.485; (b) 3.297   **10.** (a) 70.5; (b) 65.5

11. $\log_8 32 = \frac{\log 32}{\log 8} \approx 1.667$

Use the properties of logarithms to expand each expression.

1. $\log_b 5x$

2. $\log_3 7x$

3. $\log_6 \dfrac{x}{7}$

4. $\log_b \dfrac{2}{y}$

5. $\log_3 a^2$

6. $\log_5 y^4$

7. $\log_5 \sqrt{x}$

8. $\log \sqrt[3]{z}$

9. $\log_b x^2 y^4$

10. $\log_7 x^3 z^2$

11. $\log_4 y^2 \sqrt{x}$

12. $\log_b x^3 \sqrt[3]{z}$

13. $\log_b \dfrac{x^2 y}{z}$

14. $\log_5 \dfrac{3}{xy}$

15. $\log \dfrac{xy^2}{\sqrt{z}}$

16. $\log_4 \dfrac{x^3 \sqrt{y}}{z^2}$

17. $\log_5 \sqrt[3]{\dfrac{xy}{z^2}}$

18. $\log_b \sqrt[4]{\dfrac{x^2 y}{z^3}}$

Write each expression as a single logarithm.

19. $\log_b x + \log_b y$

20. $\log_5 x - \log_5 y$

21. $3 \log_5 x - 2 \log_5 y$

22. $3 \log_b x + \log_b z$

23. $\log_b x + \dfrac{1}{2} \log_b y$

24. $\dfrac{1}{2} \log_b x - 3 \log_b z$

25. $\log_b x - 2 \log_b y - \log_b z$

26. $2 \log_5 x - (3 \log_5 y + \log_5 z)$

27. $\dfrac{1}{2} \log_6 y - 3 \log_6 z$

28. $\log_b x - \dfrac{1}{3} \log_b y - 4 \log_b z$

29. $\dfrac{1}{3}(2 \log_b x + \log_b y - \log_b z)$

30. $\dfrac{1}{5}(2 \log_4 x - \log_4 y + 3 \log_4 z)$

Given that $\log 2 = 0.301$ and $\log 3 = 0.477$, find each logarithm.

31. $\log 24$

32. $\log 36$

33. $\log 8$

34. $\log 81$

35. $\log \sqrt{2}$

36. $\log \sqrt[3]{3}$

37. $\log \dfrac{1}{4}$

38. $\log \dfrac{1}{27}$

In Exercises 39 to 44, use your calculator to find each logarithm.

39. log 7.3

40. log 68

41. log 680

42. log 6,800

43. log 0.72

44. log 0.068

In Exercises 45 and 46, find the pH. You are given the hydrogen ion concentration $[H^+]$ for each solution. Use the formula pH $= -\log [H^+]$. Are the solutions acidic or basic?

45. Blood: $[H^+] = 3.8 \times 10^{-8}$

46. Lemon juice: $[H^+] = 6.4 \times 10^{-3}$

Use your calculator to approximate the antilogarithm for each logarithm.

47. 0.749

48. 1.749

49. 3.749

50. -0.251

In Exercises 51 and 52, given the pH of the solutions, approximate the hydrogen ion concentration $[H^+]$.

51. Wine: pH $= 4.7$

52. Household ammonia: pH $= 7.8$

Use your calculator to find each logarithm.

53. ln 2

54. ln 3

55. ln 10

56. ln 30

The average score on a final examination for a group of psychology students, retested after time t (in weeks), is given by

$$S = 85 - 8 \ln (t + 1)$$

Find the average score on the retests.

57. After 3 weeks

58. After 12 weeks

Estimate each logarithm by "trapping" it between consecutive integers. To estimate $\log_4 52$, we note that $4^2 = 16$ and $4^3 = 64$, so $\log_4 52$ must lie between 2 and 3.

59. $\log_3 25$

60. $\log_5 30$

61. $\log_2 70$

62. $\log_2 19$

63. log 680

64. log 6,800

Without a calculator, use the properties of logarithms to evaluate each expression.

65. log 5 + log 2

66. log 25 + log 4

67. $\log_3 45 - \log_3 5$

In Exercises 68 and 69, use the change-of-base formula to find each logarithm.

68. $\log_3 25$ **69.** $\log_5 30$

The amount of a radioactive substance remaining after time t is given by

$$A = e^{\lambda t + \ln A_0}$$

where A is the amount remaining after time t, A_0 is the original amount of the substance, and λ is the radioactive decay constant.

70. How much plutonium 239 will remain after 50,000 years if 24 kg was originally stored? Plutonium 239 has a radioactive decay constant of -0.000029.

71. How much plutonium 241 will remain after 100 years if 52 kg was originally stored? Plutonium 241 has a radioactive decay constant of -0.053319.

72. How much strontium 90 was originally stored if after 56 years it is discovered that 15 kg still remains? Strontium 90 has a radioactive decay constant of -0.024755.

73. How much cesium 137 was originally stored if after 90 years it is discovered that 20 kg still remains? Cesium 137 has a radioactive decay constant of -0.023105.

74. Which keys on your calculator are function keys and which are operation keys? What is the difference?

75. How is the pH factor relevant to your selection of a hair care product?

76. (a) Use the change-of-base formula to write $\log_2 5$ in terms of base-10 logarithms. Then use your calculator to find $\log_2 5$ rounded to three decimal places.

    (b) Use the change-of-base formula to write $\log_2 5$ in terms of base-e logarithms. Then use your calculator to find $\log_2 5$ rounded to three decimal places.

    (c) Compare your answers to parts (a) and (b).

SECTION 13.5 Logarithmic and Exponential Equations

13.5 OBJECTIVES

1. Solve a logarithmic equation
2. Solve an exponential equation
3. Solve an application involving an exponential equation

Much of the importance of the properties of logarithms developed in Section 13.4 lies in the application of those properties to the solving of equations involving logarithms and exponents. Our work in this section will consider techniques for both types of equations. Let's start with a definition.

> A **logarithmic equation** is an equation that contains a logarithmic expression.

We solved some simple examples in Section 13.3. Let's review for a moment. To solve $\log_3 x = 4$ for x, recall that we simply convert the logarithmic equation to exponential form. Here,

$$x = 3^4$$

so

$$x = 81$$

and {81} is the solution set for the given equation.

Now, what if the logarithmic equation involves more than one logarithmic term? Example 1 illustrates how the properties of logarithms must then be applied.

Example 1 Solving a Logarithmic Equation

Solve each logarithmic equation.

(a) $\log_5 x + \log_5 3 = 2$

We apply the product rule for logarithms:

$\log_b M + \log_b N = \log_b MN$

The original equation can be written as

$$\log_5 3x = 2$$

Now, since only a single logarithm is involved, we can write the equation in the equivalent exponential form:

$$3x = 5^2$$
$$3x = 25$$
$$x = \frac{25}{3}$$

and $\left\{\dfrac{25}{3}\right\}$ is the solution set.

(b) $\log x + \log (x - 3) = 1$

Write the equation as

Since no base is written, it is assumed to be 10.

$$\log x(x - 3) = 1$$

or

$$x(x - 3) = 10^1$$

Given the base of 10, this is the equivalent exponential form.

We now have

$$x^2 - 3x = 10$$
$$x^2 - 3x - 10 = 0$$
$$(x - 5)(x + 2) = 0$$

Possible solutions are $x = 5$ or $x = -2$.

Note that substitution of -2 into the original equation gives

Checking possible solutions is particularly important here.

$$\log (-2) + \log (-5) = 1$$

Since logarithms of negative numbers are *not* defined, -2 is an extraneous solution and we must reject it. The only solution for the original equation is 5. The solution set is $\{5\}$.

  CHECK YOURSELF 1

Solve $\log_2 x + \log_2 (x + 2) = 3$ for x.

The quotient property is used in a similar fashion for solving logarithmic equations. Consider Example 2.

Example 2

Solving a Logarithmic Equation

Solve each equation for x.

(a) $\log_5 x - \log_5 2 = 2$

We apply the quotient rule for logarithms:

Rewrite the original equation as

$$\log_b M - \log_b N = \log_b \frac{M}{N}$$

$$\log_5 \frac{x}{2} = 2$$

Now,
$$\frac{x}{2} = 5^2$$

$$\frac{x}{2} = 25$$

$$x = 50$$

The solution set is $\{50\}$.

(b) $\log_3 (x + 1) - \log_3 x = 3$

$$\log_3 \frac{x + 1}{x} = 3$$

$$\frac{x + 1}{x} = 27$$

$$x + 1 = 27x$$

$$1 = 26x$$

$$x = \frac{1}{26}$$

Again, you should verify that substituting $\frac{1}{26}$ for x leads to a positive value in each of the original logarithms.

and $\left\{\dfrac{1}{26}\right\}$ is the solution set.

 CHECK YOURSELF 2

Solve $\log_5 (x + 3) - \log_5 x = 2$ for x.

The solving of certain types of logarithmic equations calls for the one-to-one property of the logarithmic function.

> If $\log_b M = \log_b N$
>
> then $M = N$

Example 3   Solving a Logarithmic Equation

Solve the following equation for x.

$$\log (x + 2) - \log 2 = \log x$$

Again, we rewrite the left-hand side of the equation. So

$$\log \frac{x + 2}{2} = \log x$$

Since the logarithmic function is one-to-one, this is equivalent to

$$\frac{x+2}{2} = x$$

or

$$x = 2$$

So, {2} is the solution set.

✔ *CHECK YOURSELF 3*

Solve for x.

$$\log (x + 3) - \log 3 = \log x$$

The following algorithm summarizes our work in solving logarithmic equations.

Solving Logarithmic Equations

Step 1. Use the properties of logarithms to combine terms containing logarithmic expressions into a single term.

Step 2. Write the equation formed in step 1 in exponential form.

Step 3. Solve for the indicated variable.

Step 4. Check your solutions to make sure that possible solutions do not result in the logarithms of negative numbers.

Let's look now at **exponential equations,** which are equations in which the variable appears as an exponent.

We solved some particular exponential equations in Section 13.2. In solving an equation such as

$$3^x = 81$$

we wrote the right-hand member as a power of 3, so that

Again, we want to write both sides as a power of the same base, here 3.

$$3^x = 3^4$$

or

$$x = 4$$

The technique here will work only when both sides of the equation can be conveniently expressed as powers of the same base. If that is not the case, we must use logarithms to solve the equation, as illustrated in Example 4.

Example 4

Solving an Exponential Equation

Solve $3^x = 5$ for x, accurate to the nearest thousandth.

We begin by taking the common logarithm of both sides of the original equation.

$$\log 3^x = \log 5$$

Again:

If $M = N$

then $\log_b M = \log_b N$

Caution

This is *not* $\log 5 - \log 3$, a common error.

Now we apply the power property so that the variable becomes a coefficient on the left.

$$x \log 3 = \log 5$$

Dividing both sides of the equation by $\log 3$ will isolate x, and we have

$$x = \frac{\log 5}{\log 3}$$

$$\approx 1.465 \qquad \text{(to three decimal places)}$$

We will write the solution set as $\{1.465\}$.

Note: The solution 1.465 is not exact. You can verify the approximate solution on a calculator. Raise 3 to power 1.465. You should see a result close to 5, but not exactly 5.

✔ *CHECK YOURSELF 4*

Solve $2^x = 10$ for x, rounded to the nearest thousandth.

Example 5 shows how to solve an equation with a more complicated exponent.

Example 5

On the left, we apply

$\log_b M^p = p \log_b M$

On a graphing calculator, the sequence would be

$\boxed{(}\ \boxed{\log}\ 8\ \boxed{)}\ \boxed{\div}\ \boxed{\log}\ 5\ \boxed{)}\ \boxed{-}$
$1\ \boxed{)}\ \boxed{\div}\ 2\ \boxed{\text{ENTER}}$

A scientific calculator sequence to find x would be

$8\ \boxed{\log}\ \boxed{\div}\ 5\ \boxed{\log}\ \boxed{-}\ 1\ \boxed{=}$
$\boxed{\div}\ 2\ \boxed{=}$

Solving an Exponential Equation

Solve $5^{2x+1} = 8$ for x.

The process begins as in Example 4.

$$\log 5^{2x+1} = \log 8$$

$$(2x + 1) \log 5 = \log 8$$

$$2x + 1 = \frac{\log 8}{\log 5}$$

$$2x = \frac{\log 8}{\log 5} - 1$$

$$x = \frac{1}{2}\left(\frac{\log 8}{\log 5} - 1\right)$$

$$x \approx 0.146$$

The solution set is $\{0.146\}$.

✔ *CHECK YOURSELF 5*

Solve $3^{2x-1} = 7$ for x.

The procedure is similar if the variable appears as an exponent in more than one term of the equation.

| **Example 6** | **Solving an Exponential Equation** |

Solve $3^x = 2^{x+1}$ for x.

Use the power property to write the variables as coefficients.

$$\log 3^x = \log 2^{x+1}$$
$$x \log 3 = (x + 1) \log 2$$
$$x \log 3 = x \log 2 + \log 2$$

We now isolate x on the left.

$$x \log 3 - x \log 2 = \log 2$$
$$x(\log 3 - \log 2) = \log 2$$

To check the reasonableness of this result, use your calculator to verify that

$$x = \frac{\log 2}{\log 3 - \log 2}$$
$$\approx 1.710$$

$$3^{1.710} \approx 2^{2.710}$$

The solution set is $\{1.710\}$.

  CHECK YOURSELF 6

Solve $5^{x+1} = 3^{x+2}$ for x.

The following algorithm summarizes our work with solving exponential equations.

Solving Exponential Equations

Step 1. Try to write each side of the equation as a power of the same base. Then equate the exponents to form an equation.

Step 2. If the above procedure is not applicable, take the common logarithm of both sides of the original equation.

Step 3. Use the power rule for logarithms to write an equivalent equation with the variables as coefficients.

Step 4. Solve the resulting equation.

Step 5. Check for extraneous solutions.

There are many applications of our work with exponential equations. Consider the following.

| **Example 7** | **An Interest Application** |

If an investment of P dollars earns interest at an annual interest rate r and the interest is compounded n times per year, then the amount in the account after t years is given by

$$A = P\left(1 + \frac{r}{n}\right)^{nt} \tag{1}$$

If \$1,000 is placed in an account with an annual interest rate of 6%, find out how long it will take the money to double when interest is compounded annually and how long when compounded quarterly.

(a) Compound interest annually.

Since the interest is compounded *once* per year, $n = 1$.

Using equation (1) with $A = 2,000$ (we want the original 1,000 to double), $P = 1,000$, $r = 0.06$, and $n = 1$, we have

$$2,000 = 1,000(1 + 0.06)^t$$

Dividing both sides by 1,000 yields

$$2 = (1.06)^t$$

We now have an exponential equation that can be solved by our earlier techniques.

From accounting, we have the **rule of 72,** which states that the doubling time is approximately 72 divided by the interest rate as a percentage. Here $\dfrac{72}{6} = 12$ years.

$$\log 2 = \log (1.06)^t$$
$$= t \log 1.06$$

or

$$t = \frac{\log 2}{\log 1.06}$$
$$\approx 11.9 \text{ years}$$

It takes just a little less than 12 years for the money to double.

(b) Compound interest quarterly.

Since the interest is compounded 4 times per year, $n = 4$.

Now $n = 4$ in equation (1), so

$$2,000 = 1,000\left(1 + \frac{0.06}{4}\right)^{4t}$$
$$2 = (1.015)^{4t}$$
$$\log 2 = \log (1.015)^{4t}$$
$$\log 2 = 4t \log 1.015$$
$$\frac{\log 2}{4 \log 1.015} = t$$
$$t \approx 11.6 \text{ years}$$

Note that the doubling time is reduced by approximately 3 months by the more frequent compounding.

✔ *CHECK YOURSELF 7*

Find the doubling time in Example 7 if the interest is compounded monthly.

Problems involving rates of growth or decay can also be solved by using exponential equations.

Example 8	**A Population Application**

A town's population is presently 10,000. Given a projected growth rate of 7% per year, t years from now the population P will be given by

$$P = 10,000e^{0.07t}$$

In how many years will the town's population double?

We want the time t when P will be 20,000 (doubled in size). So

Divide both sides by 10,000.

$$20,000 = 10,000e^{0.07t}$$

or

$$2 = e^{0.07t}$$

In this case, we take the *natural logarithm* of both sides of the equation. This is because e is involved in the equation.

$$\ln 2 = \ln e^{0.07t}$$

Apply the power property.

$$\ln 2 = 0.07t \ln e$$

Note: $\ln e = 1$

$$\ln 2 = 0.07t$$

$$\frac{\ln 2}{0.07} = t$$

$$t \approx 9.9 \text{ years}$$

The population will double in approximately 9.9 years.

✔ CHECK YOURSELF 8

If \$1,000 is invested in an account with an annual interest rate of 6%, compounded continuously, the amount A in the account after t years is given by

$$A = 1,000e^{0.06t}$$

Find the time t that it will take for the amount to double ($A = 2,000$). Compare this time with the result of the Check Yourself 7 Exercise. Which is shorter? Why?

✔ CHECK YOURSELF ANSWERS

1. $\{2\}$ **2.** $\left\{\dfrac{1}{8}\right\}$ **3.** $\left\{\dfrac{3}{2}\right\}$ **4.** $\{3.322\}$ **5.** $\{1.386\}$ **6.** $\{1.151\}$

7. 11.58 years   **8.** 11.55 years. The doubling time is shorter, because interest is compounded more frequently.

Exercises · 13.5

Solve each logarithmic equation for x.

1. $\log_5 x = 3$

2. $\log_3 x = -2$

3. $\log (x + 1) = 2$

4. $\log_5 (3x + 2) = 3$

5. $\log_2 x + \log_2 8 = 6$

6. $\log 5 + \log x = 2$

7. $\log_3 x - \log_3 6 = 3$

8. $\log_4 x - \log_4 8 = 3$

9. $\log_2 x + \log_2 (x + 2) = 3$

10. $\log_3 x + \log_3 (2x + 3) = 2$

11. $\log_7 (x + 1) + \log_7 (x - 5) = 1$

12. $\log_2 (x + 2) + \log_2 (x - 5) = 3$

13. $\log x - \log (x - 2) = 1$

14. $\log_5 (x + 5) - \log_5 x = 2$

15. $\log_3 (x + 1) - \log_3 (x - 2) = 2$

16. $\log (x + 2) - \log (2x - 1) = 1$

17. $\log (x + 5) - \log (x - 2) = \log 5$

18. $\log_3 (x + 12) - \log_3 (x - 3) = \log_3 6$

19. $\log_2 (x^2 - 1) - \log_2 (x - 2) = 3$

20. $\log (x^2 + 1) - \log (x - 2) = 1$

Solve each exponential equation for x. If your solution is an approximation, give your solutions in decimal form, rounded to three decimal places.

21. $6^x = 1{,}296$

22. $4^x = 64$

23. $2^{x+1} = \dfrac{1}{8}$

24. $9^x = 3$

25. $8^x = 2$

26. $3^{2x-1} = 27$

27. $3^x = 7$

28. $5^x = 30$

29. $4^{x+1} = 12$

30. $3^{2x} = 5$

31. $7^{3x} = 50$

32. $6^{x-3} = 21$

33. $5^{3x-1} = 15$

34. $8^{2x+1} = 20$

35. $4^x = 3^{x+1}$

36. $5^x = 2^{x+2}$

37. $2^{x+1} = 3^{x-1}$

38. $3^{2x+1} = 5^{x+1}$

Use the formula

$$A = P\left(1 + \frac{r}{n}\right)^{nt}$$

to solve Exercises 39 to 42.

39. **Interest.** If \$5,000 is placed in an account with an annual interest rate of 9%, how long will it take the amount to double if the interest is compounded annually?

40. Repeat Exercise 39 if the interest is compounded semiannually.

13-63

1019

41. Repeat Exercise 39 if the interest is compounded quarterly.

42. Repeat Exercise 39 if the interest is compounded monthly.

Suppose the number of bacteria present in a culture after t hours is given by $N(t) = N_0 \cdot 2^{t/2}$, where N_0 is the initial number of bacteria. Use the formula to solve Exercises 43 to 46.

43. How long will it take the bacteria to increase from 12,000 to 20,000?

44. How long will it take the bacteria to increase from 12,000 to 50,000?

45. How long will it take the bacteria to triple? (*Hint:* Let $N = 3N_0$.)

46. How long will it take the culture to increase to 5 times its original size? (*Hint:* Let $N = 5N_0$.)

The radioactive element strontium 90 has a half-life of approximately 28 years. That is, in a 28-year period, one-half of the initial amount will have decayed into another substance. If A_0 is the initial amount of the element, then the amount A remaining after t years is given by

$$A(t) = A_0\left(\frac{1}{2}\right)^{t/28}$$

Use the formula to solve Exercises 47 to 50.

47. If the initial amount of the element is 100 g, in how many years will 60 g remain?

48. If the initial amount of the element is 100 g, in how many years will 20 g remain?

49. In how many years will 75% of the original amount remain? (*Hint:* Let $A = 0.75A_0$.)

50. In how many years will 10% of the original amount remain? (*Hint:* Let $A = 0.1A_0$.)

Given projected growth, t years from now a city's population P can be approximated by $P(t) = 25{,}000e^{0.045t}$. Use the formula to solve Exercises 51 and 52.

51. How long will it take the city's population to reach 35,000?

52. How long will it take the population to double?

The number of bacteria in a culture after t hours can be given by $N(t) = N_0e^{0.03t}$, where N_0 is the initial number of bacteria in the culture. Use the formula to solve Exercises 53 and 54.

53. In how many hours will the size of the culture double?

54. In how many hours will the culture grow to 4 times its original population?

The atmospheric pressure P, in inches of mercury (in. Hg), at an altitude h feet above sea level is approximated by $P(t) = 30e^{-0.000004h}$. Use the formula to solve Exercises 55 and 56.

55. Find the altitude if the pressure at that altitude is 25 in. Hg.

56. Find the altitude if the pressure at that altitude is 20 in. Hg.

Carbon-14 dating is used to measure the age of specimens and is based on the radioactive decay of the element carbon 14. This decay begins once a plant or animal dies. If A_0 is the initial amount of carbon 14, then the amount remaining after t years is $A(t) = A_0 e^{-0.000124t}$. Use the formula to solve Exercises 57 and 58.

57. Estimate the age of a specimen if 70% of the original amount of carbon 14 remains.

58. Estimate the age of a specimen if 20% of the original amount of carbon 14 remains.

59. In some of the earlier exercises, we talked about bacteria cultures that double in size every few minutes. Can this go on forever? Explain.

60. The population of the United States has been doubling every 45 years. Is it reasonable to assume that this rate will continue? What factors will start to limit that growth?

In Exercises 61 to 64, use your calculator to describe the graph of each equation, then explain the result.

61. $y = \log 10^x$

62. $y = 10^{\log x}$

63. $y = \ln e^x$

64. $y = e^{\ln x}$

65. In this section, we solved the equation $3^x = 2^{x+1}$ by first applying the logarithmic function, base 10, to each side of the equation. Try this again, but this time apply the natural logarithm function to each side of the equation. Compare the solutions that result from the two approaches.

Summary for Chapter 13

Example	Topic	Reference
	Inverse Relations and Functions	13.1
The inverse of the relation $$\{(1, 2), (2, 3), (4, 3)\}$$ is $$\{(2, 1), (3, 2), (3, 4)\}$$ To find the inverse of $$f(x) = 4x - 8$$ $$y = 4x - 8$$ change y to x and x to y $$x = 4y - 8$$ so $$4y = x + 8$$ $$y = \frac{1}{4}(x + 8)$$ $$y = \frac{1}{4}x + 2$$ $$f^{-1}(x) = \frac{1}{4}x + 2$$	The **inverse** of a relation is formed by interchanging the components of each ordered pair in the given relation. If a relation (or function) is specified by an equation, interchange the roles of x and y in the defining equation to form the inverse.	p. 959
If $f(x) = 4x - 8$, then $$f^{-1}(x) = \frac{1}{4}x + 2$$ *Not* one-to-one	The inverse of a function f may or may not be a function. If the inverse *is* also a function, we denote that inverse as f^{-1}, read "the inverse of f." A function f has an inverse f^{-1}, which is also a function, if and only if f is a **one-to-one** function. That is, no two ordered pairs in the function have the same second component. A function f is a one-to-one function if and only if it has an inverse f^{-1}, which is also a function. The **horizontal line test** can be used to determine whether a function is one-to-one.	p. 964 p. 965

Finding Inverse Relations and Functions

p. 961

1. Interchange the x and y components of the ordered pairs of the given relation or the roles of x and y in the defining equation.

2. If the relation was described in equation form, solve the defining equation of the inverse for y.

3. If desired, graph the relation and its inverse on the same set of axes. The two graphs will be symmetric about the line $y = x$.

Exponential Functions

13.2

An **exponential function** is any function defined by an equation of the form

p. 973

$$y = f(x) = b^x \qquad b > 0, b \neq 1$$

If b is greater than 1, the function is always increasing (a **growth function**). If b is less than 1, the function is always decreasing (a **decay function**).

In both cases, the exponential function is one-to-one. The domain is the set of all real numbers, and the range is the set of positive real numbers.

The function defined by $f(x) = e^x$, in which e is an irrational number (approximately 2.71828), is called *the* exponential function.

Graphing an Exponential Function

p. 975

1. Establish a table of values by considering the function in the form $y = b^x$.

2. Plot points from that table of values and connect them with a smooth curve to form the graph.

3. If $b > 1$, the graph increases from left to right. If $0 < b < 1$, the graph decreases from left to right.

4. All graphs will have the following in common:

    (a) The y intercept will be $(0, 1)$.

    (b) The graphs will approach, but not touch, the x axis.

    (c) The graphs will represent one-to-one functions.

	Logarithmic Functions	**13.3**

	In the expression	**p. 984**
$\log_3 9 = 2$ is in logarithmic form.	$$y = \log_b x$$	
	y is called the *logarithm of x to base b*, when $b > 0$ and $b \neq 1$.	
$3^2 = 9$ is the exponential form.	An expression such as $y = \log_b x$ is said to be in **logarithmic form.**	
$\log_3 9 = 2$ is equivalent to $3^2 = 9$.	An expression such as $x = b^y$ is said to be in **exponential form.**	
2 is the power to which we must raise 3 to get 9.	$y = \log_b x$ means the same as $x = b^y$	
	A logarithm is an exponent or a power. The logarithm of x to base b is the power to which we must raise b to get x.	
	A **logarithmic function** is any function defined by an equation of the form	
	$$f(x) = \log_b x \qquad b > 0, b \neq 1$$	
	The logarithm function is the inverse of the corresponding exponential function. The function is one-to-one with domain $\{x \mid x > 0\}$ and range composed of the set of all real numbers.	

	Properties of Logarithms	**13.4**

	If M, N, and b are positive real numbers with $b \neq 1$ and if p is any real number, then we can state the following properties of logarithms:	**p. 996**
$\log 10 = 1$	**1.** $\log_b b = 1$	
$\log_2 1 = 0$	**2.** $\log_b 1 = 0$	
$3^{\log_3 2} = 2$	**3.** $b^{\log_b x} = x$	
$\log_5 5^x = x$	**4.** $\log_b b^x = x$	

	Product Property	**p. 997**
$\log_3 x + \log_3 y = \log_3 xy$	$$\log_b MN = \log_b M + \log_b N$$	

	Quotient Property	**p. 997**
$\log_5 8 - \log_5 3 = \log_5 \dfrac{8}{3}$	$$\log_b \frac{M}{N} = \log_b M - \log_b N$$	

	Power Property	**p. 997**
$\log 3^2 = 2 \log 3$	$$\log_b M^p = p \log_b M$$	
	Common logarithms are logarithms to base 10. For convenience, we omit the base in writing common logarithms:	**p. 1000**
$\log_{10} 1{,}000 = \log 1{,}000$		
$\qquad = \log 10^3 = 3$	$$\log M = \log_{10} M$$	
	Natural logarithms are logarithms to base e. By custom we also omit the base in writing natural logarithms:	**p. 1003**
$\ln 3 = \log_e 3$		
	$$\ln M = \log_e M$$	

	Logarithmic and Exponential Equations	**13.5**
To solve $\log_2 x = 5$: Write the equation in the equivalent exponential form to solve $\quad x = 2^5 \quad$ or $\quad x = 32$	A **logarithmic equation** is an equation that contains a logarithmic expression. $$\log_2 x = 5$$ is a logarithmic equation.	**p. 1011**
To solve $\log_4 x + \log_4 (x - 6) = 2$ $\qquad \log_4 x(x - 6) = 2$ $\qquad\qquad x(x - 6) = 4^2$ $\qquad\quad x^2 - 6x - 16 = 0$ $\qquad (x - 8)(x + 2) = 0$ $x = 8 \quad$ or $\quad x = -2$ Because substituting -2 for x in the original equation results in the logarithm of a negative number, we reject that answer. The only solution is 8.	**Solving Logarithmic Equations** 1. Use the properties of logarithms to combine terms containing logarithmic expressions into a single term. 2. Write the equation formed in step 1 in exponential form. 3. Solve for the indicated variable. 4. Check your solutions to make sure that possible solutions do not result in the logarithms of negative numbers or zero.	**p. 1014**
To solve $4^x = 64$: Because $64 = 4^3$, write $\quad 4^x = 4^3 \quad$ or $\quad x = 3$	An **exponential equation** is an equation in which the variable appears as an exponent. The algorithm on the next page summarizes the steps in solving any exponential equation.	**p. 1014**

$$2^{x+3} = 5^x$$

$$\log 2^{x+3} = \log 5^x$$

$$(x + 3)\log 2 = x \log 5$$

$$x \log 2 + 3 \log 2 = x \log 5$$

$$x \log 2 - x \log 5 = -3 \log 2$$

$$x (\log 2 - \log 5) = -3 \log 2$$

$$x = \frac{-3 \log 2}{\log 2 - \log 5} \approx 2.269$$

Solving Exponential Equations   p. 1016

1. Try to write each side of the equation as a power of the same base. Then equate the exponents to form an equation.

2. If the above procedure is not applicable, take the common logarithm of both sides of the original equation.

3. Use the power rule for logarithms to write an equivalent equation with the variables as coefficients.

4. Solve the resulting equation.

This summary exercise set is provided to give you practice with each of the objectives in the chapter. Each chapter is keyed to the appropriate chapter section.

[13.1] Find the inverse function f^{-1} for the given function.

1. $f(x) = -2x + 3$ **2.** $f(x) = 4x + 5$ **3.** $f(x) = \dfrac{x - 3}{4}$ **4.** $f(x) = \dfrac{x}{4} - 3$

Find the inverse of each function. In each case, determine whether the inverse is also a function.

5.

x	$f(x)$
-3	1
-1	2
2	3
3	4

6.

x	$f(x)$
-2	5
0	4
3	5
5	4

7. $\{(1, 5), (2, 7), (3, 9)\}$ **8.** $\{(3, 1), (5, 1), (7, 1)\}$

9. $\{(2, 4), (4, 3), (6, 4)\}$ **10.** $\{(-2, 6), (0, 0), (3, 8)\}$

For each function f, find its inverse f^{-1}. Then graph both on the same set of axes.

11. $f(x) = 5x + 3$ **12.** $f(x) = -3x + 9$ **13.** $f(x) = \dfrac{x - 4}{5}$ **14.** $f(x) = \dfrac{3 - x}{2}$

Determine whether the given function is one-to-one. In each case determine if the inverse is a function.

15. $f = \{(-1, 2), (1, 3), (2, 5), (4, 7)\}$ **16.** $g = \{(-3, 2), (0, 2), (1, 2), (3, 2)\}$

17.

x	$f(x)$
-1	2
3	4
4	5
5	6

18.

x	$f(x)$
1	3
2	4
3	5
4	3

19.

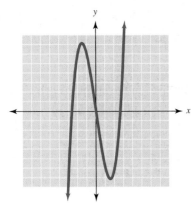

20.

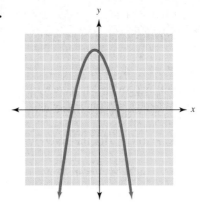

Let $f(x) = 4x + 12$ and $f^{-1}(x) = \dfrac{x}{4} - 3$. Given these two functions, in Exercises 21 to 26, find the following.

21. $f(-2)$ **22.** $f^{-1}(4)$ **23.** $f(f^{-1}(8))$ **24.** $f^{-1}(f(-4))$

25. $f(f^{-1}(x))$ **26.** $f^{-1}(f(x))$

[13.2] Graph the exponential functions defined by each of the following equations.

27. $y = 3^x$ **28.** $y = \left(\dfrac{3}{4}\right)^x$

[13.3] Solve each of the following exponential equations for x.

29. $5^x = 125$ **30.** $2^{2x+1} = 32$ **31.** $3^{x-1} = \dfrac{1}{9}$

[13.3] If it takes 2 h for the population of a certain bacteria culture to double (by dividing in half), then the number N of bacteria in the culture after t hours is given by $N = 1{,}000 \cdot 2^{t/2}$, where the initial population of the culture was 1,000. Using this formula, find the number in the culture.

32. After 4 h   **33.** After 12 h   **34.** After 15 h

[13.3] Graph the logarithmic functions defined by each of the following equations.

35. $y = \log_3 x$ **36.** $y = \log_2 (x - 1)$

[13.3] Convert each of the following statements to logarithmic form.

37. $2^5 = 32$

38. $10^3 = 1,000$

39. $5^0 = 1$

40. $5^{-2} = \dfrac{1}{25}$

41. $25^{1/2} = 5$

42. $16^{3/4} = 8$

[13.3] Convert each of the following statements to exponential form.

43. $\log_4 64 = 3$

44. $\log 100 = 2$

45. $\log_{81} 9 = \dfrac{1}{2}$

46. $\log_5 25 = 2$

47. $\log 0.001 = -3$

48. $\log_{32} \dfrac{1}{2} = -\dfrac{1}{5}$

[13.3] Solve each of the following equations for the unknown variable.

49. $y = \log_5 125$

50. $\log_b \dfrac{1}{9} = -2$

51. $\log_7 x = 2$

52. $y = \log_5 1$

53. $\log_b 3 = \dfrac{1}{2}$

54. $y = \log_9 3$

55. $y = \log_8 2$

[13.3] The decibel (dB) rating for the loudness of a sound is given by

$$L = 10 \log \dfrac{I}{I_0}$$

where I is the intensity of that sound in watts per square centimeter and I_0 is the intensity of the "threshold" sound $I_0 = 10^{-16}$ W/cm^2. Find the decibel rating of each of the given sounds.

56. A table saw in operation with intensity $I = 10^{-6}$ W/cm^2

57. The sound of a passing car horn with intensity $I = 10^{-8}$ W/cm^2

[13.3] The formula for the decibel rating of a sound can be solved for the intensity of the sound as

$$I = I_0 \cdot 10^{L/10}$$

where L is the decibel rating of the given sound.

58. What is the ratio of intensity of a 60-dB sound to one of 50 dB?

59. What is the ratio of intensity of a 60-dB sound to one of 40 dB?

[13.3] The magnitude of an earthquake on the Richter scale is given by

$$M = \log \frac{a}{a_0}$$

where a is the intensity of the shock wave of the given earthquake and a_0 is the intensity of the shock wave of a zero-level earthquake. Use that formula to solve the following.

60. The Alaskan earthquake of 1964 had an intensity of $10^{8.4}a_0$. What was its magnitude on the Richter scale?

61. Find the ratio of intensity of an earthquake of magnitude 7 to an earthquake of magnitude 6.

[13.4] Use the properties of logarithms to expand each of the following expressions.

62. $\log_b x^2 y$

63. $\log_4 \dfrac{y^3}{5}$

64. $\log_5 \dfrac{x^2 y}{z^3}$

65. $\log_5 x^3 y z^2$

66. $\log \dfrac{xy}{\sqrt{z}}$

67. $\log_b \sqrt[3]{\dfrac{x^2 y}{z}}$

[13.4] Use the properties of logarithms to write each of the following expressions as a single logarithm.

68. $\log x + 2 \log y$

69. $3 \log_b x - 2 \log_b z$

70. $\log_b x + \log_b y - \log_b z$

71. $2 \log_5 x - 3 \log_5 y - \log_5 z$

72. $\log x - \dfrac{1}{2} \log y$

73. $\dfrac{1}{3}(\log_b x - 2 \log_b y)$

[13.4] Given that $\log 2 = 0.301$ and $\log 3 = 0.477$, find each of the following logarithms. Verify your results with a calculator.

74. $\log 18$

75. $\log 16$

76. $\log \dfrac{1}{8}$

77. $\log \sqrt{3}$

[13.4] Use your calculator to find the pH of each of the following solutions, given the hydrogen ion concentration $[H^+]$ for each solution, where

$$pH = -\log [H^+]$$

Are the solutions acidic or basic?

78. Coffee: $[H^+] = 5 \times 10^{-6}$

79. Household detergent: $[H^+] = 3.2 \times 10^{-10}$

[13.4] Given the pH of the following solutions, find the hydrogen ion concentration $[H^+]$.

80. Lemonade: pH $= 3.5$

81. Ammonia: pH $= 10.2$

[13.4] The average score on a final examination for a group of chemistry students, retested after time t (in weeks), is given by

$$S(t) = 81 - 6 \ln(t + 1)$$

Find the average score on the retests after the given times.

82. After 5 weeks

83. After 10 weeks

84. After 15 weeks

85. Graph these results.

[13.4] The formula for converting from a logarithm with base a to logarithms with base b is

$$\log_a x = \frac{\log_b x}{\log_b a}$$

Use that formula to find each of the following logarithms.

86. $\log_4 20$

87. $\log_8 60$

[13.5] Solve each of the following logarithmic equations for x.

88. $\log_3 x + \log_3 5 = 3$

89. $\log_5 x - \log_5 10 = 2$

90. $\log_3 x + \log_3 (x + 6) = 3$

91. $\log_5 (x + 3) + \log_5 (x - 1) = 1$

92. $\log x - \log (x - 1) = 1$

93. $\log_2 (x + 3) - \log_2 (x - 1) = \log_2 3$

[13.5] Solve each of the following exponential equations for x. Give your results rounded to three decimal places.

94. $3^x = 243$

95. $5^x = \dfrac{1}{25}$

96. $5^x = 10$

97. $4^{x-1} = 8$

98. $6^x = 2^{2x+1}$

99. $2^{x+1} = 3^{x-1}$

[13.5] If an investment of P dollars earns interest at an annual rate of 12% and the interest is compounded n times per year, then the amount A in the account after t years is

$$A(t) = P\left(1 + \frac{0.12}{n}\right)^{nt}$$

Use that formula to solve each of the following.

100. If \$1,000 is invested and the interest is compounded quarterly, how long will it take the amount in the account to double?

101. If \$3,000 is invested and the interest is compounded monthly, how long will it take the amount in the account to reach \$8,000?

[13.5] A certain radioactive element has a half-life of 50 years. The amount A of the substance remaining after t years is given by

$$A(t) = A_0 \cdot 2^{-t/50}$$

where A_0 is the initial amount of the substance. Use this formula to solve each of the following.

102. If the initial amount of the substance is 100 milligrams (mg), after how long will 40 mg remain?

103. After how long will only 10% of the original amount of the substance remain?

[13.5] A city's population is presently 50,000. Given the projected growth, t years from now the population P will be given by $P(t) = 50{,}000e^{0.08t}$. Use this formula to solve each of the following.

104. How long will it take the population to reach 70,000?

105. How long will it take the population to double?

[13.5] The atmospheric pressure, in inches of mercury, at an altitude h miles above the surface of the earth, is approximated by $P(h) = 30e^{-0.021h}$. Use this formula to solve the following exercises.

106. Find the altitude at the top of Mt. McKinley in Alaska if the pressure is 27.7 in. Hg.

107. Find the altitude outside an airliner in flight if the pressure is 26.1 in. Hg.

The purpose of this self-test is to help you check your progress and to review for a chapter test in class. Allow yourself about 1 hour to take the test. When you are done, check your answers in the back of the book. If you missed any answers, be sure to go back and review the appropriate sections in the chapter and the exercises that are provided.

In Exercises 1 to 4, find the inverse function f^{-1} for the given function.

1. $f(x) = -4x + 5$

2. $f(x) = \dfrac{x - 3}{5}$

3.

x	$f(x)$
-3	-2
1	2
4	5
5	6

4. $\{(-3, 1), (4, 2), (5, -2)\}$

In Exercises 5 and 6, determine whether the given function is one-to-one. In each case, determine if the inverse is a function.

5. $f = \{(3, 4), (5, -1), (6, 2), (7, -1)\}$

6.

x	$f(x)$
1	3
3	8
5	11
7	5

Let $f(x) = 5x - 7$ and $f^{-1}(x) = \dfrac{x + 7}{5}$. Given those two functions, in Exercises 7 to 10, find the following.

7. $f(1)$

8. $f^{-1}(3)$

9. $f(f^{-1}(x))$

10. $f^{-1}(f(x))$

Graph the exponential functions defined by each of the following equations.

11. $y = 4^x$

12. $y = \left(\dfrac{2}{3}\right)^x$

13. Solve each of the following exponential equations for x.

(a) $5^x = \dfrac{1}{25}$

(b) $3^{2x-1} = 81$

14. Graph the logarithmic function defined by the following equation.

$$y = \log_4 x$$

Convert each of the following statements to logarithmic form.

15. $10^4 = 10,000$

16. $27^{2/3} = 9$

Convert each of the following statements to exponential form.

17. $\log_5 125 = 3$ **18.** $\log 0.01 = -2$

Solve each of the following equations for the unknown variable.

19. $y = \log_2 64$ **20.** $\log_b \dfrac{1}{16} = -2$

21. $\log_{25} x = \dfrac{1}{2}$

Use the properties of logarithms to expand each of the following expressions.

22. $\log_b x^2 y z^3$ **23.** $\log_5 \sqrt{\dfrac{xy^2}{z}}$

Use the properties of logarithms to write each of the following expressions as a single logarithm.

24. $\log x + 3 \log y$ **25.** $\dfrac{1}{3} (\log_b x - 2 \log_b z)$

Solve each of the following logarithmic equations for x.

26. $\log_6 (x + 1) + \log_6 (x - 4) = 2$ **27.** $\log (2x + 1) - \log (x - 1) = 1$

Solve each of the following exponential equations for x. Give your results rounded to three decimal places.

28. $3^{x+1} = 4$ **29.** $5^x = 3^{x+1}$

30. $2^{3x-1} = 8^{2x+1}$

Cumulative Test ▪ 0–13

This test is provided to help you in the process of reviewing the previous chapters. Answers are provided in the back of the book. If you missed any answers, be sure to go back and review the appropriate chapter section.

Solve each of the following.

1. $2x - 3(x + 2) = 4(5 - x) + 7$

2. $|3x - 7| > 5$

3. $\log x - \log (x - 1) = 1$

Graph each of the following.

4. $5x - 3y = 15$

5. $-8(2 - x) \geq y$

6. Find an equation of the line that passes through the points $(2, -1)$ and $(-3, 5)$.

7. Solve the linear inequality

$$3x - 2(x - 5) \geq 20$$

Simplify each of the following expressions.

8. $4x^2 - 3x + 8 - 2(x^2 + 5) - 3(x - 1)$

9. $(3x + 1)(2x - 5)$

Factor each of the following completely.

10. $2x^2 - x - 10$

11. $25x^3 - 16xy^2$

Perform the indicated operations.

12. $\dfrac{2}{x-4} - \dfrac{3}{x-5}$

13. $\dfrac{x^2 - x - 6}{x^2 + 2x - 15} \div \dfrac{x-2}{x+5}$

Simplify each of the following radical expressions.

14. $\sqrt{18} + \sqrt{50} - 3\sqrt{32}$

15. $(3\sqrt{2} + 2)(3\sqrt{2} + 2)$

16. $\dfrac{5}{\sqrt{5} - \sqrt{2}}$

17. Find three consecutive odd integers whose sum is 237.

Solve each of the following equations.

18. $x^2 + x - 2 = 0$

19. $2x^2 - 6x - 5 = 0$

20. Solve the following inequality:

$$2x^2 + x - 3 \le 0$$

Determinants and Cramer's Rule

A.1 OBJECTIVES

1. Find the determinant of a matrix
2. Use Cramer's rule to solve a system

Thus far, we have seen four different algebraic methods for solving systems of linear equations. We have solved these systems by substitution, by addition (sometimes called elimination), by graphing, and by using elementary row operations on a matrix that represented the system. In this section, we will examine determinants to see how they are used in solving systems of equations.

Note: The idea of a determinant dates back to Seki Kowa in Japan (1683) and Gottfried Leibniz (one of the inventors of the calculus) in Europe (1693). The name "determinant" came from the French mathematician Augustin Cauchy and was not used until 1812.

Determinant

A **determinant** is an operation on a square matrix of numbers written as shown here. The symbol | | is used to denote the determinant.

$$\begin{array}{c} \text{Column 1} \quad \text{Column 2} \\ \text{Row 1} \longrightarrow \\ \text{Row 2} \longrightarrow \end{array} \begin{vmatrix} a & c \\ b & d \end{vmatrix}$$

Note: In general, a matrix of order n has n rows and n columns.

Note: Notice that a determinant, by this definition, will be a number whenever all the matrix entries are numbers.

The *order* of a square matrix is the number of rows (or columns). The matrix in the definition is said to be of order 2. It is sometimes called a 2 × 2 matrix for convenience. The *value* of the determinant of a second-order matrix is defined as follows:

$$\begin{vmatrix} a & c \\ b & d \end{vmatrix} = ad - bc$$

Example 1 **Evaluating a Determinant**

Find the value of each determinant.

(a) $\begin{vmatrix} 2 & 3 \\ -1 & 4 \end{vmatrix} = 2(4) - (-1)(3)$

$= 8 + 3 = 11$

(b) $\begin{vmatrix} 4 & -3 \\ 5 & 0 \end{vmatrix} = 4(0) - 5(-3)$

$= 15$

The use of determinants to solve linear systems is called **Cramer's rule.** To develop that rule, consider the system of two linear equations in two unknowns:

$$a_1 x + b_1 y = c_1 \tag{1}$$

$$a_2 x + b_2 y = c_2 \tag{2}$$

To solve for x, we multiply both sides of equation (1) by b_2 and both sides of equation (2) by $-b_1$. We then have

$$a_1 b_2 x + b_1 b_2 y = c_1 b_2$$

$$-a_2 b_1 x - b_1 b_2 y = -c_2 b_1$$

Adding these new equations will eliminate the variable y from the system.

$$(a_1 b_2 - a_2 b_1)x = c_1 b_2 - c_2 b_1$$

or

$$x = \frac{c_1 b_2 - c_2 b_1}{a_1 b_2 - a_2 b_1} \tag{3}$$

We can solve for y in a similar fashion:

Note: To verify this, return to the original system. Multiply equation (1) by a_2 and equation (2) by $-a_1$. Then add to eliminate the variable x.

$$y = \frac{a_1 c_2 - a_2 c_1}{a_1 b_2 - a_2 b_1} \tag{4}$$

To see how determinants are used, we will rewrite equations (3) and (4) in a determinant form.

$$x = \frac{\begin{vmatrix} c_1 & b_1 \\ c_2 & b_2 \end{vmatrix}}{\begin{vmatrix} a_1 & b_1 \\ a_2 & b_2 \end{vmatrix}} \qquad y = \frac{\begin{vmatrix} a_1 & c_1 \\ a_2 & c_2 \end{vmatrix}}{\begin{vmatrix} a_1 & b_1 \\ a_2 & b_2 \end{vmatrix}}$$

Note: This rule was named for the Swiss mathematician Gabriel Cramer, who invented the idea of determinants independently of Kowa and Leibniz and published the rule for solving linear systems of equations in 1750.

Using the determinant definition, you should verify that these are simply restatements of equations (3) and (4). In determinant form this result is called Cramer's rule.

You do not need to "memorize" the result. It is better if you observe the following patterns. We will look at our original system again.

$$\begin{cases} a_1 x + b_1 y = c_1 \\ a_2 x + b_2 y = c_2 \end{cases}$$

$\begin{vmatrix} a_1 & b_1 \\ a_2 & b_2 \end{vmatrix}$

The determinant of the coefficients

c_1
c_2

The column of constants

Returning to our statement of Cramer's rule, we note that

1. The denominator for each variable is simply the determinant of the coefficients.
2. The numerator for x is formed from the determinant of the coefficients by replacing the x-coefficients with the column of constants.
3. The numerator for y is found by replacing the y-coefficients with the column of constants.

We will denote the determinant of the coefficients as D, the determinant in the numerator for x as D_x, and the determinant in the numerator for y as D_y. The following summarizes our discussion.

Cramer's Rule

To solve the linear system

$$a_1 x + b_1 y = c_1$$
$$a_2 x + b_2 y = c_2$$

write

$$D = \begin{vmatrix} a_1 & b_1 \\ a_2 & b_2 \end{vmatrix} \qquad \text{The determinant of the coefficients.}$$

$$D_x = \begin{vmatrix} c_1 & b_1 \\ c_2 & b_2 \end{vmatrix} \qquad \text{The } x\text{-coefficients have been replaced by the constants.}$$

$$D_y = \begin{vmatrix} a_1 & c_1 \\ a_2 & c_2 \end{vmatrix} \qquad \text{The } y\text{-coefficients have been replaced by the constants.}$$

Then

$$x = \frac{D_x}{D} \qquad \text{and} \qquad y = \frac{D_y}{D} \qquad \text{when } D \neq 0$$

Example 2

Solving a Two-Variable System

Use Cramer's rule to solve the system.

$$2x - 3y = 12$$
$$4x + 3y = 6$$

$$x = \frac{D_x}{D} = \frac{\begin{vmatrix} 12 & -3 \\ 6 & 3 \end{vmatrix}}{\begin{vmatrix} 2 & -3 \\ 4 & 3 \end{vmatrix}} = \frac{54}{18} = 3$$

$$y = \frac{D_y}{D} = \frac{\begin{vmatrix} 2 & 12 \\ 4 & 6 \end{vmatrix}}{\begin{vmatrix} 2 & -3 \\ 4 & 3 \end{vmatrix}} = \frac{-36}{18} = -2$$

Note: You should check the solution by substituting these values for x and y in the original system.

The solution set for the system is $\{(3, -2)\}$.

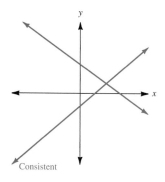

Consistent

From our earlier work with systems we know that there are three possibilities for the solution of a linear system.

1. The system is *consistent* and has a unique solution.

2. The system is *inconsistent* and has no solutions.

3. The system is *dependent* and has an infinite number of solutions.

We will see how these cases are identified when using Cramer's rule.

1. If $D \neq 0$, the system is consistent.

2. If $D = 0$ and either $D_x \neq 0$ or $D_y \neq 0$, the system is inconsistent.

3. If $D = 0$ and both $D_x = 0$ and $D_y = 0$, the system is dependent.

Example 3   Solving a Two-Variable System

Use Cramer's rule to solve the system.

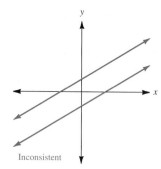

Inconsistent

$$2x - 3y = 4$$
$$-4x + 6y = 3$$

$$D = \begin{vmatrix} 2 & -3 \\ -4 & 6 \end{vmatrix} = 12 - 12 = 0$$

$$D_x = \begin{vmatrix} 4 & -3 \\ 3 & 6 \end{vmatrix} = 24 + 9 = 33$$

$$D_y = \begin{vmatrix} 2 & 4 \\ -4 & 3 \end{vmatrix} = 6 + 16 = 22$$

In applying Cramer's rule, we have

$$x = \frac{33}{0} \quad \text{and} \quad y = \frac{22}{0}$$

Because these values are undefined, no solution for the system exists and the system is inconsistent.

Example 4   Solving a Two-Variable System

Solve, using Cramer's rule.

$$x - 2y = 4$$
$$-2x + 4y = -8$$

Dependent

$$D = \begin{vmatrix} 1 & -2 \\ -2 & 4 \end{vmatrix} = 4 - 4 = 0$$

$$D_x = \begin{vmatrix} 4 & -2 \\ -8 & 4 \end{vmatrix} = 16 - 16 = 0$$

$$D_y = \begin{vmatrix} 1 & 4 \\ -2 & -8 \end{vmatrix} = -8 - (-8) = 0$$

Because $D = D_x = D_y = 0$, the system is dependent and has an infinite number of solutions, in this case all ordered pairs (x, y) satisfying the equation $x - 2y = 4$.

The use of second-order determinants to solve systems of equations is easily extendable to systems involving three unknowns, as we will now see.

To solve systems of three linear equations by determinants, we must first be able to evaluate a third-order determinant. To start, a third-order determinant is written in a similar fashion to that which we saw earlier and can be evaluated as follows:

$$\begin{vmatrix} a_1 & b_1 & c_1 \\ a_2 & b_2 & c_2 \\ a_3 & b_3 & c_3 \end{vmatrix} = a_1b_2c_3 + a_2b_3c_1 + a_3b_1c_2 - a_1b_3c_2 - a_2b_1c_3 - a_3b_2c_1 \qquad (1)$$

Even though this formula can always be used, it is rather awkward and difficult to remember. Fortunately, a more convenient pattern is available that allows us to evaluate a 3×3 determinant in terms of 2×2 determinants. This is called *expansion of the determinant by minors*.

The **minor** of an element in a 3×3 determinant is the 2×2 determinant remaining when the row and column to which the element belongs have been deleted. The diagram here will help you picture the definition.

Note: Delete the first row and the first column.

For a_1:
$$\begin{vmatrix} a_1 & b_1 & c_1 \\ a_2 & b_2 & c_2 \\ a_3 & b_3 & c_3 \end{vmatrix}$$
The minor for a_1 is $\begin{vmatrix} b_2 & c_2 \\ b_3 & c_3 \end{vmatrix}$.

Note: Delete the second row and the first column.

For a_2:
$$\begin{vmatrix} a_1 & b_1 & c_1 \\ a_2 & b_2 & c_2 \\ a_3 & b_3 & c_3 \end{vmatrix}$$
The minor for a_2 is $\begin{vmatrix} b_1 & c_1 \\ b_3 & c_3 \end{vmatrix}$.

Note: Delete the third row and the first column.

For a_3:
$$\begin{vmatrix} a_1 & b_1 & c_1 \\ a_2 & b_2 & c_2 \\ a_3 & b_3 & c_3 \end{vmatrix}$$
The minor for a_3 is $\begin{vmatrix} b_1 & c_1 \\ b_2 & c_2 \end{vmatrix}$.

We can now evaluate a 3×3 determinant by using the three minors just defined. To expand about the first column:

Note: We will come back to this sign pattern later.

1. Write the product of each element with its minor.

2. Connect those products with a $+ - +$ sign pattern.

From these steps we have

$$\begin{vmatrix} a_1 & b_1 & c_1 \\ a_2 & b_2 & c_2 \\ a_3 & b_3 & c_3 \end{vmatrix} = a_1 \begin{vmatrix} b_2 & c_2 \\ b_3 & c_3 \end{vmatrix} - a_2 \begin{vmatrix} b_1 & c_1 \\ b_3 & c_3 \end{vmatrix} + a_3 \begin{vmatrix} b_1 & c_1 \\ b_2 & c_2 \end{vmatrix} \qquad (2)$$

This is called the *expansion* of a 3 × 3 determinant about the first column. Using our earlier definition of a 2 × 2 determinant, you can verify that equations (1) and (2) are the same. However, the pattern of equation (2) is much easier to remember and is the most commonly used.

We will look at the evaluation of a 3 × 3 determinant, using this definition.

| Example 5 | **Evaluating a 3 × 3 Determinant** |

Evaluate the determinant by expanding about the first column.

$$\begin{vmatrix} 1 & -3 & 2 \\ -4 & 0 & 3 \\ 2 & -1 & 3 \end{vmatrix}$$

First we write the three minors.

Element	Minor
1	$\begin{vmatrix} 0 & 3 \\ -1 & 3 \end{vmatrix}$
-4	$\begin{vmatrix} -3 & 2 \\ -1 & 3 \end{vmatrix}$
2	$\begin{vmatrix} -3 & 2 \\ 0 & 3 \end{vmatrix}$

Now we write the product of each element and its minor, and connect those products with the + − + sign pattern.

$$+ 1 \begin{vmatrix} 0 & 3 \\ -1 & 3 \end{vmatrix} - (-4) \begin{vmatrix} -3 & 2 \\ -1 & 3 \end{vmatrix} + 2 \begin{vmatrix} -3 & 2 \\ 0 & 3 \end{vmatrix}$$

+ − + pattern

$$= 1(3) + 4(-7) + 2(-9)$$
$$= 3 - 28 - 18$$
$$= -43$$

The value of the determinant is −43.

A 3×3 determinant can actually be expanded in a similar fashion about *any row or column*. But first we need the following.

The *sign array* for a 3×3 determinant is given by

$$\begin{vmatrix} + & - & + \\ - & + & - \\ + & - & + \end{vmatrix}$$

Note: We simply place a + sign in the upper left-hand corner and then alternate signs in a checkerboard pattern.

Now to generalize our earlier approach:

1. Choose *any* row or column for the expansion.

2. Write the product of each element, in the row or column chosen, with its minor.

3. Connect those products with the appropriate sign pattern from the sign array.

How do we pick a row or column for the expansion? The value of the determinant will be the same in any case. However, the computation will be much easier if you can choose a row or column with one or more zeros.

We now will return to Example 5, expanded in a different manner in Example 6.

| Example 6 | **Evaluating a 3 × 3 Determinant** |

Evaluate the determinant by expanding about the second row.

$$\rightarrow \begin{vmatrix} 1 & -3 & 2 \\ -4 & 0 & 3 \\ 2 & -1 & 3 \end{vmatrix}$$

The expansion is now

Note: Notice the − + − from the sign array for the second row.

$$-(-4)\begin{vmatrix} -3 & 2 \\ -1 & 3 \end{vmatrix} + 0\begin{vmatrix} 1 & 2 \\ 2 & 3 \end{vmatrix} - 3\begin{vmatrix} 1 & -3 \\ 2 & -1 \end{vmatrix}$$

$$= 4(-7) + 0(-1) - 3(5)$$

$$= -28 + 0 - 15 = -43$$

Of course, the result is the same as before. Note that in practice, the 0 makes it unnecessary to evaluate (or even write) the second minor. That is why the choice of the second row for the expansion is more efficient in this case.

We are now ready to apply our work with 3×3 determinants to solving linear systems in three unknowns. Given the system

$$a_1 x + b_1 y + c_1 z = d_1$$
$$a_2 x + b_2 y + c_2 z = d_2$$
$$a_3 x + b_3 y + c_3 z = d_3$$

let D be the determinant of the coefficients.

$$D = \begin{vmatrix} a_1 & b_1 & c_1 \\ a_2 & b_2 & c_2 \\ a_3 & b_3 & c_3 \end{vmatrix}$$

We form D_x by replacing the first column of D with the constants d_1, d_2, and d_3. Similarly D_y is formed by replacing the second column with the constants and D_z by replacing the third column with the constants. Then

$$D_x = \begin{vmatrix} d_1 & b_1 & c_1 \\ d_2 & b_2 & c_2 \\ d_3 & b_3 & c_3 \end{vmatrix} \qquad D_y = \begin{vmatrix} a_1 & d_1 & c_1 \\ a_2 & d_2 & c_2 \\ a_3 & d_3 & c_3 \end{vmatrix} \quad \text{and} \quad D_z = \begin{vmatrix} a_1 & b_1 & d_1 \\ a_2 & b_2 & d_2 \\ a_3 & b_3 & d_3 \end{vmatrix}$$

With these definitions, we can now extend Cramer's rule to the case of a system of three linear equations in three unknowns.

Cramer's Rule

Given the system

$$a_1 x + b_1 y + c_1 z = d_1$$
$$a_2 x + b_2 y + c_2 z = d_2$$
$$a_3 x + b_3 y + c_3 z = d_3$$

Note: If $D = 0$, the system will have either an infinite number of solutions or no solutions.

If $D \neq 0$, the solution for the system is given by

$$x = \frac{D_x}{D} \qquad y = \frac{D_y}{D} \quad \text{and} \quad z = \frac{D_z}{D}$$

Example 7 — Solving a Three-Variable System

Use Cramer's rule to solve the system.

$$x - 2y - 3z = -8$$
$$2x + 3y + 2z = 9$$
$$-3x \qquad + 2z = 2$$

We first calculate D, D_x, D_y, and D_z.

$$D = \begin{vmatrix} 1 & -2 & -3 \\ 2 & 3 & 2 \\ -3 & 0 & 2 \end{vmatrix}$$

Note: We expand about row 3. Do you see why?

$$= -3 \begin{vmatrix} -2 & -3 \\ 3 & 2 \end{vmatrix} + 2 \begin{vmatrix} 1 & -2 \\ 2 & 3 \end{vmatrix}$$

$$= (-3)(5) + 2(7)$$

$$= -15 + 14 = -1$$

$$D_x = \begin{vmatrix} -8 & -2 & -3 \\ 9 & 3 & 2 \\ 2 & 0 & 2 \end{vmatrix}$$

Note: Again we expand about row 3.

$$= 2\begin{vmatrix} -2 & -3 \\ 3 & 2 \end{vmatrix} + 2\begin{vmatrix} -8 & -2 \\ 9 & 3 \end{vmatrix}$$

$$= 2(5) + 2(-6)$$

$$= 10 - 12 = -2$$

$$D_y = \begin{vmatrix} 1 & -8 & -3 \\ 2 & 9 & 2 \\ -3 & 2 & 2 \end{vmatrix}$$

Note: Here we expand about column 1.

$$= 1\begin{vmatrix} 9 & 2 \\ 2 & 2 \end{vmatrix} - 2\begin{vmatrix} -8 & -3 \\ 2 & 2 \end{vmatrix} + (-3)\begin{vmatrix} -8 & -3 \\ 9 & 2 \end{vmatrix}$$

$$= 1(14) - 2(-10) - 3(11)$$

$$= 14 + 20 - 33 = 1$$

$$D_z = \begin{vmatrix} 1 & -2 & -8 \\ 2 & 3 & 9 \\ -3 & 0 & 2 \end{vmatrix}$$

Note: We expand about column 2.

$$= -(-2)\begin{vmatrix} 2 & 9 \\ -3 & 2 \end{vmatrix} + 3\begin{vmatrix} 1 & -8 \\ -3 & 2 \end{vmatrix}$$

$$= 2(31) + 3(-22)$$

$$= 62 - 66 = -4$$

We can now find x, y, and z.

$$x = \frac{D_x}{D} = \frac{-2}{-1} = 2$$

$$y = \frac{D_y}{D} = \frac{1}{-1} = -1$$

Note: The solution should be verified by substitution as before.

$$z = \frac{D_z}{D} = \frac{-4}{-1} = 4$$

The solution set for the system is $\{(2, -1, 4)\}$.

Note: The technique shown in this section for evaluating a 3×3 determinant can be used to evaluate a determinant of any order. Also, there are rules for simplifying determinants *before* the expansion that are particularly helpful when working with higher-order determinants.

Answers to Odd Exercises, Self-Tests, and Cumulative Tests

Chapter 0

Section 0.1

1. $\dfrac{6}{14}, \dfrac{9}{21}, \dfrac{12}{28}$ **3.** $\dfrac{8}{18}, \dfrac{16}{36}, \dfrac{40}{90}$ **5.** $\dfrac{10}{12}, \dfrac{15}{18}, \dfrac{50}{60}$

7. $\dfrac{20}{34}, \dfrac{30}{51}, \dfrac{100}{170}$ **9.** $\dfrac{18}{32}, \dfrac{27}{48}, \dfrac{90}{160}$ **11.** $\dfrac{14}{18}, \dfrac{35}{45}, \dfrac{140}{180}$

13. $\dfrac{2}{3}$ **15.** $\dfrac{5}{7}$ **17.** $\dfrac{2}{3}$ **19.** $\dfrac{7}{8}$ **21.** $\dfrac{1}{4}$

23. $\dfrac{1}{3}$ **25.** $\dfrac{8}{9}$ **27.** $\dfrac{4}{5}$ **29.** $\dfrac{5}{7}$ **31.** $\dfrac{4}{5}$

33. $\dfrac{7}{9}$ **35.** $\dfrac{12}{35}$ **37.** $\dfrac{21}{20}$ **39.** $\dfrac{3}{7}$ **41.** $\dfrac{8}{39}$

43. $\dfrac{7}{33}$ **45.** $\dfrac{1}{9}$ **47.** $\dfrac{5}{21}$ **49.** $\dfrac{8}{15}$ **51.** $\dfrac{2}{3}$

53. $\dfrac{63}{50}$ **55.** $\dfrac{4}{3}$ **57.** $\dfrac{5}{9}$ **59.** 150   **61.** 240

63. 60   **65.** 120   **67.** 50   **69.** $\dfrac{13}{20}$ **71.** $\dfrac{13}{15}$

73. $\dfrac{19}{24}$ **75.** $\dfrac{23}{45}$ **77.** $\dfrac{107}{90}$ **79.** $\dfrac{11}{30}$ **81.** $\dfrac{5}{9}$

83. $\dfrac{4}{7}$ **85.** $\dfrac{5}{24}$ **87.** $\dfrac{7}{18}$ **89.** $\dfrac{1}{33}$ **91.** $\dfrac{23}{252}$

93. $\dfrac{7}{6}$ cups   **95.** $\dfrac{23}{12}$ mi   **97.** $\dfrac{1}{16}$ lb

Section 0.2

1. True   **3.** True   **5.** True   **7.** False   **9.** False
11. False   **13.** True   **15.** True   **17.** False
19. True   **21.** True   **23.** False   **25.** 10   **27.** 20
29. 7   **31.** -30 **33.** 6   **35.** 50   **37.** 3
39. -7 **41.** $>$ **43.** $<$ **45.** $=$ **47.** $<$
49. $-3, 2, 0$ **51.** $0, 2$ **53.** $-2, 0, 1$ **55.** $0, 1$
57.
59. (a) 3; (b) -3; (c) 3; (e) -7

Section 0.3

1. -11 **3.** 4   **5.** -2 **7.** 16   **9.** -6
11. -15 **13.** -8 **15.** 1   **17.** 0   **19.** -6
21. -3 **23.** 0   **25.** $\dfrac{9}{4}$ **27.** $\dfrac{3}{2}$ **29.** $\dfrac{23}{8}$
31. -8 **33.** -12 **35.** -7 **37.** -2 **39.** 6
41. -11 **43.** -20 **45.** 1   **47.** 0   **49.** 50
51. 10   **53.** -10 **55.** 35   **57.** 63   **59.** $\dfrac{17}{2}$
61. $\dfrac{3}{2}$ **63.** -8 **65.** -11 **67.** -24.883
69. -11.494 **71.** -5.702 **73.** 10.947   **75.** 27°F
77. 28th floor   **79.** 925 ft   **81.** 98.6°   **83.** 24 pts
85.   **87.**   **89.**

Section 0.4

1. 56   **3.** -12 **5.** -72 **7.** $-\dfrac{5}{3}$ **9.** 0
11. 64   **13.** $-\dfrac{20}{7}$ **15.** 108   **17.** -20

19. -7.8 **21.** 150   **23.** $\dfrac{1}{4}$ **25.** $-\dfrac{2}{9}$ **27.** $\dfrac{1}{6}$
29. 120   **31.** -135 **33.** -150 **35.** 144
37. 15   **39.** -48 **41.** 32   **43.** 36   **45.** -5
47. 6   **49.** -10 **51.** 8   **53.** -15 **55.** -18
57. 0   **59.** 12   **61.** Undefined   **63.** -25
65. 5   **67.** $-\dfrac{20}{9}$ **69.** $-\dfrac{5}{4}$ **71.** -5
73. -2.349 **75.** 3.464   **77.** -0.668 **79.** 14 weeks
81. $1.98   **83.** -2.4°F   **85.** -2 **87.** -2
89. 4   **91.**   **93.**

Section 0.5

1. 3^5 **3.** 7^5 **5.** 8^6 **7.** $(-2)^3$ **9.** 9
11. 16   **13.** -512 **15.** -512 **17.** -25
19. 16   **21.** 32   **23.** 1,000   **25.** 1,000,000
27. 128   **29.** 48   **31.** 9   **33.** 1,296   **35.** 19
37. 54   **39.** -14 **41.** 1   **43.** 60   **45.** 41
47. 75   **49.** 225   **51.** 34   **53.** 21   **55.** -4
57. 40   **59.** 245   **61.** 361   **63.** 80   **65.** -72
67. -25 **69.** 96   **71.** -2 **73.** -9 **75.** 6
77. 1.2   **79.** 7.8   **81.** 2^5 **83.** $36 \div (4 + 2) - 4$

Summary Exercises—Chapter 0

1. $\dfrac{10}{14}, \dfrac{20}{28}, \dfrac{50}{70}$ **3.** $\dfrac{8}{18}, \dfrac{16}{36}, \dfrac{24}{54}$ **5.** $\dfrac{1}{9}$ **7.** $\dfrac{2}{3}$

9. $\dfrac{3}{2}$ **11.** $\dfrac{7}{18}$ **13.** 12   **15.** -3 **17.** -4

19. 16   **21.** $<$ **23.** $<$ **25.** 8   **27.** -11

29. $-\dfrac{31}{39}$ **31.** 4   **33.** -5 **35.** 5   **37.** $\dfrac{13}{2}$

39. 1   **41.** 0   **43.** 36   **45.** -15 **47.** 16

49. -360 **51.** -4 **53.** -24 **55.** -4

57. Undefined   **59.** $\dfrac{7}{6}$ **61.** -2 **63.** $3 \cdot 3 \cdot 3$

65. $2 \cdot 2 \cdot 2 \cdot 2 \cdot 2 \cdot 2$ **67.** 12   **69.** 48   **71.** 41
73. 20   **75.** 324   **77.** 11   **79.** $198
81. 48 lots

Self-Test—Chapter 0

1. $\dfrac{3}{11}$ **2.** $\dfrac{25}{16}$ **3.** $\dfrac{29}{30}$ **4.** $\dfrac{21}{80}$ **5.** $\dfrac{9}{44}$

6. $\dfrac{14}{5}$ **7.** $<$ **8.** $>$ **9.** -3 **10.** -12

11. -12 **12.** 11   **13.** -35 **14.** 54   **15.** -7
16. 4   **17.** 20   **18.** 115   **19.** 11   **20.** 113

Chapter 1

Section 1.1

1. $c + d$ **3.** $w + z$ **5.** $x + 5$ **7.** $y + 10$
9. $a - b$ **11.** $7 - b$ **13.** $r - 6$ **15.** wz
17. $5t$ **19.** $8mn$ **21.** $8(m + n)$ **23.** $2(x + y)$
25. $2x + y$ **27.** $2(x - y)$ **29.** $(a + b)(a - b)$
31. $m(m - 3)$ **33.** $\dfrac{5}{x}$ **35.** $\dfrac{a + b}{7}$ **37.** $\dfrac{p - q}{4}$

39. $\dfrac{a+3}{a-3}$ **41.** $x+5$ **43.** $x-7$ **45.** $9x$

47. $3x+6$ **49.** $2(x+5)$ **51.** $(x+2)(x-2)$

53. $\dfrac{x}{7}$ **55.** $\dfrac{x+5}{8}$ **57.** $\dfrac{x+6}{x-6}$ **59.** $4s$

61. $\pi r^2 h$ **63.** $\dfrac{1}{2}h(b_1+b_2)$ **65.** Expression

67. Not an expression   **69.** Not an expression
71. Expression   **73.** $35-x$ **75.** Prt **77.** $400-x$
79. $\$375-x$

Section 1.2

1. -22 **3.** 2 **5.** -20 **7.** 12 **9.** 4
11. 83 **13.** -20 **15.** -40 **17.** 14 **19.** -9
21. 2 **23.** 2 **25.** -11 **27.** 1 **29.** 11
31. -56 **33.** -8 **35.** 91 **37.** 29 **39.** 9
41. 16 **43.** 2 **45.** -16 **47.** -72 **49.** -15.3
51. -1.1 **53.** 1.1 **55.** 14.0 **57.** -90
59. $-77{,}922$ **61.** True   **63.** False   **65.** $5.625\ \Omega$
67. 30 in.   **69.** $\$1{,}875$ **71.** $14°F$ **73.** $\$19.09$
75.   **77.** (a) 0, 80, 96, 72, 32, 0; (b) 2;
(c) 94.5, 95.744, 96.492, 96.768, 96.596, 96, 95.004, 93.632, 91.908, 89.856, 87.5; (d) 1.8; (e) 1.81

79.

Section 1.3

1. $5a$, 2 **3.** $5x^4$ **5.** $3x^2$, $3x$, -7 **7.** $5ab$, $4ab$
9. $2x^2y$, $-3x^2y$, $6x^2y$ **11.** $15p$ **13.** $17b^3$ **15.** $28xyz$
17. $6z^2$ **19.** 0 **21.** $-p^2q$ **23.** $-15p^2q$
25. $6x^2$ **27.** $-c+3d$ **29.** $-3x-y$
31. $4a-10b-1$ **33.** $-2a-3b$ **35.** $5a-2b+3c$
37. $-2x-4y$ **39.** $8p-2q$ **41.** $9a+4$
43. $-3p^2+4p$ **45.** $-2x^2$ **47.** $5x^2-2x+1$
49. $2b^2+5b+16$ **51.** $8y^3-2y$ **53.** $-3x^3-2x^2$
55. $-2x^2-x+3$ **57.** $2x-7$ **59.** m^2-3m
61. $-2y^2$ **63.** $2x^2-x+1$ **65.** $8a^2-12a-7$
67. $-4p^2+2p$ **69.** $2x^2+12$ **71.** $6b-1$
73. $-5x+4$ **75.** $2x^2+5x-12$ **77.** $-6y^2-8y$
79. $5x^2-3x+9$ **81.** 206.8 **83.** 6.5 **85.** $28x+4$
87. $-x^2+65x-150$ **89.** $(2y^2+3y+6)$ m
91. $(2x^2-x+4)$ ft   **93.** $(22y+29)$ cm
95.   **97.**   **99.**

101. 1, 3, 5, 7

Section 1.4

1. {Monday, Tuesday, Wednesday, Thursday, Friday, Saturday, Sunday}   **3.** {1, 2, 3, 6, 9, 18}
5. {2, 3, 5, 7, 11, 13, 17, 19, 23, 29}
7. {$-5, -4, -3, -2, -1$}   **9.** {0, 2, 4, 6, 8, 10, 12}
11. {3, 4, 5, 6}   **13.** {$-3, -2$}
15. {$-5, -4, -3, -2, -1, 0, 1, 2$}   **17.** {1, 3, 5, 7, ...}
19. {0, 2, 4, 6, ..., 96, 98}   **21.** {5, 10, 15, 20, ...}
23. $\{x\mid x>10\}$; $(10, \infty)$ **25.** $\{x\mid x\ge -5\}$; $[-5, \infty)$
27. $\{x\mid 2\le x\le 7\}$; $[2, 7]$ **29.** $\{x\mid -4\le x\le 4\}$; $[-4, 4]$
31.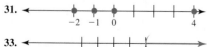

33.

35.

37.

39.

41.

43.

45.

47. $\{x\mid x\le 1\}$; $(-\infty, 1]$ **49.** $\{x\mid x>-2\}$; $(-2, \infty)$
51. $\{x\mid -2\le x\le 2\}$; $[-2, 2]$ **53.** $\{x\mid -3<x<2\}$; $(-3, 2)$
55. $\{x\mid -2\le x<4\}$; $[-2, 4)$ **57.** $\{x\mid -2<x\le 4\}$; $(-2, 4]$
59. $\{x\mid -2\le x\le 4\}$; $[-2, 4]$ **61.** {1, 2, 3, 4, 5, 6, 7, 8, 9}
63. $\varnothing$ **65.** {2, 4, 6, 8}   **67.** {1, 3, 5}   **69.** {1, 3, 5}

Summary Exercises—Chapter 1

1. $y+8$ **3.** $8a$ **5.** $5mn$ **7.** $17x+3$
9. $(x+6)(x-6)$ **11.** $\dfrac{x-4}{x+2}$ **13.** -7 **15.** 15
17. 19 **19.** 25 **21.** 1 **23.** -3 **25.** -48
27. -20 **29.** $4a^3$, $-3a^2$ **31.** $5m^2$, $-4m^2$, m^2
33. $16x$ **35.** $2a$ **37.** $3xy$ **39.** $19a+b$
41. $3x^3+9x^2$ **43.** $10a^3$ **45.** $(37-x)$ ft   **47.** $x+5$
49. $(x+4)$ m   **51.** x, $25-x$ **53.** $(2x+8)$ m
55. {1, 3}   **57.** {$-1, 0, 1, 2, 3$}   **59.** {1, 3}
61. $\{x\mid x>9\}$; $(9, \infty)$ **63.** $\{x\mid x\le -5\}$; $(-\infty, -5]$

65.

67.

69. $\{x\mid x\le 3\}$; $(-\infty, 3]$ **71.** $\{x\mid -3\le x<2\}$; $[-3, 2)$
73. $\{x\mid -2<x<4\}$; $(-2, 4)$ **75.** {1, 2, 5, 7, 9, 11, 15}
77. {2, 5, 9, 11, 15}

Self-Test—Chapter 1

1. $x+y$ **2.** $m-n$ **3.** $a\cdot b$ **4.** $\dfrac{p}{q-3}$
5. $a+7$ **6.** $c-5$ **7.** $3(2x-3y)$ **8.** $3(m-n)$
9. 4 **10.** 36 **11.** -4 **12.** 144 **13.** 5
14. $3a-b$ **15.** $2x^2+8$
16.

17. (a) $\{x\mid -3<x\le 2\}$; (b) $(-3, 2]$ **18.** $85-8\cdot 7=\$29$
19. $\dfrac{184-160}{12}=\dfrac{24}{12}=2$ lb/wk   **20.** (a) {1, 2, 3, 5, 7}; (b) {5}

Chapter 2

Section 2.1

1. Yes   **3.** No   **5.** No   **7.** No   **9.** No
11. Yes   **13.** Yes   **15.** Yes   **17.** Yes   **19.** No

21. Yes   **23.** Linear equation   **25.** Expression
27. Linear equation   **29.** {2}   **31.** {11}   **33.** {−2}
35. {6}   **37.** {4}   **39.** {−10}   **41.** {−3}
43. {2}   **45.** {4}   **47.** {9}   **49.** {6}   **51.** {6}
53. {−18}   **55.** {16}   **57.** {8}   **59.** {2}
61. {9}   **63.** {12}   **65.** $x + 3 = 7$
67. $3x − 7 = 2x$ **69.** $2(x + 5) = x + 18$ **71.** (c)
73. (a)   **75.** True   **77.** 26; $x + 7 = 33$
79. 22; $x − 15 = 7$ **81.** 1,420; $1,840 + x = 3,260$
83. $290; $x + 360 = 650$ **85.** $460; $x + 215 = 675$
87.   **89.**   **91.**

Section 2.2

1. {4}   **3.** {6}   **5.** {7}   **7.** {−4}   **9.** {−8}
11. {−9}   **13.** {3}   **15.** {−7}   **17.** {9}
19. {8}   **21.** {15}   **23.** {42}   **25.** {−20}
27. {−24}   **29.** {9}   **31.** {−20}   **33.** {−25}
35. {4}   **37.** {36}   **39.** {−1}   **41.** {4}
43. {−3}   **45.** {3}   **47.** {4}   **49.** $6x = 72$

51. $\dfrac{x}{7} = 6$ **53.** $\dfrac{1}{3}x = 8$ **55.** $\dfrac{3}{4}x = 18$ **57.** $\dfrac{2x}{5} = 12$

59. 116   **61.** 26 min   **63.** 3   **65.** 35 in.

Section 2.3

1. {4}   **3.** {3}   **5.** {−2}   **7.** {3}   **9.** {8}
11. {32}   **13.** {3}   **15.** {6}   **17.** {−2}   **19.** {2}

21. {6}   **23.** {5}   **25.** $\left\{-\dfrac{20}{3}\right\}$ **27.** {5}   **29.** {4}

31. {5}   **33.** $\left\{\dfrac{5}{2}\right\}$ **35.** {5}   **37.** $\left\{-\dfrac{4}{3}\right\}$

39. {4}   **41.** {7}   **43.** {30}   **45.** {7}

47. {15}   **49.** {3}   **51.** $\left\{\dfrac{2}{9}\right\}$ **53.** Conditional; {6}

55. Contradiction; { }   **57.** Identity; $\mathbb{R}$
59. Contradiction; { }   **61.** Identity; $\mathbb{R}$ **63.** $2x + 3 = 7$

65. $4x − 7 = 41$ **67.** $\dfrac{2}{3}x + 5 = 21$ **69.** $3x = x + 12$

71. 4   **73.** 18   **75.** 35, 36   **77.** 29, 30, 31
79. 32, 34   **81.** 25, 27   **83.** 12, 13   **85.** 1,550, 1,710
87. Washer: $520; Dryer: $430   **89.** 18, 9

91. $12,000 at $3\dfrac{1}{4}$%; $9,000 at $2\dfrac{1}{2}$%   **93.** $5,000

95.

97. A value for which the original equation is true.
99. Multiplying by 0 would always give $0 = 0$.

101. (a) $-\dfrac{3}{2}$; (b) $-\dfrac{7}{4}$; (c) $\dfrac{1}{6}$; (d) $\dfrac{2}{5}$; (e) $\dfrac{8}{3}$; (f) $-\dfrac{9}{5}$; (g) $-\dfrac{b}{a}$

Section 2.4

1. $s = \dfrac{P}{4}$ **3.** $R = \dfrac{E}{I}$ **5.** $H = \dfrac{V}{LW}$

7. $B = 180 − A − C$ **9.** $x = -\dfrac{b}{a}$ **11.** $g = \dfrac{2s}{t^2}$

13. $y = \dfrac{15 − x}{5}$ or $y = -\dfrac{1}{5}x + 3$

15. $L = \dfrac{P − 2W}{2}$ or $L = \dfrac{P}{2} − W$

17. $T = \dfrac{PV}{K}$ **19.** $b = 2x − a$

21. $C = \dfrac{5}{9}(F − 32)$ or $C = \dfrac{5(F − 32)}{9}$

23. $h = \dfrac{S − 2\pi r^2}{2\pi r}$ or $h = \dfrac{S}{2\pi r} − r$ **25.** 3 cm   **27.** 3%

29. 25°C   **31.** $2(x + 6) = 18$ **33.** $3(x − 5) = 21$
35. $2x + 3(x + 1) = 48$ **37.** 10, 18   **39.** 8, 15
41. 5, 6   **43.** 12 in., 25 in.   **45.** 8 m, 29 m
47. Legs, 13 cm; base, 10 cm
49. 150 $14 tickets; 100 $9 tickets
51. 45 37¢ stamps; 35 3¢ stamps
53. 60 coach; 40 berth; 20 sleeping room
55. 40 mi/h; 30 mi/h   **57.** 6 P.M.   **59.** 2 P.M.
61. 3 P.M.   **63.** 20 l   **65.**   **67.**

Section 2.5

1. <   **3.** >   **5.** <   **7.** >   **9.** x is less than 3
11. x is greater than or equal to −4
13. −5 is less than or equal to x

15. (number line: point at 2, open, arrow left; marks 0, 2)

17. (number line: point at 6, open, arrow left; marks 0, 6)

19. (number line: point at 1, open, arrow left; marks 0, 1)

21. (number line: point at 8, open, arrow left; marks 0, 8)

23. (number line: point at −5, open, arrow left; marks −5, 0)

25. (number line: point at 9, closed, arrow right; marks 0, 9)

27. (number line: point at 0, open, arrow left; mark 0)

29. $\{x \mid x \le 11\}$ (number line: closed at 11, arrow left; marks 0, 11)

31. $\{x \mid x \ge 2\}$ (number line: closed at 2, arrow right; marks 0, 2)

33. $\{x \mid x < 7\}$ (number line: open at 7, arrow left; marks 0, 7)

35. $\{x \mid x \le 8\}$ (number line: closed at 8, arrow left; marks 0, 8)

37. $\{x \mid x \ge 8\}$ (number line: closed at 8, arrow right; marks 0, 8)

39. $\{x \mid x < -9\}$ (number line: open at −9, arrow left; marks −9, 0)

41. $\{x \mid x \ge 12\}$ (number line: closed at 12, arrow right; marks 0, 12)

43. $\{x \mid x > -2\}$

45. $\{x \mid x \in \mathbb{R}\}$ **47.** (a)   **49.** (c)   **51.** (b)

Section 2.6

1. $\{x \mid x \le 3\}; (-\infty, 3]$

3. $\{x \mid x > -7\}; (-7, \infty)$

5. $\{x \mid x \le -3\}; (-\infty, -3]$

7. $\{x \mid x > 6\}; (6, \infty)$

9. $\{x \mid x > 20\}; (20, \infty)$

11. $\{x \mid x \le 6\}; (-\infty, 6]$

13. $\{x \mid x < 9\}; (-\infty, 9)$

15. $\{x \mid x > 4\}; (4, \infty)$

17. $\{x \mid x < 1\}; (-\infty, 1)$

19. $\{x \mid x < -1\}; (-\infty, -1)$

21. $\{x \mid x \le -6\}; (-\infty, -6]$

23. $\{x \mid x \ge -12\}; [-12, \infty)$

25. $\{x \mid x \ge -1\}; [-1, \infty)$

27. $\left\{x \mid x < \dfrac{7}{4}\right\}; \left(-\infty, \dfrac{7}{4}\right)$

29. $\{x \mid x < 8\}; (-\infty, 8)$

31. $\left\{x \mid x \ge -\dfrac{5}{2}\right\}; \left[-\dfrac{5}{2}, \infty\right)$

33. $\{x \mid x \in \mathbb{R}\}$ **35.** $\{x \mid x \in \mathbb{R}\}$ **37.** $x + 5 > 3$
39. $2x - 4 \le 7$ **41.** $4x - 15 > x$ **43.** $P < 1{,}000$
45. $x \ge 88$ **47.** Sales $> \$10{,}000$ **49.** 5 courses

51. 50 cm   **53.** \$48
55.   **57.**

Section 2.7

1. 15   **3.** 21   **5.** 5   **7.** 4   **9.** 31   **11.** 8
13. 25   **15.** -25 **17.** $\{5, -5\}$ **19.** $\{-8, 4\}$
21. $\{3.5, -2.5\}$ **23.** $\{1.5, 4.5\}$ **25.** No solution

27. $\{0, 4\}$ **29.** $\{7, 1\}$ **31.** $\left\{-\dfrac{2}{3}, 4\right\}$
33. $\left\{-\dfrac{2}{3}, \dfrac{6}{7}\right\}$ **35.** $\mathbb{R}$ **37.** $\{-14, 4\}$ **39.** $|x| = 5$
41. $|x - 3| = 4$ **43.** $|x + 3| = 4$

Section 2.8

1. $\{x \mid 3 \le x \le 6\}$

3. $\{x \mid -4 < x < 2\}$

5. $\left\{x \mid 2 \le x \le \dfrac{9}{2}\right\}$

7. $\{x \mid -2 < x < 1\}$

9. $\{x \mid x < -2 \text{ or } x > 4\}$

11. $\left\{x \mid x \le -\dfrac{5}{3} \text{ or } x \ge 3\right\}$

13. $\left\{x \mid x < -2 \text{ or } x > \dfrac{8}{3}\right\}$

15. $\{x \mid -5 < x < 5\}$

17. $\{x \mid x \le -7 \text{ or } x \ge 7\}$

19. $\{x \mid x < -2 \text{ or } x > 8\}$

21. $\{x \mid -10 \le x \le -2\}$

23. $\{x \mid x < -2 \text{ or } x > 8\}$

25. No solution
27. $\{x \mid 1 < x < 4\}$

29. $\left\{x \mid x \le -3 \text{ or } x \ge \frac{1}{3}\right\}$

31. $\left\{x \mid x < -\frac{4}{5} \text{ or } x > 2\right\}$

33. $\left\{x \mid -3 \le x \le \frac{13}{3}\right\}$

35. $|x| < 7$ **37.** $|x| \ge 5$ **39.** $|x + 2| < 7$
41. $|x + 4| \ge 3$

Summary Exercises—Chapter 2

1. Yes   **3.** Yes   **5.** No   **7.** {2}   **9.** {−5}
11. {5}   **13.** {1}   **15.** {−7}   **17.** {7}
19. {−4}   **21.** {32}   **23.** {27}   **25.** {3}
27. {−2}   **29.** $\left\{\frac{7}{2}\right\}$ **31.** {18}   **33.** {6}
35. $\left\{\frac{2}{5}\right\}$ **37.** {6}   **39.** {6}   **41.** {4}   **43.** {5}
45. $\left\{\frac{1}{2}\right\}$ **47.** $\left\{\frac{9}{2}\right\}$ **49.** $\left\{\frac{5}{3}\right\}$ **51.** {5}
53. {20}   **55.** {14}   **57.** $L = \frac{P}{2} - W$ **59.** $h = \frac{2A}{b}$
61. $p = n - mq$ **63.** 8   **65.** 17, 19, 21
67. Susan: 7 years; Larry: 9 years; Nathan: 14 years
69. 3 h at 28 mi/h; 2 h at 24 mi/h   **71.** 2 P.M.

73.

75.

77.

79. $\{x \mid x \le 11\}$ **81.** $\{x \mid x > -3\}$ **83.** $\{x \mid x > -3\}$
85. $\left\{x \mid x \ge \frac{1}{2}\right\}$ **87.** $\left\{x \mid x < -\frac{4}{3}\right\}$ **89.** $\left\{x \mid x \ge \frac{3}{2}\right\}$
91. $\mathbb{R}$ **93.** {−3, 2}   **95.** $\left\{-\frac{1}{3}, 5\right\}$ **97.** $\left\{-\frac{5}{2}, 4\right\}$
99. $\left\{-2, \frac{24}{5}\right\}$ **101.** No solution   **103.** {−5, 4}
105. {−1, 3}   **107.** $\{x \mid 1 \le x \le 2\}$
109. $\{x \mid -2 < x \le 3\}$ **111.** $\{x \mid x < -2 \text{ or } x > 6\}$
113. $\{x \mid -3 \le x \le 3\}$ **115.** $\{x \mid x < 3 \text{ or } x > 11\}$
117. $\{x \mid x \le -6 \text{ or } x \ge -1\}$ **119.** $\left\{x \mid -5 < x < \frac{7}{3}\right\}$

Self-Test—Chapter 2

1. No   **2.** Yes   **3.** {12}   **4.** {30}   **5.** $\left\{\frac{1}{3}\right\}$
6. {2}   **7.** $\left\{\frac{35}{4}\right\}$ **8.** {−2, 7}   **9.** $\left\{\frac{3V}{h}\right\}$
10. 7   **11.** 21, 22, 23   **12.** 6, 12, 17
13. 10 in. by 21 in.   **14.** 3:30 P.M.
15. $\{x \mid x \le 14\}$

16. $\{x \mid x < -4\}$

17. $\{x \mid 1 \le x \le 4\}$

18. $\{x \mid x < -5 \text{ or } x > 4\}$

19. $\{x \mid -3 < x < 5\}$

20. $\{x \mid x \le -4 \text{ or } x \ge 3\}$

Cumulative Test—Chapters 0–2

1. $\frac{7}{11}$ **2.** $\frac{6}{5}$ **3.** 2   **4.** 80   **5.** 7   **6.** 7
7. −16   **8.** 4   **9.** 63   **10.** −16   **11.** 0
12. −5   **13.** 9   **14.** $\frac{1}{3}$ **15.** $\frac{19}{12}$ **16.** $\frac{7}{10}$
17. −9   **18.** 13   **19.** 3   **20.** 7   **21.** $15x - 9y$
22. $10x^2 - 13x + 2$ **23.** {4}   **24.** {−2, 8}
25. {71}   **26.** {4}   **27.** $\{x \mid x \le -4\}$
28. $\{x \mid -3 \le x < 3\}$ **29.** $\{x \mid -3 \le x \le 5\}$
30. $\{x \mid x < -9 \text{ or } x > 7\}$ **31.** $r = \frac{I}{Pt}$ **32.** $h = \frac{2A}{b}$
33. $y = \frac{c - ax}{b}$ **34.** $W = \frac{P}{2} - L$ **35.** 13; $4x - 7 = 45$
36. 42, 43; $x + (x + 1) = 85$ **37.** 7; $3x - 12 = x + 2$
38. $420; $x + (x + 120) = 720$
39. 5 cm, 17 cm; $2x + 2(3x + 2) = 44$
40. 8 in., 13 in., 16 in.; $x + (x + 5) + 2x = 37$

Chapter 3

Section 3.1

1. (4, 2), (0, 6), (−3, 9)   **3.** (5, 2), (4, 0), (6, 4)
5. (2, 0), (1, 4)   **7.** (3, 0), (6, 2), (0, −2)
9. (4, 0), $\left(\frac{2}{3}, -5\right)$, $\left(5, \frac{3}{2}\right)$ **11.** (0, 3), $\left(1, \frac{12}{5}\right)$
13. (3, 5), (3, 0), (3, 7)   **15.** 8, 7, 12, 12   **17.** 0, 0, 4, 9
19. 3, −5, 5, 2   **21.** 4, 1, −4, 7   **23.** −3, 11, 9, 7
25. −4, 3, $\frac{4}{3}$, 1   **27.** (0, −10), (10, 0), (5, −5), (12, 2)
29. (0, −6), (3, 0), (6, 6), (9, 12)
31. (8, 0), (−4, 3), (0, 2), (4, 1)
33. (0, −5), (4, 5), (−6, −20), (2, 0)
35. (0, 3), (1, 5), (2, 7), (3, 9)
37. (−5, 0), (−5, 1), (−5, 2), (−5, 3)   **39.** 1   **41.** −6
43. 5   **45.** $9.50, $11.75, $15.50, $19.25, $23
47. 16 cm², 121 cm², 196 cm², 289 cm²   **49.**

Section 3.2

1. (4, 4)   **3.** (2, 0)   **5.** (−4, −5)   **7.** (−6, −6)
9. (−3, 5)

11.

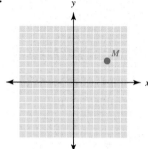

13.

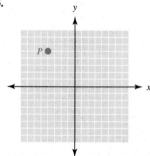

15.

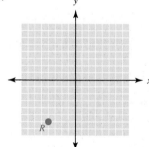

17.

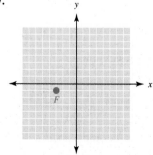

19.

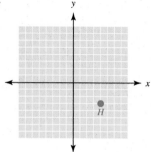

21.

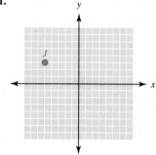

23. I   **25.** III   **27.** *x* axis   **29.** II   **31.** *y* axis
33. IV
35. (1, 30), (2, 45), (3, 60), (4, 60), (5, 75), (6, 90), (7, 95)
37. (7, 100), (15, 70), (20, 80), (30, 70), (40, 50), (50, 40), (60, 30), (70, 40), (80, 25)
39. (a) (7, 2); (b) (0, 1); (c) (1, −6); (d) $\left(\frac{3}{2}, \frac{3}{2}\right)$; (e) $\left(-4, -\frac{15}{2}\right)$; (f) $\left(-12, -\frac{1}{2}\right)$
41. The points lie on a line; (2, −6)
43.

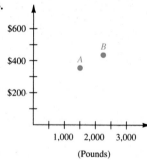

45.

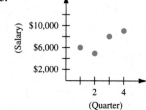

47.

49.

Section 3.3

1.

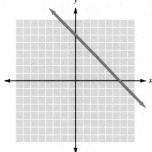

3.

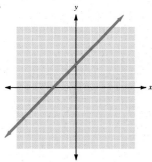

5.

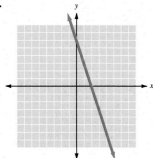

7.

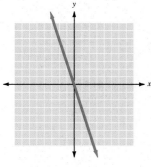

9.

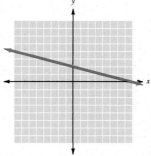

11.

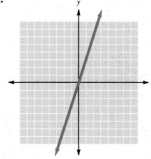

13.

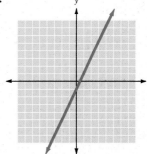

15.

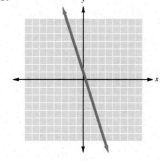

17.

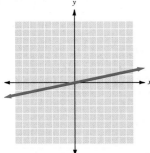

19.

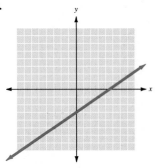

21.

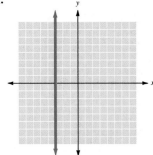

23.

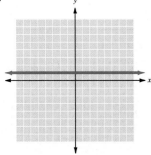

25.

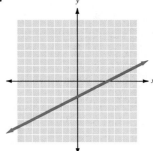

27.

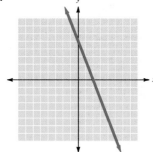

29.

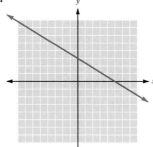

31.

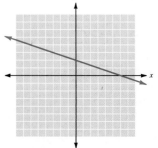

$y = -\dfrac{1}{3}x + 2$

33.

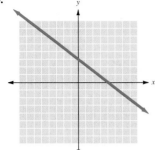

$y = -\dfrac{3}{4}x + 3$

35.

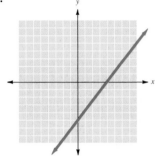

$y = \dfrac{5}{4}x - 5$

37. $y = 2x$ **39.** $y = x + 3$ **41.** $y = 3x - 3$
43. $x - 4y = 12$ **45.** $(3, 1)$

47.

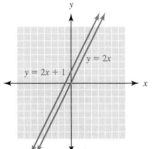

Parallel lines

$y = 2x$
$y = 2x + 1$

49.

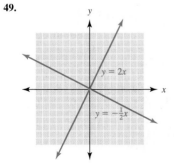

Perpendicular lines

$y = 2x$
$y = -\dfrac{1}{2}x$

51.

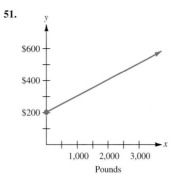

53. (a)

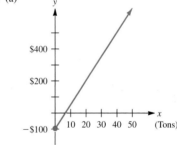

(b) $\dfrac{100}{15}$ or ≈ 7 tons; (c) \$140; (d) $y = 17x - 125$

55.

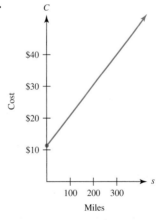

57. (a) $T = 35h + 75$; (b)

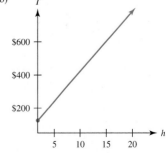

59. (a) 13, 15, 17, 19, 21;

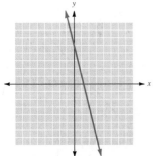

(b) Increase by 2; (c) Yes; (d) Grow by 2 units
61. (b) Increase by 3; (c) Yes; (d) Grow by 3 units
63. (b) Decrease by 4; (c) Yes; (d) Decrease by 4 units

Section 3.4

1. 1   **3.** $\dfrac{4}{5}$ **5.** $\dfrac{4}{5}$ **7.** -2 **9.** $-\dfrac{3}{2}$

11. Undefined   **13.** $\dfrac{5}{7}$ **15.** 0   **17.** $-\dfrac{4}{3}$

19. Slope 3; y intercept: $(0, 5)$
21. Slope -3; y intercept: $(0, -6)$

23. Slope $\dfrac{3}{4}$; y intercept: $(0, 1)$

25. Slope $\dfrac{2}{3}$; y intercept: $(0, 0)$

27. Slope $-\dfrac{4}{3}$; y intercept: $(0, 4)$

29. Slope 0; y intercept: $(0, 9)$

31. Slope $\dfrac{3}{2}$; y intercept: $(0, -4)$

33.

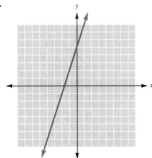

$y = 3x + 5$

35.

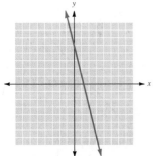

$y = -4x + 5$

37.

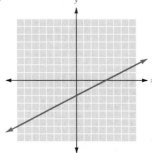

$y = \dfrac{1}{2}x - 2$

39.

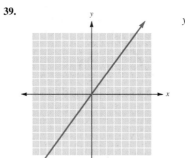

$y = \dfrac{4}{3}x$

41.

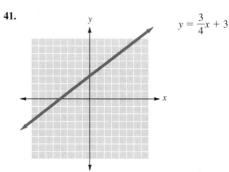

$y = \dfrac{3}{4}x + 3$

43. IV   **45.** III   **47.** I   **49.** III and IV
51. (g)   **53.** (e)   **55.** (h)   **57.** (c)

59.

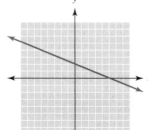

$y = -\dfrac{2}{5}x + 2$

61.

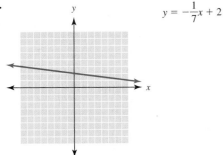

$y = -\dfrac{1}{7}x + 2$

63. 2   **65.** −2   **67.** 3   **69.** $-\dfrac{2}{5}$

71. Slope: 0.10, market price per jug; y intercept: (0, 200) minimum the $200, award
73. Slope represents price of newsprint; y intercept represents cost of the truck
75. −0.30   **77.** 3.10 ¢/yr
79.

81.

83.

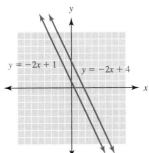

Parallel lines; no

$y = -2x + 1$ $y = -2x + 4$

85.

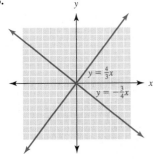

Perpendicular lines; −1

$y = \frac{4}{3}x$

$y = -\frac{3}{4}x$

Section 3.5

1. Parallel   **3.** Neither   **5.** Perpendicular   **7.** $\dfrac{1}{3}$

9. 12   **11.** $y = \dfrac{5}{4}x - 5$ **13.** $y = 5x - 2$

15. $y = -3x - 9$ **17.** $y = \dfrac{2}{5}x - 5$ **19.** $x = 1$

21. $y = -\dfrac{4}{5}x + 4$ **23.** $y = x + 1$ **25.** $y = \dfrac{3}{4}x - \dfrac{3}{2}$

27. $y = 2$ **29.** $y = \dfrac{3}{2}x - 3$ **31.** $y = \dfrac{5}{2}x + 4$

33. $y = 4x - 2$ **35.** $y = -\dfrac{1}{2}x + 2$ **37.** $y = 4$

39. $y = -2x + 7$ **41.** $y = 3x + 3$ **43.** $y = \dfrac{1}{2}x + 4$

45. $y = 3$ **47.** $y = -4x - 11$ **49.** $y = \dfrac{4}{3}x - 2$

51. $y = \dfrac{3}{2}x - \dfrac{11}{2}$ **53.** $y = -\dfrac{1}{2}x$ **55.** Yes, yes

57. Yes, no   **59.** $F = \dfrac{9}{5}C + 32$

61. (a) $C = 65x + 75$; (b) $302.50; (c) 2.65 h
63. Slope 1, y intercept (0, 3)
65. Slope 2, y intercept (0, 1)
67. Slope −3, y intercept (0, 1)
69. Slope −2, y intercept (0, −3)

Summary Exercises—Chapter 3
1. (6, 0), (3, −3), (0, −6)   **3.** (4, 1)   **5.** (−1, −5)
7.–10.

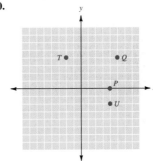

11. I   **13.** III   **15.** x axis

17.

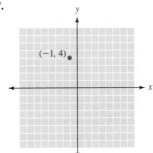

(−1, 4)

19.

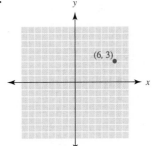

(6, 3)

21.

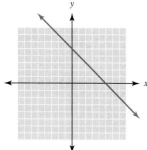

29.

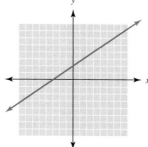

23.

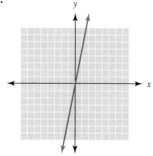

31.

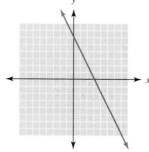

25.

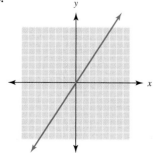

33.

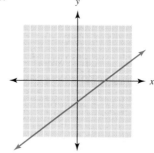

27.

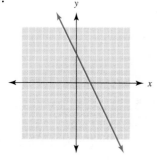

35.

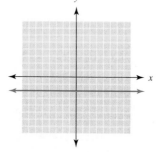

37.

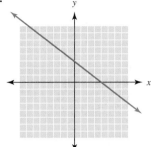

39.

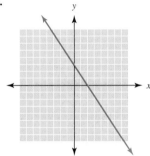

41. 2   **43.** $-\dfrac{1}{2}$ **45.** 0   **47.** $\dfrac{1}{2}$

49. Slope 2, y intercept $(0, 5)$

51. Slope $-\dfrac{3}{4}$, y intercept $(0, 0)$

53. Slope $-\dfrac{2}{3}$, y intercept $(0, 2)$

55. Slope 0, y intercept $(0, -3)$ **57.** $y = 2x + 3$

59. $y = -\dfrac{2}{3}x + 2$ **61.** Perpendicular   **63.** Parallel

65. $y = -3$ **67.** $x = 4$ **69.** $y = -3$

71. $y = -\dfrac{4}{3}x - 2$ **73.** $x = -\dfrac{5}{2}$ **75.** $y = 2x - 1$

77. $y = -\dfrac{5}{4}x + 2$ **79.** $y = -\dfrac{3}{5}x - 1$ **81.** $y = \dfrac{4}{3}x + \dfrac{14}{3}$

Self-Test—Chapter 3

1. $(4, 0), (5, 4)$ **2.** $(3, 0), (0, 4), \left(\dfrac{3}{4}, 3\right)$ **3.** $(3, 6)$

4. $(4, -2)$ **5.** $(0, -7)$

6.–8.

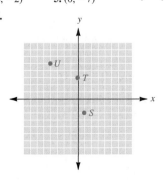

9.

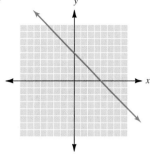

10.

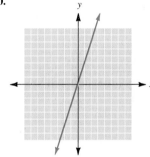

11.

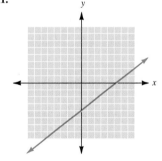

12.

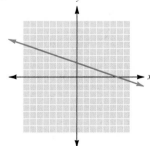

13.

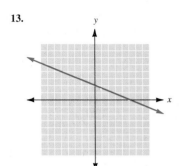

14.

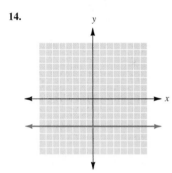

15. 1   **16.** $\dfrac{3}{7}$

17. $y = -3x + 6$

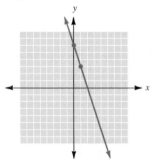

18. $y = \dfrac{2}{5}x - 3$

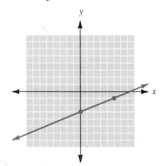

19. Slope -5, y intercept $(0, -9)$

20. Slope $-\dfrac{6}{5}$, y intercept $(0, 6)$

21. Slope 0, y intercept $(0, 5)$
22. $y = 5x - 2$ **23.** $y = -4x - 16$ **24.** $y = 4x + 3$
25. $y = -\dfrac{5}{2}x - 17$

Cumulative Test—Chapters 0–3

1. $-\dfrac{1}{3}$ **2.** $\dfrac{2}{3}$ **3.** 17   **4.** 8   **5.** 1   **6.** $-\dfrac{33}{5}$

7. $6x + 3y$ **8.** $-x - 4$ **9.** $-6x^2 + x + 2$
10. $-8x^2 - 5x + 25$ **11.** $\{-11\}$ **12.** $\{100\}$

13. $\{3, 4\}$ **14.** $\left\{-\dfrac{7}{2}\right\}$ **15.** $\left\{\dfrac{2}{3}, 4\right\}$ **16.** $\left\{-\dfrac{3}{2}\right\}$

17. $C = \dfrac{5}{9}(F - 32)$ **18.** $h = \dfrac{3V}{\pi r^2}$

19. $\{x \mid x < 4\}$

20. $\{x \mid x > -4\}$

21. $\{x \mid 5 \le x \le 9\}$

22. $\{x \mid x < 2 \text{ or } x > 7\}$

23. $\left\{x \mid -\dfrac{13}{2} \le x \le \dfrac{3}{2}\right\}$

24. $\{x \mid x < 1 \text{ or } x > 11\}$

25. -2 **26.** 0
27. Slope -4, y intercept $(0, 9)$
28. Slope $\dfrac{2}{5}$, y intercept $(0, -2)$
29. Slope 0, y intercept $(0, 9)$
30. Slope not defined; y intercept none   **31.** $y = 5x - 6$
32. $x + 10y = 86$ **33.** $y = -\dfrac{2}{3}x + 6$ **34.** $4y + 5x = 26$
35. $y = \dfrac{3}{2}x - 3$ **36.** $y = 3x + 10$ **37.** 9   **38.** 7
39. Coach \$450; first class \$900
40. Sides are 24 m; 12 m; and 28 m

Chapter 4

Section 4.1
1. x^9 **3.** x^{10} **5.** 3^7 **7.** $(-2)^8$ **9.** $4x^{13}$
11. $\left(\dfrac{1}{2}\right)^6$ **13.** $(-2)^5 x^9$ **15.** $(2x)^9$ **17.** $x^6 y^5$
19. $x^9 y^7$ **21.** $-24x^{10}$ **23.** $-30x^9$ **25.** $30x^4 y^5$
27. $x^9 y^7 z^3$ **29.** x^3 **31.** $x^3 y^8$ **33.** $x^4 y^2 z$
35. $3x^3 y^3$ **37.** 64   **39.** 81   **41.** 256
43. 384   **45.** 512   **47.** 1,536   **49.** $-15x^6$
51. $8x^3$ **53.** x^{21} **55.** $-24x^4$ **57.** $32x^{15}$
59. $-216x^{12}$ **61.** $50x^9$ **63.** $144x^{20}$ **65.** $\dfrac{4}{9}$
67. $\dfrac{a^4}{16}$ **69.** $\dfrac{a^{16}}{b^{12}}$ **71.** $\dfrac{x^{15} y^6}{z^{12}}$ **73.** $\dfrac{8x^6}{27}$

75. $-675x^{19}y^{14}$ **77.** $\dfrac{6x^{10}y^9}{5}$ **79.** $5,909.82$

81.   **83.** $(y^3)^5$ **85.** $(m^5)^4$

87. 9^4; 9^7; 9^{20}; 9^{14} **89.** $3x^3y^2z^4$ or $-3x^3y^2z^4$

Section 4.2

1. $\dfrac{1}{x^5}$ **3.** $\dfrac{1}{25}$ **5.** $\dfrac{1}{25}$ **7.** $-\dfrac{1}{8}$ **9.** $\dfrac{27}{8}$

11. $\dfrac{3}{x^2}$ **13.** $\dfrac{-5}{x^4}$ **15.** $\dfrac{1}{9x^2}$ **17.** x^3 **19.** $\dfrac{2x^3}{5}$

21. $\dfrac{y^4}{x^3}$ **23.** x^2 **25.** $\dfrac{1}{a^3}$ **27.** $\dfrac{1}{z^{10}}$ **29.** 1

31. $\dfrac{1}{x^3}$ **33.** x^{15} **35.** $2x^5$ **37.** $3a$ **39.** $x^{10}y$

41. $a^{17}b^8c^{13}$ **43.** $\dfrac{1}{x^{15}}$ **45.** b^8 **47.** $\dfrac{x^{10}}{y^6}$

49. $x^{12}y^6$ **51.** $\dfrac{x^{15}}{32}$ **53.** $\dfrac{y^4}{x^2}$ **55.** $\dfrac{y^2}{x^4}$

57. $\dfrac{48}{x^8}$ **59.** $16x^{26}$ **61.** $\dfrac{72}{x^3}$ **63.** $x^{42}y^{33}z^{25}$

65. $75x^2$ **67.** $144w^2$ **69.** $\dfrac{3x^3}{2y^4}$ **71.** $-567x^{22}y^{25}$

73. $\dfrac{6}{x^2y^5}$ **75.** $\dfrac{y^5}{x^3}$ **77.** $\dfrac{3xy^5}{4z^6}$ **79.** $\dfrac{1}{x^5y^3}$

81. $\dfrac{y^8}{x^7}$ **83.** x^{5n} **85.** x^2 **87.** y^{3n^2} **89.** x^2

91. 9.3×10^7 **93.** 1.3×10^{11} **95.** 28 **97.** 0.008
99. 0.000028 **101.** 5×10^{-4} **103.** 3.7×10^{-4}
105. 8×10^{-8} **107.** 3×10^5 **109.** 8×10^9
111. 2×10^2 **113.** 6×10^{16} **115.** 66 years
117. 1.55×10^{23}; 2.46×10^{13} l   **119.** 7.54×10^{14} l
121.   **123.**

Section 4.3

1. Polynomial   **3.** Polynomial   **5.** Polynomial
7. Not a polynomial   **9.** $2x^2$, $-3x$; 2, -3
11. $4x^3$, $-3x$, 2; 4, -3, 2 **13.** Binomial   **15.** Trinomial
17. Not classified   **19.** Monomial   **21.** Not a polynomial
23. $4x^5 - 3x^2$; 5 **25.** $-5x^9 + 7x^7 + 4x^3$; 9
27. $-9x$; 1 **29.** $x^6 - 3x^5 + 5x^2 - 7$; 6 **31.** 7, -11
33. 4, -4 **35.** 62, 30 **37.** 0, 0 **39.** Always
41. Sometimes   **43.** Sometimes   **45.** Sometimes
47. $3x + 20$; $170   **49.** $337

Section 4.4

1. $9a + 4$ **3.** $13b^2 - 18b$ **5.** $-2x^2$
7. $5x^2 - 2x + 1$ **9.** $3b^2 + 2b - 14$ **11.** $8y^3 - 2y$
13. $-a^3 + 4a^2$ **15.** $-2x^2 - x + 3$ **17.** $-4a - 5b$
19. $5a - 2b + 3c$ **21.** $6r - 5s$ **23.** $8p - 2q$
25. $x - 7$ **27.** $m^2 - 3m$ **29.** $-2y^2$
31. $2x^2 - x + 1$ **33.** $8a^2 - 12a - 7$ **35.** $-6b^2 + 8b$
37. $2x^2 + 12$ **39.** $6b - 1$ **41.** $10x - 9$
43. $2x^2 + 5x - 12$ **45.** $-6y^2 - 8y$ **47.** $6w^2 - 2w + 2$
49. $9x^2 - x$ **51.** $2a^2 + 5a$ **53.** $3x^2 + x$
55. $3x^2 + 3x - 9$ **57.** $3x^2 - 3x + 9$
59. $a = 3$; $b = 5$; $c = 0$; $d = -1$ **61.** $28x + 4$
63. $-x^2 + 65x - 150$

Section 4.5

1. $15x^5$ **3.** $-28b^{10}$ **5.** $40p^{13}$ **7.** $-12m^6$
9. $32x^5y^3$ **11.** $-6m^9n^3$ **13.** $10x + 30$
15. $12a^2 + 15a$ **17.** $12s^4 - 21s^3$ **19.** $15x^3 - 9x^2 - 3x$
21. $6x^3y^2 + 3x^2y^3 + 15x^2y^2$ **23.** $18m^4n^2 - 12m^3n^2 + 6m^3n^3$
25. $x^2 + 5x + 6$ **27.** $m^2 - 14m + 45$ **29.** $p^2 - p - 56$
31. $w^2 - w - 42$ **33.** $3x^2 - 29x + 40$
35. $6x^2 - x - 12$ **37.** $12a^2 - 31ab + 9b^2$
39. $21p^2 - 13pq - 20q^2$ **41.** $20x^2 + 23xy + 6y^2$
43. $x^2 + 6x + 9$ **45.** $w^2 - 12w + 36$
47. $z^2 + 24z + 144$ **49.** $4a^2 - 4a + 1$
51. $36m^2 + 12m + 1$ **53.** $9x^2 - 6xy + y^2$
55. $4r^2 + 20rs + 25s^2$ **57.** $36a^2 - 60ab + 25b^2$
59. $x^2 + x + \dfrac{1}{4}$ **61.** $x^2 - 36$ **63.** $m^2 - 49$
65. $x^2 - \dfrac{1}{4}$ **67.** $p^2 - 0.16$ **69.** $a^2 - 9b^2$
71. $9x^2 - 4y^2$ **73.** $64w^2 - 25z^2$ **75.** $25x^2 - 81y^2$
77. $24x^3 - 10x^2 - 4x$ **79.** $80a^3 - 45a$
81. $60s^3 - 39s^2 + 6s$ **83.** $x^3 - 4x^2 + x + 6$
85. $a^3 - 3a^2 + 3a - 1$ **87.** $\dfrac{x^2}{3} + \dfrac{11x}{45} - \dfrac{4}{15}$
89. $x^2 - y^2 + 4y - 4$ **91.** False   **93.** True
95. $(6x^2 - 11x - 35)$ cm^2 **97.** $12x - 0.05x^2$
99. $25x^2 - 40x + 16$ **101.** $x(x + 2)$ or $x^2 + 2x$
103.   **105.**   **107.**
109. $x^2 + 10x + 25$ **111.** $x^2 - 25$ **113.** 891
115. $9,996$ **117.** $3,584$

Section 4.6

1. $2x^4$ **3.** $5m^2$ **5.** $a + 2$ **7.** $3b^2 - 4$
9. $4a^2 - 6a$ **11.** $-4m - 2$ **13.** $2a^2 - 5a + 7$
15. $4x^2y - 3y^2 + 2x$ **17.** $x + 3$ **19.** $x - 5$
21. $x + 6$ **23.** $2x + 3 + \dfrac{4}{x - 3}$ **25.** $4x + 2 + \dfrac{-5}{x - 5}$
27. $2x + 3 + \dfrac{5}{3x - 5}$ **29.** $x^2 - x - 2$
31. $x^2 + 2x + 3 + \dfrac{8}{4x - 1}$ **33.** $x^2 + x + 2 + \dfrac{9}{x - 2}$
35. $7x^2 + 3x + 1 + \dfrac{3}{7x - 3}$ **37.** $x^2 + 4x + 5 + \dfrac{2}{x - 2}$
39. $x^3 + x^2 + x + 1$ **41.** $x - 3$ **43.** $x^2 - 1 + \dfrac{1}{x^2 + 3}$
45. $y^2 + y + 1$ **47.** $x^2 + 1$ **49.** $c = -2$
51.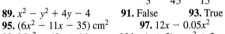

Summary Exercises—Chapter 4

1. r^{13} **3.** $\dfrac{1}{8w^3}$ **5.** $\dfrac{1}{y^3}$ **7.** $\dfrac{1}{x^3}$ **9.** $\dfrac{10}{b^3}$

11. $\dfrac{m^3}{n^{21}}$ **13.** $\dfrac{y^6}{x^9}$ **15.** $9a^8$ **17.** 4.25×10^{-5}

19. Binomial   **21.** Trinomial   **23.** Binomial
25. $13x^2$; 2 **27.** $x + 5$; 1 **29.** $7x^6 + 9x^4 - 3x$; 6
31. $9x^2 - 3x - 7$ **33.** $4x^2 + 8x$ **35.** $4x^2 - 4x - 3$
37. $10a - 5$ **39.** $3x^2 + 9x - 6$ **41.** $5x^2 - 5x + 5$
43. $9m^2 - 8m$ **45.** $6x^7$ **47.** $-21a^5b^7$
49. $10a^2 - 6a$ **51.** $21m^3n^2 - 14m^2n^3 + 35m^2n^2$

53. $w^2 - 19w + 90$ **55.** $p^2 - 6pq + 9q^2$
57. $2b^2 - 13b - 24$ **59.** $15r^2 - 24rs - 63s^2$
61. $b^3 - 2b^2 - 22b - 21$ **63.** $m^4 + 4m^2 - 21$
65. $4a^3 - 24a^2b + 35ab^2$ **67.** $a^2 - 14a + 49$
69. $9p^2 + 24p + 16$ **71.** $64x^2 - 48xy + 9y^2$
73. $y^2 - 81$ **75.** $16r^2 - 25$ **77.** $49a^2 - 9b^2$
79. $3c^3 - 75cd^2$ **81.** $3a^3$ **83.** $3a - 2$

85. $-3rs + 6r^2$ **87.** $x - 5$ **89.** $x - 3 + \dfrac{2}{x - 5}$

91. $x^2 + 2x - 1 + \dfrac{-4}{6x + 2}$ **93.** $x^2 + x + 2 + \dfrac{1}{x + 2}$

Self-Test—Chapter 4

1. $-6x^3y^4$ **2.** $\dfrac{16m^4n^{10}}{p^6}$ **3.** x^8y^{10} **4.** $\dfrac{c^2}{2d}$

5. $108x^8y^7$ **6.** $\dfrac{9y^4}{4x}$ **7.** 3 **8.** Binomial

9. Trinomial   **10.** $8x^4 - 3x^2 - 7$; $8, -3, -7$; 4
11. $10x^2 - 12x - 7$ **12.** $7a^3 + 11a^2 - 3a$
13. $3x^2 + 11x - 12$ **14.** $7a^2 - 10a$ **15.** $4x^2 + 5x - 6$
16. $2x^2 - 7x + 5$ **17.** $15a^3b^2 - 10a^2b^2 + 20a^2b^3$
18. $4x^2 + 7xy - 15y^2$ **19.** $9m^2 + 12mn + 4n^2$

20. $4x^2 + 3x + 13 + \dfrac{17}{x - 2}$

Cumulative Test—Chapters 0–4

1. -10 **2.** $\dfrac{7}{5}$ **3.** $\{7\}$ **4.** $\{-36\}$ **5.** $\{4\}$

6. $\{23\}$ **7.** $\mathbb{R}$ **8.** $\varnothing$ **9.** $\left\{-\dfrac{13}{8}\right\}$ **10.** $\{-1, 9\}$

11. $\left\{3, -\dfrac{19}{3}\right\}$ **12.** $\left\{\dfrac{2}{3}, 4\right\}$ **13.** $\{x \mid x < -15\}$

14. $\{x \mid x < 5\}$ **15.** $\{x \mid 3 < x < 13\}$
16. $\left\{x \mid x \le \dfrac{2}{3} \text{ or } x \ge 6\right\}$ **17.** slope: -4; y-intercept: $(0, 9)$
18. slope: 3 **19.** $y = -2x + 4$ **20.** $y = \dfrac{4}{5}x - \dfrac{2}{5}$

21. $y = 4x - 7$ **22.** $y = -2x$ **23.** $3x^2 + 2x - 4$
24. $x^2 - 3x - 18$ **25.** $12x^2 - 20x$ **26.** $6x^2 + x - 40$
27. $x^3 - x^2 - x + 10$ **28.** $4x^2 - 49$
29. $9x^2 - 30x + 25$ **30.** $20x^3 - 100x^2 + 125x$

31. $4xy - 2x^3 + 1$ **32.** $2x^2 + 6x + 3 + \dfrac{2}{x - 3}$ **33.** $18x^7$

34. $\dfrac{x^{16}}{y^{12}}$ **35.** $72x^4y^2$ **36.** $\dfrac{2x^5}{y^{11}}$ **37.** 2.1×10^{10}

38. 8 **39.** $65, 67$ **40.** 4 cm by 24 cm

Chapter 5

Section 5.1

1. Domain: {Colorado, Edmonton, Calgary, Vancouver};
   Range: {21, 5, 18, 17}
3. Domain: {Chamber, Testament, Rainmaker, Street Lawyer};
   Range: $\left\{\pi, 2\pi, \dfrac{1}{2}, 6\right\}$
5. Domain: {1, 3, 5, 7, 9}; Range: {2, 4, 6, 8, 10}
7. Domain: {1}; Range: {2, 3, 4, 5, 6}
9. Domain: {−3, −2, −1, 4, 5}; Range: {3, 4, 5, 6}
11. {(1, 9,274), (2, 9,096), (3, 8,814), (4, 8,801), (5, 8,684)}
13. {(−2, −8), (−1, −1), (0, 0), (1, 1), (2, 8)}; Domain: {−2, −1, 0, 1, 2}; Range: {−8, −1, 0, 1, 8}

15 {(10, 90), (20, 180), (30, 270), (40, 360)}; Domain: {10, 20, 30, 40}; Range: {90, 180, 270, 360}   **17.** Function
19. Function   **21.** Not a function   **23.** Not a function
25. Function   **27.** Not a function   **29.** Function
31. (a) -2; (b) 4; (c) -2 **33.** (a) 9; (b) -1; (c) 3
35. (a) -62; (b) -2; (c) 2 **37.** (a) 45; (b) 3; (c) -75
39. (a) -1; (b) 2; (c) 13 **41.** $f(x) = -3x + 2$

43. $f(x) = 4x - 8$ **45.** $f(x) = -\dfrac{3}{2}x + 3$

47. $f(x) = \dfrac{1}{3}x + \dfrac{3}{2}$ **49.** $f(x) = -\dfrac{5}{8}x + \dfrac{9}{8}$

51.

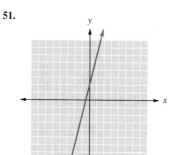

53.

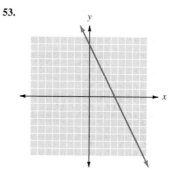

55.

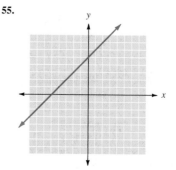

57. $5a - 1$ **59.** $5x + 4$ **61.** $5x + 5h - 1$
63. $-3m + 2$ **65.** $-3x - 4$
67. $107, 75, 49, 29, 15, 7, 5, 9, 19, 35, 57$
69. $-221, -114, -51, -20, -9, -6, 1, 24, 75$
71. \$124,400   **73.** (a) \$4.12; (b) \$2.73
75. (a) 51.425 ft; (b) 83.3 ft

Section 5.2

1.
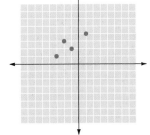
Function

3.
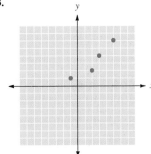
Function

5.
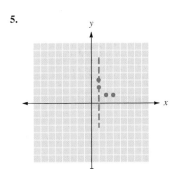
Not a Function

7. Function; $D = \{-2, -1, 0, 1, 2\}$; $R = \{-1, 0, 1, 2, 3\}$
9. Function; $D = \{-2, -1, 0, 2, 3, 5\}$; $R = \{-1, 2, 4, 5\}$
11. Function; $D = \{x \mid -4 \le x \le 3\}$; $R = \{-2\}$
13. Function; $D = \{x \mid -3 \le x \le 4\}$; $R = \{y \mid -2 \le y \le 5\}$
15. Not a function; $D = \{x \mid -3 \le x \le 3\}$; $R = \{y \mid -3 \le y \le 4\}$
17. Not a function; $D = \{-3\}$; $R = \mathbb{R}$
19. Function; $D = \mathbb{R}$; $R = \mathbb{R}$
21. Function; $D = \mathbb{R}$; $R = \{y \mid y \ge -5\}$
23. Not a function; $D = \{x \mid -6 \le x \le 6\}$; $R = \{y \mid -6 \le y \le 6\}$
25. Function; $D = \mathbb{R}$; $R = \{y \mid y \ge 0\}$
27. Function; $D = \mathbb{R}$; $R = \{y \mid y \ge 3\}$
29. Not a function; $D = \mathbb{R}$; $R = \{-4, 3\}$ **31.** -3 **33.** 5
35. -3 **37.** 0, 1   **39.** 2   **41.** 3   **43.** $-3, 6$
45. (a) 4; (b) -3; (c) 4; (d) 1
47. (a) -2; (b) 2; (c) -2, 3.7; (d) 4.5
49. (a) -5; (b) 4; (c) -2, 2; (d) -2.5, 2.5

Section 5.3

1. 7   **3.** -2 **5.** Does not exist   **7.** $\{0, 2\}$
9. $\{0, 2\}$ **11.** (a) $3x + 1$; (b) $-11x + 9$; (c) 10; (d) -13

13. (a) $3x + 4$; (b) $13x - 8$; (c) 13; (d) 18
15. (a) $-2x^2 - x + 4$; (b) $4x^2 + 3x - 6$; (c) -17; (d) 16
17. (a) $-x^3 + 2x^2 - 2x + 4$; (b) $-x^3 - 2x^2 - 8x + 12$; (c) -11;
(d) -20
19. (a) $6x + 4$; (b) $\mathbb{R}$ **21.** (a) $3x + 2 + \dfrac{1}{x-2}$; (b) $\{x \mid x \ne 2\}$
23. (a) $x^2 + x - 5 + \dfrac{2}{3x+1}$; (b) $\left\{x \mid x \ne -\dfrac{1}{3}\right\}$ **25.** -15
27. 0   **29.** -5 **31.** Undefined   **33.** $\{0, 2\}$
35. (a) $(2x-1)(x-3) = 2x^2 - 7x + 3$; (b) $\dfrac{2x-1}{x-3}$; (c) $\{x \mid x \ne 3\}$
37. (a) $6x^2 + x - 2$; (b) $\dfrac{3x+2}{2x-1}$; (c) $\left\{x \mid x \ne \dfrac{1}{2}\right\}$
39. (a) $-2x^2 - x + 10$; (b) $\dfrac{2-x}{5+2x}$; (c) $\left\{x \mid x \ne -\dfrac{5}{2}\right\}$
41. $P(x) = -x^2 + 21x - 50$ **43.** $V(t) = 10 - 9.8t$
45. $R(x) = 119x - 6x^2$

Section 5.4

1. -3 **3.** 5   **5.** 2   **7.** Does not exist
9. Does not exist   **11.** -4
13. (a) -2; (b) -6; (c) 4; (d) $2x - 2$
15. (a) -8; (b) -32; (c) 37; (d) $12x + 1$
17. (a) 9; (b) 1; (c) 12; (d) $x^2 + 3$
19. (a) 2; (b) -14; (c) 71; (d) $8x^2 - 1$
21. $h(x) = (g \circ f)(x)$ **23.** $h(x) = (g \circ f)(x)$
25. $h(x) = (g \circ f)(x)$ **27.** $h(x) = (g \circ f)(x)$
29. $h(x) = (f \circ g)(x)$
31. (a) $N(v) = 0.8v$; (b) $C(N) = 40N + 100$; (c) $C(v) = 32v + 1000$
33. \$118.78

Summary Exercises—Chapter 5

1. Domain: {Maine, Massachusetts, Vermont, Connecticut};
Range: {5, 13, 7, 11}
3. Domain: {Dean Smith, John Wooden, Denny Crum, Bob
Knight}; Range: {65, 47, 42, 41}
5. Domain: {1, 3, 4, 7, 8}; Range: {1, 2, 3, 5, 6}
7. Domain: {1}; Range: {3, 5, 7, 9, 10}
9. Function   **11.** Not a function   **13.** Function
15. Not a function   **17.** (a) 5; (b) 9; (c) 3
19. (a) 5; (b) 5; (c) 5   **21.** (a) 9; (b) 1; (c) 3
23. $f(x) = -2x + 5$ **25.** $f(x) = -\dfrac{2}{3}x + 2$
27. $f(x) = \dfrac{3}{4}x + 3$

29.
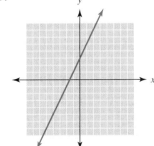

31.

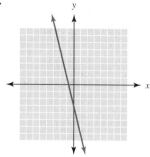

33. $3a - 2$ **35.** $3x + 3h - 2$

37.

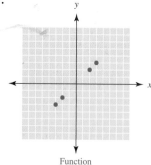

Function

39.

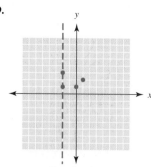

Not a function

41. Function; $D = \mathbb{R}$; $R = \{y \mid y \geq -4\}$
43. Not a function; $D = \{x \mid -6 \leq x \leq 6\}$; $R = \{y \mid -6 \leq y \leq 6\}$
45. (a) 0; (b) -3; (c) -5; (d) 5; (e) $-3.3, -2, 3.4$; (f) $-4, 1$
47. (a) -2; (b) 12; (c) $-5, 4$; (d) -2; (e) 8; (f) 12
49. 2   **51.** Does not exist   **53.** Does not exist
55. $\{-2, 3, 7\}$ **57.** $2x^2 + 6x - 8$
59. $2x^4 + x^3 - x^2 + 6x + 5$
61. $5x^2 + 3x + 10$ **63.** $4x^2 + 8x$
65. $6x^2 - 10x$ **67.** $3x^3$ **69.** $\dfrac{2x}{x - 3}$; $D = \{x \mid x \neq 3\}$
71. $\dfrac{3}{x}$; $D = \{x \mid x \neq 0\}$ **73.** 8   **75.** 2   **77.** -3
79. (a) -2; (b) -8; (c) 7; (d) $3x - 2$
81. (a) 25; (b) 49; (c) 4; (d) $x^2 - 5$
83. $(f \circ g)(x)$

Self-Test—Chapter 5

1. (a) $D = \{-3, 1, 2, 3, 4\}$; $R = \{-2, 0, 1, 5, 6\}$; (b) $D = \{$United States, Germany, Russia, China$\}$; $R = \{101, 65, 63, 50\}$
2. (a) Function; $D = \{-4, -1, 0, 2\}$; $R = \{-2, 5, 6\}$; (b) Not a function; $D = \{-3, 0, 1, 2\}$; $R = \{0, 1, 2, 4, 7\}$
3. (a) 6; (b) 12; (c) 2   **4.** (a) 2; (b) -5

5.

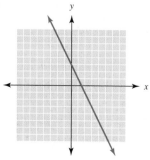

6. 6

7.

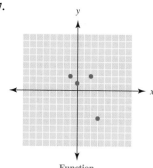

Function

8. Function   **9.** Not a function
10. (a) 3; (b) 1; (c) 2; (d) $-5, 5$
11. Domain: $\{x \mid -7 \leq x \leq 7\}$; Range: $\{y \mid -7 \leq y \leq 7\}$
12. Domain: $\{x \mid x \in \mathbb{R}\}$; Range: $\{y \mid y \leq 9\}$
13. (a) $6x^2 - 3x + 7$; (b) $2x^2 - 3x + 7$
14. (a) $-3x^3 + 3x^2 + 5x - 9$; (b) $-3x^3 + 7x^2 - 9x - 5$
15. $(f \cdot g)(x) = 3x^3 - 2x^2 - 3x + 2$; $D = \mathbb{R}$
16. $(f \div g)(x) = \dfrac{x^2 - 1}{3x - 2}$; $D = \left\{x \mid x \neq \dfrac{2}{3}\right\}$
17. $9x^2 - 12x + 3$ **18.** $3x^2 - 5$ **19.** 48   **20.** 22

Cumulative Test—Chapters 0–5

1. 29   **2.** 9   **3.** $\dfrac{20}{3}$ **4.** -3 **5.** $2x^2 - x + 11$
6. $12x^2 - 30x + 21$ **7.** $\{15\}$ **8.** $\{-108\}$
9. $\{-28\}$ **10.** $\left\{-\dfrac{5}{3}, 5\right\}$ **11.** $\{-5, 6\}$
12. $\{-3, 1\}$ **13.** $R = \dfrac{R_1 R_2}{R_1 + R_2}$

14. $\{x \mid x \leq 1\}$

15. $\{x \mid x < -2\}$

16. $\{x \mid -3 < x < 6\}$

17. $\{x \mid x > 13 \text{ or } x < -3\}$

18. $\left\{x \mid x < \dfrac{1}{3} \text{ or } x \geq 4\right\}$

19. -2 **20.** 0 **21.** Slope 3; y intercept $(0, 6)$

22. Slope $\dfrac{3}{2}$; y intercept $(0, -6)$ **23.** $y = 2x - 3$

24. $y = \dfrac{2}{3}x + \dfrac{7}{3}$ **25.** $y = -\dfrac{5}{4}x - 2$ **26.** $y = -x - 2$

27. 22 **28.** -3

29.

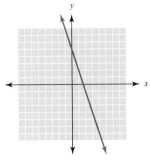

30. Function   **31.** Not a function

32. (a) Not a function; (b) Function   **33.** (a) -5; (b) -2; (c) 5

34. (a) $x^2 - 4x - 1$; (b) $-x^2 + 10x - 9$

35. (a) $-2x + 2$; (b) $4x^2 + 12x - 20$

36. (a) $-12x^2 + 15x$; (b) $\dfrac{-3x}{4x - 5}$

37. (a) $2x^3 - 5x^2$; (b) $\dfrac{2x - 5}{x^2}$

38. (a) -3; (b) $-6x + 3$; (c) -1; (d) $-6x + 5$

39. (a) 16; (b) $16x^2$; (c) 4; (d) $4x^2$ **40.** 5 cm $\times$ 17 cm

Chapter 6

Section 6.1

1. 2 **3.** 8 **5.** x^2 **7.** a^3 **9.** $5x^4$

11. $2a^4$ **13.** $3xy$ **15.** $5b$ **17.** $3abc^2$

19. $(x + y)^2$ **21.** $5(2x + 1)$ **23.** $8(3m - 4n)$

25. $4m(3m + 2)$ **27.** $5s(2s + 1)$ **29.** $12x(2x - 5)$

31. $5a^2(3a - 5)$ **33.** $6pq(1 + 3p)$ **35.** $7mn(m^2 - 3n^2)$

37. $6(x^2 - 3x + 5)$ **39.** $5a(a^2 - 3a + 1)$

41. $4x(3 + 2y - 7y^2)$ **43.** $5xy(2x + 3 - y)$

45. $5r^2s^2(2r + 5 - 3s)$ **47.** $3a(3a^4 - 5a^3 + 7a^2 - 9)$

49. $5mn(3m^2n - 4m + 7n^2 - 2)$ **51.** $(x - 9)(x + 5)$

53. $(p - 2q)(p - q)$ **55.** $(y - z)(x + 3)$

57. $(b - c)(a + b)$ **59.** $(r + 2s)(6r - 1)$

61. $(a - 2)(b^2 + 3)$ **63.** $(x + 3)(x - 4y)$

65. $(m - 3n)(m + 2n^2)$ **67.** Correct   **69.** Incorrect

71. Correct   **73.** 10 **75.** $6x$

77.   **79.** $33 - t$ **81.**

83.

Section 6.2

1. No   **3.** Yes   **5.** No   **7.** No   **9.** Yes

11. $(m + n)(m - n)$ **13.** $(x + 13)(x - 13)$

15. $(7 + y)(7 - y)$ **17.** $(3b + 4)(3b - 4)$

19. $(4w + 7)(4w - 7)$ **21.** $(2s + 3r)(2s - 3r)$

23. $(3w + 7z)(3w - 7z)$ **25.** $(7a + 3b)(7a - 3b)$

27. $(x^2 + 6)(x^2 - 6)$ **29.** $(xy + 4)(xy - 4)$

31. $(5 + ab)(5 - ab)$ **33.** $(r^2 + 2s)(r^2 - 2s)$

35. $(9a + 10b^3)(9a - 10b^3)$ **37.** $2x(3x + y)(3x - y)$

39. $3mn(2m + 5n)(2m - 5n)$

41. $(4a^2 + 9b^2)(2a + 3b)(2a - 3b)$

43. $(y + 5)(y^2 - 5y + 25)$ **45.** $(m - 5)(m^2 + 5m + 25)$

47. $(ab - 3)(a^2b^2 + 3ab + 9)$

49. $(2w + z)(4w^2 - 2wz + z^2)$

51. $(r - 4s)(r^2 + 4rs + 16s^2)$

53. $(2x - 3y)(4x^2 + 6xy + 9y^2)$

55. $3(a + 3b)(a^2 - 3ab + 9b^2)$

57. $4(x - 2y)(x^2 + 2xy + 4y^2)$

59. (a) -16; (b) $-2x(3x^2 + 5)$; (c) -16

61. (a) -30; (b) $5x^3(x^2 - 7)$; (c) -30

63. (a) -10; (b) $2x^5(3x - 8)$; (c) -10

65. $(b - c)(a + 4b)(a - 4b)$

67. $3a(2a + b)(a + 3b)(a - 3b)$ **69.** 49 **71.** 45

73.   **75.** $3x - 2y$; $27x^3 - 8y^3$

77. No; $4(x + 2)(x - 2)$

Section 6.3

1. True   **3.** False   **5.** True   **7.** False   **9.** True

11. $a = 1$; $b = 7$; $c = -5$ **13.** $a = 1$; $b = -3$; $c = 8$

15. $a = 3$; $b = 5$; $c = -8$ **17.** $a = 4$; $b = 8$; $c = 11$

19. $a = -7$; $b = -5$; $c = 2$ **21.** Factorable; 3, -2

23. Not factorable   **25.** Factorable; -3, -2

27. Factorable; 6, -1 **29.** Factorable; -15, -4

31. $2x + 4x$; $(x + 2)(x + 4)$ **33.** $-5x - 4x$; $(x - 5)(x - 4)$

35. $-9x + 7x$; $(x - 9)(x + 7)$ **37.** $(x + 4)(x + 6)$

39. $(x - 4)(x - 7)$ **41.** $(s + 10)(s + 3)$

43. $(a - 8)(a + 6)$ **45.** $(x - 1)(x - 7)$

47. $(x - 9)(x + 2)$ **49.** $(x - 7)(x - 7)$

51. $(p - 12)(p + 2)$ **53.** $(x + 11)(x - 6)$

55. $(c + 4)(c + 15)$ **57.** $(n + 10)(n - 5)$

59. $(x + 2y)(x + 5y)$ **61.** $(x - 3y)(x + 4y)$

63. $(x - 5y)(x - 8y)$ **65.** $(3x + 2)(2x + 5)$

67. $(5x - 3)(3x + 2)$ **69.** $(6m - 5)(m + 5)$

71. $(3x - 2)(3x - 2)$ **73.** $(6x + 5)(2x - 3)$

75. $(3y - 2)(y + 3)$ **77.** $(8x + 5)(x - 4)$

79. $(2x + y)(x + y)$ **81.** $(5a + 2b)(a - 2b)$

83. $(9x - 5y)(x + y)$ **85.** $(3m - 4n)(2m - 3n)$

87. $(12a - 5b)(3a + b)$ **89.** $(x + 2y)^2$

91. $5(2x - 3)(2x + 1)$ **93.** $4(2m + 1)(m + 1)$

95. $3(5r - 2s)(r - s)$ **97.** $2x(x - 2)(x + 1)$

99. $y^2(2y + 3)(y + 1)$ **101.** $6a(3a - 1)(2a - 3)$

103. $3(p + q)(3p + 7q)$ **105.** 6 or 9   **107.** 8 or 10 or 17

109. 4 **111.** 2 **113.** 3, 8, 15, 24, . . .

115. (a) $(x - 4)(x + 2)$; (b) 4, -2

117. (a) $(2x - 3)(x + 1)$; (b) $\dfrac{3}{2}$, -1

Section 6.3*

1. True   **3.** False   **5.** True   **7.** False

9. False   **11.** $a = 1$; $b = 7$; $c = -5$

13. $a = 1$; $b = -3$; $c = 8$

15. $a = 3$; $b = 5$; $c = -8$ **17.** $a = 4$; $b = 8$; $c = 11$

19. $a = -7$; $b = -5$; $c = 2$ **21.** $(x + 4)(x + 6)$

23. $(x - 5)(x - 4)$ **25.** $(x - 9)(x + 7)$

27. Not factorable   **29.** $(x - 4)(x - 7)$

31. $(s + 10)(s + 3)$ **33.** $(a - 8)(a + 6)$

35. $(x - 1)(x - 7)$ **37.** $(x + 3)(x + 8)$

39. $(x - 7)(x - 7)$ **41.** Not factorable

43. $(x + 11)(x - 6)$ **45.** $(c + 4)(c + 15)$

47. $(n + 10)(n - 5)$ **49.** $(x + 2y)(x + 5y)$

51. $(x - 3y)(x + 4y)$ **53.** $(x - 5y)(x - 8y)$
55. $(3x + 2)(2x + 5)$ **57.** $(5x - 3)(3x + 2)$
59. $(6m - 5)(m + 5)$ **61.** $(3x - 2)(3x - 2)$
63. $(6x + 5)(2x - 3)$ **65.** $(3y - 2)(y + 3)$
67. $(8x + 5)(x - 4)$ **69.** $(2x + y)(x + y)$
71. $(5a + 2b)(a - 2b)$ **73.** $(9x - 5y)(x + y)$
75. $(3m - 4n)(2m - 3n)$ **77.** $(12a - 5b)(3a + b)$
79. $(x + 2y)^2$ **81.** $5(2x - 3)(2x + 1)$
83. $4(2m + 1)(m + 1)$ **85.** $3(5r - 2s)(r - s)$
87. $2x(x - 2)(x + 1)$ **89.** $y^2(2y + 3)(y + 1)$
91. $6a(3a - 1)(2a - 3)$ **93.** $3(p + q)(3p + 7q)$
95. 6 or 9   **97.** 8 or 10 or 17   **99.** 4   **101.** 2
103. 3, 8, 15, 24, . . .   **105.** (a) $(x - 4)(x + 2)$; (b) 4, -2

107. (a) $(2x - 3)(x + 1)$; (b) $\dfrac{3}{2}$, -1

Section 6.4

1. $\{-2, 5\}$ **3.** $\{-3, 5\}$ **5.** $\{5, 6\}$ **7.** $\{-3, 7\}$
9. $\{-5, 10\}$ **11.** $\{-7, 12\}$ **13.** $\{0, 8\}$
15. $\{-10, 0\}$ **17.** $\{0, 5\}$ **19.** $\{-5, 5\}$
21. $\left\{-\dfrac{5}{3}, \dfrac{5}{3}\right\}$ **23.** $\left\{-\dfrac{3}{2}\right\}$ **25.** $\left\{4, \dfrac{9}{2}\right\}$ **27.** $\left\{-3, \dfrac{6}{5}\right\}$
29. $\left\{\dfrac{1}{2}, \dfrac{2}{3}\right\}$ **31.** $\{-3, 9\}$ **33.** $\{0, 9\}$ **35.** $\{0, 3\}$
37. $\{-5, 3\}$ **39.** $\left\{-\dfrac{3}{2}, 3\right\}$ **41.** $\left\{-\dfrac{7}{3}, 2\right\}$
43. $\{-2, 6\}$ **45.** $\{-4, 7\}$ **47.** $\{2, 6\}$ **49.** $\{-5, 1\}$
51.   **53.** $x^2 - x - 20 = 0$

55. $x^2 - 8x + 12 = 0$ **57.** $\{-2, 0, 5\}$ **59.** $\{-3, 0, 3\}$
61. $\{-2, -1, 2\}$ **63.** $\{-3, -1, 1, 3\}$
65. $\{0 \text{ cm}^3, 100 \text{ cm}^3\}$ **67.** $\{0 \text{ cm}^3, 132 \text{ cm}^3\}$ **69.** $\{30\}$
71. 2, 6; these are the zeros of Exercise 47
73.

Section 6.5

1. 5, 13   **3.** 4, 6   **5.** -9, -8 or 8, 9   **7.** 5, 6
9. 4, 5   **11.** -5, -4, -3 or 3, 4, 5   **13.** 4
15. 7 ft by 10 ft   **17.** 5 cm by 17 cm   **19.** 5 cm by 6 cm
21. 30 cm by 60 cm   **23.** 6 ft by 10 ft   **25.** 40 cm
27. 4 s   **29.** 6 s   **31.** 6 s   **33.** 6 s   **35.** 7 s
37. 40 mi/h   **39.** 70 mi/h   **41.** 50   **43.** $7 or $10
45. 30 or 90

Section 6.6

1. $3(x + 5)^2$ **3.** $5(n^2 + 4n + 6)$ **5.** $3a(y - 4)(y + 4)$
7. $(x + 6)(y - 2)$ **9.** $2x(a - 4)(b - 7)$
11. Not factorable   **13.** $3(y - 3)(y^2 + 3y + 9)$
15. $3m(3n - 4)(2n + 1)$ **17.** $x(3x - 2)(3x + 2)$
19. $(x - 3)(x + 3)(y - 2)(y + 2)$ **21.** $2(2x + 1)^2$
23. $(x + 5)(x - 2)(x + 2)$

Summary Exercises—Chapter 6

1. $7(2a - 5)$ **3.** $8s^2(3t - 2)$ **5.** $9s^3(3s + 2)$
7. $9m^2n(2n - 3 + 4n^2)$ **9.** $8ab(a + 3 - 2b)$
11. $(3x + 4y)(2x - y)$ **13.** $(p + 7)(p - 7)$
15. $(3n + 5m)(3n - 5m)$ **17.** $(5 + z)(5 - z)$
19. $(5a + 6b)(5a - 6b)$ **21.** $3w(w + 2z)(w - 2z)$
23. $2(m + 6n^2)(m - 6n^2)$ **25.** $(x - 4)(x + 5)$
27. $(4x + 3)(2x - 5)$ **29.** $x(2x + 3)(3x - 2)$
31. (a) 18; (b) $3x^2(x + 5)$; (c) 18
33. (a) 4; (b) $4x^4(3x^2 - 2)$; (c) 4   **35.** $(x + 9)(x - 9)$

37. $(4m + 7n)(4m - 7n)$ **39.** $(a^2 + 4b^2)(a + 2b)(a - 2b)$
41. $(2x + 1)(4x^2 - 2x + 1)$
43. $(5m + 4n)(25m^2 - 20mn + 16n^2)$
45. Factorable; $m = -6$, $n = 5$
47. Factorable; $m = -3$, $n = -8$
49. $(x + 10)(x + 2)$ **51.** $(w - 6)(w - 9)$
53. $(x - 12y)(x + 4y)$ **55.** $(7x - 3)(2x + 7)$
57. $(x + 4)(x + 5)$ **59.** $(a - 4)(a - 3)$
61. $(x + 8)(x + 8)$ **63.** $(b - 7c)(b + 3c)$
65. $m(m + 7)(m - 5)$ **67.** $3y(y - 7)(y - 9)$
69. $(3x + 5)(x + 1)$ **71.** $(2b - 3)(b - 3)$
73. $(5x - 3)(2x - 1)$ **75.** $(4y - 5x)(4y + 3x)$
77. $4x(2x + 1)(x - 5)$ **79.** $3x(2x - 3)(x + 1)$
81. $\{-6, 1\}$ **83.** $\{-10, 3\}$ **85.** $\{-5, 4\}$
87. $\{0, 10\}$ **89.** $\{-5, 5\}$ **91.** $\left\{-1, \dfrac{3}{2}\right\}$
93. $\{-5, 2\}$ **95.** $\{-4, 9\}$ **97.** $\{0, -3, 5\}$
99. -6 and 9; -9 and 6   **101.** Width 12 ft; length 18 ft
103. 4 s   **105.** 50   **107.** $5x^2(x - 7)(x + 6)$
109. $2y(6 - y)(6 + y)$ **111.** $10x^3(x + 4)^2$
113. $(x - 5)(x + 5)(x + 2)$

Self-Test—Chapter 6

1. $7(b + 6)$ **2.** $5(x^2 - 2x + 4)$ **3.** $(4y + 7x)(4y - 7x)$
4. $2b(4a + 5b)(4a - 5b)$ **5.** $(3y - 2x)(9y^2 + 6xy + 4x^2)$
6. $(a - 7)(a + 2)$ **7.** $(y + 10z)(y + 2z)$
8. $(3x - 2y)(3x - 2y)$ **9.** $(x + 2)(x - 5)$
10. $(3w + 7)(w + 1)$ **11.** $(3x + 2)(2x - 5)$
12. $(4x - 3y)(2x + y)$ **13.** $3x(2x + 5)(x - 2)$
14. $(x^2 + 9)(x + 3)(x - 3)$ **15.** $\{-1, 3\}$ **16.** $\{5, 6\}$
17. $\left\{-\dfrac{1}{3}, \dfrac{3}{2}\right\}$ **18.** $\{-5, -3\}$ **19.** 20 cm   **20.** 2 s

Cumulative Test—Chapters 0–6

1. 9   **2.** $\{11\}$ **3.** $\{-33\}$ **4.** $\{-3, 15\}$
5. $\{-10, 1\}$ **6.** $\{x \mid x > -6\}$ **7.** $\{x \mid x \geq -6\}$
8. $\left\{x \mid \dfrac{1}{2} < x < \dfrac{13}{2}\right\}$ **9.** $\{x \mid x < -8 \text{ or } x > -2\}$
10. Slope $\dfrac{-2}{5}$, y intercept $(0, 2)$ **11.** $y = 2x + 5$
12. $y = -3x - 2$ **13.** $2x^2y + 5xy$ **14.** $-x^2 + 2x - 1$
15. $x^2 + 9x - 2$ **16.** $3x^3 - 6x^2 + 6x$
17. $2x^2 - 3x - 5$ **18.** $x^3 + 27$ **19.** $8x^2 + 14x - 15$
20. $a^2 - 9b^2$ **21.** $x^2 - 4xy + 4y^2$ **22.** $50x^3 - 18xy^2$
23. $-\dfrac{8y^9}{x^6}$ **24.** $3a^4b^6$ **25.** $\dfrac{y^3}{4x^9}$ **26.** $\dfrac{2x^3z}{y^5}$
27. $-x^2 + 3x + 5$ **28.** $-3x^3 - 5x^2$ **29.** $-3x^2 + 5$
30. -121 **31.** $4(3x + 5)$ **32.** $(5x + 7y)(5x - 7y)$
33. $(4x - 5)(3x + 2)$ **34.** $(x - 5)(2x - 3)$
35. $\{-5, 3\}$ **36.** $\{-3, 2\}$ **37.** $8xy^4 - 5x^3y + 3$
38. $x + 4 + \dfrac{6}{x - 1}$ **39.** 17; $3x - 5 = 46$
40. $43; $x + (x - 5) = 81$ **41.** 60 mi/h   **42.** 10 or 50

Chapter R

Section R.1

1. $x - 5$ **3.** $2x - 5$ **5.** $p(6 + p)$ **7.** $\dfrac{4}{3}\pi r^3$
9. $\dfrac{1}{2}h(b_1 + b_2)$ **11.** 4   **13.** 2   **15.** 12
17. $18x^2$ **19.** $-3xy + 8y$ **21.** $2x - 8$

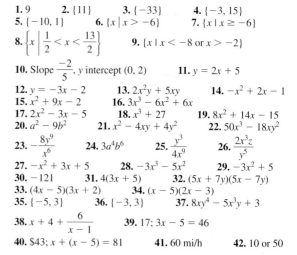

23. {1, 2, 3, 4, 6, 8, 12, 24}   **25.** {. . . , −4, −2, 0, 2, 4}

27. {x | −2 ≤ x ≤ 5}; [−2, 5]

29. {x | −4 < x < 4}; (−4, 4)

31.
$$-5 \quad -1\,0 \quad 2\ 3 \quad 5$$

33.
$$-5 \qquad 0 \quad 2$$

35. {x | −3 ≤ x ≤ 4}; [−3, 4]

37. {x | −4 < x ≤ 4}; (−4, 4]

39. $A \cup B = \{1, 2, 3, 4, 7, 10\}$; $A \cap B = \{7, 10\}$

Section R.2

1. No   **3.** Yes   **5.** {24}

7. {3}   **9.** $\left\{\dfrac{3}{2}\right\}$

11. $x < -12$
$$-12 \qquad 0$$

13. $x > 2$
$$0 \quad 2$$

15. $x < -2$
$$-2 \quad 0$$

17. $x < \dfrac{1}{2}$
$$0 \quad \dfrac{1}{2}$$

19. {−5, 11}   **21.** {−10, 5}   **23.** {−5, 6}

25. No solution

27. $-1 \le x \le 11$
$$-1\,0 \qquad 11$$

29. $-4 < x < 3$
$$-4 \quad 0 \quad 3$$

31. $-\dfrac{9}{2} \le x \le 3$
$$-\dfrac{9}{2} \quad 0 \quad 3$$

33. 19   **35.** 7, 15   **37.** $295; $11.80

39. 4 cm by 19 cm

Section R.3

1. (1, −2); (2, 0)   **3.** (4, 0); (4, −1)

5. (4, 5); (0, 9), (9, 0), (10, −1)

7. (0, −8), (1, −6), (4, 0), (3, −2)   **9.** A(1, 3)

11. C(2, −4)

13.–16.

17.

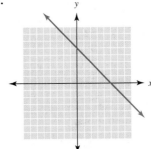

19.

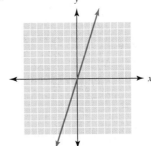

21.

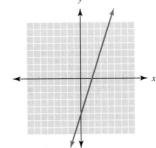

23.

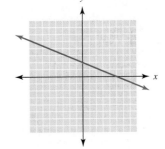

25.

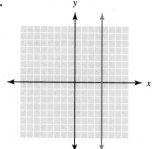

27. −1   **29.** 0

31. Slope 3, y intercept $(0, 2)$

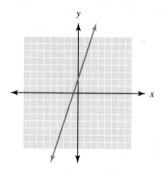

33. Slope −5, y intercept $(0, 3)$

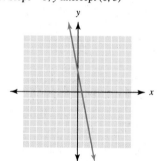

35. Perpendicular   **37.** $y = -\frac{1}{2}x - 1$ **39.** $y = 3x + 6$
41. $(7, 0)$ **43.** $(2, -6)$

Section R.4

1. x^{18} **3.** $\frac{y^2}{x}$ **5.** $\frac{5x^2y^3}{7z}$ **7.** $\frac{1}{2y^{13}}$ **9.** $4x^8y^3$

11. 5.07×10^{-3} **13.** $42{,}800{,}000$ **15.** Monomial, -3
17. Binomial, 6   **19.** $6x^2 - 3$ **21.** $27n^2 - 32n - 6$
23. $-15x^2 - 15x - 13$ **25.** $84y^3 + 108y^2$
27. $21x^2 + 20x + 4$ **29.** $12r^2 - 11rt + 2t^2$
31. $9x^2 - 16y^2$ **33.** $x^2 - 6x + 9$ **35.** $3x^3 - 6x^2 - 24x$
37. $25x^3 - 4xy^2$ **39.** $-16x^8 - 40x^7 + 24x^6$

41. $3n^2 - n + 4$ **43.** $y - 7 + \frac{8}{y + 2}$

Section R.5

1. D: $\{-1, 0, 2, 5\}$; R: $\{3, 5, 4, 7\}$ **3.** D: $\{-1, 0, 2, 4\}$; R: $\{3\}$
5. D: $\{-3, 0, 1, 2\}$; R: $\{1, 5, 3, -4\}$ **7.** Function

9. Not a function   **11.** Function   **13.** −3   **15.** −18
17. Function; D: $\{-3, 0, 2, 3\}$; R: $\{2, 1, 3, -5\}$
19. Function; D: $\mathbb{R}$; R: $\{y \mid y \geq -4\}$
21. (a) 3; (b) −3; (c) −1; (d) 3
23. (a) −6; (b) 0; (c) −2, 3; (d) −1, 2
25. (a) $4x + 3$; (b) $-10x + 7$; (c) 11; (d) 17
27. (a) $-x^2 - 2x + 5$; (b) $3x^2 + 8x - 9$; (c) −3; (d) −14

29. $5x - 2 + \dfrac{1}{x - 3}$; D: $\{x \mid x \neq 3\}$

31. (a) $2x^2 - x - 10$; (b) $\dfrac{-x + 2}{2x - 5}$; (c) −4; (d) 0

33. (a) $-5x^2 + 17x - 6$; (b) $\dfrac{2 - 5x}{x - 3}$; (c) 8; (d) $-\dfrac{12}{5}$ **35.** 5

37. (a) −1; (b) −4; (c) 1; (d) −8
39. (a) 3; (b) 39; (c) 36; (d) $3(x^2 + 3)$

Section R.6

1. $x(x - y)$ **3.** $2xy(18x^3y - 9x^2y^2 - 2)$
5. $y(y - 1)(3 + y)$ **7.** $(3x - 2)(5x + 3)$
9. $4(3m - 5)(m - 2)$ **11.** $(2y + 5)(2y - 5)$
13. $(x - 3 + y)(x - 3 - y)$ **15.** $2(5x + 3)^2$
17. $(3x - 2y)(9x^2 + 6xy + 4y^2)$ **19.** Yes; $m = -4$, $n = 2$
21. $(x - 2)(5x - 4)$ **23.** $2(3m + 5)(2m - 3)$

25. $(7w + 2)(9w - 8)$ **27.** $\{0, 5\}$ **29.** $\left\{\dfrac{3}{2}, -\dfrac{5}{3}\right\}$

31. $\{-5, 2\}$ **33.** −2 and 7, or −7 and 2   **35.** 3 s

Final Exam 0–6

1. 2   **2.** 10   **3.** $\dfrac{5}{8}$ **4.** $\dfrac{4}{7}$ **5.** 180

6. 118   **7.** −16   **8.** −285

9.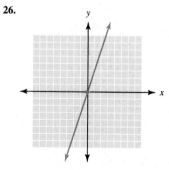

10. $\{x \mid -4 < x \leq 2\}$ **11.** $2x^2y + 3x + 2xy$

12. $x^2 + 9x$ **13.** $\{11\}$ **14.** $\{-33\}$ **15.** $\left\{\dfrac{8}{5}\right\}$

16. $\{1\}$ **17.** $\{-3, 15\}$ **18.** $\{-10, 1\}$

19. $\left\{-\dfrac{31}{2}\right\}$ **20.** $\{x \mid x > -6\}$

21. $\left\{x \mid \dfrac{1}{2} < x < \dfrac{13}{2}\right\}$ **22.** $\{x \mid x < -8 \text{ or } x > -2\}$

23. $\left\{x \mid x \geq \dfrac{14}{3} \text{ or } x \leq -2\right\}$ **24.** $\{x \mid -4 < x < 2\}$

25. $\{x \mid x < -3 \text{ or } x > 2\}$

26.

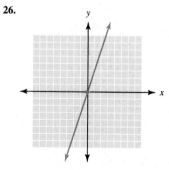

27.

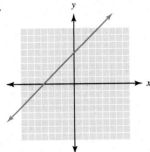

28.

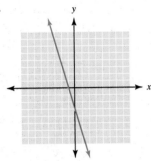

29.

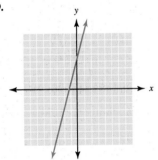

30. -2 **31.** Slope -4, y intercept $(0, 9)$
32. $y = -3x - 1$ **33.** $y = 2x + -1$ **34.** Function
35. Not a function   **36.** Not a function   **37.** Function
38. $-x + 2$ **39.** $-7x + 12$ **40.** $8x^3 - 14x^2$

41. $\dfrac{1}{8}$ **42.** $8x^2 + 7$ **43.** -242 **44.** $a^2 - 9b^2$

45. $x^2 - 4xy + 4y^2$ **46.** $x^2 + 3x - 10$
47. $a^2 + a - 12$ **48.** $3x + 2$ **49.** $3x + 3 + \dfrac{1}{x - 1}$
50. $4(3x + 5)$ **51.** $(5x + 6y)(5x - 6y)$
52. $(4x - 5)(3x + 2)$ **53.** $(2x - 3)(x - 5)$

54. $\{-4, 7\}$ **55.** $\{-2, 5\}$ **56.** $\left\{-\dfrac{3}{7}, \dfrac{3}{7}\right\}$

57. $\left\{-8, \dfrac{1}{2}\right\}$ **58.** 30.5 in. by 31.5 in.

59. Biology \$73; math \$68   **60.** \$43.50
61. -3 and 5, or -5 and 3   **62.** Width 5 ft; length 13 ft
63. 1 s; 4 s   **64.** 25   **65.** $(2m - n)(4m^2 + 2mn + n^2)$
66. $3x(3 - x)(3 + x)$ **67.** $5y(y + 2)(x + 3)$
68. $3m(m + 4)^2$

Chapter 7

Section 7.1

1. 3   **3.** Never undefined   **5.** $\dfrac{1}{2}$ **7.** 0

9. -2 **11.** 0   **13.** 2   **15.** 2   **17.** $\dfrac{2}{3}$

19. $\dfrac{2x^3}{3}$ **21.** $\dfrac{2xy^3}{5}$ **23.** $\dfrac{-12x^3}{7y^2}$ **25.** $\dfrac{a^3b^2}{3c^2}$

27. $\dfrac{6}{x + 4}$ **29.** $\dfrac{x + 1}{6}$ **31.** $\dfrac{x - 4}{x + 9}$ **33.** $3b + 2$

35. $\dfrac{y - z}{y + 3z}$ **37.** $\dfrac{x^2 + 4x + 16}{x + 4}$ **39.** $\dfrac{(a^2 + 9)(a - 3)}{a + 2}$

41. $\dfrac{y - 2}{x + 5}$ **43.** $\dfrac{x + 6}{x^2 - 2}$ **45.** $\dfrac{-2}{m + 5}$ **47.** $\dfrac{-11 - x}{2x + 1}$

49. Rational   **51.** Rational
53. Not rational   **55.** (a) $f(x) = x - 2$; (b) $(-1, -3)$, $x \neq -1$
57. (a) $f(x) = 3x - 1$; (b) $(-2, -7)$, $x \neq -2$

59. (a) $f(x) = \dfrac{x + 2}{5}$; (b) $(-2, 0)$, $x \neq -2$

61.

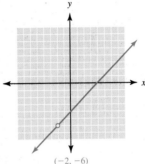

$(-2, -6)$

63.

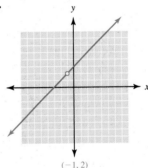

$(-1, 2)$

65.

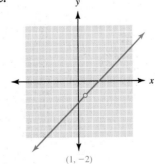

$(1, -2)$

67.   **69.**

71. 2   **73.** 3   **75.** $2x + h$ **77.** $x + 5$

79. (a) $R(x) = \dfrac{3{,}500 + 8.75x}{x}$; (b) \$78.75

81. (a) 3; (b)

x	$f(x)$	(d) x	$f(x)$
4	1	2	-1
3.1	10	2.9	-10
3.01	100	2.99	-100
3.001	1,000	2.999	$-1{,}000$
3.0001	10,000	2.9999	$-10{,}000$

(f)

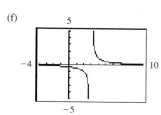

Section 7.2

1. $\dfrac{2}{x}$ **3.** $\dfrac{3}{x^8}$ **5.** $\dfrac{5}{12x}$ **7.** $\dfrac{16b^3}{3a}$ **9.** $5mn$

11. $8x^2$ **13.** $\dfrac{9b}{8}$ **15.** $x^2 + 2x$ **17.** $\dfrac{3(c-2)}{5}$

19. $\dfrac{5x^3}{x-3}$ **21.** $\dfrac{5d}{4(d-3)}$ **23.** $\dfrac{x-5}{2x+3}$

25. $\dfrac{2a+1}{2a}$ **27.** $\dfrac{-6}{w+2}$ **29.** $\dfrac{-a}{6}$ **31.** $\dfrac{2}{x}$

33. $\dfrac{3}{m}$ **35.** $\dfrac{5}{x}$

37. (a) -4; (b) undefined; (c) $h(x) = (x+1)(x-4)$, $x \neq -2$, $x \neq 4$; (d) -4; (e) undefined

39. (a) -6; (b) undefined; (c) $h(x) = (2x-5)(3x-1)$, $x \neq -2$, $x \neq -1$; (d) -6; (e) undefined

41. (a) $\dfrac{-4}{5}$; (b) $\dfrac{5}{4}$; (c) $h(x) = \dfrac{(3x-2)(x+4)}{(x-2)(x-5)}$; (d) $-4, -1, 2, 5$

43. $\dfrac{x^2+2}{4}$ **45.** $\dfrac{x+2}{x(x+3)}$

Section 7.3

1. $\dfrac{3}{x^3}$ **3.** $\dfrac{7}{3a+7}$ **5.** 2   **7.** $\dfrac{y-1}{2}$ **9.** 2

11. $\dfrac{3}{x+2}$ **13.** $\dfrac{23}{10x}$ **15.** $\dfrac{3(2a+1)}{a^2}$

17. $\dfrac{2(n-m)}{mn}$ **19.** $\dfrac{9b-20}{12b^3}$ **21.** $\dfrac{2b-9}{b(b-3)}$

23. $\dfrac{5x+7}{(x+1)(x+2)}$ **25.** $\dfrac{4(y+2)}{(y-3)(y+1)}$

27. $\dfrac{w(7w-30)}{(w-6)(w-2)}$ **29.** $\dfrac{7x}{(3x-2)(2x+1)}$

31. $\dfrac{1}{x-9}$ **33.** $\dfrac{2x+11}{(x+4)(x-4)}$ **35.** $\dfrac{3m+1}{(m-1)(m-2)}$

37. $\dfrac{15}{y-5}$ **39.** $\dfrac{-12}{x-4}$ **41.** $\dfrac{5z+14}{(z+2)(z-2)(z+4)}$

43. (a) $\dfrac{1}{2}$; (b) $h(x) = \dfrac{5x^2-7x}{(x+1)(x-3)}$; (c) $\left(1, \dfrac{1}{2}\right)$

45. (a) $\dfrac{3}{4}$; (b) $h(x) = \dfrac{x^2+x+1}{(x+1)^2}$; (c) $\left(1, \dfrac{3}{4}\right)$

47. (a) $\dfrac{-5}{6}$; (b) $h(x) = \dfrac{20x}{(x-5)(x+5)}$; (c) $\left(1, \dfrac{5}{6}\right)$

49. (a) $\dfrac{-3}{16}$; (b) $h(x) = \dfrac{(x+5)}{4(x-9)}$; (c) $\left(1, \dfrac{-3}{16}\right)$

51. $\dfrac{-16}{5}$ **53.** $\dfrac{49}{25}$ **55.** 2   **57.** 8

59. Undefined

Section 7.4

1. $\dfrac{8}{9}$ **3.** $\dfrac{13}{11}$ **5.** $\dfrac{5}{6}$ **7.** $\dfrac{1}{2x}$ **9.** $\dfrac{m}{2}$

11. $\dfrac{2(y+1)}{y-1}$ **13.** $\dfrac{3b}{a^2}$ **15.** $\dfrac{x}{(x-5)(x+4)}$

17. $\dfrac{2x-1}{2x+1}$ **19.** $y - x$ **21.** $\dfrac{x-y}{y}$ **23.** $\dfrac{a+4}{a+3}$

25. $\dfrac{x^2 y(x+y)}{(x-y)}$ **27.** $\dfrac{x}{x+2}$ **29.** $\dfrac{y+2}{(y-1)(y+4)}$

31. $\dfrac{x}{3}$ **33.** 1   **35.** $\dfrac{2a}{a^2+1}$ **37.** $\dfrac{2x+1}{x+1}$

39. $\dfrac{3x+2}{2x+1}$ **41.** $\dfrac{5x+3}{3x+2}$ **43.**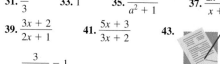

45. $\dfrac{\dfrac{3}{3+h} - 1}{h} = \dfrac{-1}{3+h}$ **47.**

49. $44\dfrac{4}{9}$ mi/h   **51.** $54\dfrac{6}{11}$ mi/h   **53.**

55.

x	-3	-2	-1	0	1	2	3
$\dfrac{-8 + \dfrac{20}{x}}{4 - \dfrac{25}{x^2}}$	-12	8	1.3333	Error	-0.5714	-0.8889	-1.091
$\dfrac{-4x}{2x+5}$	-12	8	1.3333	0	-0.5714	-0.8899	-1.091

Section 7.5

1. Equation, $\{40\}$ **3.** Expression, $\dfrac{3x}{10}$ **5.** Equation, $\{5\}$

7. Equation, $\{3\}$ **9.** $\{5\}$ **11.** $\{8\}$ **13.** $\left\{\dfrac{3}{2}\right\}$

15. $\{3\}$ **17.** $\left\{-\dfrac{9}{5}\right\}$ **19.** $\left\{-\dfrac{2}{3}\right\}$

21. No solution or $\varnothing$ **23.** $\{-23\}$ **25.** $\{9\}$

27. $\{4\}$ **29.** $\{8\}$ **31.** $\{4\}$ **33.** $\left\{\dfrac{3}{2}\right\}$

35. No solution or $\varnothing$ **37.** $\{7\}$ **39.** $\{5\}$

41. $\left\{-\dfrac{5}{2}\right\}$ **43.** $\left\{-\dfrac{1}{2}, 6\right\}$ **45.** $\left\{-\dfrac{1}{2}\right\}$

47. $\left\{-\dfrac{1}{3}, 7\right\}$ **49.** $\{-8, 9\}$

51. Sides are 3 and 9 on the first triangle and 8 and 24 on the second.

53. $\dfrac{ab}{b-a}$ **55.** $\dfrac{RR_2}{R_2-R}$ **57.** $\dfrac{y+1}{y-1}$ **59.** 6, 24

61. 9   **63.** 4 mi/h   **65.** 150 mi/h   **67.** 36 min

69.

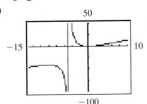

Section 7.6

1. 24   **3.** $\dfrac{25}{7}$ **5.** $\dfrac{3}{4}$ **7.** 3, 13   **9.** -5, 2

11. $x<-2$ or $x>3$ **13.** $x<-3$ or $x>\dfrac{5}{3}$

15. $\{x \mid -1 < x < 2\}$

17. $\{x \mid x < 2 \text{ or } x > 4\}$

19. $\{x \mid -3 < x \le 5\}$

21. $\left\{x \mid x < -3 \text{ or } x \ge \dfrac{1}{2}\right\}$

23. $\{x \mid -2 \le x < 3\}$

25. $\{x \mid -3 < x \le 4\}$

27. $\{x \mid x < 2 \text{ or } x > 3\}$

29.

Section 7.7

1. (a) $\dfrac{5}{2}$, -8; (b) $\{x \mid x \in \mathbb{R}, x \ne 1\}$; (c) -2, 4;

(d)

3. (a) $-\dfrac{1}{3}$, $\dfrac{3}{2}$; (b) $\{x \mid x \in \mathbb{R}, x \ne -5\}$; (c) -1, $\dfrac{3}{2}$;

(d)

5. (a) 8, 0; (b) $\{x \mid x \in \mathbb{R}, x \ne -5\}$; (c) 2, 4;

(d)

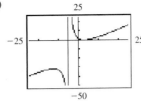

7. (a) $-\dfrac{5}{2}$, 0; (b) $\{x \mid x \in \mathbb{R}, x \ne 1\}$; (c) -2, $\dfrac{1}{3}$;

(d)

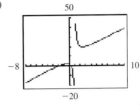

9. (a) 0, 18; (b) $\{x \mid x \in \mathbb{R}, x \ne 3\}$; (c) -2, 1;

(d)

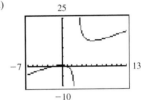

Summary Exercises—Chapter 7

1. Never undefined   **3.** $x \ne 5$ **5.** $\dfrac{3x^2}{4}$ **7.** 7

9. $\dfrac{-(x+3)}{x+5}$ **11.** $\dfrac{2a-b}{3a-b}$

13.

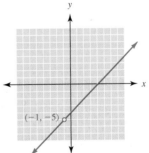

15. $\dfrac{2x^3}{3}$

17. $\dfrac{4}{3y}$ **19.** $\dfrac{3}{2a}$ **21.** $\dfrac{x+2y}{x-5y}$

23. (a) -8; (b) x^2+x-20; (c) -8 **25.** $\dfrac{3x^2-16x-8}{(x+4)(x-4)}$

27. $\dfrac{x+5}{x(x-5)}$ **29.** $\dfrac{-11}{6(m-1)}$ **31.** $\dfrac{4(x-4)}{5(x+1)(x-1)}$

33. $\dfrac{-4}{s+2}$ **35.** $\dfrac{3x-1}{x+2}$

37. (a) 8; (b) $\dfrac{3x^2-8x}{x^2-5x+6}$; (c) (4, 8)   **39.** $\dfrac{2}{3x^2}$

41. $\dfrac{b + a}{b - a}$ **43.** $\dfrac{s + r}{rs}$ **45.** $\dfrac{x - 4}{(x - 1)(x - 1)}$

47. $-y + 2$ **49.** $\dfrac{x + 2}{(x - 1)(x + 4)}$ **51.** $\{5\}$

53. $\{6\}$ **55.** $\{3\}$ **57.** $\{-1\}$ **59.** $\{4\}$

61. $\{0, 7\}$

63. Sides are 4 and 11 on the first triangle and 8 and 22 on the second triangle.

65. 5, 6   **67.** 12 h   **69.** 0   **71.** -2

73. $x < -2$ or $x > 4$ **75.** $\{x \mid -1 < x < 2\}$

77. $\{x \mid x < 2$ or $x > 4\}$ **79.** $\{x \mid -3 < x \le 5\}$

81. $\left\{x \mid x < -3 \text{ or } x \ge \dfrac{1}{2}\right\}$ **83.** $\{x \mid -2 \le x < 3\}$

85. $\{x \mid -3 < x \le 4\}$ **87.** $\{x \mid x < 2$ or $x > 3\}$

89. (a) 8, 8; (b) D: $\{x \mid x \in \mathbb{R}, x \ne -1\}$; (c) -4, 2; (d)

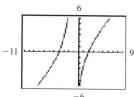

Self-Test—Chapter 7

1. $\dfrac{-3x^4}{4y^2}$ **2.** $\dfrac{w + 1}{w - 2}$ **3.** $\dfrac{x + 3}{x - 2}$

4.

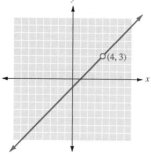

5. $\dfrac{4a^2}{7b}$

6. $\dfrac{m - 4}{4}$ **7.** $\dfrac{2}{x - 1}$ **8.** $\dfrac{2x + y}{x(x + 3y)}$

9. $\dfrac{-5(3x + 1)}{3x - 1}$ **10.** $\dfrac{2(2x + 1)}{x(x - 2)}$ **11.** $\dfrac{2}{x - 3}$

12. $\dfrac{4}{x - 2}$ **13.** $\dfrac{8x + 17}{(x - 4)(x + 1)(x + 4)}$ **14.** $\dfrac{y}{3y + x}$

15. $\dfrac{z - 1}{2(z + 3)}$ **16.** $\{2, 6\}$

17. $\{x \mid -4 \le x < 3\}$

18. $\{x \mid -3 < x \le -1\}$

19. 3   **20.** (a) $\dfrac{1}{4}$; (b) $x < -\dfrac{2}{3}$ or $x > \dfrac{1}{4}$

Cumulative Test—Chapters 0–7

1. $\{12\}$ **2.** -14 **3.** $6x + 7y = -21$

4. x intercept $(-6, 0)$; y intercept $(0, 7)$

5. $f(x) = 6x^2 - 5x - 2$ **6.** $f(x) = 2x^3 + 5x^2 - 3x$

7. Domain: $\{x \mid x = 2\}$ **8.** 16

9. $x(2x + 3)(3x - 1)$ **10.** $(4x^8 + 3y^4)(4x^8 - 3y^4)$

11. $\dfrac{3}{x - 1}$ **12.** $\dfrac{1}{(x + 1)(x - 1)}$ **13.** $x^2 - 3x$

14. (a) $\dfrac{8x^2 - 34x - 29}{x - 5}$; (b) $\dfrac{1}{2(4x + 3)(x - 5)}$;

(c) $\left\{x \mid x \ne -\dfrac{3}{4} \text{ or } x \ne 5\right\}$

15. $\left\{\dfrac{7}{2}\right\}$ **16.** $\left\{-\dfrac{8}{9}, -\dfrac{4}{9}\right\}$ **17.** $\left\{-\dfrac{30}{17}\right\}$

18. $\{x \mid x > -2\}$ **19.** $\left\{x \mid \dfrac{1}{5} < x < \dfrac{7}{5}\right\}$

20. $\{x \mid x \le -4$ or $x \ge -2\}$ **21.** $\{-5\}$

22. $\{3\}$ **23.** $\dfrac{b^6}{a^{10}}$ **24.** Don 6 h, Barry 3 h

25. Length 18 cm, width 10 cm

Chapter 8

Section 8.1

1.

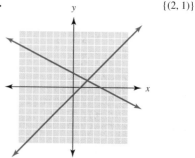

$\{(5, 1)\}$

3.

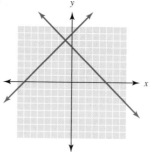

$\{(-1, 6)\}$

5.

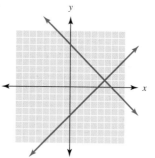

$\{(2, 1)\}$

7. 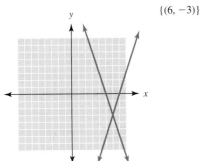 $\{(6, -3)\}$

9. 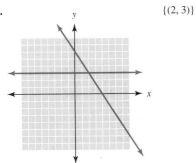 $\{(6, 2)\}$

11. 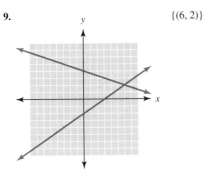 $\{(2, 3)\}$

13. 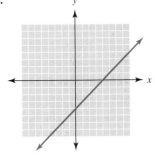 Infinite number of solutions, dependent system

15. 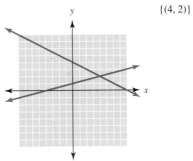 $\{(4, 2)\}$

17. 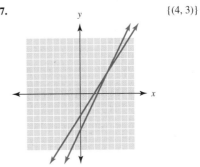 $\{(4, 3)\}$

19. 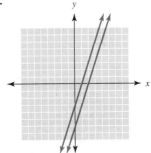 No solutions, inconsistent system

21. 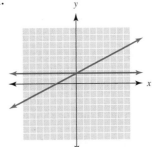 $\left\{\left(0, \dfrac{3}{2}\right)\right\}$

23. 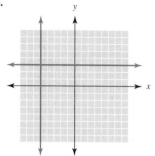 $\{(-5, 3)\}$

25. $\{(3.18, 28.58)\}$ **27.** $\{(-445.35, -253.01)\}$
29. $\{(29.31, -18.03)\}$ **31.** Consistent
33. Inconsistent   **35.** Consistent   **37.** Dependent
39. $m = 2; b = 5$ **41.**   **43.**

45. (a) $\dfrac{-A}{B}$; (b) $\dfrac{-D}{E}$; (c) $AE - BD \neq 0$

Section 8.2

1. $\{(2, 3)\}$ **3.** $\left\{\left(-3, \dfrac{7}{3}\right)\right\}$

5. $\{(2, 1)\}$ **7.** Infinite number of solutions, dependent system
9. $\{(5, -3)\}$ **11.** No solutions, inconsistent system
13. $\{(-8, -2)\}$ **15.** $\{(5, -2)\}$ **17.** $\{(-4, -3)\}$
19. Infinite number of solutions, dependent system

21. $\left\{\left(\dfrac{1}{3}, 2\right)\right\}$ **23.** $\{(10, 1)\}$

25. No solutions, inconsistent system   **27.** $\left\{\left(-2, -\dfrac{8}{3}\right)\right\}$

29. $\{(3, 4)\}$ **31.** $\{(4, -5)\}$ **33.** $\left\{\left(\dfrac{20}{3}, -2\right)\right\}$

35. $\{(9, -15)\}$ **37.** (d)   **39.** (g)   **41.** (h)
43. (c)   **45.** 550 adult tickets; 200 student tickets
47. 27 in. by 15 in.   **49.** Mulch $1.80; fertilizer $3.20
51. 105 lb of $4 beans; 45 lb of $6.50 beans
53. $8,000 bond; $4,000 time deposit
55. 100 ml of 10%; 300 ml of 50%
57. 15 mi/h boat; 3 mi/h current   **59.** 26
61. 15 battery-powered calculators; 20 solar models
63. (15, 150)   **65.** 100 mi

67. $\left\{\left(\dfrac{2}{3}, \dfrac{2}{5}\right)\right\}$ **69.** $\left\{\left(\dfrac{1}{3}, -\dfrac{3}{2}\right)\right\}$ **71.** $y = \dfrac{3}{2}x - 2$

73. $\{(1.3, -0.5)\}$ **75.** $\{(5.8, 1.7)\}$ **77.**

79. (a) $y = \dfrac{AF - CD}{AE - BD}, AE - BD \neq 0$;

(b) $x = \dfrac{CE - BF}{AE - BD}, AE - BD \neq 0$

Section 8.3

1. $\{(1, 2, 4)\}$ **3.** $\{(-2, 1, 2)\}$ **5.** $\{(-4, 3, 2)\}$
7. Infinite number of solutions, dependent system

9. $\left\{\left(3, \dfrac{1}{2}, -\dfrac{7}{2}\right)\right\}$ **11.** $\{(3, 2, -5)\}$ **13.** $\{(2, 0, -3)\}$

15. $\left\{\left(4, -\dfrac{1}{2}, \dfrac{3}{2}\right)\right\}$ **17.** No solutions, inconsistent system

19. $\left\{\left(2, \dfrac{5}{2}, -\dfrac{3}{2}\right)\right\}$ **21.** 3, 5, 8

23. 3 nickels; 5 dimes; 17 quarters   **25.** 4 cm, 7 cm, 8 cm
27. $2,000 savings; $4,000 bond; $6,000 money market
29. 243   **31.** Roy 8 mi; Sally 16 mi; Jeff 26 mi
33. $\{(1, 2, -1, -2)\}$ **35.** $\{(7, 2)\}$
37. $T = 20, C = 25, B = 40$
39. (a) $y = 2x^2 - x + 4$; (b) $y = 3x^2 - 2x + 1$

Section 8.4

1. $\begin{bmatrix} 2 & -3 & \vdots & 5 \\ 1 & 4 & \vdots & 2 \end{bmatrix}$ **3.** $\begin{bmatrix} 1 & -5 & \vdots & 6 \\ 0 & 1 & \vdots & 2 \end{bmatrix}$

5. $\begin{bmatrix} 1 & 2 & -1 & \vdots & 3 \\ 1 & 0 & 3 & \vdots & 1 \\ 0 & 1 & -2 & \vdots & 4 \end{bmatrix}$ **7.** $\begin{aligned} x + 2y &= 3 \\ x + 5y &= -6 \end{aligned}$

9. $\begin{aligned} x + 3y &= 5 \\ y &= 2 \end{aligned}$ **11.** $\begin{aligned} x + 2y &= 4 \\ y + 5z &= 3 \\ x + y + z &= 1 \end{aligned}$ **13.** $\{(1, -2)\}$

15. $\{(5, -3)\}$ **17.** $\{(2, -4)\}$ **19.** $\{(4, -3)\}$
21. Inconsistent, no solutions   **23.** $\{(1, -2, 0)\}$

25. $\{(-1, -8, -3)\}$ **27.** $\left\{\left(\dfrac{5}{2}, -3, -\dfrac{7}{2}\right)\right\}$

29. $\{(-1, 1, 0)\}$

Section 8.5

1.

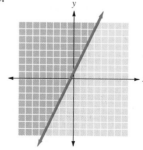

3.

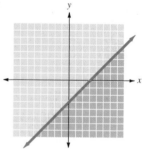

5.

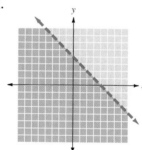

7.

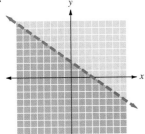

9.

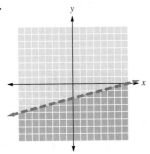

11.

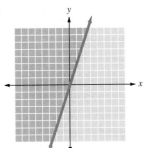

13.

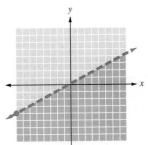

15.

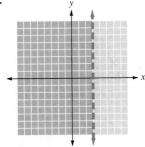

17.

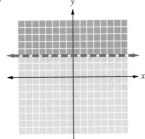

19.

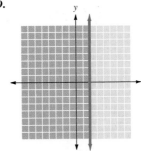

21.

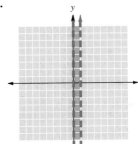

23.

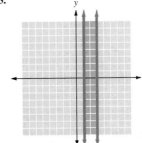

25.

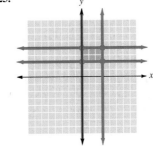

27.

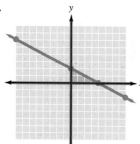

29.

5.

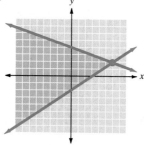

31.

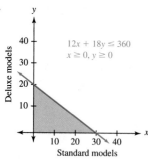

$12x + 18y \leq 360$
$x \geq 0, y \geq 0$

Deluxe models

Standard models

7.

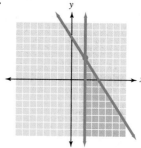

33.

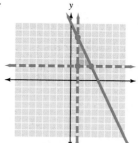

$3x + 4y \leq 1,000$
$x \geq 0$
$y \geq 0$

$(0, 250)$

$(333\frac{1}{3}, 0)$

Special diet

Normal diet

35. $y \geq -x + 4$

37. $y < \dfrac{1}{2}x - 3$

9.

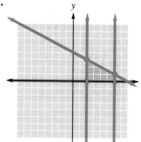

Section 8.6

1.

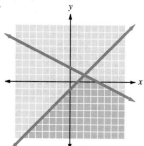

11.

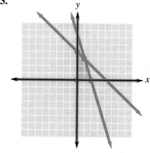

3.

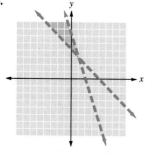

13.

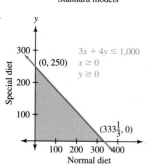

15.

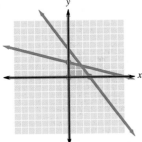

17.

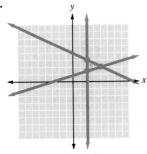

19.

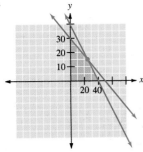

21.

23. $y \le 2x + 3$
$y \le -3x + 5$
$y \ge -x - 1$

Summary Exercises—Chapter 8

1. $\{(6, 2)\}$

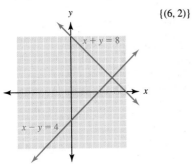

3. $\{(3, 2)\}$

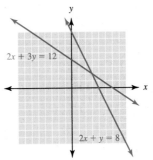

5. $\{(38.05, -15.98)\}$ **7.** $\{(3, 2)\}$ **9.** $\{(0, -1)\}$
11. $\{(-4, 3)\}$ **13.** $\{(-3, 2)\}$ **15.** $\{(6, -4)\}$
17. $\{(9, 5)\}$ **19.** No solutions, inconsistent system

21. $\{(2, -1)\}$ **23.** $\left\{\left(-\frac{2}{3}, 5\right)\right\}$ **25.** 7, 23

27. 800 adult tickets; 400 student tickets   **29.** 20 cm; 12 cm
31. \$8,000 in savings; \$9,000 in time deposit

33. Jet 500 mi/h; wind 50 mi/h   **35.** 15 of $5\frac{1}{4}$ in.; 8 of $3\frac{1}{2}$ in.

37. 300 mi   **39.** $\{(6, 1, -2)\}$ **41.** $\{(5, 2, 1)\}$

43. $\left\{\left(\frac{1}{2}, -\frac{3}{2}, 0\right)\right\}$ **45.** 2, 5, 8

47. 200 orchestra; 40 balcony; 120 box seats
49. \$6,000 savings; \$2,000 stock; \$4,000 mutual fund

51. $\begin{bmatrix} 5 & 3 & | & 11 \\ -1 & -2 & | & 2 \end{bmatrix}$ **53.** $\{(4, -3)\}$

55.

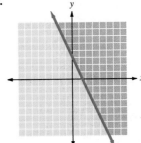

57.

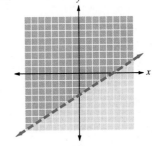

59.

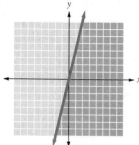

61.

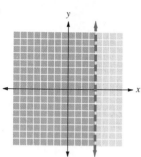

63.

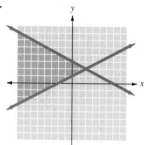

65.

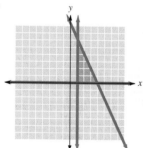

67.

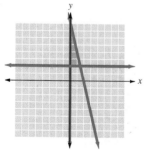

69.

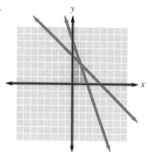

Self-Test—Chapter 8

1. $\{(-3, 4)\}$

2. Infinite number of solutions, dependent system

3. No solutions, inconsistent system   **4.** $\{(-2, -5)\}$

5. $\{(5, 0)\}$ **6.** $\left\{\left(3, -\dfrac{5}{3}\right)\right\}$ **7.** $\{(-1, 2, 4)\}$

8. $\left\{\left(2, -3, -\dfrac{1}{2}\right)\right\}$ **9.** Disks $2.50; ribbons $6

10. 60 lb jawbreakers; 40 lb licorice

11. Four 5-in. sets; six 12-in. sets

12. $3,000 savings; $5,000 bond; $8,000 mutual fund

13. 50 ft by 80 ft

14.

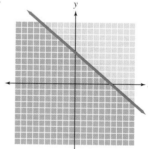

15.

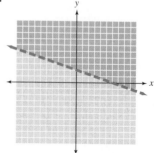

16.

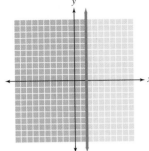

17.

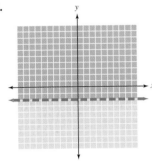

18.

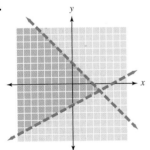

19.

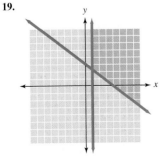

20.

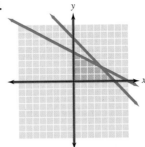

Cumulative Test—Chapters 0–8

1. $\left\{\dfrac{11}{2}\right\}$ **2.** $x > -2$ **3.** $\{-1, 4\}$

4. $-4 \le x \le \dfrac{2}{3}$ **5.** $x < -\dfrac{17}{5}$ or $x > 5$ **6.** $\{2\}$

7.

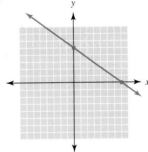

8.

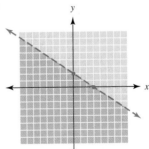

9. $\dfrac{P - P_0}{IT}$ **10.** 7   **11.** $y = -x + 3$

12. $2x^2 - 5x - 3$ **13.** $9x^2 - 12x + 4$

14. $(x - 3)(x^2 - 5)$ **15.** $y = \dfrac{4}{5}x - \dfrac{2}{5}$

16. 90   **17.** $\left\{\left(-10, \dfrac{26}{3}\right)\right\}$ **18.** $\{(2, 2, -1)\}$

19. 8 cm by 19 cm   **20.** 73   **21.** $\dfrac{y^9}{x^{15}}$

22. $\dfrac{8x - 4}{x(x - 2)}$ **23.** x **24.** $\dfrac{y(y - 3)}{8}$

25. $\dfrac{x(3 - xy)}{4}$

Chapter 9

Section 9.1

1.

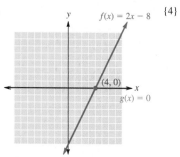

$f(x) = 2x - 8$ {4}

(4, 0) $g(x) = 0$

3.

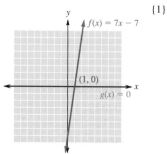

$f(x) = 7x - 7$ {1}

(1, 0) $g(x) = 0$

5.

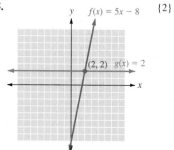

$f(x) = 5x - 8$ {2}

(2, 2) $g(x) = 2$

7.

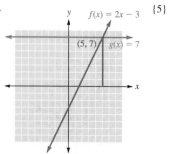

$f(x) = 2x - 3$ {5}

(5, 7) $g(x) = 7$

9.

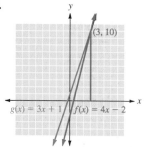

{3}

(3, 10)

$g(x) = 3x + 1$ $f(x) = 4x - 2$

11.

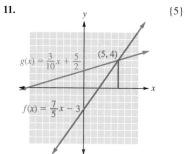

{5}

$g(x) = \frac{3}{10}x + \frac{5}{2}$ (5, 4)

$f(x) = \frac{7}{5}x - 3$

13.

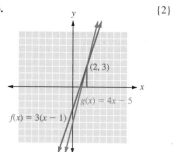

{2}

(2, 3)

$g(x) = 4x - 5$

$f(x) = 3(x - 1)$

15.

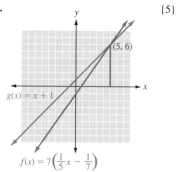

{5}

(5, 6)

$g(x) = x + 1$

$f(x) = 7\left(\frac{1}{5}x - \frac{1}{7}\right)$

17.

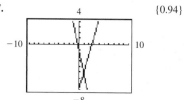

{0.94}

19.

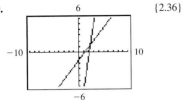

{2.36}

21. For a given number of miles use the lower graph to determine the cheaper cost.

23. 250 sets   **25.**

Section 9.2

1.

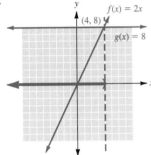

Solution set $\{x \mid x < 4\}$

3.

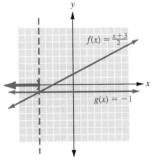

Solution set $\{x \mid x < -5\}$

5.

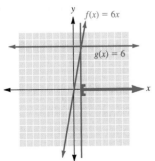

Solution set $\{x \mid x \geq 1\}$

7.

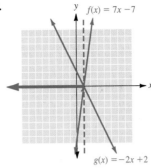

Solution set $\{x \mid x < 1\}$

9.

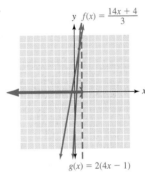

Solution set $\{x \mid x < 1\}$

11.

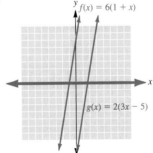

Solution set $\{x \mid x \in R\}$

13.

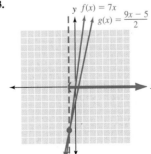

Solution set $\{x \mid x > -1\}$

15.

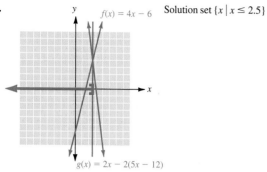

$f(x) = 4x - 6$

Solution set $\{x \mid x \le 2.5\}$

$g(x) = 2x - 2(5x - 12)$

17. $x \ge 500$

19. If miles are under 700, Downtown Edsel; if over 700, Wheels, Inc.; $W = 28 \times 7 = 196$; $DE = 98 + 0.14x$ (x is number of miles)

21. 110 people   **23.**

Section 9.3

1.

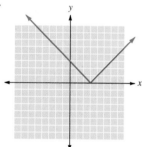

3.

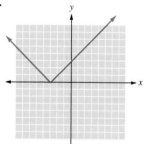

5.

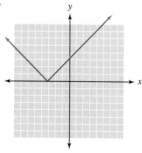

7.

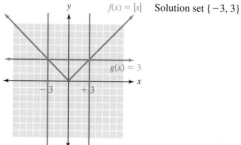

$f(x) = |x|$ Solution set $\{-3, 3\}$

$g(x) = 3$

9.

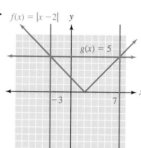

$f(x) = |x - 2|$ Solution set $\{-3, 7\}$

$g(x) = 5$

11.

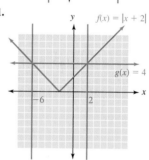

$f(x) = |x + 2|$ Solution set $\{-6, 2\}$

$g(x) = 4$

13.

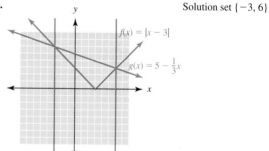

Solution set $\{-3, 6\}$

$f(x) = |x - 3|$

$g(x) = 5 - \frac{1}{3}x$

15.

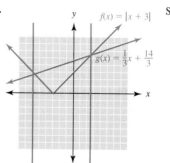

$f(x) = |x + 3|$ Solution set $\{-\frac{23}{4}, \frac{5}{2}\}$

$g(x) = \frac{1}{3}x + \frac{14}{3}$

17. $f(x) = |x - 2|$ **19.** $f(x) = |x - 3|$

21.

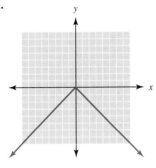

23.

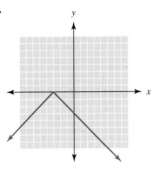

25.

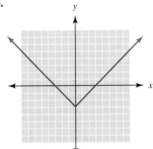

27.

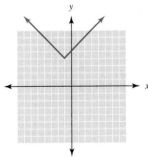

29.

31.

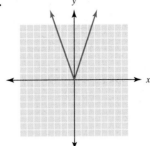

33.

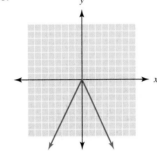

35. Answers will vary
37. $\{-1\}$
39. $\{1\}$

Section 9.4

1. $\{x \mid -5 < x < 5\}$

3. $\{x \mid x \leq -7 \text{ or } x \geq 7\}$

5. $\{x \mid x < 2 \text{ or } x > 6\}$

7. $\{x \mid -10 \leq x \leq -2\}$

9. $\{x \mid x < -2 \text{ or } x > 8\}$

11. No solution
13. $\{x \mid 1 < x < 4\}$

15. $\left\{x \mid x \leq -3 \text{ or } x \geq \dfrac{1}{3}\right\}$

17.

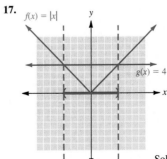

$f(x) = |x|$ $g(x) = 4$

Solution set $\{x \mid -4 < x < 4\}$

19.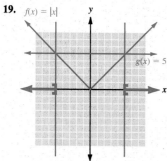

$f(x) = |x|$

y

$g(x) = 5$

x

Solution set $\{x \mid x \le -5 \text{ or } x \ge 5\}$

21.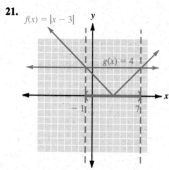

$f(x) = |x - 3|$

y

$g(x) = 4$

x

-1 7

Solution set $\{x \mid -1 < x < 7\}$

23.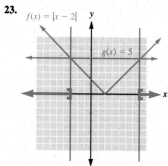

$f(x) = |x - 2|$

y

$g(x) = 5$

x

Solution set $\{x \mid x \le -3 \text{ or } x \ge 7\}$

25.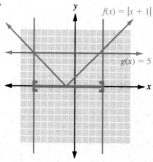

y

$f(x) = |x + 1|$

$g(x) = 5$

x

Solution set $\{x \mid -6 \le x \le 4\}$

27.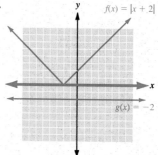

y

$f(x) = |x + 2|$

x

$g(x) = -2$

Solution set: All real numbers

29.

Summary Exercises—Chapter 9

1.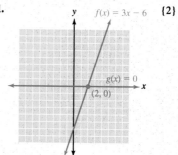

y $f(x) = 3x - 6$ $\{2\}$

$g(x) = 0$

x

$(2, 0)$

3.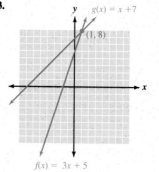

y $g(x) = x + 7$ $\{1\}$

$(1, 8)$

x

$f(x) = 3x + 5$

5.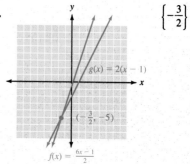

y $\left\{-\dfrac{3}{2}\right\}$

$g(x) = 2(x - 1)$

x

$\left(-\dfrac{3}{2}, -5\right)$

$f(x) = \dfrac{6x - 1}{2}$

7.

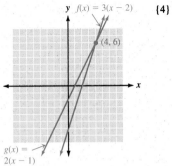

$f(x) = 3(x - 2)$ {4}

(4, 6)

$g(x) = 2(x - 1)$

15.

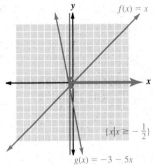

$f(x) = x$ **17.** $x \geq 15$

$\{x | x \geq -\frac{1}{2}\}$

$g(x) = -3 - 5x$

9.

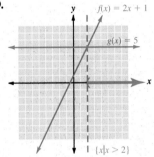

$f(x) = 2x + 1$

$g(x) = 5$

$\{x | x > 2\}$

19.

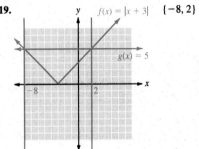

$f(x) = |x + 3|$ $\{-8, 2\}$

$g(x) = 5$

-8 2

11.

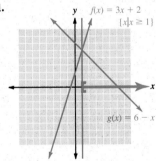

$f(x) = 3x + 2$

$\{x | x \geq 1\}$

$g(x) = 6 - x$

21.

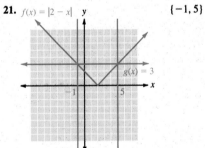

$f(x) = |2 - x|$ $\{-1, 5\}$

$g(x) = 3$

-1 5

13.

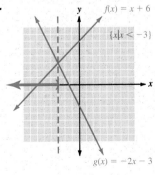

$f(x) = x + 6$

$\{x | x < -3\}$

$g(x) = -2x - 3$

23.

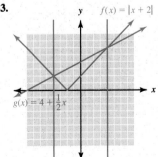

$f(x) = |x + 2|$ $\{-4, 4\}$

$g(x) = 4 + \frac{1}{2}x$

25. 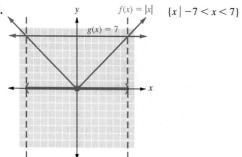 $\{x \mid -7 < x < 7\}$

27. 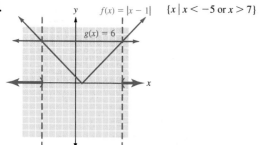 $\{x \mid x < -5 \text{ or } x > 7\}$

29. 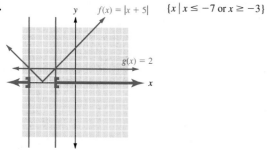 $\{x \mid x \le -7 \text{ or } x \ge -3\}$

31. $\{x \mid -4 \le x \le 4\}$

33. $\{x \mid -1 \le x \le 3\}$

35. No solution

Self-Test—Chapter 9

1. 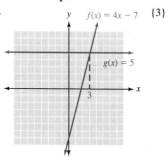 $\{3\}$

2. 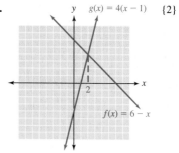 $\{2\}$

3. 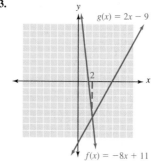 $\{2\}$

4. 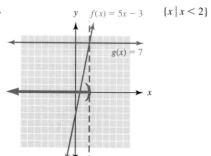 $\{2\}$

5. 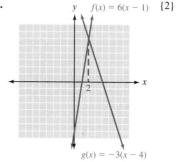 $\{x \mid x < 2\}$

6.

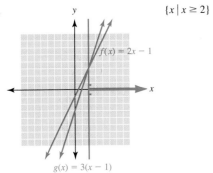

$f(x) = 2x - 1$

$g(x) = 3(x - 1)$

$\{x \mid x \geq 2\}$

7.

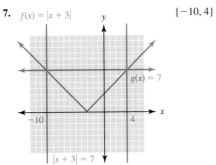

$f(x) = |x + 3|$

$g(x) = 7$

$|x + 3| = 7$

-10 4

$\{-10, 4\}$

8.

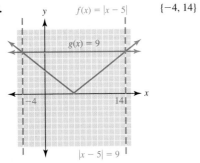

$f(x) = |x - 5|$

$g(x) = 9$

-4 141

$|x - 5| = 9$

$\{-4, 14\}$

9.

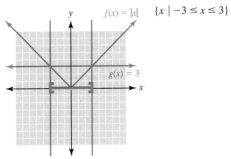

$f(x) = |x|$

$g(x) = 3$

$\{x \mid -3 \leq x \leq 3\}$

10.

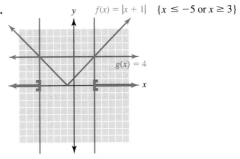

$f(x) = |x + 1|$ $\{x \leq -5 \text{ or } x \geq 3\}$

$g(x) = 4$

11.

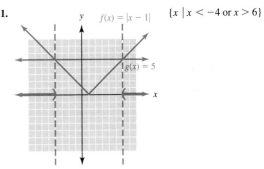

$f(x) = |x - 1|$ $\{x \mid x < -4 \text{ or } x > 6\}$

$g(x) = 5$

12.

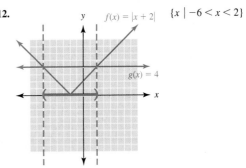

$f(x) = |x + 2|$ $\{x \mid -6 < x < 2\}$

$g(x) = 4$

13.

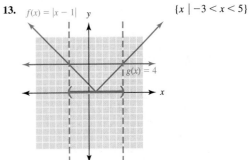

$f(x) = |x - 1|$

$g(x) = 4$

$\{x \mid -3 < x < 5\}$

14.

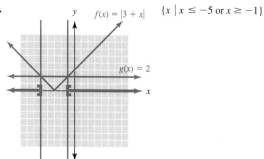

$\{x \mid x \le -5 \text{ or } x \ge -1\}$

19.

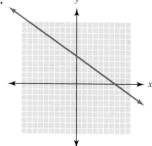

15.

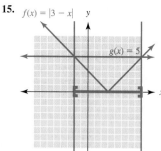

$\{x \mid -2 \le x \le 8\}$

20.

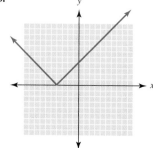

16.

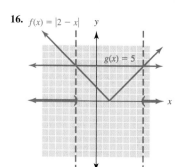

$\{x \mid x < -3 \text{ or } x > 7\}$

21.   Solution set $\{-3\}$

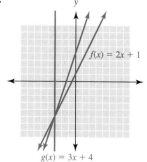

17. 150 mi at 30 mi/h and 80 mi at 40 mi/h
18. 20,000 boxes
19. $\left\{x \mid -\dfrac{9}{2} < x < \dfrac{7}{2}\right\}$ **20.** $\left\{x \mid x \le -\dfrac{7}{3} \text{ or } x \ge \dfrac{11}{3}\right\}$

Cumulative Test—Chapters 0–9

1. $\left\{-\dfrac{4}{3}\right\}$ **2.** $\{-2, 8\}$ **3.** -21 **4.** $y = \dfrac{2}{3}x - 2$

5. $4x^2 - 5x - 3$ **6.** $x(x+3)(x-3)$ **7.** $(2x+1)(x-3)$

8. $(x-2)(x+5)$ **9.** $\dfrac{1}{x+1}$ **10.** $\dfrac{-x^2 + x + 5}{(x+3)(x-2)}$

11. $\dfrac{1}{(x-4)(x+1)}$ **12.** $(2, -2)$

13. x intercept 3, y intercept 6   **14.** $\{x \mid x \ge -8\}$

15. $\{x \mid -3 \le x \le 9\}$ **16.** $\{x \mid x < -3 \text{ or } x > 1\}$

17. $\dfrac{1}{x-5} + x + 6$ **18.** $\dfrac{1}{(x-5)(x+6)}$

22.   Solution set $\{-9, 1\}$

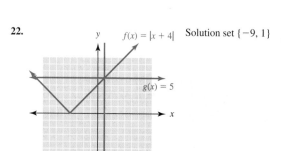

23.

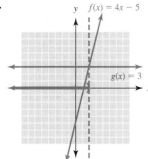

$f(x) = 4x - 5$

$g(x) = 3$

Solution set $\{x \mid x < 2\}$

24.

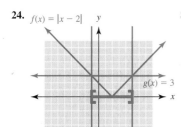

$f(x) = |x - 2|$

$g(x) = 3$

Solution set $\{x \mid -1 \le x \le 5\}$

25.

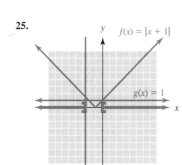

$f(x) = |x + 1|$

$g(x) = 1$

Solution set $\{x \mid x \le -2 \text{ or } x \ge 0\}$

Chapter 10

Section 10.1

1. 7 **3.** -6 **5.** ± 9 **7.** Not a real number
9. 3 **11.** -4 **13.** -6 **15.** 3 **17.** -2

19. -2 **21.** Not a real number **23.** -3 **25.** $\dfrac{2}{3}$

27. $\dfrac{2}{3}$ **29.** 6 **31.** 3 **33.** 4 **35.** 3

37. $|x|$ **39.** y **41.** $3|x|$ **43.** $|a^2 b^3|$ **45.** $4x^2$
47. $|y^5|$ **49.** $|m^2 n^3|$ **51.** $5a$ **53.** $2xy^3$
55. 3.873 **57.** 12.490 **59.** Not a real number
61. 4.362 **63.** 2.618 **65.** -2.466 **67.** False
69. True **71.** False **73.** False **75.** No
77. **79.** **81.** (a) 6; (b) 6; (c) 12;
(d) 12; (e) 10;
(f) Not possible

Section 10.2

1. $2\sqrt{3}$ **3.** $5\sqrt{2}$ **5.** $-6\sqrt{3}$ **7.** $2\sqrt{13}$
9. $2\sqrt{15}$ **11.** $-5\sqrt{5}$ **13.** $2\sqrt[3]{2}$ **15.** $-2\sqrt[3]{6}$
17. $3\sqrt[3]{5}$ **19.** $2\sqrt[4]{2}$ **21.** $3\sqrt[4]{3}$ **23.** $3a\sqrt{5}$
25. $3w^2\sqrt{6}$ **27.** $5a^2\sqrt{3a}$ **29.** $6p^2 q\sqrt{3p}$
31. $2x\sqrt[3]{2}$ **33.** $-2a^2\sqrt[3]{10}$ **35.** $5x^4\sqrt[3]{2x}$
37. $-5ab^5 c^3\sqrt[3]{2a}$ **39.** $2wz^3\sqrt[4]{6wz}$ **41.** $2w^2\sqrt[5]{2}$
43. $\dfrac{\sqrt{5}}{4}$ **45.** $\dfrac{\sqrt{5}}{3y^2}$ **47.** $\dfrac{\sqrt[3]{5}}{2}$ **49.** 3 units
51. $\sqrt{10}$ units **53.** $4\sqrt{2}$ units **55.** $\dfrac{5\sqrt{7}}{7}$
57. $\dfrac{7\sqrt{3}}{6}$ **59.** $\dfrac{\sqrt{30}}{5}$ **61.** $\dfrac{\sqrt[3]{14}}{2}$ **63.** $\dfrac{5\sqrt[3]{4}}{4}$
65. $\dfrac{2\sqrt{3w}}{w}$ **67.** $\dfrac{2m\sqrt{10mn}}{5n}$ **69.** $\dfrac{\sqrt[3]{5y^2}}{y}$
71. $\dfrac{3\sqrt[3]{4x^2}}{2x}$ **73.** $\dfrac{\sqrt[3]{50x}}{5x}$ **75.** $\dfrac{\sqrt[3]{10a}}{2a}$
77. $\dfrac{a\sqrt[3]{a^2 b^2}}{b^3}$ **79.** True **81.** True **83.** True
85. $x^5 y^5$ **87.**   **89.** (a) 15.65; (b) 15.65;
(c) 12.25; (d) 12.25

Section 10.3

1. $7\sqrt{5}$ **3.** $3\sqrt{3a}$ **5.** Cannot be simplified
7. $9\sqrt[3]{2}$ **9.** $9\sqrt{6}$ **11.** $3\sqrt{5}$ **13.** $5\sqrt{7}$
15. $6\sqrt{2}$ **17.** $4\sqrt[3]{3}$ **19.** $\sqrt{6w}$ **21.** $5x\sqrt{2x}$
23. $3w\sqrt[3]{2w^2}$ **25.** $\dfrac{4}{3}\sqrt{3}$ **27.** $\dfrac{3}{2}\sqrt[3]{6}$ **29.** $\dfrac{2}{3}\sqrt{6}$
31. $\dfrac{1}{3}\sqrt{3}$ **33.** $\sqrt{11a}$ **35.** $\sqrt{42}$ **37.** $\sqrt[3]{36}$
39. 6 **41.** $3p\sqrt[3]{2}$ **43.** $2xy\sqrt[3]{5y}$ **45.** $\sqrt{6} + 5\sqrt{2}$
47. $2\sqrt{6}$ **49.** $4x\sqrt{3}$ **51.** $6\sqrt[3]{2}$ **53.** $-10 - \sqrt{2}$
55. $-28 + \sqrt{10}$ **57.** -4 **59.** $a - 3$
61. $28 - 10\sqrt{3}$ **63.** $a + 6\sqrt{a} + 9$ **65.** $x + 2\sqrt{xy} + y$
67. $\dfrac{\sqrt{21}}{7}$ **69.** $\dfrac{\sqrt{6ab}}{3b}$ **71.** $\dfrac{3\sqrt[3]{2}}{2}$ **73.** $2 - \sqrt{3}$
75. $6 + 2\sqrt{5}$ **77.** $3 + 2\sqrt{2}$ **79.** $\dfrac{w + 6\sqrt{w} + 9}{w - 9}$
81. $\dfrac{x - 2\sqrt{xy} + y}{x - y}$ **83.** $14x^2\sqrt[3]{x}$ **85.** $\sqrt{2x + 3}$

Section 10.4

1. $\{4\}$ **3.** $\left\{\dfrac{1}{4}\right\}$ **5.** $\{4\}$ **7.** $\{6\}$
9. No solution **11.** $\{1\}$ **13.** $\{w \mid w \ge -3\}$
15. $\left\{\dfrac{7}{2}\right\}$ **17.** $\left\{\dfrac{7}{3}\right\}$ **19.** $\{3\}$ **21.** $\{4\}$
23. $\{6\}$ **25.** $\{7\}$ **27.** $\left\{0, \dfrac{1}{2}\right\}$ **29.** $\{32\}$
31. $\{3, -3\}$ **33.** $\{1\}$ **35.** $\{1\}$ **37.** $\left\{-\dfrac{7}{4}\right\}$
39. $\{3, 7\}$ **41.** $\{0\}$ **43.** $\{6\}$ **45.** $\{-15, 3\}$
47. $\{7\}$ **49.** $\{16\}$ **51.** $\{x \mid x \ge 1\}$ **53.** $q = \dfrac{h^2}{p}$
55. $R = \dfrac{v^2}{2g}$ **57.** $S = 2\pi r^2$ **59.** $V = \dfrac{\pi h r^2}{2}$
61. $x = 1 \pm \sqrt{d^2 - (y - 2)^2}$ **63.** 1.88 s **65.** $l = \dfrac{t^2 g}{4\pi^2}$

67. 9 **69.** 16 **71.** ≈ 245 mi **73.** $\dfrac{d^2}{2}$

75. ≈ 36 ft **77.** ≈ 28 ft **79.** ≈ 61 mi/h **81.** $\{-2\}$
83. $\{0.44\}$

Section 10.5

1. 7 **3.** -5 **5.** Not a real number **7.** 3

9. 3 **11.** $\dfrac{2}{3}$ **13.** 9 **15.** 16 **17.** 4

19. 729 **21.** $\dfrac{4}{9}$ **23.** $\dfrac{1}{7}$ **25.** $\dfrac{1}{3}$ **27.** $\dfrac{1}{27}$

29. $\dfrac{1}{32}$ **31.** $\dfrac{5}{2}$ **33.** x **35.** $y^{6/7}$ **37.** $b^{13/6}$

39. $x^{1/3}$ **41.** s **43.** $w^{2/3}$ **45.** x **47.** $a^{3/5}$

49. $\dfrac{1}{y^4}$ **51.** a^4b^9 **53.** x^2y^3 **55.** $st^{1/3}$

57. $4pq^{5/3}$ **59.** $x^{2/5}y^{1/2}z$ **61.** $a^{1/2}b^{1/4}$ **63.** $\dfrac{s^2}{r^4}$

65. $\dfrac{x^3}{y^2}$ **67.** $\dfrac{1}{mn^2}$ **69.** $\dfrac{s^3}{r^2t}$ **71.** $\dfrac{2xz^3}{y^2}$

73. $\dfrac{2n}{m^{1/5}}$ **75.** $xy^{3/4}$ **77.** $\sqrt[4]{a^3}$ **79.** $2\sqrt[3]{x^2}$

81. $3\sqrt[5]{x^2}$ **83.** $\sqrt[5]{9x^2}$ **85.** $(7a)^{1/2}$ **87.** $2m^2n^3$

89. 9.946 **91.** 0.370 **93.** **95.** $a^2 + a^{5/4}$

97. $a - 4$ **99.** $m - n$ **101.** $x + 4x^{1/2} + 4$
103. $r + 2r^{1/2}s^{1/2} + s$ **105.** $(x^{1/3} + 1)(x^{1/3} + 3)$
107. $(a^{2/5} - 3)(a^{2/5} - 4)$ **109.** $(x^{2/3} - 2)(x^{2/3} + 2)$
111. x^{5n} **113.** y^{4n} **115.** r^2 **117.** $a^{6n}b^{4n}$
119. x **121.** $\sqrt[4]{x}$ **123.** $\sqrt[8]{y}$ **125.** 2×10^4
127. 2×10^{-3} **129.** 4×10^{-4} **131.** 40

133. (a) $\dfrac{3}{2}$; (b) $-\dfrac{6}{5}$; (c) $-\dfrac{3}{4}$; (d) 3 **135.** (a) 81; (b) not defined

Section 10.6

1. $4i$ **3.** $-11i$ **5.** $i\sqrt{21}$ **7.** $2i\sqrt{3}$

9. $-6i\sqrt{3}$ **11.** $-1 - i$ **13.** $-1 + i$ **15.** $2 + 2i$
17. $5 - 7i$ **19.** $7 - i$ **21.** $3 + 11i$ **23.** $6 - 3i$
25. $0 + 0i$ or 0 **27.** $-15 + 9i$ **29.** $30 + 12i$
31. $-6 - 8i$ **33.** $-5 + 4i$ **35.** 25 **37.** $23 + 14i$
39. $18 + i$ **41.** $40 - 42i$ **43.** $3 + 2i; 13$
45. $3 - 2i; 13$ **47.** $-3 + 2i; 13$ **49.** $-5i; 25$

51. $2 - 3i$ **53.** $-\dfrac{2}{3} - \dfrac{5}{3}i$ **55.** $\dfrac{6}{29} - \dfrac{15}{29}i$

57. $2 - 3i$ **59.** $\dfrac{17}{25} + \dfrac{6}{25}i$ **61.** $-\dfrac{7}{25} - \dfrac{24}{25}i$

63. **65.** $-\sqrt{35}$ **67.** -6 **69.** $-3\sqrt{10}$
71. -10 **73.** -1 **75.** 1
77. -1 **79.** $-i$

81.

Summary Exercises—Chapter 10

1. 11 **3.** Not a real number **5.** -4 **7.** $\dfrac{3}{4}$

9. 8 **11.** a^2 **13.** $7w^2z^3$ **15.** $-3b^2$ **17.** $2xy^2$

19. $3\sqrt{5}$ **21.** $2x\sqrt{15}$ **23.** $2\sqrt[3]{4}$

25. $\dfrac{3}{4}$ **27.** $\dfrac{y^2}{7}$ **29.** $\dfrac{\sqrt{5}}{4x}$ **31.** $11\sqrt{10}$

33. $10\sqrt[3]{2x}$ **35.** $11\sqrt{2}$ **37.** $3\sqrt{7}$ **39.** $13\sqrt[3]{2}$

41. $10a\sqrt[3]{2a^2}$ **43.** $\dfrac{11}{2}\sqrt{2x}$ **45.** $\dfrac{4}{15}\sqrt{15}$

47. $6x\sqrt{3}$ **49.** $\sqrt{15} + 2\sqrt{5}$ **51.** $6a\sqrt{5}$
53. $7 + \sqrt{21} - \sqrt{14} - \sqrt{6}$ **55.** 4 **57.** $7 - 2\sqrt{10}$

59. $\dfrac{2\sqrt{3x}}{x}$ **61.** $\dfrac{\sqrt[3]{3a}}{a}$ **63.** $\dfrac{\sqrt[3]{x^2y}}{y^2}$

65. $\dfrac{5 + \sqrt{3}}{2}$ **67.** $\dfrac{x - 6\sqrt{x} + 9}{x - 9}$ **69.** $\left\{\dfrac{11}{3}\right\}$

71. $\{5\}$ **73.** $\{\pm 5\}$ **75.** $\{1, 5\}$ **77.** -10

79. 2 **81.** 125 **83.** $\dfrac{1}{7}$ **85.** x^4 **87.** r

89. $x^{2/5}$ **91.** x^8y^{15} **93.** $\dfrac{x^6}{y^{7/2}}$ **95.** $\sqrt[4]{x^3}$

97. $3\sqrt[3]{a^2}$ **99.** $(7x)^{1/5}$ **101.** $3pq^3$ **103.** $7i$
105. $-2i\sqrt{15}$ **107.** $4 - 5i$ **109.** $-3 + 5i$

111. $23 + 14i$ **113.** 13 **115.** $\dfrac{6}{5} + \dfrac{8}{5}i$

117. $4 + 3i$

Self-Test—Chapter 10

1. $7a^2$ **2.** $-3w^2z^3$ **3.** $p^2q\sqrt[3]{9pq^2}$ **4.** $\dfrac{7x}{8y}$

5. $\dfrac{\sqrt{10xy}}{4y}$ **6.** $\dfrac{\sqrt[3]{3x^2}}{x}$ **7.** $3x\sqrt{3x}$

8. $5m\sqrt[3]{2m}$ **9.** $4x\sqrt{3}$ **10.** $3 - 2\sqrt{2}$
11. $x^3\sqrt{14x}$ **12.** $2x^3\sqrt[3]{4x^2}$ **13.** $\{11\}$

14. $\{4\}$ **15.** $64x^6$ **16.** $\dfrac{9m}{n^4}$ **17.** $\dfrac{8s^3}{r}$

18. $\dfrac{32x^4}{y}$ **19.** 6.6 **20.** 1.4 **21.** $a^2b\sqrt[5]{a^4b}$

22. $(125p^9q^6)^{1/3} = 5p^3q^2$ **23.** $3 + 10i$
24. $-14 + 22i$ **25.** $5 - 5i$

Cumulative Test—Chapters 0–10

1. $\{-15\}$ **2.** 5 **3.** $y = \dfrac{3}{2}x - 6$ **4.** $\left\{3, \dfrac{1}{3}\right\}$

5. $10x^2 - 2x$ **6.** $10x^2 - 39x - 27$ **7.** $x(x - 1)(2x + 3)$
8. $9(x^2 + 2y^2)(x^2 - 2y^2)$ **9.** $(x + 2y)(4x - 5)$

10. $\dfrac{x + 5}{3x - 1}$ **11.** $\dfrac{x + 7}{(x - 5)(x - 1)}$ **12.** $\dfrac{a + 2}{2a}$

13. 1 **14.** $2x^4y^3\sqrt{3y}$ **15.** $\sqrt{6} - 5\sqrt{2} + 3\sqrt{3} - 15$

16.

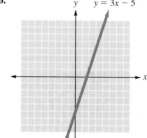

17. $x = -5$

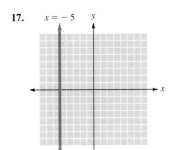

18.

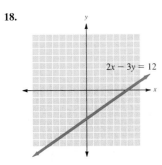

$2x - 3y = 12$

19. $\{(3, -1)\}$ **20.** (a) $\{x \mid x \leq -4\}$; (b) $\{x \mid -6 < x < 10\}$

21. $x^6 y^{14}$ **22.** $216x^4 y^5$ **23.** $\dfrac{4xy^3}{3}$ **24.** $\{1\}$

25. $\left\{-\dfrac{1}{3}, 1\right\}$

Chapter 11

Section 11.1

1. $\{-7, -2\}$ **3.** $\{-5, 7\}$ **5.** $\left\{-\dfrac{1}{2}, 3\right\}$

7. $\left\{-\dfrac{1}{2}, \dfrac{2}{3}\right\}$ **9.** $\{-11, 11\}$ **11.** $\{-\sqrt{7}, \sqrt{7}\}$

13. $\{-\sqrt{6}, \sqrt{6}\}$ **15.** $\{\pm 2i\}$ **17.** $\{-1 \pm 2\sqrt{3}\}$

19. $\left\{\dfrac{-1 \pm \sqrt{5}}{3}\right\}$ **21.** 81   **23.** 16   **25.** $\dfrac{9}{4}$

27. $\dfrac{1}{4}$ **29.** $\dfrac{1}{100}$ **31.** $\dfrac{1}{144}$ **33.** $\{-5 \pm \sqrt{29}\}$

35. $\{-2, 4\}$ **37.** $\{1 \pm \sqrt{6}\}$ **39.** $\{-5 \pm 2\sqrt{3}\}$

41. $\left\{\dfrac{5 \pm \sqrt{53}}{2}\right\}$ **43.** $\left\{\dfrac{3 \pm \sqrt{29}}{2}\right\}$ **45.** $\left\{\dfrac{-1 \pm \sqrt{17}}{4}\right\}$

47. $\left\{\dfrac{-1 \pm \sqrt{3}}{2}\right\}$ **49.** $\left\{\dfrac{4 \pm \sqrt{22}}{3}\right\}$ **51.** $\left\{\dfrac{1 \pm i\sqrt{35}}{3}\right\}$

53. $\{-5 \pm i\sqrt{3}\}$ **55.** 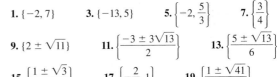   **57.** a^2

59. $\dfrac{9}{4}a^2$ **61.** 1   **63.** $\{-a \pm \sqrt{4 + a^2}\}$

65. 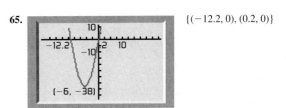 $\{(-12.2, 0), (0.2, 0)\}$

67. 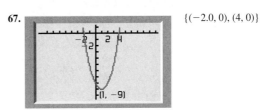 $\{(-2.0, 0), (4, 0)\}$

69. (a) There are none; (b) $\pm i$; (c) If the graph of $f(x)$ has no x intercepts, the zeros of the function are not real.
71. (a) $x^2 + 8x + 8x$; (b) 64; (c) $x^2 + 16x + 64 = (x + 8)^2$

Section 11.2

1. $\{-2, 7\}$ **3.** $\{-13, 5\}$ **5.** $\left\{-2, \dfrac{5}{3}\right\}$ **7.** $\left\{\dfrac{3}{4}\right\}$

9. $\{2 \pm \sqrt{11}\}$ **11.** $\left\{\dfrac{-3 \pm 3\sqrt{13}}{2}\right\}$ **13.** $\left\{\dfrac{5 \pm \sqrt{13}}{6}\right\}$

15. $\left\{\dfrac{1 \pm \sqrt{3}}{2}\right\}$ **17.** $\left\{-\dfrac{2}{3}, 1\right\}$ **19.** $\left\{\dfrac{1 \pm \sqrt{41}}{4}\right\}$

21. $\{1, 3\}$ **23.** $\{4\}$ **25.** $\left\{\dfrac{-3 \pm \sqrt{29}}{-2}\right\}$

27. $\left\{\dfrac{1 \pm i\sqrt{2}}{3}\right\}$ **29.** $\left\{\dfrac{1 \pm \sqrt{161}}{8}\right\}$ **31.** $\left\{\dfrac{3 \pm \sqrt{39}}{15}\right\}$

33. $\left\{\dfrac{-1 \pm \sqrt{41}}{2}\right\}$ **35.** $\left\{\dfrac{1 \pm \sqrt{23}}{2}\right\}$ **37.** $\left\{\dfrac{7 \pm \sqrt{73}}{4}\right\}$

39. $\left\{\dfrac{3 \pm \sqrt{33}}{4}\right\}$ **41.** $\{1 \pm \sqrt{6}\}$ **43.** 25, two

45. 0, one   **47.** 37, two   **49.** -63, none   **51.** $\{4\}$

53. $\left\{\dfrac{7 \pm \sqrt{37}}{6}\right\}$ **55.** $\left\{0, \dfrac{2}{5}\right\}$ **57.** $\left\{\dfrac{-5 \pm \sqrt{33}}{4}\right\}$

59. $\{-3i, 3i\}$ **61.** $\{-2, 5\}$
63. (a) 1.2 or 5.8 s; (b) 1.7 or 5.3 s   **65.** -9, -8 or 8, 9
67. 7 ft by 10 ft   **69.** 2.7 cm and 5.4 cm
71. (a) 4 s; (b) 1 s   **73.** 50 chairs
75. 5.5 in., 6.5 in., 8.5 in.   **77.** 63 or 27
79. $0.94 or $17.06   **81.** (a) 5.5 s; (b) 1 s, 2 s
83. (a) $h(t) = 100 + 20t - 5t^2$; (b) 6.9 s; (c) 5 s; (d) no
85. 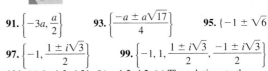   **87.** $\{\pm\sqrt{z^2 - y^2}\}$ **89.** $\{-6a, 6a\}$

91. $\left\{-3a, \dfrac{a}{2}\right\}$ **93.** $\left\{\dfrac{-a \pm a\sqrt{17}}{4}\right\}$ **95.** $\{-1 \pm \sqrt{6}\}$

97. $\left\{-1, \dfrac{1 \pm i\sqrt{3}}{2}\right\}$ **99.** $\left\{-1, 1, \dfrac{1 \pm i\sqrt{3}}{2}, \dfrac{-1 \pm i\sqrt{3}}{2}\right\}$

101. (a) $\{-1.2, 4.2\}$; (b) -1.2, 4.2; (c) The solutions to the quadratic equation are the x intercepts of the graph.

Section 11.3

1. (f)   **3.** (g)   **5.** (h)   **7.** (c)   **9.** (a), (c)
11. (a), (c)   **13.** (b), (c)

15.

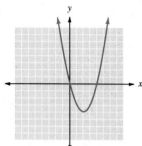

$x = 2$; vertex $(2, -4)$;
$(0, 0)$ and $(4, 0)$

25.

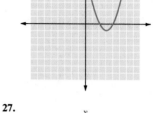

$x = 3$; vertex $(3, -1)$;
$(2, 0)$ and $(4, 0)$

17.

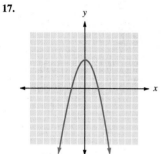

$x = 0$; vertex $(0, 4)$;
$(-2, 0)$ and $(2, 0)$

27.

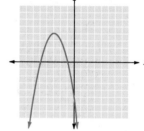

$x = -3$; vertex $(-3, 4)$;
$(-5, 0)$ and $(-1, 0)$

19.

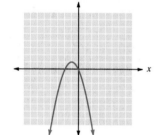

$x = -1$; vertex $(-1, 1)$;
$(-2, 0)$ and $(0, 0)$

29.

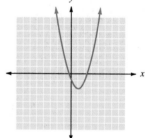

$x = 1$; vertex $(1, -2)$;
$(0, -1)$ and $(2, -1)$

21.

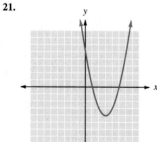

$x = 3$; vertex $(3, -4)$;
$(1, 0)$ and $(5, 0)$

31.

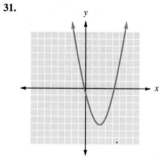

$x = 2$; vertex $(2, -5)$;
$(0, -1)$ and $(4, -1)$

23.

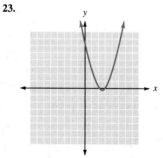

$x = \dfrac{5}{2}$; vertex $\left(\dfrac{5}{2}, -\dfrac{1}{4}\right)$;
$(3, 0)$ and $(2, 0)$

33.

$x = \dfrac{3}{2}$; vertex $\left(\dfrac{3}{2}, -\dfrac{3}{4}\right)$;
$(0, -3)$ and $(3, -3)$

35.

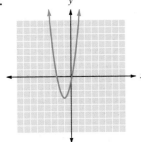

$x = -1$; vertex $(-1, -3)$; $(0, -1)$ and $(-2, -1)$

37.

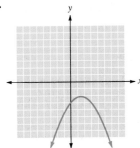

$x = \dfrac{3}{2}$; vertex $\left(\dfrac{3}{2}, -\dfrac{9}{4}\right)$; $(0, -3)$ and $(3, -3)$

39.

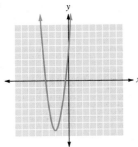

$x = -2$; vertex $(-2, -7)$; $(0, 5)$ and $(-4, 5)$

41. $\{-4, 3\}$ **43.** $\{0, 3.167\}$ **45.** $\{-1.772, 0.439\}$
47. $\{-3.828, 1.828\}$ **49.** 100 items; \$2,600
51. 500 ft by 500 ft; 250,000 ft^2 **53.** 144 ft   **55.** 88 ft/s
57. (a) $(-4, 2)$; (b) $y \geq 2$ **59.** (a) $(-2, -1)$; (b) $y \leq -1$
61. (a) $(4, 0)$; (b) $y \leq 0$ **63.** $-3 \leq x \leq 3$; $-25 \leq y \leq 0$
65. $-2 \leq x \leq 4$; $-10 \leq y \leq 0$

Section 11.4

1. $\{x \mid -4 < x < 3\}$

3. $\{x \mid x < -4 \text{ or } x > 3\}$

5. $\{x \mid -4 \leq x \leq 3\}$

7. $\{x \mid x \leq -4 \text{ or } x \geq 3\}$

9. $\{x \mid x < -1 \text{ or } x > 4\}$

11. $\{x \mid -4 \leq x \leq 3\}$

13. $\{x \mid x \leq 2 \text{ or } x \geq 3\}$

15. $\{x \mid -6 \leq x \leq 4\}$

17. $\{x \mid x < -9 \text{ or } x > 3\}$

19. $\left\{x \mid -2 \leq x \leq \dfrac{3}{2}\right\}$

21. $\left\{x \mid -1 < x < \dfrac{3}{4}\right\}$

23. $\{x \mid -4 \leq x \leq 4\}$

25. $\{x \mid x \leq -5 \text{ or } x \geq 5\}$

27. $\{x \mid x < -2 \text{ or } x > 2\}$

29. $\{x \mid 0 \leq x \leq 4\}$

31. $\{x \mid x \leq 0 \text{ or } x \geq 6\}$

33. $\{x \mid 0 < x < 4\}$

35. $\{x \mid x = 2\}$

37. $\{x \mid -4 \leq x \leq 7\}$

39.

41. $\{x \mid x < -1 \text{ or } 0 < x < 2\}$

43. $\{x \mid -2 \leq x \leq 1 \text{ or } x \geq 3\}$

45. $\{x \mid x \leq -3 \text{ or } 0 \leq x \leq 5\}$

47. $\{x \mid 50 \leq x \leq 60\}$ **49.** $\{x \mid 2 \leq t \leq 3\}$

51.

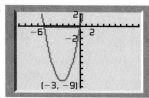
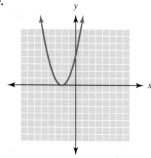

$\{x \mid -6 < x < 0\}$

53.

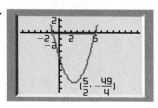

$\{x \mid -1 \le x \le 6\}$

Summary Exercises—Chapter 11

1. $\{\pm 2\sqrt{3}\}$ **3.** $\{2 \pm 2\sqrt{5}\}$ **5.** 49   **7.** $\{-1, 5\}$

9. $\{5 \pm 2\sqrt{7}\}$ **11.** $\left\{\dfrac{3 \pm \sqrt{7}}{2}\right\}$ **13.** $\{-3, 8\}$

15. $\left\{\dfrac{5 \pm \sqrt{17}}{2}\right\}$ **17.** $\left\{\dfrac{-2 \pm \sqrt{7}}{3}\right\}$ **19.** $\{1 \pm \sqrt{29}\}$

21. $\left\{\dfrac{-1 \pm i\sqrt{14}}{3}\right\}$ **23.** None   **25.** One   **27.** 4, 8

29. 5   **31.** Width 5 cm; length 7 cm   **33.** $30 \le x \le 50$
35. Width 4 cm; length 9 cm   **37.** Width 8 ft; length 15 ft
39. $-1 + \sqrt{31}, 1 + \sqrt{31}$ or 4.6, 6.6 m   **41.** $\{(2, 0), (-1, 0)\}$

43. $\left\{\left(-\dfrac{3}{2}, 0\right), (-2, 0)\right\}$ **45.** $x = 0; (0, 0)$
47. $x = 0; (0, -5)$ **49.** $x = -2; (-2, 0)$
51. $x = -3; (-3, 1)$ **53.** $x = 5; (5, -2)$

55. $x = 1; (1, 1)$ **57.** $x = -\dfrac{1}{2}; \left(-\dfrac{1}{2}, \dfrac{25}{4}\right)$

59. $x = -3; (-3, 13)$ **61.** $\{-3.712, 1.212\}$
63. $\{0.807, 6.192\}$
65.

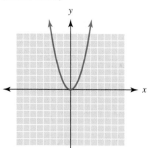

67.

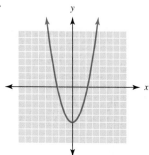

69.

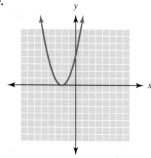

71.

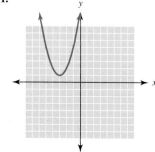

73.

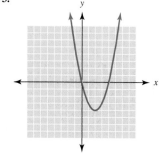

75.

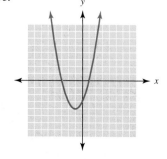

77.

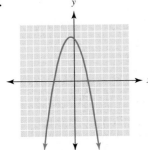

79.

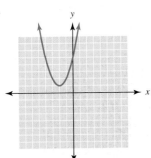

81.

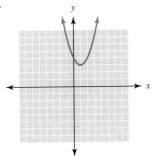

83.

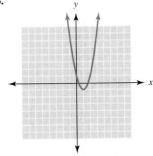

85. $\{x \mid x < -5 \text{ or } x > 2\}$

87. $\{x \mid -3 \le x \le -1\}$

89. $\{x \mid -3 \le x \le 8\}$

91. $\{x \mid x \le -8 \text{ or } x \ge 8\}$

93. $\{x \mid -3 < x < 7\}$

95. $1 \le t \le 3$

Self-Test—Chapter 11

1. $\left\{-3, -\dfrac{1}{2}\right\}$ **2.** $\left\{-\dfrac{5}{2}, \dfrac{2}{3}\right\}$ **3.** $\left\{0, \dfrac{3}{2}, -\dfrac{3}{2}\right\}$

4. $\{\pm\sqrt{5}\}$ **5.** $\{1 \pm \sqrt{10}\}$ **6.** $\left\{\dfrac{2 \pm \sqrt{23}}{2}\right\}$

7. $\left\{\dfrac{-3 \pm \sqrt{13}}{2}\right\}$ **8.** $\left\{\dfrac{5 \pm \sqrt{19}}{2}\right\}$ **9.** $\left\{\dfrac{5 \pm \sqrt{37}}{2}\right\}$

10. $\{-2 \pm \sqrt{11}\}$ **11.** $\left\{-\dfrac{2}{3}, 4\right\}$ **12.** 7, 9

13. 2.7 s   **14.** $x = -2; (-2, 1)$ **15.** $x = 2; (2, -9)$

16. $x = \dfrac{3}{2}; \left(\dfrac{3}{2}, \dfrac{3}{2}\right)$ **17.** $x = 3; (3, -2)$

18. $x = 3; (3, -7)$

19.

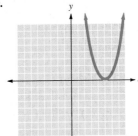

20.

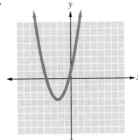

21.

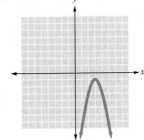

22.

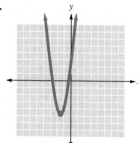

23. $\{-4.541, 1.541\}$ **24.** $\{-1.396, 0.896\}$

25.

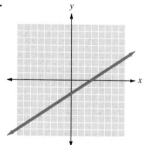

$\{x \mid x < -3 \text{ or } x > 1\}$

26. $\{x \mid -7 < x < 2\}$

27. $\{x \mid x \le -3 \text{ or } x \ge 6\}$

28. $\left\{x \mid -2 \le x \le \dfrac{5}{3}\right\}$

29. $\{x \mid 0 \le x \le 3\}$

30.

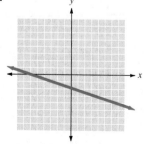

$\{x \mid x < -5 \text{ or } x > 0\}$

Cumulative Test—Chapters 0–11

1.

2.

3.

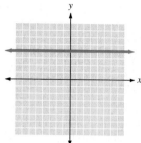

4. -3 **5.** $\dfrac{4}{3}$ **6.** 35 **7.** $x^3 + 3x^2 - x - 3$

8. $x(x + 3)(x - 2)$ **9.** $\dfrac{-x - 4}{(x - 3)(x + 2)}$ **10.** $6xy^2\sqrt{2xy}$

11. $\left\{\dfrac{7}{2}\right\}$ **12.** $\{-4\}$ **13.** $\{-5, 3\}$ **14.** $\{1, 2\}$

15. $\{-10, 3\}$ **16.** $\left\{\dfrac{3 \pm \sqrt{21}}{2}\right\}$ **17.** $\{3 \pm \sqrt{5}\}$

18. $\{-2, 6\}$ **19.** $\left\{\dfrac{13}{6}\right\}$ **20.** $\{x \mid x \le 9\}$

21. $\{x \mid x \le -7 \text{ or } x \ge 9\}$ **22.** $\{x \mid -2 \le x \le 3\}$

23. -13 **24.** $-2 + 2\sqrt{97}, 2 + 2\sqrt{97}$ or 17.7 ft, 21.7 ft

25. Between 21 and 59 units

Chapter 12

Section 12.1

1. $y = 3x^2 + 2x - 5$ **3.** $y = \dfrac{1}{2}x^2 - \dfrac{3}{2}x - \dfrac{5}{2}$

5. $y = -x^2 + 2x + \dfrac{3}{2}$

7.

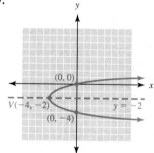

9.

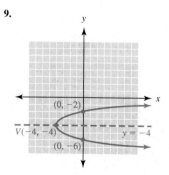

11.

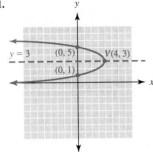

13.

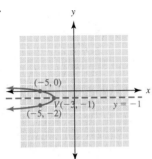

15.

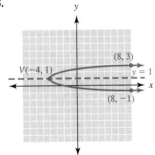

17. $a > 0$ and $h > 0$ or $a < 0$ and $h < 0$

19.

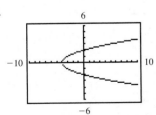

21.

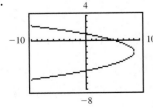

23.

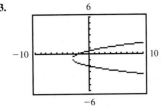

25. $\{x \mid x \geq -9\}$ **27.** $\{x \mid x \leq 2\}$

Section 12.2

1. Parabola   **3.** Line   **5.** Circle   **7.** Circle
9. None of these   **11.** Parabola
13. Center $(0, 0)$; radius 6   **15.** Center $(3, -1)$; radius 4
17. Center $(-2, 0)$; radius 5   **19.** Center $(3, -4)$; radius $\sqrt{41}$

21.

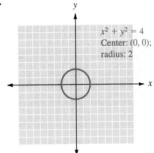

23.

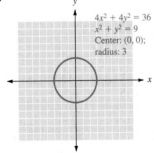

25.

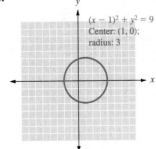

27.

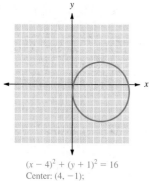

$(x - 4)^2 + (y + 1)^2 = 16$
Center: $(4, -1)$;
radius: 4

29.

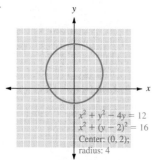

$x^2 + y^2 - 4y = 12$
$x^2 + (y - 2)^2 = 16$
Center: $(0, 2)$;
radius: 4

31.

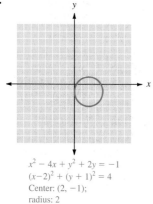

$x^2 - 4x + y^2 + 2y = -1$
$(x-2)^2 + (y + 1)^2 = 4$
Center: $(2, -1)$;
radius: 2

33. $x^2 + y^2 = 25$ **35.** $(x - 3)^2 + (y + 2)^2 = 25$
37. Circle with center $(1, 2)$ and radius zero
39. $10\sqrt{5}$ cm; ≈ 22.4 cm   **41.** $x^2 + y^2 = \dfrac{4}{9}$
43. $y = \sqrt{36 - x^2}$
$\quad y = -\sqrt{36 - x^2}$

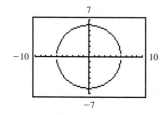

45. $y = \sqrt{36 - (x + 5)^2}$
$\quad y = -\sqrt{36 - (x + 5)^2}$

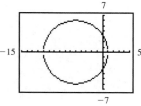

47. Domain $\{x \mid -7 \le x \le 1\}$, range $\{y \mid -2 \le y \le 6\}$
49. Domain $\{x \mid -5 \le x \le 5\}$, range $\{y \mid -2 \le y \le 8\}$

Section 12.3

1.

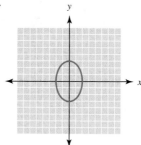

3.

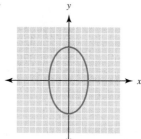

5.

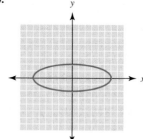

7.

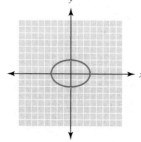

9.

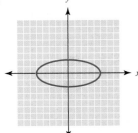

11.

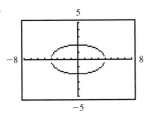

13. Yes

15.

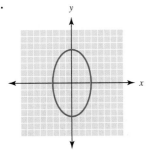

17.

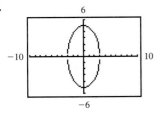

19. Domain $\{x \mid -3 \le x \le 3\}$, range $\{y \mid -4 \le y \le 4\}$

21. $\dfrac{x^2}{16} + \dfrac{y^2}{9} = 1$

Section 12.4

1. Circle   **3.** Parabola   **5.** Hyperbola
7. Hyperbola   **9.** Circle   **11.** Hyperbola   **13.** (d)
15. (h)   **17.** (b)   **19.** (c)
21.

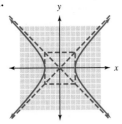

23.

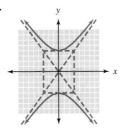

25.

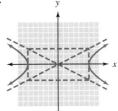

27.

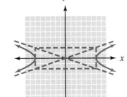

29.

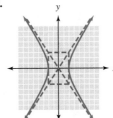

31.

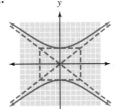

33.

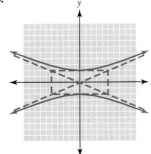

35.

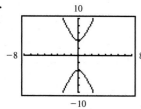

37.

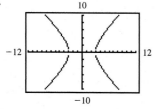

Summary Exercises—Chapter 12

1. $y = \frac{2}{5}x^2 + \frac{3}{5}x - \frac{1}{5}$

3.

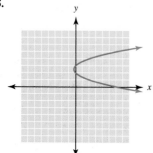

5.

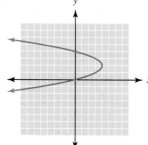

7.

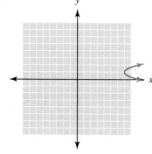

9. Center $(0, 0)$; radius 9   **11.** Center $(0, 0)$; radius 3
13. Center $(3, 0)$; radius 6   **15.** Center $(-2, 1)$; radius 5
17. Center $(-4, -5)$; radius 8

19.

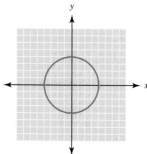

21.

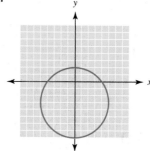

23.

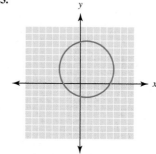

25.

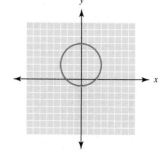

27.

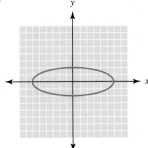

29.

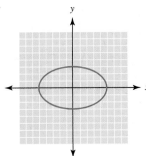

31.

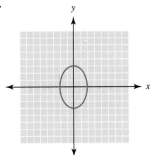

33.

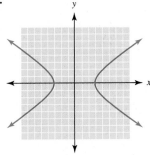

35.

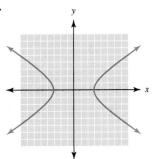

37. Line   **39.** Circle   **41.** Parabola
43. Parabola   **45.** Ellipse   **47.** Hyperbola
49. Ellipse   **51.** Hyperbola   **53.** Line
55. Domain $\{x \mid -11 \le x \le 11\}$, range $\{y \mid -11 \le y \le 11\}$
57. Domain $\{x \mid -4 \le x \le 4\}$, range $\{y \mid -3 \le y \le 3\}$
59. $x^2 + y^2 = 25$

Self-Test—Chapter 12

1. $y = -x^2 + 3x - \dfrac{3}{2}$

2.

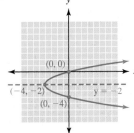

3.

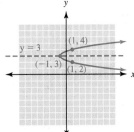

4.

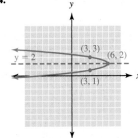

5.

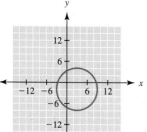

Center $(3, -2)$; radius 6

6.

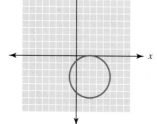

Center $(-1, 2)$; radius 5

7.

Center $(-3, 0)$; radius 2

8.

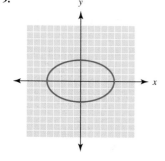

Center $(2, -3)$; radius 3

9.

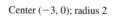

10.

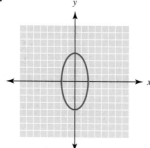

11.

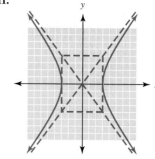

12.

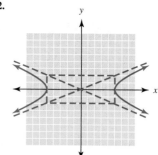

13. Line   **14.** Parabola   **15.** Circle   **16.** Circle
17. Parabola   **18.** Ellipse   **19.** Hyperbola   **20.** Circle

Cumulative Test—Chapters 0–12

1. $\left\{\dfrac{11}{2}\right\}$ **2.** $\{x \mid x > -2\}$ **3.** $\{-1, 4\}$

4. $\left\{x \mid -4 \leq x \leq \dfrac{2}{3}\right\}$ **5.** $\left\{x \mid x < -\dfrac{17}{5} \text{ or } x > 5\right\}$

6. $\{-3, 8\}$

7.

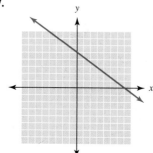

8.

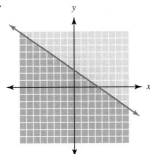

9. $\sqrt{601} \approx 24.5$ **10.** 7   **11.** $f(x) = -x + 3$
12. $2x^2 - 5x - 3$ **13.** $9x^2 - 12x + 4$
14. $f(x) = (x - 3)(x^2 - 5)$

15.

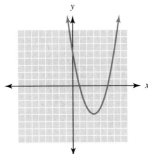

16.

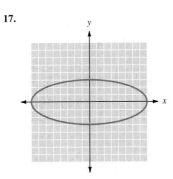

17.

18. $\left\{\left(-10, \dfrac{26}{3}\right)\right\}$ **19.** 8 cm by 19 cm   **20.** 73

Chapter 13

Section 13.1

1. $f^{-1}(x) = \dfrac{x - 5}{3}$ **3.** $f^{-1}(x) = 2x + 1$

5. $f^{-1}(x) = \dfrac{x + 3}{2}$ **7.** $f^{-1}(x) = 3x - 4$

9. $f^{-1}(x) = 3x - 15$
11. $\{(-5, -2), (6, 2), (4, 3), (3, 4)\}$; a function
13. $\{(3, -4), (7, -2), (5, 3), (-4, 7)\}$; a function
15. $\{(3, 2), (4, 3), (5, 4)\}$; a function
17. $\{(5, 1), (5, 2), (5, 3)\}$; not a function
19. $\{(3, 2), (5, 3), (7, 4)\}$; a function

21. $f^{-1}(x) = \dfrac{6 - 3x}{5}$ **23.** $f^{-1}(x) = \dfrac{5x - 10}{11}$

25. $f^{-1}(x) = \dfrac{x + 6}{3} = \dfrac{1}{3}x + 2$

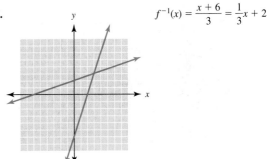

27. $f^{-1}(x) = \dfrac{6 - x}{2} = -\dfrac{1}{2}x + 3$

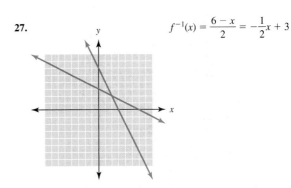

29. Not one-to-one; inverse is not a function
31. One-to-one; inverse is a function
33. Not one-to-one; inverse is not a function
35. One-to-one; inverse is a function   **37.**
39. 12   **41.** 6   **43.** x
45. 2   **47.** 3   **49.** x
51. 16   **53.** 4   **55.** x
57. 5   **59.** $f^{-1}(x) = \dfrac{x - b}{m}$ **61.** $\dfrac{5}{2}$

Section 13.2

1. (c)   **3.** (b)   **5.** (h)   **7.** (f)   **9.** 1

11. 16   **13.** 1   **15.** 64   **17.** $\dfrac{5}{4}$ **19.** 17

21. $\dfrac{1}{4}$ **23.** 1

25.

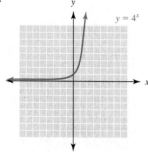

$y = 4^x$

33.

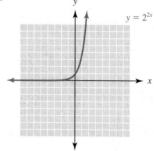

$y = 2^{2x}$

27.

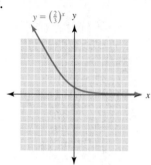

$y = \left(\frac{2}{3}\right)^x$

35.

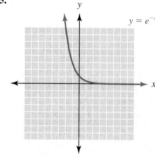

$y = e^{-x}$

29.

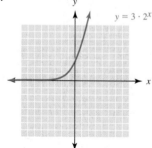

$y = 3 \cdot 2^x$

37.

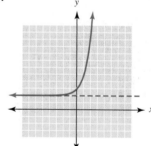

31.

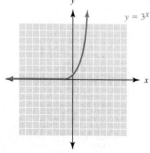

$y = 3^x$

39.

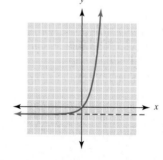

41. {3}   **43.** {4}   **45.** {−4}

47. {2}   **49.** {3}   **51.** {−1}   **53.** 800

55.

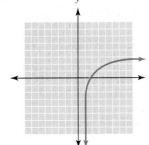

57. 16 g

59.

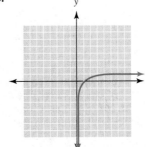

61. $1,469.33

63.

65.

67.

69. 2.7169   **71.** 2.71823   **73.**

75.

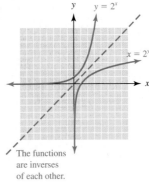

The functions
are inverses
of each other.

77. (−0.77, 0.59); (2, 4); (4, 16)

Section 13.3

1.

3.

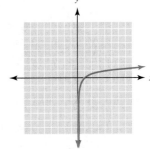

5.

7. $\log_2 32 = 5$ **9.** $\log_{10} 100 = 2$ **11.** $\log_3 1 = 0$

13. $\log_4 \dfrac{1}{16} = -2$ **15.** $\log_{10} \dfrac{1}{100} = -2$

17. $\log_{16} 4 = \dfrac{1}{2}$ **19.** $\log_{64} \dfrac{1}{4} = -\dfrac{1}{3}$ **21.** $\log_{27} 9 = \dfrac{2}{3}$

23. $\log_{27} \dfrac{1}{9} = -\dfrac{2}{3}$ **25.** $2^4 = 16$ **27.** $5^0 = 1$

29. $10^1 = 10$ **31.** $5^3 = 125$ **33.** $3^{-3} = \dfrac{1}{27}$

35. $10^{-3} = 0.001$ **37.** $16^{1/2} = 4$ **39.** $8^{2/3} = 4$

41. $25^{-1/2} = \dfrac{1}{5}$ **43.** 6   **45.** 3   **47.** -4

49. -2 **51.** $\dfrac{1}{2}$ **53.** $\{2\}$ **55.** $\{4\}$ **57.** $\{100\}$

59. $\{1\}$ **61.** $\left\{\dfrac{27}{8}\right\}$ **63.** $\{5\}$ **65.** $\left\{\dfrac{1}{1{,}000}\right\}$

67. $\{-2\}$ **69.** $\{3\}$ **71.** $\{25\}$ **73.** $\left\{-\dfrac{2}{3}\right\}$

75. 50 dB   **77.** 70 dB   **79.** 10^{-8} W/cm^2 **81.** 10
83. 1,000   **85.** 6   **87.** $10^5 \cdot a_0$ **89.**
91. 24,000 yr   **93.** 77,000 yr
95. Pu-239: 240,000 yr; SR-90: 280 yr;
      Th-230: 770,000 yr; Cs-135: 20,000,000 yr
97. (a) 6; (b) 2, 4; (c) $\log_3 (9 \cdot 81) = \log_3 9 + \log_3 81$

Section 13.4

1. $\log_b 5 + \log_b x$ **3.** $\log_6 x - \log_6 7$ **5.** $2 \log_3 a$

7. $\dfrac{1}{2} \log_5 x$ **9.** $2 \log_b x + 4 \log_b y$

11. $2 \log_4 y + \dfrac{1}{2} \log_4 x$ **13.** $2 \log_b x + \log_b y - \log_b z$

15. $\log x + 2 \log y - \dfrac{1}{2} \log z$ **17.** $\dfrac{1}{3}(\log_5 x + \log_5 y - 2 \log_5 z)$

19. $\log_b xy$ **21.** $\log_5 \dfrac{x^3}{y^2}$ **23.** $\log_b x \sqrt{y}$

25. $\log_b \dfrac{x}{y^2 z}$ **27.** $\log_6 \dfrac{\sqrt{y}}{z^3}$ **29.** $\log_b \sqrt[3]{\dfrac{x^2 y}{z}}$
31. 1.380   **33.** 0.903   **35.** 0.151   **37.** -0.602
39. 0.863   **41.** 2.833   **43.** -0.143
45. 7.42, basic   **47.** 5.61   **49.** 5610   **51.** 2×10^{-5}
53. 0.693   **55.** 2.303   **57.** 74   **59.** Between 2 and 3
61. Between 6 and 7   **63.** Between 2 and 3   **65.** 1
67. 2   **69.** 2.113   **71.** 0.25 kg   **73.** 160 kg
75.

Section 13.5

1. $\{125\}$ **3.** $\{99\}$ **5.** $\{8\}$ **7.** $\{162\}$ **9.** $\{2\}$

11. $\{6\}$ **13.** $\left\{\dfrac{20}{9}\right\}$ **15.** $\left\{\dfrac{19}{8}\right\}$ **17.** $\left\{\dfrac{15}{4}\right\}$

19. $\{3, 5\}$ **21.** $\{4\}$ **23.** $\{-4\}$ **25.** $\left\{\dfrac{1}{3}\right\}$

27. $\{1.771\}$ **29.** $\{0.792\}$ **31.** $\{0.670\}$ **33.** $\{0.894\}$
35. $\{3.819\}$ **37.** $\{4.419\}$ **39.** 8.04 yr   **41.** 7.79 yr
43. 1.47 h   **45.** 3.17 h   **47.** 20.6 yr   **49.** 11.6 yr
51. 7.5 yr   **53.** 23.1 h   **55.** 4,558 ft   **57.** 2,876 yr
59.   **61.** The graph is that of $y = x$. The two
      operations cancel each other.
      63. The graph is that of $y = x$. The two
      operations cancel each other.
65.

Summary Exercises—Chapter 13

1. $f^{-1}(x) = \dfrac{3 - x}{2}$ **3.** $f^{-1}(x) = 4x + 3$

5. $\{(1, -3), (2, -1), (3, 2), (4, 3)\}$; function
7. $\{(5, 1), (7, 2), (9, 3)\}$; function
9. $\{(4, 2), (3, 4), (4, 6)\}$; not a function
11.

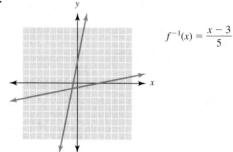

$f^{-1}(x) = \dfrac{x - 3}{5}$

13.

$f^{-1}(x) = 5x + 4$

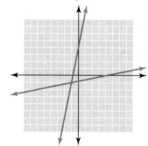

15. One-to-one; f^{-1}: $\{(2, -1), (3, 1), (5, 2), (7, 4)\}$; function
17. One-to-one; f^{-1}: $\{(2, -1), (4, 3), (5, 4), (6, 5)\}$; function
19. Not one-to-one; f^{-1} is not a function   **21.** 4
23. 8   **25.** x
27.

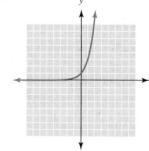

29. $\{3\}$
31. $\{-1\}$
33. 64,000

35.

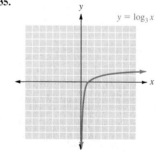

$y = \log_3 x$

37. $\log_2 32 = 5$ **39.** $\log_5 1 = 0$ **41.** $\log_{25} 5 = \dfrac{1}{2}$

43. $4^3 = 64$ **45.** $81^{1/2} = 9$ **47.** $10^{-3} = 0.001$

49. $\{3\}$ **51.** $\{49\}$ **53.** $\{9\}$ **55.** $\left\{\dfrac{1}{3}\right\}$

57. 80 dB   **59.** 100   **61.** 10   **63.** $3 \log_4 y - \log_4 5$

65. $3 \log_5 x + \log_5 y + 2 \log_5 z$

67. $\dfrac{2}{3}\log_b x + \dfrac{1}{3}\log_b y - \dfrac{1}{3}\log_b z$ **69.** $\log_b \dfrac{x^3}{z^2}$

71. $\log_5 \dfrac{x^2}{y^3 z}$ **73.** $\log_b \sqrt[3]{\dfrac{x}{y^2}}$ **75.** 1.204   **77.** 0.239

79. 9.495, basic   **81.** 6.3×10^{-11} **83.** 66.6

85.

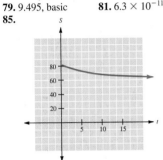

87. 1.969   **89.** $\{250\}$ **91.** $\{2\}$ **93.** $\{3\}$

95. $\{-2\}$ **97.** $\{2.500\}$ **99.** $\{4.419\}$ **101.** 8.21 yr

103. 166 yr   **105.** 8.7 yr   **107.** 6.6 mi

Self-Test—Chapter 13

1. $f^{-1}(x) = \dfrac{5 - x}{4}$ **2.** $f^{-1}(x) = 5x + 3$

3. $f^{-1}(x) = \{(-2, -3), (2, 1), (5, 4), (6, 5)\}$

4. $f^{-1}(x) = \{(1, -3), (2, 4), (-2, 5)\}$

5. Not one-to-one; $f^{-1}(x)$: $\{(4, 3), (-1, 5), (2, 6), (1, 7)\}$; f^{-1} is *not* a function

6. One-to-one; f^{-1}: $\{(3, 1), (8, 3), (11, 5), (5, 7)\}$; f^{-1} is one-to-one

7. -2 **8.** 2   **9.** x **10.** x

11.

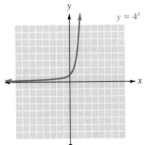

12.

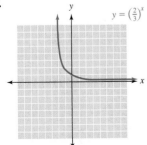

13. (a) $\{-2\}$; (b) $\left\{\dfrac{5}{2}\right\}$

14.

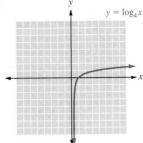

15. $\log 10{,}000 = 4$ **16.** $\log_{27} 9 = \dfrac{2}{3}$ **17.** $5^3 = 125$

18. $10^{-2} = 0.01$ **19.** $\{6\}$ **20.** $\{4\}$ **21.** $\{5\}$

22. $2 \log_b x + \log_b y + 3 \log_b z$

23. $\dfrac{1}{2}(\log_5 x + 2 \log_5 y - \log_5 z)$ **24.** $\log (xy^3)$

25. $\log_b \sqrt[3]{\dfrac{x}{z^2}}$ **26.** $\{8\}$ **27.** $\left\{\dfrac{11}{8}\right\}$ **28.** $\{0.262\}$

29. $\{2.151\}$ **30.** $\{-1.333\}$

Cumulative Test—Chapters 0–13

1. $\{11\}$ **2.** $\left\{x \mid x < \dfrac{2}{3} \text{ or } x > 4\right\}$ **3.** $\left\{\dfrac{10}{9}\right\}$

4.

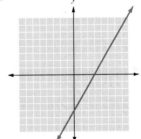

5.

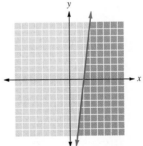

6. $6x + 5y = 7$ **7.** $\{x \mid x \geq 10\}$

8. $2x^2 - 6x + 1$ **9.** $6x^2 - 13x - 5$

10. $(2x - 5)(x + 2)$ **11.** $x(5x + 4y)(5x - 4y)$

12. $\dfrac{-x + 2}{(x - 4)(x - 5)}$ **13.** $\dfrac{x + 2}{x - 2}$ **14.** $-4\sqrt{2}$

15. $22 + 12\sqrt{2}$ **16.** $\dfrac{5}{3}(\sqrt{5} + \sqrt{2})$ **17.** 77, 79, 81

18. $\{-2, 1\}$ **19.** $\left\{\dfrac{3 \pm \sqrt{19}}{2}\right\}$ **20.** $\left\{x \mid -\dfrac{3}{2} \leq x \leq 1\right\}$

Index